Food Engineering Series

Springer's Food Engineering Series is essential to the Food Engineering profession, providing exceptional texts in areas that are necessary for the understanding and development of this constantly evolving discipline. The titles are primarily reference-oriented, targeted to a wide audience including food, mechanical, chemical, and electrical engineers, as well as food scientists and technologists working in the food industry, academia, regulatory industry, or in the design of food manufacturing plants or specialized equipment.

More information about this series at http://www.springer.com/series/5996

Amauri Rosenthal
Rosires Deliza
Jorge Welti-Chanes
Gustavo V. Barbosa-Cánovas
Editors

Fruit Preservation

Novel and Conventional Technologies

Springer

Editors
Amauri Rosenthal
Embrapa Agroindústria de Alimentos
Rio de Janeiro, RJ, Brazil

Rosires Deliza
Embrapa Agroindústria de Alimentos
Rio de Janeiro, RJ, Brazil

Jorge Welti-Chanes
Tecnológico de Monterrey
Monterrey, México

Gustavo V. Barbosa-Cánovas
Washington State University
Pullman, WA, USA

ISSN 1571-0297
Food Engineering Series
ISBN 978-1-4939-3309-9 ISBN 978-1-4939-3311-2 (eBook)
https://doi.org/10.1007/978-1-4939-3311-2

Library of Congress Control Number: 2017958452

This Springer imprint is published by the registered company Springer Science+Business Media, LLC, part of Springer Nature.
The registered company address is: 233 Spring Street, New York, NY 10013, U.S.A.

Preface

Fruits and fruit-based products are, in most cases, associated with very good sensory characteristics, health, well-being, and perishability and are relatively easy to mix with food products of diverse origin and amenable to be processed by conventional and novel technologies. Given the multiplicity of aspects whenever ***fruit preservation*** is considered, we took the challenge of covering in a thorough, comprehensive manner most aspects dealing with this topic.

To accomplish our goals, we invited well-known colleagues with expertise in specific disciplines associated with fruit preservation to contribute chapters to this book. Eighteen chapters were assembled in a sequence that would facilitate, like building blocks, to have at the same time, a bird's-eye view and an in-depth coverage of traditional and novel technologies to preserve fruits.

Even though processing took center stage in this book, ample space was dedicated to other relevant and timely topics on fruit preservation such as safety, consumer perception, sensory and health aspects. A brief summary of each chapter is provided in the following paragraphs.

Chapter 1 is centered on consumer perceptions of fruit products manufactured using novel technologies and how to increase consumer acceptance of these products. Among other topics, relevance is given to the processes involved in the formation of consumer attitudes towards new technologies, factors that influence these attitudes, strategies to change them, and the need to increase consumers' awareness and knowledge about new food processing technologies.

Chapter 2 addresses fruit and fruit products safety issues in terms of microbial contamination by the main pathogenic microorganisms as well as those contributing to product deterioration. Control measures that should be taken to prevent high levels of contamination of these undesirable microorganisms are also included.

Chapter 3 focuses on why the consumption of fruit products is widely recommended, i.e., rich source of micronutrients like vitamins; bioactive phytochemicals such as carotenoids, flavonoids. Quality changes of these valuable fruit components due to processing and storage are discussed, mainly those related to vitamin C, carotenoids, flavonoids, and folates.

In order to significantly reduce undesirable changes due to processing and overprocessing, minimal processing of foods is receiving considerable attention in the last few years. This approach, as applied to fruits, is presented and discussed in Chap. 4. Special attention is given to disinfection techniques followed by careful handling of the fruits to extend their shelf lives while keeping their freshness.

Chapter 5 covers in depth the combination of two or more technologies to treat fruits and fruit-based products, i.e., the use of the hurdle technology approach in this kind of products. This comprehensive chapter starts highlighting and justifying the increasing demand for fresh-like, minimally processed foods. This section is followed by an exhaustive analysis of the hurdle concept which exploits synergistic and additive interactions between sublethal stress factors to warrant safety and quality while reducing energy input and treatment intensity. Then, the most used hurdle combinations are presented, discussed, and summarized in a number of tables. High pressure, pulsed electric fields, ultraviolet, and high power ultrasound are some of the technologies that are combined.

Chapter 6 presents a thorough, in-depth review of a number of strategies to extend the shelf life of fruits and food products by exposing them to different temperatures. The authors are giving relevance to below room temperatures, i.e., cooling and freezing, but they are also covering controlled and modified atmosphere; selected thermal treatments, UV-C irradiation. All aspects of freezing are covered with great degree of detail including ice formation, fruit quality changes (during freezing and storage), and available equipment.

Chapter 7 covers systematically and in great detail different approaches to thermally dry fruits and food products. The author describes in a masterful way the fundamentals of a number of drying approaches and later on presents a good number of relevant applications which include two comprehensive tables and many meaningful illustrations.

Chapter 8 is devoted to reviewing in great detail the fundamentals and applications of the processing of fruit juices by membrane technologies. It includes a very extensive description of the best known ones like microfiltration, ultrafiltration, nanofiltration, reverse osmosis, electrodialysis, and pervaporation to later on emphasize those that are applicable to fruit juices, indicating which ones are in use and those that could be used in the near future. This chapter includes numerous and very useful tables summarizing and highlighting a number of relevant aspects in the usage of these technologies as applied to fruit juices.

Chapter 9 is devoted to analyze factors that need to be taking into account to develop reliable decision-making tools that will lead to optimize the modified atmosphere packaging (MAP) of fruits and food products. These tools need to take into account the needs of the produce as well as constraints and wishes of the stakeholders such as biodegradability of the packaging material and costs. The authors present in detail one of those tools that rely on the creation of a database on fresh fruits (optimal storage conditions, respiration, transpiration, etc.) to feed mathematical models in the MAP optimization step and a database on packaging materials (gases and vapor transfer rates, permeance, permeability, etc.) that are coupled with stakeholder's requests. The chapter includes how to construct the databases, how to optimize MAP, and what computing and statistical methods are needed to process all the gathered information.

Chapter 10 is an in-depth review of the most popular frying of foods approaches including classic frying, deep-fat frying, and vacuum frying. Special attention is given to snack manufacturing, oil absorption kinetics, effect of processing conditions on food and oil quality, acrylamide formation, structure and sensory characteristics of fried products, and industrial equipment. Many examples of a variety of fried foods including fruits and vegetables are presented and analyzed. Similar products fried by different techniques are compared in terms of overall quality.

Chapter 11 is focused on the use of ultrasound, more specifically on power ultrasound (20–100 kHz) which has different applications to high-frequency ultrasound (20–100 MHz). The latter is used for nondestructive inspection and identification of food composition. The first part of this chapter is dedicated to presenting how ultrasound is generated and why it is an effective technique suitable for preservation. It is followed by sections detailing how it is used in juice processing, surface decontamination, postharvest quality enhancement, as a drying aid, extraction of selected and valuable compounds, blanching, and pest control. The combination of ultrasound with thermal and moderate pressures is also described and analyzed.

Chapter 12 deals with vacuum impregnation, a very challenging and good alternative to preserve fruits, to enhance nutritional value, and to develop creative fruit-based products. Among the topics covered, there is an analysis of the mass transfer taking place in this process, which includes the role of the physical properties and characteristics of the fruits as well as processing pressure and temperature. The incorporation of a number of impregnants such as salts, sugars, minerals, phenolic compounds, vitamins, and microorganisms and their impact on the treated fruits is thoroughly discussed.

Chapter 13 is devoted to high pressure processing (HPP) of fruits and fruit products, a technology that is receiving significant attention from the food industry to offer high-quality products. To the best of our knowledge, this chapter is one of the most comprehensive review on the subject. It includes a brief *racconto* on high pressure since it was introduced in 1898 as a possible technology to process foods followed by a thorough description of the most common HHP equipment used at the industrial level. Then, there is an extensive review on how HPP is used to preserve a number of fruit products such as fresh-cut, dried fruits, juices, nectars, pastes, and purées. Finally, there is an exhaustive analysis of the effects of HPP on selected microorganisms, enzymes, and bioactive compounds.

The safety and quality of irradiated fruits and vegetables is the subject of Chap. 14. Irradiation is a controversial technology which is gaining acceptance in some parts of the world. The three most used technologies—Electron Beam, Gamma Ray, and X-ray—are extensively covered providing an excellent picture of where this technology currently stands. Applications, advantages and disadvantages, and treatment dose ranges for a number of fruits and vegetables are listed into very comprehensive tables. Other topics covered include the mode of operation, shelf life extension, sensory aspects, regulations, and packaging.

Chapter 15 is dedicated to the use of microwaves to treat fruits and fruit products in a number of unit operations such as blanching, drying, and thermal processing. At the beginning, the authors cover in great detail the fundamentals of this technology and its advantages and disadvantages, and they include as well the dielectric properties of a great number of fruits. At the end, there are case studies on selected fruits treated by this very promising technology which is rapidly growing where applications such as pasteurization and sterilization are taking center stage.

Ohmic heating and pulsed electric fields are two technologies that have been adopted by the food industry for some very relevant applications. At the same time, R&D efforts to facilitate implementation at large scale are quite intense. The fundamentals of these two technologies and their use in the processing of fruits are the subject of Chap. 16. The current status of ohmic heating and pulsed electric fields, advantages and disadvantages of these technologies, as well as potential new applications to treat fruits are thoroughly discussed.

Chapter 17 deals with continuous and pulsed UV light for processing fresh fruits and fruit products. Fundamentals and features of UV light generation, propagation, and evaluation of UV light parameters are reviewed as well as the latest applications. Good part of the chapter is dedicated to analyzing the effects of UV light on the survival of pathogenic and spoilage microorganisms that are typically present in fruits.

Chapter 18 describes and analyzes how ozone can be applied at almost any step in the fruit supply chain. The authors state that in addition to improving fruit safety and extending product shelf life, ozone treatments may also be selected to enhance the nutritional quality of fruits as well as remove residues of pesticides. They also mention that the potential benefits of ozone in fruit processing seem very promising; therefore, the adoption of ozone is likely to continue growing within the fresh fruit industry. This chapter depicts in detail the state of the art of ozone as applied to fruit processing covering a vast number of aspects such as when and how to apply it and how it could be used in the fruit supply chain. The application of this technology in a variety of fruits such as pomes, berries, melons, oranges, tangerines, kiwi fruit, and figs is meticulously analyzed.

Fruit Preservation will serve as an excellent text or reference book to graduate and undergraduate students to learn the state of the art of this challenging, relevant topic. At the same time, since the book covers a vast area of research, development, and applications, it will also serve as a good reference to food industry professionals and practitioners, in particular to those involved in the processing of fruits and fruit products. The book will be equally important to food safety specialists and process authorities in both the government and food industry. Moreover, it will be a valuable reference for authorities involved in the import and export of fruits and fruit products.

The editors are very thankful to the 49 authors for sharing their expertise, experience, and vision to come up with very valuable chapters to make the whole book project an excellent reference on ***Fruit Preservation***. The editors are aware of some overlaps between a few chapters, and this is inevitable in a book of this magnitude, but this will help to visualize basic concepts from different angles for the benefit of the readers in this rapidly evolving field. Gratitude is also extended to all the reviewers who contribute their time and expertise to make better each chapter.

We hope this book will become a worthy addition to the body of knowledge on ***Fruit Preservation*** and readers will find in it balanced, systematic, and harmonized information.

Rio de Janeiro, RJ, Brazil	Amauri Rosenthal
Rio de Janeiro, RJ, Brazil	Rosires Deliza
Monterrey, México	Jorge Welti-Chanes
Pullman, WA, USA	Gustavo V. Barbosa-Cánovas

Contents

Contributors

Jaime Amaya-Farfan University of Campinas, Campinas, SP, Brazil

Stella Maris Alzamora Universidad de Buenos Aires, Buenos Aires, Argentina

Gastón Ares Universidad de la República, Pando, Uruguay

Gustavo V. Barbosa-Cánovas Washington State University, Pullman, WA, USA

Marie-Pierre Belleville Université de Montpellier, Montpellier, France

Pedro Bouchon Pontificia Universidad Católica de Chile, Santiago, Chile

Patrice Buche INRA, Montpellier, France

Alicia Chaves CIDCA and Universidad Nacional de La Plata—CONICET, La Plata, Buenos Aires, Argentina

Rosires Deliza Embrapa Agroindústria de Alimentos, Rio de Janeiro, RJ, Brazil

Cynthia Ditchfield University of São Paulo, Pirassununga, SP, Brazil

Manuel Dornier Montpellier SupAgro, Montpellier, France

Verónica Dueik Pontificia Universidad Católica de Chile, Santiago, Chile

Zamantha Escobedo-Avellaneda Tecnológico de Monterrey, Monterrey, Mexico

Hao Feng University of Illinois at Urbana—Champaign, Champaign, IL, USA

Rébeca García-García Tecnológico de Monterrey, Monterrey, Mexico

Nathalie Gontard Université Montpellier 2, Montpellier, France

Barbara Gouble INRA, Avignon, France

Université d'Avignon et des Pays de Vaucluse, Avignon, France

Sandra Norma Guerrero Universidad de Buenos Aires, Buenos Aires, Argentina

José Ángel Guerrero-Beltrán Universidad de las Américas-Puebla, Cholula, Puebla, Mexico

Valérie Guillard Université Montpellier 2, Montpellier, France

Carole Guillaume Université Montpellier 2, Montpellier, France

David R. Kasler The Ohio State University, Columbus, OH, USA

Tatiana Koutchma Agriculture and Agri-Food Canada, Guelph, ON, Canada

Hyoungill Lee University of Illinois at Urbana—Champaign, Champaign, IL, USA

Aurelio López-Malo Universidad de las Americas—Puebla, Cholula, Puebla, Mexico

Olga Martín-Belloso Department of Food Technology, University of Lleida, Lleida, Spain

Antonio Martínez Instituto de Agroquímica y Tecnología de Alimentos (IATA-CSIC), Valencia, Spain

Katia Nicolau Matsui Federal University of Rio Grande do Norte, Natal, Brazil

Mariana Morales-de la Peña Tecnológico de Monterrey, Monterrey, Mexico

Hugo Mújica-Paz Tecnológico de Monterrey, Monterrey, Mexico

Brendan A. Niemira USDA-ARS Eastern Regional Research Center, Wyndmoor, PA, USA

Marta Orlowska Agriculture and Agri-Food Canada, Guelph, ON, Canada

Prashant Raj Pokhrel Washington State University, Pullman, WA, USA

Elisa Helena da Rocha Ferreira Federal Rural University of Rio de Janeiro, Seropédica, RJ, Brazil

Dolores Rodrigo Instituto de Agroquímica y Tecnología de Alimentos (IATA-CSIC), Valencia, Spain

Delia B. Rodriguez-Amaya University of Campinas, Campinas, SP, Brazil

Amauri Rosenthal Embrapa Agroindústria de Alimentos, Rio de Janeiro, RJ, Brazil

Henry T. Sabarez CSIRO, Werribee, VIC, Australia

Carmen Cecilia Tadini University of São Paulo, São Paulo, SP, Brazil

María Soledad Tapia Universidad Central de Venezuela, Caracas, Venezuela

Julia Hauck Tiburski Federal Rural University of Rio de Janeiro, Seropédica, RJ, Brazil

Fabrice Vaillant CIRAD, Montpellier, France

Aurora Valdez-Fragoso Tecnológico de Monterrey, Monterrey, Mexico

Jorge Welti-Chanes Tecnológico de Monterrey, Monterrey, Mexico

Ahmed E. Yousef The Ohio State University, Columbus, OH, USA

Surama F. Zanini Espirito Santo Federal University, Alegre, ES, Brazil

Noemí Zaritzky CIDCA and Universidad Nacional de La Plata—CONICET, La Plata, Buenos Aires, Argentina

Bin Zhou University of Illinois at Urbana—Champaign, Champaign, IL, USA

Yan Zhu Agriculture and Agri-Food Canada, Guelph, ON, Canada

Chapter 1
Consumer Perception of Novel Technologies

Rosires Deliza and Gastón Ares

1.1 Introduction

Consumption of fruits is recommended as part of a healthy diet and has been associated with positive health effects, such as decreased risk of heart diseases and some types of cancer (FAO/WHO 2004; Habauzit et al. 2013; Wang et al. 2011; Wooton-Beard and Ryan 2011). However, consumption of these products is still below recommendations (Pomerleau et al. 2004; Shaikh et al. 2008; World Health Organization 2003). Lack of availability, effort of preparation, price, quality deterioration in the supply chain, as well as pesticide residues are among the barriers for fruit consumption (Yeh et al. 2010, 85–98; Williams and Hammit 2001). In this context, the fruit industry faces opportunities for the development of new products that address consumer concerns and meet their demands, which can also help to differentiate products in a highly competitive marketplace (Jaeger et al. 2011).

New technologies provide the fruit sector opportunities for the development of new products that meet consumer demands (Onwezen and Bartels 2011). These technologies have several advantages over conventional thermal processes, providing safer, healthier, and more nutritious food, with a minimal modification of their sensory characteristics, while using less energy, water, and chemicals and producing less waste (Knorr 1999). However, these advantages do not assure consumer acceptance, which is the main determinant of the success of new technologies (Siegrist 2008).

R. Deliza (✉)
Embrapa Agroindústria de Alimentos, Rio de Janeiro, RJ, Brazil
e-mail: rosires.deliza@embrapa.br

G. Ares
Universidad de la República, Pando, Uruguay

A. Rosenthal et al. (eds.), *Fruit Preservation*, Food Engineering Series,
https://doi.org/10.1007/978-1-4939-3311-2_1

Food choice is a complex phenomenon, which depends on several interrelated factors (Köster 2009). Food not only provide the necessary nutrients but are part of a wider psychological, social, and cultural setting (Rozin 2005). Consumer decisions on what food to eat depend on characteristics of the food product (sensory and non-sensory characteristics), characteristics of the person making the choice (physiology, attitudes, experiences, knowledge, etc.), and characteristics of the context in which the choice is made (place, time, social context, culture, etc.) (Furst et al. 1996). For this reason, consumer acceptance of new technologies is far from being unconditional, being dependent on a wide range of interrelated factors.

Throughout history, consumers have been suspicious about several new food technologies, being reluctant to accept canned, irradiated, and genetically modified food when first introduced to the marketplace (Young 2003). Consumers reject new technologies for different reasons, including safety, ethical, and environmental concerns (Ronteltap et al. 2007).

Lack of insight into consumer perception has been identified as one of the main determinants of the high failure rate of a large proportion of the new food products launched into the market and their withdrawal within short times (Stewart-Knox and Mitchell 2003; van Trijp and Steenkamp 2005, 87–124). Therefore, understanding consumer perception of new technologies is crucial for their success, as it can contribute to the identification of potential barriers to consumer acceptance and the design of strategies to encourage their adoption (Ronteltap et al. 2007).

In this context, the aims of the present chapter are to provide an overview of the main determinants of consumer perception of fruit products manufactured using new technologies and to discuss strategies to increase consumer awareness and acceptance of these products.

1.2 Consumer Attitudes Towards New Technologies

According to the theory of planned behavior, given availability of opportunities and resources, people's decision to perform a behavior is determined by their intentions (i.e., how much effort they are willing to invest to perform such behavior) and their perceived behavioral control (people's perception of how easy it would be for them to perform the behavior) (Ajzen 1988, 1991).

Intentions are determined by attitudes, subjective norms, and perceived behavioral control (Ajzen and Madden 1986). Attitudes are related to the degree to which people have positive or negative appraisals of the behavior, whereas subjective norms refer to perceived social pressure to perform or avoid the behavior (Ajzen 1988). The more positive the attitudes and subjective norms towards a behavior, the stronger will be a person's intention to perform that particular behavior. A graphical representation of the theory of planned behavior is shown in Fig. 1.1.

Several studies have relied on the theory of planned behavior to understand consumer acceptance of new technologies (Bredahl 2001; Frewer et al. 1997; Olsen et al. 2010; Siegrist 2000). These studies have shown that consumer willingness to accept new food technologies is strongly influenced by their attitudes.

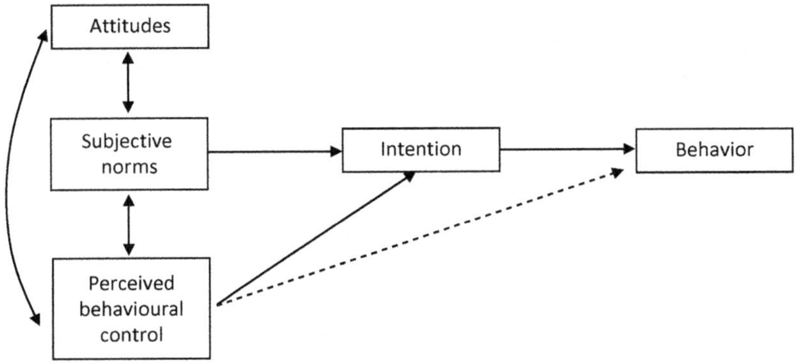

Fig. 1.1 Graphical representation of the theory of planned behavior (adapted from Ajzen 1991)

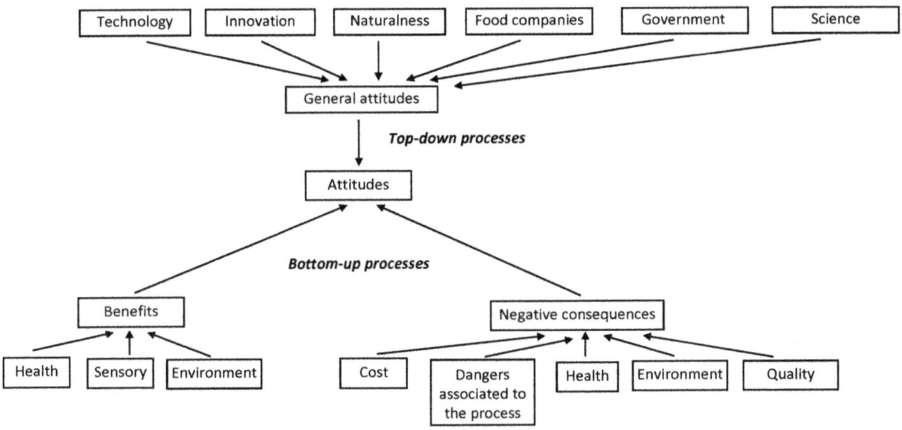

Fig. 1.2 Schematic representation of the processes involved in the formation of attitudes towards new technologies

Attitudes are formed through the evaluation of available information using both existing schemas stored in long-term memory and schemas created from contextualization of information (Greehy et al. 2013). Two main processes, which are simultaneous and not mutually exclusive, have been identified: bottom-up and top-down (Eagly and Chaiken 1993; Scholderer and Frewer 2003). Figure 1.2 presents a summary of the characteristics that have been reported to determine attitudes towards new technologies through bottom-up and top-down processes.

Bottom-up processes imply that attitudes towards new technologies are formed through a cognitive and affective evaluation of their characteristics (Fishbein 1963). Consumers use their prior knowledge and available information to estimate the benefits and negative consequences of new technologies, considering how they differ from conventional production methods (Grunert et al. 2004b). Although new technologies have several different characteristics, only a limited number of benefits and potential consequences are salient in determining people's attitudes (Miller 1956). Thus, consumers perform a weighted evaluation of these salient characteristics to form their attitudes (Ajzen and Fishbein 1970; Fishbein 1963). Therefore, consumers will hold positive attitudes towards new technologies if they are mainly associated with benefits and positive consequences, and negative attitudes if they are associated with negative or undesirable consequences (Ajzen 1991).

Three main benefits have been regarded as relevant for the formation of attitudes towards new technologies: health, sensory, and environmental issues (Cardello 2003; Cardello et al. 2007; Olsen et al. 2011; Onwezen and Bartels 2011; Sonne et al. 2010).

One of the most relevant benefits of new technologies, such as high-pressure processing (HPP) and pulsed electric fields (PEF), is that they preserve the products' nutritional value, particularly their vitamin content (Nielsen et al. 2009; Sonne et al. 2010). The fact that products are manufactured with HPP and do not contain added preservatives also contributes to healthfulness perception (Deliza et al. 2005; Sonne et al. 2010). According to Butz et al. (2003), European consumers may be willing to buy products manufactured with new technologies if they have an associated health benefit. Laboissiere et al. (2007b) also reported that Brazilian consumers expected to like more a passion fruit juice when the information on preserving the vitamins and maintaining fresh-fruit like flavor of the product were presented on the package. Similarly, Abadio-Finco et al. (2010) found that consumers stated a higher intention to purchase for pineapple juice processed by HPP when the package contained the following claim: *Nutritious and with more flavor. The high pressure technology keeps the flavor and preserves the vitamins.*

These above-mentioned results are in accordance with the fact that consumers have also highlighted the sensory characteristics of the products as a major benefit of new technologies (Cardello et al. 2007). Sonne et al. (2010) reported that consumers imagined that HPP and PEF fruit juices would taste like fresh fruits, which led to more enjoyment and contributed to having more pleasure in life. According to Nielsen et al. (2009), sensory quality is a key determinant of consumer acceptance and commercial marketability of food products manufactured using new technologies. In this sense, several tropical fruit juices such as pineapple, *pitanga* (Brazilian cherry, *Eugenia uniflora* L.), passion fruit, mango, papaya, and yellow mombin (*Spondias mombin* L.), all of them processed by HPP, were evaluated by Brazilian consumers and achieved higher liking scores than their thermally treated counterparts (commercial products available in the market) (Barros et al. 2007; Ferreira 2013; Laboissiere et al. 2007a; Pontes et al. 2008; Shinagawa et al. 2013; Tiburski et al. 2009). Besides, the sensory description of the pressurized juices revealed products with characteristics closer to the fresh fruits, suggesting that this particular attribute has driven consumer liking. Figure 1.3 shows an example of these findings. Six pineapple juices were considered: four thermally processed commercial brands available in the market (C, D, E, F), one pressurized (HHP), and one from the "in natura" pulp (IN). The juices were evaluated by a trained panel using Quantitative Descriptive Analysis, i.e., the assessors evaluated the intensity of a series of sensory attributes using scales. Data were analyzed using Principal Component Analysis (PCA) on the average scores of the evaluated attributes. Results revealed that the sensory characteristics of samples IN and HHP were similar, as they are close in the first two principal components of the PCA (Fig. 1.3a). They were mainly characterized by the sensory attributes that were correlated with the

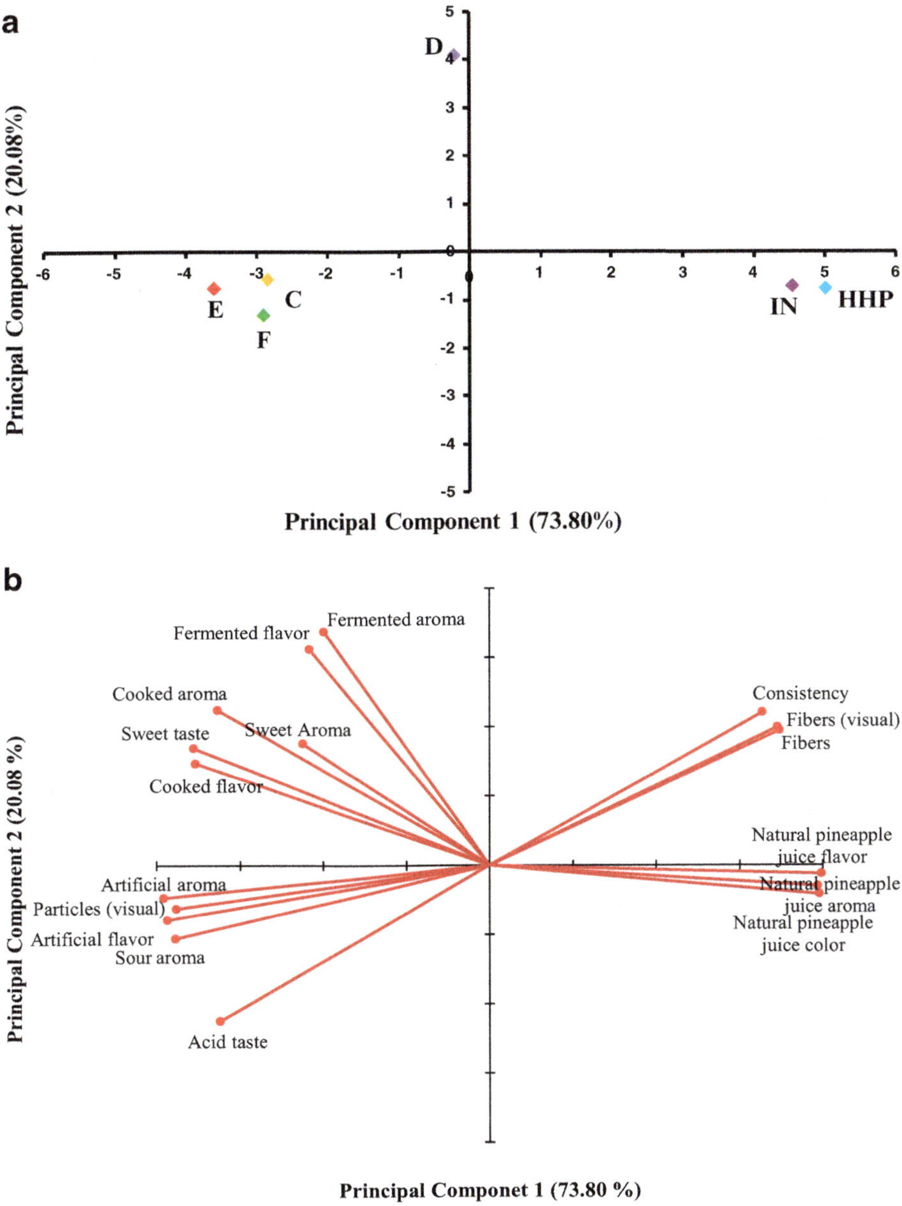

Fig. 1.3 Quantitative descriptive analysis of pineapple juices with different processing conditions. Results of the principal component analysis (PCA) performed on average scores for the evaluated sensory attributes. (**a**) Representation of the samples; and (**b**) Sensory attributes in the first two principal components of the PCA. *IN: in natura* pineapple juice, *HHP:* pressurized pineapple juice, *C – F:* thermally treated commercial juices (Barros et al. 2007)

first principal component of the PCA: *natural pineapple juice flavor, aroma,* and *color* (Fig. 1.3b). On the other hand, the thermally processed juices C, E and F, located at negative values of the first principal component (Fig. 1.3a), were characterized by their *sour taste, sour aroma, artificial pineapple juice aroma, particle presence,* and *artificial pineapple flavor* (Fig. 1.3b). Finally, commercial juice D was located at positive values of the second component and was characterized by its fermented flavor and aroma, its sweet taste, and cooked flavor (Barros et al. 2007).

New technologies are perceived as environmentally friendly due to the fact that they use less energy and produce less waste than conventional processes (Cardello et al. 2007; Nielsen et al. 2009). Environmental issues are relevant for consumers when choosing products manufactured using new technologies over conventional products because they feel responsible for nature, future generation, and mankind in general (Sonne et al. 2010).

Lack of perceived benefits has been associated with concerns about the need for and usefulness of novel technologies, as well as an increase in perceived negative consequences (Gaskell 2000). Negative consequences of new technologies are mainly related to their increased price, dangers in the processes, as well as potential negative effects on the environment, quality, and health (Fig. 1.2).

An increase in price has been reported to be a potential barrier for consumer acceptance of new technologies. Consumers are usually not willing to pay a premium price for these products, as highlighted by Butz et al. (2003), who reported that British and German consumers were reluctant to buy products manufactured with new technologies if they were more expensive than conventional products. However, Cardello et al. (2007) reported that cost was the least important barrier for the acceptance of new food technologies. In this sense, it is important to highlight that price has been reported to influence consumer purchase intention in two opposite ways: it could reduce purchase intention due to a greater monetary sacrifice, or it could encourage purchase intention of an increase in perceived product quality (Jaeger 2006). Some studies have shown a positive effect of price on consumers' willingness to consume products produced using new technologies. French participants were willing to pay more for products manufactured with new technologies if they are associated with an increased quality (Butz et al. 2003). Also, Nielsen et al. (2009) reported that higher price was perceived as a benefit of new technologies when applied to baby food, due to the fact that parents want high quality products for their children. Laboissiere et al. (2007b) identified two segments of consumers, which gave different importance to price. One of them expected to like more a low priced HPP-passion fruit juice in relation to the second segment that preferred the package with a more expensive price.

According to Cardello et al. (2007), the potential risks of new technologies are the most relevant factors in shaping consumers' attitudes and their intention to use these types of products. Risks perceived as most threatening are those that are perceived as unknown, involuntary, unobservable, out of consumers' control, and that are associated with delayed and potentially fatal health effects (Slovic 1987). This is usually the case of new food technologies (Cardello 2003; Olsen et al. 2010; Ronteltap et al. 2007). Consumers are usually not aware of the processes applied to food products and cannot reverse their effect once applied, which increases risk perception of new technologies (Cardello 2003), although they have low risk from a technical standpoint (Fischoff et al. 1978).

Consumers have been extensively reported to be concerned about the potential health risks of irradiation, mainly because of fear that it can make food radioactive or led to the formation of harmful compounds (Frenzen et al. 2001; Gunes and Tekin 2006). In this sense, He et al. (2005) reported that a considerable proportion of consumers might try to avoid products labeled as irradiated because they would consider it a health warning.

Consumers also report concerns related to loss of quality, perception of inherent risks in the processes, and environmental risks, although they are less relevant than potential health risks (Cardello et al. 2007; Nielsen et al. 2009; Olsen et al. 2010).

As it has been said before, the acceptance of a new technology depends on several factors and the consumer himself—with his/her cultural, emotional, psychological background—plays an important role on the process. As an example, a choice-based conjoint study that has been carried out on irradiated papayas with consumers in Rio de Janeiro (Brazil) is presented. The findings demonstrated that the most important factor for consumer's intention to purchase was the appearance of the fruits (Deliza et al. 2009). This result suggests that for the Brazilian individuals who participated in the study, the use of irradiation was not a relevant determinant of their selection, when choosing papaya. However, another scenario appeared when these consumers were asked about *How concerned are you about using irradiation in food processing?* (Concern); and how much they agreed on three statements. The three statements were as follows: *Eating irradiated food is not a safe thing to do* (Irradiate is not safe); *Eating irradiated food is probably safer than eating non-irradiated food* (Irradiated safer than non-irradiated); *Eating irradiated food will increase my likelihood of experiencing health problems later* (Health problems later). Finally, participants were provided with information about food irradiation, which stated: "The irradiation is an efficient method for preserving the quality of the food and was approved by the Ministry of Health. When applied under controlled conditions it brings benefits to the consumer." After reading information about food irradiation, subjects rated their opinion about the technology (Opinion after information). Again, different perceptions were noticed, which were revealed by the existence of two consumer segments. Figure 1.4b shows the representation of the participants in the PCA performed on their responses. As it can be seen, respondents were widely distributed along the first two dimensions of the PCA, suggesting high heterogeneity in their responses. Two main segments were identified: Segment 1 and Segment 2. People in Segment 1 were more concerned than those in Segment 2. These participants perceived irradiation as more risky and agreed on the statement *Eating irradiated food will increase my likelihood of experiencing health problems later* (Fig. 1.4a).

Bottom-up processes fail to explain how consumers form attitudes towards new technologies when their knowledge and awareness is scarce (Olsen et al. 2010). For this reason, top-down processes have been reported to be the main determinant of consumers' attitudes towards new technologies (Scholderer and Frewer 2003;

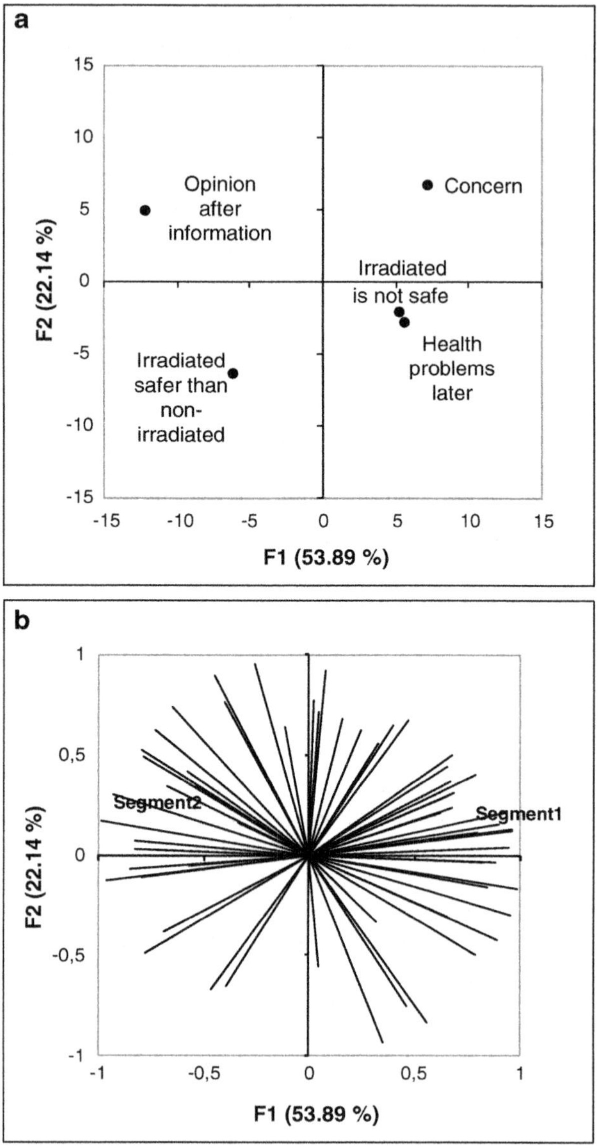

Fig. 1.4 Results of a PCA performed on consumers' responses to a set of questions about food irradiation: (**a**) representation of the questions (participants' concern, perceived health risks associated with eating irradiated products, and opinion) on the first two principal components; (**b**) representation of the consumers in the first two principal components (Deliza et al. 2009)

Siegrist and Cvetkovich 2000; Søndeargaard et al. 2005). Top-down processes imply that attitudes towards objects and behaviors are formed by classifying them with higher-order attitudes and values (Prislin et al. 1998). This means that in the absence of any factual or experiential knowledge, attitudes towards new technologies are formed by associating them with other general concepts. Thus, the mental representation of new technologies may be embedded into a multidimensional structure composed by a large number of interrelated concepts, which may determine attitudes (Olsen et al. 2010; Scholderer and Frewer 2003; Søndeargaard et al. 2005).

Attitudes towards food companies have been reported to be relevant in shaping consumers' attitudes towards new technologies. Consumers have been reported to be skeptical to the benefits of new technologies as they believe that they have been developed by the food industry to increase their profits and not to provide benefits to consumers (Nielsen et al. 2009).

Consumers' attitudes towards the role of science and technology in society have been regarded as relevant determinants of attitudes towards new food technologies (Matin et al. 2012). Consumers who have faith in science and technology and think that they contribute to improve standard of living are usually more willing to consume products manufactured with these technologies. On the contrary, consumers who are more concern with nature and the environment usually reject new technologies and prefer natural fresh products (Mireaux et al. 2007; Olsen et al. 2010).

Trust in scientists, policy makers, and governmental organizations provides sense of protection against potential risks and decrease negative associations (Greehy et al. 2013).

Attitudes towards Innovation have been positively correlated to acceptance of new technologies. Consumers who have a positive impression of innovation and regard it as positive for development and well-being are usually more willing to accept new technologies than those that bear negative associations with innovation and technology (Nielsen et al. 2009). As a consequence, it is relevant to stress that great efforts should be concentrated at the initial development steps. Knowing consumer needs and offering products that meet their expectations is an important factor for the success of product development (Costa and Jongen 2006; Saguy and Moskowitz 1999; Urban and Hauser 1993; van Kleef et al. 2005).

Top-down formation of attitudes may explain differences in the acceptance of different technologies. New food technologies may be linked to other technologies. Food irradiation raises associations with the application of radioactive materials in other fields, which may explain the negative attitude towards this technology (Cardello 2003; Frenzen et al. 2001; Frewer et al. 2011; Gunes and Tekin 2006). PEF are usually associated with electricity, which raise negative attitudes related to danger and fear (Nielsen et al. 2009). However, HPP does not usually raise negative associations related to other technologies, being the technology most easily accepted by consumers (Cardello et al. 2007; Mireaux et al. 2007; Sonne et al. 2010).

In summary, research has shown that consumer acceptance of new technologies depends on whether consumers perceive that their benefits outweigh their negative associations, consequences, and potential risks. The trade-offs between perceived risks and negative consequences have been regarded as an important cue for the diffusion of innovations and may determine acceptance or rejection of new technologies (Frewer 2003; Greehy et al. 2013; Rogers 2003). Butz et al. (2003) reported that 90 % of the potential buyers of new HPP products perceived personal advantages, while 60 % of the non-buyers did not perceived personal advantages.

1.3 Factors That Influence Consumer Attitudes Towards New Technologies

Several individual variables have been reported to affect consumer attitudes towards new technologies. It has been shown that certain consumer groups are more willing to accept new food technologies (Cardello 2003).

Females have been reported to be more concerned, less likely to perceive benefits, and less willing to accept new food technologies than males (Cardello 2003; Cardello et al. 2007; Ronteltap et al. 2007). This difference can be related to the fact that females usually assign different meanings and values to food than males, due to their earlier involvement in food-related activities, which is enhanced by their active role in the provision of adequate food for their family (Rozin et al. 1999).

Regarding age, research has shown that older age groups are more concerned about technology-related food safety issues and new technologies than younger groups (He et al. 2005; Miles et al. 2004). Butz et al. (2003) reported that younger UK consumers were more willing to buy HPP orange juice than older consumers. However, other researchers have identified no differences in attitudes towards new technologies between age groups (Frenzen et al. 2001; Gunes and Tekin 2006).

Other demographic characteristics have not been reported to have a clear effect on attitudes towards new food technologies (Lyndhurst 2009). In a study conducted in Canada, Henson et al. (2007) reported that highly educated people and those with higher income tended to be more concerned about new technologies than the rest. Similarly, Frenzen et al. (2001) reported that higher education and income level positively influenced willingness to consume irradiated food. However, Butz et al. (2003) reported that European consumers with higher education qualification were more positive about buying HPP orange juice.

Consumer differences in general attitudes and values have been reported to have a larger influence than socio-demographic variables on acceptance of new food technologies (Cardello 2003; Lyndhurst 2009; Ronteltap et al. 2007). Attitudes towards technology and social trust vary across countries and have been related to differences in attitudes towards new food technologies (Bruhn et al. 1987; Lampila and Lahteenmaki 2007; Poppe and Kjaernes 2003; Siegrist and Cvetkovich 2000). Trust in the food industry, science, and governmental organizations has also been associated with a higher likelihood of accepting new food technologies (Bord and Conner 1990). Furthermore, consumers with strong pro-environmental values are usually more concerned about new food technologies (Bruhn et al. 1987; Cardello 2003).

Knowledge is also expected to affect attitudes towards new technologies, decrease risk perception, and increase willingness to consume products manufactured with these technologies (Bouyer et al. 2001; Ronteltap et al. 2007; Siegrist 1998). According to Rimal et al. (2004), knowledge about safety was positively correlated to consumer intention to consume irradiated products (Rimal et al. 2004).

Other more enduring psychological attitudes, such as universalism and hedonism, could also potentially affect attitudes and willingness to consume food produced using new technologies (Honkanen and Verplanken 2004). Furthermore, consumers' adoption of new food technologies has been related to specific personality traits such as food neophobia, which can be regarded as the reluctance to try novel food (Pliner and Hobden 1992) and has been related to adoption of innovations (Cox and Evans 2008; Ronteltap et al. 2007; Schnettler et al. 2013).

Finally, consumer attitudes towards new technologies are also dependent on the specific technology and the food product to which the technologies are applied. Consumers have been extensively reported to have negative attitudes towards genetically modified organisms and irradiation, whereas attitudes towards HPP and PEF have been reported to range from neutral to slightly positive (Cardello 2003; Lyndhurst 2009; Cardello et al. 2007; Sonne et al. 2010). Besides, HPP products are generally easier to accept than PEF products (Butz et al. 2003; Mireaux et al. 2007; Nielsen et al. 2009; Olsen et al. 2011). Regarding the product to which new technologies are applied, consumers are, in general, more negative towards the application of new technologies to animal products than to fruit and vegetables (Cardello et al. 2007; Funcane and Holup 2005; Lyndhurst 2009; Moses 1999). This suggests that the application of new technologies to fruit products holds a great potential.

1.4 Strategies for Changing Consumer Attitudes Towards New Technologies

Consumer knowledge and awareness of new technologies is very limited; in fact a large percentage of consumers have never heard about them (Nielsen et al. 2009). According to Lampila and Lahteenmaki (2007) and Frenzen et al. (2000), lack of information is one of the main determinants of rejection to consume products manufactured with new technologies, suggesting that providing consumers with trustable information may be a valuable strategy to overcome any lack of confidence on the new technology.

Attitudes towards new technologies are mainly determined by top-down processes, which are highly resistant to change (Eagly and Chaiken 1993; Scholderer and Frewer 2003). However, attitudes towards new technologies may change when new beliefs about the technology are formed (Olsen et al. 2010). This emphasizes the importance of providing clear, understandable, and trustable information, as well as making it easy for consumers to try products manufactured with new technologies, before the introduction of products into the marketplace.

According to Tversky and Kahneman (1974), people employ simplified cognitive strategies for making judgments and taking decisions. For this reason, information processing and judgment are usually guided by affective reactions towards a stimulus (Zajonc 1980). Affective heuristics implies that if feelings generated by a behavior are positive, people will try to reproduce the behavior, whereas if it generates negative feelings actions and thoughts will try to avoid that behavior. Therefore, according to Finucane et al. (2000), if consumers first receive positive information about a new food technology, they will have a positive attitude and decrease their perceived risk, compared to a situation in which they receive information about its potential negative consequences.

Information about the benefits and safety of new technologies has been reported to have a positive influence on consumer acceptance and likelihood of purchase of food products manufactured using these processes (Bruhn 1995; Cardello et al. 2007; Frewer et al. 1996, 1997; Rollin et al. 2011; Schutz et al. 1989). Positive attitude towards HHP was observed among participants of several studies focusing on different fruit juices when information on the benefits was provided. Consumers valued the benefits HHP would bring (nutritional and sensory) to the product and stated a higher intention to purchase pressurized juices with such information (Abadio-Finco et al. 2010; Pontes et al. 2009). However, research has shown that information can have a negative influence on consumer perception when they have a stable negative attitude towards a new technology. O'Fallon et al. (2007) and Scholderer and Frewer (2003) reported that information about genetically modified products had a negative influence on consumer perception by making negative associations more salient in their mind. These results stress that information about new technologies should be positively framed and communicated from the early stages of their introduction into the marketplace.

Health and environmental benefits of new technologies are credence attributes, which involve a high level of uncertainty (Darby and Karni 1973). Consumers cannot directly evaluate these characteristics of products produced using new technologies and therefore they have to trust the information provided by producers or other governmental and non-governmental organizations (Jahn et al. 2005). For this reason, these benefits of new technologies can only become relevant in shaping consumers' attitudes if they are provided with accurate, understandable, and trustable information. In this sense, information sources have a crucial role in determining the efficacy of communication strategies. Public confidence in adequate risk assessment prior to the implementation of new technologies regulations is essential for consumer acceptance (Greehy et al. 2013).

According to Cardello et al. (2007), the method selected to communicate product information should be determined according to the consumer segment being targeted. Trust in the information source is a key factor that mediates the assimilation of new information (Bruhn 2008). In general, consumers tend to rely on the information provided by friends, family, or other people with personal significance (Mellman Group 2006).

The most trusted sources of information vary with culture. The food industry tends to be the least trusted information source because consumers are suspicious of the main drivers for the development of new technologies (Nielsen et al. 2009). For this reason, Deliza et al. (2003) stated that food companies in the UK could potentially risk their reputation by including unknown new technologies into the marketplace. On the contrary, Gunes and Tekin (2006) reported that Turkish consumers would be willing to consume food products manufactured using new technologies if well-known and trusted companies would produce them. Further studies in specific countries where new technologies may be available are recommended.

The media tends to be the most trusted source of information about new food technologies for Asian consumers; consumers in the USA rely on governmental organizations and policy makers, whereas European consumers tend to rely on independent non-governmental organizations and consumer groups (Chen and Li 2007; Lyndhurst 2009). According to Hayes et al. (2002), messages about the potential risks of new technologies sent by independent non-governmental organizations can have a larger effect on consumer attitudes than official information sources.

The media has been recognized as consumers' primary source of information about science and technology (Allan 2002; National Science Board 2010). It increases consumer's awareness of scientific issues and influence the formation of attitudes (Dudo et al. 2011). Frequent coverage in news media can make new technologies more easily accessible in consumers' mind and increase the relevance of specific issues in the formation of attitudes (Scheufele and Tewksbury 2007). Frenzen et al. (2000) attributed a decrease in consumers' willingness to buy irradiated food to the increase in media attention during the late 1990s.

Framing of the messages has been reported to have a large influence on consumer perception of some new technologies. How new technologies are presented in the media can potentially affect how they are understood and perceived by consumers (Scheufele 1999). Cardello (2003) reported that the term "ionizing energy" generates less negative associations and raises less concern than the word "irradiation." This result suggests that the term "ionizing energy" can be an interesting alternative for the design of information and marketing campaigns aimed at increasing consumer acceptance of irradiated food.

Including information about new technologies on the product label can raise awareness, increase perception of personal control over the consumption of products manufactured with these technologies, and improve consumer acceptance (Costa-Font et al. 2008; Rollin et al. 2011). Labels play a key role in attracting consumers' attention and providing information that influence consumer expectations and could largely determine their purchase intention (Moskowitz et al. 2009). When consumers have difficulty in selecting among several options of a specific product, they can use specific information from labels to make up their mind (Imm et al. 2012). In general, consumers and stakeholders support the inclusion of information about new technologies on food labels (Frewer et al. 2004; Landmark Europe 2009). Deliza et al. (2005) reported that including information on the benefits offered by high pressure technology on fruit juice packages had a positive influence on consumers' purchase intention. However, when consumers have negative attitudes towards a product, including salient and extensive information on food labels can negatively affect their purchase intent. Including extensive information about irradiation or genetically modified products has been regarded as a warning sign that encourages consumers to avoid consuming the product (O'Fallon et al. 2007).

One of the most relevant benefits that can be stressed in marketing campaigns of products manufactured using new technologies is to inform consumers about the improved sensory characteristics (Cardello et al. 2007). Therefore, letting consumers try products in shops or other places can contribute to the generation of a positive association and increase acceptance of new technologies through evaluative conditioning (Nielsen et al. 2009). Evaluative conditioning refers to changes in consumer acceptance of a product due to the pairing with a stimulus with positive or negative valence (De Houwer 2007). Thus, by pairing the concept of a new technology with a product that imparted a positive sensory experience, the positive affective reaction resulting from tasting the product can be transferred to the technology (Walther et al. 2011). In this sense, previous research has shown that attitudes to a new technology become more positive after trying products manufactured using that technology (Cardello 2003; Grunert et al. 2004a; Olsen et al. 2011; Terry and Tabor 1988). A qualitative in-home exploratory study showed that when a group of housewives were allowed to drink mango juices, pressurized juice was perceived as more natural and having more mango flavor than the market leader brand (Pontes et al. 2009).

1.5 Conclusions and Remaining Challenges

New technologies offer food companies several opportunities to deliver new fruit products that meet many of consumers' unmet demands. However, consumers' awareness and knowledge about new food technologies is limited. As a consequence, attitudes towards new technologies are in many instances negative due to negative associations with other higher-order concepts, which make consumers skeptical about their benefits. Hence, providing information about new technologies seems to be a key strategy for increasing consumers' acceptance of fruit products manufactured by means of these technologies. Information and communication campaigns should be performed at the early phases of the introduction of new technologies into the marketplace. Further research on the influence of information sources and framing of messages is necessary to assure successful communication of the benefits of new technologies. In this sense, building consumer trust in new technologies requires an integrated action from food producers and governmental and non-governmental organizations.

Consumer-oriented innovation is a key strategy that should be considered by the fruit industry when developing fruit products manufactured using new technologies. Considering that consumer perception depends on a wide range of sensory and non-sensory variables, research aiming at identifying the trade-offs that determine consumer choice of products manufactured using new technologies over conventional products could contribute to the development and marketing of successful products. This type of approach can be useful to identify barriers to consumer adoption of products produced using new technologies prior to substantial time and monetary investments.

References

Abadio-Finco, F.D.B., R. Deliza, A. Rosenthal, and C.H.O. Silva. 2010. The effect of extrinsic product attributes of pineapple juice on consumer intention to purchase. *Journal of International Food & Agribusiness Marketing* 22: 125–142.

Ajzen, I. 1988. *Attitudes, personality, and behavior*. Chicago: Dorsey Press.

———. 1991. The theory of planned behavior. *Organizational Behavior and Human Decision Processes* 50: 179–211.

Ajzen, I., and M. Fishbein. 1970. The prediction of behavior from attitudinal and normative variables. *Journal of Experimental Social Psychology* 6: 466–487.

Ajzen, I., and T.J. Madden. 1986. Prediction of goal-directed behavior: Attitudes, intentions, and perceived behavioral control. *Journal of Experimental Social Psychology* 22: 453–474.

Allan, S. 2002. *Media risk and society*. Philadelphia, PA: Open University Press.

Barros, A.M., R. Deliza, L.H.E.S. Laboissiére, L.M.A.Q. Camargo, and A. Rosenthal. 2007. Does high pressure really preserve pineapple juice sensory attributes? *High Pressure Bioscience and Biotechnology* 1: 302–307.

Bord, R.J., and R.E. Conner. 1990. Risk communication, knowledge and attitudes: Explaining reactions to a technology perceived as risky. *Risk Analysis* 10: 499–506.

Bouyer, M., S. Bagdassarian, S. Chaabanne, and E. Mullet. 2001. Personality correlates of risk perception. *Risk Analysis* 21: 457–465.

Bredahl, L. 2001. Determinants of consumer attitudes and purchase intentions with regard to genetically modified foods: Results of a cross-national survey. *Journal of Consumer Policy* 24: 23–61.

Bruhn, C.M. 1995. Strategies for communicating the facts on food irradiation to consumers. *Journal of Food Protection* 58: 213–216.

———. 2008. Consumer acceptance of food innovations. *Innovation: Management, Policy and Practice* 10: 91–95.

Bruhn, C.M., H.G. Schutz, and R. Sommer. 1987. Food irradiation and consumer values. *Ecology of Food and Nutrition* 21: 219–235.

Butz, P., E.C. Needs, A. Baron, O. Bayer, B. Geisel, B. Gupta, U. Oltersdorf, and B. Tauscher. 2003. Consumer attitudes to high pressure food processing. *Food, Agriculture & Environment* 1: 30–34.

Cardello, A.V. 2003. Consumer concerns and expectations about novel food processing technologies: Effects on product liking. *Appetite* 40: 217–233.

Cardello, A.V., H.G. Schutz, and L.L. Lesher. 2007. Consumer perceptions of foods processed by innovative and emerging technologies: A conjoint analytic study. *Innovative Food Science & Emerging Technologies* 8: 73–83.

Chen, M.-F., and H.-L. Li. 2007. The consumer's attitude toward genetically modified foods in Taiwan. *Food Quality and Preference* 18: 662–674.

Costa, A.I.A., and W.M.F. Jongen. 2006. New insights into consumer-led food product development. *Trends in Food Science & Technology* 17: 457–465.

Costa-Font, M., J.M. Gil, and B.W. Traill. 2008. Consumer acceptance, valuation of and attitudes towards genetically modified food: Review and implications for food policy. *Food Policy* 33: 99–111.

Cox, D.N., and G. Evans. 2008. Construction and validation of a psychometric scale to measure consumers' fears of novel food technologies: The food technology neophobia scale. *Food Quality and Preference* 19: 704–710.

Darby, M., and E. Karni. 1973. Free competition and the optimal amount of fraud. *Journal of Law and Economics* 16: 67–88.

De Houwer, J. 2007. A conceptual and theoretical analysis of evaluative conditioning. *Spanish Journal of Psychology* 10: 230–241.

Deliza, R., M.C. da Costa, A. Rosenthal, D. Hedderley, and L. Frewer. 2003. Acceptance of novel food technologies—Comparison between Brazilian and British consumers. FoodInfo Online Features, June 17, 2003. www.foodsciencecentral.com/fsc/ixid12294.

Deliza, R., A. Rosenthal, F.B.D. Abadio, C.H.O. Silva, and C. Castillo. 2005. Application of high pressure technology in the fruit juice processing: Benefits perceived by consumers. *Journal of Food Engineering* 67: 241–246.

Deliza, R., A. Rosenthal, D. Hedderley, and S.R. Jaeger. 2009. Consumer perception of irradiated fruit: A case study using choice-based conjoint analysis. *Journal of Sensory Studies* 25: 184–200.

Dudo, Anthony, Doo-Hun Choi, and Dietram A. Scheufele. 2011. Food nanotechnology in the news. Coverage patterns and thematic emphases during the last decade. *Appetite* 56: 78–89.

Eagly, A.H., and S. Chaiken. 1993. *The psychology of attitudes*. Fort Worth, TX: Harcourt Brace Jovanovitch.

FAO/WHO. 2004. *Fruit and vegetables for health. Joint FAO/WHO workshop on fruit and vegetables for health*. Kobe, Japan: FAO/WHO.

Ferreira, E.A.S. 2013. [Development of tropical *pitanga* (*Eugenia uniflora* L.) nectar using the high hydrostatic pressure: Microbiological, nutritional and sensory aspects]. Master (Food Sci. and Technology)—Federal Rural University of Rio de Janeiro.

Finucane, M.L., A. Alhakami, P. Slovic, and S.M. Johnson. 2000. The affect heuristic in judgments of risks and benefits. *Journal of Behavioral Decision Making* 13: 1–17.

Fischoff, B., P. Slovic, S. Lichenstein, S. Read, and B. Combs. 1978. How safe is safe enough? A psychometric study of attitudes towards technological risks and benefits. *Policy Studies* 9: 127–152.

Fishbein, M. 1963. An investigation of the relationships between beliefs about an object and the attitude toward that object. *Human Relations* 26: 233–240.

Frenzen, P., A. Majchrowicz, B. Buzby, B. Imhoff, and The FoodNet Working Group. 2000. Consumer acceptance of irradiated meat and poultry products. *Agriculture Information Bulletin* 757: 1–8.

Frenzen, P.D., E.E. DeBess, K.E. Hechemy, H. Kassenborg, M. Kennedy, K. McCombs, A. McNees, and The FoodNet Working Group. 2001. Consumer acceptance of irradiated meat and poultry in the United States. *Journal of Food Protection* 64: 2020–2026.

Frewer, L.J. 2003. Societal issues and public attitudes towards genetically modified foods. *Trends in Food Science and Technology* 14: 319–332.

Frewer, L.J., C. Howard, and R. Shepherd. 1996. The influence of realistic product exposure on attitudes towards genetic engineering foodstuffs. *Food Quality and Preference* 7: 61–67.

Frewer, L.J., C. Howard, D. Hedderley, and R. Shepherd. 1997. Consumer attitudes towards different food-processing technologies used in cheese production: The influence on consumer benefit. *Food Quality and Preference* 8: 271–280.

Frewer, L., J. Lassen, B. Kettlitz, J. Scholderer, V. Beekman, and K.G. Berdal. 2004. Societal aspects of genetically modified foods. *Food and Chemical Toxicology* 42: 1181–1193.

Frewer, L.J., K. Bergmann, M. Brennan, R. Lion, R. Meertens, G. Rowe, M. Siegrist, and C. Vereijken. 2011. Consumer response to novel agri-food technologies. Implications for predicting consumer acceptance of emerging food technologies. *Trends in Food Science & Technology* 22: 442–456.

Funcane, M.L., and J.L. Holup. 2005. Psychosocial and cultural factors affecting perceived risk of genetically modified food: An overview of the literature. *Social Science and Medicine* 60: 1603–1612.

Furst, T., M. Connors, C.A. Bisogni, J. Sobal, and L. Winter Falk. 1996. Food choice: A conceptual model of the process. *Appetite* 26: 247–266.

Gaskell, G. 2000. Agricultural biotechnology and public attitudes in the European Union. *AgBioforum* 3(2&3): 87–96.

Greehy, M.G., M.B. McCarthy, M.M. Henchion, E.J. Dillon, and S.N. McCarthy. 2013. Complexity and conundrums. Citizens' evaluations of potentially contentious novel food technologies using a deliberative discourse approach. *Appetite* 70: 37–46.

Grunert, K., T. Bech-Larsen, L. Lähteemäki, Ø. Ueland, and A. Åström. 2004a. Attitudes towards the use of GMOs in food production and their impact on buying intention: The role of positive sensory experience. *Agribusiness* 20: 95–107.

Grunert, K., H. Sondergaard, and J. Scholderer. 2004b. How can we know what we like when we don't understand it? Consumer attitude formation towards complex technical issues. Crossenz: QLK1-CT-2002-02208.

Gunes, G., and M.D. Tekin. 2006. Consumer awareness and acceptance of irradiated foods: Results of a survey conducted on Turkish consumers. *LWT - Food Science and Technology* 39: 443–444.

Habauzit, V., D. Milenkovic, and C. Morand. 2013. Vascular protective effects of fruit polyphenols. In *Polyphenols in human health and disease*, ed. R. Ross Watson, V. Preedy, and S. Zibadi. New York: Academic.

Hayes, D.J., J.A. Fox, and J.F. Shogren. 2002. Experts and activists: How information affects the demand for food irradiation. *Food Policy* 27: 185–193.

He, S., S. Fletcher, and A. Rimal. 2005. Unwillingness to consume irradiated beef. *Journal of Food Distribution Research* 36: 71–78.

Henson, S., M. Annou, J. Cranfield, J. Ryks, and D. Herath. 2007. Understanding consumer attitudes towards food technologies in Canada. International Food Economy Research Group working paper no. 13. Guelph, Canada: Department of Food, Agricultural and Resource Economics, University of Guelph.

Honkanen, P., and B. Verplanken. 2004. Understanding attitudes towards genetically modified food: The role of values and attitude strength. *Journal of Consumer Policy* 27: 401–420.

Imm, B.-Y., J.H. Lee, and S.H. Lee. 2012. Effects of sensory labels on taste acceptance of commercial food products. *Food Quality and Preference* 25: 135–139.

Jaeger, S.R. 2006. Non-sensory factors in sensory science research. *Food Quality and Preference* 17: 132–144.

Jaeger, S.R., L.G. Axten, A.G. Paisley, M.W. Wohlers, K.B. Marsh, F. Sullivan, and F.R. Harker. 2011. Developing models systems for testing the sensory properties and consumer acceptance of new fruit cultivars: The example of kiwifruit. *Food Quality and Preference* 22: 521–531.

Jahn, G., M. Schramm, and A. Spiller. 2005. The reliability of certification: Quality labels as a consumer policy tool. *Journal of Consumer Policy* 28: 53–73.

Knorr, D. 1999. Novel approaches in food-processing technology: New technologies for preserving foods and modifying function. *Current Opinion in Biotechnology* 10: 485–491.

Köster, E.P. 2009. Diversity in the determinants of food choice: A psychological perspective. *Food Quality and Preference* 20: 70–82.

Laboissiere, L.H.E.S., R. Deliza, A.M. Barros-Marcellini, A. Rosenthal, L.M.A.Q. Camargo, and R.G. Junqueira. 2007a. Effects of high hydrostatic pressure (HHP) on sensory characteristics of yellow passion fruit juice. *Innovative Food Science & Emerging Technologies* 8: 469–477.

———. 2007b. Food processing innovation: A case study with pressurized passion fruit juice. *Journal of Technology Management & Innovation* 2: 108–123.

Lampila, P., and L. Lahteenmaki. 2007. Consumers' attitudes towards high pressure freezing of food. *British Food Journal* 109: 838–851.

Landmark Europe. 2009. *Nanotechnology in consumer products: Challenges and opportunities for Europe*. Brussels, Belgium: Landmark Public Policy Advisers Europe.

Lyndhurst, B. 2009. *An evidence review of public attitudes to emerging food technologies. Social Science Research Unit*. London: Food Standards Agency.

Matin, A.H., E. Goddard, F. Vandermoere, S. Blanchemanche, A. Bieberstein, S. Marette, and J. Roosen. 2012. Do environmental attitudes and food technology neophobia affect perceptions of the benefits of nanotechnology? *International Journal of Consumer Studies* 362: 149–157.

Mellman Group. 2006. *Public sentiment about genetically modified food*. Washington, DC: Pew Initiative on Food and Biotechnology.

Miles, S., M. Brennan, S. Kuznesof, M. Ness, C. Ritson, and L.J. Frewer. 2004. Public worry about specific food safety issues. *British Food Journal* 106: 9–22.

Miller, O.A. 1956. The magical number seven plus or minus two: Some limits on our capacity for processing information. *Psychological Review* 63: 81–97.

Mireaux, M., D.N. Cox, A. Cotton, and G. Evans. 2007. An adaptation of repertory grid methodology to evaluate Australian consumers' perceptions of food products produced by novel technologies. *Food Quality and Preference* 18: 834–848.

Moses, V. 1999. Biotechnology products and European consumers. *Biotechnology Advances* 17: 647–678.

Moskowitz, H.R., M. Reisner, J.B. Lawlor, and R. Deliza. 2009. *Packaging research in food product design and development*. Ames, IA: Wiley-Blackwell.

National Science Board. 2010. Science and Engineering Indicators 2010 (Chapter 7). National Science Foundation. http://www.nsf.gov/statistics/seind10/c7/c7h.htm (Accessed April 29, 2010).

Nielsen, H., A.M. Sonne, K.G. Grunert, D. Banati, A. Pollák-Tóth, Z. Lakner, N.V. Olsen, T.P. Zontar, and M. Peterman. 2009. Consumer perception of the use of high pressure processing and pulsed electric field technologies in food production. *Appetite* 52: 115–126.

O'Fallon, M.J., D. Gursoy, and N. Swanger. 2007. To buy or not to buy: Impact of labeling on purchasing intentions of genetically modified foods. *International Journal of Hospitality Management* 26: 117–130.

Olsen, N.V., K.G. Grunert, and A.M. Sonne. 2010. Consumer acceptance of high-pressure processing and pulsed-electric field: A review. *Trends in Food Science & Technology* 21: 464–472.

Olsen, N.V., E. Menichelli, K.G. Grunert, A.M. Sonne, E. Szabó, D. Bánáti, and T. Næs. 2011. Choice probability for apple juice based on novel processing techniques: Investigating the choice relevance of mean-end-chains. *Food Quality and Preference* 22: 48–59.

Onwezen, M.C., and J. Bartels. 2011. Which perceived characteristics make product innovations appealing to the consumer? A study on the acceptance of fruit innovations using cross-cultural consumer segmentation. *Appetite* 57: 50–58.

Pliner, P., and K. Hobden. 1992. Development of a scale to measure the trait of food neophobia. *Appetite* 19: 105–120.

Pomerleau, J., K. Lock, M. McKee, and D.R. Altmann. 2004. The challenge of measuring global fruit and vegetable intake. *Journal of Nutrition* 134: 1175–1180.

Pontes, M.M.M., R. Deliza, J.G. Santos, and A. Rosenthal. 2008. Effect of high hydrostatic pressure on mango (*Mangifera indica*, L. (Ubá)) juice sensory characteristics. In *XXI Congresso Brasileiro e Ciência e Tecnologia de Alimentos, XV Seminário Latino Americano e do Caribe de Ciência e Tecnologia de Alimentos, Belo Horizonte, Brazil*.

Pontes, M.M.M., R. Deliza, A. Rosenthal, and L. Casotti. 2009. Consumption habits, preference and labeling of ready-to-drink juice: An exploratory study on pressurized mango juice. In *8th Pangborn Sensory Science Symposium, Florence, Italy*.

Poppe, C., and U. Kjaernes. 2003. *Trust in food in Europe—A comparative analysis*. Professional report no. 5–2003. Oslo, Norway: National Institute for Consumer Research.

Prislin, R., W. Wood, and G.J. Pool. 1998. Structural consistency and the deduction of novel from existing attitudes. *Journal of Experimental Social Psychology* 34: 66–89.

Rimal, A.P., K.H. McWatters, I.B. Hashim, and S.M. Fletcher. 2004. Intended vs. actual purchase behavior for irradiated beef: A simulated supermarket setup (sss) experiment. *Journal of Food Products Marketing* 10: 1–15.

Rogers, E.M. 2003. *Diffusion of innovations*. New York: Free Press.

Rollin, F., J. Kennedy, and J. Wills. 2011. Consumers and new food technologies. *Trends in Food Science & Technology* 22: 99–111.

Ronteltap, A., J.C.M. van Trijp, R.J. Renes, and L.J. Frewer. 2007. Consumer acceptance of technology-based food innovations: Lessons for the future of nutrigenomics. *Appetite* 49: 1–17.

Rozin, P. 2005. The meaning of food in our lives. A cross-cultural perspective on eating and well-being. *Journal of Nutrition Education and Behavior* 27: S107–S112.
Rozin, P., C. Fischler, S. Imada, A. Sarubin, and A. Wrzesniewski. 1999. Attitudes to food and the role of food in life in the U.S.A., Japan, Flemish Belgium and France. Possible implications for the diet-health debate. *Appetite* 33: 163–180.
Saguy, I.S., and H.R. Moskowitz. 1999. Integrating the consumer into new product development. *Food Technology* 53: 68–73.
Scheufele, D.A. 1999. Framing as a theory of media effects. *Journal of Communication* 49: 103–122.
Scheufele, D.A., and D. Tewksbury. 2007. Framing, agenda setting, and priming. The evolution of three media effects models. *Journal of Communication* 57: 9–20.
Schnettler, B., G. Crisóstomo, J. Sepúlveda, M. Mora, G. Lobos, H. Miranda, and K.G. Grunert. 2013. Food neophobia, nanotechnology and satisfaction with life. *Appetite* 69: 71–79.
Scholderer, J., and L. Frewer. 2003. The biotechnology communication paradox: Experimental evidence and the need for a new strategy. *Journal of Consumer Policy* 26: 125–157.
Schutz, H.G., C.M. Bruhn, and V. Diaz-Knauf. 1989. Consumer attitude toward irradiated foods: Effects of labeling and benefit information. *Food Technology* 43: 80–86.
Shaikh, A.R., A.L. Yaroch, L. Nebeling, M.C. Yeh, and K. Resnicow. 2008. Psychosocial predictors of fruit and vegetable consumption in adults: A review of the literature. *American Journal of Preventive Medicine* 34: 535–543.
Shinagawa, F.B., R. Deliza, A. Rosenthal, and M.A. Zarur. 2013. High pressure effects on sensory attributes of papaya nectar. *Ciência Rural* 43: 1898–1904.
Siegrist, M. 1998. Belief in gene technology: The influence of environmental attitudes and gender. *Personality and Individual Differences* 24: 861–866.
———. 2000. The influence of trust and perceptions of risks and benefits on the acceptance of gene technology. *Risk Analysis* 20: 195–203.
———. 2008. Factors influencing public acceptance of innovative food technologies and products. *Trends in Food Science & Technology* 19: 603–608.
Siegrist, M., and G. Cvetkovich. 2000. Perception of hazards: The role of social trust and knowledge. *Risk Analysis* 20: 713–720.
Slovic, P. 1987. Perception of risk. *Science* 236: 280–285.
Søndeargaard, H.A., K.G. Grunert, and J. Scholderer. 2005. Consumer attitudes to enzymes in food production. *Trends in Food Science and Technology* 16: 466–474.
Sonne, A.M., K.G. Grunert, N.V. Olsen, B.S. Granli, E. Szabó, and D. Banati. 2010. Consumers' perceptions of HPP and PEF food products. *British Food Journal* 114: 85–107.
Stewart-Knox, B.J., and P. Mitchell. 2003. What separates the winners from the losers in new food product development? *Trends in Food Science & Technology* 14: 58–64.
Terry, D.E., and R.L. Tabor. 1988. Consumer acceptance of irradiated produce. *Journal of Food Distribution Research* 19: 73–89.
Tiburski, J.H., R. Deliza, and A. Rosenthal. 2009. Preference mapping of yellow mombin nectar submitted to high pressure processing. In *8th Pangborn Sensory Science Symposium, Florence*.
Tversky, A., and D. Kahneman. 1974. Judgement under uncertainty: Heuristics and biases. *Science* 185: 1124–1131.
Urban, G.L., and J.R. Hauser. 1993. *Design and marketing of new products*. Englewood Cliffs, NJ: Prentice-Hall.
van Kleef, E., H.C.M. van Trip, and P. Luning. 2005. Consumer research in the early stages of new product development: A critical review of methods and techniques. *Food Quality and Preference* 16: 181–201.
van Trijp, J.C.M., and J.B.E.M. Steenkamp. 2005. Consumer-oriented new product development. Principles and practice. In *Innovation in agri-food systems*, ed. W. Jongen and M. Meulenberg. Wageningen, the Netherlands: Academic.
Walther, E., R. Weil, and J. Dusing. 2011. The role of evaluative conditioning in attitude formation. *Current Directions in Psychological Science* 20: 192–196.
Wang, S., J.P. Melnyk, R. Rsao, and M.F. Marcone. 2011. How natural dietary antioxidants in fruits, vegetables and legumes promote vascular health. *Food Research International* 44: 14–22.

Williams, P.R.D., and J.K. Hammit. 2001. Perceived risks of conventional and organic produce. Pesticides, pathogens, and natural toxins. *Risk Analysis* 21: 319–330.

Wooton-Beard, P.C., and L. Ryan. 2011. Improving public health?: The role of antioxidant-rich fruit and vegetable beverages. *Food Research International* 44: 3135–3148.

World Health Organization. 2003. *Diet, nutrition and the prevention of chronic diseases: Report of a joint WHO/FAO expert consultation.* Geneva, Switzerland: World Health Organization.

Yeh, M.C., J. Obenchain, and A. Viladrich. 2010. Barriers and facilitating factors affecting fruit and vegetable consumption. In *Bioactive foods in promoting health*, ed. Ronald Ross Watson and Victor R. Preedy. New York: Academic.

Young, A.L. 2003. Food irradiation—After 35 years, have we made progress: A government perspective. *Environmental Science and Pollution Research* 10: 82–88.

Zajonc, R.B. 1980. Feeling and thinking: Preferences need no inferences. *American Psychologist* 35: 151–175.

Chapter 2
Safety Issues on the Preservation of Fruits and Vegetables

Antonio Martínez, Dolores Rodrigo, and Surama F. Zanini

2.1 State of the Art

In the recent years, there has been a large increase in the consumption of fruits and vegetables (De Roever 1999; Regmi et al. 2004). One main reason is the advice from nutritional experts that at least five portions of fruit and vegetables should be consumed per day. A balanced diet which is low in fat, high in fibre, and includes plenty of fruits and vegetables has been shown to protect against heart disease and many cancers (De Roever 1999; Johnson et al. 2004). Fresh fruits and vegetables as well as their derivatives are an important part of the food chain having differential characteristics of meat and fish foodstuffs, containing a huge range of bioactive compounds. Consequently, fresh produce has become one of our most desirable food because today's consumer perceives it as being healthy, tasty, and convenient.

Despite the credit that these products have with their image as healthy food, risks are associated to their consumption that could influence the consumer's perception once they become aware of the potential microbiological hazards. It is now commonly accepted that fruit and vegetable consumption is a risk factor for infection with enteric pathogens. Thus, food is essential to life but, if contaminated, can cause illness and even death.

This has already been observed regarding the consumption of pesticide-free fresh fruits and vegetables where consumers have shown willingness to pay more for these products, or to accept cosmetically imperfect produce as a trade-off for

A. Martínez (✉) • D. Rodrigo
Instituto de Agroquímica y Tecnología de Alimentos (IATA-CSIC), Valencia, Spain
e-mail: conaml@iata.csic.es

S.F. Zanini
Espirito Santo Federal University, Alegre, ES, Brazil

A. Rosenthal et al. (eds.), *Fruit Preservation*, Food Engineering Series,
https://doi.org/10.1007/978-1-4939-3311-2_2

lower pesticide residue levels (Bunn et al. 1990; Lynch 1991; Eom 1994; Boccaletti and Nardella 2000). Also due to their positive image, they are also eaten at all population levels including high-risk groups as young, old, pregnant, and immune-compromised individuals (YOPI's). This fact increases the likelihood of fresh produce being involved in severe foodborne illness. However, consumers of fresh fruits and vegetables are interested in the food quality and raise no concerns about the food safety (Worsfold et al. 2004a, 2004b).

Fruit and vegetables can become contaminated with pathogenic microorganisms while growing in fields or orchards, during harvesting, at postharvest handling, and during processing. All fruits and vegetables allow the proliferation of pathogenic and spoilage microorganisms (bacteria, yeast, and fungi) due to their physicochemical and nutritional characteristics. The low pH level of fruits and some vegetables is due to the high concentration of organic acids as citric, malic, or tartaric; additionally they contain vitamins and minerals (Holland et al. 1992). The distribution of fresh produce through the various markets is variable and diverse. The diversification of this distribution may have an impact on food safety of produce because there are many steps involved, thereby increasing the opportunity for potential contamination by pathogenic microorganisms. Today, the fresh produce industry is focusing much of its efforts on training employees in the importance of traceback plans, as well as developing and implementing traceback systems. This could also help to identify which part of the food chain is responsible for a pathogenic contamination. Moreover, after approval of Regulations (EC) No 178/2002 and 852/2004, traceability is mandatory for all food producers since January 1st 2005 and primary production will also be included in the HACCP systems shortly.

There are surveys demonstrating the presence of human pathogens on fruit and vegetables (FDA 2001, 2003; Kärenlampi and Hänninen 2004), indicating several factors contributing to the potential adherence and growth. These include fertilization with manure or waste water contaminated with pathogens, some harvest practices (Beuchat 1996), not always refrigerated storage, long-term transportation as well as risks associated to preparation and eating habits. The convenient raw produce sold by the traders can be already trimmed, peeled, or cut. These pre-prepared food presumably are responsible for the growth of the fresh produce market. At the consumer's home, fresh pre-prepared fruits and vegetables are often consumed without washing or additional processing. However, the cut plant surface provides a source of nutrition and the damaged tissue can allow penetration of bacteria. A temperature abuse during transportation, storage, and distribution can then result in an accelerated growth of the pathogenic bacteria. Minimal processing could, therefore, contribute to the risk of food poisoning microorganisms being present on or within fresh produce. In addition, the increase in importation of exotic or "out of season" products from countries with different or lower standards, in "Agricultural of Hygienic Practices", could also contribute to an increased risk of food poisoning organisms being present on imported produce (FDA 2001, 2003).

The most common and important pathogenic bacteria that can be found on fresh produce are *Listeria monocytogenes, Clostridium botulinum, Shigella, Salmonella, Aeromonas, Staphylococcus aureus, Bacillus cereus, Vibrio cholerae,*

Campylobacter, E. coli 0157:H7 (Everis 2004; FDA 2001, 2003). Moreover, fresh produce has been associated with a number of documented outbreaks of foodborne illness, particularly in Europe, Japan, United States, and Canada, including cases of *Escherichia coli* O157:H7 (spinach, lettuce), *Salmonella Typhimurium* and *Salmonella Newport* (tomatoes, lettuce), *Salmonella Thompson* (rocket), and hepatitis A (spring onion) by Anon (2001a, 2001b). Human diseases originating from raw consumed fruits and vegetables have not been regarded extensively in the past. However, they play an important role in the effort of governments, producers, and traders for a high-level protection of consumer health. Some sources for contamination of fresh produce with pathogenic bacteria have been identified, but until now in Europe there has been no broad survey considering all steps in the production chain as well as variable systems in different countries.

Other source of concern in the consumption of fruits and vegetables is the presence of chemical contaminants. This type of contamination can be due to the presence of heavy metals or pesticide residuals over the allowed limit.

In this chapter, fruit and fruit products safety issues in relation to the microbiological contamination will be considered. Information on main pathogenic microorganisms hazards will be presented considering the intrinsic and extrinsic factors contributing to product deterioration, as well as control measures that should be taken in place to prevent high levels of microorganisms.

2.2 Microbial Hazards

One of the major growth segments in the food retail industry is fresh and minimally processed fruits and vegetables that were stimulated largely by consumer demand for fresh, healthy, convenient, and additive-free food which are safe and nutritious. Thus, there is a consumer preference for food that is less processed, contains fewer preservatives, is convenient, and perceived as fresh (De Roever 1999).

Fruits and vegetables are unique food in that they are often consumed raw or with minimal preparation. Fresh-cut products are fruits or vegetables, initially called minimally processed or lightly processed products, that can be defined as any fresh fruit or vegetable that have been trimmed, peeled, and/or cut into a fully usable product, to obtain 100 % edible product which is subsequently packaged to offer consumers high nutrition, convenience, and flavour while maintaining freshness and kept in refrigerated storage (Martin-Belloso and Soliva-Fortuny 2010).

On the market, fresh-cut tropical fruits include melons, cantaloupe, watermelon, mangoes, mangosteen, rambutan, jackfruit, pummelo, papaya, durian, grapefruit, pineapples, and fruit mixes. Fresh-cut salads include shredded leafy vegetables and salad mixes. Fresh-cut vegetables for cooking include peeled baby carrots, baby corn, broccoli and cauliflower florets, cut celery stalks, shredded cabbage, cut asparagus, stir-fry mixes, and cut sweet potatoes. Fresh-cut herbs are also marketed widely (James and Ngarmsak 2011).

Inevitably, with the growing demand for fresh fruits and vegetables the Centers for Disease Control and Prevention (CDC) reported an increase in the frequency of produce-associated foodborne disease outbreaks (Bean et al. 1997; Mead et al. 1999). The potential for contamination increases as the fresh produce moves from farm to table, i.e. irrigation water, improperly composted manure, wash water systems, soiled equipment, unsanitary practices, etc.

In the USA, between 1990 and 2002, 56 outbreaks with 6762 cases were linked to fruits. Four per cent of the cases were caused by berries, 2 % by melon, and 94 % by other fruits. In fruits, *Salmonella*, Noroviruses, and *E. coli* represented the most significant hazards (Smith DeWaal and Barlow 2002). Salmonellosis outbreaks have been associated with contaminated fresh fruits (CDC 1991) such as strawberries (Niu et al. 1992; CDC 1997) and raspberries (Ramsay and Upton 1989; Herwaldt and Ackers 1997).

Minimally processed fruits are more perishable than raw materials and preparations from vegetables. Thus, the marketing of fresh cut fruits has been limited to 5–7 days if compared to 15–20 days of vegetable-based products (Watada 1997a, b).

Ready-to-eat vegetables must be stored under appropriate temperatures in order to inhibit the growth of pathogens (Francis et al. 1999). However, most microorganisms found in fresh produce are psychotrophic (Nguz et al. 2005), e.g., *Listeria monocytogenes*, which may grow in fresh produce stored under refrigeration (Beuchat 1996), and *Clostridium* spp that are the most important spoilage groups for fruits and vegetables.

Slicing, dicing, and shredding procedures, as well as temperature abuse during storage, could result in increases in populations of mesophilic aerobic microorganisms (Brackett 1992; Nguyen-the and Carlin 1994) associated with fresh-cut products. Mesophilic bacteria from plate count studies typically ranged from 10^3 to 10^8 CFU/g (Beuchat 1996). Total counts on products after processing ranged from 10^3 to 10^6 CFU/g (Nguyen-the and Carlin 1994). Thus, the slicing step may increase the risk of contamination because the cut of surfaces exude nutrients, which become available to the microorganisms naturally present in the produce, i.e., growth of pathogens. So, all this processing contributes to microorganism multiplication and eventually increases bacterial counts (Berbari et al. 2001; Gleeson and O'Beirne 2005). Therefore, the presence of cut surfaces, with a consequent release of nutrients, the absence of treatments able to ensure the microbial stability, the active metabolism of fruits or vegetables, and the confinement of the final product can all increase the growth extent of the naturally occurring microbial population (Nguyen-the and Carlin 1994).

Minimally processed food may represent a microbiological risk because the processes for limiting pathogens proliferation that include storage at low temperature and/or the packaging in modified atmospheres are stringent conditions only to some microorganisms and may promote the prevalence of anaerobic and psychrotrophic bacteria (De Martinis et al. 2002). Also, the fresh-cut products do not generally contain preservatives or antimicrobial substances and rarely undergo any heat processing before consumption.

Temperature is another environmental factor that affects the activity and microbial growth. This is mainly due to the influence of temperature on the activity of microbial enzymes. The minimum and maximum temperatures for growth of a microorganism depend on factors such as pH and water activity (Aw). If these

Table 2.1 Major environmental conditions for microbial growth

Pathogens	Parameters					
	T_{min} (°C)	T_{max} (°C)	pH_{min}	pH_{min}	$A_{w\,min}$	$NaCl_{max}$
C. jejuni	32	45	4.9	9.0	0.98	2
***C. botulinum* type A or B proteolytic**	10	50	4.6	8.5	0.93	10
***C. botulinum* type E nonproteolytic**	3	45	4.6	8.5	0.97	5
E. coli	7	46	4.4	9.0	0.95	6.5
L. monocytogenes	0	45	4.39	9.4	0.92	10
***Salmonella* spp.**	5	47	4.2	9.5	0.94	8
***Shigella* spp.**	7	47	4.9	9.3	0.97	5.2
Y. enterocolitica	-1	42	4.2	9.6	0.97	7

FDA (2001) and ICMSF (1996)

Table 2.2 Doses of some pathogenic microorganisms necessary to cause illness in healthy adults

Microorganisms	Doses (cells)
Shigella dysenteriae	10^1-10^4
Shigella flexneri	10^2-10^9
Salmonella typhi	10^4-10^9
Others Salmonella	10^5-10^{10}
E. coli (pathogenic)	10^6-10^{10}
Y. enterocolitica	10^9

FDA (2001)

environmental factors (pH and Aw) are outside the optimum range, the minimum temperature increases and the maximum decreases, thus narrowing the range of growth (Garbutt 1997), as showed in Table 2.1.

The Table 2.2 shows, for some pathogenic microorganisms, values found in literature concerning infectious doses liable to cause disease in healthy adults.

Even after some processing, ready-to-eat vegetables retain much of their original microbiota; this is a serious health issue, because pathogens may be part of that microbiota. Fruits and vegetables have an important pathogenic and non-pathogenic microbiological load coming from soil, water, insects, and handling by man. In general, the common microbial flora is composed by *Pseudomonas spp, Erwinia herbicola, and Enterobacter aglomerans.*

However, the fruits and vegetables are protected from microbial invasion by the skin and thus they are expected to be able to retain high quality longer than the cut products (Hurst 1995). Thus, the growth of foodborne pathogens is not common on intact surfaces because they do not produce the enzymes necessary to break down the protective outer barriers. This restricts the availability of nutrients and moisture. One exception is the reported growth of *E. coli* O157:H7 on the surface of watermelon and cantaloupe rinds (Del Rosario and Beuchat 1995) as well as *E. coli* O157:H7 and *Listeria monocytogenes* have been shown to attach to the cut surfaces of lettuce leaves and penetrate the internal tissue, indicating protection from chemical sanitizers. Moreover, several pathogenic microorganisms can grow and survive in many fresh produce, like lettuce, broccoli, asparagus, because these food have nutrients necessary for their rapid growth (Abdul-Raouf et al. 1993).

The microorganisms normally present on the surface of raw fruits and vegetables may consist of chance contaminants from the soil or dust, or bacteria or fungi that have grown and colonized by utilizing nutrients exuded from plant tissues. Therefore, the consumption of raw food or salads may endanger the health of consumers. Several outbreaks of gastroenteritis have been attributed to the consumption of contaminated fresh vegetables. Toxinfections associated with ingestion of vegetables, which revealed a lack of hygiene during manipulation, have been recorded (De Roever 1998; Michino et al. 1999). Thus, the precise establishment of the origin of a disease outbreak is of crucial importance in planning the strategies and interventions to minimize future health hazards.

Pathogens within soil may contaminate crops directly when heavy rain or water gun irrigation causes leaf splash. The ability of the pathogen to survive in the environment will impact on the likelihood of crop contamination and pathogen viability at harvest and through to consumption. Table 2.3 shows survival times for each enteropathogen in the environment.

Table 2.3 Pathogens survival (day) in environment

Pathogens	Environment	Survival (day)	References
E. coli **O157:H7**	Soil + animal manure	99	Nicholson et al. (2005)
E. coli **O157:H7**	Animal manure	60	Avery et al. (2005)
E. coli **O157:H7**	Slurries	60	Avery et al. (2005)
E. coli **O157:H7**	Nonaerated ovine manure	>365	Kudva et al. (1998)
E. coli **O157:H7**	Aerated ovine manure	120	Kudva et al. (1998)
E. coli **O157:H7**	Nonaerated slurry	600	Kudva et al. (1998)
Salmonella	Soil	968	Nicholson et al. (2005)
Salmonella	Soil + bovine slurry	300	Nicholson et al. (2005)
Salmonella	Slurry + durty water	90	Nicholson et al. (2005)
Campilobacter	Slurry + durty water	90	Nicholson et al. (2005)
Listeria	Soil + animal manure	30	Nicholson et al. (2005)
Listeria	Slurry + durty water	180	Nicholson et al. (2005)
Hepatitis A	Water	>365	Seymour and Appleton (2001)
Hepatitis A	Soil	96	

The contamination of fruits and vegetables can occur in any stage of food processing. It has long been known that the improper use of manure can transfer pathogens onto crops, resulting in human disease. Raw manure should not be applied to crops. In addition to the hazard of pathogen transmission, it is well recognized that salt injury to sensitive vegetable crops and transfer of viable weed seed may result unless the manure is subjected, at least, to a period of undisturbed (no thorough mixing or pile inversion) composting. In places where noncomposted animal manure as fertilizer and untreated water are used to wash fresh products or to irrigate vegetable crops, *Salmonella, Shigella, Bacillus cereus,* and *Clostridium botulinum* are easily observed. Health problems due to the consumption of swiss chard, lettuce, cabbage and water-cress contaminated with *Salmonella* and *Shigella* have been reported. According to Frank and Takeushi (1999), fresh produce, especially lettuce, was identified as carriers of pathogenic bacteria relevant to human health, such as *Salmonella, Shigella, Listeria monocytogenes, Yersinia enterocolitica, E. coli* enteropathogenic, *E. coli* enterotoxigenic, and *E. coli* enterohaemorrhagic (O157:H7), also protozoa, parasites, and hepatitis A virus (Nascimento et al. 2003).

The National Advisory Committee on Microbiological Criteria for Foods (NACMCF) lists 11 agents associated with produce-borne outbreaks. Foremost among them are *E. coli* O157:H7 and various *Salmonella* serotypes (Tauxe 1997). Health officials at a national food safety meeting disclosed preliminary data, which demonstrated that foodborne illnesses associated with fresh produce in the United States are related predominantly to pathogens of animal origin. Illnesses attributed to imported produce predominantly align with human sources of contamination. The prevalence of *E. coli* O157:H7 and *Salmonella* spp. in manure also varies with the source animal. *Escherichia coli* O157:H7 colonizes cattle and other ruminants but generally not poultry. The prevalence of cattle pathogen shedding varies among different studies. Cassin et al. (1998) projected that the number of *E. coli* O157:H7 shedding animals varies from 0.3 to 0.8 %, but may be considerably higher in a population consisting exclusively of young or stressed animals. In addition, survey results may be strongly influenced by regional and seasonal variation.

Bacteria such as *Clostridium botulinum, Bacillus cereus,* and *Listeria monocytogenes*, all capable of causing illness, are normal inhabitants of many soils, whereas *Salmonella, Shigella, Escherichia coli,* and *Campylobacter* reside in the intestinal tracts of animals, including humans, and are more likely to contaminate raw fruits and vegetables through contact with faeces, sewage, untreated irrigation water, or surface water.

Therefore, the prevention of risks of contamination by pathogens may occur from the realization of good agricultural practices ranging from planting to harvesting and other important aspects, even in the process of farming, such as the water quality used for irrigation and employment of adequate sanitation practices by producers in the handle and care of plants in the farmer's field.

Bacterial pathogens continue to be a major contributor to produce-associated foodborne illnesses. In a review of produce-associated outbreaks in the USA from 1973 to 1997, bacteria were responsible for 60 % of outbreaks in which an etiologic agent was identified (Sivaplasingham et al. 2004). Salmonella was the most commonly reported bacterial pathogen, accounting for nearly half of the outbreaks due to bacteria (Sivaplasingham et al. 2004).

A description of pathogens of most concern and that have been isolated from fresh products or raw products, including fruit and vegetables, with emphasis on their association with foodborne outbreaks is given below.

Listeria monocytogenes: The bacterium is Gram-positive, non-sporulating and rod-shaped, psychrotrophic, facultative anaerobic. Six species of the genus *Listeria* have been recognized (ICMSF 1996). Two are considered non-pathogenic; *L. innocua* and *L. murrayi* (syn. L. grayi), while *L. seeligeri, L. ivanovii*, and *L. welshimeri* rarely cause human infection. This leaves *L. monocytogenes* as the most important species with respect to human health which can cause severe infection mainly in immunocompromised persons and in pregnant women. The predominant *L. monocytogenes* serotype isolated from salad vegetables has been shown to be serogroup 1 (Harvey and Gilmour 1993; Heisick et al. 1989).

Listeria spp. grows optimally under microaerophilic conditions, but grows well both aerobically and anaerobically (anaerobic incubation has been shown to be more favourable to *Listeria* growth or survival than aerobic incubation). It can grow in food packaged under vacuum or nitrogen gas (AIFST 2003). Growth of the organism was not retarded by a 5–10 % CO_2 atmosphere and it can also grow in relatively high (e.g. 30 %) CO_2, but growth is inhibited under 75 % CO_2.

Most literature reports on modified atmosphere packaging have studied *L. monocytogenes* and the data suggest that modified atmospheres containing approximately 75 % CO_2 and no oxygen will inhibit this organism according to Hudson et al. (1994). This bacterium can survive, grow, and multiply in different environmental conditions, on refrigeration or warm temperatures, low pH, and high salt concentrations (Gandhi and Chikindas 2007), therefore this bacterium has wide environmental distribution. Thus, *L. monocytogenes* is relatively resistant to freezing, drying, high salts (growth at 10 %; survival at 20–30 %), and $pH < 5.0$. The risk of listeriosis increases when these vegetables are stored for longer periods before consumption because *L. monocytogenes* has a greater opportunity to grow. Controlled atmosphere storage has been shown to extend the shelf-life of broccoli and asparagus, but does not influence the rate of growth of *L. monocytogenes* (Berrang et al. 1989).

The organism also exists in nature as a saprophyte, growing on decaying plant materials, so its presence on raw fruits and vegetables is not rare (Beuchat 1992, 1996; Beuchat et al. 1990).

Listeria monocytogenes has been isolated from pre-packaged mixed vegetable products, chicory, endive and fresh-cut lettuce, sliced cucumber and fruits such as tomatoes and cantaloupe. It has also been implicated in foodborne disease outbreaks across the globe. In USA, it was isolated on cucumbers, potatoes, and radishes by Heisick et al. (1989). Beuchat (1998) reported a number of surveys documenting the presence of *L. monocytogenes* on cucumber, peppers, potato, radish, leafy vegetables, beansprout, broccoli, tomato, and cabbage. Thus, *L. monocytogenes* can grow in plant tissue under refrigeration temperature even if the initial concentrations of *Listeria* are not very high (NACMCF 1991), although it can increase during storage in refrigeration temperature.

Generally, contamination is higher on root vegetables and Heisick et al. (1989) suggested that this is due to increased contact with soil. Crepet et al. (2007) analysed 165 studies and reported that prevalence on salad vegetables is usually under 5 %, with lower numbers isolated from leafy salad vegetables than from sprouted seeds and other vegetables (e.g. carrots, cabbage, celery, and spinach).

Beuchat and Brackett (1990) showed that *L. monocytogenes* is capable of growth on lettuce when exposed to processing conditions, although carrot juice seemed inhibitory. Farber et al. (1998) demonstrated that *L. monocytogenes* populations declined on grated carrot by 2-logs over 9 days.

Listeria outbreaks linked to fresh produce are infrequent and tend to be limited to vulnerable groups. The two documented outbreaks which have occurred, in 1979 and 1981 respectively, were attributed to cabbage (in coleslaw) and salad items (celery, lettuce, and tomatoes) served as part of hospital meals (Anonymous 2001a, 2001b).

Thus, studies have demonstrated that the growth of *L. monocytogenes,* when inoculated onto asparagus and broccoli, increased by 3 log at 15 °C and 0.5 log at 4 °C (Berrang et al. 1989). On cabbage, a 2.1 and 4 log increase of *L. monocytogenes* at 25 °C and 5 °C was demonstrated, respectively, by Kallander et al. (1991) and Beuchat et al. (1986). Increased levels of *L. monocytogenes* on cauliflower (Berrang et al. 1989; Beuchat et al. 1990), broccoli (Beuchat et al. 1990), asparagus (Beuchat et al. 1990), endive (Carlin et al. 1995), and lettuce (Carlin and Nguyen-the 1994) has also been documented.

Recently, Conway et al. (2000) determined that *Listeria monocytogenes* survived and proliferated on *Delicious* apple slices stored at 10 or 20 °C (50 or 68 °F) in air or controlled atmosphere (0.5 % O_2 + 15 % CO_2), but did not grow at 5 °C (41 °F). Controlled atmosphere had no significant effect on the survival or growth of *L. monocytogenes* at elevated temperatures.

Yersinia enterocolitica: Swine are the predominant natural reservoir for *Y. enterocolitica*, although the pathogen has been found in a variety of terrestrial and freshwater ecosystems, including soil, vegetation, and water in lakes, rivers, wells, and streams (Kapperud 1991) and also isolated from raw vegetables. Certainly, application of improperly composted pig manure to vegetable fields should be avoided to reduce the possibility of pathogenic strains being present on produce when it reaches the consumer. The pathogen can grow at refrigeration temperatures commonly used during transport and storage of fresh products.

The outbreaks of *Y. enterocolitica* have been documented by contamination of mung bean sprouts (Harris et al. 2003), carrots (Catteau et al. 1985), or grated carrots (Darbas et al. 1985). The incidence of *Y. enterocolitica* was higher in fresh produce as on root and leafy vegetables than in tomatoes and cucumbers (Darbas et al. 1985).

Salmonella: is a Gram-negative rod-shaped bacterium, facultative anaerobic, nonlactose fermenting, nonspore forming, mesophile, and most are motile. This bacterium is a member of the family Enterobacteriaceae. Complete inhibition of growth occurs at pH $<$ 3.8 and $>$9.0, temperature $<$7 °C, or water activity $<$0.94 (Ray 1996; Jay 2000; Gray and Fedorka-Cray 2002). Optimum growth occurs at pH near neutrality and temperatures between 35 and 37 °C (Ray 1996). *Salmonella* spp. may be detected in both cattle and poultry manure. The prevalence among dairy herds may range from 57 to 84 % (Smith et al. 1993). Thus, animals and birds are the natural reservoirs. *Salmonella* reaches food directly by infected workers or indirectly by animals and humans waste, water polluted with waste, or in soil fertilization. It is more probable that the disease occur when a large number of microorganisms are taken after multiplication in food that were exposed to room temperature for several hours. Symptoms of the disease appear in 6–36 h or more after the ingestion of contaminated food. The disease duration is 1–7 days or more (Hobbs 1998).

Outbreaks and sporadic cases of infection have been associated mainly to poultry, pork, beef, and vegetables (Price 1997). Although fresh fruits and vegetables are less frequently related to salmonellosis outbreaks, they have been associated with the consumption of sprouted seeds, cut cantaloupe, watermelon and honeydew (Golden et al. 1993), oranges (Pao et al. 1998), lettuce, cauliflower, mustard, cress, endive and spinach (Thunberg et al. 2002), and mushrooms (Doran et al. 2005).

Thus, this bacterium is one of the pathogens involved in most cases associated with fresh produce-related infection, isolated in 48 % of cases between 1973 and 1997 in the USA (Sivaplasingham et al. 2004) and in 41 % of cases during 1992–2000 in the UK by Health Protection Agency.

Salmonella does not grow in food stored at temperatures lower than 7 °C; therefore, it is not a risk to public health in fresh-cut products, provided these products are maintained at or below 7 °C. But, improper storage temperature combined with the favorable conditions for growth on the surface of cut melons or cantaloupe were factors that could contribute to the outbreak.

A wide spectrum of produce vehicles have been associated with *Salmonella* infections. Prepared melon salad has been responsible for outbreaks of *Salmonella poona* enteritis in the US (CDC 1991; Madden 1992). It has been shown that the relatively high pH of the melon flesh (pH 5.9–6.7) permits rapid multiplication of *Salmonella* when the melon is kept above refrigeration temperature (Golden et al. 1993). In papaya, growth was observed during 6 h at 25–27 °C at a pH of 5.7 and in apples at pH 4.1 *Salmonella* spp. can survive during 66 h at 8 °C. Studies showed that the pathogen can rapidly grow in damaged, chopped, or sliced tomatoes (pH 4.0–4.5) stored at 20–30 °C over a 1–3 day period (FDA 2001). At temperatures lower than 5 °C, there was a gradual decline over a 12-day storage period.

In 2008, jalapeño and serrano peppers were vehicles for a large multistate outbreak of *Salmonella* serovar Saintpaul infections by CDC (2008). Examples of other outbreaks of Salmonella enterica linked to ready-to-eat plant produce include an outbreak in Scandinavia and the UK of serovar Thompson infections associated with consumption of rocket leaves (Nygard et al. 2008).

Escherichia coli included ***Escherichia coli* O157:H7**: *E. coli* is a normal inhabitant of the intestinal tract of animals; however, their occurrence indicates poor handling of food during processing, use of equipment in poor sanitary conditions, or use of contaminated raw material (ICMSF 1978). The major groups of *E. coli* are designated as enterotoxigenic, enterohaemorrhagic, enteropathogenic, enteroinvasive, diffuse-adhering, and enteroaggregative (Doyle et al. 1997). *E. coli* is a Gram-negative, motile, nonsporulating, rod-shaped, facultative anaerobic bacterium, mesophile.

Some *E. coli* strains as *E. coli* O157:H7 and the recently involved in the German outbreak *E. coli* O104:H5 are pathogens that produce verotoxins or shiga-like enterotoxins (VTI/STEC).

E. coli O157:H7 grows rapidly at 30–42 °C, poorly at 44–45 °C, and does not grow at <10 °C (Ray 1996). This pathogen was first recognized as a pathogen in 1982, when it was associated with two foodborne outbreaks of hemorrhagic colitis (Doyle et al. 1997).

Strains of *E. coli* O157:H7 belong to the group enterohaemorrhagic being the most frequently associated with outbreaks of hemorrhagic colitis that may progress to hemolytic uremic syndrome. The pathogenicity of enterohaemorrhagic strains seems to be associated with several factors, including the production of cytotoxins called verotoxins or "shiga-like" toxins similar to the toxin produced by the bacterium *Shigella dysenteriae* type I (Desmarchelier and Grau 1997). The infectious dose is unknown but appears to be in the range of 10 cells per gram or milliliter of food consumed (FDA/CFSAN 2001). The main characteristics that distinguish *E. coli* O157: H7 from other *E. coli* strains are poor or no growth at 44 °C and the inability to use the sorbitol and produce the enzyme ß-glucuronidase (March and Ratnam 1986; Meng et al. 1994). Therefore, they are not detected in analyzes of fecal coliforms by the most probable number method which uses the fermentation of lactose at 44.5 °C as a confirmatory test, or the direct analysis of *E. coli* using substrates for the enzyme ß-glucuronidase.

E. coli O157:H7 occurs in the intestinal tract of humans and other warm-blooded animals, including cattle, deer, horses, goats, sheep, cats, dogs, rabbits and poultry, with prevalence rates of up to 5.2 % (Knight 1993; WHO 1998; Fratamico et al. 2002) as well as in the faeces of wild birds, for example, starlings (Moller Nielsen et al. 2004) and gulls (Wallace et al. 1997). The incidence in the faeces of the animals ranges from 0 to 10 % (Desmarchelier and Grau 1997). A survey of cattle herds indicated that the prevalence of *E. coli* O157:H7 among feedlot animals was as high as 36.8 % (Chapman et al. 1997). Wang et al. (1996) revealed that *E. coli* O157:H7 survived in bovine faeces for 42–49 days at 37 °C, for 49–56 days at 22 °C, and for 63–70 days at 5 °C. Therefore, the researchers concluded that regulations requiring the ageing of bovine manure for 60 days before using it as a fertilizer were inadequate.

Houseflies can also serve as a vector of dissemination as they carry the pathogen in their intestine and other parts of their body. Since cattle appear to be a natural reservoir for the pathogen, with prevalence rates of 1.8–28 % (Fratamico et al. 2002), contamination of raw fruits and vegetables may occur when cattle inadvertently enter fields, or improperly composted cow manure is applied as fertilizer (WHO 1998).

Since cattle appear to be a natural reservoir for the pathogen, most illness outbreaks have been associated with the consumption of contaminated, undercooked beef and dairy products. The potential for contamination may be enhanced when fruits or vegetables have fallen from the plant to the ground and are then picked and placed into the handling and processing chain. Also, because contaminated manure may become airborne dust particles, it is possible that fruits on trees and vines become contaminated. Workers on farms and in packing houses may also be a source of *E. coli* O157:H7. Therefore, livestock grazing in orchards may contaminate fallen apples with faeces and, as *E. coli* O157:H7 can proliferate in damaged apple tissue (Stopforth et al. 2004), this can result in the contamination of unpasteurized fruit juices⁄ciders.

In accordance with the data from the Center for Disease Control and Prevention (Olsen et al. 2000), fruits, vegetables, and salads represent about 20 % of food most often implicated in outbreaks caused by enteropathogenic strains of *E. coli*, in the period of 1993–1997. Thus, in recent years an increase in the number of outbreaks of *E. coli* O157: H7 associated with fruit, fruit juices, vegetables, and salads prepared with vegetables or fresh produce has been observed. The salad in question contained as ingredients onions, carrots, zucchini, peppers, broccoli, mushrooms, and tomatoes (Beuchat 1996).

Studies have shown the survivability of *E. coli* O157: H7 in acidic and alcoholic environment (Molina et al. 2003) and also that it can remain viable in different food like lettuce, cucumber and carrots (Abdul-Raouf et al. 1993), cider (Semanchek and Golden 1996), and commercial mayonnaise (Zhao and Doyle 1994).

In 1991, there was an outbreak of *E. coli* O157: H7 in the US by the consumption of unpasteurized apple cider. It was suggested that the cider had been produced with cider apples collected from the ground and contaminated with cattle manure. In 1993, an outbreak occurred by consumption of melons that were probably cross-contaminated with meat products handled in the kitchen (Feng 1995). Therefore, some outbreaks associated with melon consumption were a result of an infected food handler or cross-contamination from raw beef via knives, cutting boards, or hands.

It was found by Del Rosario and Beuchat (1995) that growth of *E. coli* O157:H7 was observed on the rind of melons stored under high relative humidity at 25 °C for 14–22 days. The pathogen rapidly died on the rind surface of melons stored at 5 °C. Cut cantaloupe is considered a potentially hazardous food in the FDA Food Code because it is capable of supporting the growth of pathogens due to its low acidity, pH 5.2–6.7, and high water activity from 0.97 to 0.99.

In 1997, outbreaks of *E. coli* by consumption of alfalfa sprouts (CDC 1997) were documented. The association with alfafa sprouts may be due to the volume consumed, as these are the most popular type of sprouted seed commonly eaten raw.

At 25 °C, Red Delicious apples supported survival of *E. coli* O157:H7. Winesap apples were the least favorable for survival of *E. coli* O157:H7 at 25 °C. At 10 °C, survival of *E. coli* O157:H7 was poorest in ground Red Delicious apples. When stored at 4 °C, Golden Delicious and Rome apples were not statistically different in supporting survival of the pathogen (Fisher and Golden 1998).

According to US FDA review (2001), growth and survival of *E. coli* O157:H7 in unpasteurized juices or apple ciders over the range 3.5–4.2 was reported. In refrigeration temperatures (<8 °C), there appeared to be a decrease in levels of *E. coli* O157:H7. But at ambient temperatures, there was an increase of *E. coli* O157:H7 levels.

Thus, it would appear from these studies that the conditions for survival and/or growth of *E. coli* O157:H7 on fresh produce (raw fruits and vegetables) are influenced by storage temperature. Levels of this organism decreased at refrigeration temperature but increased at nonrefrigeration temperatures (>12 °C). Packing under modified atmosphere has little or no effect on the survival or growth *E. coli* O157:H7 (Abdul-Raouf et al. 1993).

Shigella: This bacterium is isolated from the intestinal tract of humans and belongs to the family Enterobacteriaceae as do *Salmonella* and *Escherichia coli*. There are four serological subgroups under the genus *Shigella*: *S. dysenteriae, S. flexneri, S. boydii,* and *S. sonnei* are pathogenic to humans at low dose infection. The infective dose of this microorganism is as low as 200 cells, although Lampel and Maurelli (2002) reported that ten cells are enough to cause disease.

These bacteria are gram-negative, facultative anaerobic, non-motile, rod-shaped bacteria. The strain can grow between 7 and 46 °C, with an optimum at 37 °C (Ray 1996; Jay 2000). The *Shigella* spp reaches the food through contamination with human fecal matter, either through water or through the hands of manipulators. The presence of the agent in various food is directly related to the role played by man as shedders of the bacteria, especially when the personal hygiene is limited.

This bacterium is rapidly destroyed at temperatures above 65 °C and supports changes in pH from 4.9 to 9.3 (*S. sonnei*), but does not resist to values lower than 4.5. It does not survive pasteurization, is sensitive to ionizing radiation, but it is not affected by reducing the water activity. Disinfectants like chlorine, iodine, and quaternary ammonium are effective in the destruction of this microorganism (Germano and Germano 2001).

Outbreaks of *Shigella* are related to the consumption of contaminated raw fruits and vegetables. *Shigella* can contaminate food by several routes, including insects and the hands of persons who handle the produce, although shigellosis is more often transmitted from person to person.

The outbreaks of *Shigella* have been associated with the consumption of lettuce and fresh green onions (Beuchat 1996). *Shigella sonnei* can survive on lettuce at 5 °C for 3 days without decreasing in number and it can increase by more than 1000-fold at 22 °C. *Shigella* can grow in shredded cabbage and chopped parsley stored at 24 °C. Populations of *S. sonnei, S fleneri,* and *S. dysenteriae* inoculated onto the surface of cut cubes of papaya, jicama, and watermelon increased substantially within 4–6 h at 22–27 °C.

S. sonnei survived on refrigerated, shredded lettuce for 3 days without decreasing in number and increased when held at 22 °C (Satchell et al. 1990).

Clostridium botulinum: This is a spore-forming bacterium. The spores of *Clostridium botulinum* are found in soil and sediments and on the surfaces of fruits and vegetables. *C. botulinum* spores are capable of growing on fresh-cut vegetables under conditions of low O_2 and high temperature (Sugiyama and Yang 1975). They grow in anaerobic vacuum packaging and modified atmosphere environments, commonly used for packaging vegetables.

Botulism has been more associated to consumption of cooked vegetables than fresh produce. Growth and toxin production often lag behind spoilage in fresh vegetables.

Outbreaks of this pathogen have been documented for cabbage (Solomon et al. 1990) and garlic in oil (St. Louis et al. 1988), coleslaw prepared from packaged, shredded cabbage mixed with coleslaw dressing. The botulinum toxin can be formed in shredded cabbage when the cabbage is packaged under an atmosphere containing reduced oxygen and stored at 22–25 °C for 4–6 days (Solomon et al. 1990). Botulism toxin was produced in polyvinyl chloride film-packaged mushrooms held at 20 °C for 3–4 days (Sugiyama and Yang 1975). Proteolytic strains of *C. botulinum* grew and produced toxin in vacuum-packaged mushrooms held at 15–27 °C (Malizio and Johnson 1991).

Growth of *C. botulinum* and toxin production tend to occur more quickly at higher temperatures (>15 °C), as demonstrated in lettuce (Austin et al. 1998).

Botulinun toxin was not recovered from fresh-cut cantaloupe or honeydew melons inoculated with a 10-strain mixture of proteolytic and non-proteolytic *Clostridium botulinum* after 21 days at 7 °C (44.6 °F). However, toxin was recovered in some inoculated honeydew samples stored 9 days at 15 °C (59 °F) in hermetically sealed packages (Larsen and Johnson 1999).

Campylobacter jejuni: Thermophilic campylobacter are widespread in the environment, where they are a sign of recent contamination with animal and avian faeces, agricultural run-off, and sewage effluent. Although intestinal carriage of campylobacters is ubiquitous in livestock, domestic animals, wild animals, wild birds, and poultry, contamination of the environment with the bacteria in faeces is sporadic and varies seasonally, depending on factors such as stress and changes in diet. Wild birds, and not sewage effluent, are the source of campylobacters in some coastal waters. The density of *Campylobacter* spp. in sewage effluent depends on the source of the sewage and the type of treatment. There is a qualitative, but not a quantitative, correlation between campylobacter and faecal indicators in environmental samples. The marked seasonal pattern of campylobacter in temperate, aquatic environments is a result of variations in *Campylobacter* die-off rates at different times of the year.

Campylobacter is a leading cause of bacterial enteritis in many countries. Epidemiological evidence suggests that salad vegetables are the second-highest risk factor for *Campylobacter* infection after poultry (Evans et al. 2003). Reservoirs of this pathogen include several wild animals as well as poultry, cows, pigs, and domestic pets (Nachamkin 1997). The *Campylobacter* cannot grow outside of a warm-blooded host or at temperatures below 30 °C, so the survival on fresh produce is limited, especially if not protected from UV light (Obiri-Danso et al. 2001). This bacterium is sensitive to acid pH, but it can survive on cut fruits for sufficient time to be a risk to the consumer (Castillo and Escartin 1994). According to the same authors, *C. jejuni* can survive on sliced watermelon and papaya for up to 6 h at a pH values from 3.0 to 5.5 (watermelon) and pH of 5.0 for papaya slices.

Although the modes of transmission are not well known (Vierikko et al. 2004), handling or consumption of undercooked or raw chicken has been regarded as the major risk factor for infection with *C. jejuni* (WHO 2002; Sopwith et al. 2003). Cross-contamination of other food items, especially those eaten raw (e.g. salads), is also a significant risk factor. Commercial catering establishments like restaurants provide opportunities for outbreaks of foodborne disease because large quantities of different food are handled in the same kitchen. This is the reason why eating chicken in the restaurants is associated with increased risk of infection (Altekruse and Tollefson 2003).

Thus, *Campilobacter* outbreaks are most commonly linked to the consumption of poultry or with the consumption of raw fruits and vegetables (Bean and Griffin 1990; Harris et al. 1986) that were contaminated with poultry products (Gillespie et al. 2003). Outbreaks of campylobacter have been linked to lettuce, sweet potatoes, cucumber, melon, and strawberries (Kirk et al. 1997; Brandl et al. 2004). In outbreaks of this pathogen associated with lettuce or salads by cross-contamination from poultry meat cases, cross-contamination during food preparation was likely, and consequently, the contamination control should focus on reducing cross-contamination during food storage and preparation. According to Kumar et al. (2001), *C. jejuni* was isolated from spinach, fenugreek, lettuce, radish, parsley, green onions, potatoes, and mushrooms. Park and Sanders (1992) reported *Campylobacter* spp. on 1.6–3.3 % of the vegetables tested, but other extensive studies of raw organic and prepack salad vegetables failed to isolate *Campylobacter* (Evans et al. 2003). It is suggested that outbreaks linked to fresh produce may be due to cross-contamination in the kitchen (Evans et al. 2003).

Virus: There are many different ways in which food can become contaminated with viruses including by infected food handlers, contaminated food preparation surfaces, irrigation and fertilization of crops with animal and human waste, and sewage contamination of shellfish growing waters (Calder et al. 2003; Koopmans and Duizer 2004). Thus, use of contaminated water and handling of produce by infected individuals at the farm or at post-harvesting can also contaminate food and since viruses can remain viable on inanimate objects and surfaces, poor hygiene practices can lead to food contamination. Shellfish taken from waters contaminated with human faeces have been the vehicle in most outbreaks, but any food handled by an infected person may become contaminated and transmit the infection (Cliver 1985).

Although viruses will not grow on food, raw fruits and vegetables may serve as vehicles for infection. The extent to which hepatitis A and other viruses are removed from the surface of fruits and vegetables upon treatment with chemical disinfectants is not known. According to Calder et al. (2003), the outbreaks of viral gastroenteritis are known to be mainly caused by norovirus (NoV) and outbreaks of viral hepatitis are caused by Hepatitis A virus (HAV).

Enteric viruses have a low infective dose and remain active even after exposure to low pH like as <3 (Seymour and Appleton 2001) and temperature extremes. Freezing, for example, increases virus persistence (Le Guyader et al. 2004).

Noroviruses are probably the most common cause of epidemic non-bacterial acute gastroenteritis and can be transmitted by faecal-contaminated food, water, surfaces, and hands. Noroviruses are pathogens of particular concern for produce quality. Ice made from contaminated water has been implicated as the vehicle in more than one outbreak, but salad items have also been linked to Norwalk-like gastroenteritis (Kuritsky et al. 1984). Workers who have prepared salads linked to viral gastroenteritis have been shown to have high antibody titers to Norwalk virus (Gross et al. 1989). Studies have shown that viruses may persist for weeks or even months on vegetable crops and in soils that have been irrigated or fertilized with sewage wastes (Larkin et al. 1978).

Several multi-country outbreaks of norovirus associated with raspberries from China or Eastern Europe that were irrigated with contaminated agricultural waters have been described (Hjertqvist et al. 2006). Outbreaks associated with food consumption, particularly fresh fruit (or frozen fresh fruit), vegetables and oysters, and ready-to-eat food, are frequent (Le Guyader et al. 2004, 2006; Hjertqvist et al. 2006).

Multiple outbreaks of norovirus gastroenteritis associated with raspberries have been reported, especially in Europe (Korsager et al. 2005; Hjertqvist et al. 2006). Similarly, outbreaks of hepatitis A infection have been reported as associated with raspberries, strawberries, green onions and lettuce (Hutin et al. 1999; Dentinger et al. 2001), diced tomatoes (Williams et al. 1995), raspberries (Ramsay and Upton 1989) and strawberries (Niu et al. 1992). Hernandez et al. (1997) suggested that lettuce contaminated with sewage could be a vehicle for hepatitis A virus and rotavirus.

References

Abdul-Raouf, U.M., L.R. Beuchat, and M.S. Ammar. 1993. Survival and growth of Escherichia coli O157:H7 on salad vegetables. *Applied and Environmental Microbiology* 59: 1999–2006.

AIFST. 2003. *Foodborne microorganisms of public health significance*, 6th ed. Waterloo, NSW: Australian Institute of Food Science and Technology Incorporated, NSW Branch, Food Biology Group. Chapter 13.

Altekruse, S.F., and L.K. Tollefson. 2003. Human campylobacteriosis: A challenge for the veterinary profession. *Journal of the American Veterinary Medical Association* 223: 445–452.

Anon. 2001a. Analysis and Evaluation of Preventive Control Measures for the Control and Reduction/Elimination of Microbial Hazards on Fresh and Fresh-Cut Produce, Chapter IV, Outbreaks Associated with Fresh Produce: Incidence, Growth, and Survival of Pathogens in Fresh and Fresh-Cut Produce. Rockville, MD: US Food and Drug Administration. Accessed November 26 2006. http://www.cfsan.fda.gov/~comm/ift3-4o.html.

Anon 2001b. *The Safe Sludge Matrix*. 3rd ed. Wolverhampton, UK: ADAS. Accessed January 18 2007. Downloadable from http://www.adas.co.uk/media_files/Publications/SSM.pdf.

Austin, J.W., K.L. Dodds, B. Blanchfield, and J.M. Farber. 1998. Growth and toxin production by *Clostridium botulinum* on inoculated fresh-cut packaged vegetables. *Journal of Food Protection* 61: 324–328.

Avery, L.M., K. Killham, and D.L. Jones. 2005. Survival of E. coli O157: H7 in organic wastes destined for land application. *Journal of Applied Microbiology* 98: 814–822.

Bean, N.H., and P.M. Griffin. 1990. Foodborne disease outbreaks in the United States, 1973-1987: Pathogens, vehicles and trends. *Journal of Food Protection* 53: 807–814.
Bean, N.H., J.S. Goulding, M.T. Daniels, and F.J. Angulo. 1997. Surveillance for foodborne disease outbreaks—United States, 1998–1992. *Journal of Food Protection* 60: 1265–1286.
Berbari, S.A.G., J.E. Paschoalino, and N.F.A. Silveira. 2001. Efeito do cloro na água de lavagem para desinfecção de alface minimamente processada. *Ciência e Tecnologia de Alimentos* 21: 197–201.
Berrang, M.E., R. Brackett, and L. Beuchat. 1989. Growth of *Listeria monocytogenes* on fresh vegetables stored under controlled atmosphere. *Journal of Food Protection* 52(10): 702–705.
Beuchat, L.R. 1992. Surface disinfection of raw produce. *Dairy, Food and Environmental Sanitation* 12: 6–9.
———. 1996. Pathogenic microorganisms associated with fresh produce. *Journal of Food Protection* 59: 204–261.
———. 1998. *Surface decontamination of fruits and vegetables eaten raw: A review*. Geneva, Switzerland: World Health Organisation, Food Safety Unit. WHO/FSF/FOS/98.2.
Beuchat, L.R., and R.E. Brackett. 1990. Survival and growth of Listeria monocytogenes on lettuce as influenced by shredding, chlorine treatment, modified atmosphere packaging and temperature. *Journal of Food Science* 55: 755–758.
Beuchat, L.R., R.E. Brackett, D.Y. Hao, and D.E. Conner. 1986. Growth and thermal inactivation of Listeria monocytogenes in cabbage and cabbage juice. *Canadian Journal of Microbiology* 32: 791–795.
Beuchat, L.R., M.E. Berrang, and R.E. Brackett. 1990. Presence and public health implications of Listeria monocytogenes on vegetables. In *Foodborne listeriosis*, ed. A.L. Miller, J.L. Smith, and G.A. Somkuti, 175–181. Amsterdam: Elsevier.
Boccaletti, S., and M. Nardella. 2000. Consumer willingness to pay for pesticide-free fresh fruits and vegetables in Italy. *International Food and Agribusiness Management Review* 3: 297–310.
Brackett, R.E. 1992. Microbiological safety of chilled foods current issues. *Trends in Food Science & Technology* 3(4): 81–85.
Brandl, M.T., A.F. Haxo, A.H. Bates, and R.E. Mandrell. 2004. Comparison of survival of Campylobacter jejuni in the phyllosphere with that in the rhizosphere of spinach and radish plants. *Applied and Environmental Microbiology* 70: 1182–1189.
Bunn, D., G.W. Feenstra, L. Lynch, and R. Sommer. 1990. Consumer acceptance of cosmetically imperfect produce. *Journal of Consumer Affairs* 24(2): 268–279.
Calder, L., G. Simmons, C. Thornley, P. Taylor, K. Pritchard, et al. 2003. An outbreak of hepatitis A associated with consumption of raw blueberries. *Epidemiology and Infection* 131: 745–751.
Carlin, F., and C. Nguyen-the. 1994. Fate of *Listeria monocytogenes* on four types of minimally processed green salads. *Letters in Applied Microbiology* 18: 222–226.
Carlin, F., C. Nguyen-the, and A. Abreu Da Silva. 1995. Factors affecting the growth of *Listeria monocytogenes* on minimally processed fresh endive. *Journal of Applied Bacteriology* 78: 636–646.
Cassin, M.H., A.M. Lammerding, E.C.D. Todd, W. Ross, and R.S. McColl. 1998. Quantitative risk assessment for Escherichia coli 0157:H7 in ground beef hamburgers. *International Journal of Food Microbiology* 41: 21–44.
Castillo, A., and E.F. Escartin. 1994. Survival of *Campylobacter jejuni* on sliced watermelon and papaya [a research note]. *Journal of Food Protection* 57(2): 166–168.
Catteau, M., C. Krembel, and G. Wauters. 1985. *Yersinia enterocolitica* in raw vegetables. *Sciences des Aliments* 5: 103–106. Hors Serie IV.
CDC. 1991. Multistate outbreak of *Salmonella poona* infections—United States and Canada, 1991. *Morbidity and Mortality Weekly Report* 40: 549–552.
———. 1997. Outbreak of *Escherichia coli* O157:H7 infections associated with eating alfafa sprouts—Michigan and Virginia, June–July 1997. *Morbidity and Mortality Weekly Report* 46(32): 741–744.
———. 2008. Outbreak of salmonella serotype Saintpaul infections associated with multiple raw produce items—United States, 2008. *MMWR. Morbidity and Mortality Weekly Report* 57: 929–934.

Chapman, P.A., C.A. Siddons, A.T. Cerdan Malo, and M.A. Harkin. 1997. A 1-year study of Escherichia coli O157 in cattle, sheep, pigs and poultry. *Epidemiology and Infection* 119: 245–250.

Cliver, D.O. 1985. Vehicular transmission of hepatitis A. *Public Health Reviews* 13: 235–292.

Conway, W.S., B. Leverentz, R.A. Saftner, W.J. Janisiewicz, C.E. Sams, and E. Leblanc. 2000. Survival and growth of Listeria monocytogenes on fresh-cut apple slices and its interaction with Glomerella cingulata and Penicillium expansum. *Plant Disease* 84: 177–181.

Crepet, A., I. Albert, C. Dervin, and F. Carlin. 2007. Estimation of microbial contamination of food from prevalence and concentration data: Application to Listeria monocytogenes in fresh vegetables. *Applied and Environmental Microbiology* 73: 250–258.

Darbas, H., M. Riviere, and J. Obertic. 1985. Yersinia in refrigerated vegetables. *Sciences des Aliments* 5: 81–84.

De Martinis, E.C.P., V.F. Alves, and B.D.G.M. Franco. 2002. Fundamentals and perspectives for the use of bacteriocins produced by lactic acid bacteria in meat products. *Food Reviews International* 18(2–3): 191–208.

De Roever, C. 1998. Microbiological safety evaluations and recommendations on fresh produce. Review. *Food Control* 69: 321–347.

———. 1999. Microbiological safety evaluations and recommendations on fresh produce. *Food Control* 9: 321–347.

Del Rosario, Brenda A., and Larry R. Beuchat. 1995. Survival and growth of enterohemorrhagic *Escherichia coli* O157:H7 in cantaloupe and watermelon. *Journal of Food Protection* 58(1): 105–107.

Dentinger, C.M., W.A. Bower, O.V. Nainan, S.M. Cotter, G. Myers, L.M. Dubusky, et al. 2001. An outbreak of hepatitis A associated with green onions. *Journal of Infectious Diseases* 183: 1273–1276.

Desmarchelier, P.M., and F.H. Grau. 1997. Escherichia coli. In *Foodborne microorganisms of public health significance*, ed. A.D. Hocking, G. Arnold, I. Jenson, K. Newton, and P. Sutherland, 231–264. Sydney, NSW: Australian Institute of Food Science and Technology. Chapter 7.

Doran, G., F. Sheridan, N. Delappe, C. O'Hare, W. Anderson, G. Corbett-Feeney, and M. Cormican. 2005. Salmonella enterica serovar Kedougou contamination of commercially grown mushrooms. *Diagnostic Microbiology and Infectious Disease* 51: 73–76.

Doyle, M.P., T. Zhao, J. Meng, and S. Zhao. 1997. Escherichia coli O157:H7. In *Food microbiology: Fundamentals and frontiers*, ed. M.P. Doyle, L.R. Beuchat, and T.J. Montville, 171–191. Washington, DC: ASM Press.

Eom, Y.S. 1994. Pesticide residue risk and food safety valuation: A random utility approach. *American Journal of Agricultural Economics* 76: 760–771.

Evans, M.R., C.D. Ribeiro, and R.L. Salmon. 2003. Hazards of healthy living: Bottled water and salad vegetables as risk factors for Campylobacter infection. *Emerging Infectious Diseases* 9: 1219–1225.

Everis, L. 2004. *Risks of pathogens in ready-to-eat fruits, vegetables, and salads through the production process*. Chipping Campden, England: Campden and Chorleywood Food Research Association Group. Review no. 44.

Farber, J.M., S.L. Wang, Y. Cai, and S. Zhang. 1998. Changes in populations of Listeria monocytogenes inoculated onto packaged fresh cut vegetables. *Journal of Food Protection* 61: 192–195.

FDA. 2001. FDA survey of imported fresh produce. U.S. Food and Drug Administration. http://www.cfsan.fda.gov/~dms/prodsur6.html.

———. 2003. FDA survey of domestic fresh produce. U.S. Food and Drug Administration. http://www.cfsan.fda.gov/~dms/prodsu10.html.

FDA/CFSAN. 2001. Escherichia coli O157:H7. In *Foodborne pathogenic microorganisms and natural toxins handbook (the "Bad Bug Book")*. New Hampshire, MD: Food and Drug Administration, Center for Food Safety & Applied Nutrition. Chap. 15. Disponível no site http://vm.cfsan.fda.gov/~mow/badbug.zip.

Feng, P. 1995. *Escherichia coli* serotype O157:H7: Novel vehicles of infections and emergence of phenotypic variants. *Emerging Infectious Diseases* 1(2): 1–9. Disponível no site http://www.cfsan.fda.gov/~mow/feng.htm.

Fisher, Tomeka L., and David A. Golden. 1998. Fate of *Escherichia coli* O157:H7 in ground apples used in cider production. *Journal of Food Protection* 61(10): 1372–1374.

Francis, G.A., C. Thomas, and D. O'Beirne. 1999. The microbiological safety of minimally processed vegetables. Review article. *International Journal of Food Science and Technology* 34: 1–22.

Frank, J.F, and K. Takeushi. 1999. Direct observation of *E. coli* O157:H7 inactivation on lettuce leaf using confocal scanning laser microscopy. In *Proceedings of International Conference of International Committee on Food Microbiology and Hygiene, Veldhoven, The Netherlands*, 795–797.

Fratamico, P.M., J.L. Smith, and R.L. Buchanan. 2002. Escherichia coli. In *Foodborne diseases*, ed. D.O. Cliver and H.P. Riemann, 79–101. San Diego, CA: Academic.

Gandhi, M., and M.L. Chikindas. 2007. *Listeria*: A foodborne pathogen that knows how to survive. *International Journal of Food Microbiology* 113: 1–15.

Garbutt, J. 1997. *Essentials of food microbiology*. Londres, Reino Unido: Arnold.

Germano, Manuel Leal, and Maria Isabel Simões Germano. 2001. *Higiene e Vigilâncias Sanitárias de Alimentos*. São Paulo, Brazil: Livraria Varela.

Gillespie, I.A., S.J. O'Brien, G.K. Adak, C.C. Tam, J.A. Frost, and F.J. Bolton. 2003. Point source outbreaks of Campylobacter jejuni infection—Are they more common than we think and what might cause them? *Epidemiology and Infection* 130: 367–375.

Gleeson, E., and D. O'Beirne. 2005. Effects of process severity on survival and growth of Escherichia coli and Listeria innocua on minimally processed vegetables. *Food Control* 16: 677–685.

Golden, D.A., E.J. Rhodehamel, and D.A. Kautter. 1993. Growth of *Salmonella* spp. in cantaloupe, watermelon, and honeydew melon. *Journal of Food Protection* 56: 194–196.

Gray, J.T., and P.J. Fedorka-Cray. 2002. Salmonella. In *Foodborne diseases*, ed. D.O. Cliver and H.P. Riemann, 55–68. San Diego, CA: Academic.

Gross, T.P., G. Ceade, G.W. Gary, D. Harting, D. Goeller, and E. Israel. 1989. An outbreak of acute infectious nonbacterial gastroenteritis in a high school in Maryland. *Public Health Reports* 104: 164–169.

Harris, M.V., T. Kimball, N.S. Weiss, and C. Nolan. 1986. Dairy products, produce and other non-meat foods as possible sources of Campylobacter jejuni and Campylobacter coli enteritis. *Journal of Food Protection* 49: 347–351.

Harris, L.J., J.N. Farber, L.R. Beuchat, M.E. Parish, T.V. Suslow, E.H. Garrett, and F.F. Busta. 2003. Outbreaks associated with fresh produce: Incidence, growth, and survival of pathogens in fresh and fresh-cut produce chapter III. *Comprehensive Reviews in Food Science and Food Safety* 2(Suppl s1): 78–141.

Harvey, J., and A. Gilmour. 1993. Occurrence and characteristics of Listeria in food produced in Northern Ireland. *International Journal of Food Microbiology* 19: 193–205.

Heisick, J.E., D.E. Wagner, M.L. Neierman, and J.T. Peeler. 1989. Listeria spp. found on fresh market produce. *Applied and Environmental Microbiology* 55: 1925–1927.

Hernandez, F., R. Monge, C. Jimenez, and L. Taylor. 1997. Rotavirus and hepatitis A virus in market lettuce (*Latuca sativa*) in Costa Rica. *International Journal of Food Microbiology* 37: 221–223.

Herwaldt, B.L., and M.L. Ackers. 1997. An outbreak in 1996 of cyclosporiasis associated with imported raspberries. The Cyclospora Working Group. New England Journal of Medicine 336 (22): 1548–1556.

Hjertqvist, M., A. Johansson, N. Svensson, P.E. Abom, C. Magnusson, et al. 2006. Four outbreaks of norovirus gastroenteritis after consuming raspberries, Sweden, June–August 2006. *Euro Surveillance* 11: E060907.

Hobbs, B.C. 1998. *Toxinfecções e controle higiênico-sanitário de alimentos*. São Paulo, Brazil: Livraria Varela.

Holland, B., I.D. Unwin, and D.H. Buss. 1992. *Fruit and nuts. First supplement to 5th edition of McCance and Widdowson's The composition of foods*. Cambridge, England: Royal Society of Chemistry.

Hudson, J.A., J.S. Mott, and N. Penney. (1994). Growth of Listeria monocytogenes, Aeromonas hydrophilla, and Yersinia enterocolitica on vacuum and saturated carbon dioxide controlled atmosphere packaged sliced roast beef. *Journal of Food Protection* 57: 204–208.

Hurst, W.C. 1995. Sanitation of lightly processed fruits and vegetables. *HortScience* 30(1): 22–24.

Hutin, Y.J., V. Pool, E.H. Cramer, O.V. Nainan, J. Weth, I.T. Williams, et al. 1999. A multistate, foodborne outbreak of hepatitis A. National Hepatitis A Investigation Team. *New England Journal of Medicine* 340: 595–602.

International Commission on Microbiological Specifications for Foods (ICMSF). 1978. *Microorganisms in foods: Their significance and methods of enumeration*, vol. 1, 2nd ed. Toronto, Canada: University of Toronto Press.

———. 1996. *Microorganisms in foods: Characteristics of microbial pathogens*, vol. 5, 513. London: Blackie Academic & Professional.

James, J.B., and T. Ngarmsak. 2011. Processing oh-cut tropical fruits and vegetables: A technical guide. In ed. Rosa S. Rolle, 86p.

Jay, J.M. 2000. *Modern food microbiology*, 6th ed, 511–525. Gaithersburg, MD: Aspen Publishers.

Johnson, D.B., S. Beaudouin, L.T. Smith, S.A. Beresford, and J.P. Logerfo. 2004. Increasing fruit and vegetable intake in home-bound elders: The Seattle Senior Farmers' Market Nutrition Pilot Program. *Preventing Chronic Disease* 1: A03.

Kallander, K.D., A.D. Hitchins, G.A. Lancette, J.A. Schmieg, G.R. Garcia, H.M. Solomon, and J.N. Sofos. 1991. Fate of *Listeria monocytogenes* in shredded cabbage stored at 5 and 25°C under a modified atmosphere. *Journal of Food Protection* 54: 302–304.

Kapperud, G. 1991. Yersinia enterocolitica in food hygiene. *International Journal of Food Microbiology* 12: 53–66.

Kärenlampi, R., and M.L. Hänninen. 2004. Survival of *Campylobacter jejuni* on various fresh produce. *International Journal of Food Microbiology* 97: 187–195.

Karitsky, J.N., M.T. Osterholm, H.B. Greenberg, J.A. Keriath, J.R. Godes, C.W. Hedberg, J.C. Forfang, A.Z. Kapikian, J.C. McCullough, and K.E. White. 1984. Norwalk gastroenteritis: A community outbreak associated with bakery product consumption. *Annals of Internal Medicine* 100: 519–521.

Kirk, M., et al. 1997. A prolonged outbreak of Campylobacter infection at a training facility. *Communicable Diseases Intelligence* 21(12): 57–61.

Knight, P. 1993. Hemorragic *Escherichia coli:* The danger increases. *ASM News* 59(5): 247–250.

Koopmans, M., and E. Duizer. 2004. Foodborne viruses: An emerging problem. *International Journal of Food Microbiology* 90: 23–41.

Korsager, B., S. Hede, H. Boggild, B.E. Bottiger, and K. Molbak. 2005. Two outbreaks of norovirus infections associated with the consumption of imported frozen raspberries, Denmark. *Euro Surveillance* 10. pii=2729.

Kudva, I.T., K. Blanch, and C.J. Hovde. 1998. Analysis of Escherichia coli O157:H7 survival in ovine or bovine manure and manure slurry. *Applied and Environmental Microbiology* 64: 3166–3174.

Kuritsky, J.N., M.T. Osterholm, H.B. Greenberg, J.A. Korlath, J.R. Godes, C.W. Hedberg, et al. 1984. Norwalk gastroenteritis: A community outbreak associated with bakery product consumption. *Annals of Internal Medicine* 100: 519–521. doi: https://doi.org/10.7326/0003-4819-100-4-519.

Kumar, A., R.K. Agarwal, K.N. Bhilegaonkar, B.R. Shome, and V.N. Bachhil. 2001. Occurrence of Campylobacter jejuni in vegetables. *International Journal of Food Microbiology* 67: 153–155.

Lampel, K.A., and A.T. Maurelli. 2002. Shigella. In *Foodborne diseases*, 2nd ed, ed. O.D. Cliver and H.P. Riemann, 69–77. London: Academic Press.

Larkin, E.P., J.T. Tierney, J. Lovett, D. Van Donsel, and D.W. Francis (1978). Land application of sewage wastes: potential for contamination of foodstuffs and agricultural soils by viruses, bacterial pathogens and parasites. In *State of Knowledge in Land Treatment of Wastewater, International Symposium*, vol. 2. Hanover, NH: U.S. Army Corps of Engineers.

Larsen, A.E., and E.A. Johnson. 1999. Evaluation of botulinal toxin production in packaged fresh-cut cantaloupe and honeydew melons. *Journal of Food Protection* 62: 948–952.

Le Guyader, F.S., C. Mittelholzer, L. Haugarreau, K.-O. Hedlund, R. Alsterlund, M. Pommepuy, and L. Svensson. 2004. Detection of noroviruses in raspberries associated with a gastroenteritis outbreak. *International Journal of Food Microbiology* 97: 179–186.

Le Guyader, F.S., F. Bon, D. Demedici, S. Parnaudeau, A. Bertone, et al. 2006. Detection of multiple noroviruses associated with an international gastroenteritis outbreak linked to oyster consumption. *Journal of Clinical Microbiology* 44: 3878–3882.

Lynch, L. 1991. Consumers choose lower pesticide use over picture perfect products. *Food Review* 14(1): 9–11.

Madden, J.M. 1992. Microbial pathogens in fresh produce—The regulatory perspective. *Journal of Food Protection* 55: 821–823.

Malizio, C.J., and E.A. Johnson. 1991. Evaluation of the botulism hazard from vacuum packed Enoki mushrooms (Flammulina velutipes). *Journal of Food Protection* 54: 20–21.

March, S.B., and S. Ratnam. 1986. Sorbitol MacConkey medium for detection of *Escherichia coli* O157:H7 associated with haemorragic colitis. *Journal of Clinical Microbiology* 23: 869–872.

Martin-Belloso, O., and R.S. Soliva-Fortuny. 2010. *Advances in Fresh-Cut Fruits and Vegetables Processing*. Boca Raton, FL: CRC Press. 424 p.

Mead, P.S., L. Slutsker, V. Dietz, L.F. McCaig, J.S. Bresee, C. Shapiro, P.M. Griffin, and R.V. Tauxe. 1999. Food-related illness and death in the United States. *Emerging Infectious Diseases* 5(5): 607–625.

Meng, J., M.P. Doyle, T. Zhao, and S. Zhao. 1994. Detection and control of *Escherichia coli* O157:H7 in foods. *Trends in Food Science & Technology* 5: 179–185.

Michino, H., K. Araki, S. Minami, S. Takaya, N. Sakai, M. Miyazaki, A. Ono, and H. Yanagawa. 1999. Massive outbreak of Escherichia coli O157:H7 infection in schoolchildren in Sakai City, Japan, associated with consumption of white radish sprouts. *American Journal of Epidemiology* 150: 787–796.

Moller Nielsen, E., M.N. Skov, J.J. Madsen, J. Lodal, J. Brochner Jespersen, and D.L. Baggesen. 2004. Verocytotoxin-producing *Escherichia coli* in wild birds and rodents in close proximity to farms. *Applied and Environmental Microbiology* 70: 6944–6947.

Molina, P.M., A.E. Parma, and M.E. Sanz. 2003. Survival in acid and alcoholic medium of Shiga toxin-producing *Escherichia coli* O157:H7 and non-O157:H7 isolated in Argentina. *BMC Microbiology* 3: 17.

Nachamkin, I. 1997. Campylobacter jejuni. In *Food microbiology: Fundamentals and frontiers*, ed. M.P. Doyle, L.R. Beuchat, and T.J. Montville, 159–170. Washington, DC: American Society for Microbiology Press.

NACMCF. 1991. Listeria monocytogenes. Recommendations by the National Advisory Committee on microbiological criteria for foods. *International Journal of Food Microbiology* 14: 185–246.

Nascimento, M.S., N. Silva, and M.M. Okazaki. 2003. Avaliação comparativa da eficácia de cloro, vinagre, ácido acético e ácido peracético na redução da população de microrganismos aeróbios mesófilos em verduras e frutas. *REV NET- DTA* 3(6). 3 de novembro de 2003.

Nguyen-the, C., and F. Carlin. 1994. The microbiology of minimally processed fresh fruits and vegetables. *Critical Reviews in Food Science and Nutrition* 34(4): 371–401.

Nguz, K., J. Shindano, S. Samapundo, and A. Huyghebaert. 2005. Microbiological evaluation of fresh-cut organic vegetables produced in Zambia. *Food Control* 16: 623–628.

Nicholson, F.A., S.J. Groves, and B.J. Chambers. 2005. Pathogen survival during livestock manure storage and following land application. *Bioresource Technology* 96: 135–143.

Niu, M.T., L.B. Polish, B.H. Robertson, B.K. Khanna, B.A. Woodruff, C.N. Shapiro, M.A. Miller, J.D. Smith, J.K. Gedrose, M.J. Alter, and H.S. Margoles. 1992. Multistate outbreak of hepatitis A associated with frozen strawberries. *Journal of Infectious Disease* 166: 518–524.

Nygard, K., J. Lassen, L. Vold, Y. Andersson, I. Fisher, S. Lofdahl, et al. 2008. Outbreak of Salmonella Thompson infections linked to imported rucola lettuce. *Foodborne Pathogens and Disease* 5: 165–173.

Obiri-Danso, K., N.D. Paul, and K. Jones. 2001. The effects of UVB and temperature on the survival of natural populations and pure cultures of Campylobacter jejuni, Camp. coli, Camp. lari and urease-positive thermophilic campylobacters (UPTC) in surface waters. *Journal of Applied Microbiology* 90: 256–267.

Olsen, S.J., L.C. MacKinnon, J.S. Goulding, N.H. Bean, and L. Slutsker. 2000. Surveillance for foodborne-disease outbreaks United States, 1993–1997. *MMWR CDC Surveillance Summary* 49 (1): 1–62.

Pao, S., G.E. Brown, and K.R. Schneider. 1998. Challenge studies with selected pathogenic bacteria on freshly peeled Hamlin orange. *Journal of Food Science* 63: 359–362.

Park, C.E., and G.W. Sanders. 1992. Occurrence of thermotolerant Campylobacters in fresh vegetables sold at farmer's outdoor markets and supermarkets. *Canadian Journal of Microbiology* 38: 313–316.

Price, R.J. 1997. *Compendium of fish and fishery products processing methods*, 45. Davis, CA: HACCP.

Ramsay, C.N., and P.A. Upton. 1989. Hepatitis A and frozen raspberries. *Lancet* 1: 43–44.

Ray, B. 1996. *Fundamental food microbiology*. Boca Raton, FL: CRC Press.

Regmi, A., N. Ballenger, and J. Putman. 2004. Globalisation and income growth promote the Mediterranean diet. *Public Health Nutrition* 7: 977–983.

Satchell, F.B., P. Stephenson, W.H. Andrews, L. Estela, and G. Allen. 1990. The survival of Shigella sonnei in shredded cabbage. *Journal of Food Protection* 53: 558–562. 624.

Semanchek, J.J., and D.A. Golden. 1996. Survival of *Escherichia coli* O157:H7 during fermentation of apple cider. *Journal of Food Protection* 59: 1256–1259.

Seymour, I.J., and H. Appleton. 2001. A review, foodborne viruses and fresh produce. *Journal of Applied Microbiology* 91: 759–773.

Sivaplasingham, S., C.R. Friedman, L. Cohen, and R.V. Tauxe. 2004. Fresh produce: A growing cause of outbreaks of foodborne illness in the United States, 1973 through 1997. *Journal of Food Protection* 67: 2342–2353.

Smith, B.P., D.W. Dilling, and M. Thurmand. 1993. How common is Salmonella on dairies? In *Proceedings of the 97th Annual Meeting of the US Animal Health Association, October 23–29, Las Vegas, NV*, 460–462. Richmond, VA: U.S. Animal Health Association.

Smith DeWaal, C., and K. Barlow. 2002. *Outbreak alert! Closing the gaps in our federal food safety net*, 5th ed. Washington, DC: Center for Science in the Public Interest.

Solomon, H.M., D.A. Kautter, T. Lilly, and E.J. Rhodehamel. 1990. Outgrowth of Clostridium botulinum in shredded cabbage at room temperature under modified atmosphere. *Journal of Food Protection* 53: 831–833.

Sopwith, W., M. Ashton, J.A. Frost, K. Tocque, S. O'Brien, M. Regan, and Q. Syed. 2003. Enhanced surveillance of Campylobacter infection in the North West of England 1997–1999. *Journal of Infection* 46: 35–45.

St. Louis, M.E., S.H.S. Peck, D. Bowering, C.B. Morgan, J. Blatherwick, S. Banerjee, G.D.M. Kettyls, W.A. Black, M.E. Milling, A.H.W. Hauschild, R.V. Tauxe, and P.A. Blake. 1988. Botulism from chopped garlic: Delayed recognition of a major outbreak. *Annals of Internal Medicine* 108: 363–368.

Stopforth, J.D., J.S. Ikeda, P.A. Kendall, and J.N. Stofos. 2004. Survival of acid-adapted or non-adapted Escherichia coli O157:H7 in apple wounds and surrounding tissue following chemical treatments and storage. *International Journal of Food Microbiology* 90: 51–61.

Sugiyama, H., and K.H. Yang. 1975. Growth potential of Clostridium botulinum in fresh mushrooms packaged in semipermeable plastic film. *Journal of Applied Microbiology* 30: 964–969.

Tauxe, R.V. 1997. Emerging foodborne diseases: An evolving public health challenge. *Dairy, Food and Environmental Sanitation* 17(12): 788–795.

Thunberg, R.L., T.T. Tran, R.W. Bennett, and R.N. Matthews. 2002. Research note: Microbial evaluation of selected fresh produce obtained at retail markets. *Journal of Food Protection* 65: 677–682.

Vierikko, A., M. Haarmann, A. Siitonen, P. Ruutu, and H. Rautelin. 2004. Domestically acquired campylobacter infections in Finland. *Emerging Infectious Diseases* 10: 127–130.

Wallace, J.S., T. Cheasty, and K. Jones. 1997. Isolation of verocytotoxigenic producing *Escherichia coli* O157 from wild birds. *Journal of Applied Microbiology* 82: 399–404.

Wang, G., T. Zhao, and M.P. Doyle. 1996. Fate of enterohemorrhagic Escherichia coli O157:H7 in bovine feces. *Applied and Environmental Microbiology* 62: 2567–2570.

Watada, A.E. 1997a. Quality maintenance of freshcut fruits and vegetables. *Food Biotechnology* 6: 229–233.

———. 1997b. Quality maintenance of fresh-cut fruits and vegetables. *Food Biotechnology* 6(4): 224–233.

Williams, I.T., B. Bell, D. Berry, and C. Shapiro. 1995. Foodborne outbreak of hepatitis A, Arkansas. In *Epidemic Intelligence Service 44th Annual Conference, March 27–31, 1994*, 19. Atlanta, GA: Centers for Disease Control and Prevention.

World Health Organization. 1998. Zoonotic non-O157 Shiga toxin producing Escherichia coli (STEC). Report of a WHO Scientific Working Group Meeting. http://www.who.int/emcdocuments/zoonoses/whocsraph988c.html.

———. 2002. Fact sheets: Campylobacter. WHO. http://www.who.int/inf-fs/en/fact255.html (accessed August 25, 2004).

Worsfold, D., C.J. Griffith, and P. Worsfold. 2004a. A survey of environmental health officers' views of food hygiene training. *British Food Journal* 106: 51–64.

Worsfold, D., P.M. Worsfold, and C.J. Griffith. 2004b. An assessment of food hygiene and safety at farmers' markets. *International Journal of Environmental Health Research* 14: 109–119.

Zhao, T., and M.P. Doyle. 1994. Fate of enterohemorrhagic *Escherichia coli* O157:H7 in commercial mayonnaise. *Journal of Food Protection* 57: 780–783.

Chapter 3
Nutritional and Functional Attributes of Fruit Products

Delia B. Rodriguez-Amaya and Jaime Amaya-Farfan

3.1 Introduction

The consumption of fruits is widely recommended because they are rich sources of micronutrients (e.g., vitamins) and bioactive phytochemicals (e.g., carotenoids, flavonoids). These highly important fruit constituents, however, may suffer alterations during processing and storage of food. Investigation of the changes and the influencing factors is necessary so as to diminish undesirable consequences, while promoting desirable effects. Processing and storage conditions should be optimized for maximum retention of valuable compounds for human health. This chapter focuses on ascorbic acid (vitamin C), carotenoids, flavonoids, and folates, principal health-promoting components of fruits.

In the last two decades, nonthermal technologies have been introduced. These technologies have been reported to inactivate microorganisms and enzymes, without provoking adverse thermal effects on sensory and nutritional properties. Both thermal and nonthermal processing effects will be discussed in this chapter.

3.2 Carotenoids

Carotenoids are fat-soluble natural pigments that confer yellow, orange, or red color to fruits, especially tropical fruits. Some carotenoids (e.g., β-carotene, α-carotene, β-cryptoxanthin) have provitamin A activity, and both vitamin A-active and vitamin A-inactive carotenoids have been credited with other health benefits: enhancement of immune function and reduction of the risk of developing degenerative diseases

---

D.B. Rodriguez-Amaya (✉) • J. Amaya-Farfan
University of Campinas, Campinas, SP, Brazil

A. Rosenthal et al. (eds.), *Fruit Preservation*, Food Engineering Series,
https://doi.org/10.1007/978-1-4939-3311-2_3

such as cancer, cardiovascular diseases, cataract, and macular degeneration (Krinsky and Johnson 2005; Tapiero et al. 2004; Voutilainen et al. 2006). These physiological activities have been widely attributed to antioxidant activity, but other mechanisms of the carotenoid action against chronic diseases have been increasingly cited: modulation of carcinogen metabolism, regulation of cell growth, inhibition of cell proliferation, enhancement of cell differentiation, stimulation of cell-to-cell gap junctional communication, retinoid-dependent signaling, modulation of DNA repair mechanism, induction of detoxifying enzymes, and filtering of blue light.

The carotenoids most commonly encountered in food are β-carotene, α-carotene, β-cryptoxanthin, lycopene, lutein, and violaxanthin (Rodriguez-Amaya et al. 2008a). Except violaxanthin, these are also the principal carotenoids encountered in human blood, and together with zeaxanthin, the most investigated in terms of human health. The carotenoids of fruits had been recently reviewed (Rodriguez-Amaya et al. 2008b).

3.2.1 Effects of Thermal Processing

The major alterations of the highly unsaturated carotenoid molecule during processing and storage are geometric isomerization and enzymatic or non-enzymatic oxidation (Rodriguez-Amaya 1997, 1999, 2002) (Fig. 3.1).

Isomerization of all-*E*-carotenoids (*trans*-carotenoids), the usual configuration in nature, to the Z-isomers (*cis*-isomers) is well documented. It is promoted by acids, heat, and light. The release of organic acids during slicing, pulping, or juicing of fruits can be sufficient to provoke *E–Z* isomerization, but this isomerization occurs to a greater

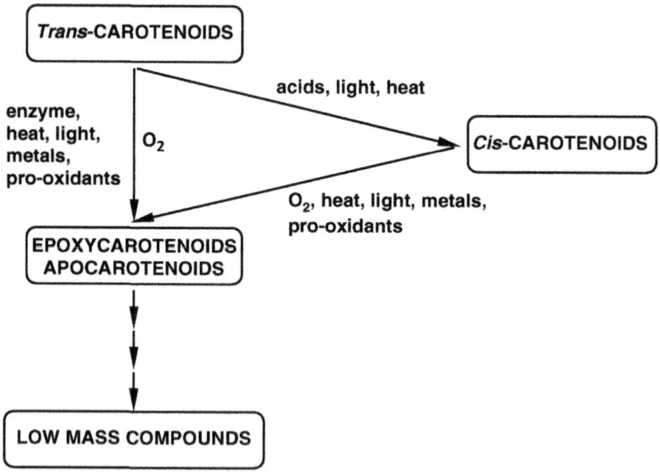

Fig. 3.1 Proposed scheme for the degradation of carotenoids. Taken from Rodriguez-Amaya (1999)

extent during thermal treatment. Consequently, the color turns paler, bioavailability is affected, and bioconversion of provitamin A carotenoids to vitamin A is reduced.

Z-provitamins A have long been known to have lower provitamin A activity than their all-*E*-isomers (Zechmeister 1962). Moreover, all-*E*-β-carotene was found to preferentially accumulate in human chylomicrons (Stahl et al. 1995) and human serum (Gaziano et al. 1995; Ben-Amotz and Levy 1996) compared to 9-*Z*-β-carotene. On the other hand, the human plasma was found to contain high levels of *Z*-lycopene isomers (Stahl et al. 1992; Schierle et al. 1997), contrary to the absolute predominance of all-*E*-lycopene in food. This shift to the *Z*-isomers of lycopene could not be accounted for by geometric isomerization during thermal processing of food and isomerization in the stomach provoked by the low pH; it was attributed to preferential absorption of these isomers (Boileau et al. 2002).

Enzymatic or nonenzymatic oxidation depends on the availability of oxygen and on the structure of the carotenoid (Rodriguez-Amaya 1999, 2002). It is promoted by light, heat, metals, enzymes, and peroxides and is inhibited by antioxidants.

Enzyme-catalyzed oxidation takes place prior to heat treatment, during peeling, slicing, and pulping; it can also occur in minimally processed food and in unblanched frozen food during thawing. Typically, carotenoid loss happens rapidly, immediately after tissue disruption, after which the carotenoid concentrations stabilize.

Nonenzymatic oxidation, often called autoxidation, during and after thermal processing is increased by destruction of the cellular structure, increase of surface area or porosity, duration and severity of processing, duration and conditions of storage, permeability of the packaging material to oxygen, and exposure to light.

Autoxidation during storage of processed food is usually characterized by a lag phase, followed by rapid decrease of the carotenoid content, coherent with a free radical mechanism. In thermally processed guava juice (Padula and Rodriguez-Amaya 1987), mango puree and slices (Godoy and Rodriguez-Amaya 1987), and papaya puree (Godoy and Rodriguez-Amaya 1991), nonenzymatic oxidation began only after 10 months of storage at room temperature.

Carotenoid oxidation is often accompanied by isomerization, and both the *Z*- and *E*-isomers are oxidized (Rodriguez and Rodriguez-Amaya 2007, 2009). The initial stages of oxidation involve epoxidation and cleavage to apocarotenals (Fig. 3.1). Subsequent fragmentations result in compounds with low molecular masses, similar to those produced in fatty acid oxidation. Now devoid of the color and biological activity of carotenoids, these volatile compounds could give rise to desirable flavor or off-flavor.

Numerous papers have reported variable, often considerable, losses of carotenoids during thermal processing. Earlier studies published up to 1996 were reviewed by Rodriguez-Amaya (1997), leading to the following conclusions:

- The stability of carotenoids differs in different food (different matrices), even when the same processing and storage conditions are used. Carotenoids per se have different susceptibilities to degradation.
- The main cause of carotenoid destruction during processing and storage of food is enzymatic or non-enzymatic oxidation. Enzymatic degradation of carotenoids may be a more serious problem than thermal decomposition in some food.

- Carotenoids are naturally protected in plant tissues; peeling, cutting, or pulping of fruits increases exposure to oxygen and brings together carotenoids and enzymes, which catalyze carotenoid oxidation. Thus, thermal processing should be carried out immediately after tissue disruption.
- Peeling and juicing physically remove substantial amounts of carotenoids, the losses often surpassing those of heat treatment.
- Whatever the processing method chosen, retention of carotenoids decreases with longer processing time, higher processing temperatures, and cutting or maceration of the food.
- The heat treatment in blanching may provoke some losses of carotenoids, but the inactivation of oxidative enzymes will prevent further and greater losses during slow processing (as in drying) and storage.
- Exclusion of oxygen, such as through vacuum or hot filling, oxygen-impermeable packaging or inert atmosphere, protection from light and storage at low temperatures all protect carotenoids from decomposition.

Carotenoid losses during thermal processing continue to be observed in more recent work, essentially confirming the findings of the earlier studies. In an orange juice-milk beverage, pasteurization (90 °C, 30 s) provoked significant reductions of 22.8 % and 22.5 %, respectively, in the xanthophylls lutein and zeaxanthin, but changes in provitamin A carotenoids (α-carotene, β-carotene, β-cryptoxanthin) were not significant (Zulueta et al. 2010). Thermal pasteurization (90 °C for 30 s) of orange juice, followed by rapid chilling in an ice water-bath, resulted in reduction of 46 % in violaxanthin and 25 % in antheraxanthin (Lee and Coates 2003). Microwave heating did not have a significant effect on the total xanthophyll and β-carotene contents of papaya puree at 285, 570, and 850 W for 30 s, but at 475 W for 45 and 60 s, there was a small but significant reduction of the xanthophyll level (De Ancos et al. 1999).

Drying slices of mango "Kent" and "Tommy Atkins" in an over-flow tray dryer for 3.5 h, the fruit temperature rising within 2 h from ambient to more than 50 °C in the dark, resulted in losses of 7 % and 31 %, respectively, of all-*E*-β-carotene (Pott et al. 2003). Freeze-drying significantly decreased the lycopene content of the grapefruit "Rio Red" pulp, the loss being greater in the control compared to irradiated samples (300 Gy) (Vanamala et al. 2005). In terms of β-carotene, reduction was significant only in the control sample.

3.2.2 Effects of Minimal Processing

Carotenoid biosynthesis may continue in fruits even after harvest, provided they are kept intact and biosynthetic enzymes are not inactivated (Rodriguez-Amaya 1997). This was more recently shown in nine Thai mango cultivars, in which accumulation of β-carotene occurred during post-harvest ripening (Vásquez-Caicedo et al. 2005). Loss of cellular integrity and of cell compartmentalization in fresh-cut fruits due to cutting operations destroys the natural protection of carotenoids and exposes them to oxidative enzymes. Thus, carotenoid losses can be observed, due to enzymatic oxidation prevailing over continued biosynthesis.

In fresh-cut fruits, no losses in carotenoids were observed in kiwi slices and watermelon cubes, but a 25 % loss was found in pineapple pieces and 10–15 % losses in cantaloupe cubes, mango cubes, and strawberry slices after 6 days at 5 °C (Gil et al. 2006). Except for watermelon, in which lycopene content was similar for the whole and the fresh-cut fruits, and kiwi, in which the total, lutein, and β-carotene levels in the fresh-cut fruits were slightly higher after 9 days, the carotenoid contents of whole fruits were generally higher than those of fresh-cut fruits. The carotenoid contents of the whole fruits increased in strawberry throughout the 9 days of storage and initially (up to 3 days) in the pineapple and kiwi whole fruits; in mango the total carotenoid content decreased in 3 days, but increased thereafter. In fresh-cut cubes of watermelon "Summer Flavor 800" and "Sugar Shack," there was slight decrease in lycopene after 7 days of storage at 2 °C, which was attributed to senescence, since losses did not occur before this period (Perkins-Veazie and Collins 2004). In orange, the total carotenoid content showed a significant increase for the whole fruits, whereas no significant change was observed in segments or peeled fruits during storage at 4 °C for 12 days (Plaza et al. 2011a).

Modified atmosphere packaging and low temperature storage are used to extend the shelf-life of minimally processed fruits. The effect of controlled-atmospheres of 2 % O_2, 12 % CO_2 in air, and 2 % O_2 + 12 % CO_2 on the quality and carotenoid content was assessed in sliced persimmon held for 8 days and peach held for 7 days at 5 °C (Wright and Kader 1997). Peach slices stored in air + 12 % CO_2 had lower β-carotene and β-cryptoxanthin contents. The different carotenoids in persimmon did not show a definite trend. In fresh-cut papaya "Maradol" stored at 5, 10, or 20 °C, loss of β-carotene was greater with higher temperature in both cubes and slices (Rivera-López et al. 2005). In both studies, the shelf-life based on the visual quality ended before significant loss of carotenoids occurred.

During 1 year of frozen storage of watermelon chunks or puree, lycopene suffered losses of ~30–40 % at −20 °C and ~5–10 % at −80 °C (Fish and Davis 2003). Lycopene was slightly more stable in the puree than in diced watermelon at −20 °C but not at −80 °C. β-carotene appeared to be more stable than lycopene during storage at −20 °C. Similarly, Lee and Coates (2002) observed greater than 20 % loss of lycopene and a 7 % loss in β-carotene in pink grapefruit stored at −23 °C for 12 months.

3.2.3 *Effects of High Pressure Processing*

High pressure (HP) treatments of orange juice at 100 MPa/5 min/60 °C, 350 MPa/2.5 min/30 °C, and 400 MPa/1 min/40 °C led to equal or higher concentrations of individual carotenoids. After 10 days of refrigerated storage, no significant change in total carotenoid content was observed with the first treatment while small losses were observed with the other two treatments (Sánchez-Moreno et al. 2003).

Moderate decrease (<11 %) of the carotenoids occurred in orange juice subjected to 400 MPa/1 min/40 °C towards the end of 20 days of storage at 4 °C (Plaza et al. 2011b).

HP-treated orange juice had the highest carotenoid content compared with untreated orange juice and those subjected to traditional thermal processing and pulsed electric field processing (Sánchez-Moreno et al. 2005). HP treatment at 350 MPa produced significant increases (20–43 %) in the carotenoid content of orange juice (de Ancos et al. 2002). An increase in time (beyond 5 min) or temperature (above 30 °C) did not improve the amount of carotenoids extracted.

Reported higher concentrations of carotenoids in HP-processed fruit products are not due to increased formation of carotenoids but to greater extractability, brought about by HP's effect on macromolecular structures, such as proteins and polymer carbohydrates, facilitating the release of carotenoids from the food matrix.

HP treatments at 50, 150, 300, and 400 MPa/15 min/25 °C of "Rojo Brillante" and "Sharon" persimmon fruit purees did not significantly lower the concentrations of individual carotenoids, but increased extractable amounts of some carotenoids (9–27 %) (de Ancos et al. 2000).

In watermelon juice, HP treatments at 300–900 MPa, 5–60 min, 60 °C, had a small impact on the all-*E*-lycopene, total *Z*-lycopene, and total lycopene concentrations (Zhang et al. 2011). It had the least effect on the carotenoid levels, compared with thermal and ultraviolet-C treatments, for which the concentrations were significantly lower with higher temperature and higher UV-C dose.

3.2.4 Effects of High-Intensity Pulsed Electric Field Processing

With high-intensity pulsed electric field (HIPEF) (30 kV/cm, 100 μs) processing, there was no significant decrease in the concentration of any carotenoid in comparison with the untreated orange juice (Cortés et al. 2006). Total carotenoids decreased 6.7 % when the juice was treated with HIPEF and 12.6 % when the juice was pasteurized (90 °C, 20 s). During refrigerated storage, carotenoids were maintained for a longer time in juice treated with HIPEF than in pasteurized juice.

HIPEF processing (at 25, 30, 35 and 40 kV/cm, 30–340 μs) of an orange-carrot juice mixture caused a significant increase in the concentrations of most of the carotenoids as treatment time increased (Torregrosa et al. 2005).

Watermelon juice showed high retention of lycopene when high electric field strengths (30–35 kV/cm), frequencies (50–250 Hz), and pulse widths (1–7 μs) were applied (treatment time of 50–2050 μs) (Oms-Oliu et al. 2009). Maximum relative lycopene content was obtained when HIPEF treatments were set at 35 kV/cm for 50 μs using 7 μs bipolar pulses at 200 Hz.

Pulsed electric fields (15, 25, 35, 40 kV/cm, from 40 to 700 μs) influenced the concentration of extracted carotenoids in an orange juice-milk beverage, producing a slight increase at 15 kV/cm and a slight decrease at 40 kV/cm (Zulueta et al. 2010).

3.3 Vitamin C

Vitamin C is a water-soluble micronutrient that is essential for the growth and repair of all tissues. It acts as an antioxidant and a cofactor in enzymatic and hormonal processes (Institute of Medicine 2006). It participates in the biosynthesis of carnitine, neurotransmitters, collagen (needed for healthy bones, teeth, gums, and blood vessels), and other component of connective tissue and modulates the absorption, transport, and storage of iron. As an effective scavenger of reactive oxygen species, it minimizes oxidative stress, thereby controlling inflammation and tissue damage associated with immune responses (Johnston 2006). It has been linked with the prevention of atherosclerosis and certain types of cancer.

In their review of preharvest and postharvest factors influencing vitamin C content of horticultural crops, Lee and Kader (2000) noted that:

- The retention of vitamin C is lowered by bruising and other mechanical injuries, and by excessive trimming.
- Irradiation at low doses (1 kGy or lower) has no significant effects on vitamin C content of fruits and vegetables.
- The loss of vitamin C after harvest can be reduced by storing fruits and vegetables in reduced O_2 and/or up to 10 % CO_2 atm; higher CO_2 levels can accelerate vitamin C loss.
- Vitamin C is subject to degradation during processing and cooking.
- Blanching reduces the vitamin C content, but limits further decreases during frozen storage by preventing the action of enzymes, particularly ascorbic acid oxidase.

3.3.1 Effects of Thermal Processing

Ascorbic acid is highly susceptible to oxidation, catalyzed by transition metals such as Cu^{2+} and Fe^{3+} and accelerated by heat and light. Degradation of ascorbic acid primarily involves oxidation to dehydroascorbic acid, followed by hydrolysis to 2,3-diketogulonic acid (Fig. 3.2). Further oxidation, dehydration, and polymerization take place, forming a wide array of nutritionally inactive products (Gregory 2008). Uncatalyzed oxidation is essentially negligible; oxidation catalyzed by trace metals in food accounts for much of the oxidative degradation of ascorbic acid. Highly soluble in aqueous solutions, ascorbic acid can also be lost significantly by leaching from cut or bruised surfaces of fruits.

Both ascorbic acid and dehydroascorbic acid have vitamin C activity. Loss of this vitamin activity occurs with the hydrolysis of dehydroascorbic acid to form 2,3-diketogulonic acid, this hydrolysis being favored by alkaline conditions. Dehydroascorbic acid is most stable at pH 2.5–5.5; its stability decreases as pH increases (Gregory 2008).

Vitamin C decreased during production of strawberry nectar and juice, the biggest losses being caused by pressing and pasteurization (Klopotek et al. 2005). The

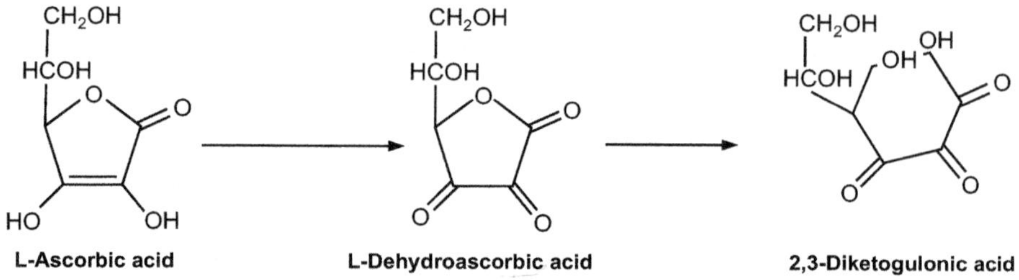

Fig. 3.2 Initial steps in the oxidative degradation of ascorbic acid

pressing process led to a vitamin C loss of about 22 %; the heat treatment (85 °C, ~5 min) caused a decrease of 35 % in the juice and 28 % in the nectar as compared to the filtered juice. Processing to puree resulted in a 12 % loss of this vitamin as compared to raw strawberries.

HP (400 MPa/40 °C/1 min), PEF (35 kV/cm, 750 μs), high pasteurization (90 °C, 1 min), and high pasteurization (70 °C, 30 s) + freezing (−38 °C/15 min) caused significant decreases (~7.79 %) in vitamin C of orange juice; low pasteurization and freezing did not change the vitamin C content (Sánchez-Moreno et al. 2005). In relation to total vitamin C, PEF, high pasteurization, high pasteurization + freezing, and freezing resulted in decreases (8.2 %), whereas HP and low pasteurization did not provoke any change.

Kiwi fruits were dried under varying conditions: air temperatures at 35, 45, 55, and 65 °C; mean velocities at 0.3, 0.6, and 0.9 ms^{-1}; relative humidity at 40, 55, 70, and 85 % (Kaya et al. 2010). The degradation of vitamin C was considerably affected by the drying conditions. Increasing drying air temperature increased decomposition of vitamin C in dried fruits. Conversely, loss of this vitamin was reduced with increasing relative humidity of the drying air.

3.3.2 *Effects of Minimal Processing*

Citrus fruits of different species and cultivars ("Red blush" grapefruit, "Palazzetti" mandarin-type fruit, "Minneola" tangelo and "Salustiana" and "Shamouti" orange) were minimally processed and cold-stored (4 °C) for up to 12 or 15 days (del Caro et al. 2004a). Ascorbic acid decreased significantly in "Minneola" and "Salustiana" segments, although only in the last samples. This vitamin also decreased in "Salustiana" juice.

Some losses were observed in ascorbic acid (19–24 %) and total vitamin C (ascorbic acid plus dehydroascorbic acid) contents (15–23 %) in whole fruits, hand-peeled fruits, and manually separated segments of oranges at the end of refrigerated storage (4 °C for 12 days), although no significant differences were found among the different samples (Plaza et al. 2011a).

In fresh-cut strawberries and persimmons, the post-cutting life based on visual quality ended before significant losses of ascorbic acid occurred (Wright and Kader 1997). Controlled atmospheres of 2 % O_2, air + 12 % CO_2, or 2 % O_2 + 12 % CO_2 had no significant effect on the total ascorbate content for either fruit.

3.3.3 Effects of High Pressure Processing

A high pressure treatment of 600 MPs at 40 °C for 4 min led to better retention of ascorbic acid during post-processing storage of orange juice, compared to conventional thermal pasteurization (80 °C, 60 s) (Polydera et al. 2005). Based on ascorbic acid retention, shelf-life was increased from 49 % (storage at 15 °C) to 112 % (storage at 0 °C), compared to the thermally pasteurized juice. This vitamin was not significantly reduced in orange and in an orange-carrot-lemon juice product processed at 500 and 800 MPa for 5 min (Fernandez García et al. 2001). During subsequent storage for 21 days at 4 °C, vitamin C reductions of 6–10 % in non-treated samples and 5–23 % in pressurized samples occurred, losses in 500 MPa treated samples being smaller than those treated at 800 MPa. In another study, orange juice processed at 800 MPa/25 °C/1 min suffered less than 20 % loss of ascorbic acid after storage for 3 months at 4 °C or 2 months at 15 °C (Nienaber and Shellhammer 2001).

Ascorbic acid degradation rates were lower in high pressurized (500 MPa, 35 °C, 5 min) reconstituted orange juice, resulting in an extension of its shelf-life compared to conventionally pasteurized juice (80 °C, 30 s) (Polydera et al. 2003). Based on ascorbic acid retention, the increase of shelf-life of high pressurized juice stored in bottles compared to the thermally pasteurized juice ranged from 11 % (storage at 15 °C) to 65 % (storage at 0 °C).

Orange juices subjected to 100 MPa/60 °C/5 min and 400 MPa/40 °C/1 min had 10.6 % and 6.9 %, respectively, of ascorbic acid reduction (10.3 % and 8.1 %, respectively, in total vitamin C, the sum of ascorbic acid and dehydroascorbic acid) just after HP treatment, whereas the juice submitted to 350 MPa/30 °C/2.5 min maintained the same level of this vitamin as the untreated juice (Sánchez-Moreno et al. 2003). This was taken to mean that loss of ascorbic acid was due to thermal degradation since it occurred at treatments with higher temperature.

3.3.4 Effects of High-Intensity Pulsed Electric Field Processing

High-intensity electric field pulses using varying field strengths (0.5, 1.0, and 2.0 kV/cm corresponding to 12, 48, and 192 J/kg, per pulse, respectively, with a pulse duration of 400 μs) and pulse numbers (2–50) were applied to apple slices as a pretreatment for osmotic dehydration (Taiwo et al. 2003). The vitamin C content of samples treated with 2 kV/cm was about 50–60 % lower than that of the untreated samples or of those treated with 0.5 and 1.0 kV/cm. Vitamin C also decreased with longer immersion time. Loss of ascorbic acid was attributed to diffusion of fruit solutes (e.g., ascorbic acid) toward the dehydration solution and to first-order deterioration.

Watermelon juice was subjected to HIPEF, at different electric field strengths (30–35 kV/cm), pulse frequencies (50–250 HZ), treatment times (20–2050 μs), and pulse polarity (monopolar/bipolar) (Oms-Oliu et al. 2009). Juices treated at 35 kV/cm for 50 μs at 50 Hz using mono- or bipolar 1-μs pulses exhibited the highest vitamin C retention. On the other hand, vitamin C loss was higher than 50 % when HIPEF treatment was set at 35 kV/cm, for 2050 μs at 250 Hz applying mono- or bipolar 7-μs pulses, thus severe HIPEF treatments led to greater vitamin C reduction.

3.4 Flavonoids

Flavonoids are widely distributed in plants. They can be classified into seven groups: flavones, flavanones, flavonols, flavanonols, isoflavones, flavanols (catechins), and anthocyanidins. Common biological and chemical properties of almost all flavonoids are antioxidant activity, ability to scavenge active oxygen species, ability to scavenge electrophiles, ability to inhibit nitrosation, ability to chelate metals, potential to produce hydrogen peroxide in the presence of certain metals, and capability to modulate certain cellular enzyme activities (Ho et al. 2008).

Flavonoids have been reported to have antioxidant, anticarcinogenic, antithrombotic, hepatoprotective, antiviral and anti-inflammatory activities (Duthie et al. 2000; Middleton et al. 2000; Reed 2002; Ross and Kasum 2002; Lule and Xia 2005; Chong et al. 2010).

Amarowicz et al. (2009) reviewed extensively the influence of postharvest processing and storage on the contents of phenolic acids (hydroxycinnamic and hydroxybenzoic acids) and flavonoids (flavanone, flavones, flavonols, monomeric flavanols, and anthocyanins), involving mostly vegetables with a few papers on fruits. They concluded that for most of the subclasses in question, the effect of storage and processing on the polyphenol content is negligible in comparison with the differences between different varieties of plants.

Among the flavonoids, anthocyanins play a dual role in fruits. They impart red, blue, and purple hues to fruits, especially temperate fruits, and are also associated with reduced risk for cardiovascular and neuronal illnesses, cancer, diabetes, and other diseases (Wu et al. 2002; Cho et al. 2003; Galvano et al. 2004; Stintzing and Carle 2004; Nichenametla et al. 2006). On the other hand, they are highly susceptible to degradation. Thus, most studies on processing effects on flavonoids focus on these pigments. The most common anthocyanins in food are pelargonidin-3-*O*-glucoside, cyanidin-3-*O*-glucoside, delphinidin-3-*O*-glucoside, peonidin-3-*O*-glucoside, petunidin-3-*O*-glucoside, and malvidin-3-*O*-glucoside.

Anthocyanin stability has been shown to be affected by various physical and chemical factors, such as the chemical structure and concentration of the anthocyanin, temperature, pH, light, oxygen, presence of enzymes, proteins, metallic ions, and other flavonoids and phenolics (Bridle and Timberlake 1997; Schwartz et al. 2008; Castañeda-Ovando et al. 2009; Patras et al. 2010). Generally, increased hydroxylation decreases stability, whereas increased methylation increases stability. Glycosylation of the anthocyanin, especially in C-3, increases stability and solubility in water. Acylation of sugar residues with acids also increases stability.

Anthocyanin molecules with complex patterns of glycosylation and acylation exhibit remarkable stability to pH changes, heat treatment, and light exposure. The stabilization has been attributed to intramolecular and intermolecular co-pigmentation, self-association, metal complexing, and presence of inorganic salts.

Anthocyanins change color reversibly with pH. These compounds are encountered in different forms and show diverse colors at different pHs. In aqueous medium, including food, anthocyanins can exist in four possible forms, depending on the pH: the red cation favilium ($pH<3$), the colorless pseudo base carbinol, or a colorless or pale yellow chalcone (pH 3–6), and the purple blue quinoidal base (pH 7–8).

Anthocyanins and ascorbic acid disappear simultaneously in fruit juices; this was initially attributed to a direct interaction between the two molecules. It is now believed that it is the hydrogen peroxide formed from the oxidation of ascorbic acid that attacks the C-2 position of the anthocyanin, provoking the cleavage of the pyrylium ring and producing colorless esters and coumarin derivatives (Schwartz et al. 2008).

3.4.1 Effects of Thermal Processing

Anthocyanins are highly reactive; they easily degrade or react with other constituents, forming colorless or brown compounds. The mechanisms of anthocyanin degradation during thermal processing are not well established. Suggested pathways involve polymerization (browning) or cleavage (loss of color) (Schwartz et al. 2008). In alkaline or acid medium, in the presence of the enzyme β-glucosidase, anthocyanin is hydrolyzed, releasing the anthocyanidin, which can be transformed first into the carbinol base and subsequently into α-diketone (Fig. 3.3). The latter can polymerize, forming brown products, or fragment into an aldehyde and a derivative of benzoic acid, both of which are colorless. On the other hand, anthocyanin can also be converted to carbinol and later into chalcone, and finally cleaved into a coumarin derivative and a compound corresponding to the B ring. Coumarin 3,5-diglucoside is a common product of the degradation of anthocyanins (3,5-diglycosides) The degradation depends on the anthocyanin present and the temperature.

In heat-treated (over 6 h at 95 °C, pH 3.5) elderberry and strawberry pigment isolates, acylated anthocyanins initially underwent hydrolysis, splitting the acyl-glycoside moieties from the flavylium backbone, forming anthocyanidin (aglycone) (Sadilova et al. 2006, 2007). Pentoses were more readily split off than hexoses. Opening of the pyrylium ring then followed. Finally, phenolic acids (4-hydroxybenzoic acid from

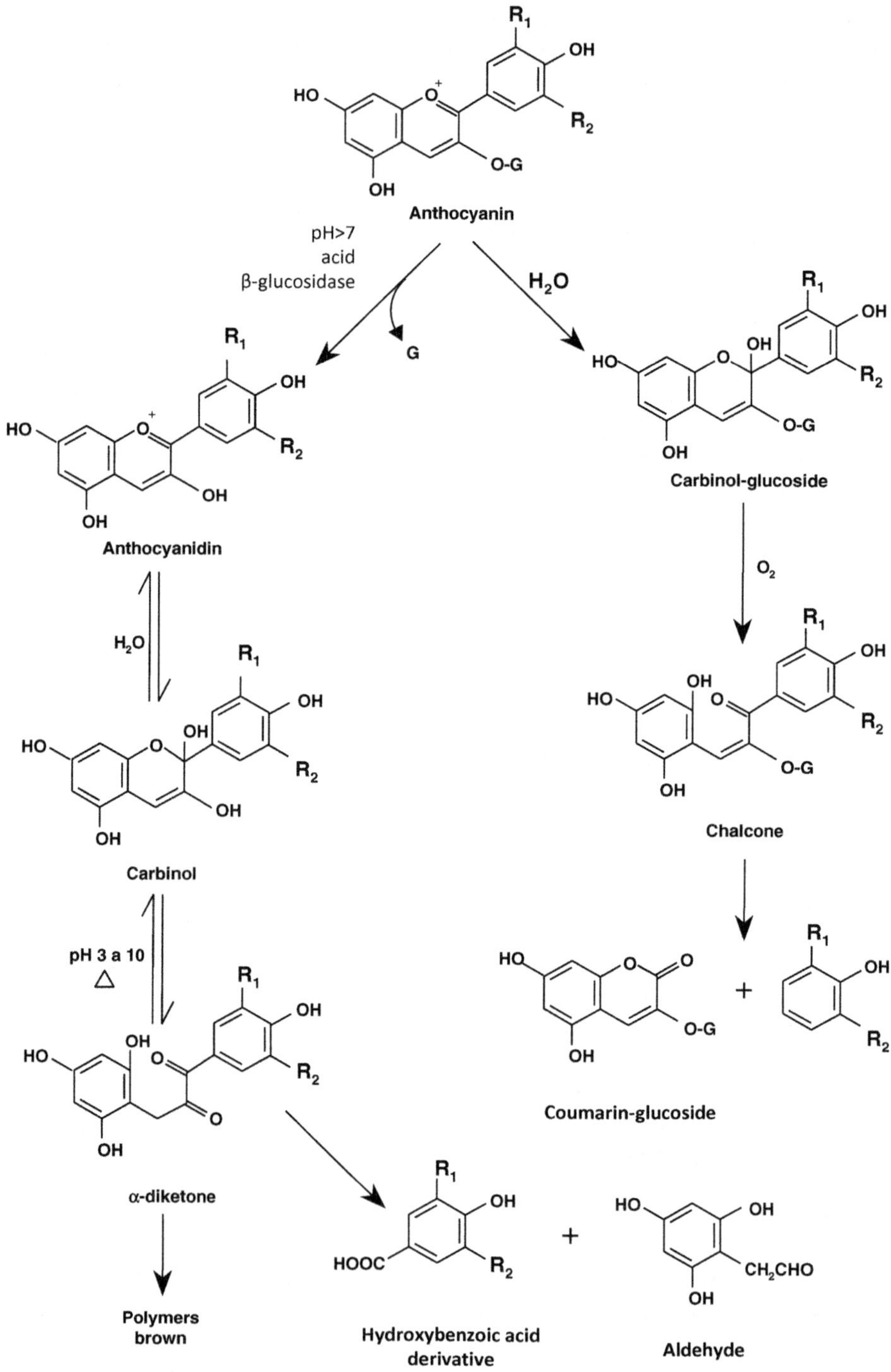

Fig. 3.3 Transformations of anthocyanins during thermal processing. Based on Stintzing and Carle (2004), with modifications

pelargonidin and protocatechuic acid from cyanidin) and phloroglucinaldehyde (form both cyanidin and pelargonidin) were formed as terminal degradation products, residues of the B- and A-ring, respectively.

Pasteurization (85 °C, ~5 min) led to a decrease (related to the filtered juice) in total anthocyanins of 27 % in strawberry juice and 39 % in strawberry nectar (Klopotek et al. 2005). Alteration of the total anthocyanin content was modest during puree production.

The use of frozen strawberries significantly improved color stability of the nectar, allowing production of this product with a shelf-life of up to 12 months (Gössinger et al. 2009). The half-life of anthocyanin monomers increased significantly.

Considerable losses of anthocyanins occurred during the processing of canned strawberry and jam (Ngo et al. 2007). Compared to the fresh samples, the total combined anthocyanin in canned strawberries and in the syrup decreased as much as 68.8 %. The pronounced loss and leaching of anthocyanin explained the yellow-colored appearance of the canned berries. Based on anthocyanin differences between jam and frozen fruits, approximately 70 % loss of anthocyanin occurred during manufacture of jam from frozen berries.

Substantial losses of anthocyanins and polyphenolics occurred when blueberries were processed into juice and concentrate; different classes of compounds had varying susceptibility to degradation (Skrede et al. 2000). The highest losses occurred with milling and depectinization, aggravated by native polyphenol oxidase. Inactivation of this enzyme with a steam blanching step induced a significant recovery increase of anthocyanins and cinnamates (Rossi et al. 2003).

In the processing of frozen blueberries into juice and concentrate, only 13–23 % of the anthocyanins and 36–39 % of the polyphenolics were recovered in the pasteurized juice. There was considerable loss of total (60–65 %) and individual polyphenolics (cinnamic acids and flavonol-glycosides) in the initial steps (thawing, crushing, depectinization, and pressing) (Lee et al. 2002). The press-cakes held 15–20 % of the frozen berries' polyphenolics. Forty-two to 45 % of frozen berries' polyphenolics was lost during juice processing and not accounted in the final pasteurized juices and press cakes. Overall anthocyanin levels were higher in the treated samples (heat and SO_2), but polyphenolic levels remained similar to those in the control.

Processing blueberries into various forms (canned in syrup, canned in water, pureed, and juiced clarified or nonclarified) resulted in losses of 28–59 % of monomeric anthocyanins, the greatest losses occurring in clarified juices and the least in nonclarified juices (Brownmiller et al. 2008). Storage at 25 °C for 6 months resulted in dramatic losses in all the thermally processed products, ranging from 62 % in canned blueberries in water and 85 % in clarified juice. In canned products, significant amounts of monomeric anthocyanins (14–25 %) leached out of the berries into the liquid canning medium.

Only 12–27 % of the anthocyanins and 20–32 % of the polyphenolics (hydrobenzoic acids and flavonol glycosides) were recovered in ultrahigh-temperature bayberry juices (Fang et al. 2006). Fifty-two to 58 % of the anthocyanins and 30–35 % of the polyphenolics were left in the centrifuged cakes. Substantial losses in total and individual polyphenolics occurred in the initial steps of blanching, crushing, pasteurization, and depectinization. Total and monomeric anthocyanins were significantly higher in the pasteurized and blanched samples than in the SO_2-treated samples, which had higher levels than the control juice. Overall, polyphenolic levels were significantly higher in the pasteurized and blanched samples than in the SO_2-treated and control samples. Flavonol deoxyhexosides were more stable than the flavonol hexosides during bayberry juice processing.

In grape juice, the concentration of flavan-3-ols (catechins) was influenced, in decreasing order of importance, by pressing method, cultivar, pasteurization, and vintage (Fuleki and Ricardo-da-Silva 2003). Cold pressing without maceration was the least and hot pressing after maceration at 60 °C for 60 min the most effective method for extracting flavan-3-ols. Pasteurization increased the concentration of catechins in cold-pressed juices, but decreased it in hot-pressed juices. The concentration of most proanthocyanidins increased with pasteurization.

Five processes at industrial scale (squeezing, mild pasteurization, standard pasteurization, concentration, and freezing) were compared (Gil-Izquierdo et al. 2002). The concentration process caused a mild precipitation of phenolics to the juice cloud. In the pulp, pasteurization led to degradation of several phenolic compounds, caffeic acid derivatives, and vicenin 2 and narirutin, with losses of 34.5 %, 30.7 %, and 28 %, respectively.

In comparison with fresh samples, the total anthocyanins in cabinet-dried blueberries and in cabinet-dried blueberries with osmotic pretreatment were significantly reduced, the loss being 41 % and 49 %, respectively (Lohachoompol et al. 2004). There was no significant decrease in the anthocyanin level in frozen samples during 3 months of storage.

The total phenolic levels in air-dried marionberries were 15.6–21.1 % lower than those found in frozen fruits, the difference between frozen and freeze-dried marionberries not being significant (Asami et al. 2003). On the other hand, in strawberries, the total phenolic levels in freeze-dried and air-dried samples were 36.5–42.6 % lower than those of frozen strawberries, no statistical difference being found between freeze-dried and air-dried strawberries.

Anthocyanin content was higher in low-temperature (60 °C) dried President prunes than in the high-temperature (85 + 70 °C) dried prunes (del Caro et al. 2004b). These pigments disappeared completely after 8 months of storage at 20 °C.

Microwave-vacuum and the combination of hot-air and microwave-vacuum dried red raspberries had higher retention of ellagic acid, quercetin, and kaempferol (glycoside and aglycone forms) as well as total polyphenols (Mejia-Meza et al. 2010). Freeze drying and microwave-vacuum drying resulted in higher retention of anthocyanins (aglycone form); anthocyanins appeared to be less heat-stable than polyphenols.

The following independent variables in açai powder production by spray drying were investigated: inlet air temperature (138–202 °C), feed flow rate (5–25 g/min), maltodextrin concentration (10–30 %) (Tonon et al. 2008). Anthocyanin retention was affected only by the inlet air temperature, due to the high sensitivity of these pigments to high temperature.

3.4.2 *Effects of Minimal Processing*

Several enzymes linked to anthocyanin degradation can enter into action once compartmentalization is lost in fresh-cut fruits. Glycosidases (anthocynases) may hydrolyze the glycosidic bonds of anthocyanins to yield the much more unstable anthocyanidins which undergo spontaneous degradation and decolorization. Both blueberry polyphenol oxidase (Skrede et al. 2000) and peroxidase (Kader et al. 2002) have been shown to be involved in anthocyanin degradation. It has also been proposed that polyphenol oxidase oxidizes polyphenolics to quinones, which subsequently react with anthocyanins to produce brown pigments (Kader et al. 1999; Jiménez and García-Carmona 1999).

Minimally processed segments and juices of citrus fruits of different species and cultivars showed different behavior with regard to the flavonoid content during storage at 4 °C (del Caro et al. 2004a). Total flavonoid (mainly hesperidin) increased significantly in the segments, but decreased in the juice during storage. In another study, the flavanone content showed a significant increase throughout refrigerated storage (4 °C for 12 days), but no significant differences were found among the samples (segments, peeled and whole) (Plaza et al. 2011a).

3.4.3 *Effects of High Pressure Processing*

HP treatment (400 MPa, 40 °C, 1 min) increased the extractability of flavanones in orange juice (Plaza et al. 2011b), while PEF (35 kV/cm, 750 μs) and low pasteurization (LP) (70 °C, 30 s) treated juices had similar levels to those of untreated juice. The flavanone content decreased significantly (around 50 %) during the first 20 days of storage at 4 °C for all treated juices, remaining higher in the HP juice than in the PEF and LP juices. An earlier work of the same group (Sánchez-Moreno et al. 2005) showed that HP treatment led to increased naringenin (20.16 %) and hesperetin (39.88 %) contents, whereas PEF treatment did not modify the flavanone content. In general, pasteurization plus freezing led to diminished naringenin content (16.04 %), with no modification in hesperetin.

3.5 Folates

Folate is a water-soluble B-complex vitamin that functions as a coenzyme in the acceptance and transfer of one carbon moieties involved in the synthesis, interconversion, and modification of nucleotides, amino acids, and other cellular components (Bailey and Gregory 2006; Institute of Medicine 2006). It has a specific role in the prevention of neural tube defects and its deficiency may cause megaloblastic anaemia. Epidemiological studies suggest that folate may protect against vascular disease, cancer, and mental disorders.

The term folate refers to two forms: naturally occurring folates in food and folic acid in supplements and fortified food. Folic acid (pteroylmonoglutamic acid) is found naturally only in trace quantities. The major naturally occurring forms of folate are polyglutamyl species of 5,6,7,8-tetrahydrofolates (H_4 folates); folate in fruits and vegetables mostly exist as 5-methyl-H_4folate,

3.5.1 Effects of Processing

Considering their reactivity and solubility in water, folates can potentially suffer extensive losses during food processing and preparation. Aside from oxidative degradation, considerable leaching to water used for washing, blanching, canning, or cooking can occur. However, few studies on processing effects on folates have been published, and these few papers are mostly on vegetables. Investigation of the processing effects on fruit folates is lacking.

Oxidative cleavage of H_4folates, H_2folates and, to a lesser extent, folic acid yields nutritionally inactive products (*p*-aminobenzoylglutamate and a pterin) (Gregory 2008). The mechanism of oxidation and the nature of the pterin produced during oxidative cleavage of H_4folate vary with pH.

5-Methyl-H_4folate can degrade to at least two products (Fig. 3.4) (Gregory 2008). The first has been tentatively identified as 5-methyl-H_2folate, which retains vitamin activity because it can be readily reduced back to 5-methyl-H_4folate by weak reductants such as thiols or ascorbate.

Stability of folates has been studied in buffer solutions, showing the effects of pH, oxygen, and temperature. Stability of H_4folate is maximum at pH 8–12 and 1–2 and minimum at pH 4–6 (Gregory 2008).

Based on detailed kinetic studies on HP stability of natural folates such as 5-formyl- and 5-methyl-H_4folate in buffer solutions, it was observed that different folates have different pressure and temperature stability (Nguyen et al. 2003, 2006; Indrawati et al. 2005). Folate degradation is enhanced by increasing pressure at constant temperature (above 40 °C) and by increasing temperature at constant pressure.

Mild technologies, e.g., mild pasteurization and high pressure processing (400 MPa or 600 MPa), resulted in good folate retentions, particularly for red oranges (>90 %). Traditional processing such as sterilization led to 25 % loss in red orange juice, but similar folate retention was obtained with mild processing and sterilization for grapefruit juice and pineapple juice. There was no decrease in folate content during the shelf-life of grapefruit juice; in pineapple juice, a slight decrease in folate content was seen (Jägerstad et al. 2005).

Temperature (70–120 °C) and pressure (from 50 to 200 MPa/25 °C and 500 MPa/60 °C) stabilities of 5-methyltetrahydrofolate were studied by Indrawati et al. (2004) in model systems and in some food products. 5-Methyltetrahydsrofolate in orange juice and kiwi puree was relatively temperature- (up to 120 °C) and pressure- (up to 500 MPa/60 °C) stable in contrast to carrot juice and asparagus. Endogenous ascorbic acid was found to play an important and positive role in increasing folate stability and seemed to be more important than the pH of the food medium.

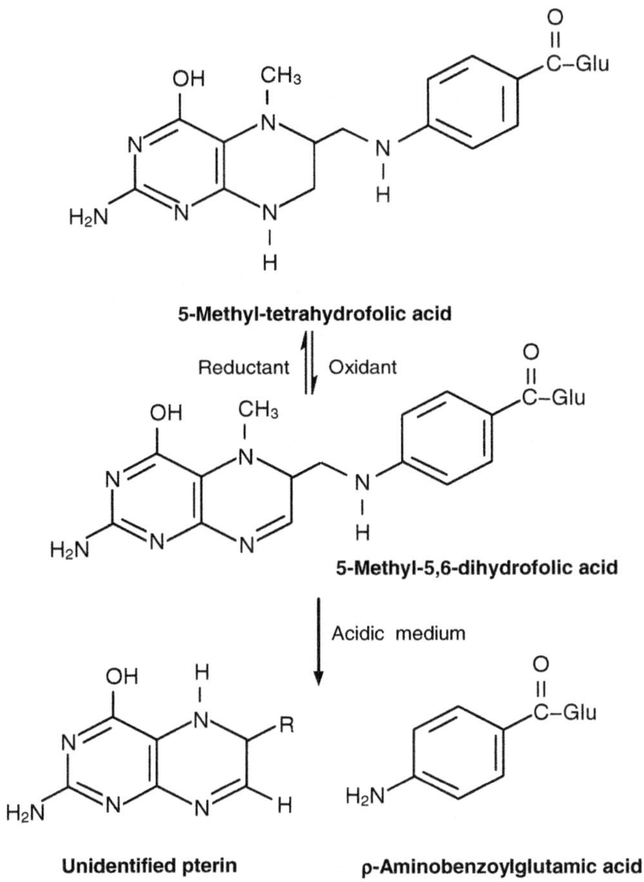

Fig. 3.4 Proposed mechanism for oxidative degradation of 5-methyl-tetra-hydrofolate

In a model orange juice (folates of orange juice were extracted and converted to folate monoglutamates), excess ascorbic acid strongly protected folates against pressure and heat (Butz et al. 2004). Pressurization at 600 MPa and 80 °C affected the model juice folates synergistically, and pressure increased the formation of 5,10-methenylfolate. In fresh orange juice, components other than ascorbate appeared to additionally stabilize folates, thus pressure preservation and pressure sterilization appeared very feasible.

References

Amarowicz, R., R. Carle, G. Dongowski, A. Durazzo, R. Galensa, D. Kammerer, G. Maiani, and M.K. Piskula. 2009. Influence of postharvest processing and storage on the content of phenolic acids and flavonoids in foods. *Molecular Nutrition & Food Research* 53: S151–S183.

Asami, D.K., Y.J. Hong, D.M. Barrett, and A.E. Mitchell. 2003. Comparison of the total phenolic and ascorbic acid content of freeze-dried and air-dried marionberry, strawberry, and corn grown using conventional, organic, and sustainable agricultural practices. *Journal of Agricultural and Food Chemistry* 51: 1237–1241.

Bailey, L. B., and J. F. Gregory. 2006. Folate. In *Present knowledge in nutrition*, ed. B.A. Bowman and R.M. Russell, 9th ed., 278–301. Washington, DC: ILSI Press.

Ben-Amotz, A., and Y. Levy. 1996. Bioavailability of a natural mixture compared with synthetic all-*trans*-β-carotene in human serum. *American Journal of Clinical Nutrition* 63: 729–734.

Boileau, T.W.M., A.C. Boileau, and J.W. Erdman Jr. 2002. Bioavailability of all-trans and cis-isomers of lycopene. *Experimental Biology and Medicine* 227: 914–919.

Bridle, P., and C.F. Timberlake. 1997. Anthocyanins as natural food colours—Selected aspects. *Food Chemistry* 58: 103–109.

Brownmiller, C., L.R. Howard, and R.I. Prior. 2008. Processing and storage effects on monomeric anthocyanins, percent polymeric color, and antioxidant capacity of processed blueberry products. *Journal of Food Science* 73: H72–H79.

Butz, P., Y. Serfert, A. Fernandez Garcia, S. Dietrich, R. Lindauer, A. Bognar, and B. Tauscher. 2004. Influence of high-pressure treatment at 25 °C and 80 °C on folates in orange juice and model media. *Journal of Food Science* 69: SNQ117–SNQ121.

Castañeda-Ovando, A., M.L. Pacheco-Hernández, M.E. Páez-Hernández, J.A. Rodríguez, and C.A. Galàn-Vidal. 2009. Chemical studies of anthocyanins: A review. *Food Chemistry* 113: 859–871.

Cho, J., J.S. Kang, P.H. Long, J. Jing, Y. Back, and K.S. Chung. 2003. Antioxidant and memory enhancing effects of purple sweet potato anthocyanin and *Cordyceps* mushroom extract. *Archives of Pharmacal Research* 26:821–825.

Chong, M.F.F., R. Macdonald, and J.A. Lovegrove. 2010. Fruit polyphenols and CVD risk: A review of human intervention studies. *British Journal of Nutrition* 104: S28–S39.

Cortés, C., F. Torregrosa, M.J. Esteve, and A. Frígola. 2006. Carotenoid profile modification during refrigerated storage in untreated and pasteurized orange juice and orange juice treated with high-intensity pulsed electric fields. *Journal of Agricultural and Food Chemistry* 54: 6247–6254.

de Ancos, B., M.P. Cano, A. Hernandez, and M. Monreal. 1999. Effects of microwave heating on pigment composition and colour of fruit purees. *Journal of the Science of Food and Agriculture* 79: 663–670.

de Ancos, B., E. Gonzalez, and M.P. Cano. 2000. Effect of high-pressure treatment on the carotenoid composition and the radical scavenging activity of persimmon fruit purees. *Journal of Agricultural and Food Chemistry* 48: 3542–3548.

de Ancos, B., S. Sgroppo, L. Plaza, and M.P. Cano. 2002. Possible nutritional and health-related value promotion in orange juice preserved by high-pressure treatment. *Journal of the Science of Food and Agriculture* 82: 790–796.

del Caro, A., A. Piga, V. Vacca, and M. Agabbio. 2004a. Changes of flavonoids, vitamin C and antioxidant capacity in minimally processed citrus segments and juices during storage. *Food Chemistry* 84: 99–105.

del Caro, A., A. Piga, I. Pinna, P.M. Fenu, and M. Agabbio. 2004b. Effect of drying conditions and storage period on polyphenolic content, antioxidant capacity, and ascorbic acid of prunes. *Journal of Agricultural and Food Chemistry* 52: 4780–4784.

Duthie, G.G., S.J. Duthie, and J.A.M. Kyle. 2000. Plant polyphenol in cancer and heart disease: Implications as nutritional antioxidants. *Nutrition Research Reviews* 13: 79–106.

Fang, Z., M. Zhang, Y. Sun, and J. Sun. 2006. How to improve bayberry (*Myrica rubra* Sieb. et Zucc.) juice color quality: Effect of processing on bayberry anthocyanins and polyphenolics. *Journal of Agricultural and Food Chemistry* 54: 99–106.

Fernandez García, A., P. Butz, A. Bògnar, and B. Tauscher. 2001. Antioxidative capacity, nutrient content and sensory quality of orange juice and an orange-lemon-carrot juice product after high pressure treatment and storage in different packaging. *European Food Research and Technology* 213: 290–296.

Fish, W.W., and A.R. Davis. 2003. The effects of frozen storage conditions on lycopene stability in watermelon tissue. *Journal of Agricultural and Food Chemistry* 51: 3582–3585.
Fuleki, T., and J.M. Ricardo-da-Silva. 2003. Effects of cultivar and processing method on the contents of catechins and procyanidins in grape juice. *Journal of Agricultural and Food Chemistry* 51: 640–646.
Galvano F., L. la Fauci, G. Lazzarino, V. Fogliano, A. Ritieni, S. Ciappellano, N. C. Battistini, B. Tavazzi, and G. Galvano. 2004. Cyanidins: metabolism and biological properties. *The Journal of Nutritional Biochemistry* 15: 2–11.
Gaziano, J.M., E.J. Johnson, R.M. Russell, J.E. Manson, M.J. Stampfer, P.M. Ridker, B. Frei, C.H. Hennekens, and N.I. Krinsky. 1995. Discrimination in absorption or transport of β-carotene isomers after oral supplementation with either all-*trans*- or 9-*cis*-β-carotene. *American Journal of Clinical Nutrition* 61: 1248–1252.
Gil, M.I., E. Aguayo, and A.A. Kader. 2006. Quality changes and nutrient retention in fresh-cut versus whole fruits during storage. *Journal of Agricultural and Food Chemistry* 54: 4284–4296.
Gil-Izquierdo, A., M.I. Gil, and Federico Ferreres. 2002. Effect of processing techniques at industrial scale on orange juice antioxidant and beneficial health compounds. *Journal of Agricultural and Food Chemistry* 50: 5107–5114.
Godoy, H.T., and D.B. Rodriguez-Amaya. 1987. Changes in individual carotenoids on processing and storage of mango (*Mangifera indica*) slices and purée. *International Journal of Food Science and Technology* 22: 451–460.
———. 1991. Comportamento dos carotenóides de purê de mamão (*Carica papaya*) sob processamento e estocagem. *Ciência e Tecnologia de Alimentos* 11: 210–220.
Gössinger, M., S. Moritz, M. Hermes, S. Wendelin, H. Scherbichler, H. Halbwirth, K. Stich, and E. Bedrghofer. 2009. Effects of processing parameters on color stability of strawberry nectar from puree. *Journal of Food Engineering* 90: 171–178.
Gregory III, J.F. 2008. Vitamins. In *Fennema's food chemistry*, ed. S. Damodaran, K.L. Parkin, and O.R. Fennema, 439–521. Boca Raton, FL: CRC.
Ho, C.T., M.M. Rafi, and G. Ghai. 2008. Bioactive substances: Nutraceuticals and toxicants. In *Fennema's food chemistry*, ed. S. Damodaran, K.L. Parkin, and O.R. Fennema, 751–779. Boca Raton, FL: CRC.
Indrawati, C. Arroqui, I. Messagie, M.T. Nguyen, A. van Loey, and M. Hendrickx. 2004. Comparative study on pressure and temperature stability of 5-methyltetrahydrofolic acid in model systems and in food products. *Journal of Agricultural and Food Chemistry* 52: 485–492.
Indrawati, I., A. van Loey, and M. Hendrickx. 2005. Pressure and temperature stability of 5-methyltetrahydrofolic acid: A kinetic study. *Journal of Agricultural and Food Chemistry* 53: 3081–3087.
Institute of Medicine. 2006. *Dietary DRI reference intakes: The essential guide to nutrient requirements*. Washington, DC: National Academy of Sciences.
Jägerstad, M., V. Piironen, C. Walker, G. Ros, E. Carnovale, M. Holasova, and H. Nau. 2005. Increasing natural food folates through bioprocessing and biotechnology. *Trends in Food Science and Technology* 16: 298–306.
Jiménez, M., and F. García-Carmona. 1999. Oxidation of the flavonol quercetin by polyphenol oxidase. *Journal of Agricultural and Food Chemistry* 47: 56–60.
Johnston, C.S. 2006. Vitamin C. In *Present knowledge in nutrition*, vol. 1, 9th ed, ed. B.A. Bowman and R.M. Russell, 233–241. Washington, DC: International Life Sciences Institute Press.
Kader, F., J. Nicolas, and M. Metche. 1999. Degradation of pelargonidin-3-glucoside in the presence of chlorogenic acid and blueberry polyphenol oxidase. *Journal of the Science of Food and Agriculture* 79: 517–522.
Kader, F., M. Irmouli, J.P. Nicolas, and M. Metche. 2002. Involvement of blueberry peroxidases in the mechanisms of anthocyanin degradation in blueberry juice. *Journal of Food Science* 67: 910–915.
Kaya, A., O. Aydin, and S. Kolayli. 2010. Effect of different drying conditions on the vitamin C (ascorbic acid) content of Haward kiwifruits (*Actinidia deliciosa* Planch). *Food and Bioproducts Processing* 88: 165–173.

Klopotek, Y., K. Otto, and V. Böhm. 2005. Processing strawberries to different products alters contents of vitamin C, total phenolics, total anthocyanins and antioxidant capacity. *Journal of Agricultural and Food Chemistry* 53: 5640–5646.
Krinsky, N.I., and E.J. Johnson. 2005. Carotenoid actions and their relation to health and disease. *Molecular Aspects of Medicine* 26: 459–516.
Lee, H. S., and G. A. Coates. 2002. Characterization of color fade during frozen storage of red grapefruit juice concentrates. *Journal of Agricultural and Food Chemistry* 50: 3988–3991.
Lee, H.S., and G.A. Coates. 2003. Effect of thermal pasteurization on Valencia orange juice color and pigments. *Lebensmittel-Wissenchaft und Technologie* 36: 153–156.
Lee, S.K., and A.A. Kader. 2000. Preharvest and postharvest factors influencing vitamin C content of horticultural crops. *Postharvest Biology and Technology* 20: 207–220.
Lee, J., R.W. Durst, and R.E. Wrolstad. 2002. Impact of juice processing on blueberry anthocyanins and polyphenolics: Comparison of two pretreatments. *Journal of Food Science* 67: 1660–1667.
Lohachoompol, V., G. Srzednicki, and J. Craske. 2004. The change of total anthocyanins in blueberries and their antioxidant effect after drying and freezing. *Journal of Biomedicine and Biotechnology* 5: 248–252.
Lule, S.U., and W. Xia. 2005. Food phenolics, pros and cons: A review. *Food Reviews International* 21: 367–388.
Mejia-Meza, E.I., J.A. Yáñez, C.M. Remsberg, J.K. Takemoto, N.M. Davies, B. Rasco, and C. Clary. 2010. Effect of dehydration on raspberries: Polyphenol and anthocyanin retention, antioxidant capacity, and antiadipogenic activity. *Journal of Food Science* 75: 5–12.
Middleton Jr., E., C. Kandaswami, and T.C. Theoharides. 2000. The effects of plant flavonoids on mammalian cells: Implications for inflammation, heart disease, and cancer. *Pharmacological Reviews* 52: 673–751.
Ngo, T., R.E. Wrolstad, and Y. Zhao. 2007. Color quality of Oregon strawberries—Impact of genotype, composition, and processing. *Journal of Food Science* 72: 25–32.
Nguyen, M.T., Indrawati, and M. Hendrickx. 2003. Model studies on the stability of folic acid and 5-methyltetrahydrofolic acid degradation during the thermal treatment in combination with high hydrostatic pressure. *Journal of Agricultural and Food Chemistry* 51: 3352–3357.
Nguyen, M.T., I. Oey, M. Hendrickx, and A. van Loey. 2006. Kinetics of (*6R,S*) 5-formyltetrahydrofolic acid isobaric-isothermal degradation in a model system. *European Food Research and Technology* 223: 325–331.
Nichenametla, S.N., T.G. Taruscio, D.L. Barney, and J.H. Exon. 2006. A review of the effects and mechanisms of polyphenolics in cancer. *Critical Reviews in Food Science and Nutrition* 46: 161–183.
Nienaber, U., and T.H. Shellhammer. 2001. High-pressure processing of orange juice: Combination treatments and shelf-life study. *Journal of Food Science* 66: 332–336.
Oms-Oliu, G., I. Odriozola-Serrano, R. Soliva-Fortuny, and O. Martín-Belloso. 2009. Effects of high-intensity pulsed electric field processing conditions on lycopene, vitamin C and antioxidant capacity of watermelon juice. *Food Chemistry* 115: 1312–1319.
Padula, M., and D.B. Rodriguez-Amaya. 1987. Changes in individual carotenoids and vitamin C on processing and storage on processing and storage of guava juice. *Acta Alimentaria* 16: 209–216.
Patras, A., N.P. Brunton, C. O'Donnell, and B.K. Tiwari. 2010. Effect of thermal processing on anthocyanin stability in foods; mechanisms and kinetics of degradation. *Trends in Food Science & Technology* 21: 3–11.
Perkins-Veazie, P., and J.K. Collins. 2004. Flesh quality and lycopene stability of fresh-cut water melon. *Postharvest Biology and Technology* 31: 159–166.
Plaza, L., I. Crespo, S. de Pascual-Teresa, B. de Ancos, C. Sánches-Moreno, M. Muñoz, and M.P. Cano. 2011a. Impact of minimal processing on orange bioactive compounds during refrigerated storage. *Food Chemistry* 124: 646–651.
Plaza, L., C. Sánchez-Moreno, B. de Ancos, P. Elez-Martínez, O. Martín-Belloso, and M.P. Cano. 2011b. Carotenoid and flavanone content during refrigerated storage of orange juice processed by high-pressure, pulsed electric fields and low pasteurization. *LWT--Food Science and Technology* 44: 834–839.

Polydera, A.C., N.G. Stoforos, and P.S. Taoukis. 2003. Comparative shelf life study and vitamin C loss kinetics in pasteurized and high pressure processed reconstituted orange juice. *Journal of Food Engineering* 60: 21–29.

———. Quality degradation kinetics of pasteurized and high pressure processed fresh Navel orange juice: Nutritional parameters and shelf life. *Innovative Food Science and Emerging Technologies* 6: 1–9.

Pott, I., M. Marx, S. Neidhart, W. Mühlbauer, and R. Carle. 2003. Quantitative determination of β-carotene stereoisomers in fresh, dried, and solar-dried mangoes (*Mangifera indica* L.). *Journal of Agricultural and Food Chemistry* 51: 4527–4531.

Reed, J. 2002. Cranberry flavonoids, atherosclerosis and cardiovascular health. *Critical Reviews in Food Science and Nutrition* 42: 301–316.

Rivera-López, J., F.A. Vásquez-Ortiz, J.F. Ayala-Zavala, R.R. Sotelo-Mundo, and G.A. González-Aguilar. 2005. Cutting shape and storage temperature affect overall quality of fresh-cut papaya cv. 'Maradol'. *Journal of Food Science* 70: 482–489.

Rodriguez, E.B., and D.B. Rodriguez-Amaya. 2007. Formation of apocarotenals and epoxycarotenoids from β-carotene by chemical reactions and autoxidation in model systems and processed foods. *Food Chemistry* 101: 563–572.

———. 2009. Lycopene epoxides and apo-lycopenals formed by chemical reactions and autoxidation in model systems and processed foods. *Journal of Food Science* 74: 674–682.

Rodriguez-Amaya, D.B. 1997. *Carotenoids and food preparation: The retention of provitamin carotenoids in prepared, processed, and stored foods*. Washington, DC: USAID-OMNI.

———. 1999. Changes in carotenoids during processing and storage of foods. *Archivos Latinoamericanos de Nutrición* 49: 38S–47S.

———. 2002. Effects of processing and storage on food carotenoids. *Sight and Life Newsletter* 3(Special Issue): 25–35.

Rodriguez-Amaya, D.B., M. Kimura, H.T. Godoy, and J. Amaya-Farfan. 2008a. Updated Brazilian database on food carotenoids: Factors affecting carotenoid composition. *Journal of Food Composition and Analysis* 21: 445–463.

Rodriguez-Amaya, D.B., J. Amaya-Farfan, and E.B. Rodriguez. 2008b. Carotenoids in fruits: Biology, chemistry, technology and health benefits. In *Current trends in phytochemistry*, ed. F. Epifano, 167–188. Kerala, India: Research Signpost.

Ross, J.A., and C.M. Kasum. 2002. Dietary flavonoids: Bioavailability, metabolic effects, and safety. *Annual Reviews of Nutrition* 22: 19–34.

Rossi, M., E. Giussani, R. Morelli, R. lo Scalzo, R.C. Nani, and D. Torreggiani. 2003. Effect of fruit blanching on phenolics and radical scavenging activity of highbush blueberry juice. *Food Research International* 36: 999–1005.

Sadilova, E., F.C. Stintzing, and R. Carle. 2006. Thermal degradation of acylated and nonacylated anthocyanins. *Journal of Food Science* 71: 504–512.

Sadilova, E., R. Carle, and F.C. Stintzing. 2007. Thermal degradation of anthocyanins and its impact on color and in vitro antioxidant capacity. *Molecular Nutrition & Food Research* 51: 1461–1471.

Sánchez-Moreno, C., L. Plaza, B. de Ancos, and M.P. Cano. 2003. Vitamin C, provitamin A carotenoids, and other carotenoids in high-pressurized orange juice during refrigerated storage. *Journal of Agricultural and Food Chemistry* 51: 647–653.

Sánchez-Moreno, C., L. Plaza, P. Elez-Martínez, B. de Ancos, O. Martín-Belloso, and M.P. Cano. 2005. Impact of high pressure and pulsed electric fields on bioactive compounds and antioxidant activity of orange juice in comparison with traditional thermal processing. *Journal of Agricultural and Food Chemistry* 53: 4403–4409.

Schierle, J., W. Bretzel, I. Büchler, N. Faccin, D. Hess, K. Steiner, and W. Schüep. 1997. Content and isomeric ratio of lycopene in food and human blood plasma. *Food Chemistry* 59: 459–465.

Schwartz, S.J., J.H. von Elbe, and M.M. Giusti. 2008. Colorants. In *Fennema's food chemistry*, ed. S. Damodaran, K.L. Parkin, and O.R. Fennema, 571–638. Boca Raton, FL: CRC.

Skrede, G., R.E. Wrolstad, and R.W. Durst. 2000. Changes in anthocyanins and polyphenolics during juice processing of highbush blueberries (*Vaccinium corymbosum* L.). *Journal of Food Science* 65: 357–364.

Stahl, W., W. Schwarz, A.R. Sundquist, and H. Sies. 1992. Cis-trans isomers of lycopene and β-carotene in human serum and tissues. *Archives of Biochemistry and Biophysics* 294: 173–177.

Stahl, W., W. Schwarz, J. von Laar, and H. Sies. 1995. All-*trans*-β-carotene preferentially accumulates in human chylomicrons and very low density lipoproteins compared with the 9-*cis* geometrical isomer. *The Journal of Nutrition* 125: 2128–2133.

Stintzing, F.C., and R. Carle. 2004. Functional properties of anthocyanins and betalains in plants, food, and in human nutrition. *Trends in Food Science & Technology* 15: 19–38.

Taiwo, K.A., A. Angersbach, and D. Knorr. 2003. Effects of pulsed electric field on quality factors and mass transfer during osmotic dehydration of apples. *Journal of Food Process Engineering* 26: 31–48.

Tapiero, H., D.M. Townsend, and K.D. Tew. 2004. The role of carotenoids in the prevention of human pathologies. *Biomedicine & Pharmacotherapy* 58: 100–110.

Tonon, R.V., C. Brabet, and M.D. Hubinger. 2008. Influence of process conditions on the physicochemical properties of acai (*Euterpe oleraceae* Mart.) powder produced by spray drying. *Journal of Food Engineering* 88: 411–418.

Torregrosa, F., C. Cortés, M.J. Esteve, and A. Frígola. 2005. Effect of high-intensity pulsed electric fields processing and conventional heat treatment on orange-carrot juice carotenoids. *Journal of Agricultural and Food Chemistry* 53: 9519–9525.

Vanamala, J., G. Cobb, N.D. Turner, J.R. Lupton, K.S. Yoo, L.M. Pike, and B.S. Patil. 2005. Bioactive compounds of grapefruit (*Citrus paradise* Cv. Rio Red) respond differently to postharvest irradiation, storage, and freeze drying. *Journal of Agricultural and Food Chemistry* 53: 3980–3985.

Vásquez-Caicedo, A.L., P. Sruamsiri, R. Carle, and S. Neidhart. 2005. *Journal of Agricultural and Food Chemistry* 53: 4827–4835.

Voutilainen, S., T. Nurmi, J. Mursu, and T.H. Rissanen. 2006. Carotenoids and cardiovascular health. *American Journal of Clinical Nutrition* 83: 1265–1271.

Wright, K.P., and A.A. Kader. 1997. Effect of controlled-atmosphere storage on the quality and carotenoid content of sliced persimmons and peaches. *Postharvest Biology and Technology* 10: 89–97.

Wu, X., G. Cao, and R.L. Prior. 2002. Absorption and metabolism of anthocyanins in human subjects following consumption of elderberry or blueberry. *Journal of Nutrition* 132: 1865–1871.

Zechmeister, L. 1962. *Cis-trans isomeric carotenoids, vitamins A and arylpolyenes*. Vienna: Springer.

Zhang, C., B. Trierweiler, W. Li, P. Butz, Y. Xu, C.E. Rüfer, Y. Ma, and X. Zhao. 2011. Comparison of thermal, ultraviolet-c, and high pressure treatments on quality parameters of watermelon juice. *Food Chemistry* 126: 254–260.

Zulueta, A., F.J. Barba, M.J. Esteve, and A. Frígola. 2010. Effects on the carotenoids pattern and vitamin A of a pulsed electric field-treated orange juice-milk beverage and behavior during storage. *European Food Research and Technology* 231: 525–534.

Chapter 4
Minimal Processing of Fruits

Zamantha Escobedo-Avellaneda, José Ángel Guerrero-Beltrán, María Soledad Tapia, Gustavo V. Barbosa-Cánovas, and Jorge Welti-Chanes

4.1 Introduction

Food are biological systems in which microbial growth and enzymatic and chemical reactions are constantly taking place causing deterioration and quality losses; thus, an adequate handling and processing is necessary, after harvesting, to avoid undesirable reactions to extend their shelf-life.

Fruits are very perishable food. The postharvest losses are estimated in 5–25 % in developed countries and 20–50 % in the developing ones. They show much more susceptibility to deterioration than cereals (Chakraverty et al. 2003). The higher rate of deterioration is mainly due to their high water activity and chemical composition. Once harvested, metabolic processes of fruits can continue conducting them to ripening and/or senescence; the rate of these reactions depends on the climacteric or non-climacteric behavior. Once reached the desired maturity level, non-climacteric fruits are harvested and stored under adequate conditions to delay senescence.

Z. Escobedo-Avellaneda • J. Welti-Chanes (✉)
Tecnológico de Monterrey, Monterrey, Mexico
e-mail: jwelti@itesm.mx

J.Á. Guerrero-Beltrán
Universidad de las Américas-Puebla, Cholula, Puebla, Mexico

M.S. Tapia
Universidad Central de Venezuela, Caracas, Venezuela

G.V. Barbosa-Cánovas
Washington State University, Pullman, WA, USA

A. Rosenthal et al. (eds.), *Fruit Preservation*, Food Engineering Series,
https://doi.org/10.1007/978-1-4939-3311-2_4

Climacteric fruits on the other hand continue ripening; therefore, the physiological changes must be adequately controlled to avoid unpleasant color, flavor, and texture throughout the postharvest shelf-life, period in which the consumption of the food is considered acceptable according to sensorial, microbial, chemical, and nutritional characteristics (Singh and Cadwallader 2004).

The knowledge of the health benefits of fruits has increased their consumption as fresh products. However, the consumption of fresh products could be a vehicle for foodborne diseases. Thus, the utilization of adequate posthandling techniques to assure the safety of produce is necessary. Washing and sanitizing are the first procedures applied to decrease the microbial loads of fresh fruits, and in addition, they remove undesirable residues.

Fresh-cut fruits are products obtained from entire fruits that after washing and/or sanitizing are trimmed, peeled, sliced, diced, or cut and that are not exposed to other transformation method for preserving their nutritional properties and freshness (**International Fresh-Cut Produce Association, IFPA**). These minimally processed fruits are ready-to-eat products highly convenient to consumers. Nevertheless, all these simple operations can contribute to the contamination of fruits, increasing their susceptibility to deterioration; thus, the use of any preservation technique, such as refrigeration, controlled or modified atmospheres, chemical or natural antimicrobial compounds, UV-light, pulsed light, irradiation, or high hydrostatic pressure (HHP), is required to increase their shelf-life (Artés et al. 2009; Rico et al. 2007). The minimal processing of fruits including the methods already mentioned improves safety while preserving quality. The combination of more than one preservation method is known as "hurdle technology"; each method has an additive or synergetic effect on microbial inactivation or other deterioration phenomena.

In this chapter, the disinfection techniques for minimally processed fruits and the subsequent methods used to preserve their quality and to extend their shelf-life while preserving their freshness will be reviewed.

4.2 Physiological Aspects Affecting the Postharvest Life of Fruits

After harvesting, fruits continue their metabolic processes. Fruit postharvest shelf-life depends completely on their composition and physiological characteristics as well as environmental conditions involved during storing. Respiration, transpiration, and ethylene production are the three physiological processes responsible for ripening and senescence of fruits; therefore, the control of these processes, through adequate postharvest handing and minimal processing, is necessary to delay senescence and maintain the best possible quality (Kader 2002).

Fruit respiration involves the oxidation of carbohydrates by atmospheric oxygen-generating energy (ATP), CO_2, and water. The ATP produced is then used in many ripening processes; therefore, a way for controlling the rate of senescence is by controlling the respiration rate. The respiration process can also be controlled using the adequate temperature and packaging system for storing; thus, low temperatures will decrease the activity of enzymes involved in ripening. The use of the adequate package may also diminish the available oxygen needed for deteriorative enzymes such as polyphenol oxidase (PPO) and peroxidase (POD). An adequate package material should be selected based on its permeability to water, CO_2, and oxygen; consequently, the fruit respiration rate could be modified due to changes in the consumption of O_2 and the releasing of CO_2. Generally, it is desirable to reduce the O_2 concentration in the package; thus, the minimal exchange amount of O_2 in the system should be controlled in order to prevent fermentation.

Transpiration is the process of water loss that can conduct to wilting, shriveling, shrinkage, softening, flaccidity, limpness as well as loss of crispness, juiciness, and nutritional quality of fruits. Internal fruit biological processes and surrounding fruit characteristics are implicated in the rate of water loss. The epidermal system of fruits (including cuticle, epidermal cells, stoma, lenticels, and trichomes) regulates the water loss of the product, while temperature and relative humidity control the rate of water diffusion. Transpiration of fruits can be reduced by using physical barriers around the fruit such as edible films or by controlling the storage conditions.

Ethylene is a hormone produced by higher plants. It regulates the fruit characteristics during growth and senescence. Ethylene can be active at very low concentrations; however, its production can be controlled by using adequate storage conditions. Based on the respiration rate and ethylene production during ripening, fruits are classified as climacteric or non-climacteric. Climacteric fruits have a great CO_2 and ethylene production rate which coincides with the ripening rate. Non-climacteric fruits show low changes in the CO_2 and ethylene production rates during ripening. Some examples of climacteric and non-climacteric fruits are shown in Table 4.1.

4.3 Minimal Processing Technologies Used in Fruit Preservation

Microbiological, enzymatic, and chemical reactions are responsible for fruit deterioration. These unwanted phenomena must be controlled in order to preserve fruits for longer times. Fruits can be polluted with deteriorative and pathogenic microorganisms as well as with microorganisms indicating bad manufacturing practices such as fecal coliforms. They are also affected by endogenous or exogenous enzymes (released by microorganisms) that can breakdown proteins, lipids, carbohydrates, pigments, vitamins, and other compounds into smaller molecules that can cause off flavors and/or engender toxic compounds. In addition, polyphenol oxidase (PPO) and peroxidase (POD) enzymes can cause enzymatic browning and affect phenolic compounds, decreasing the sensorial and nutritional quality of food. Chemical compounds in food can also react among them, or with atmospheric oxygen, affecting color, flavor, and nutrients. The enzymatic activity, microbiological contamination, and nutrient loss are high in products that were chopped, cut, sliced, or peeled (Chakraverty et al. 2003). Adequate cleaning, sanitizing, and preservation

Table 4.1 Climacteric and non-climacteric fruits (Adapted from Kader 2002)

Climacteric		Non Climacteric	
Common name	Scientific name	Common name	Scientific name
Apricot	*Prunus armeniaca*	Black raspberry	*Rubus occidentalis*
Avocado	*Persea americana*	Blueberry	*Vaccinium uliginosum*
Banana	*Musa×paradisiaca*	Cranberry	*Vaccinium oxycoccus*
Black sapote	*Diospyros digyna*	Cucumber	*Cucumis sativus*
Kiwi	*Actinidia deliciosa*	Grape	*Vitis vinifera*
Mango	*Mangifera indica*	Grapefruit	*Citrus×paradisi*
Papaya	*Carica papaya*	Melon	*Cucumis melo*
Peach	*Prunus persica*	Orange	*Citrus× sinensis*
Pear	*Pyrus communis*	Prickle pear	*Opuntia ficus-indica*
Pineapple	*Ananas comosus*	Raspberry	*Rubus ideas*
Plum	*Prunus domestica*	Strawberry	*Fragaria×anamosa*
Prickle pear	*Opuntia ficus-indicia*	Tangerine	*Citrus tangerina*
Tomato	*Solanum lycopersicum*	Watermelon	*Citrullus lanatus*
White sapote	*Casimiroa edulis*	Blackberry	*Rubus fruticosus*

methods must be applied during fresh fruit production to decrease microbiological loads and to avoid all these unwanted reactions. Preservation approaches such as mild irradiation, adequate packaging, and HHP, among others, can be applied to preserve the product quality (Alzamora et al. 2000). These preservation methods can delay or eliminate the microbial growth and other deteriorative reactions; therefore, increasing the shelf-life of the product (Ramos et al. 2013).

The increase of the respiration rate and ethylene production of fresh-cut fruits depends on the degree of injury during processing as well as the subsequent storage conditions; for example, when stored at 5 °C whole kiwi has a respiration rate of 4.6 mg CO_2/kg-h, while sliced kiwi has a respiration rate of 11.6 mg CO_2/kg-h; whole peach, on the other hand, has a respiration rate of 8.1 mg CO_2/kg-h and the sliced one delivers 10 mg CO_2/kg-h (Garcia and Barrett 2005).

4.3.1 Washing and Sanitizing of Fruits

Every step in the chain production of fresh-cut fruits contributes to their final microbial load. Microorganisms are responsible for foodborne diseases; therefore, an adequate handing of the food to prevent and diminish the microbial populations is essential (Kader and Barrett 2005). Washing and sanitizing are two important strategies used in the industry to improve safety of food (Artés et al. 2009).

The microflora of fresh-cut fruits include bacteria, yeast, and molds (Rico et al. 2007). Bacteria such as *Leuconostoc mesenteroides, Lactobacillus* spp.,

Pseudimonas spp., *Erwinia herbicola, Flavobacterium, Xanthomonas,* and *Enterobacter agglomerans* as well as various molds such as *Alternaria, Penicillium, Fusarium,* and *Aspergillus* and yeast such as *Torulopsis, Saccharomyces,* and *Candida* can be found in fruits (Ramos et al. 2013; Zagory 1999).

The washing of fruits with water is not enough to eliminate microbial contamination; however, the addition of sanitizing agents may increase the antimicrobial effectiveness. Disinfection is one of the most important processing steps applied to improve quality, safety, and shelf-life of fresh-cut fruits (Gil et al. 2009). Before disinfection, dirt, pesticide residues, plant debris, and other possible contaminants must be removed from fruit surfaces (Gil et al. 2009; Soliva-Fortuny and Martın-Belloso 2003). The washing, disinfestation, and rinsing procedures depend on each product and type of disinfectant used. Chlorine, chlorine dioxide, acidified sodium chlorite, hydrogen peroxide, peracetic acid, peroxyacetic acid, trisodium phosphate, electrolyzed water, and ozone are among the most common sanitizers used to sanitize fruits.

4.3.1.1 Chlorine

Although their sanitizing effect could be lowered when in contact with organic matter, chlorine and sodium hypochlorite (NaClO) are widely used to disinfect food. The common concentrations used for sanitizing fruits are of 50–200 ppm with an exposure time of up to 5 min (Gonzalez et al. 2004; Tapia and Welti-Chanes 2012). Their effectiveness is also pH-dependent, having a more antimicrobial effect at acid pHs; however, due to the corrosion of metallic materials, they are preferably used at pH between 6 and 7.5 (Beuchat 2000; Rico et al. 2007). It is necessity to eliminate residual chlorine on the fruit surface by rinsing. Due to the loss of activity at high pHs and to the production of carcinogenic halogenated byproducts such as trihalomethanes, haloacetic acids, and chloramines, in some European countries the use of chlorine in fresh-cut products is forbidden (Artés et al. 2009; Ölmez and Kretzschmar 2009; Rico et al. 2007). Nevertheless, it will be necessary to consume many kilograms of products treated with chlorine per day to have toxic effects (Russell 2005).

The level of microbial reduction obtained during the rinsing of fruits in chlorine solution depends on the chlorine concentration, amount of free available chlorine (in the form of hypochlorous acid, HOCl), dipping time, and type of microorganism (Ramos et al. 2013). Beuchat et al. (1998) showed that the spray application of chlorine solutions to raw produce could be a suitable and more convenient alternative than dipping them. Inoculated (*Salmonella, Escherichia coli* O157:H7 and *Listeria monocytogenes*) whole apples, tomatoes, and lettuce leaves were treated (sprayed and then dipped) with 200 or 2000 ppm of chlorine solution for 0, 1, 3, 5, or 10 min and then rinsed with sterile water. Pathogens were reduced from 0.35 to 2.30 log cycles. In fresh-cut apple, sodium hypochlorite (100 mg/L, pH 6.5) was adequate to eliminate *E. coli* O157:H7, *Salmonella* spp. and *Listeria* spp. (inoculated at 10^6 CFU/mL) (Abadias et al. 2011).

Although there are studies favoring the use of chlorine for sanitizing fruits, most of the investigations showed that it is not effective on reducing microbial loads; therefore, no safe products are obtained. Dipping watermelon cubes in chlorine solution (40 μL/L) was not an effective procedure for reducing microbial populations (Fonseca and Rushing 2006). In Golden delicious apples treated with 200 ppm Cl_2, 2.00 log cycles reductions of *E. coli* were observed (Sapers et al. 1999); nevertheless, this reduction in the population of *E. coli* is not considered enough to assure safety (Annous et al. 2001). The rinsing of strawberries with 100 and 200 ppm NaClO was also ineffective to reduce the *E. coli* O157:H7 (initial inoculum of 10^8 CFU/mL) (Yu et al. 2001) and total mesophiles (Alexandre et al. 2012) populations. Similar results were obtained when using 150 ppm sodium hypochlorite to inactivate aerobic mesophiles, coliform bacteria, and fungi from the surface of strawberries (Koseki et al. 2004).

Besides all its disadvantages, chlorine is still widely used as a disinfectant; however, it is necessary to develop and/or apply new sanitizers to obtain microbiologically safe fruits.

4.3.1.2 Chlorine Dioxide

Chlorine dioxide (ClO_2), obtained by the reaction of either an acid or chlorine gas with sodium chlorite, has about 2.5 times more oxidation capacity than chlorine; it does not form chloramines and it has been approved by the FDA (Goodburn and Wallace 2013; Rico et al. 2007). ClO_2 is more effective for inactivating spores than chlorine. It has shown effectively against *L. monocytogenes, S. Typhimurium, E. coli, Legionella, Amoebal cysts, Giardia cysts, Cryptosporidium*, and viruses (Xie 2003; Artés et al. 2009).

A concentration of up to 3 ppm of ClO_2 is commonly used to treat whole fruits and vegetables. ClO_2 gas has a greater penetration capacity in the fruit than liquid ClO_2. Factors such as gas concentration, time of exposure (7–135 min), relative humidity (55–95 %), and temperature (5–25 °C) can influence the effectiveness of the sanitation process. It is necessary to produce ClO_2 gas on-site to avoid explosion, and as well as for chorine, products sanitized with ClO_2 should be rinsed with water (Artés et al. 2009; Gómez-López et al. 2009; Goodburn and Wallace 2013; Ölmez and Kretzschmar 2009).

Mari et al. (1999) showed the effectiveness of ClO_2 in the germination of *Monilinia laxa* conidia. They inoculated nectarine and plums adding 20 μL (6×10^3 conidia/mL). Fruits were dipped in a 10 μg/mL ClO_2 solution for 20 min; conidia were completely inhibited and no brown rot was observed.

Apples, strawberries, and cantaloupes were treated with 3 and 5 ppm of ClO_2, by dipping, for up to 5 min. The exposure of produce to 5 ppm reduced the microbial populations in about 5.60 log cycles, while a concentration of 3 ppm caused a decreasing of about 4.90 log. Populations of *E. coli* O157:H7 and *L. monocytogenes* remained nearly unchanged during 9 days of storage at 4 °C. Mesophilic bacteria increased in about 2.00–3.00 log cycles and molds plus yeasts were significantly higher than the initial counts (Rodgers et al. 2004).

The application of gaseous ClO_2 from sachets was effective in reducing viable *Alicyclobacillus acidoterrestris* spores on apple surfaces. Spores exposed to high and medium ClO_2 concentrations, released from sachets for 1 h, were reduced 5.00 log cycles; however, visual quality was decreased. After 1, 2, and 3 h of ClO_2 exposure to low release, the spores were reduced 2.70, 3.70, and 4.50 log cycles, respectively. After 7 days of storage, no significant visual quality differences were observed between apples exposed to low release sachet and non-treated apples (Lee et al. 2006).

Results about the use of gaseous ClO_2 indicate its effectiveness as sanitizer for fruits such as blueberries, strawberries, and red raspberries. The exposure of blueberries to 8.0 mg ClO_2/L for 120 min significantly reduced the population of *Salmonella* in 2.40–3.70 log (CFU/g); for strawberries, the microbial population was reduced in 3.80–4.40 log (CFU/g); and in raspberries, a reduction of 1.50 log (CFU/g) was achieved. Treatments with 4.1/30, 6.2/60, and 8.0/120 mg ClO_2 per L/min caused reductions in the yeast plus molds population on blueberries, strawberries, and raspberries of 1.4–2.5, 1.4–4.2, and 2.6–3.0 log (CFU/g), respectively. Treatment with 4.1 mg ClO_2/L did not significantly affect the sensory quality of fruits stored for up to 10 days at 8 °C (Sy et al. 2005).

A study in blueberry showed that 1, 3, 5, 10, and 15 ppm of ClO_2 at exposure times of 10 s, 1, 5, 10, 20, 30 min, 1 h, and 2 h was more effective in reducing *L. monocytogenes* (4.88 log, CFU/g) than for *P. aeruginosa, S. typhimurium, Staphylococcus aureus,* and *Yersinia enterocolitica. P. aeruginosa* was reduced 2.20 log cycles after 5 min treatment with 15 ppm of ClO_2. Shorter treatment times were more effective in reducing *S. typhimurium* than longer times. The highest reduction (4.60 log CFU/g) of *S. aureus* was achieved with 15 ppm of ClO_2 for 30 min. *Y. enterocolitica* was reduced by 3.50 log (CFU/g) after 2 h of exposure to 5 ppm of ClO_2. Fifteen ppm of ClO_2 reduced foodborne yeasts plus molds by 2.80 log CFU/g after 1 h of treatment (Wu and Kim 2007).

Jin et al. (2007) pointed out that the microbial load of strawberries sanitized with 50 ppm ClO_2 increased after 7 days storage at 4 °C, but the increase was lower than in no-sanitized strawberries. Total aerobic bacteria increased from 1.40 to 2.10 log (CFU/g) and from 2.75 to 4.32 log (CFU/g) in sanitized and no-sanitized strawberries. Yeast plus molds in sanitized strawberries increased from 1.10 to 1.97 log (CFU/g) and from 2.55 to 4.50 log (CFU/g) in the no-sanitized fruit.

4.3.1.3 Acidified Sodium Chlorite

Acidified sodium chlorite is used in the range of 500–1200 ppm for sanitizing fruits (Ramos et al. 2013; Lukasik et al. 2003). *E. coli* O157:H7, *S.* Montevideo, poliovirus 1, and the bacteriophages PRD1, X174, and MS2 were significantly reduced in strawberry with 100 or 200 ppm of acidified sodium chlorite washes, in which 200 ppm was the most effective concentration (Lukasik et al. 2003).

4.3.1.4 Hydrogen Peroxide

Although not yet approved by the FDA as a sanitizing agent in fresh produce, hydrogen peroxide (H_2O_2) does not leave residues on the product because it is reduced completely into water and oxygen (Ölmez and Kretzschmar 2009; Soliva-Fortuny and Martın-Belloso 2003). It is regularly used to treat some food surfaces and packaging materials during aseptic packaging. Its use as an antimicrobial is restricted to products such as milk, dried eggs, starch, tea, and wine in the range of 0.04–1.25 %. For sanitizing effects, concentrations in the range 2–4 % are recommended (Ramos et al. 2013).

There are many studies about the application of H_2O_2 to sanitize fruit products. Cantaloupe samples rinsed with H_2O_2 at 50 °C yield a fresh-cut product with a shelf-life larger than 2 weeks (Sapers et al. 2001). *E. coli* ATCC 25922 inoculated in apples was reduced in about 3.00 log cycles when rinsed with 1 % H_2O_2 at 20 or 40 °C for 15 or 30 min (Sapers et al. 1999). Apple inoculated by dipping in a mix (10^6 CFU/mL) of *E. coli* O157:H7, *Salmonella* spp., and *Listeria* spp. and then treated with H_2O_2 (5, 10, 20 mL/L) showed bacterial reductions similar (or higher) than those obtained with hypochlorite solutions (100 mg/L, pH 6.5); in addition, bacteria populations were maintained low throughout storage (Abadias et al. 2011). The use of 5 % H_2O_2 provided high reductions of total mesophilic bacteria in strawberries (Alexandre et al. 2012).

Some researchers have shown that H_2O_2 is not effective to produce safe fresh-fruits. The apple washing with 5 % H_2O_2, at 20 and 50 °C, was not effective to inactivate *E. coli* (Annous et al. 2001). Similarly, 1 % H_2O_2 solution was ineffective to destroy *E. coli* 766 inoculated on cantaloupes (Sapers and Sites 2003). Three percent of H_2O_2 reduced the *E. coli* O157:H7 population (10^8 CFU/g) on strawberries in about 2.20 log (CFU/g) (Yu et al. 2001). H_2O_2 solution (0.5 %) was slightly less effective than free chlorine to reduce loads of *E. coli* O157:H7, *S.* Montevideo, poliovirus 1, and the bacteriophages PRD1, 174, and MS2 in strawberries; however, it caused slight fruit discoloration (Lukasik et al. 2003).

4.3.1.5 Peracetic Acid

Peracetic acid (PAA) is a strong oxidizing agent formed by the reaction of acetic acid and hydrogen peroxide. It has been used to clean surfaces in contact with food and its efficacy depends on concentration and time of exposure.

Sweet cherries, apricots, peaches, and nectarines treated with PAA reduced the incidence of brown rot caused by *Monilinia laxa* and soft rot caused by *Rhizopus stolonifer* (Mari et al. 2004). Mari et al. (1999) pointed out that brown rot was totally controlled when conidia remained in contact with 250 mg/L PAA for 5 min. Fruit, neither wounded nor inoculated, dipped for 1 min in a 125 mg/L PAA solution showed a significant reduction of *Monilinia* rots compared with control. Significant inhibition of *R. stolonifer* (inoculated) was also observed on wounded fruit after treatment with 250 mg/L PAA solution for 1 min. 1000 μg/mL PAA treatment was effective in reducing the decay incidence of plums by 50 % after 1 h of contact with conidia (6×10^3 conidia per mL) (Mari et al. 1999).

4.3.1.6 Peroxyacetic Acid

Liquid peroxyacetic acid, obtained by combining peracetic acid and hydrogen peroxide, is a strong oxidizing agent and it is more effective than chlorine to inactivate pathogens. It should be used in a concentration up to 80 ppm in fruits and vegetables. It is adequate to sanitize fruits due to its resistance to several conditions such as high temperature, pH range (1–8), hardness, and soil contamination (Artés et al. 2009; Ölmez and Kretzschmar 2009).

Peroxyacetic acid at concentrations of 80 and 120 mg/L eliminates *E. coli* O157:H7, *Salmonella* spp. and *Listeria* spp. (10^6 CFU/mL) on fresh-cut apple (Abadias et al. 2011). It is also effective in reducing *E. coli* O157:H7 on cantaloupe rinds when compared with water (Wang et al. 2006). Significant reductions of *E. coli* O157:H7, *S. Montevideo* and poliovirus one were obtained in strawberries treated with 100 ppm of peroxyacetic acid (Lukasik et al. 2003). Apples, strawberries, and cantaloupes were inoculated by dipping them until containing about 10^6 CFU/g of *E. coli* O157:H7 or *L. monocytogenes.* After some time, they were submerged in 80 ppm peroxyacetic acid for up to 5 min observing about 4.40 log reductions. The population of both *E. coli* O157:H7 or *L. monocytogenes* remained relatively unchanged during the storage at 4 °C for 9 days (Rodgers et al. 2004).

4.3.1.7 Trisodium Phosphate

Trisodium phosphate (TSP) solutions have a pH of 11–12 that limits its use in fruits and vegetables (Ramos et al. 2013). A study performed by Annous et al. (2001) showed that washing apples (contaminated with *E. coli*) with 8 % TSP at 20 and 50 °C was not effective to inactivate the microorganism. Nevertheless, the efficiency of this compound increases when combined with chlorine. Rodgers et al. (2004) showed that chlorinated TSP (200 ppm chlorine) reduced in about 4.90 log cycles the population of *E. coli* O157:H7 and *L. monocytogenes* inoculated on apples, strawberries, and cantaloupes.

4.3.1.8 Electrolyzed Water

Electrolyzed water (EW) is produced by electrolysis of aqueous sodium chloride. An electrolyzed basic aqueous solution and an electrolyzed acid solution are obtained at the cathode and anode, respectively. According to this, the EW with sanitizing properties can be divided in acid electrolyzed water (AEW), also called electrolyzed oxidizing water, and neutral electrolyzed water (NEW). The AEW has strong effect on pathogens and spoilage microorganisms attributed to its pH (2.1–4.5), high oxidation-reduction potential (higher than 1000 mV), and the presence of hypochlorous acid. NEW (pH 5.0–8.5) has also strong bacterial effect due to its oxidation–reduction potential (500–700 Mv) (Ramos et al. 2013; Rico et al. 2007).

Apple and cantaloupe cylinders were inoculated with *E. coli* O157:H7 by dipping, and then treated with AEW for 8–15 min; higher levels of microbial inactivation were achieved than when washing fruits only with water (Wang et al. 2006). Wounded pear fruit was inoculated with 5×10^5 conidia of *Botryosphaeria berengeriana*/mL and then immersed in AEW water; the treatment suppressed the incidence and disease severity showing its effectiveness as surface sanitizer (Al-Haq et al. 2002). AEW (30 ppm free available chlorine) was unable to completely inactivate aerobic mesophiles, coliform bacteria, and fungi from the surface of strawberries (Koseki et al. 2004).

AEW has potential as an alternative to chlorine disinfectants for controlling infection of *Penicillium expansum* in apples during handling and processing. One hundred and 50 % AEW decreased spores of *P. expansum* as much as 4.00 and 2.00 log cycles, respectively, on wounded apples (Okull and Laborde 2004).

AEW and NEW showed similar efficiency to sodium hypochlorite at the same free chlorine concentration in the disinfection of apple slices inoculated with *E. coli, L. innocua,* or *S. choleraesuis*. AEW at 100 mg free chlorine/L was the treatment with the highest bactericidal activity. The microbial reductions ranged from 1.20 to 2.40 log cycles for NEW and AEW at 100 and 50 mg of free chlorine/L, respectively. In general, these treatments were equal or more effective than sodium hypochlorite at 100 mg of free chlorine/L used for washing (Graca et al. 2011).

More than 2.00 log cycles reductions of aerobic mesophiles were obtained in strawberries washed for 10 or 15 min with AEW, prepared with 0.10 % (w/v) NaCl solution. Washing fruit surfaces with distilled water resulted in 1.90 and 1.30 log cycles reductions of *L. monocytogenes* and *E. coli* O157:H7, respectively, in the rinsing water. On the other hand, up to 2.35 log cycles of rinse fluid reduction of *E. coli* O157:H7 were observed on fruit surfaces washed with AEW water. However, the use of AEW, as a sanitizer, did not show better antimicrobial effects than water during refrigeration storage (Udompijitkul et al. 2007).

4.3.1.9 Ozone

Ozone (O_3), a strong antimicrobial agent approved by the FDA, has higher antimicrobial activity than chlorine and it is not pH-dependent. It must be generated on site because of its instability, in contact with organic matter forms aldehydes, ketones, and carboxylic acids. Ozone is used in the range 0.03–20 ppm, but in the gas form concentrations higher than 20,000 ppm can be used. Ozone (3 ppm) was used for reduce microbial loads of E. *coli* O157:H7 and *L. monocytogenes* on inoculated apples, strawberries, and cantaloupes. Produce were inoculated, by dipping, to contain about 10^6 CFU/g of *E. coli* O157:H7 or *L. monocytogenes*, let stand overnight, and later on submerged in ozone for up to 5 min. Ozone reduced the microbial populations in about 5.60 log units. Populations of both pathogens remained relatively unchanged during 9 days storage at 4 °C. Mold plus yeast, on the other hand, were significantly higher than initial counts; therefore, this sanitizer was ineffective to stop the growing of these microorganisms (Rodgers et al. 2004). Similarly, Pérez et al. (1999) showed that a treatment with 0.35 ppm ozone was ineffective in preventing fungal decay in strawberries after 4 days at 20 °C. Experiments done by Fonseca and Rushing (2006) also demonstrated that the dipping of watermelon cubes in ozone (0.4 μL/L) was not effective in reducing microbial populations. Koseki et al. (2004) demonstrated that 5 ppm ozone was unable to completely inactivate aerobic mesophilic bacteria, coliform bacteria, and fungi from the surface of strawberries. Smilanick et al. (2002) found that the immersion of citrus fruits and peaches in ozonized water did not control the postharvest decay at concentrations that actually should control fruits decay. Green mold and sour rot on citrus fruit, caused by *Penicillium digitatum* and *Geotrichum citriurantii*, respectively, were not reduced in 10 ppm O_3 for 20 min of immersion. On five peach varieties, the average natural incidence of brown rot, caused by *Monilinia fructicola*, was reduced from 10.9 to 5.4 % by immersion in 1.5 ppm O_3 for 1 min. A treatment with 5 ppm O_3 for 15 min further reduced the fruits decay to 1.7 %; however, the control of brown rot is associated with the formation of some hollows on the fruit.

4.3.2 Minimal Processing Methods to Extend Shelf-Life of Fresh-Fruits

The use of sanitizers in fresh-cut fruits is not enough to obtain safe products. A reduction of microbial load is obtained, but an additional preservation factor or preservation method is required to extend their shelf-life. In this way, the application of minimal processing technologies to extend shelf-life of produce, while maintaining the fresh-like state, is relevant. Methods such as refrigeration, the use of natural preservatives, edible coating, irradiation, UV, pulsed light, ultrasound, HHP, controlled (CA) and modified atmospheres (MA) may help to maintain the quality of fresh-cut fruits after disinfection.

4.3.2.1 Refrigeration

Temperature is the factor that most affects the rate of deterioration of food. The rate of deterioration is proportional to the respiration rate, so a reduction of 10 °C may reduce the respiration rate in about 2–4 times. The selection of the adequate temperature to store each fresh product is essential, not only to reduce the respiration rate but also to avoid chilling injuries that can cause surface lesions, water soaking of tissues, internal discoloration, breakdown of tissues, compositional changes, acceleration of senescence, and in general greater susceptibility to decay. At 4 °C, most of minimally processed products may extend their shelf-life by 5–7 times (Rennie et al. 2003).

Cooling, which consist in removing heat from food, can be achieved by using different methods. The adequate system to remove heat will depend on factors such as respiration rate, amount of product, time of storage, and cooling. Some of the main methods used to cooling fruits are shown in Table 4.2.

Table 4.2 Cooling methods used in fruit preservation (Rennie et al. 2003; Thompson et al. 2002)

Method	Characteristics
Hydrocooling	Products are submerged in water or sprayed with it. One of the advantages of this method is that products do not loss moisture as compared with other cooling methods. Immersion systems work in continuous flow and are more useful for products with density higher than water. The water used must be sanitized with chlorine
Contact icing	Fruits are put into direct contact with the ice. This method is not expensive but it requires high times for cooling. Ice must be 38 % of the product weight to reduce temperature from 35 to 2 °C
Vacuum cooling	It is a rapid and adequate cooling method for products of large surfaces that require loss water easily. The main disadvantage of this method is its limited capacity, which is due to the room and vacuum requirements
Cooling in room	Products are placed in cold rooms. It is one of the most widely used cooling methods because it works also as storage. It is not recommended for vegetables with high respiration rates because due to the long time required for cooling products will deteriorate rapidly. It is adequate for long-term storage products
Forced-air cooling	This method consists in flowing air through the product that is placed in tunnels of other adequate system. The time of cooling is increased from around 4 to 10 times as compared with cooling in rooms

4.3.2.2 Natural Preservatives

The use of natural preservatives may be effective to retain quality of fresh-cut fruits by inhibiting spoilage (Rico et al. 2007). Examples of natural preservatives are organic acids, essential oils, and chitosan; these compounds can be added to food directly or indirectly as part of a coating of the food.

4.3.2.2.1 Organic Acids

Organic acids such as lactic, citric, acetic, and tartaric acids are considered antimicrobial agents to lessen microorganism in fresh-cut fruits. Their antimicrobial activity is due to pH reduction of the product, changes in the membrane permeability and transport, among other effects (Rico et al. 2007). Lukasik et al. (2003) pointed out that the use of vinegar at concentration of 10 % reduces significantly the numbers of *E. coli* O157:H7 and *S.* Montevideo (by about 90 %) as well as the numbers of poliovirus 1 in strawberries. Yu et al. (2001) showed that the use of low acetic acid concentrations (2 and 5 %) was not effective to reduce the *E. coli* O157:H7 population (initial inoculum of 10^8 CFU/mL) on strawberries. Venkitanarayanan et al. (2002), on the other hand, showed that the treatment of apples and oranges with 1.5 % lactic acid plus 1.5 % hydrogen peroxide for 15 min at 40 °C reduced more than 5.00 log cycles of *E. coli* O157:H7, *S. enteritidis*, and *L. monocytogenes,* and in addition the quality of apples was maintained.

4.3.2.2.2 Essential Oils

Essential oils are used as natural preservatives in food due to their antimicrobial properties and natural origin. Essential oils are secondary metabolites produced as part of the natural defense mechanism of plants against insects and pathogens. Essential oils may have antibacterial, antiparasitic, and antifungal properties. They can also be used as insecticides and antioxidants. Cloves, thyme, cinnamon, oregano, and vanilla are some common examples used to obtain essential oils. The cloves essential oil is made up of eugenol (64 %), eugenyl acetate (16.3 %), and caryophyllene (14.5 %); it has been used to reduce the microbial load of some microorganism: *Campylobacter jejuni, E. aerogenes, E. coli, L. monocytogenes, S. enteritidis*, and *S. aereus*. The thyme essential oil is made up of thymol (63.8 %), α-phellandrene (13.3 %), and cis-sabinene hydroxide (8.1 %); it has been used to inhibit the growth of *L. monocytogenes* (Hyldgaard et al. 2002). Abadias et al. (2011) demonstrated that populations (10^6 CFU/g) of *E. coli* O157:H7, *Salmonella* spp., and *Listeria* spp. in apple were reduced with vanillin (12 g/L), a natural essential oil obtained from vanilla; the bacterial populations were maintained at low counts throughout 6 days storage at 10 °C. These microbial reductions were similar or higher than those obtained by sodium hypochlorite (100 mg/L, pH 6.5).

4.3.2.3 Blanching

Blanching is a thermal treatment used in fruits and vegetables to reduce the enzymatic activity and microbiological load. Fruits and vegetables are immersed in hot (91–99 °C) or boiling water for 1–10 min, depending on the type of fruit or vegetable (FAO 2003). Siegel et al. (1971) blanched red tart cherries (*Prunus cerasus* L. cv. Montmorency) at 100 °C using steam for 0, 30, 45, and 60 s, followed by freezing at −20 °C. No color change was observed in either blanched or non-blanched unfrozen cherries. Around 14 to 25 % of color was lost when frozen and non-blanched cherries were thawed at room temperature, while in cherries blanched for 45 or 60 s no significant color change was observed. When cherries were disintegrated, the non-blanched cherries lost 70 % of their color after 30 min in contact with atmospheric conditions, while those blanched maintained their color. Results of this research show the importance of blanching for maintaining the color quality of cherries. Similar results were obtained by Rossi et al. (2003) for blueberries.

4.3.2.4 Ultraviolet Light

Nonionizing ultraviolet (UV) light at wavelength of 190–280 nm is used to treat food disinfecting fruit surfaces. UV-C light at 254 nm does not leave chemical residues. It is easy to use and effective against most types of microorganisms.

Lu et al. (1991) showed that the UV-C light treatment at doses of 0.84–40 kJ/m^2 reduced the mold rots and delayed ripening of peaches and apples during storage.

The amount of rotten decreased with increasing UV dose. Treated peaches were firmer; however, pH and soluble solids were lower in UV-C light-treated peaches than for non-treated fruits. For apples, pH was lower and ascorbic acid was higher in UV-C light-treated than in non-treated products. Peeled pear slices irradiated (254 nm) at doses between 0 and 15 kJ/m^2 improved the inactivation of *L. innocua* ATCC 33090, *L. monocytogenes* ATCC 19114 D, *E. coli* ATCC 11229, and *Zygosaccharomyces bailli* NRRL 7256 (Schenk et al. 2008).

UV light treatment has potential as surface disinfection method in ready-to-eat food. In cantaloupe exposed to UV-C light, 2.00 log reductions were achieved for both the total viable count and enterobacteria. In addition, microbial loads remained lower in treated melon than in untreated fruits during storage and the general quality of the product was maintained (Manzocco et al. 2011). For cut cantaloupe, the UV-C treatment was effective in reducing the yeast, mold, and *Pseudomonas* spp. populations. Cutting of melon under UV-C light delivers melon pieces with low aerobic mesophilic and lactic acid bacteria in comparison to microbial population in untreated melon or pieces of melon obtained after UV-C light treatment (Lamikanra et al. 2005). The exposition of packaged watermelons cubes to UV-C light at 4.1 kJ/m^2 reduced more than 1.00 log cycles, the microbial load of the product at the end of its shelf-life without changes in general quality. UV-C light doses lower than 1.4 kJ/m^2 reduced few microbial populations. Higher UV doses did not show any difference in microbial populations (6.3 kJ/m^2) or result in quality deterioration (13.7 kJ/m^2) (Fonseca and Rushing 2006).

4.3.2.5 Irradiation

Although irradiation of food appears in 1945, it has not been widely used due to consumer concern. In 1958, it was classified as a food additive rather than preservation process (Hawkes 2000). This technique involves the exposure of food to a field of ionizing energy for controlling spoilage and eliminates foodborne pathogens by breaking chemical bonds in molecules implicated in the cell growth. It has also been used for pest control, for inducing mutation in plants, and for preventing sprouting of tubers (Borsa 2000). Spices and dry vegetables comprise about 46 % of irradiated food in the world, garlic and potato 22 %, grain and fruits 20 %, meat and seafood 8 %, and others food 4 % (Lado and Yousef 2002).

Irradiation is a non-thermal method used for pasteurizing many food in a safe, efficient, and environmentally clean way (Arvanitoyannis et al. 2009). The type of food and the irradiation purpose will determine the conditions used for each particular food. The FDA recommends 1 kGy for the radiation of fruits (Farkas 1998). In the USA, irradiated food must be labeled with the international irradiation logo; either legends "Treated by irradiation" or "Do not irradiate again" must appear in the label.

The irradiation of food can be performed using gamma rays, from radionucleus such as Co^{60} or Cs^{137}, electron beam generators, and X-ray machines. Packaged fresh or frozen products can be processed directly. Co^{60} emits gamma rays with energies of 1.17 and 1.33 MeV in the form of intense energy; up to 95 % of the energy is available for use. Gamma rays penetrate deeply, yielding a uniform dosage in the food product, but the use of Co^{60} requires frequent replacement and its irradiation is relatively slow. The energy delivered by Cs^{137} is less penetrating than that of Co^{60}, but it has longer half-life. The use of electron beams for treating food is restricted due to its limited penetration; 8 cm is the maximum depth reached by the maximum permitted energy (10 MeV). X-rays are generated by bombarding a metal with electrons. Part of the energy is absorbed and part is converted to X-rays. X-rays penetrates deeply, thus food can be processed in its container (http://www.epa.gov/rpdweb00/sources/food_irrad.html; Kilcast 1995).

Pasteurized by irradiation can cause some quality changes, such as softening in food. Therefore, it is recommended to use this technology in combination with other treatments. Irradiation with mild heat is a promising combination of treatments for decreasing the dose of irradiation needed for preserving food quality. Treatments at 55 °C for 5 min and 50 °C for 10 min have been used for irradiating mangoes and papayas, respectively (Arvanitoyannis et al. 2009).

Table 4.3 shows some applications of irradiation, alone or combined with mild heat and other preservation factors.

4.3.2.6 Pulsed Light

Pulsed light, high intensity pulsed light, pulsed white light, or pulsed UV-C light destroy microorganisms rapidly in both solid and liquid food, by causing structural changes at the microbial DNA level. Wavelengths in the range of 100–1100 nm are reached by UV-C light (Gómez-López et al. 2007; Ramos et al. 2013).

Bialka and Demirci (2008) showed the potential of pulsed UV-C light for decontamination of raspberries and strawberries. For raspberries, *E. coli* O157:H7 and *Salmonella* were reduced by 3.90 and 3.40 log cycles (CFU/g) after treatment at 72 and 59.2 J/cm^2, respectively. For strawberries, the maximum reductions were 2.10 and 2.80 log units (CFU/g) after treatment at 25.7 and 34.2 J/cm^2, respectively. Visually, fruits were not damaged by the treatments. Sauer and Moraru (2009) informed that pulsed light, at doses of up to 13.10 J/cm^2, deduced *E. coli* ATCC 25922 and *E. coli* O157:H7 in apple juice and cider treated in static and turbulent modes. For static treatments, in apple juice, inactivation of 2.70 log units was obtained for *E. coli* ATCC 25922 and 2.5 log for *E. coli* O157:H7. In cider, inactivation levels of 2.30 log and 3.20 log cycles were obtained. Turbulent flows resulted in 5.80 log reduction in cider and 7.20 log reductions in juice. Lagunas-Solar et al. (2006) used pulsed UV-C light as an alternative to pesticides to disinfect the surface of apple, kiwi, lemon, nectarine, orange, peach, pear, raspberry, and grape. Plant pathogens (fungal) were efficiently inhibited (more than 5.00 log) in less than 10 s of exposition time. However, in naturally infected or inoculated (sprayed) fruits, only partial disinfection was observed.

Table 4.3 Application of irradiation alone or combined with other technologies for the treatment of fruits

Product	Type of irradiation	Conditions	Main findings	Reference
Papaya harvested at three stages of maturation	Gamma radiation	0.75 kGy were applied and the color, firmness, and physicochemical characteristics of papaya were analyzed after different times of storage at 11 and 24 °C	Papaya maturation degree at harvest did not influence the effect of irradiation. Irradiation maintained firmness of papaya, by delaying ripening. There was no effect of irradiation in papaya weight loss, occurrence of diseases, croma of flesh color, pH and total soluble solids contents	De Azevedo Pimentel and Melges Walder (2004)
Papayas, rambutans, and Kau oranges	X-irradiation	0.75 kGy were applied and samples were stored for 2 and 9 days to determine the effect of X-irradiation on sensory quality attributes	While aroma and flavor tended to be more intense in the irradiated fruits, the firmness decreased as a result of irradiation and storage, but with significant effect only in rambutans. Irradiation did not significantly change the ascorbic acid and carotenoid contents, pH, titratable acidity, and total soluble solids	Boylston et al. (2002)
Early and late season 'Rio Red' grapefruit	Gamma irradiation (Cs^{137})	Fruit was exposed to 0, 70, 200, 400, and 700 Gy with a centerline absorbed dose of about 40 Gy/min. After the irradiation, fruits were stored for 4 weeks at 10 °C followed by 1 week at 20 °C with 90–95 % relative humidity	The response of fruit to irradiation is harvest time-dependent. Lower doses (below 200 Gy) of irradiation after 35 days of storage were useful in enhancing health-promoting compounds in early season grapefruit. Higher doses of irradiation (400 and 700 Gy) and 35 days of storage had detrimental effects on quality of early season grapefruit; however, no significant effect was observed on the quality of the late season fruit	Patil et al. (2004)
Apple slices	Gamma irradiation (Cobalt 60) combined or not with calcium treatments in different O_2 atmospheres concentrations	The effect on firmness of apple slices treated at doses up to 5 kGy at 0.4 and 2 kGy/h dose rates in 0, 21, and 100 % O_2 atmospheres were evaluated	At irradiation doses higher than 0.34 kGy, the firmness of apples decreased. High dose rates had a positive effect on softening, but the dose rate become insignificant during storage	Gunes et al. (2001)

4.3.2.7 Ultrasound

Ultrasound uses frequencies from 20 to 100 kHz and requires the presence of a liquid medium for power transmission. This technology is often combined with aqueous sanitizer solutions (Ramos et al. 2013). Cao et al. (2010a) pointed out that an ultrasonic treatment of 250 W for 9.8 min was effective in inhibiting decaying of strawberries as well as maintaining their quality. Cao et al. (2010b) showed that the application of ultrasound to strawberries at 40 kHz significantly reduced the decay of fruit maintaining its firmness, total soluble solids, acidity, and vitamin C contents. Chen and Zhu (2011) demonstrated that the combination of ClO_2 (40 mg/L ClO_2 for 10 min) and ultrasound (100 W for 10 min) could be a promising approach to maintain the postharvest storage quality of plums. Huang et al. (2006) informed that ultrasound combined with ClO_2 water solutions has a potential to improve the antimicrobial efficacy of ClO_2. About 2.50 and 4.30 log reductions of *Salmonella* and *E. coli* O157:H7 were achieved in apples when combining ultrasound (170 kHz) and ClO_2 (20–40 ppm) (Ölmez and Kretzschmar 2009).

4.3.2.8 High Hydrostatic Pressure

HHP is a non-thermal preservation method used to extend the shelf-life of food without altering their nutritional and sensory properties. Thus, this technology could be used to preserve fresh-cut fruits. In the HHP processing, products are placed on vessels filled frequently with water. A uniform pressure might cause microbiological injury that can lead to death and enzyme inactivation without altering the sensory and nutritional properties of the food. Due to the adiabatic compression, temperature of water increases about 3 °C per each 100 MPa. Commercially, pressures from 400 to 700 MPa are commonly used at temperatures up to 50 °C; however, when pressurization is performed at higher temperatures, the procedure is known as pressure-assisted thermal processing (PATP). This combination of pressure and temperature could have some negative effects on nutritional compounds of food (Escobedo-Avellaneda et al. 2011).

HHP processing has been widely used to preserve fruit juices due to their acidity. Its application in whole or fresh-cut fruits has not been widely investigated. Alemán et al. (1994) showed the potential application of HHP on whole fruits by treating fresh cut pineapple pieces at 340 MPa for 15 min. They attained reductions of bacteria, molds, and yeasts of 3.00 log (~4 °C), 3.10 log (21 °C), and more than 2.50 log cycles (38 °C), respectively. There are some studies showing that the HHP processing can cause negative effect on some physicochemical characteristics of fruits which depends on the fruit variety. Wolbang et al. (2008) showed that the HHP processing of melon did not have an effect on total soluble solids, but color was adversely affected. Vitamin C concentrations decreased, while the levels of β-carotene were significantly increased.

4.3.2.9 Food Packaging

Atmospheric air is composed of about 78, 21, and 0.03 % of N_2, O_2, and CO_2, respectively. The concentration of these gases in the surrounding atmosphere of food can influence their quality and shelf-life. Oxygen is responsible for many undesirable changes such as oxidations, loss of color, vitamins, and some amino acids; it also influences the microbial growth. Packaging is the simplest method to protect food from the surrounding conditions. Although the package acts as a barrier to environmental conditions to extend the shelf-life of food, it is necessary to alter the gas concentration inside the package. Modification of the atmospheric gas concentrations may reduce the respiration rate of fresh produce as well as control ethylene production retarding ripening. The alteration of gas conditions in a system includes the use of vacuum, controlled atmosphere (CA), and modified atmosphere (MA). In order to establish the optimum environmental conditions, it is necessary to know the gas exchange rate of each particular food and some material packaging properties like their permeability to water vapor, CO_2, and O_2. CA and MA are obtained by changing the concentration of at least one of these gases: N_2, O_2, CO_2, and ethylene. In MA, the gas concentrations are not continually controlled and they change depending on the respiration rate of the product, package permeability, and storage conditions, while in CA the gas concentrations are constantly monitored and adjusted. Another method to reduce deterioration rates of food is by using edible coatings. This method does not modify the concentration of gases surrounding the product, but rather avoids the contact of the product with the air, decreasing some deterioration reactions.

4.3.2.9.1 Controlled Atmospheres

During the CA storage of fruits and vegetables, the concentration of O_2 is generally lowered and the level of CO_2 is often increased to diminish the action of ethylene and oxidative enzymes and to reduce the loss of O_2 sensitive compounds like vitamin C and carotenoids. Controlled atmospheres have also been used for pest control, including insects, rats, and mice (Vroom and C.J.M. Zuyderwijk 2006). There are many methods to control the CO_2, O_2, and ethylene concentrations for storing samples at CA. For controlling oxygen, active systems such as external burners, liquid or gaseous nitrogen, gas separator systems, or hypobaric systems can be used. Scrubbing systems for removal of CO_2 are based on NaOH, hydrated lime, water, activated charcoal, and molecular sieves. Commercial ethylene scrubbers include the heated catalyst scrubber, ethylene absorbing beads, and UV light.

Suitable conditions, in terms of gases concentration, may differ for each fruit or vegetable. These conditions depend on each particular species, stage of maturity, and grown conditions. Undesirable changes associated with controlled atmospheres have been observed; for example, carbohydrates could ferment in apples and pears stored at very low concentrations of O_2; this may generate a change in flavor (from a sweet taste to alcoholic taste) as well as turning color from purple to red or yellow to brown. Undesirable changes also may occur at high concentrations of CO_2; at these conditions, the skin becomes hard and darkening occurs on the skin and core (Kupferman 2001). CA conditions recommended for storing some products are shown in Table 4.4.

Some products are classified according to the minimum concentrations of O_2 and the maximum concentration of CO_2 required for storing (Table 4.5).

4.3.2.9.2 Modified Atmospheres

In MA, the food is placed in packages that act as semipermeable membranes to regulate the exchange of gasses inside and outside the film. In this system, the slowing of respiration rates reduces microbial growth and enzymatic reactions. The concentrations of O_2 and CO_2 inside the package changes as a result of the respiration rate of the product, the permeability of the film in which the product

Table 4.4 Oxygen concentration recommended to storage selected fruits under controlled atmospheres (Ke and Kader 1990, 1992; Ke et al. 1990, 1991)

Fruit	O_2 (%)	Shelf-life (days)
Apple	0.02	12
Blueberry	0.00	14–21
Cherry	0.02	25
Mango	0.10–0.20	5
Nectarine	0.02	14
Orange	0.02	16
Papaya	0.20–0.40	3
Peach	0.02	40
Pear	0.02	14
Strawberry	0.25	10

Table 4.5 Classification of fruits according to the minimum concentrations of O_2 and the maximum concentration of CO_2 required in controlled atmospheres (Kader and Ke 1994)

O_2 (%)	Examples
2	Apples, pears, kiwi, peaches, cherries, nectarines, papaya, strawberry, pineapple, melon
5	Citrus fruits
CO_2 (%)	*Examples*
2	Apple (Golden delicious), pear, grapes, tomatoes, peppers, and lettuce
5	Apple, peach, nectarine, orange, avocado, banana, mango, kiwi, and papaya
10	Grapefruit, lemon, lime, pineapple, cucumber, zucchini, broccoli, potatoes, onions, and garlic
15	Strawberry, raspberry, blueberry, cherry, mushrooms, and spinach

is packaged, and the storage conditions. It is desirable that the rates of permeation of O_2 and CO_2 be equal to the respiration rate of the product. Permeability of the film to O_2, CO_2, and water vapor is one of the most important factors to consider in MA, because it will have a direct effect on product quality. Films with high permeability to O_2 are not recommended because the product will produce ethylene and will ripen shortly. On the other hand, films with too low permeability to O_2 will ferment rapidly. In addition to permeability, temperature, membrane surface area, and pressure difference inside and outside the package will affect the rate of diffusion. Food packaged in MA are regularly stored under refrigeration.

The atmosphere modification can be passive or active. The passive MA is designed for packaging food with selected concentrations of gases (O_2, CO_2, and N_2), which will be adjusted with the gases of the respiration process of the product: CO_2 liberated and O_2 consumed. In the active MA, air is removed from the package containing a particular product, and subsequently replaced by the selected mixture of gases (Kader 1985; Kader et al. 1989). Qi et al. (1999) reported that an atmosphere with low O_2 concentrations reduced the ethylene production dramatically in honeydew melon cubes. Rattanapanone et al. (2001) pointed out that the marketable period of mango cubes could be extended by 1–2 days compared to samples kept at atmospheric conditions. Martínez-Ferrer et al. (2002) reported that a gas mixture consisting of 4 % O_2, 10 % CO_2, and 86 % N_2 extended the shelf-life of pineapple and mango in comparison with fruits packaged under vacuum, samples stored in 100 % oxygen, and control (non-storage). Serrano et al. (2005) showed that the use of MA (2–3 % of CO_2 and 11–12 % of O_2) in combination with eugenol, thymol, or menthol was effective for maintaining the cherry fruit quality; the combination reduced the incidence of decay because essential oils reduced the molds plus yeasts and total aerobic mesophiles populations by 4.00 and 2.00 log cycles, respectively.

4.3.2.9.3 Edible Films and Coatings

An edible film is made up of edible materials such as polysaccharides, proteins, and lipids. They have a thickness of about 0.3 mm and work as a barrier to face adverse environmental conditions, microorganisms as well as avoid quality loss due to transfer of water, gases, and flavors (Pavlaths and Orts 2009). Edible films are degraded much faster than synthetic films due to the nature of the materials from which they are made. They are generally used to pack individual food (fruits for example); they can also be found covering ingredients for preventing migration of flavor from one part to another (Guilbert 1986). They can also be used as carriers of antimicrobials and antioxidants (Pavlaths and Orts 2009).

Coatings of food can be obtained by dipping the product in the solution containing the polymer or using foams or brushes or sprinkling for coating the product. The excess of solution is removed with a rubber band (Donhowe and Fennema 1993; Grant and Burns 1994).

Some products which are coated to form a film on the surface are melons to increase their shelf-life; apples, avocados, and carrots to give brightness; grapefruit to maintain color and prevent water losses; mangoes, pineapples, bananas, and avocados in order to give a shine, delay ripening, and prevent weight loss; fresh-cut papaya in order to improve firmness; fresh-cut Fuji apples to incorporate antibrowning agents; fresh-cut apple and papaya for probiotic coating (Dalal et al. 1970, 1971; Mathur and Srivastava 1955; Rojas-Graü et al. 2007; Tapia et al. 2007, 2008).

As mentioned before, edible coating can be used to incorporate antimicrobial compounds to fresh fruits to improve safety. Antimicrobials in edible films and coatings may provide inhibitory effects against spoilage and pathogenic bacteria by maintaining effective concentrations of the active compound on food surfaces (Rojas-Graü et al. 2009). Lee et al. (2003) reported that apple slices coated with carrageenan containing ascorbic acid, citric acid, and oxalic acid extend their shelf-life by 2 weeks when packed in trays and stored at 3 °C. Tangor (or Murcott tangor produced with tangerine and orange) coated with 0.1 % low molecular weight chitosan (LMWC, Mw = 15 kDa) decreased its decay when stored at 15 °C, and this decrease was reduced about 20 % as compared with the fungicide TBZ. A concentration of 0.2 % LMWC was more effective in controlling the growth of *Penicillium digitatum* and *P. italicum* on citrus fruits (Chien et al. 2007a). Slices of mango coated with 0.5–2 % chitosan inhibited the growth of microorganisms (Chien et al. 2007b). The application of alginate and gellan-based edible coatings to fresh-cut apples retarded the microbiological deterioration of the fruits (Rojas-Graü et al. 2008).

4.4 Final Remarks

Microbiological, enzymatic, and chemical reactions are continually taking place in fruits; the rate of these reactions increases due to operation such as peeling, cutting, and slicing during the fresh-cut products production. Washing and sanitizing are two very important steps to be taken into account during fresh-cut fruit production to reduce the microbiological load.

Although chlorine is one of the main sanitizers used for food decontamination, its low effectiveness, depending on the range of pH used, decreases more when in contact with water delivering toxic compound. It is necessary to look for alternative sanitizers with greater oxidizing characteristics and nontoxic residues. In general, the effectiveness of each sanitizer depends on their concentration, type of food, and type and load of microorganism. Thus, each system must be studied individually to determine the best sanitizer to be used.

To maintain the overall quality of food during storage, it is necessary to use other preservation methods, based on the concept of minimal processing, such as mild irradiation, packaging, and HHP, among many other approaches. In general, the effectiveness of minimal processing techniques depends on type of food, type and load of microorganism, and level of each technology. Thus, individual studies for each food are necessary to determine the more suitable preservation method.

References

Abadias, M., I. Alegre, J. Usall, R. Torres, and I. Viñas. 2011. Evaluation of alternative sanitizers to chlorine disinfection for reducing foodborne pathogens in fresh-cut apple. *Postharvest Biology and Technology* 59: 289–297.

Al-Haq, M.I., Y. Seo, S. Oshita, and Y. Kawagoe. 2002. Disinfection effects of electrolyzed oxidizing water on suppressing fruit rot of pear caused by Botryosphaeria berengeriana. *Food Research International* 35(7): 657–664.

Alemán, G., F. Farkas Dl, J.A. Torres, E. Wilhelmsen, and S. McIntyre. 1994. Ultra-high pressure pasteurization of fresh cut pineapple. *Journal of Food Protection* 10(4): 931–934.

Alexandre, E.M.C., T.R.S. Brandão, and C.L.M. Silva. 2012. Assessment of the impact of hydrogen peroxide solutions on microbial loads and quality factors of red bell peppers, strawberries and watercress. *Food Control* 27: 362–368.

Alzamora, S.M., A. López-Malo, and M.S. Tapia. 2000. Overview. In *Minimally processed fruits and vegetables*, Fundamentals and applications, ed. S.M. Alzamora, M.S. Tapia, and A. Lopez-Malo, 1–9. Gaithersburg, Maryland: Aspen.

Annous, B.A., G.M. Sapers, A.M. Mattrazzo, and D.C.R. Riordan. 2001. Efficacy of washing with a commercial flatbed brush washer, using conventional and experimental washing agents, in reducing populations of *Escherichia coli* on artificially inoculated apples. *Journal of Food Protection* 64(2): 159–163.

Artés, F., P. Gómez, E. Aguayo, V. Escalona, and F. Artés-Hernández. 2009. Sustainable sanitation techniques for keeping quality and safety of fresh-cut plant commodities. *Postharvest Biology and Technology* 51: 287–296.

Arvanitoyannis, I., A.C. Stratakos, and P. Tsarouhas. 2009. Irradiation applications in vegetables and fruits: A review. *Critical Reviews in Food Science and Nutrition* 49: 427–462.

Beuchat, L.R., B.V. Nail, B.B. Adler, and M.R.S. Clavero. 1998. Efficacy of spray application of chlorinated water in killing pathogenic bacteria on raw apples, tomatoes and lettuce. *Journal of Food Protection* 61: 1305–1311.

Beuchat, L.R. 2000. Use of sanitizers in raw fruit and vegetable processing. In *Minimally processed fruits and vegetables*, Fundamental aspects and applications, ed. S.M. Alzamora, M.S. Tapia, and A. Lopez-Malo, 63–79. Gaithersburg, Maryland: Aspen.

Bialka, K.L., and A. Demirci. 2008. Efficacy of pulsed UV-Light for the decontamination of *Escherichia coli* O157:H7 and *Salmonella* spp. on raspberries and strawberries. *Journal of Food Science* 73(5): M201–M207.

Borsa, J. 2000. Irradiation of foods. In *Wiley encyclopedia of food science and technology*, vol. 1, ed. F.J. Francis, 1428–1436. New York: Wiley.

Boylston, T.D., C.A. Reitmeier, J.H. Moy, G.A. Mosher, and L. Taladriz. 2002. Sensory quality and nutrient composition of three Hawaiian fruits treated by X-irradiation. *Journal of Food Quality* 25(5): 419–433.

Cao, S., Z. Hu, and B. Pang. 2010a. Optimization of postharvest ultrasonic treatment of strawberry fruit. *Postharvest Biology and Technology* 55: 150–153.

Cao, S., Z. Hu, B. Pang, H. Wang, H. Xie, and F. Wu. 2010b. Effect of ultrasound treatment on fruit decay and quality maintenance in strawberry after harvest. *Food Control* 21: 529–532.

Chakraverty, A., A.S. Mujumdar, G.S. Raghavan, and H.S. Ramaswamy. 2003. Introduction: Production, trade, losses, causes, and preservation. In *Handbook of postharvest technology: Cereals, fruits, vegetables, tea and species*, ed. A. Chakraverty, A.S. Mujumdar, G.S. Raghavan, and H.S. Ramaswamy. New York: Marcel Dekker.

Chen, Z., and C. Zhu. 2011. Combined effects of aqueous chlorine dioxide and ultrasonic treatments on postharvest storage quality of plum fruit (*Prunus salicina* L.). *Postharvest Biology and Technology* 61(2): 117–123.

Chien, P.J., F. Sheu, and H.R. Lin. 2007a. Coating citrus (*Murcott tangor*) fruit with low molecular weight chitosan increases postharvest quality and shelf life. *Food Chemistry* 100: 1160–1164.

Chien, P.J., F. Sheu, and F.H. Yang. 2007b. Effects of edible chitosan coating on quality and shelf life of sliced mango fruit. *Journal of Food Engineering* 78: 225–229.
Dalal, V.B., W.E. Eipeson, and N.S. Singh. 1971. Wax Emulsion for fresh fruits and vegetables to extend their storage life. *Indian Food Packer* 25: 9.
Dalal, V.B., P. Thomas, N. Nagaraia, G.R. Shah, and B.C. Arnla. 1970. Effect of wax coating on bananas of varying maturity. *Indian Food Packer* 24(2): 36–39.
De Azevedo Pimentel, R.M., and J.M. Melges Walder. 2004. Gamma radiation in papaya harvested at three stages of maturation. *Scientific Agriculture* 61(2): 146–150.
Zagory, D. 1999. Effects of post-processing handling and packaging on microbial populations. *Postharvest Biology and Technology* 15: 313–321.
Donhowe, I.G., and O.R. Fennema. 1993. The effects of plasticizers on crystallinity, permeability, and mechanical properties of methylcellulose films. *Journal of Food Processing and Preservation* 17: 247–257.
Escobedo-Avellaneda, Z., M. Pateiro Moure, N. Chotyakul, J.A. Torres, J. Welti-Chanes, and C. Pérez Lamela. 2011. Benefits and limitations of food processing by high-pressure technologies: Effects on functional compounds and nonbiotic contaminants. *CyTA—Journal of Food* 9(4): 352–365.
FAO. 2003. Handling and preservation of fruits and vegetables by combined methods for rural areas. Technical manual. FAO Agricultural Services Bulletin 149, ed. G.V. Barbosa-Cánovas, J.J. Fernández-Molina, S.M. Alzamora, M.S. Tapia, A. López-Malo, and J. Welti-Chanes, 3–18. Rome, Italy: FAO.
Farkas, J. 1998. Irradiation as a method for decontaminating food: A review. *International Journal of Food Microbiology* 44(3): 189–204.
Fonseca, J.M., and J.W. Rushing. 2006. Effect of ultraviolet-C light on quality and microbial population of fresh-cut watermelon. *Postharvest Biology and Technology* 40: 256–261.
Garcia, E., and D. Barrett. 2005. Fresh-cut fruits. In *Processing fruits science and technology*, 2nd ed, ed. D.M. Barrett, L. Somogyi, and H. Ramaswamy, 53–72. Boca Raton, FL: CRC.
Grant, L., and J.K. Burns. 1994. Application of coatings. In *Edible coatings and films to improve food quality*, ed. J.M. Krochta, E.A. Baldwin, and M.O. Nisperos-Carriedo, 89. Boca Raton, FL: CRC.
Gil, M.I., M.V. Selma, F. López-Gálvez, and A. Allende. 2009. Fresh-cut product sanitation and wash water disinfection: Problems and solutions. *International Journal of Food Microbiology* 134: 37–45.
Gómez-López, V.M., P. Ragaert, J. Debevere, and F. Devlieghere. 2007. Pulsed light for food decontamination: A review. *Trends in Food Science & Technology* 18: 464–473.
Gómez-López, V.M., A. Rajkovic, P. Ragaert, N. Smigic, and F. Devlieghere. 2009. Chlorine dioxide for minimally processed produce preservation: A review. *Trends in Food Science & Technology* 20: 17–26.
Gonzalez, R.J., Y. Luo, S. Ruiz-Cruz, and J.L. McEvoy. 2004. Efficacy of sanitizers to inactivate *Escherichia coli* O157:H7 on fresh-cut carrot shreds under simulated process water conditions. *Journal of Food Protection* 67: 2375–2380.
Goodburn, C., and C.A. Wallace. 2013. The microbiological efficacy of decontamination methodologies for fresh produce: A review. *Food Control* 32: 418–427.
Graca, A., M. Abadias, M. Salazar, and C. Nunes. 2011. The use of electrolyzed water as a disinfectant for minimally processed apples. *Postharvest Biology and Technology* 61: 172–177.
Guilbert, S. 1986. Technology and application of edible protective films. In *Food packaging and preservation: Theory and practice*, ed. M. Mathlouthi, 371–399. London: Elsevier.
Gunes, G., J.H. Hotchkiss, and C.B. Watkins. 2001. Effects of gamma irradiation on the texture of minimally processed apple slices. *Journal of Food Science* 66(1): 63–67.
Hawkes, J.G. 2000. Food processing: Effect on nutritional quality. In *Wiley encyclopedia of food science and technology*, vol. 1, ed. F.J. Francis, 969–986. New York: Wiley.
Huang, T.S., C.L. Xu, K. Walker, P. West, S.Q. Zhang, and J. Weese. 2006. Decontamination efficacy of combined chlorine dioxide with ultrasonication on apples and lettuce. *Journal of Food Science* 71: M134–M139.
Hyldgaard, M., T. Mygind, and R.L. Meyer. 2002. Essential oils in food preservation: Mode of action, synergies, and interactions with food matrix components. *Frontiers in Microbiology* 3: 1–24.

Jin, Y.Y., Y.J. Kim, K.S. Chung, M. Won, and B. Song. 2007. Effect of aqueous chlorine treatment on the microbial growth and qualities of strawberries during storage. *Food Science and Biotechnology* 16: 1018–1022.

Kader, A., and D.M. Barrett. 2005. Classification, composition of fruits and postharvest maintenance of quality. In *Processing fruits Science and technology*, Secondth ed, ed. D.M. Barrett, L. Somogyi, and H. Ramaswamy, 3–22. Boca Raton, FL: CRC.

Kader, A.A., and D. Ke. 1994. Controlled atmospheres. In *Insect pests and fresh horticultural products*, ed. R.E. Paull and J.W. Armstrong, 223–236. Wallingford, England: CAB International.

Kader, A.A. 1985. Modified atmosphere and low-pressure systems during transport and storage. In *Postharvest technology of horticultural crops*, ed. A.A. Kader, R.F. Kasmire, F.G. Mitchell, M.S. Reid, W.S. Sommer, and J.F. Thompson, 458–467. Davis, CA: Special Publication 3311. University of California.

Kader, A. 2002. Quality parameters of fresh-cut fruit and vegetable products. In *Fresh-cut fruits and vegetables: Science, technology and market*, ed. O. Lamikanra. Boca Raton, FL: CRC.

Kader, A.A., D. Zagory, and E.V. Kerbel. 1989. Modified atmosphere packaging of fruit and vegetables. *Critical Reviews in Food Science and Nutrition* 28(1): 1–30.

Ke, D., and A.A. Kader. 1990. Tolerance of 'Valencia' oranges to controlled atmospheres as determined by physiological responses and quality attributes. *Journal of American Society of Horticultural and Science* 115(5): 779–783.

Ke, D., and A.A. Kader. 1992. Potential of controlled atmospheres for postharvest insect disinfestation of fruits and vegetables. *Postharvest News and Information* 3(2): 31N–37N.

Ke, D., L. Goldstein, M. O'Mahony, and A.A. Kader. 1991. Effects of short term exposure to low O_2 and high CO_2 atmosphere on quality attributes of strawberries. *Journal of Food Science* 56: 50–54.

Ke, D., H. Van Gorsel, and A.A. Kader. 1990. Physiological and quality responses of 'Bartlett' pears to reduced O_2 and enhanced-CO_2 atmospheres and storage temperatures. *Journal of the American Science for Horticultural Science* 155: 435–439.

Kilcast, D. 1995. Food irradiation: Current problems and future potential. *International Biodeterioration & Biodegradation* 36(3–4): 279–296.

Koseki, S., K. Yoshida, S. Isobe, and K. Itoh. 2004. Efficacy of acidic electrolyzed water for microbial decontamination of cucumbers and strawberries. *Journal of Food Protection* 67(6): 1247–1251.

Kupferman, E. 2001. Controlled atmosphere storage of apples and pears. Washington State University. Tree Fruit Research and Extension Center. http://postharvest.tfrec.wsu.edu/EMK2001D.pdf.

Lado, B.H., and A.E. Yousef. 2002. Alternative food-preservation technologies: Efficacy and mechanisms. *Microbes and Infection* 4: 433–440.

Lagunas-Solar, M.C., C. Piña, J.D. MacDonald, and L. Bolkan. 2006. Development of pulsed UV light processes for surface fungal disinfection of fresh fruits. *Journal of Food Protection* 2(9): 376–384.

Lamikanra, O., D. Kueneman, D. Ukuku, and K.L. Bett-Garber. 2005. Effect of processing under ultraviolet light on the shelf life of fresh-cut cantaloupe melon. *Food Science* 70(9): C534–C539.

Lee, J.Y., H.J. Park, C.Y. Lee, and W.Y. Choi. 2003. Extending shelf-life of minimally processed apples with edible coatings and antibrowning agents. *LWT--Food Science and Technology* 36(3): 323–329.

Lee, S.Y., G.I. Dancer, S.S. Chang, M.K. Rhee, and D.H. Kang. 2006. Efficacy of chlorine dioxide gas against *Alicyclobacillus acidoterrestris* spores on apple surfaces. *International Journal of Food Microbiology* 108: 364–368.

Lu, J.Y., C. Stevens, V.A. Khan, M. Kabwe, and C.L. Wilson. 1991. The effect of ultraviolet irradiation on shelf-life and ripening of peaches and apples. *Journal of Food Quality* 14(4): 299–305.

Lukasik, J., M.L. Bradley, T.M. Scott, M. Dea, A. Koo, W.Y. Hsu, et al. 2003. Reduction of poliovirus 1, bacteriophages, *Salmonella* Montevideo, and *Escherichia coli* O157:H7 on strawberries by physical and disinfectant washes. *Journal of Food Protection* 66(2): 188–193.

Manzocco, L., S. Da Pieve, and M. Maifreni. 2011. Impact of UV-C light on safety and quality of fresh-cut melon. *Innovative Food Science and Emerging Technologies* 12: 13–17.

Mari, M., T. Cembali, E. Baraldi, and L. Casalini. 1999. Peracetic acid and chlorine dioxide for postharvest control of *Monilinia laxa* in stone fruits. *Plant Disease* 83(8): 773–776.

Mari, M., R. Gregori, and I. Donati. 2004. Postharvest control of Monilinia laxa and Rhizopus stolonifer in stone fruit by peracetic acid. *Postharvest Biology and Technology* 33(3): 319–325.

Mathur, P.B., and H.C. Srivastava. 1955. Effect of akin coatings on the storage behaviour of mangoes. *Journal of Food Science* 20(6): 559–566.

Martínez-Ferrer, M., C. Harper, F. Pérez-Muñoz, and M. Chaparro. 2002. Modified atmosphere packaging of minimally processed mango and pineapple fruits. *Journal of Food Science* 67(9): 3365–3371.

Okull, D.O., and L.F. Laborde. 2004. Activity of electrolyzed oxidizing water against *Penicillium expansum* in suspension and on wounded apples. *Journal of Food Science* 69(1): FMS23–FMS27.

Ölmez, H., and U. Kretzschmar. 2009. Potential alternative disinfection methods for organic fresh-cut industry for minimizing water consumption and environmental impact. *LWT—Food Science and Technology* 42: 686–693.

Patil, B.S., J. Vanamala, and G. Hallman. 2004. Irradiation and storage influence on bioactive components and quality of early and late season 'Rio Red' grapefruit (*Citrus paradisi* Macf.). *Postharvest Biology and Technology* 34: 53–64.

Pavlaths, A.E., and W. Orts. 2009. Edible films: Why, what and how! In *Edible films and coatings for food and other applications*, ed. M.E. Embuscado and K.C. Huber, 1–24. New York: Springer.

Pérez, A.G., C. Sanz, J.J. Ríos, R. Olıas, and J.M. Olıas. 1999. Effects of ozone treatment on postharvest strawberry. *The Journal of Agricultural and Food Chemistry* 47: 1652–1656.

Qi, L., T. Wu, and A.E. Watada. 1999. Quality changes of fresh-cut honeydew melons during controlled atmosphere storage. *Journal of Food Quality* 22(5): 513–521.

Ramos, B., F.A. Miller, T.R.S. Brandão, P. Teixeira, and C.L.M. Silva. 2013. Fresh fruits and vegetables—An overview on applied methodologies to improve its quality and safety. *Innovative Food Science and Emerging Technologies* 20: 1–15.

Rattanapanone, N., Y. Lee, T. Wu, and A.E. Watada. 2001. Quality and microbial changes of fresh-cut mango cubes held in controlled atmosphere. *HortScience* 36(6): 1091–1095.

Rennie, T.J., C. Vigneault, J.R. Deell, and G.S.V. Raghavan. 2003. Cooling and storage. In *Handbook of postharvest technology: Cereals, fruits, vegetables, tea and species*, ed. A. Chakraverty, A.S. Mujumdar, G.S. Raghavan, and H.S. Ramaswamy, 505–538. New York: Marcel Dekker.

Rico, D., A.B. Martín-Diana, J.M. Barat, and C. Barry-Ryan. 2007. Extending and measuring the quality of fresh-cut fruit and vegetables: A review. *Trends in Food Science and Technology* 18: 373–386.

Rodgers, S.L., J.N. Cash, M. Siddiq, and E.T. Ryser. 2004. A comparison of different chemical sanitizers for inactivating *Escherichia coli* O157:H7 and *Listeria monocytogenes* in solution and on apples, lettuce, strawberries and cantaloupe. *Journal of Food Protection* 67: 721–731.

Rojas-Graü, M.A., R. Soliva-Fortuny, and O. Martín-Belloso. 2009. Edible coatings to incorporate active ingredients to fresh-cut fruits: A review. *Trends in Food Science and Technology* 20: 438–447.

Rojas-Graü, M.A., M.A. Tapia, and O. Martín-Belloso. 2008. Using polysaccharide-based edible coatings to maintain quality of fresh-cut Fuji apples. *LWT—Food Science and Technology* 41: 139–147.

Rojas-Graü, M.A., M.S. Tapia, F.J. Rodríguez, A.J. Carmona, and O. Martin-Belloso. 2007. Alginate and gellan-based edible coatings as carriers of antibrowning agents applied on fresh-cut Fuji apples. *Food Hydrocolloids* 21(1): 118–127.

Rossi, M., E. Giussani, R. Morelli, R. Lo Scalzo, R.C. Nani, and D. Torreggiani. 2003. Effect of fruit blanching on phenolics and radical scavenging activity of highbush blueberry juice. *Food Research International* 36(9–10): 999–1005.

Russell, R. 2005. The disinfection of raw vegetables and salads-focus on mixed oxidants or anolyte. Washing and decontamination of fresh produce forum. Gloucestershire, England: Campden and Chorleywood Food Research Association Group. Newsletter Issue 6.

Sapers, G.M., and J.E. Sites. 2003. Efficacy of 1% hydrogen peroxide wash in decontaminating apples and cantaloupe melons. *Journal of Food Science* 68(5): 1793–1797.

Sapers, G.M., R.L. Miller, and A.M. Mattrazzo. 1999. Effectiveness of sanitizing agents in inactivating *Escherichia coli* in Golden Delicious apples. *Journal of Food Science* 64(4): 734–737.

Sapers, G.M., R.L. Miller, V. Pilizota, and A.M. Mattrazzo. 2001. Antimicrobial treatments for minimally processed cantaloupe melon. *Journal of Food Science* 66(2): 345–349.

Sauer, A., and C.I. Moraru. 2009. Inactivation of *Escherichia coli* ATCC 25922 and *Escherichia coli* O157:H7 in apple juice and apple cider, using pulsed light treatment. *Journal of Food Protection* 5(8): 937–944.

Serrano, M., D. Martınez-Romero, S. Castillo, F. Guillen, and D. Valero. 2005. The use of natural antifungal compounds improves the beneficial effect of MAP in sweet cherry storage. *Innovative Food Science and Emerging Technologies* 6: 115–123.

Schenk, M., S. Guerrero, and S.M. Alzamora. 2008. Response of some microorganisms to ultraviolet treatment on fresh-cut pear. *Food and Bioprocess Technology* 1(4): 384–392.

Siegel, A., P. Markakis, and C.L. Bedford. 1971. Stabilization of anthocyanins in frozen tart cherries by blanching. *Journal of Food Science* 36(6): 962–963.

Singh, T.K., and K.R. Cadwallader. 2004. Ways of measuring shelf-life and spoilage. In *Understanding and measuring the shelf-life of food*, ed. R. Steele. Cambridge, England: Woodhead.

Soliva-Fortuny, and O. Martın-Belloso. 2003. New advances in extending the shelf life of fresh-cut fruits: A review. *Trends in Food Science & Technology* 14: 341–353.

Smilanick, J.L., D.M. Margosan, and F. Mlikota Gabler. 2002. Impact of ozonated water on the quality and shelf-life of fresh citrus fruit, stone fruit, and table grapes. *Ozone: Science and Engineering* 24(5): 343–356.

Sy, K.V., K.H. McWatters, and L.R. Beuchat. 2005. Efficacy of gaseous chlorine dioxide as a sanitizer for killing *Salmonella*, yeasts, and molds on blueberries, strawberries, and raspberries. *Journal of Food Protection* 68(6): 1165–1175.

Tapia, M.S., M.A. Rojas-Graü, A. Carmona, F.J. Rodríguez, R. Soliva-Fortuny, and O. Martin-Belloso. 2008. Use of alginate- and gellan-based coatings for improving barrier, texture and nutritional properties of fresh-cut papaya. *Food Hydrocolloids* 22(8): 1493–1503.

Tapia, M.S., M.A. Rojas-Graü, F.J. Rodríguez, J. Ramírez, A. Carmona, and O. Martin-Belloso. 2007. Alginate- and gellan-based edible films for probiotic coatings on fresh-cut fruits. *Journal of Food Science* 72(4): E190–E196.

Tapia, M.S., and J. Welti-Chanes. 2012. Hurdle technology principles applied in decontamination of whole and fresh-cut produce. In *Decontamination of fresh and minimally processed produce*, 1st ed, ed. V.M. Gómez-López, 417–444. Oxford: Wiley.

Thompson, J.F., F.G. Mitchell, and R.F. Kasmire. 2002. Cooling horticultural commodities. In *Postharvest technology of horticultural crops*, 3rd ed, ed. A.A. Kader, 97–112. Oakland, CA: Regents of the University of California.

Udompijitkul, P., M.A. Daeschel, and Y. Zhao. 2007. Antimicrobial effect of electrolyzed oxidizing water against *Escherichia coli* O157:H7 and *Listeria monocytogenes* on fresh strawberries (Fragaria × ananassa). *Journal of Food Science* 72(9): M397–M406.

Venkitanarayanan, K.S., C.M. Lin, H. Bailey, and M.P. Doyle. 2002. Inactivation of *Escherichia coli* O157:H7, *Salmonella enteritidis*, and *Listeria monocytogenes* on apples, oranges and tomatoes by lactic acid with hydrogen peroxide. *Journal of Food Protection* 65: 100–105.

Vroom, N., and C.J.M. Zuyderwijk. 2006. Improvement treatments for controlled atmospheres in stored product pest control. In Proceedings of the Annual Conference on Methyl Bromide Alternatives and Emissions Reductions. November 6–9. Orlando, FL: The Double Tree Hotel.

Wang, H., H. Feng, and Y. Luo. 2006. Dual-phasic inactivation of *Escherichia coli* O157:H7 with peroxyacetic acid, acidic electrolyzed water and chlorine on cantaloupes and fresh-cut apples. *Journal of Food Safety* 26: 335–347.

Wolbang, C.M., J.L. Fitos, and M.T. Treeby. 2008. The effect of high pressure processing on nutritional value and quality attributes of *Cucumis melo* L. *Innovative Food Science & Emerging Technologies* 9(2): 196–200.

Wu, V.C., and B. Kim. 2007. Effect of a simple chlorine dioxide method for controlling five food-borne pathogens, yeasts and molds on blueberries. *Food Microbiology* 24: 794–800.

Xie, F. 2003. *Disinfection byproducts in drinking water: Form, analysis and control*, 176. Boca Raton, FL: CRC/Technology and Industrial Arts.

Yu, K., M.C. Newman, D.D. Archbold, and T. Hamilton-Kemp. 2001. Survival of *Escherichia coli* O157:H7 on strawberry fruit and reduction of the pathogen population by chemical agents. *Journal of Food Protection* 64(9): 1334–1340.

Chapter 5
The Hurdle Concept in Fruit Processing

Stella Maris Alzamora, Aurelio López-Malo, Sandra Norma Guerrero, and María Soledad Tapia

5.1 Introduction

Fruits and vegetables (F&V) are important elements of nutrition in most world cultures (Schwentesius and Gómez 2000). Today, and increasingly more often, humans tend to eat healthy diets that include an increase in the intake of fruits and vegetables as their beneficial effects on health have been evidenced. Comparisons made between populations who consume small amounts of F&V and those that eat larger amounts as part of a balanced diet, show that the latter have a reduced risk of chronic diseases, including some types of cardiovascular disorders (Dauchet et al. 2004), type II diabetes (Montonen et al. 2005; Venn and Mann 2004) and certain types of cancer (Chang et al. 2005; Nkondjock et al. 2005; Rashidkhani et al. 2005; Ray 2005; Zhang et al. 2005; De Stefani et al. 2005). Based on these facts it is not surprising that authorities of most countries in the world advice to increase the intake of F&V as part of a healthy diet. The World Health Organization (WHO) and the Food and Agriculture Organization of the United Nations (FAO) recommend consumption of at least 400 g of fruit and vegetable (five servings per day) and this has been included into dietary guidelines such as is the case of the United States (WHO/FAO 2003; USDA 2005).

S.M. Alzamora (✉) • S.N. Guerrero
Universidad de Buenos Aires, Buenos Aires, Argentina

A. López-Malo
Universidad de las Americas—Puebla, Cholula, Puebla, Mexico

M.S. Tapia
Universidad Central de Venezuela, Caracas, Venezuela

A. Rosenthal et al. (eds.), *Fruit Preservation*, Food Engineering Series,
https://doi.org/10.1007/978-1-4939-3311-2_5

Another aspect that affects consumption positively is the variety and availability of products in the current scenario of globalized trade that allows a significant flow of fruits considered exotic. This has resulted in major advances in technologies that have improved postharvest management and control of the cold chain, a condition for long-distance transportation of highly perishable commodities. One of the main problems of developing countries is the absence of an efficient integration between production and consumption, which is reflected in the lack of modern distribution channels. Moreover, the seasonality of production imposes an accelerated time for processing the fruit, and an accumulation of high-price stocks during the harvest months. Because of this, it is necessary to develop expedite, simple, low-cost, preservation techniques that may act as a regulatory factor of the offer. These techniques in their majority have to be used in combination so as to be alternatives to concentration, freezing, and other techniques intensive in energy expenditure, even refrigeration, with the enormous challenge of not compromising safety and quality during prolonged storage, and ideally be simple enough to be applied in the production sites in small agro-industries, which may in turn generate an important social and economic impact. The incorporation of a hurdle approach that utilizes individual sublethal physical, chemical, and biological stressors in a sequential application with a cumulative inhibitory effect on the microbiological load present and quality loss reactions in a food matrix is very adequate for the fruit industry, especially for the minimal processing sector with all its variants. This approach allows tailoring practices to individual operations considering process and final products (Tapia and Welti-Chanes 2012).

The produce industry is faced with an ever-increasing demand for fresh-likeness, convenience, and "health" in foods with the minimally processed fruit sector becoming one of its fastest growing segments. To harmonize these trends without compromising safety, a number of milder fruit preservation factors (some already used commercially) had been explored (i.e., "nonthermal" physical agents such as high hydrostatic pressure, pulsed electric fields, ultrasound, pulsed light, ultraviolet light; chemical agents, such as natural antimicrobials, ozone, hydrogen peroxide, among others, and minimal heating) (Gómez et al. 2011c). The alternative "nonthermal" physical agents, intensely investigated in the last two decades, can cause inactivation of microorganisms at ambient or sublethal temperatures avoiding the deleterious effects that severe heating has on quality, but most preservation systems would still require refrigeration to assure safety and keeping properties. The trend for mildly preserved fruits comes with the combined preservation/hurdle technology as the principle in designing the overall treatment. In general, the processes are less robust and need to be well controlled through adequate product and process design and proper implementation and monitoring through HACCP (Havelaar et al. 2010).

Assuring safety is essential to accessing the market, and global recognition of standardized protocols to eliminate risk at every step from "farm to fork" has translated into food safety policy for many countries. Although the Hazard Analysis and Critical Control Points (HACCP) system is not yet mandatory and not currently required by law for an important sector of the minimal processed industry like fresh-cut produce, the whole and fresh and industry itself has been encouraging their fresh-cut processors to voluntarily implement HACCP programs in their facilities, as well as to demand safe practices by partners throughout the supply chain. Many segments of the fresh-cut produce industry have adopted HACCP principles. In produce operations, however, HACCP systems have limited application since specific critical limits cannot be established and monitored to ensure that the hazard is reduced to acceptable levels. Instead, Good Agricultural Practices (GAPs)—voluntary—, Good Manufacturing Practices (GMPs)—mandatory—, and Sanitation Standard Operating Procedures (SSOPs)—voluntary—, provide the primary levels of risk management for operations in the field that will directly impact the fresh-cut processing plant (Tapia et al. 2009).

Most fruits contain between 10 % and 25 % carbohydrate, less than 1 % protein and less than 0.5 % lipid (USDA 2004). There are fruits that are exception in terms of lipid content, such as avocados and olives which may contain up to 20 % or 30 % of total fat. Carbohydrates include cellulose, hemicelluloses, lignin, and pectic substances that contribute as a source of dietary fiber, and sugars that give them a sweet taste. The relative acidity of the fruit is due to the presence of acids; mainly malic and citric. The acidity of the fruits is also influenced by the crop, the soil conditions, and the degree of fruit maturity, but in general is always within a certain range. Fruits may differ in their composition and structure, which determines the kind of deterioration and how easily it can be attacked by microorganisms. The more acidic pH of fruits and the presence of carbohydrates promote the deterioration due to the growth of molds, yeasts, and some acid-tolerant bacteria in a greater extent. Water is the major component of fruits and fruit–water activity (a_w) is determined by the nature and concentration of dissolved naturally occurring chemicals, such as sugars, organic acids, inorganic salts, and other soluble substances. As the concentration of solutes (nonionic or ionizable) naturally present in the aqueous phase of fresh fruits is relatively small, a_w is close to unity (Chirife and Ferro Fontán 1982). This high value facilitates the growth of microbial populations that have access to these foods, as is evident by observing the natural occurrence of numerous deteriorative genera of bacteria, molds, and yeasts as well as occasional pathogenic bacteria such as *Listeria monocytogenes*, *Salmonella*, *Escherichia coli* O157:H7, *Clostridium botulinum,* and others. Fruits have also become increasingly important identified vehicles for microorganisms capable of causing disease, which is found in the many documented outbreaks associated with fresh fruits and fresh juices in recent years (Lee et al. 2001; Sewell and Farber 2001; Sivapalasingam et al. 2004).

In the case of fruits, commercial quality and storage life depends on certain conditions that occur before, during, and after harvest. During growth, fruit diseases can be produced by: molds, yeasts, bacteria, mycoplasma, and viruses (Messiaen et al. 1994). Levels of disease vary greatly from season to season. Crop losses can be the result of the attack by more than one pathogen to the plant. The microbial load of fruits grown in the field reflects the land on which they are grown. Therefore, during preharvest, important factors are the land, irrigation water, the presence of human or animal fecal material, the type of fertilizer, air and people who care for the crops (Fernández Escartín 2000; Jay 2002). These elements may constitute a source of various microorganisms that are normally deposited on the surface of fruits and vegetables, which remain in place while maintaining the structural integrity of the first layers of cells. Fruit spoilage usually occurs during storage and transport and/or during postharvest handling and waiting time for processing. As soon as fruits are collected in boxes, baskets, or trucks during the harvest, are exposed to contamination with organisms from spoiled fruits (collected earlier) and/or containers. Mechanical damage can increase susceptibility to decay and microbial growth. When the epicarp is broken, the microorganisms on the surface may enter and find the right conditions to cause deterioration. The microbial load expected in plants will be greater in those crops that have more contact with the soil and its irregular surface also allows for greater adhesion of contamination such as strawberries. Commodities that are not directly in contact with the soil and have smooth surfaces will have lower microbial loads. In postharvest handling major contamination sources are machinery and equipment, containers, pets and wildlife, workers, vehicles and the atmosphere (Beuchat 2000; Bracket 2001; Jay 2002). Fruit sold in markets are primarily contaminated during handling. Another mechanism of contamination that occurs when losing the integrity of the cuticles, shells, or tests covering fruits, is that insects can be deposited in these wounds with microorganisms attached and can even lay their own eggs (Jones and Widemo 2005).

Internalization is another process of contamination of fruits. Microorganisms can enter into plant tissue through structures such as stomata, lenticels, and stem vascular tissues. Although washing is a primary step to reduce microbial load, it can be a source of contamination and a vehicle for dissemination of microorganisms. The temperature differential between the fruit and the wash water favors the entry of microorganisms. The extent of infiltration of water in fruits generally depends on factors such as time of exposure, magnitude of the temperature differential, depth of immersion, agitation, viscosity of the external environment, and the size and number of points of entry that allow access to the inner space, but also on the adherence of microorganisms to the porous structure of the tissue. This phenomenon reinforces the need for proper management to change the temperature of the fruit after harvest and to use water with good microbiological quality in the washing step (Richards and Beuchat 2004; Bartz 1982; Penteado et al. 2004). The way by which microorganisms penetrate plant tissues have not been clearly established. Products in the field can carry heavy microbial loads on their surfaces, but sometimes the internal tissues contain certain bacteria too, which by unknown mechanisms penetrate the interior. In addition, procedures for decontamination, reducing the initial microbial load, and methods to prevent post-process contamination, change the original microflora ecosystem and therefore affect the microbial associations. The microbial association is specific to each type of food and is affected by intrinsic and extrinsic factors that play an important ecological role in the establishment of the saprophytic microflora and colonization of the food by pathogens (Martínez et al. 2000). All operations undergone by the precut fresh products may damage plant tissues and can increase microbial populations. The use of equipment improperly sanitized may allow microorganisms transferred by contact, as in the case of *Geotrichum candidum*, which can accumulate in the equipment and contaminate fruits (Bracket 2001). In general, the specific effects of pre- and postharvest and storage on the microflora of fresh produce is difficult to predict, given the complexity of ecosystems and the variety of products.

In an attempt to satisfy consumers and industry, a number of combinations of several antimicrobial factors in a multifactorial fruit preservation approach ("hurdle" technology) have been developed in the past 20 years. Targeted application of the hurdle concept has aimed to improving quality and safety of fruit products at the farm level, and in the whole and fresh-cut minimally processed fruits industry (Alzamora et al. 2000; Leistner and Gould 2002; Raso and Barbosa-Cánovas 2003; Ross et al. 2003; Raso et al. 2005; Allende et al. 2006; Rico et al. 2007; Raybaudi-Massilia et al. 2009; Gómez et al. 2011c; Tapia and Welti-Chanes 2012).

The objective of this chapter was to present a panoramic view of recent promising hurdle combinations explored for the conservation of tropical and subtropical fruits and their subproducts, and point out some areas of study to fully exploit the potential of the hurdle concept in the design and optimization of preservation techniques.

5.2 The Hurdle Concept

5.2.1 Basic Aspects

Microorganisms have evolved different mechanisms to resist the adverse effects of the stresses provided by preservation factors. As internal media stability (composition and volume of fluids) is vital for survival and growth, these mechanisms, called "homeostatic mechanisms", act to ensure that key physiological activities and parameters in the cells remain relatively unchanged, even when the environment around the cell is different and greatly perturbed (Leistner and Gould 2002; Gould 1995). When a stress is sensed by the microorganism, signals that induce mechanisms to cope with the stressor are developed. These mechanisms involve modifications in gene expression and protein activities (Capozzi et al. 2009). Homeostatic mechanisms that vegetative cells have evolved in order to survive extreme environmental stresses are energy dependent and allow microorganisms to keep functioning. In contrast, homeostasis in spores is passive, and adaptive mechanisms are built into the bacterial spore prior to the environmental stress being imposed (Leistner and Gould 2002). They act to keep the central protoplast in a constant low water level environment, this being the prime reason for the extreme metabolic inertness or dormancy and resistance of these cells to high temperature, high hydrostatic pressure, ultrasonication, and other hostile environments. For instance, one of the most mechanisms studied in more depth is osmoregulation. When a microorganism is put into an environment of reduced a_w, water is extracted from the cytoplasm of the cell (in a passive way or possibly mediated by water channels) and membrane turgor is lost. The homeostasis (or internal equilibrium) is disturbed and the organism will not multiply but will remain in the lag-phase until the equilibrium is reestablished. Although the specific details of how each organism responds to an hyperosmotic shock are different and the organisms differ widely in the range of osmolarity over which they will grow, several common features, both physiologically and genetic, have arisen (Tapia et al. 2007; O'Byrne and Booth 2002). Bacterial response to hyperosmolarity encloses two aspects. The first one has to do with the ability of bacteria to accumulate osmoprotective low molecular weight compounds (by synthesis and/or by active transport) in their cytoplasm at concentrations sufficient to just exceed the osmolality of the external medium for turgor and growth restoration. The second one refers to the osmotic regulation of the expression of a number of genes to optimize growth under the stress condition, allowing cells to modulate the rate of acquisition of compatible solutes. Many of these genes are under the control of alternative stress and stationary-phase sigma factors, σ^S in the Gram-negative and σ^B in the Gram-positive species (Pichereau et al. 2000). For example, the growth of *E. coli* in absence of other compatible solutes from the growth medium occurs by the accumulation via its synthesis of trehalose. Trehalose synthetic enzymes are under the control of the *rpoS* sigma factor, which accumulates when cells are grown at high osmolarity (O'Byrne and Booth 2002). In *E. coli* the σ^S regulon includes over 50 different genes and the products of these genes confer resistance to wide range of stress conditions.

The second one concerns the osmotic induction of general stress systems, with the consequent development of multitolerances towards other environmental stresses when subjected to hypertonic environments (Pichereau et al. 2000; O'Byrne and Booth 2002).

As presently known, there is a general response mechanism (the so-called "global response") underlying many of the apparent distinct responses of microorganisms to different stresses imposed on them in foods (e.g., low a_w, low pH, low or high temperature, oxidative stress, starvation, etc.). This global response is mediated by the stationary-phase regulator RpoS, that regulates the expression of many important stationary-phase stress resistance genes linked to survival under starvation conditions and to survival in the stationary-phase. As Gould (2000) stated, this overlapping stress responses would explain the cross-resistances to different stresses that have usually been found to occur in response to a single stress. Reduced a_w causes an increase in maintenance metabolism and a reduction in yield and growth rate because the solute accumulation process is energy-dependent. If the osmoregulatory capacity of the cell is exceeded (by a severe reduction in a_w), the cell ceases growth.

Other example is acid stress response. The maintenance of intracellular pH within a narrow range is essential for microorganism growth. Lowering the external pH by strong acids causes denaturation of enzymes present on cell surface and lowering of the cytoplasmic pH due to proton permeation through membrane when the pH gradient is very large. When weak acids are used, undissociated acids act as "proton ionophores" and permeate through the membrane, increasing the rate at which protons enter the cytoplasm, but also the acid anion may have specific effects on metabolism amplifying the action of the low pH. Proton influx could lead to a complete dissipation of the proton motive force (Capozzi et al. 2009). Major adaptive mechanism to regulate the cytoplasmic pH is the energy-dependent proton extrusion, which acts to keep the cytoplasmic pH higher than that of the environment, and, sometimes to extrude the organic acid (Booth and Kroll 1989; Leistner and Gould 2002). When the stress severity increases and the microorganisms' capacity for generating energy is not enough to prevent the net proton influx, the cytoplasmic pH falls, growth ceases, and cells may die.

Oxidative stress by reactive oxygen species (ozone, chlorine dioxide, hydrogen peroxide, electrolyzed water, peroxyacetic acid), and nitrogen species caused an imbalance between intracellular oxidant concentration, cellular antioxidant protection, and oxidative change of lipids of membrane, proteins, and DNA repair enzymes. A number of protection systems (catalases, peroxidases, superoxide dismutases, superoxide reductases, etc.) acts in cellular defense against oxidative stresses (Capozzi et al. 2009).

Temperature downshift decreases membrane fluidity and stabilizes secondary structures of nucleic acids (reducing the efficiency of mRNA translation and transcription). Homeoviscous adaptation involves among others the incorporation of fatty acids with lower melting points into lipids to reestablish the optimal fluidity of the membrane and the production of cold shock inducible proteins to prevent mRNA secondary structures (Capozzi et al. 2009).

Preservation procedures are effective when they overcome, temporally or permanently, the various homeostatic reactions that microorganisms have evolved in order to resist stresses. The degree of change in environmental conditions will determine whether the microorganism lose their viability, become injured, or express adaptive mechanisms that would allow them to survive or even to growth during stress (Capozzi et al. 2009).

In foods preserved by combined methods, the active homeostasis of vegetative microorganisms and the passive refractory homeostasis of spores are disturbed by a combination of gentle antimicrobial factors at a number of sites ("targets") or in a cooperative manner (Gould and Jones 1989). Low levels of different stresses are employed rather a single intensive stress. Moreover, a more effective preservation (i.e., synergistic effects of hurdles or preservation factors) is obtained if small stresses with different targets ("multitarget preservation") are selected to inhibit microorganisms' growth instead of small stresses with the same target (i.e., additive effect of hurdles). For example, for vegetative cells (where homeostasis is energy dependent), the goal is to reduce the availability of energy (for instance by limiting the amount of oxygen available for facultative organisms) and/or to increase the demand for energy (by imposing some other stresses). Placing a number of sublethal stresses (i.e., hurdles or preserving factors) and/or increasing the intensity of a particular sublethal hurdle on a vegetative microbial cell increases the expenditure of energy, and energy is completely used up for repairing the homeostasis. The microorganism becomes metabolically exhausted and dies (Leistner and Gould 2002). On the contrary, when preservation factors are used at high intensity, metabolic exhaustion does not occur because the initiation of homeostatic mechanisms is prevented, and survival of cells is actually enhanced. The metabolic exhaustion is of enormous practical significance in hurdle-preserved fruits, since the microbiological status of such fruits improves with time of storage (Alzamora et al. 1995; Tapia de Daza et al. 1996). Sublethal treatments also may result in an increased sensitivity to adverse environmental factors, such as so longer lag-phase of sublethally damaged cells, when the cell resumes growth after treatment (Smelt et al. 2002). For spores (where homeostasis is nonenergetic and depends on the structure of the organism), the goal is to damage key structures (i.e., by chemical, enzymatic, or physical attack on coats, cortex, etc.) or to release spores from dormancy (i.e., initiating germination with natural germinants or with false triggers, or applying high pressures).

The hurdle concept exploits synergistic and/or additive interactions between sublethal stressors, helping to reduce the detrimental effects on product quality, the energy input, and the treatment intensities required while guarantying safety.

5.2.2 Most Commonly Used Hurdle Combinations

The combination of physical and chemical methods to decontaminate whole and fresh-cut F&V produce is the way forward to follow for the industry and different approaches have been explored for obtaining stability and fresh-likeness. One of these approaches involved the use of combinations of traditional stressors in simple and inexpensive processing methods. Over the past decades, Alzamora et al. (1995, 2000) developed innovative technologies for obtaining shelf-stable "high-moisture fruit products" with shelf-life of 3–8 months without need of refrigeration. Thus, for various fruit products (pieces, purées), the stress factors employed were: blanching and/or a mild heat treatment—applied without affecting the sensory and nutritional characteristics—, a_w and pH reductions, and antimicrobial agents added to prevent potential microbial spoilage. The a_w reduction (a_w 0.94–0.98, usually adjusted with glucose, sucrose, fructose, maltodextrins, corn syrups, and/or some polyols), control of pH (pH 3.0–4.1, usually adjusted with citric or phosphoric acid), addition of antimicrobials (in doses legally approved and/or sensory compatible, usually weak acids), and depending on the fruit type, the addition of antibrowning agents, were the factors selected to formulate fruit preservation procedures (Alzamora et al. 1989, 1993, 1995; Guerrero et al. 1994; Cerrutti et al. 1997; Argaiz et al. 1995; Tapia de Daza et al. 1995). Analyzing the role of each hurdle in the combined-technique system, the blanching step applied with saturated vapor is a critical operation in the decontamination of fruits. Although its primary objective is enzymes inactivation, heating during blanching also inactivates yeasts, most molds, and aerobic natural flora and sensitizes remaining microorganisms to other hurdles. Reductions in the microbial load from 60 % to 99 % have been reported after blanching of papaya, mango, pineapple, and strawberry (Alzamora et al. 1995, 2000; Tapia de Daza et al. 1995, 1996). The a_w factor was selected in the range 0.93–0.98, accomplishing emergent interest for "fresh-likeness" and low-sugar in foods. The pH was maintained equal or near the pH value of fresh fruit. In less acidic fruits, pH was adjusted to the lower value that was sensory compatible with the natural flavor of the fruit. Foods with high a_w are suitable for the growth of bacteria, molds, and yeasts, but high acidity determines an unsuitable environment for the growth of most bacteria. So, the low pH determines a potential type of spoilage by fungi and acid-tolerant bacteria. Considering that a slight reduction of pH increases the lower limit of a_w for bacterial growth and, vice versa, a slight reduction of a_w diminishes the range of pH that permits growth, it is expected that interaction pH—a_w in those ranges will be enough to suppress the growth of most bacteria of concern in fruit preservation (Alzamora et al. 1993, 2003). Ability of fungi to tolerate reduced a_w and pH, on the contrary, demands the incorporation of antifungal (e.g., sorbic or benzoic acid) in moderate amounts (400–1000 ppm potassium sorbate or sodium benzoate). The major goal for the design of these combined techniques was the development of simple and inexpensive techniques for bulk storage without refrigeration that were energy efficient and suitable to preserve fruits "in situ", that helped overcoming seasonal production constraints and reduce postharvest losses (Alzamora et al. 1995; Argaiz et al. 1995; Tapia de Daza et al. 1995). To optimize the level of the stress factors at high-moisture contents, the microbial response to stress factors was addressed using different approaches: studies in laboratory media, studies of evolution of native flora in fruit products and microbial challenge tests with microorganisms of concern. The results of such studies have been reported (Alzamora et al. 1989, 2000) and demonstrated that the appropriate selection of stress factors and their levels, can lead to fruit products with longer shelf-life, stable at room temperature. The use of antimicrobials of natural origin as replacement (total or partial) of sorbates, benzoates, and other synthetic additives to meet consumer's expectations about chemicals, was the other aspect considered to improve combined techniques (Alzamora et al. 2003).

A second approach for improving control of foodborne and spoilage microorganisms with promising results involves combinations of emerging "nonthermal" factors or combinations of these stressors with traditional ones. Table 5.1 presents some selected hurdles or stressors (already used industrially or still in development or testing) along with their mode of action, their advantages and disadvantages and the combined processes in which they had been applied to preserve fruits.

Emerging nonthermal factors reported herein have not broad-spectrum inactivation processes like thermal treatment, but represent pasteurization techniques that allow minimizing the disadvantages of severe thermal processing. High hydrostatic pressure, pulsed electric field, ultraviolet radiation, pulsed light, and ultrasound are gaining commercial uses most quickly with fruit-derived products, probably due the low pH that naturally exists in this type of food materials, and are considered "hurdles" that cooperate well in an overall preservation strategy. On the other hand, it should be mentioned that it is probable that the acid adaptation of contaminant flora could adversely affect the microorganism resistance to these technologies, fact that promotes the intelligent combination with other hurdles.

5.3 Research and Commercial Application: Examples of Combined Traditional and Novel Stressors

Targeted application of the hurdle concept has been used/suggested in fruit preservation in different arrangements: (a) using two or more stressors simultaneously to prevent growth or inactivation of spoilage and pathogenic microorganisms; (b) using one or more stressors to inactivate/injure or physically remove some microorganisms, and

Table 5.1 Selected emerging stressors applied in fruit preservation

Factor and mechanism of action	Advantages	Limitations and drawbacks	Potential application/products on the market	Hurdles investigated in combination
High hydrostatic pressure (HHP)				
Application of 100–800 MPa, below 0–100 °C, from seconds to about 20 min, instantaneously and uniformly throughout food, independent of size, shape and food composition Mechanism: Multitarget and dependent on pressure level: membrane damage, protein denaturation, leakage of cell contents, and dissociation of ribosomes	Inactivation of some enzymes according to HHP dose Little change in vitamins, pigments, flavor and antioxidant activity, although effects depend on fruit matrix, pressure, and temperature	High cost of equipment, increased metal fatigue, long cycle times High resistance of browning enzymes and PME to HHP Undesirable sensory changes at high doses (color, appearance, skin loss, structural/texture changes)	Jam, jellies, fruit juices and purées, guacamole, fruit yogurts, dairy-based fruit smoothie, sauces (in use since 1990)	Low pH Natural and synthetic antimicrobials Temperatures below or above room temperature Vacuum packaging and refrigerated storage Mild heating
Short-wave ultraviolet light (UV-C)				
Radiation from the short-wave ultraviolet region of the electromagnetic spectrum (200–280 nm) Mechanism: Damage to DNA, membranes and enzyme activity induced by UV-C light absorption. Hormetic effects in agricultural produce	Moderate to low cost of equipments Little effect on color, vitamin C and taste of fruit juices Little changes in tissue darkening, color, texture, and visual quality of cut fruits at low doses	Low penetration into solids and opaque juices, long treatment times in solids Enzymatic browning of cut fruit surfaces at high doses, more notorious as storage time increase	Pasteurization of apple cider and clear juices (in use since 2000). Surface decontamination of whole and cut fruit surfaces. Reduction of fruit decay and softening	Refrigerated storage MAP Mild thermal treatment US Sanitizers

Pulse light (PL)				
Few flashes applied in a fraction of a second of intense pulses of broad spectrum light (ultraviolet to the near-infrared region) Mechanism: Damage to DNA and destruction of cellular components by the high peak power and the photothermal effects of visible and near-infrared portions of the flash spectrum	Very short treatment times (≤60 s) Little effect on color, texture, antioxidant, and sensory properties at low doses	Low penetration into solids and opaque juices Engineering solutions needed for juice treatment Thermal damage of product at high doses Browning and dehydration of cut fruit surfaces, more notorious as storage time and PL dose increase	Reduction of microbial load on surfaces of whole and cut fruits and in clear juices	Refrigerated storage UV-C Mild thermal treatment
Pulsed electric fields (PEF)				
Application of oscillating, bipolar, exponentially decaying or square wave electric pulses of high voltage (20–70 kv/cm, pulse duration 1–30 s) Mechanism: perturbation of cell membrane and loss of membrane permeability	Little effect on food quality attributes Very short treatment times (≤30 s)	Restricted to foods that can sustain high electric fields, have low electrical conductivity and do not have/produced bubbles	Limited to pasteurization of fruit juices (in use since 2005) and fruit smoothies	Bio-preservatives, essential oils, other antimicrobials Moderate temperatures UV-C PL
High power ultrasound (US)				
Energy generated by sound waves of 20 kHz or more and intensities higher than 1 W/cm^2 Mechanism: Disruption of cellular structures (wall, membranes, organelles, DNA) and cell lysis attributed to cavitation	Inactivation of enzymes when US is combined with heat, and pressure Little change in color of juices and cut fruits	High energy consumption, intensity of industrial-scale equipments limited, long treatment times Heating of the product. Undesirable sensory changes and rupture of skin in berries at high doses	No commercial fruit products; suggested for juice pasteurization. Actual applications limited to product modification and process efficiency improvements (enhancement of mass and heat transfer, degassing of liquids, cleaning of surfaces)	Moderate temperature, pressure Sanitizers Natural antimicrobials UV-C PEFs

then, in sequential mode, one or more stressors to prevent survival/proliferation of remaining refractory or sublethally damaged cells (these last with greater sensitivity to adverse agents); (c) using two or more stressors in sequence for inactivating microorganisms. The order in which they are applied may affect the effectiveness of inactivation. Type, number, and intensity of hurdles had been proposed according to the type of fruit, the impact of the hurdles on quality, the shelf-life required, and/or the available processing/storage infrastructure. These processes have their own set of limitations and advantages. It must be emphasized that no single process will allow obtaining high quality and safety for every fruit product. Some applications to exemplify the different possibilities are addressed next.

5.3.1 Cut and Whole Fruits

UV-C (short-wave ultraviolet) radiation can be considered a promising tool for keeping overall quality of fresh-cut fruits. Direct inactivation by UV-C of microorganisms is limited solely to those associated to the surface of the fruit as UV-C has extremely low penetration into solids, but inactivation can occur in the entire fruit at the dose levels used to induced hormesis (0.5–9 kJ/m^2 for optimal effects according to the type of fruit), with production of antifungal enzymes and phytoalexins (Shama and Anderson 2005). However, semilogarithmic survival curves of inoculated microorganisms on tomatoes, pears, and apple surfaces showed upward concavity and pronounced tailing effect attributed not only to the heterogeneity in the resistances of the population to UV-C irradiation, but to the shielding or physical protection of microorganisms on the solid surface from incident UV-C (effect of surface topography) and/or the internalization of microorganisms into fruit porous tissues (Schenk et al. 2008; Yaun et al. 2004; Gómez et al. 2010). This inactivation pattern reinforced the need for a hurdle approach to reach microbiological stability.

UV-C disinfection has been extensively studied as a postharvest treatment for reducing the number of microorganisms on the surface of fresh and cut fruits (Shama 2006; Allende and Artes 2003; Yaun et al. 2004; Fonseca and Rushing 2006), combined with posterior chilling or MAP to preserve quality. Doses of UV-C light up to 6.9 kJ/m^2 were a satisfactory sanitizing treatment ($\approx$1–1.5 log reduction) for fresh-cut watermelon without causing deterioration of quality in terms of juice leakage, flesh darkening, visual quality, and color values compared to controls after 7 days of storage at 3 °C (Fonseca and Rushing 2006). Gómez et al. (2010, 2011a) examined the effect of UV-C irradiation at different doses on native flora and inoculated microorganisms, surface color, and rheological characteristics of cut-apple disks stored in refrigeration for 7 days. They also explored the use of some pretreatments (hot water blanching, dipping into a solution containing ascorbic acid and calcium chloride) to minimize browning of UV-C irradiated apple slices. Color and compression parameters were found to be dependent on UV-C dose, storage time, and type of pretreatment. Changes in structural features, color, and viscoelastic parameters were mainly evidenced after refrigerated storage. At the end of storage, samples exposed to only UV-C light turned darker (lower L^* values) and less green (higher a^* value) when compared to fresh-cut apple slices or to samples on day 0 and this effect was more pronounced at the greatest UV-C dose. Light microscopic images showed breakage of cellular membranes in UV-C treated samples which may explain the increase in browning of irradiated apples. Both pretreatments helped in maintaining the original color of apple slices after UV-C light exposure. Natural microflora counts were higher in untreated UV-C than in UV-C treated samples along the whole storage. All samples showed a viscoelastic solid behavior with the storage modulus (G') dominating the viscoelastic response. Overall, both dynamic moduli decreased, and creep instantaneous compliance (J_0), decay compliances (J_1 and J_2), and fluidity significantly increased after treatments and storage at 5 °C. However, a texture-trained panel only significantly differentiated stored untreated apple from the other samples regarding fracturability and juiciness, showing the potential of this technology to maintain texture characteristics in preserved fruit.

Recently, Gómez et al. (2011b) performed a similar study but using pulsed light (PL) instead of continuous UV-C. They examined the dose effect of PL irradiation on surface color, microstructure, and microbial stability of cut-apples stored under refrigeration. An increase in surface browning was noticed when increasing PL doses were applied. Again, light microscopy observations indicated that the modifications on color of treated apples could be at least partially ascribed to the breakage of cellular membranes, which would cause a loss of functional cell compartmentalization, increasing enzyme-substrate contact with the consequent increase in tissue browning. But increases in temperature during PL irradiation at high doses could also cause non-enzymatic browning.

The combination of UV-C with mild heat treatment (sequential hurdles) had been suggested by Marquenie et al. (2002) and Pan et al. (2004) for controlling postharvest decay of berries (strawberries and sweet cherries) stored at room temperature. Previous irradiation with UV-C (4.1 kJ/m^2) enhanced the benefits of heat treatment (45 °C, 3 h in air) and further reduced decay, softening, and reddening of the strawberry fruit (Pan et al. 2004).

An important group of combined treatments recently developed were oriented to hurdle strategies that consider the use of traditional or novel sanitizers (ozone, acidified sodium chlorite, hydrogen peroxide, electrolyzed water–acidic, or neutral, peroxiacetic acid, organic acids, etc.) combined among themselves and/or combined with physical inactivation stressors (ultrasound; light-based techniques; mild heat, low temperature, etc.). In many cases, the result is the on-site destruction of even refractory organisms without the generation of residues. Contemporary strategies based on sanitizers to provide the whole and fresh-cut produce industry have been deeply discussed in the comprehensive review by Tapia and Welti-Chanes (2012). One of these approaches is the employment of UV-C in combination with hydrogen peroxide for decontaminating fresh fruits. Schenk et al. (2012) investigated the simultaneous or serial combined use of UV-C light (7.5 min; 3.7 kJ/m^2) and H_2O_2 treatment (3 % w/v; pH 3.0; 25 °C) to preserve fresh-cut pear discs under 10-day refrigerated storage (5 °C). This combination was successful preventing the proliferation of the residual native flora population during refrigerated storage, being the browning effect less pronounced compared with the individual treatments. Additionally, the product processed by the combined H_2O_2/UV-C treatment was well accepted by the consumers.

Among ultrasound-based approaches, Jang and Moon (2011) proposed the combined use of ultrasound (40 kHz) and ascorbic acid (1 %) to extend the shelf-life of refrigerated (10 °C) fresh-cut "Fuji" apples by inhibiting enzymes related to enzymatic browning (polyphenol oxidase and peroxidase). These authors observed that the simultaneous treatment has synergistic inhibitory effects on PPO activity without deformation of enzyme protein. They inferred that the combined treatment allowed the ascorbic acid to act inside the cell disrupted by ultrasound treatment.

Chen and Zhu (2011) designed a combined preservation method for plum fruit (*Prunus salicina* L.) storage based on the simultaneous application of the hurdles ClO_2 (40 mg/L; 10 min) and ultrasound (100 W, 10 min) with the purpose of reducing respiration rates and maintaining firmness of the fruits, which were packaged after the treatment into aseptic polyethylene bags and stored at 4 °C for 60 days. The ClO_2 and ultrasound-treated fruit showed significantly higher flesh firmness than the untreated plum. The activities of ripening-related enzymes in plum fruit might be retarded by the ClO_2 and ultrasonic treatments and thus softening inhibited. The effect of combined treatments on maintaining contents of total flavonoids, ascorbic acid, reducing sugars, and titratable acids were similar but were more beneficial than the individual treatments and the untreated control. This confirmed that ascorbic acid is insufficient in controlling browning and maintaining the commercial value of fresh-cut products (its inhibitory enzyme effect is reversible and temporary). The simultaneous mode was effective in reducing the initial microflora, leaving no detectable chemical residues and retaining sensory qualities of plum fruit, and fruit shelf-life could be extended to 60 days compared to 35 days for the control.

HHP is gaining popularity in the fruit industry because its ability to destroy microorganisms and to significantly reduce the enzymatic activity on acid fruit juices and fresh fruits without greatly affecting vitamins, pigments, and flavor and antioxidant activity, probably due to the stability of covalent unions to high pressure. Because of inherent low pH, most fruits can be easily stabilized by HHP, since yeasts, molds, and vegetative cells of bacteria can be inactivated by pressures in the range of 200–700 MPa near room temperature. But the presence of HHP-resistant enzymes requires a careful selection of the operative/storage conditions for best quality retention. HHP affects enzymes and there is an optimum temperature range at which proteins are more resistant to pressure. As an example, good storage stability of strawberries was obtained over at least 3 months when HHP at temperatures between 20 and 40 °C was combined with vacuum packaging and refrigerated storage since polyphenol oxidase was highly resistant to high-pressure inactivation (Terefe et al. 2009).

The combination of HHP and plant-essential oils had been suggested as an alternative control for fruit diseases. *Colletotrichum gloeosporioides* spores, which cause anthracnose in papaya, were efficiently inhibited by a 350 MPa-30 min treatment or by a combination of 150 MPa-30 min and 0.75 mg/ml of citral or lemongrass oil. An explanation for the enhanced effect of pressure plus lemongrass essential oil is that pressure facilitates the uptake of the oil constituents into the spore, increasing the number of targets affected (Palhano et al. 2004).

Krebbers et al. (2003) analyzed the effects of combined HHP—thermal treatments on consistency, viscosity, color, lycopene content, enzyme activity, and microorganisms of tomato purée as a model of tomato-based products. Single HHP (700 MPa, 2 min, 20 °C) resulted in inactivation of natural flora to a level below the detection limit and partial inactivation of galacturonase but activation of pectin methylesterase. The product appeared stable during storage at 4 °C at least 8 weeks. After high-pressure sterilization treatments combined with elevated starting temperatures (≥80 °C, one or two pulses), an ambient stable product was obtained, with more than 99 % inactivation of polygalacturonase and pectin methylesterase, with better color, greater lycopene content, and improved water-binding capacity than the conventional sterilized tomato product.

Several chemical compounds have been used in combination to reduce bacterial populations on fruit and they are still the most widely used treatments, either before processing or during pre- and post-cutting operations. Many efforts are being made to replace chlorine-based chemicals used as fruit sanitizers since in many European countries their use is prohibited due to their potential toxicity (Gil et al. 2009). Electrolyzed oxidizing water (EOW) has been shown to be a promising alternative decontamination technique with a strong bactericidal effect. This technique has been suggested as a valuable disinfection tool for wash water sanitation in the minimally processed fruit industry (Gil et al. 2009). An evidence is provided by Wang et al. (2007), who proposed a sequential washing treatment (5 min) of acidic EOW (pH 2.7; oxidation-reduction potential 1150 mV; free chlorine 45 mg/L) followed by calcium ascorbate solution (5 %; pH 5.0) to effectively delay bacterial growth, browning, and firmness loss of fresh-cut apple wedges under passive modified atmosphere storage conditions (polypropylene bags; 4 °C; 11 days).

An interesting and more recent approach was proposed by Silveira et al. (2011), who examined the effect of hot water immersion dipping (60 °C; 90 and 120 s) followed by immersion in peroxiacetic acid (80 mg/L; 60 s) on metabolic activity and microbial and sensory quality changes of fresh-cut Galia melon pieces packaged into oriented polypropylene heat-sealed trays (generation of a passive modified atmosphere packaging, 7.4 kPa O_2 and 7.4 kPa CO_2) and stored for 10 days at 5 °C. This combined treatment, using an eco-friendly sanitizer, had positive effect on melon overall quality and reduced its metabolism helping to maintain fruit firmness.

5.3.2 *Juices*

The nonthermal factors applied to fruit juices that have been mostly investigated were pulsed electric field (PEF), high hydrostatic pressure (HHP), short-wave ultraviolet irradiation (UV-C), and ultrasound (US) (Bermúdez-Aguirre and Barbosa-Cánovas 2012; Keenan et al. 2010; Liang et al. 2006).

Table 5.2 resumed some investigations on PEF combined with many other hurdles in PEF-based strategies. Several studies, mainly in apple juice, reported that microbial PEF inactivation enhanced by its combination with UV-C irradiation. Walkling-Ribeiro et al. (2008) observed an *Staphylococcus aureus* reduction of 9.5 $\log_{10}$ cycles in apple juice reconstituted from concentrate processed by batch UV-C treatment (30 min;

Table 5.2 Efficacy of PEF-based hurdle strategies for fruit juice preservation

PEF-based strategy/storage conditions	Evaluated conditions	Parameters of evaluation	Product	Conclusions on hurdle strategy	Reference
Continuous flow PEF system (PEF; 30 kV; 1 μs; 200 pulses/s)/moderate temperature (*T*)/antimicrobials	PEF (27 and 33 kV/cm; 3, 6, and 10 L/h); *T* (45 and 50 °C); (2) Antimicrobials: nisin (27.5 U/ml)/lysozyme (690 U/ml), clove oil (3 ml/100 ml)	Natural microflora; PPO	Apple cider	PEF: 3.1 $\log_{10}$ red (3 L/h; 50 °C) PEF/T/nisin/lisozyme: Less than additive PEF/T/clove oil: No significant additional effect 33 % PPO activity decrease (PEF/T)	Liang et al. (2006)
Continuous flow lab scale PEF system (40 kV/cm, 100 μs; 15.75 ml/min)/ Batch UV-C system (254 nm, 30 min, 30 W)	(1) UV-C; (2) PEF; (3) UV-C + PEF; (4) PEF + UV-C; (5) Heat processing (26 s, 94 °C) and (6) Heat processing (26 s, 72 °C)	Natural microflora; NEBI; color; pH; Antioxidant activity; polyphenol content; PPO; POD	Fresh apple juice	(3) UV-C + PEF or (4) PEF + UV-C: Less than additive antimicrobial effect; similar to (5) No significant differences in chemical parameters with respect to (1) and (2) No significant differences in % enzyme activity with respect to (2): 47.2 % and 49.5 % POD activity; 42.8 % and 41.3 % PPO activity	Noci et al. (2008)
Continuous flow lab scale PEF system (3.5 μs)/continuous flow UV-C system (254 nm; 25 W)	(1) UV-C (10, 20, 30, 40 and 50 cm length tube; 8–20 ml/min); (2) PEF (40, 50 and 60 kV/cm; 8, 14 and 20 ml/min); (3) UV-C + PEF and (4) PEF + UV	*E. coli* ATCC 23472	Pasteurized apple juice	UV-C + PEF or PEF + UV-C: additive effect (5.3 $\log_{10}$ red.; 60 kV/cm; 8 ml/min; 30 cm UV-C length)	Gachovska et al. (2008)

Batch UV-C system (254 nm; 26 cm distance; 30 W; 30 min; 20 °C)/ continuous mild preheating (60 s)/ continuous flow lab scale PEF system (1.0 μs pulse width; 15 Hz)	(1) Combined treatment, orthogonal design: UV-C/ Preheating (T_{inlet} 35, 43, 46 and 50 °C)/PEF (28, 32, 36 and 40 kV/cm; 25, 50, 75 and 100 μs); (2) HTST pasteurization (94 °C, 26 s)	*Staphylococcus aureus* SST 2.4; color; pH; conductivity; NEBI; polyphenol content; AA content	Apple juice reconstituted from concentrate	(1) UV-C/preheating/PEF: 9.5 $\log_{10}$ red. (30 min-UV-C/46 °C preheating/40 kV-75 μs PEF). No significant changes in physical and chemical properties (2) HTST: 8.2 $\log_{10}$ red.	Walkling-Ribeiro et al. (2008)
Continuous flow lab scale pulsed light system (PL; 360 μs pulse width; 200–1100 nm; 3 Hz; 1.213 J/cm^2/ pulse)/continuous flow lab scale PEF system (PEF; 1.0 μs pulse width; 18 Hz; 20.8 ml/min; T_{inlet}:20 °C; 34 kV/cm; 93 μs) Continuous flow lab scale UV system (UV; 254 nm; 30 W; 176 ml/min; 5.3 J/cm^2/30 s)	(1) UV-C; (2) PL; (3) PL+PEF; (4) UV-C+PEF; (5) Heat processing (26 s, 72 °C) (H72)	*E. coli* K12 DSM 1607 *Pichia fermentans* DSM 70090; Shelf-life (natural microflora at 4 °C)	90:10 Fresh apple and cranberry juice blend	*E. coli* and *P. fermentans*: (1) and (2): less than 5 $\log_{10}$ red. (3) and (4): more than 6 $\log_{10}$ red. (5): 6 $\log_{10}$ red. Shelf-life (3): 21–28 days Shelf-life (4): 14–21 days	Palgan et al. (2011)
Continuous flow lab scale PL system (360 μs pulse width; 200–1100 nm; 3 Hz; 17 ml/min)/continuous flow lab scale PEF system (1.0 μs pulse width; T_{inlet}: 20 °C; 89 μs)	(1) PL (4.0 and 5.1 J/cm^2); (2) PEF (24 and 34 kV/cm; 13.4 and 17.0 ml/min; 15 and 12 Hz); (3) PEF+PL; (4) PL+PEF	*E. coli* K12 DSM 1607; color; pH; polyphenol content; NEBI; sensory consumer tests	Apple juice reconstituted from concentrate (1:7.8 v/ water v)	(1) PL: 3.3–3.9 $\log_{10}$ red.; (2) 1.8–3.5 $\log_{10}$ red. (3) Additive or synergistic effect (more than 6.5 $\log_{10}$ red.) (4) 4.5–6.2 $\log_{10}$ red. (~additive effect) No significant changes in physical; sensory; and chemical properties (4) Most acceptable treatment	Caminiti et al. (2011)

(continued)

Table 5.2 (continued)

PEF-based strategy/storage conditions	Evaluated conditions	Parameters of evaluation	Product	Conclusions on hurdle strategy	Reference
Continuous flow lab scale PEF system (40 kV/cm; 1.0 μs pulse width; 15 Hz; 100 μs; max T_{outlet}: 56 °C)/bio-preservatives (BP; 2.5 ppm nisin; 500 ppm lactic acid; 100 ppm benzoic acid or 10 ppm natamycin)	(1) PEF; (2) BP; (3) BP+PEF	*E. coli* K12 HB101; *Listeria innocua* IMD 11288; *P. fermentans* CBS	Pulp-free orange juice	(1) PEF: 4.8 $\log_{10}$ red. *P. fermentans*; 6.0 $\log_{10}$ red. *E. coli* and 4.2 $\log_{10}$ red. *L. innocua*. (2) BP: 0.5–2.0 $\log_{10}$ red. (3) Synergistic effect (7.8 $\log_{10}$ red.; *P. fermentans*; lactic acid); additive effect (5.5 $\log_{10}$ red.; *P. fermentans*; benzoic acid); without effect (4.2 $\log_{10}$ red. *P. fermentans*; natamycin); synergistic effect (5.6 and 7.9 $\log_{10}$ red.; *L. innocua* and *E. coli*; nisin); synergistic effect (6.1 $\log_{10}$ red.; *L. innocua*; lactic acid); without effect (5.8 $\log_{10}$ red.; *E. coli*; lactic acid)	McNamee et al. (2010)
Continuous flow bench-scale PEF system (35 kV/cm; 4 μs pulse width; $T<38$ °C)/citric acid (CA) or cinnamon bark oil (CO)	(1) PEF (200, 600 and 1000 μs; 100, 150 and 200 Hz); (2) CA (0.5 %, 1.0 %, 1.5 % and 2.0 % w/v); (3) CO (0.05 %, 0.10 %; 0.20 % and 0.30 % v/v); (4) CA (1 h)/PEF (1000 μs; 100 Hz); (5) CO (1 h)/PEF (1000 μs; 100 Hz)	*Salmonella enterica* Ser. Enteritidis 1.82	Tomato juice	(1) PEF (4.184 $\log_{10}$ red., 1000 us, 100 Hz) (2) CA: little effect (<1.0 $\log_{10}$ red.) (3) CO: 1.09–4.11 $\log_{10}$ red. (CO: 0.1–0.3 %) (4) CA/PEF: synergistic effect (CA>1 %); 5 $\log_{10}$ red. (CA: 2 %) (5) CO/PEF: synergistic effect (CO>0.1 %); total inactivation (CO>0.2 %)	Mosqueda-Melgar et al. (2008a)

Continuous flow bench-scale PEF system (35 kV/cm; 4 μs pulse width; *T*<38 °C)/citric acid (TA) or cinnamon bark oil (TO)	(1) PEF (500, 1250 and 2000 μs; 100, 1750 and 250 Hz); (2) CA (0.5 %, 1.0 %, 1.5 % and 2.0 % w/v); (3) CO (0.05 %, 0.10 %; 0.20 % and 0.30 % v/v); (4) CA (1 h)/PEF (1575–1700 μs; 100–235 Hz); (5) CO (1 h)/ PEF (1575–1700 μs; 100–235 Hz)	*Salmonella enterica* Ser. Enteritidis 1.82; *E. coli* O157:H7	Apple, pear, orange, and strawberry juices	(1) PEF: 4.34 (apple)–5.15 (orange) $\log_{10}$ red. *S.* Enteritidis; 4.28 (apple)–5.46 (strawberry) $\log_{10}$ red. *E. coli* (4) CA/PEF: synergistic effect (CA: 1.0–1.5 % *S.* Enteritidis; orange and strawberry); (CA: 1.5–2.0 %, *E. coli*, orange and strawberry); (CA: 1.5 %, *S.* Enteritidis and *E. coli* above 5 $\log_{10}$ red., apple and pear) (5) CO/PEF: additive effect (CO>0.1 % *S.* Enteritidis, apple, pear and orange; CO>0.05 % *E. coli*, strawberry)	Mosqueda-Melgar et al. (2008b)
Continuous flow PEF system (35 kV/cm; 4 μs pulse width; 150 Hz; 90 ml/min)/bacteriocin (B; enterocin AS-48; 3.5 AU/μg protein)	(1) PEF (100–1000 μs; 20 and 40 °C); (2) B (30; 45 and 60 μg/ml) (3) PEF/B	*Salmonella enterica* CECT 915	Apple juice	(1) PEF: 3.04 $\log_{10}$ red. (2) B: no effect (3) PEF (1000 μs; 40 °C)/B (60 μg/ml): synergistic effect, 4.5 $\log_{10}$ red.	Martinez Viedma et al. (2008)

AA ascorbic acid, *NEBI* non-enzymatic browning index, *POD* peroxidase, *PPO* polyphenoloxidase, *red.* reductions

20 °C) followed by mild preheating (35–50 °C) and PEF treatment (25–100 μs; 28–40 kV/cm). This approach introduced an advantage with respect to the sole PEF treatment from the point of view of energy consumption, requiring shorter treatment time to reach a desired inactivation. In a subsequent investigation, these researchers studied the influence of this hurdle strategy on selected quality parameters (color, pH; Brix; non-enzymatic browning index; antioxidant capacity) and enzymatic activity (polyphenoloxidase and peroxidase) of freshly squeezed apple juice. They applied these hurdles as stand-alone treatments (UV-C or PEF) or using two types of combinations (UV-C+PEF or PEF+UV-C). Additionally, they processed apple juice by using traditional thermal treatments in a heat exchanger (72 or 94 °C) (Noci et al. 2008). The application of the combination UV-C+PEF to freshly squeezed apple juice resulted in a similar total microbial reduction compared to the severe heat treatment (7.1 and 6.7 log cycle reduction, respectively), being the inactivation effect for both combinations less than additive. However, the quality attributes measured in juice processed by PEF or by the combined approach were similar to those observed in juice treated by the milder heat process (72 °C) and consistently superior when compared to the severe heat treatment (94 °C). With regard to the effects on enzyme activity, juices processed by PEF; UV-C+PEF or PEF+UV-C did not show any significant difference (47.5 %; 47.2 %; or 49.5 % respectively). Gachovska et al. (2008) found an additive effect on the inactivation of *E. coli* ATCC 23472 in pasteurized commercial apple juice processed in a continuous flow chamber using a combination of UV-C and PEF applied sequentially. Regardless of the order of treatments, a maximum *E. coli* reduction of 5.35 log CFU/ml was achieved using PEF (60 kV/cm, 11.3 pulses) and UV-C (length of 50 cm, treatment time of 2.94 s, and flow rate of 8 ml/min). Palgan et al. (2011) studied the potential of the nonthermal hurdles UV-C (5.3 J/cm^2), high-intensity light pulses (PL) (3.3 J/cm^2), PEF (34 kV/cm, 18 Hz, 93 μs) or manothermosonication (MTS) (4 bar, 43 °C, 750 W, 20 kHz) applied individually or in paired sequences, to inactivate *E. coli* and *Pichia fermentans* inoculated in a fresh blend of apple and cranberry juice. Selected sequential combined treatments (PL+PEF; PL+MTS; UV-C+PEF; UV-C+MTS) gave comparable reductions ($p \geq 0.05$) for each of the organisms examined to those observed in thermally pasteurized samples (approx. 6 log CFU/ml) while none of the individual hurdles was capable of achieving the 5 log reduction required by FDA for fruit juices. As all combinations led to reductions below detection levels, authors could not determine the presence or absence of synergistic or additive effects. This work showed the potential for combinations of a light-based technology (UV-C or PL) followed by either PEF or MTS; however more investigation would be needed since the shelf-life (4 °C) of UV-C+PEF and PL+PEF-treated samples was 14 and 21 days respectively, shorter than that corresponding to thermally treated samples (35 days). In a recent study, Caminiti et al. (2011) also proposed the PEF (24 or 34 kV/cm; 89 μs)/PL (360 μs, 3 Hz; 4.0 or 5.1 J/cm^2) hurdle combination and the reverse sequence as an alternative of thermal pasteurization for bacterial control in apple juice. They reported a synergistic interaction on *E. coli* K12 (more than 6 log reductions) in reconstituted apple juice but did not find any difference in pH, °Brix, total phenolics and sensory parameters between apple juice processed by the PEF/PL combination and the pasteurized product.

Other authors reported an increase in the killing effect of PEF on microorganisms in various fruit juices (apple, pear, tomato, orange, and strawberry) or apple cider in combination with natural antimicrobials such us bacteriocins (McNamee et al. 2010; Martinez Viedma et al. 2008); essential oils and organic acids (McNamee et al. 2010; Mosqueda-Melgar et al. 2008a, b; Liang et al. 2006). While the mechanism of synergy is not fully understood, the additional stress of PEF probably facilitates the income of antimicrobials to the cytoplasmic membrane, improving the efficacy of the antimicrobial compounds, which allows proposing very low doses of antimicrobials.

As it can be seen, although the combination of different nonthermal technologies have shown a significant microorganism inactivation effect, strong differences were found between the different nonthermal hurdle strategies depending on type of microorganism; operation mode (continuous or batch) and equipment; order of sequence; matrix; etc. Thermal treatment may still be required to achieve the desired level of inactivation for practical uses, but the combination of different hurdles with heating allows the use of significantly reduced temperature in fruit juices with the consequent less detrimental effect on quality parameters (Bermúdez-Aguirre and Barbosa-Cánovas 2012; Liang et al. 2006). Aronsson and Rönner (2001) gave a detailed discussion on thermally influenced PEF inactivation and drew conclusions about synergy of PEF and thermal treatment.

UV-C radiation can be effectively used for pasteurization of different kinds of fruit juices, without in general affecting in a severe way color profiles, vitamin C content, and taste (Tran and Farid 2004; Keyser et al. 2008). The combination of UV-C treatment and low temperature storage allowed a shelf-life extension from ≈ 2 days to more than 5 days. However, the use of UV-C is still limited due to the low UV transmittance of fruit juices. The penetration of UV-C radiation depends on the type of liquid, its absorptivity, soluble solids, and suspended matter. Thus, different UV-C reactors (thin film, turbulent, laminar Taylor-Couette, Dean flow reactors) are being studied to ensure effective radiation penetration (Koutchma 2009).

HHP can induce changes in fruit structure and texture which, along with taste, are the most important sensory attributes for consumer acceptability. For this reason, most of the commercialized and/or investigated HP processed products of plant origin are in the form of purée or juice (e. g. guacamole, fruit jams, and juices). In juices, control of PME is crucial for assure cloud stability, since demethylation of pectin results in the separation of a clear serum and a sediment constituted by complexes of low methoxyl pectin and calcium ions. In general quality-related enzymes are rather pressure-stable and pressure treatments were usually combined with mild heating to obtain juices or fruits of high quality. For example, a synergistic effect of HHP and moderate temperature on orange PME inactivation was found by Polydera et al. (2004) except in the high-temperature–low-pressure region where an antagonistic interaction was noted. Buchow et al. (2009) also reported a synergism between pressure and temperature on the inactivation of apple polyphenol oxidase (PPO) above 300 MPa where an antagonistic effect was found at lower pressures. The stability interfaces of apple juice in order to optimize a HHP treatment when considering different types of quality targets (PPO, PME, and vitamin C) were determined by Valdramidis et al. (2009) for a given number of design variables (level of pressure, treatment time/temperature, storage time/temperature). Previously, apple juice spoilage constraints of *Issatchenkia orientalis* and other vegetative microorganisms were defined. The two enzymes appeared to be much more resistant than vegetative microorganisms with respect to pressure–temperature treatments. The application of 750 MPa and 50 °C or higher was required to ensure both microbial stability and quality of apple juice, while only 350 MPa for 10 min would be necessary to minimize the probability of spoilage during storage at 8 °C.

Sonication applied alone (at room temperature and atmospheric pressure) is not very effective inactivating microorganisms in juices. However, the combination of ultrasound with other preservation factors and/or the selection of operative conditions that enhance the per se effect of high-power sonication shows considerable promise (Piyasena et al. 2003; Knorr et al. 2004; Alzamora et al. 2011). López-Malo et al. (2006) analyzed the response of *L. monocytogenes* and *S. cerevisiae* to the single and combined effects of high-intensity ultrasound (20 kHz, 400 W, 95.2 μm, *T*: 35 °C) and UV-C light (continuous flow system; 90 cm long glass tube with a 100 W Hg lamp, 1100 μW/cm^2) in clarified apple juice. The effect of the US/UV-C combination was additive and led to a great inactivation (≈4–5 log cycles reduction after 5 min treatment), with the majority of the population dead in the first minutes of treatment. Char et al. (2010) studied the use of continuous flow (0.2 L/min) US treatment (20 kHz, 95 μm, 40 °C) combined with UV-C light (100 W) to inhibit *E. coli* ATCC 35218 in orange juice (pH 3.5; 9° Brix). The poor single effect of UV-C light in orange juice (1.7 log reduction), probably due to the opaque nature of matrix, was enhanced by the combination with US achieving approximately 4 log reductions (additive effect). Combined treatment was more effective in simultaneous rather than in a series of US–UV-C arrangement.

Thermoultrasonic treatment caused a higher killing effect than only sonication treatment. Raising the temperature and hence membrane fluidity (i.e., weakening the intermolecular forces) would enhance the disruption by US (Russell 2002). However, as temperature increased toward lethal values, the benefits of ultrasound application are reduced probably as a result of an increased thermal effect and a reduced intensity of cavitation (López-Malo et al. 1999). Inactivation studies of *Listeria monocytogenes* 10403S, an ultrasound resistant strain, were conducted at sublethal (20–40 °C) and lethal (50–60 °C) temperatures in apple cider (pH 3.4) with and without application of US (20 kHz, 750 W, 99 ml sample) (Baumann et al. 2005). Ultrasound increased the inactivation rate at both lethal and sublethal temperatures. The bactericidal effect of the combined process was additive. After a 5-min of thermoultrasonic treatment at 60 °C, cells of *L. monocytogenes* 10403S died during a 6 h period at room temperature. This treatment conditions could provide a solution for apple cider industries to achieve the required 5 log reduction in pathogenic populations.

Ferrante et al. (2007) applied a ultrasound-based hurdle strategy (20 kHz; 95 μm) for *Listeria monocytogenes* inactivation in fresh-squeezed orange juice which included the use of moderate temperature (45 °C) and addition of naturally occurring antimicrobials vanillin (1000 or 1500 ppm) and citral (100 ppm). The bacteria inactivation was total in 10 min treatment of 1500 ppm vanillin and 100 ppm citral containing orange juice, being the product pleasant for the consumers according to a sensory study. Phenolic compounds have lipophilic nature and could accumulate in the lipid bilayer of the cell, disturbing and sensitizing the membrane to ultrasound (Brul and Coote 1999). In a more recent study using pineapple, grape, and cranberry juices, under 24 kHz, 400 W, 120 μm and 60 °C, between 5 and 7 log-reduction of *S. cerevisiae* were achieved (depending on the juice) after 10 min of continuous ultrasonic treatment but significant changes in color and pH of juices were detected after processing. The use of lower temperatures (40 or 50 °C) or pulsed mode delayed yeast inactivation (Bermúdez-Aguirre and Barbosa-Cánovas 2012). Gómez-López et al. (2010) reported a non-negligible effect of ultrasound -and refrigeration- on shelf-life in orange juice. They applied ultrasonic treatments at a frequency of 20 kHz and three-wave amplitudes for 2, 4, 6, 8, and 10 min, to orange juice with added calcium. Wave amplitude of 89.25 mm for 8 min was selected for final treatment of the juice, and storage studies were performed at 4 and 10 °C. The treatments decreased mesophilic counts by 1.38 log CFU/ml, and yeast and molds counts by 0.56 log CFU/ml. The sensory quality of the juice was slightly deteriorated after treatment, but during storage, control samples degraded earlier than sonicated samples. Controls were rejected by the sensory panel after 6 days storage at 4 °C due to off-flavor, while ultrasonicated juice was rejected after 10 days due to off-odor. Consequently, a shelf-life extension of 4 days was achieved. Sonication also affected color and decreased ascorbic acid content. This study shows that ultrasonication may be useful to extend the shelf-life of orange juice and could be further enhanced by the use of other hurdles.

5.4 Recommendations

Examples described above present the promising use of combined processes for the production of mildly preserved fruit products. Knowledge of the fundamentals of traditional and emerging factors and the number of reports on preservation of foods in general, and fruits in particular, have dramatically increased during the last 15 years. But more research is needed to fully exploit the potential of the hurdle concept in the design and optimization of fruit preservation techniques and facilitate/increase their adoption by the industry. Vital areas for further studies include the following points.

5.4.1 Microbial Behavior in Response to Stressors

New food preservation strategies can be developed on a sound scientific base if combination of hurdles is thought taking into account their different modes of action over microbial cells and the mechanisms mediating microbial adaptive responses (Ross et al. 2003; Gould 2000). Our knowledge of these subjects is far from complete. Nowadays, the advent of new methodic developments may provide a significant conceptual advance in the understanding of responses in microorganisms to a variety of environmental stresses.

Microbial cells react upon stress by producing specific stress transcriptors, proteins, and metabolites, to bring cellular metabolism to homeostasis (Brul et al. 2002, 2006). Booming "genomics" technologies (genomics, transcriptomics, proteomics, and metabolomics) contribute to the understanding of cellular behavior by a simultaneous approach in which the whole set of cellular biomolecules is studied in a given experimental setup. Cellular response at molecular level can then be used to study cellular physiology and/or to generate databases of cellular reactions to environmental conditions, supporting the development of effective food preservation processes.

Changes in plasma membrane potential and/or ion flux modulation are some of the earliest cellular modifications occurring when the environments become hostile, and are linked to cell viability and physiological processes. The noninvasive microelectrode ion flux estimation technique allows calculating net fluxes of H^+, Ca^{2+}, K^+, Na^+, Cl^-, Mg^{2+}, and other ions in real time from the measured voltage gradient at the surface from molds, yeasts, and biofilms (Shabala et al. 2006). The study of the kinetics of membrane-transport processes across cellular membranes has shown to be a valuable tool to investigate various aspects of adaptive responses to temperature, osmolality, and acid stress. For instance, evaluation of temperature-induced changes in net H^+ fluxes in *Listeria monocytogenes* allowed quantifying the critical temperature at which membrane-transport activity undergoes dramatic changes, thus enabling to assess the chilling resistance of the bacteria (Shabala et al. 2006). Shabala et al. (2002) used in combination noninvasive ion flux measurements with fluorescence microscopy to quantify intracellular pH when studying *L. monocytogenes* response to acidic stress. The kinetics of pH_i change and net H^+ fluxes indicated that plasma membrane H^+ transporters play a central role in the bacteria pH homeostasis and the acid tolerance response.

Multiparameter flow cytometry (FC) is another modern tool for monitoring the stress response of microorganisms. By means of both scattering and fluorescence signal measurements, information on cell parameters (physiological state, size, surface roughness, and granularity) at single cell level and their distribution within cell populations is provided with a relatively high degree of statistical resolution ($\cong$50,000 cells in minute) enabling assessment of population heterogeneity (Hewitt and Nebe-Von-Caron 2004; Díaz et al. 2010). Cell viability evaluation is an issue of particular concern in preservation technologies design. Resistance to different stressors has been attributed to an entry into viable but non-culturable (VBNC) states. VBNC, dormant, or damaged cells are not able to grow on culture media but generally maintain their metabolic machinery active and remain alive, provoking undesirable effects such as food spoilage, accumulation of toxins, or transfer of genes. FC allows identifying functionally homogeneous subpopulations including VBNC. FC was used to monitor pressure-induced changes in *Lactobacillus rhamnosus* and *Listeria monocytogenes*, and the studies showed that pressurized inactivated bacteria were still in possession of enzyme activity and not completely membrane compromised (Ananta et al. 2004; Ritz et al. 2001). Ananta et al. (2005) showed that high-intensity ultrasound provoked the rupture of the lipopolysaccharide layer of the outer membrane of

Gram-negative bacteria, suggesting the ultrasound-assisted physical disruption of the outer membrane to facilitate the entry of bacteriocins. Based on FC studies, Mathys et al. (2007) suggested a three-step inactivation model of pressure–heat-treated spores of *Bacillus licheniformis* involving a germination step following hydrolysis of spore cortex, an unknown step and finally an inactivation step with physical compromise of the spore's inner membrane. Schenk et al. (2011) demonstrated that mechanisms of UV-C induced cellular damage differed according to exposure time and the organism tested: a significant damage in the cytoplasmic membrane integrity and in the enzyme activity was found in *E. coli* and *S. cerevisiae* in the first minutes of UV-C treatment while a VBNC subpopulation of *L. innocua* was detected at near all UV-C doses assayed.

The introduction of any alternative technology as well as its optimization requires quantitative data about microbial response (Alzamora et al. 2010). In particular, kinetic parameters and models are essential to develop food preservation processes that ensure safety (McMeekin 2007; Alzamora and López-Malo 2002; Alzamora et al. 2010). The parameters also allow comparison of the ability of different process technologies to reduce microbial populations. Accurate model prediction of survival curves would be beneficial to the F&V industry in selecting the optimum combinations of lethal agents and environmental factors as well as exposure times to obtain desired levels of inactivation while minimizing production costs and maintaining a maximum degree of sensory and nutrient quality. Potential combinations of preservation factors are numerous, but until now not much quantitative microbiological information is available about combination of alternative processes with traditional constraints. Predictive microbiology provides the tools to compare the impact of different factors on reduction of microbial population (Alzamora et al. 2010). For instance, when *L. monocytogenes* and *S. cerevisiae* undergone US, UV-C and US/UV-C treatments and inactivation data were fitted using the cumulative Weibull distribution function, the correspondent frequency distributions of resistances showed that the combined action of US/UV-C light not only increased the microbicidal effect of sonication but changed the distribution of inactivation times. When both physical inactivation agents were applied together, narrowest frequency shapes, skewed to the right, with low dead time means and a very substantial decrease in its overall spread, were in general obtained, indicating a more homogeneous response of the microbial population to the combined treatment (López-Malo et al. 2006).

5.4.2 Engineering Solutions

Process uniformity is an important factor that affects the effectiveness of the treatment and attempts with its successful commercialization (Heldman et al. 2008). Engineering solutions are required to design new equipments which assure that all product pieces or volume elements receive the same stressor dose For example, in continuous UV-C and pulsed light processes, the distance and the relative position of the sample with respect to the Hg and Xenon lamps significantly influence the received dose or fluence as well as the increase in temperature with the PL dose (Gómez et al. 2010, 2011b).

5.4.3 Support Studies for the Design of Preservation Techniques

While much relevant information is available in the scientific literature concerning stressors/interaction of stressors that influence microbial activities in fruits, it is not often practical in formulating combined preservation techniques. In many cases the information provided involves only data from traditional challenge testing with pathogenic microorganisms in particular conditions, microbial presence or absence tests. These isolated results do not allow us to compare quantitatively what happens when the levels of the independent preservation factors are changed. Neither can the sensitivity of the key microorganisms to the different factors be inferred (Alzamora et al. 2010). There are no systematic in vivo studies that analyze the response of microorganisms to different doses of a stressor employed alone or in combination and/or the influence of critical extrinsic and intrinsic parameters. In the case of novel inactivation technologies, it has also been found that the principle of equi-effectiveness of the product of stressor intensity and exposure time is not always valid and the germicidal efficiency would depend on stressor intensity at similar doses (Gómez-López et al. 2007; Raffellini et al. 2011).

Studies that consider the effect that combined antimicrobial strategies inflict on the population and type of native microorganisms, and their dynamic and influence on pathogenic organisms along storage are rarely found in scientific literature. According to reported findings, mild processing combined stressors could extend the shelf-life, increase the destruction of inoculated pathogens without increasing the shelf-life, or increase both the safety and the shelf-life of the fruit product (Ferrario et al. 2011). Therefore, survival studies and spoilage studies during different storage conditions are of high interest for evaluating the shelf-life of these products.

Studies documenting the impact of mild preservation techniques on structure and quality attributes of fruits generally only consider the effect of processing per se, but major changes in quality attributes occur during storage (Gómez et al. 2010, 2011b; Welti-Chanes et al. 2009). It is also important to mark that interventions for sensory and functional stability used along with combined stressors in the preservation technologies could also impair microbial inactivation. Gómez et al. (2010, 2011b) reported that microbial inactivation by UV-C or PL decreased when cut-apples were previously immersed into an antibrowning calcium chloride–ascorbic acid aqueous solution.

Finally, physiology and kinetics studies to select preservation factors have focused on microbial cells in aqueous planktonic phase or in fruit pieces just inoculated, but association of microorganisms with surface forming biofilms is the prevailing microbial lifestyle and planktonic cell studies constitute a biased view of microbial life (Lindsay and von Holy 2006). Biofilm cells in fruit surfaces or microorganism aggregates in juices are immensely heterogeneous and much more resistant to stressors than the planktonic, freely suspended cells. This fact highlights the importance of performing experiments with attached cells for evaluating the effectiveness of the combined hurdles.

5.5 Future Trends

Hurdle concept-based techniques offer many advantages in meeting consumer and industry demands of freshness, safety, and/or convenience. However, the cumulative knowledge base should be increased. Key points for their design and commercialization include a more deep knowledge on mechanisms of stressors mode of action, the availability and interpretation of systematic kinetic data on microbial and quality attributes behavior (with special relevance to dose response and the influence of critical parameters), and the optimization of equipment. Only in this way it would be possible to address the scientific bases for: (a) comparing the relative level of protection afforded by different combinations of stressors/critical process parameters; (b) selecting the best combined strategy that ensures high quality and safe fruit products; and (c) establishing food safety objectives and performance and process criteria that expedite the commercialization of the fruit products (Stewart et al. 2002).

Acknowledgments The authors want to thank the support from Universidad de Buenos Aires, CONICET and ANPCyT–BID from Argentina, from the Universidad de Las Americas, Puebla, and CONACyT from Mexico, and from Universidad Central de Venezuela from Venezuela.

References

Allende, A., and F. Artes. 2003. UV-C radiation as a novel technique for keeping quality of fresh processed "Lollo Rosso" lettuce. *Food Research International* 36: 739–746.

Allende, A., F.A. Tomás-Barberán, and M.I. Gil. 2006. Minimal processing for healthy traditional foods. *Trends in Food Science and Technology* 17: 513–519.

Alzamora, S.M., and A. López-Malo. 2002. Microbial behavior modeling as a tool in the design and control of processed foods. In *Engineering and food for the 21st century*, ed. J. Welti-Chanes, G. Barbosa-Cánovas, and J.M. Aguilera, 631–650. Boca Raton: CRC Press.

Alzamora, S.M., L.N. Gerschenson, P. Cerrutti, and A.M. Rojas. 1989. Shelf stable pineapple for long-term non refrigerated storage. *Lebensmittel Wissenschaft und Technologie* 22: 233–236.

Alzamora, S.M., M.S. Tapia, A. Argaiz, and J. Welti. 1993. Application of combined methods technology in minimally processed fruits. *Food Research International* 26: 125–130.

Alzamora, S.M., P. Cerrutti, S. Guerrero, and A. López-Malo. 1995. Minimally processed fruits by combined methods. In *Food preservation by moisture control—Fundamentals and applications (International symposium on the properties of water—ISOPOW Practicum II, June 19–24, 1994, México)*, ed. J. Welti-Chanes and G. Barbosa-Cánovas, 463–492. Lancaster: Technomic Pub. Co.

Alzamora, S.M., A. López-Malo, and M.S. Tapia. 2000. *Minimally processed fruits and vegetables: Fundamentals and applications*. Gaithersburg: Aspen Publishers, Inc.

Alzamora, S.M., S. Guerrero, A. López-Malo, and E. Palou. 2003. Plant antimicrobials combined with conventional preservatives for fruit products. In *Natural antimicrobials for the minimal processing of foods*, ed. S. Roller, 235–249. Boca Raton: CRC Press/Woodhead Publishing Limited.

Alzamora, S.M., S.N. Guerrero, A. López-Malo, E. Palou, C.D. Char, and S. Raffellini. 2010. Models for microbial inactivation: Application in food preservation design. In *Processing effects on safety and quality of foods*, ed. E. Ortega-Rivas, 87–115. Boca Raton: CRC Press.

Alzamora, S.M., S. Guerrero, M. Schenk, S. Raffellini, and A. López-Malo. 2011. Inactivation of microorganisms. In *Ultrasound technologies for food and bioprocessing*, ed. H. Feng, G. Barbosa Cánovas, and J. Weiss, 321–344. New York: Springer.

Ananta, E., V. Heinz, and D. Knorr. 2004. Assessment of high pressure induced damage on *Lactobacillus rhamnosus* GG by flow cytometry. *Food Microbiology* 21: 567–577.

Ananta, E., D. Voight, M. Zenker, V. Heinz, and D. Knorr. 2005. Cellular injuries upon exposure of *Escherichia coli* and *Lactobacillus rhamnosus* to high-intensity ultrasound. *Journal of Applied Microbiology* 99: 271–278.

Argaiz, A., A. López-Malo, and J. Welti-Chanes. 1995. Considerations for the development and the stability of high moisture fruit products during storage. In *Food preservation by moisture control—Fundamentals and applications (International symposium on the properties of water—ISOPOW Practicum II, June 19–24, 1994, México)*, ed. J. Welti-Chanes and G. Barbosa-Cánovas, 761–792. Lancaster: Technomic Pub. Co.

Aronsson, K., and U. Rönner. 2001. Influence of pH, water activity and temperature on the inactivation of *Escherichia coli* and *Saccharomyces cerevisiae* by pulsed electric fields. *Innovative Food Science & Emerging Technologies* 2: 105–112.

Bartz, J.A. 1982. Infiltration of tomatoes immersed at different temperatures to different depths in suspensions of *Erwinia caratovora* subsp. *Caratovora*. *Plant Disease* 66: 302–305.

Baumann, A.R., S.E. Martín, and H. Feng. 2005. Power ultrasound treatment of *Listeria monocytogenes* in apple cider. *Journal of Food Protection* 68: 2333–2340.

Bermúdez-Aguirre, D., and G.V. Barbosa-Cánovas. 2012. Inactivation of *Saccharomyces cerevisiae* in pineapple, grape and cranberry juices under pulsed and continuous thermo-sonication treatments. *Journal of Food Engineering* 108: 383–392.

Beuchat, L. 2000. Use of sanitizers in raw fruit and vegetable processing. In *Minimally processed fruits and vegetables. Fundamental aspects and applications*, ed. S.M. Alzamora, M.S. Tapia, and A. López-Malo, 63–75. Gaithesburg: Aspen Publishers, Inc.

Booth, I.R., and R.G. Kroll. 1989. The preservation of foods by low pH. In *Mechanisms of action of food preservation procedures*, ed. G.W. Gould, 119–160. New York: Elsevier Applied Science.

Bracket, R.E. 2001. Fruits, vegetables, and grains. In *Food microbiology: Fundamentals and frontiers*, 2nd ed, ed. M.P. Doyle, L.R. Beuchat, and T.J. Montville. Washington: ASM Press.

Brul, S., and P. Coote. 1999. Preservative agents in foods. Mode of action and microbial resistance mechanisms. *International Journal of Food Microbiology* 50: 1–17.

Brul, S., F.M. Klis, S.J.C.M. Oomes, R.C. Montijn, F.H.J. Schuren, P. Coote, and K.J. Hellingwerf. 2002. Detailed process design based on genomics of survivors of food preservation processes. *Trends in Food Science and Technology* 13: 325–333.

Brul, S., F. Schuren, R. Montijn, B.J.F. Keijser, H. van der Spek, and S.J.C.M. Oomes. 2006. The impact of functional genomics on microbiological food quality and safety. *International Journal of Food Microbiology* 112: 195–199.

Buchow, R., U. Weiss, and D. Knorr. 2009. Inactivation kinetics of apple polyphenol oxidase in different pressure-temperature domains. *Innovative Food Science & Emerging Technologies* 10: 441–448.

Caminiti, I.M., I. Palgan, F. Noci, A. Muñoz, P. Whyte, D.A. Cronin, D.J. Morgan, and J.G. Lyng. 2011. The effect of pulsed electric fields (PEF) in combination with high intensity light pulses (HILP) on *Escherichia coli* inactivation and quality attributes in apple juice. *Innovative Food Science & Emerging Technologies* 12: 118–123.

Capozzi, V., D. Fiocco, M.L. Amodio, A. Gallone, and G. Spano. 2009. Bacterial stressors in minimally processed food. *International Journal of Molecular Sciences* 10: 3076–3105.

Cerrutti, P., S.M. Alzamora, and S.L. Vidales. 1997. Vanillin as antimicrobial for producing shelf-stable strawberry purée. *Journal of Food Science* 62: 608–610.

Chang, E.T., K.E. Smedby, S.M. Zhang, H. Hjalgrim, M. Melbye, A. Ost, B. Glimelius, A. Wolk, and H.O. Adami. 2005. Dietary factors and risk of non-Hodgkin lymphoma in men and women. *Cancer Epidemiology, Biomarkers & Prevention* 14: 512–520.

Char, C., E. Mitilinaki, S. Guerrero, and S.M. Alzamora. 2010. Use of high-intensity ultrasound and UV-C light to inactivate some microorganisms in fruit juices. *Food and Bioprocess Technology* 3: 797–803.

Chen, Z., and C. Zhu. 2011. Combined effects of aqueous chlorine dioxide and ultrasonic treatments on postharvest storage quality of plum fruit (*Prunus salicina* L.). *Postharvest Biology and Technology* 61: 117–123.

Chirife, J., and C. Ferro Fontán. 1982. Water activity of fresh foods. *Journal of Food Science* 47: 661–663.

Dauchet, L., J. Ferrières, D. Arveiler, J.W. Yarnell, F. Gey, P. Ducimetière, J.-B. Ruidavets, B. Haas, A. Evans, A. Bingham, P. Amouyel, and J. Dallongeville. 2004. Frequency of fruit and vegetable consumption and coronary heart disease in France and Northern Ireland: The PRIME study. *British Journal of Nutrition* 92: 963–972.

De Stefani, E., P. Boffetta, H. Deneo-Pellegrini, A.L. Ronco, P. Correa, and M. Mendilaharsu. 2005. The role of vegetable and fruit consumption in the etiology of squamous cell carcinoma of the esophagus: A case-control study in Uruguay. *International Journal of Cancer* 116: 130–135.

Díaz, M., M. Herrero, L.A. García, and C. Quirós. 2010. Application of flow cytometry to industrial microbial bioprocesses. *Biochemical Engineering Journal* 48: 385–407.

Fernández Escartín, E. 2000. *Microbiología e Inocuidad de los Alimentos*, 555–557. México: Universidad Autónoma de Querétaro.

Ferrante, S., S. Guerrero, and S.M. Alzamora. 2007. Combined use of ultrasound and natural antimicrobials to inactivate *Listeria monocytogenes* in orange juice. *Journal of Food Protection* 70: 1850–1856.

Ferrario, M., S.M. Alzamora, and S.N. Guerrero. Use of Weibull and modified Gompertz models to characterize microbial inactivation in orange juice processed with ultraviolet light and natural antimicrobials. Presented at VIII Congreso Iberoamericano de Ingeniería de Alimentos (CIBIA 8). Lima, Perú, 23–26 de octubre de 2011.

Fonseca, J.M., and J.W. Rushing. 2006. Effect of ultraviolet-C light on quality and microbial population of fresh-cut watermelon. *Postharvest Biology and Technology* 40: 256–261.

Gachovska, T., S. Kumar, H. Thippareddi, J. Sibbiah, and F. Williams. 2008. Ultraviolet and pulsed electric field treatments have additive effect on inactivation of *E. coli* in apple juice. *Journal of Food Science* 73: 412–441.

Gil, M.I., M.V. Selma, F. Lopez-Galvez, and A. Allende. 2009. Fresh-cut product sanitation and wash water disinfection: Problems and solutions. *International Journal of Food Microbiology* 134: 37–45.

Gómez, P.L., S.M. Alzamora, M.A. Castro, and D.M. Salvatori. 2010. Effect of ultraviolet-C light dose on quality of cut-apple: Microorganism, color and compression behaviour. *Journal of Food Engineering* 98: 60–70.

Gómez, P.L., A. García Loredo, D. Salvatori, S. Guerrero, and S.M. Alzamora. 2011a. Viscoelasticity, texture and ultrastructure of cut apple as affected by sequential anti-browning and ultraviolet-C light treatments. *Journal of Food Engineering* 107: 214–225.

Gómez, P.L., D. Salvatori, A. García Loredo, and S.M. Alzamora. 2011b. Pulsed light treatment of cut apple: Dose effect on color, structure and microbiological stability. *Food and Bioprocess Technology* 5: 2311–2322. https://doi.org/10.1007/s11947-011-0598-3.

Gómez, P., S.M. Alzamora, and J. Welti-Chanes. 2011c. Hurdle technology in fruit processing. *Annual Reviews of Food Science and Technology* 2: 447–465.

Gómez-López, V.M., P. Ragaert, J. Debevere, and F. Devlieghere. 2007. Pulsed light for food decontamination: A review. *Trends in Food Science and Technology* 18: 464–473.

Gómez-López, V.M., L. Orsolani, A. Martínez-Yépez, and M.S. Tapia. 2010. Microbiological and sensorial quality of sonicated calcium-added orange juice. *LWT – Food Science and Technology* 43: 808–813.
Gould, G.W. 1995. *New methods of food preservation*. Glasgow: Blackie Academic & Professional.
———. 2000. Induced tolerance of microorganisms to stress factors. In *Minimally processed fruits and vegetables—Fundamental aspects and applications*, ed. S.M. Alzamora, M.S. Tapia, and A. López-Malo, 29–37. Gaithersbug: Aspen Publishers Inc.
Gould, G.W., and M.V. Jones. 1989. Combination and synergistic effect. In *Mechanisms of action of food preservation procedures*, ed. G.W. Gould, 401–421. London: Elsevier Applied Science.
Guerrero, S., S.M. Alzamora, and L.N. Gerschenson. 1994. Development of a shelf-stable banana purée by combined factors: Microbial stability. *Journal of Food Protection* 57: 902–907.
Havelaar, A.H., S. Brul, A. de Jong, R. de Jonge, M.H. Zwietering, and B.H. ter Kuile. 2010. Future challenges to microbial food safety. *International Journal of Food Microbiology* 139(Suppl.): S79–S94.
Heldman, D.R., D.B. Lund, and A. Husain. 2008. Cross-process issues impacting innovative food processing technologies. *Food Science and Technology International* 14: 411–412.
Hewitt, C.J., and G. Nebe-Von-Caron. 2004. The application of multi-parameter flow cytometry to monitor individual microbial cell physiological state. *Advances in Biochemical Engineering/ Biotechnology* 89: 197–223.
Jang, J.-H., and K.-D. Moon. 2011. Inhibition of polyphenol oxidase and peroxidase activities on fresh-cut apple by simultaneous treatment of ultrasound and ascorbic acid. *Food Chemistry* 124: 444–449.
Jay, J.M. 2002. *Microbiología Moderna de los Alimentos*, 4th ed, 123, 283–288, 521–530. España: Editorial Acribia.
Jones, T.M., and F. Widemo. 2005. Survival and reproduction when food is scarce: Implications for a lekking Hawaiian Drosophila. *Ecological Entomology* 30: 397–405.
Keenan, D.F., N.P. Brunton, T.R. Gormley, F. Butler, B.K. Tiwari, and A. Patras. 2010. Effect of thermal and high hydrostatic pressure processing on antioxidant activity and colour of fruit smoothies. *Innovative Food Science & Emerging Technologies* 11: 551–556.
Keyser, M., I.A. Müller, F.P. Cilliers, W. Nel, and P.A. Gouws. 2008. Ultraviolet radiation as a non-thermal treatment for the inactivation of microorganisms in fruit juice. *Innovative Food Science & Emerging Technologies* 9: 348–354.
Knorr, D., M. Zenker, V. Heinz, and D.U. Lee. 2004. Applications and potential of ultrasonics in food processing. *Trends in Food Science and Technology* 15: 261–266.
Koutchma, T. 2009. Advances in ultraviolet light technology for non-thermal processing of liquid foods. *Food and Bioprocess Technology* 2: 138–155.
Krebbers, B., A.M. Matser, S.W. Hoogerwerf, R. Moezelaar, M.M. Tomassen, and R.W. van den Berg. 2003. Combined high-pressure and thermal treatments for processing of tomato puree: Evaluation of microbial inactivation and quality parameters. *Innovative Food Science & Emerging Technologies* 4: 377–385.
Lee, W.C., M.J. Lee, and J.S. Kim. 2001. Foodborne illness outbreaks in Korea and Japan studied retrospectively. *Journal of Food Protection* 64: 899–902.
Leistner, L., and G.W. Gould. 2002. *Hurdle technologies. Combination treatments for food stability, safety and quality*. New York: Kluwer Academic/Plenum Publishers.
Liang, Z., Z. Cheng, and G.S. Mittal. 2006. Inactivation of spoilage microorganisms in apple cider using a continuous flow pulsed electric field system. *LWT – Food Science and Technology* 39: 351–357.
Lindsay, D., and A. von Holy. 2006. What food safety professionals should know about bacterial biofilms. *British Food Journal* 108: 27–37.
López-Malo, A., S. Guerrero, and S.M. Alzamora. 1999. *Saccharomyces cerevisiae* thermal inactivation kinetics combined with ultrasound. *Journal of Food Protection* 62: 10–13.
López-Malo, A., S. Guerrero, A. Santiesteban, and S.M. Alzamora. Inactivation kinetics of *Saccharomyces cerevisiae* and *Listeria monocytogenes* in apple juice processed by novel technologies. Paper presented at ENPROMER 2005, Rio das Pedras—Brazil, August 14–18, 2006.

Marquenie, D., J. Lammertyn, A.H. Geeraerd, C. Soontjens, J.F. Van Impe, B.M. Nicolaï, and C.W. Michiels. 2002. Inactivation of conidia of *Botrytis cinerea* and *Monilinia fructigena* using UV-C and heat treatment. *International Journal of Food Microbiology* 74: 27–35.
Martinez Viedma, P., A. Sobrino Lopez, N. Ben Omar, H. Abriouel, R. Lucas Lopez, E. Valdivia, O. Martin Belloso, and A. Galvez. 2008. Enhanced bactericidal effect of enterocin AS-48 in combination with high-intensity pulsed-electric field treatment against *Salmonella enterica* in apple juice. *International Journal of Food Microbiology* 128: 244–249.
Martínez, A., R.V. Díaz, and M.S. Tapia. 2000. Microbial ecology of spoilage and pathogenic flora associated to fruit and vegetables. In *Minimally processed fruits and vegetables: Fundamental aspects and applications*, ed. S.M. Alzamora, M.S. Tapia, and A. López-Malo, 43–62. Gaithersburg: Aspen Publishers, Inc.
Mathys, A., B. Chapman, M. Bull, V. Heinz, and D. Knorr. 2007. Flow cytometric assessment of *Bacillus* spore response to high pressure and heat. *Innovative Food Science and Emerging Technologies* 8: 519–527.
McMeekin, T.A. 2007. Predictive microbiology: Quantitative science delivering quantifiable benefits to the meat industry and other food industries. *Meat Science* 77: 17–27.
McNamee, C., F. Noci, D.A. Cronin, J.G. Lyng, D.J. Morgan, and A.G. Scannell. 2010. PEF based hurdle strategy to control *Pichia fermentans*, *Listeria innocua* and *Escherichia coli* k12 in orange juice. *International Journal of Food Microbiology* 138: 13–18.
Messiaen, C.M., D. Blancard, F. Rouxel, and R. Lafon. 1994. *Enfermedades de las Hortalizas*, 3rd ed, 17. España: Ediciones Mundi-Prensa.
Montonen, J., P. Knekt, T. Härkänen, R. Järvinen, M. Heliövaara, A. Aromaa, and A. Reunanen. 2005. Dietary patterns and the incidence of type 2 diabetes. *American Journal of Epidemiology* 161: 219–227.
Mosqueda-Melgar, J., R.M. Raybaudi-Massilia, and O. Martin-Belloso. 2008a. Inactivation of *Salmonella enterica* Ser. Enteritidis in tomato juice by combining of high-intensity pulsed electric fields with natural antimicrobials. *Journal of Food Science* 73: M47–M53.
———. 2008b. Non-thermal pasteurization of fruit juices by combining high-intensity pulsed electric fields with natural antimicrobials. *Innovative Food Science & Emerging Technologies* 9: 328–340.
Nkondjock, A., D. Krewski, K.C. Johnson, and P. Ghadirian. 2005. Dietary patterns and risk of pancreatic cancer. *International Journal of Cancer* 114: 817–823.
Noci, F., J. Riener, M. Walkling-Ribeiro, D.A. Cronin, D.J. Morgan, and J.G. Lyng. 2008. Ultraviolet irradiation and pulsed electric fields (PEF) in a hurdle strategy for the preservation of fresh apple juice. *Journal of Food Engineering* 85: 141–146.
O'Byrne, C., and I.R. Booth. 2002. Osmoregulation and its importance to food–borne microorganisms. *International Journal of Food Microbiology* 74: 203–216.
Palgan, I., I.M. Caminiti, A. Muñoz, F. Noci, P. Whyte, D.J. Morgan, D.A. Cronin, and J.G. Lyng. 2011. Combined effect of selected non-thermal technologies on *Escherichia coli* and *Pichia fermentans* inactivation in an apple and cranberry juice blend and on product shelf life. *International Journal of Food Microbiology* 151: 1–6.
Palhano, F.K., T.T.B. Vilches, R.B. Santos, M.T.D. Orlando, J.A. Ventura, and P.M.B. Fernandes. 2004. Inactivation of *Colletotrichum gloeosporioides* spores by high hydrostatic pressure combined with citral or lemongrass essential oil. *International Journal of Food Microbiology* 96: 61–66.
Pan, J., A.R. Vicente, G.A. Martínez, A.R. Chaves, and P.M. Civello. 2004. Combined use of UV-C irradiation and heat treatment to improve postharvest life of strawberry fruit. *Journal of the Science of Food and Agriculture* 84: 1831–1838.
Penteado, A.L., B.S. Eblen, and A.J. Millar. 2004. Evidence of Salmonella internalization into fresh mangoes during simulated postharvest insect disinfestations procedures. *Journal of Food Protection* 67: 181–184.
Pichereau, V., A. Hartke, and Y. Auffray. 2000. Starvation and osmotic stress induced multiresistances influence of extracellular compounds. *International Journal of Food Microbiology* 55: 19–25.

Piyasena, P., E. Mohareb, and R.C. McKellar. 2003. Inactivation of microbes using ultrasound: A review. *International Journal of Food Microbiology* 87: 207–216.

Polydera, A.C., E. Galanou, N.G. Stoforos, and P.S. Taoukis. 2004. Inactivation kinetics of pectin methylesterase of Greek Navel orange juice as a function of high hydrostatic pressure and temperature process conditions. *Journal of Food Engineering* 62: 291–298.

Raffellini, S., M. Schenk, S. Guerrero, and S.M. Alzamora. 2011. Kinetics of *Escherichia coli* inactivation employing hydrogen peroxide at varying temperatures, pH and concentrations. *Food Control* 22: 920–932.

Rashidkhani, B., P. Lindblad, and A. Wolk. 2005. Fruits, vegetables and risk of renal cell carcinoma: A prospective study of Swedish women. *International Journal of Cancer* 113: 451–455.

Raso, J., and G.V. Barbosa-Cánovas. 2003. Nonthermal preservation of foods using combined processing techniques. *Critical Reviews in Food Science and Nutrition* 43: 265–285.

Raso, J., R. Pagán, and S. Condón. 2005. Nonthermal technologies in combination with other preservation factors. In *Novel food processing technologies*, ed. G.V. Barbosa-Cánovas, M.S. Tapia, and M.P. Cano, 453–475. Boca Raton: CRC Press.

Ray, A. 2005. Cancer preventive role of selected dietary factors. *Indian Journal of Cancer* 42: 15–24.

Raybaudi-Massilia, R.M., J. Mosqueda-Melgar, R. Soliva-Fortuny, and O. Martín-Belloso. 2009. Control of pathogenic and spoilage microorganisms in fresh-cut fruits and fruit juices by traditional and alternative natural antimicrobials. *CRFSFS* 8: 157–180.

Richards, G., and L.R. Beuchat. 2004. Attachment of *Salmonella* Poona to cantaloupe rind and stem scar tissues as affected by temperature of fruit and inoculum. *Journal of Food Protection* 67: 1359–1364.

Rico, D., A.B. Martín-Diana, J.M. Barat, and C. Barry-Ryan. 2007. Extending and measuring the quality of fresh-cut fruit and vegetables: A review. *Trends in Food Science and Technology* 18(2007): 373–386.

Ritz, M., J.L. Tholozan, M. Federighi, and M.F. Pilet. 2001. Morphological and physiological characterization of *Listeria monocytogenes* subjected to high hydrostatic pressure. *Applied and Environmental Microbiology* 67: 2240–2247.

Ross, A.I.V., M.W. Griffiths, G.S. Mittal, and H.C. Deeth. 2003. Combining nonthermal technologies to control foodborne microorganisms. *International Journal of Food Microbiology* 89: 125–138.

Russell, N.J. 2002. Bacterial membranes: The effect of chill storage and food processing. An overview. *International Journal of Food Microbiology* 79: 27–34.

Schenk, M., S. Guerrero, and S.M. Alzamora. 2008. Response of some microorganisms to ultraviolet treatment on fresh-cut pear. *Food and Bioprocess Technology* 1: 384–392.

Schenk, M., S. Raffellini, S. Guerrero, G. Blanco, and S.M. Alzamora. 2011. Inactivation of *Escherichia coli, Listeria innocua* and *Saccharomyces cerevisiae* by UV-C light: Study of cell injury by flow cytometry. *LWT – Food Science and Technology* 44: 191–198.

Schenk, M., A. García Loredo, S. Raffellini, S.M. Alzamora, and S. Guerrero. 2012. The effect of UV-C in combination with H_2O_2 treatments on microbial response and quality parameters of fresh cut pear discs. *International Journal of Food Science and Technology* 47: 1842–1851.

Schwentesius, R., and M.A. Gómez. 2000. *Internacionalización de la Horticultura*, 1–10, 32–35. México: Mundi-Prensa México.

Sewell, A.M., and J.M. Farber. 2001. Foodborne outbreaks in Canada linked to produce. *Journal of Food Protection* 64: 1863–1877.

Shabala, L., B. Budde, T. Ross, H. Siegumfeldt, M. Jakobsen, and T. McMeekin. 2002. Responses of *Listeria monocytogenes* to acid stress and glucose availability revealed a novel combination of fluorescence microscopy and microelectrode ion-selective techniques. *Applied and Environmental Microbiology* 68: 1794–1802.

Shabala, L., T. Ross, T. McMeekin, and S. Shabala. 2006. Non-invasive microelectrode ion flux measurements to study adaptive responses of microorganisms to the environment. *FEMS Microbiology Reviews* 30: 472–486.

Shama, G. 2006. Ultraviolet light. In *Handbook of food science, technology and engineering*, vol. 3, ed. Y.H. Hui, 122-1–122-14. Boca Raton: CRC/Taylor & Francis.

Shama, G., and P. Anderson. 2005. UV hormesis in fruits: A concept ripe for commercialization. *Trends in Food Science and Technology* 16: 128–136.

Silveira, A.C., E. Aguayo, V.H. Escalona, and F. Artés. 2011. Hot water treatment and peracetic acid to maintain fresh-cut Galia melon quality. *Innovative Food Science & Emerging Technologies* 12: 569–576.

Sivapalasingam, S., C.R. Friedman, L. Cohen, and R.V. Tauxe. 2004. Fresh produce: A growing cause of outbreaks of foodborne illness in the United States, 1973 through 1997. *Journal of Food Protection* 67: 2342–2353.

Smelt, J.P.P.M., J.C. Hellemons, P.C. Wouters, and S.J.C. van Gerwen. 2002. Physiological and mathematical aspects in setting criteria for decontamination of foods by physical means. *International Journal of Food Microbiology* 78: 57–77.

Stewart, C.M., R.B. Tompkin, and M.B. Cole. 2002. Food safety: New concepts for the new millennium. *Innovative Food Science & Emerging Technologies* 3: 105–112.

Tapia de Daza, M.S., A. Argaiz, A. López-Malo, and R.V. Díaz. 1995. Microbial stability assessment in high and intermediate moisture foods: Special emphasis on fruit products. In *Food preservation by moisture control—Fundamentals and applications (International symposium on the properties of water—ISOPOW Practicum II, June 19–24, 1994, México)*, ed. J. Welti-Chanes and G. Barbosa-Cánovas, 575–602. Lancaster: Technomic Pub. Co.

Tapia de Daza, M.S., S.M. Alzamora, and J. Welti Chanes. 1996. Combination of preservation factors applied to minimal processing of foods. *Critical Reviews in Food Science and Nutrition* 36: 629–659.

Tapia, M.S., and J. Welti-Chanes. 2012. Hurdle technology principles applied in decontamination of whole and fresh-cut produce. In *Decontamination of fresh and minimally processed produce*, 1st ed, ed. Vicente M. Gómez-López, 417–449. New York: John Wiley & Sons, Inc.

Tapia, M.S., J. Chirife, and S.M. Alzamora. 2007. Effects of water activity (a_w) on microbial stability: As a hurdle in food preservation. In *Water activity in foods: Fundamental and application*, ed. G.B. Barbosa-Cánovas, A.J. Fontana Jr., S.J. Schmidt, and T.P. Labuza. Ames: IFT Press–Blackwell Publishing Professional.

Tapia, M.S., V.M. Gómez-López, and C. Olaizola. 2009. HACCP implementation in the production of fresh-cut fruits and vegetables. *Stewart Postharvest Review* 4: 1–7.

Terefe, N.S., K. Matthies, L. Simons, and C. Versteeg. 2009. Combined high pressure-mild temperature processing for optimal retention of physical and nutritional quality of strawberries (*Fragaria × ananassa*). *Innovative Food Science & Emerging Technologies* 10: 297–307.

Tran, M.T.T., and M. Farid. 2004. Ultraviolet treatment of orange juice. *Innovative Food Science & Emerging Technologies* 5: 495–502.

USDA. 2004. National Nutrient Database for Standard Reference. www.nal.usda.gov/fnic/foodcomp/Data/SR18/reports/sr18page.

———. 2005. Dietary Guidelines for Americans. www.usda.gov/cnpp/DG2005/.

Valdramidis, V.P., W.D. Gram, A. Beattie, M. Linton, A. McKay, A.M. Fearon, and M.F. Patterson. 2009. Defining the stability interfaces of apple juice: Implications on the optimisation and design of High Hydrostatic Pressure treatment. *Innovative Food Science & Emerging Technologies* 10: 396–404.

Venn, B.J., and J.I. Mann. 2004. Cereal grains, legumes and diabetes. *European Journal of Clinical Nutrition* 58: 1443–1461.

Walkling-Ribeiro, M., F. Noci, D.A. Cronin, J. Riener, J.G. Lyng, and D.J. Morgan. 2008. Reduction of *Staphylococcus aureus* and quality changes in apple juice processed by ultraviolet irradiation, pre-heating and pulsed electric fields. *Journal of Food Engineering* 89: 267–273.

Wang, H., H. Feng, and Y. Luo. 2007. Control of browning and microbial growth on fresh-cut apples by sequential treatment of sanitizers and calcium ascorbate. *Journal of Food Science* 72: M001–M007.

Welti-Chanes, J., C.E. Ochoa-Velasco, and J.A. Guerrero-Beltrán. 2009. High pressure homogenization of orange juice to inactivate pectinmethylesterase. *Innovative Food Science & Emerging Technologies* 10: 457–462.
World Health Organization. Food and Agriculture Organization of the United Nations (WHO/FAO). 2003. *Diet, nutrition and the prevention of chronic diseases*. WHO Technical Report Series 916. Report of a Joint WHO/FAO Expert Consultation, 148.
Yaun, B.R., S.S. Summer, J.D. Eifert, and J.E. Marcy. 2004. Inhibition of pathogens on fresh produce by ultraviolet energy. *International Journal of Food Microbiology* 90: 1–8.
Zhang, Y., S.K. Vareed, and M.G. Nair. 2005. Human tumor cell growth inhibition by nontoxic anthocyanidins, the pigments in *fruits* and *vegetables*. *Life Science* 76: 1465–1472.

Chapter 6
Cooling and Freezing of Fruits and Fruit Products

Alicia Chaves and Noemí Zaritzky

6.1 Cooling of Fruits

6.1.1 Introduction

Temperature is an important factor for maintaining postharvest quality of fruits and greatly influences the rate of deterioration.

Some of the processes associated with deterioration (respiration, ethylene production, etc.) increase with an increase in temperature. For each 10 °C increase in temperature, the rate of chemical reactions increases 2–3 times (Q_{10}Van't Hoff factor). Then, lowering the temperature of fruits decreases their rate of deterioration, reducing the respiration, ethylene production, and ethylene sensitivity consequently extending their shelf-life. Generally the greatest reduction in processes related to fruit deterioration is obtained at temperatures just above the fruit-freezing point, in the case of non-chilling sensitive fruits, or above chilling injury threshold for chilling sensitive fruits.

Fruits could be classified as non-chilling sensitive and chilling sensitive. Chilling sensitive fruits (generally, products of tropical or subtropical origin), are damaged when they are kept at temperatures above freezing point and below 5–15 °C depending on their chilling sensitivity. Table 6.1 shows fruits arranged according to their sensitivity to chilling injury. Symptoms of chilling injury generally are browning, surface lesions, pitting, water-soaked areas, failure in ripen normally, development of off-flavors, incidence of fungal attack (Kader 2007a, b).

A. Chaves • N. Zaritzky (✉)
CIDCA and Universidad Nacional de La Plata—CONICET,
La Plata, Buenos Aires, Argentina

A. Rosenthal et al. (eds.), *Fruit Preservation*, Food Engineering Series,
https://doi.org/10.1007/978-1-4939-3311-2_6

Table 6.1 Fruits arranged according to their sensitivity to chilling injury (Adapted of Kader 2007a, b)

Non-chilling sensitive	Chilling sensitive
Apple[a]	Avocado
Apricot	Banana
Cherry	Lemon
Fig	Guava
Pear	Orange
Plum	Melon (cantaloupe, honey dew)
Strawberry	Papaya
Blueberry	Pineapple
Raspberry	Passion fruit
Peach[a]	Watermelon
	Lychee
	Limes

[a]Some varieties are chilling sensitive

At low temperatures, freezing injury could also occur when fruits are exposed to temperatures below their freezing point. There is, usually, a breakdown of tissue by intercellular/intracellular ice formation and loss of product.

Moreover, high temperatures (above 27–30 °C) also cause fruit injuries. More common symptoms are surface scalds, loss of pigments, tissue with watery or translucent appearance, and heterogeneous ripening.

6.1.2 Precooling Treatments and Refrigeration

6.1.2.1 Precooling

Precooling is an important postharvest operation. Precooling involves the rapid removal of field heat after harvest. It is specially employed in fruits with a short postharvest life (i.e., tropical fruits). There are different methods for fast cooling, such as: cold air (i.e., room cooling, forced air cooling), hydrocooling, icing, vacuum cooling, and evaporative cooling.

The choice of the precooling method depends on the physiology of fruits, its temperature at harvest, and the desired postharvest life. Also, it is necessary to consider the cost–benefit ratio (Wills et al. 1998).

Moreover, the rate of cooling of a product depends on the difference in temperature between the fruit and the refrigerant, the nature of the cooling medium, the rate of heat transfer from the product to the medium and the thermal conductivity of the product. As the difference in temperature between the fruit and the refrigerant decreases, the cooling rate also decreases. Usually in practice the "half cooling time" and the "7/8 cooling time" are used as references for the refrigeration process. The "half time" cooling ($t_{1/2}$) is the time required to reduce to half the initial temperature difference between product and cooling medium (Thompson et al. 2007). Considering the exponential characteristic of the refrigeration curve, seven-eighths cooling time ($t_{7/8}$) is three times the "half time". Both cooling times are independent of the initial product and medium temperatures, and remain constant throughout the cooling period for certain conditions (type of package, cooling system). $t_{7/8}$ is of more practical use, because the product temperature is close to the storage or transport temperature.

Room cooling: It is the most common precooling system. Fruits (in bulk or in boxes, other packages) are exposed to cold air in a normal cooling room. The product may be cooled and stored in the same room. Room cooling has some disadvantages, as that a large area is necessary, the refrigeration rate is slow with respect to other cooling systems and fruit dehydration often occurs. For an efficient operation it is necessary that the cold air be in contact with all the surfaces of the fruit containers with a flux of 0.3 m^3/min per ton (0.005×10^{-3} m^3 s^{-1} kg^{-1}) of stored product in order to achieve an adequate heat elimination.

Forced air cooling: It is more efficient (faster) than room cooling, because cold air circulates through the containers, so the cold air has a direct contact with the fruits. Water losses are low because the product is cooled relatively fast and dehydration varies according to the sensitivity of each product. The cooling rate is controlled by the cold air flux that circulates through the product; usually, $1–2 \times 10^{-3}$ m^3 s^{-1} kg^{-1} is used (Thompson et al. 2007).

Hydrocooling: In this method cold water is the refrigeration medium. It is an effective system for rapid cooling for different fruits, either in bulk or packaged. Thus, both (packages and product) must be tolerant to wetting. Since packaging restricts water movement and reduces cooling efficiency, products are usually hydrocooled in bulk bins.

Water has a greater heat capacity than air, and then hydrocooling is more rapid than air cooling. Usually water temperature is close to storage or transport temperature. Cold water should be properly disinfected to prevent microbiological contamination.

Other precooling methods such as icing, vacuum cooling, and evaporative cooling are less used in fruits than in vegetables. Icing uses crushed ice as cooling medium and it is usually employed for temporary cooling (transport). Vacuum cooling involves a decrease in the pressure (660 Pa) around the product, and then evaporation of water occurs. The water evaporation absorbs heat. Water loss may be diminished by spraying water on the product.

In evaporative cooling, dry air is cooled by blowing it across a wet surface. The water evaporation causes the cooling of product.

Table 6.2 shows the precooling methods recommended for the most common fruits.

Table 6.2 Recommended precooling methods for the most common fruits

Precooling method	Fruits
Cooling room	Apple, pears, lemons, oranges, peaches, mango
Forced air cooling	Apple, pears, lemons, oranges, cherries, peaches, avocado, mango, kiwi, plums, melons, strawberries
Hydrocooling	Peaches, plums, cherries, mango, papaya, melons

6.1.2.2 Refrigeration

Following precooling, cold chain should be maintained throughout the fruit's life. So transport and storage should be under refrigeration. The objective of refrigerated storage is to prolong the commercial life of perishable products, by decreasing metabolic activity without causing chilling or freezing injuries, reduction of microbial growth, and water loss of the product between others. There are tables with recommended temperatures for fruits. Fruits non-sensitive to chilling injury have recommended storage temperatures close to 0 °C (apples, pears, peaches). For the moderately sensitive to chilling, the recommended temperatures range between 7 and 10 °C (avocado, citric, pineapple) and, for the highly sensitive the storage temperatures are in the range of 12–15 °C. Storage temperatures should be kept within a range of variation of ±1 °C (or less) with respect to the recommended values. The temperature fluctuations are undesirable and should be reduced by a good air circulation.

In addition to the temperature control, the humidity inside de chamber must also be controlled. A high relative humidity (RH) is required to minimize water losses. Recommended RH values result from a compromise between reducing water losses and microbial growth prevention; usually recommended RH is between 85 and 95 %.

Ethylene should also be controlled; the presence of ethylene can accelerate ripening and senescence. High ethylene levels are undesirable for a long storage. For fruits which produce low levels of ethylene, adequate ventilation (air renovation) is sufficient to keep ethylene at safe levels. If ventilation is not sufficient to control ethylene concentration, it can be destroyed by oxidation with ozone, UV light, or potassium permanganate.

In some storage facilities, different fruits are kept in the same refrigeration place for short times. In this case, it is necessary to consider the requirements of temperature, RH, respiration rate, ethylene production, ethylene sensitivity, and possible odor absorption (Thompson et al. 2007).

Other gases, such as carbon dioxide, oxygen, and carbon monoxide could be controlled by adequate ventilation.

6.1.3 Controlled and Modified Atmosphere

6.1.3.1 Carbon Dioxide and Oxygen

The purpose of the storage in Controlled Atmosphere is to extend the shelf-life of fruits, maintaining its sensory and nutritional quality, beyond just allowing for the cold storage.

The terms modified atmosphere storage (MA) and controlled atmosphere storage (CA) implies the addition or removal of gases resulting in an atmosphere whose composition is different from that of normal air. CA or MA is used in conjunction with low temperature storage. Usually it involves the reduction in the oxygen level and the increase in carbon dioxide concentration, although there are other alternatives (Kader 2007a, b; Wills et al. 1998; Rojas-Graü et al. 2009).

In CA storage gas levels are precisely controlled while in MA the composition of the atmosphere is not closely controlled. In MA the gas levels during storage are established as a result of the balance between CO_2 production and O_2 consumption due to respiration of the product and the exchange through semi-permeable plastic film of packages (Wills et al. 1998).

It is known that the increase in CO_2 and the decrease in O_2 levels exert effects in respiration rate and other metabolic reactions (ethylene production, softening, etc.). Generally, CA uses oxygen and carbon dioxide concentrations of about 1–5 kPa. As the O_2 levels are low, care must be taken to avoid anaerobic respiration. Also, if CO_2 concentrations are too high, physiological damages to fruits then can be produced. Tolerance to low O_2 and high CO_2 concentrations varies among fruits (Table 6.3).

If properly used CA can provide certain benefits, such as delayed senescence, reduced sensitivity of the fruit to ethylene action, lower ethylene production, reduction of some physiological damages, reduction of the activity of several microorganisms, and control of certain insects.

However, if not properly used or if the storage periods extended too much CA can lead to some undesirable effects such as induction of fermentative reactions leading to the formation of undesired metabolites such as acetaldehyde and ethanol. Moreover, in certain cases some physiological disorders (browning, irregular ripening between others) are induced (Kader 2007a, b).

Table 6.3 Tolerance of fruits to high CO_2 and low O_2 concentrations at recommended storage temperatures (Kader 2007a, b)

Level O_2 (kPa)	Fruits	Levels CO_2 (kPa)	Fruits
<1	Some apples and pears, dried fruits	2	Pears, apples (Delicious), apricots, grape,
2	Apples, pears, peaches Persimmons, apricots, plums, strawberries, kiwi, cherries, papaya, nectarines	5	Apples, peaches, nectarines, avocados, mangoes, bananas, papaya, kiwi, plums, oranges
3	Persimmons, avocados, cherries	10	Lemon, lime, per Simons, pineapples,
5	Citric fruits	15–20	Strawberries, blueberries, figs, raspberries, cherries

6.1.3.2 Other Atmospheres

6.1.3.2.1 Atmospheres with O_2 at Super-Atmospheric Concentrations

Oxygen concentrations higher than 21 kPa can influence the physiology and conservation of postharvest quality of perishable fruits and vegetables. Ripening of mature-green, climacteric fruits was slightly enhanced by exposure to 30–80 kPa O_2, but levels above 80 kPa retarded their ripening and caused O_2 toxicity disorders on some fruits. Also, super-atmospheric O_2 concentrations could inhibit the growth of some bacteria and fungi. They are much more effective if combined with elevated CO_2 (Kader and Ben-Yehoshua 2000). Wszelaky and Mitchan (2000) studied the effects of elevated O_2 alone or in combination with elevated CO_2 atmospheres for postharvest decay control on strawberry fruit (*Fragaria* × *ananassa* Duch.). They found that the 100 kPa O_2 treatment reduced decay but increased production of fermentative metabolites. Beyond the potential benefits of using super-atmospheric O_2 concentrations, its use in the industry is difficult because of its cost and the dangers associated with storing gas mixtures with high oxygen concentrations.

6.1.3.2.2 Hypobaric Storage

In hypobaric storage the product is stored at partial vacuum. In these systems the total pressure inside the chamber is about 40 kPa. Thus the oxygen partial pressure is reduced as well as ethylene and other volatile compounds. Such storage system is not widespread because it is expensive. Its use is more limited due to the low pressure at which the product is subjected, which promotes fluid loss and deterioration (Kader 2007a, b; Rodríguez-Félix et al. 2005).

6.1.3.2.3 Use of Other Gases

Modified atmospheres using other gases have been investigated. Among these gases carbon monoxide, ozone, nitric oxide, and sulfur dioxide, helium and argon are recommended; however these systems are not widely used.

Carbon monoxide (CO) was used in the transport of lettuce to reduce browning. Also, CO (5–10 %) added to the MA inhibits growth of various postharvest pathogens (Sandhya 2010; Kader 2007a, b). However this method is not frequently used in fruits.

Ozone (O_3) is used to reduce the microbial load of the products. Rodoni et al. 2010 found that a short-term ozone treatment might be useful to reduce damage and excessive softening in tomato, without negatively affecting other quality attributes.

Nitric oxide application has been favorable in strawberry and other crops storage (Zhu and Zhou 2007).

Sulfur dioxide (SO_2) is a gas used in grapes to prevent decay caused by *Botrytis cinerea*. However, the phytotoxicity of SO_2 has limited its use in postharvest.

6.1.4 Novel Technologies: Thermal Treatments, UV-C Irradiation

6.1.4.1 Thermal Treatments (Heat Treatment, Heat Shock)

Heat treatments have been used in fruit postharvest technology for insect disinfestations, decay control, ripening delay, and modification of fruit responses to other stresses (Lurie 1998; Paull and Chen 2000). These treatments have been used for disinfestation and disinfection of various tropical fruits, such as mango, papaya, and citrus fruits (Jacobi et al. 2001).

In fruits and vegetables, heat treatments (heat shock) have been used to improve the capacity of conservation, but without affecting cell viability or suppression of physiological processes (Lurie 1998). These treatments generally used temperatures between 45 and 70 °C, which are applied during periods ranging from seconds to hours. This is based on the fact that the inactivation of microorganisms primarily depends on the temperature applied, whereas many undesirable quality changes depend on the duration time of the heat treatment (Paull and Chen 2000). So, strawberries heat-treated in an air oven (45 °C, 3 h) showed lower decay and less tissue damage than the non-treated fruit (Vicente et al. 2006).

Heat may be applied to fruits utilizing hot water by dipping or spray, water vapor or hot air (static or flowing air) with humidity control or not. Water is desired for many applications because it has higher heat transfer capacity than air, which can reduce exposure times. Hot water treatments are easy to apply and to control product and water temperature. However, can only be used in products that can tolerate contact with water (Fallik 2004; Paull and Chen 2000; Lurie 1998).

Ripening of most climacteric fruits is characterized by increased softening, respiratory activity, sugar content, color development, and ethylene production, meanwhile decreasing organic acids content. Heat treatment affects several aspects of fruit ripening, such as ethylene production and cell-wall degradation, probably through changes in gene expression and protein synthesis (Lurie 1998). As an example, strawberries treated at 45 °C for 3 h showed a delay in the softening and in the expression and activity of various enzymes (endoglucanase, and β-galactosidasepoligalactosidase) related to the degradation of the cell wall (Civello and Martínez 2008).

During application of a thermal treatment, the expression of most ripening-related genes decreases noticeably while the expression of genes corresponding to heat shock proteins (HSP) increases (Lurie et al. 1996).

In addition, heat shock may induce reactive oxygen species (ROS), followed by the production of oxygen radical scavengers such as superoxide dismutases (SOD), peroxidases (POD), and catalases (Mittler et al. 2004). Heat treatments also induce an increase in the activity of ascorbate-glutathione cycle enzymes (González Aguilar et al. 2010). The increase in the activity of these enzymes, in addition to the induction of carotenoids, phenolic and other compounds, is very important in maintaining and prolonging the postharvest life of fruits preventing oxidative damage in cell tissue acting as antioxidants. These antioxidants can help in controlling oxidative reactions caused by ROS (Cisneros-Zevallos 2003). Industrial implementation of heat treatments is still incipient.

6.1.4.2 UV-C Irradiation

The germicidal effect of ultraviolet (UV) radiation has been long recognized and a UV radiation sterilization application has been adopted (Guerrero-Beltrán and Barbosa-Cánovas 2004).

The UV portion of the electromagnetic spectrum ranges from approximately 200–400 nm (Bintsis et al. 2000). The UV spectrum can be subdivided into three sub-regions, that is UV-A (400–320 nm), UV-B (320–280 nm), and UV-C (280–200 nm) (Maverakis et al. 2010). UV-C light is a non-ionizing radiation; it can be absorbed and produce chemical changes in different cellular components. In particular the region between 250 and 260 nm is lethal to bacteria, viruses, fungi, protozoa, yeast, and algae (Civello et al. 2006).

UV-C irradiation can be applied at lethal and sub-lethal doses. The detrimental effect of UV-C includes tissue structural damage, changes in cytomorphology, and water permeability of inner epidermal cells (Lichtscheidl-Schultz 1985). Nevertheless, low doses of UV-C irradiation stimulated beneficial reactions in biological organs, a phenomenon known as hormesis (Shama 2007). It has been reported that hormetic doses of UV-C can prolong the postharvest life and maintain the quality of fruits. These effects include delay of senescence process and fruit ripening (Gonzalez-Aguilar et al. 2007a, b), decay reduction (Baka et al. 1999; Pan et al. 2004), induction of natural defense and elicitors against fungi and bacteria (Alothman et al. 2009).

Resistance to infection by pathogen is related with deoxyribonucleic acid (DNA) damage and the induction of plant defense mechanisms. The DNA is particularly sensitive to UV-C radiation; the absorption causes the formation of dimers, mainly between adjacent pyrimidine bases (thymine, cytosine). Because DNA and RNA polymerases cannot decode these products, when the DNA repair system capacity is exceeded, replication and transcription are compromised, resulting in the collapse and death of the affected cells (Sinha and Häder 2002).

The induction of plant defense mechanisms is manifested through the stimulation of antifungal chemical species such as phytoalexins (scoparone and resveratrol), flavonoids, and degrading fungal cell-wall enzymes (chitinases, glucanases) (El-Ghaouth et al. 1998). The induction of plant defense system can also trigger the accumulation of these compounds and other phytochemicals such as carotenoids and vitamin C which exhibit antioxidant potential, improving the nutritional status of the fruit (Alothman et al. 2009; Gonzalez-Aguilar et al. 2007a, b).

Postharvest UV-C treatments consist in exposing the commodities for a certain period of time under a bank of UV-C lamps with a maximal emission at 254 nm, which is the most efficient wavelength for damaging DNA. Similarly to heat shock treatment, UV-C irradiation must be applied before cold storage. The efficiency of UV-C radiation is directly related to the UV-C dose expressed in Joules per square meter (J/m^2). A relatively broad spectrum of treatments has been used in fruits but in most cases the doses used (Table 6.4) ranged from 0.2 to 20 kJ/m^2 (Civello et al. 2006). The UV-C dose needs to be optimized for each fruit, and even for each new variety or cultivar assayed. In some cases, the dose required to achieve the desired results changes with the ripening stage and during the harvest season (Civello et al. 2006). Exposure to very low doses do not result in substantial benefits, while excessive treatment can damage the membrane lipids and alter other biomolecules involved in the maintenance of homeostasis and increased susceptibility to microbial attack (Allende et al. 2006; Artés et al. 2009).

The effectiveness of UV-C treatment does not seem to depend on temperature in the range of 5–37 °C, but it depends on the incidence of radiation on the product and hence by the shape and surface properties (Bintsis et al. 2000; Ben-Yehoshua and Mercier 2005). UV radiation has extremely limited penetration into solids (Gardner and Shama 2000).

Effects of postharvest UV-C treatment on fruits and vegetables: In general, UV-C radiation has been shown to be effective in controlling storage rots. The effect of UV-C treatments on decay control has been assayed both in isolated microorganisms and in fruit or vegetables either carrying their normal microflora or in controlled inoculation conditions with a particular pathogen. The treatments have been reported to be effective to reduce decay caused by the most common postharvest pathogens, including *Penicillium digitatum, Botrytis cinerea, Rhizopu sstolonifer, Alternaria citri* and *Alternaria alternata, Monilinia fructigena,* and *Colletotrichum gloeosporioides*. Effective control of postharvest decay by UV-C treatments has been reported in several products including apple, peach, tangerine, grapefruit, strawberry, boysenberry, mango, and grape (Civello et al. 2006).

Beyond the known germicidal effect of UV-C radiation, several studies have shown that fruits exposed to low doses of UV-C increased the accumulation of

Table 6.4 UV-C doses effective to decrease decay in fruits

Fruit	Dose (kJ/m^2)	Reference
Apple cv Golden Delicious	4.8–7.5	Stevens et al. (1996)
cv Red Delicious	7.5	De Capdeville et al. (2002)
Boysenberry	9.2	Vicente et al. (2004)
Grapefruit	0.5–1.5	D'hallewin et al. (2000)
Mango	4.9–9.8	González-Aguilar et al. (2001)
Peach	7.5	Stevens et al. (1996)
Strawberry	4.6	Pan et al. (2004)
Satsuma mandarin	0.25–3.0	Shen et al. (2013)

compounds having antimicrobial properties (El Ghaouth et al. 2003). In citrus, the UV-C treatments increased the scoparone (phytoalexin) concentration, increasing the resistance against *Penicillium digitatum* (D'hallewin et al. 1999).

Futhermore, UV-C radiation can delay ripening. UV-C treatment can cause changes in the content of pigments, such as chlorophyll, carotenoids, and anthocyanins. Generally, UV-C treatments reduce chlorophyll degradation and produce an increase in total carotenoids. Diverse effects of UV-C radiation on anthocyanins have been reported. Pan et al. (2004) found that strawberry fruit irradiated with 4.1 or 6.9 kJ/m^2 showed less surface reddening and lower anthocyanin content than control fruits. In the case of grapes and pomegranates, UV-C treatments did not affect anthocyanin content (Cantos et al. 2000) Also, UV-C irradiation reduced the softening in raspberry (Vicente et al. 2004), mango (González-Aguilar et al. 2001), and strawberry (Baka et al. 1999), among other fruits.

Postharvest stress-type treatments, such UV treatments, have been developed to preserve fruits. The stress can activate some enzymatic and non-enzymatic antioxidant systems of the fruits (Lim et al. 2007). The effect of UV-C treatments on the synthesis of antioxidant compounds and antioxidant enzymes can vary depending on the hormetic doses, time of exposure, and treated fruit. Alothman et al. (2009) found an increase in phenols and flavonoids in guava and banana after 30 min exposure to UV-C light. Erkan et al. (2008) found that UV-C irradiation induced superoxide dismutase activity in strawberry.

In addition, UV-C irradiation may induce reactive oxygen species (ROS), followed by the production of oxygen radical scavengers such as superoxide dismutases(SOD),peroxidases(POD),catalases(CAT)monodehydroascorbatereductase (MDAR), glutathione (GSH), and oxidized glutathione (GSSG) maintaining redox homeostasis.

Moreover, it has been reported that exposure for short periods to UV radiation can reduce physiological disorders such as chilling injury (Gonzalez-Aguilar et al. 2004). UV-C treatments reduced the incidence of chilling injury in peaches and pepper fruits (Vicente et al. 2005; Andrade Cuvi et al. 2011).

UV-C radiation has several advantages that encourage their use in postharvest treatments. They are simple, fast, dry, and cold processes, which require less space than other methods, low maintenance and have low cost. Another advantage of UV-C treatments is that they could be applied in combination with other postharvest technologies. Because exposure to UV-C radiation can be harmful against biological systems, special caution with exposure of the eyes or skin of the operators should be taken. Then the section where treatments are performed should not allow radiation leaks outside, and must be just open when the lamps are off. It is also important to consider that the radiation with wavelengths below 260 nm produces ozone in the presence of oxygen. If the lamps remain lit for long periods, the level of this gas can be high and therefore must be monitored and removed (Bintsis et al. 2000). Commercial application of UV-C treatment is still limited.

6.1.5 Minimally Processed Fruits

Fresh-cut fruits are generally much more perishable than intact products since they have been subjected to severe physical stress, such as peeling, cutting, slicing, shredding, or trimming.

During minimal processing, mechanical injury results in cellular delocalization of enzymes and their substrates, leading to biochemical deteriorations such as enzymatic browning, off-flavors and texture breakdown, as well as increased respiration rate and ethylene synthesis. Enzymatic browning, caused mainly by the action of polyphenol oxidase (PPO), is a major factor limiting the shelf-life of minimally processed fruits (Lamikanra 2002). Ascorbic acid as a reducing agent has been applied in combination with organic acids or calcium salts to prevent enzymatic browning and maintain firmness of fruits (Wang et al. 2007). Since these products are very perishable, cooling is required. Low temperatures are required to reduce respiration rates, retard microbial growth, and deteriorating reactions.

The distribution of fresh-cuts is normally done at normal refrigeration temperatures at the retail and household levels (4–6 °C) and a relative humidity of 90–95 %. Storage temperatures usually recommended are between 0 and 5 °C. Generally, fresh-cut fruits are packaged in MAP (modified atmosphere packaging). Responses to MAP differ markedly depending on the product concerned.

Moreover, it is important to note that there may be differences in the optimal atmosphere for a whole or a cut fruit (Rodríguez-Félix et al. 2005; Kader 2007b). This may be due to increased metabolic activity of the fresh-cut, and its shorter storage time. For example, many fresh-cuts manifest decay before the damage by high CO_2 or low O_2 appears. Although MA composition must be determined for each product, usually, low O_2 (0.5–5 kPa) and high CO_2 (5–10 kPa) levels are employed (Cantwell and Suslow 2007).

Also, in cut fruits the stress treatments were effective. The main objective of applying an UV-C treatment in processed fruits is the destruction of pathogens to increase the shelf-life of the products. Microbial activity was effectively reduced after treated cubes of watermelon. In other cases such as in processed arils of pomegranate, unclear results were obtained on the effect of the UV-C radiation on microbial growth and no benefits in shelf-life and overall quality were found after treatments at doses between 0.5 and 13.6 kJ/m^2. Gómez et al. (2010) examined the effect of UV-C irradiation at different doses on the surface color of apple slices stored in refrigeration for 7 days. They also explored the use of some pretreatments (hot water blanching, dipping in a solution containing ascorbic acid and calcium chloride) to minimize apple browning caused by UV-C light. Color parameters were found to be dependent on UV-C dose, storage time, and the type of pretreatment. Both pretreatments contributed to maintain the original color of apple slices after UV-C light exposure.

Otherwise, heat treatments, alone or in combination with other sanitizing methods, could be a viable alternative to preserve the sensory and microbial quality of the fresh-cut fruits. Usually heat treatments are applied to minimally processed fruits as hot water dips. Silveira et al. (2011) found that a hot water dip treatment (60 °C; 90 and 120 s) followed by a peroxiacetic acid dip, are effective to control the microbial growth and maintain the overall quality in fresh-cut Galia melon.

6.1.6 *Edible Coatings*

Edible films and coatings are thin layers of edible materials applied on food products that play an important role on their conservation, distribution, and marketing. Some of their functions are to protect the product from mechanical damage, physical, chemical, and microbiological activities.

An edible coating is a thin layer of edible material formed as a coating on a food product. The edible coating is applied in liquid form on the fruit, usually by immersing the product in a solution-generating substance formed by the structural matrix (carbohydrate, protein, lipid, or multicomponent mixture) (Falguera et al. 2011).

Edible coatings create a passive modified atmosphere, which can influence various changes in fresh and minimally processed fruits, such as: antioxidant properties, color, firmness, sensory quality, microbial growth inhibition, ethylene production, and volatile compounds (Oms-Oliu et al. 2008).

Edible coatings have a high potential to carry active ingredients such as anti-browning agents, colorants, flavors, nutrients, spices, antimicrobial compounds, and active nanoencapsulated compounds that can extend product shelf-life and reduce the risk of pathogen growth (Rojas-Graü et al. 2009). Lee et al. (2003) reported that apple slices coated with carrageenan containing ascorbic acid, citric acid, and oxalic acid extended shelf-life by 2 weeks when packaged in trays at 3 °C. Garcia et al. (2001) reduced microbial growth below 6 log CFU/g at the maximum storage time assayed (28 days) and extended storage life of fresh strawberries using a starch-based coating containing potassium sorbate and citric acid. However, specific studies on fresh-cut fruits are limited and their industrial implementation is still incipient.

6.2 Freezing of Fruits

Freezing has been successfully employed for the long-term preservation of many foods, extending significantly their shelf-life. The process involves lowering the product temperature, below the freezing point, changing most of the water in the foodstuff to ice. During the freezing process, water is removed from the food matrix through the formation of ice crystals, and the concentration of dissolved substances in the unfrozen regions increases, lowering the water activity of the product. Therefore freezing is a very efficient method for food preservation, not only because the low temperatures inhibit the growth of microorganisms and retard biochemical and enzymatic reactions, but also due to the decrease in water activity (Zaritzky

2000, 2008). Freezing is one of the methods usually used for preserving food because the product tends to maintain its original attributes, keeping the nutritional properties as close as possible to those of the fresh products (Fennema et al. 1973; Canet and Alvarez 2006).

Freezing preservation retains the quality of agricultural products over long storage periods. It is considered one of the best methods of long-term preservation for fruits with respect to retention of sensory attributes and nutritional properties; it retards microorganisms' growth and slows down changes that affect quality or cause spoilage in food. Frozen fruits will retain much of their fresh flavor and nutritive value; however, their texture may be somewhat different than that of fresh fruits.

The preservation of fruits by freezing is one the most important preservation methods. The safety and nutrition quality of frozen products are enhanced when high-quality raw materials are used, good manufacturing practices are applied, and the products are maintained at the correct subzero temperatures.

Commercial freezing of small fruits and berries started in the eastern part of the United States in about 1905 (Desrosier and Tressler 1977; Barbosa Canovas et al. 2005). One of the advantages of freezing preservation of fruits is that they may be more extensively used during the off-season fruits; in addition frozen fruits can be transported to distant markets that could not be accessed with fresh fruit. Freezing preservation makes possible further processing of fruit products, such as, juice, jams, and syrups from frozen whole fruit, slices, or pulps.

The market for frozen foods is continuously growing throughout the world. The production is driven by changes in lifestyle, a higher demand for prepared food together with an increase in the availability of technology to produce and store frozen food (Kennedy 2000). The market of frozen fruits includes whole frozen fruit (conventional market) and fruit pieces to be incorporated as ingredients, into prepared foods. Many fruit-based products have appeared on the market, for example fruit pieces pre-dipped in sugar and pre-dried frozen fruits. They are widely used as basic materials, or as additional components, in many food formulations; examples include cooked dishes, pastry and confectionery products, ice cream, frozen desserts and sweets, fruit salads, and yogurt (Torreggiani et al. 2000).

Although freezing of fruits results in improved effects with respect to shelf-life and availability throughout the year, various undesirable changes occur as a result of this process due to ice formation.

6.2.1 The Freezing Process: Ice Formation

Water is a very important component in food, affecting its quality attributes and shelf-life. The formation of three-dimensional hydrogen bonds explains many of the "anomalous" properties of water, such as high boiling and freezing points, high specific heat, high latent heats of fusion and vaporization, high surface tension, high polarity, and unusual density changes (Fennema et al. 1973). One of the important "anomalies" of water is the expansion in volume of ice on freezing; when water freezes at 0 °C at atmospheric pressure, its volume increases by about 9 %. This behavior is in contrast to normal liquids, which usually contract on freezing and expand on melting. In the case of water, the volume expansion when going from liquid to solid (freezing) under ambient pressure may cause tissue damage in biological organisms (Zaritzky 2010).

Freezing involves thermodynamic factors, which define the characteristics of the system under equilibrium conditions, and kinetic factors, that describe the rates at which equilibrium might be approached as water is converted into ice. The freezing process includes two successive stages: the formation of ice crystals (nucleation), and the subsequent increase in crystal size (growth).

6.2.2 Homogeneous and Heterogeneous Nucleation

When ice and water coexist at atmospheric pressure, the temperature of the system reaches the freezing point of pure water ($T_f=0$ °C) as long as both liquid and solid are present. The amount of ice remains constant and no energy is either added or removed from the mixture. The freezing point for water (or the melting point for ice, T_m) is an equilibrium point. Thermodynamic theory indicates that below 0 °C, ice is the stable form of pure water. However, if water is cooled to 0 °C it will not freeze; it is necessary to reach a temperature (T) lower than the freezing point (T_f) before ice begins to form. Supercooling (or undercooling) is necessary to overcome the free energy that accompanies the formation of a new phase (an ordered solid ice particle) from the melted phase. Supercooling is defined as $\Delta T_s = T_f - T$; in the case of pure water, when the temperature of the refrigerated medium is −10 °C the maximum supercooling that could be reached is 10 °C.

For a crystalline phase to grow there must be an initiation seed or nucleus upon which the crystal can grow. Nucleation is the arrangement of molecules into an ordered particle of a sufficient size to survive and serve as a site for further crystal growth. An energy barrier (activation energy) must be surpassed before nucleation can occur. Nucleation serves as the initial process of freezing, and can be considered as the critical step that results in a complete phase change (Zaritzky 2010, 2011).

The nucleation rate is highly dependent on the temperature of the freezing medium, the viscosity of the system, and the sample volume. Nucleation is a statistical phenomenon and the probability of nucleation depends on the volume of the sample; when the sample volume is small the probability of nucleation is low and high supercooling is required (Hartel 2001).

Ice nucleation can be homogeneous or heterogeneous. Homogeneous nucleation occurs in pure water free from all impurities. The homogeneous nucleation temperature of pure water is about −45 °C, which is the minimum temperature that pure water can be cooled to before freezing occurs spontaneously. Heterogeneous ice nucleation takes place when water molecules aggregate in a crystalline arrangement on nucleating agents such as active surfaces. This type of nucleation predominates in solutions and in food systems (Fennema et al. 1973). The supercooling necessary for heterogeneous nucleation is lower than that for homogeneous nucleation.

Figure 6.1a represents a schematic curve showing the effect of initial supercooling on the rate of heterogeneous nucleation per unit volume of a food sample. At low supercooling values (high subfreezing temperatures), the rate of ice nucleation is low, but it increases with supercooling. Figure 6.1a also shows that at very low temperatures, close to the glass transition temperature of pure water (which is approximately −135 °C), nucleation no longer occurs. This is due to the increase

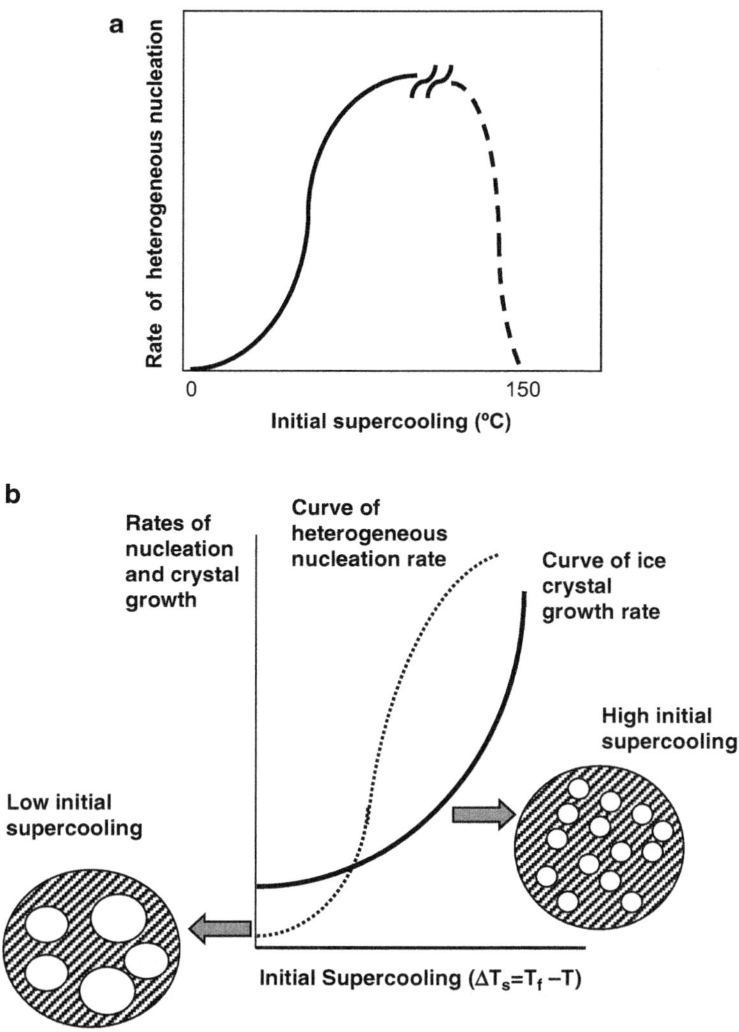

Fig. 6.1 (**a**) Schematic curve showing the effect of supercooling on the rate of heterogeneous nucleation of ice. (**b**) Rates of nucleation and crystal growth curves. Effect of supercooling on ice crystal sizes

in viscosity, a process known as vitrification, which is produced when a very small water specimen is exposed to extremely low temperatures and very high cooling rates. Ice formation is avoided and the liquid becomes a glass; at high solute concentrations or high viscosities, the viscous barrier to nucleation becomes more important than the supercooling and ice nucleation is inhibited. The glassy state is characterized by reduced molecular mobility and a very high viscosity (higher than 10^{14} Pa s).

6.2.3 Crystal Growth

Once a stable ice nucleus is formed, further growth by adding molecules to the solid–liquid interphase is possible. Growth is not instantaneous and it is controlled by the rate of removal of the latent heat released during the phase change, as well as by the rate of mass transfer, in the case of solutions. Heat transfer is not the only factor that governs crystal growth or the rate of ice propagation. If ice is crystallizing from a solution, water molecules diffuse from the surrounding solution to the surface of the ice crystals and are incorporated into the growing solid phase. At the same time, solute molecules (salts or sugars) must be rejected from the region occupied by the pure ice crystals and diffuse away from the solid surface.

The number and size of ice crystals formed in a given frozen system (small enough to avoid thermal gradients) is the result of the nucleation and growth rate curves. Crystal size varies inversely with the number of nuclei formed (Fig. 6.1b). At high supercooling values, which lead to high freezing rates, the nucleation rate is higher than the crystal growth rate; as a consequence a large number of nuclei are formed and the mass of ice is distributed in many small crystals. In contrast at low initial supercooling and low freezing rate, fewer nuclei are formed, which leads to large crystals (Fig. 6.1b).

6.2.4 Freezing Curves

Figure 6.2 shows schematic time–temperature curves for the freezing of small liquid specimens (without thermal gradients) of pure water (upper curve) and of an aqueous solution representing a simplified food model system (lower curve). The first stage in the cooling of pure water involves the removal of sensible heat (4.18 kJ/kg °C); nucleation is necessary for freezing to initiate, and the temperature can fall below 0 °C without the formation of ice crystals. Point S indicates the supercooling of the water before crystallization begins (Fennema et al. 1973, 1996; Zaritzky 2000, 2011). Once the critical mass of nuclei is reached, the system nucleates at point S and releases the latent heat of solidification faster than the heat that is being removed from the system.

The increase in temperature (point S to point A) due to the release of the latent heat of solidification after initial supercooling represents the onset of ice crystallization.

Fig. 6.2 Schematic time–temperature curves during freezing of small liquid specimens (without thermal gradients) of pure water and an aqueous solution representing a simplified food model system (lower curve)

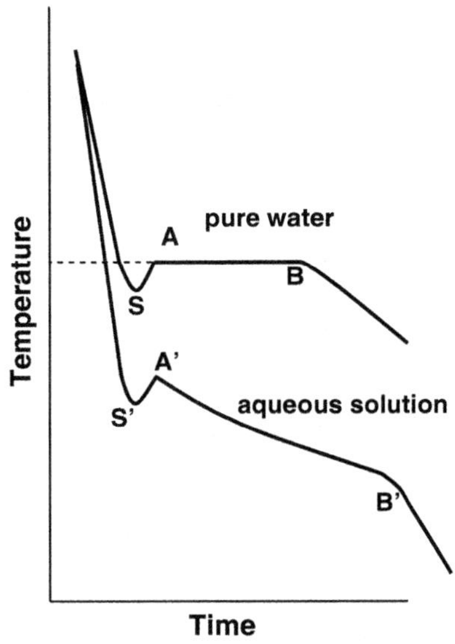

Once crystallization begins, the temperature reaches point A, the freezing point of pure water ($T_f = 0\,°C$). While ice and water are in equilibrium, the temperature remains at the freezing point until all the water has been converted into ice (point B). In pure water, the plateau from A to B represents the time during which crystal growth is occurring. Once solidification is completed, further removal of heat results in a decrease in temperature towards the temperature of the freezing medium.

The freezing of a food system can be analyzed in a simplified manner by considering the behavior of an aqueous solution (Fig. 6.2, lower curve). The cooling curve also shows a supercooling point (S′); nucleation occurs at point S′ and the released heat raises the temperature from S′ to A′. Point A′ represents the initial freezing point of the solution, which is lower than the freezing point of pure water. The freezing point depression is determined by the number of dissolved solute molecules. In aqueous solutions, the degree of supercooling is generally lower than in pure water, since the added solute promotes heterogeneous nucleation, accelerating the nucleation process. In very concentrated solutions, it is sometimes difficult to observe supercooling; further cooling from A′ to B′ results in the growth of ice crystals and substantial ice formation. The declining freezing point (negative slope of A′–B′) reflects the gradual increase in solute concentration as water is separated out in the form of ice crystals.

6.2.5 *Initial Freezing Point*

The initial freezing point of a system containing solutes (sugars, salts etc.) is lower than the freezing point of pure water due to the presence of dissolved substances in the tissue. The freezing points of fruits are directly related to their soluble solid contents. At the initial freezing point, a fraction of the water within the food crystallizes and the remaining solution becomes more concentrated in solutes, thus, the freezing point of the unfrozen portion of the food is further reduced. As the temperature continues to decrease, the formation of ice crystals increases the concentration of the solutes in solution depressing the freezing point. This is a colligative property that depends on the concentration of solutes, which lowers the effective number of solvent molecules that can undergo the phase transition from liquid to solid. Freezing point depression is directly proportional to the concentration of solutes. Using basic thermodynamic principles of aqueous solutions, it is possible to predict the freezing point depression of different foodstuffs (T_{fs}). Heldman (1974), proposed an equation (valid for dilute solutions) to calculate the freezing point depression ($\Delta T_f = T_f - T_{fs}$) with reference to the freezing point (T_f) of the pure solvent (water):

$$\Delta T_{\mathrm{f}} = \frac{R T_{\mathrm{Kf}}^{2} M_{\mathrm{A}} m}{1000 \lambda_{\mathrm{A}}} = 1.86m \tag{6.1}$$

where λ_A is the latent heat of melting for pure water (6003 kJ/mol); M_A is the molecular weight of water (18 g/mol); m is the molality of the solution representing the food system (number of moles of solute/1000 g of solvent); and T_{Kf} is the freezing point of pure water (273 K). Foodstuffs with higher solute content show a lower initial freezing point. A typical range of initial freezing points in fruits is −0.8 to −2.7 °C (Fennema et al. 1973). The freezing points of fruit tissues are governed by the concentration of water-soluble solutes (soluble solids) especially sugars, salts, and acids. Sugars are the most predominant soluble solids in fruits; the total soluble solids in fruits generally range between 4 and 23 % (Fennema et al. 1973). Table 6.5 shows ranges of initial freezing points of different fruits, based on data reported by Wang et al. (2003).

Table 6.5 Ranges of the initial freezing point of different fruits adapted from data reported by Wang et al. (2003)

Fruit	Ranges of initial freezing point (°C)
Pear	−2 to −1.5
Apple	−2 to −1.10
Orange	−2.2 to −0.8
Cherry	−1.8 to −1.7
Banana	−1.7 to −0.77
Strawberry	−0.80 to −0.77
Plum	−0.8 to −0.82
Persimmon	−2.20 to −2.16
Peach	−0.88 to −0.90
Lemon	−1.4 to −1.33
Grape	−2.10 to −1.17
Fig	−2.44 to −2.4

6.2.6 State Diagram

A phase diagram only indicates the conditions in which equilibrium phase transformation can occur (Roos 1995). In contrast, a state diagram is a map of different states of a food as function of water or solid content and temperature (Roos and Karel 1991a; Rahman 2006, 2007, 2009). State diagrams describe the physical state as a function of temperature including data on equilibrium, non-equilibrium, and metastable equilibrium states such as glass transition conditions. A schematic temperature–composition state diagram for an aqueous system with a single solute is shown in Fig 6.3.

The equilibrium thermodynamic freezing process of a food system can be represented by the equilibrium liquid–solid curve (line 1–2 in Fig. 6.3), which gives the melting temperature as a function of solute concentration. At slow rate of cooling when temperature decreases, water is removed as it forms the solid phase (ice crystals) and the solute in the unfrozen phase is freeze-concentrated. An equilibrium freezing temperature exists for each ice/unfrozen phase ratio, and is a function of solute concentration. The mass of ice in equilibrium with the unfrozen solution is characteristic of the type of foodstuff, depends on water content, and changes with temperature.

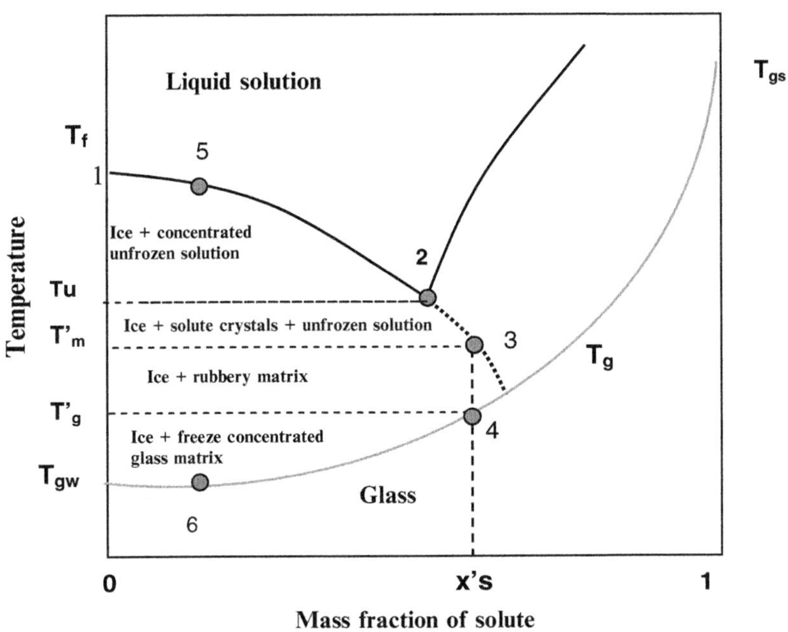

Fig. 6.3 Schematic state diagram for an aqueous system with a single solute. T_m = melting point curve; T_{eu} = eutectic point; T_g = glass transition temperature; T'_g = glass transition temperature of the maximally freeze-concentrated solution in equilibrium with ice crystals; below T'_g the freeze-concentrated unfrozen matrix becomes a glass; x'_s = concentration of solute within the glass (maximally freeze-concentrated matrix); T'_m = onset of ice melting

In the case of eutectic solutions, the equilibrium curve extends from the freezing temperature (T_f) of pure water (0 °C) (point 1) to the eutectic temperature of the solute represented by point 2 in Fig. 6.3 (Rahman 2009). Solute crystallization is mandatory to sustain thermodynamics equilibrium. Further cooling results in crystallization of ice and solute in constant proportion, leaving the unfrozen phase unchanged in composition and freezing point. This dual crystallization process continues at constant temperature until crystallization of water and solute is as complete as possible. Further cooling will decrease the sample temperature without changes in the physical state (Fennema 1995, 1996).

If water or a diluted solution is cooled very fast (point 5–6 in Fig. 6.3) the viscosity of the liquid phase rises and may reach such high values that molecular rearrangements in the liquid become extremely slow, thus avoiding ice crystallization. The liquid is in a metastable state until it is below the glass transition temperature (T_g), where the system is an amorphous solid or glass (Fig. 6.3).A glass (vitreous state) is defined as a non-equilibrium, metastable, amorphous, disordered solid of very high viscosity (Le Meste et al. 2002). The glass transition occurs when a supercooled melt is converted into a glass. Both the supercooled melt and the glass are non-crystalline states; the glass is an out-of-equilibrium state where the liquid-like structure of the melt is maintained as a rigid solid, and the supercooled melt (observed between the glass–liquid transition and the melting point) can be considered a viscoelastic high mobility "rubber" in the case of a polymeric material, or a viscous liquid for low molecular weight materials, where diffusion-controlled changes occur (Fennema et al. 1973; Franks 1985; Slade and Levine 1995). A system is considered to be vitrified if its viscosity is extremely high (10^{10}–10^{14} Pa). Figure 6.3 shows the glass transition temperature (T_g) curve as a function of solute (sugar) concentration; this curve extends from the glass transition temperature (T_g) of pure water (close to −135 °C) to T_{gs} the glass transition of the pure solute. The effect of water as a plasticizer of food systems is manifested as a depression of the glass transition temperature of the amorphous components. In aqueous solutions, as the concentration of solutes increases, the temperature T_g at which vitrification occurs also increases, and the cooling rate necessary to achieve vitrification decreases with respect to pure water. The pathway indicated by 5–6 involves a very rapid removal of heat from very small samples resulting in the vitrification of the entire sample without ice formation. Vitrification has no commercial significance in frozen foods but is important for the cryopreservation of biological material (Fennema 1995).

At intermediate cooling rates, non-equilibrium cooling occurs; this situation is quite different from the thermodynamic equilibrium freezing. Under non-equilibrium freezing the solute eventually attains its saturation temperature (eutectic point), but further cooling does not result in nucleation of solute crystals, rather the solution becomes supersaturated with solute and this condition is metastable. Ice formation at temperatures above the eutectic temperature occurs only to the extent defined by the freezing temperature, however continued cooling will cause the supersaturated unfrozen phase to convert into a metastable amorphous solid (glassy matrix) with very high viscosity (about 10^{12} Pa s).

At temperatures below the eutectic temperature if the solutes remain in the unfrozen solution, ice formation may proceed according to the extended T_m curve (dotted line in Fig. 6.3). Such solutions contain ice and a supersaturated unfrozen phase.

Therefore during non-equilibrium rapid cooling, freeze concentration may continue beyond the eutectic point into a non-equilibrium state through a viscoelastic liquid/solid glass state transition, due to reduction in molecular motion and diffusion kinetics (Goff 1994, 1997; Roos and Karel 1991a, b). Continuing freeze concentration at lower temperatures increases the viscosity of the unfrozen phase until this concentrated solution becomes a glass (Roos 1995).

The formation of an amorphous state is time-dependent since the limiting factor of the process (water removal in the form of ice) becomes more difficult as concentration increases. The marked effect of viscosity on mass transfer properties acts as the limiting factor for ice growth; in addition, under conditions where heat removal is rapid, a high level of supercooling at the interface decreases the propagation rate and freezing becomes progressively slower as ice crystallization is hindered, consequently more time is required for crystal growth at each temperature.

T'_m (Fig. 6.3) is the onset melting temperature of ice in contact with the maximally concentrated solution; ice formation above T'_m proceeds to an equilibrium extent determined by the equilibrium freezing curve.

T'_g is defined as the maximally freeze-concentrated glass transition temperature of the frozen system, where the unfrozen water in the matrix is unable to freeze and ice formation ceases within the time-scale of normal measurement (Franks 1985; Levine and Slade 1991; Roos and Karel 1991b; Goff 1994, 1997; Le Meste et al. 2002; Rahman 2009). T'_g is the temperature at which the maximum amount of ice is formed, leading to a maximally freeze-concentrated solution.

Below T'_g the unfrozen matrix takes on solid properties (glass) because of reduced molecular motion, which is responsible for the marked reduction in translational, non-rotational mobility (Roos and Karel 1991a; Slade and Levine 1995). At this temperature, the concentration of solute within the glass is x'_s (Fig. 6.3). The point 4 (T'_g, x'_s) is a characteristic transition (maximal freeze concentration transition). T'_g is practically an invariant value which depends only on solute composition. The water content at points 3 or 4 is the unfrozen water $(1-x'_s)$, that is, the water that remains unfrozen even at very low temperatures; it includes both uncrystallized free water and bound water attached to the solids matrix (Rahman 2009). The viscosity at the onset of glass transition indicated by T'_g is probably close to that of a glass (10^{12} Pa s).

Different values of T'_g were reported for sugars (Hartel 2001): fructose (−57 °C); galactose (−56 °C); glucose (−57 °C); lactose (−41 °C); sucrose (−46 °C); and trehalose (−40 °C). Values of T'_g for fruits such as apple (−40 °C); tomato (−41 °C); strawberry (−41 °C) were reported by Fennema (1996).

Roos (1995) reported the following average values of T'_g (°C) in fruits: apple (−42 to −41), banana (−35), blueberry (−41), peach (−36.5); strawberry (−41 to −33.5).

In the case of fruit juices the average T'_g values are: apple (−40.5 °C); lemon (−43), orange (−37.5), pear (−40), pineapple (−37.5), prune (−41), strawberry (−41), white grape (−42.5).

The cryostabilization of frozen foods is related to the possibility of maintaining the products below the glass transition temperature of the freeze-concentrated matrix (T'_g), or modifying the formulation of the food to increase the glass transition temperature above normal storage temperatures. If a product is stored at a temperature below T'_g, it may be expected to be composed of ice and a freeze-concentrated phase in the glassy state, and should exhibit long-term stability. If the storage temperature is between T'_g and T_m, the freeze-concentrated phase is not in the glassy state, it is more diluted (rubbery) and processes governed by diffusion are not inhibited; these processes can lead to deterioration during storage (Fennema 1996). Frozen foods stored below T'_g are considered stable to physical changes governed by diffusion, such as ice recrystallization, etc.

6.2.7 Freezing Rate

A simple definition of freezing rate is a rate of temperature change. In large specimens, temperature gradients are established along the sample and freezing rate is position dependent. The surface is cooled faster than the center and supercooling may be observed before nucleation. High freezing rates are commonly observed on the surface in contact with the refrigerant, decreasing towards the thermal center (Zaritzky 2000). In very small specimens, temperature gradients can be neglected and all the points in the sample will have similar freezing rates.

Freezing rate is one of the main factors that affect food quality because it determines ice crystal sizes. The higher the freezing rate, the greater the nucleation velocity, and thus the greater the number of smaller crystals. For a given position in a specimen, the freezing rate can be represented by the characteristic local freezing time (t_c), the time needed to change the temperature from the initial freezing point to a temperature at which, for example, 80 % of the total water content is converted into ice (Bevilacqua et al. 1979; Bevilacqua and Zaritzky 1980). However, this definition is of limited application because this rate varies with the position in the system; it is large at the surface near the refrigerated border and lower inside the product.

A better definition of freezing rate might be to consider the average rate of ice formation, or the rate of advance of the freezing interface, which is related to the rate of heat removal. The freezing process is for practical purposes complete when most of the freezable water at the thermal center of the product has been converted into ice. According to the International Institute of Refrigeration (1986), the freezing rate of a food can be defined as the ratio between the minimum distance from the surface to the thermal center, and the time elapsed between the surface reaching 0 °C and the thermal center 10 °C colder than the temperature of initial ice formation.

6.2.8 Structure of Vegetable Tissue

The epidermis of plant tissues is structurally adapted to provide protection against biological and physical stress and consists of tightly packed cells containing waxy material. The parenchymatous tissue performs much of the metabolical activity of the plant and is constituted of semirigid, polyhedral cells with cellulose walls bounded by pectinaceous middle lamella and often including a network of air spaces. Mature plant cells contain a number of organelles, such as chloroplasts, chromoplasts, large vacuoles, protein bodies, amyloplasts, and starch granules. All protoplasts of higher plant cells are separated from the cell walls by a cell membrane of granular structure, the plasmalemma of about 100 Å thick. The functions of the plasmalemma are to control the passage of water and solutes in and out of the cell and to catalyze the degradation of specific substrates. The protoplast is composed of protoplasmic components (cytoplasm, nucleus, plastids, and other organelles), and non-protoplasmic components such as vacuoles, crystals, starch granules, and oil droplets (Fennema et al. 1973; Zaritzky 2011).

The cytoplasm is a continuous viscous fluid or a gel acting as a matrix for organelles and other particulate matter. The solutes in solution are proteins, sugars, polysaccharides, and inorganic ions. In vacuolated mature cells, the cytoplasm is confined as a thin layer next to the inside surface of the cell wall. In actively dividing cells, the cytoplasm occupies most of the cell volume; the water content of active cytoplasm is about 85–90 %. The cells of fleshy parenchyma tissue of fruit often are not in perfect contact, thus leading to intercellular gas spaces. For example in mature apple tissue, 20–25 % of the total tissue volume is constituted by intercellular gas spaces; in mature peach tissue this value is of 15 %.

The vacuole, which may comprise most of the mature plant cells, contains organic acids, phenols, and hydrolytic enzymes that can be released when the fragile membranes are disrupted by freezing. Firmness and crispiness (textural properties associated with fruits) are attributed to the osmotic pressure developed within the cell when pressure is exerted on the rigid cell walls.

6.2.9 Intracellular and Extracellular Ice Formation

In food tissues, barriers to water movement such as cell membranes introduce complexity to the freezing mechanism because the internal and external environments have to be considered (intracellular and extracellular regions of the tissue) (Reid 1996).

Freezing food tissues can lead to extracellular ice formation and also, under some conditions, to intracellular ice formation. Membrane permeability and the internal properties of the cell are important factors that affect ice formation. The location of ice crystals in food tissues is a function of the nature of the cells, the freezing rate, and the temperature of the specimen. It is generally accepted that crystallization, regardless of freezing rate, starts in the extracellular fluid (Fennema et al. 1973; Mazur 1970). Intracellular nucleation of ice in the tissue will therefore not occur until a high degree of supercooling is reached. However, intracellular nucleation is produced when the rate of heat removal is high enough to eliminate the heat of crystallization of the extracellular ice nuclei and to produce a high degree of supercooling inside the cells.

Slow freezing generally causes ice crystals to form exclusively in extracellular areas. As cells contain a higher concentration of non-diffusible ions than the surrounding fluid, the total concentration of ionic particles will be greater inside the cell than in the extracellular space, and a lower freezing point would be expected for the intracellular space. Supercooling in intracellular spaces is then minimized, decreasing the probability of intracellular nucleation (Zaritzky 2011).

The rate of ice crystal growth depends on the rate of heat removal and the diffusion of water to the surface of the growing crystal. When ice starts to be formed in the extracellular space, solute concentration increases and water activity decreases in the unfrozen external region. Since water activity for the intracellular fluid at any given temperature is higher than that of the extracellular fluid, water diffuses from the cells and is deposited on the extracellular ice crystals in order to equilibrate the chemical potential in both fluids. Slow freezing results in considerable shrinkage of the cells and the formation of large extracellular ice crystals. In contrast, tissues and cellular suspensions that are frozen rapidly at very low temperature show both intracellular and extracellular ice crystals with a uniform distribution. Rapid freezing produces intracellular crystallization and results in numerous small ice crystals, minimum dislocation of water, and in the case of food systems, an appearance which is similar to the original unfrozen system (Bevilacqua et al. 1979; Bevilacqua and Zaritzky 1980).

The formation of intracellular ice is affected by several factors. One factor is cell permeability, which controls the loss of water through the membrane to the external environment when the osmotic gradient is established. The migration of water from the intracellular space increases the internal solute concentration, reducing the internal freezing point and the degree of intracellular supercooling. A membrane with high water permeability prevents intracellular freezing and sustains high supercooling.

At low freezing rates, the rate of change in concentration in the external unfrozen matrix is slow and water can migrate from the interior of the cell; under these conditions, the cell dehydrates and water is deposited on the external ice crystals (Reid 1997).

When the freezing rate is high and water permeability is low, the solute concentration of the extracellular unfrozen matrix increases rapidly. However, since water cannot be transferred rapidly, the intracellular region becomes increasingly supercooled. At some critical supercooling point, the internal contents will freeze, with low water transfer from the intracellular space. When freezing is fast and water permeability high, water migrates from intracellular to extracellular spaces. In this case, the cell dehydrates and intracellular freezing is not observed (Reid 1997).

Intracellular freezing is favored by rapid cooling to a low temperature so that the opportunity for cellular dehydration is minimized. Under these conditions, there is a high probability of intracellular ice nucleation or the growth of extracellular ice crystals through the cell membrane. Cell membranes act as effective barriers to crystal growth at high subfreezing temperatures, such as those encountered during slow freezing, whereas during rapid cooling to some critical low temperature (in the neighborhood of −10 °C), the barrier properties of membranes tend to disappear (Mazur 1970).

In fruit tissues, external water (surrounding metabolically active cells) freezes before the internal water of the protoplast. An intact plasmalemma restricts nucleation of intracellular ice at temperatures above −6 °C. At temperatures below −6 °C, the critical radius of ice may be smaller than the radius of the pores in the plasmalemma, this situation promotes the formation of intracellular ice because extracellular ice is enabled to grow through the intact plasmalemma (Fennema et al. 1973).

As mentioned, freezing of large pieces of tissue involves thermal gradients; freezing rates are higher near the surfaces in contact with the refrigerant medium and decrease towards the thermal center of the sample (Bevilacqua et al. 1979; Bevilacqua and Zaritzky 1980). Therefore, intracellular ice is only produced in a narrow zone adjacent to the area in contact with the cooling medium, which experienced high freezing rates; supercooling and nucleation takes place only in the near-surface layers of the food; in the inner zones, freezing proceeds more slowly (Zaritzky 2010).

The existence of intracellular ice constitutes an index of high freezing rates. Ice crystals nucleated in the refrigerated surface can grow towards the thermal center of the product. As the freezing rate decreases, intracellular ice disappears, and only the growth of extracellular crystals can be observed, at the expense of the intracellular water (Bevilacqua and Zaritzky 1980). Due to this dehydration process, the shape of the cells becomes irregular and distorted.

6.2.10 Freezing Times

Freezing time is often defined as the time necessary to cool the slowest cooling location from the initial temperature to a defined final temperature. Accurate predictions of freezing time are necessary to assess food quality, processing requirements, such as the minimum time the product should remain in a continuous freezer, and economic aspects of the freezing process. Since freezing is an unsteady state heat transfer process in which latent heat is released over a range of temperatures due to change of phase, it does not occur at a unique temperature.

The freezing time depends on factors directly related to the object to be frozen (size, geometry, surface area, initial and final temperatures of the product, thermophysical properties, type of packaging), and factors which are characteristic of the freezing system, such as the temperature of the cooling medium and the heat transfer coefficient in the freezing equipment.

Freezing rate depends strongly on thermophysical properties of the fruit: density, apparent specific heat (that includes the phase change) and thermal conductivity.

Changes in the frozen water fraction with temperature affect all thermophysical properties of the frozen product. Knowledge of properties such as thermal conductivity, density, and specific heat is essential when designing a freezing process. The principal feature of thermophysical properties in the frozen range is their strong dependence on temperature; this is due to the varying proportion of water converted into ice and the large difference between the properties of ice and liquid water.

6.2.11 Freezing Equipment

Freezing systems can be grouped according to the heat transfer medium (IIF 1986): Cold Air (Still or forced air freezers); Metal Contact freezing (Plate freezers); Liquid (Immersion freezers); Liquefied gases: nitrogen or carbon dioxide (Cryogenic freezers).

Cold air: Still or forced air is used as the freezing medium. It is one of the commonly used freezing equipment due to the versatility for several product types. Freezing is accomplished by placing the food in freezing rooms. Still air freezing is the slowest method of freezing due to the low surface heat transfer coefficient of circulating air inside the room (ASHRAE 2010).

The forced air freezers (blast freezers) consist of air circulation by convection inside the freezing room. There are different arrangements for air-blast freezers, primarily grouped in batch or continuous operation. Batch air-blast freezer is the simplest common form of forced convection freezer. It consists of an insulated room or cabinet containing electric fans that force the air over refrigerant evaporator coils and then circulate it over the food products that are stacked in racks to ensure that cold air pass over the surface of the products. It is important to assure a good contact between the product and the air. In batch blast freezers the air can be directed by the use of turning vanes or slotted ceilings which spread the air flow (North and Lovatt 2012).

The operating conditions for air-blast freezers are between −18 and −40 °C and air velocities up to 20 m/s. Continuous air-blast freezers include belt-type freezers, tunnel freezers, and spiral freezers. A moving belt or moving shelf system transports the products through an environment with cool air circulating at high velocity. In tunnel freezers, the products on trays are placed in racks or trolleys; trays with food are stacked on trolleys or the food is moved through a freezing tunnel by a series of stainless steel mesh belts. In belt freezers the product usually travels horizontally through a long enclosed space (freezing tunnel) and coldest air is circulated upward through the product. The moving belt is formed by a continuous flexible mesh that may be linear or spiral. Sometimes multiphase tunnels are applied with a number of belts. The product falls from one belt onto another; this also breaks up clumps of frozen food. The thickness of the food layer on the belts can vary from 25 to 125 mm. In blast freezers, large volumes of air are recycled, this can cause freezer burn and oxidative changes to unpackaged food. Moisture from the food is transferred, via the air, to the refrigeration coils, which makes frequent defrosting necessary (Miller and Butcher 2000).

Impingement freezer: it is a type of blast freezer designed with high air velocities to give quick freezing in cases where internal heat transfer is not the limiting factor. This design gives high air velocities at right angles to the product surface, on both sides of the belt (Miller and Butcher 2000). Impingement freezers use numerous jet nozzles to direct air onto the surface of food products at very high velocities. The process is usually continuous with air jets positioned above and below a mesh conveyor system (North and Lovatt 2012). The air flow direction is usually perpendicular to the product surface, disrupting the boundary layer surrounding the product and increasing the surface heat transfer coefficient. Cost-effective applications of this method are limited to thin products (less than 25 mm thick) (ASHRAE 2010). Overall heat transfer coefficients can reach 175 W/m^2 K.

Fluidized bed freezer: it is a continuous process that forces cold air up under the product at a given velocity that can fluidize small pieces of food. It consists of a perforated tray or belt through which air at temperatures ranging between of −25 and −40 °C is blown vertically upwards. The air acts as both the cooling and the transport medium; the forced cold air from beneath the conveyor belt, causes the products to suspend or float in the cold air stream. The use of high air velocity is very effective for freezing unpacked foods, especially when they can be completely surrounded by flowing air.

Fluidized beds are characterized by high heat transfer coefficients and good mixing, which ensures uniform temperature distribution and prevents frozen product clumping together. Commercial fluidized bed freezers tend to use a moving belt to control residence time. Product is loaded onto one end of a perforated belt and fluidized by an upward flow of cold air (Miller and Butcher 2000).

The use of fluidization has several advantages compared with other methods of freezing since the product is individually quick frozen (IQF), which is convenient for particles with a tendency to stick together. Individual Quick Freezing is appropriate for small pieces of food, uniform in size, and not prone to damage caused by the high-velocity mixing that occurs in a fluidized bed. This results in a better texture, there is no lump/block formation and the product is free flowing; therefore it is not necessary to thaw or defrost the whole pack to take out only a portion and the rest remains frozen. It is a freezing method of choice for seasonal products such as fruits and vegetables; each piece is frozen individually resulting in freezing times of only 10–12 min which otherwise takes at least 3–4 h or even more in the blast freezer. In this system, food comes in more extensive contact with the air than in blast freezers, so that all surfaces are frozen simultaneously and uniformly.

Indirect contact freezers (*Plate freezers*): The product to be frozen is pressed between hallow metal plates, either horizontally or vertically, with a refrigerant circulating inside the plates at a temperature of about −40 °C. Food packaged or unpackaged is placed between the plates; the plates are pressed slightly together;

this improves the contact between the food and the freezing plates. Heat is extracted by direct conduction through the surfaces which are cooled by a circulating refrigerated medium. They are considered indirect contact freezers because the product is indirectly exposed to the freezing medium; materials being frozen are separated from the refrigerant by a conducting material, usually a steel plate. One advantage of such freezers is that little dehydration of the food takes place, thus reducing the frequency of defrosting. Plate freezers operate by mechanical refrigeration and generally provide an efficient medium for heat transfer, although the system has some limitations, especially when used for packaged foods due to resistance of package to heat transfer. Besides, this type of freezing system is only limited to regular-shaped materials or block-shaped packaged products (Miller and Butcher 2000).

Immersion freezers: The immersion freezer consists of a tank with a cooled freezing media, such as glycol, glycerol, sodium chloride, calcium chloride, and mixtures of salt and sugar. The product is immersed in this solution while being conveyed through the freezer, resulting in a direct heat exchange. The solute used in the freezing system should be safe without taste, odor, color, or flavor, and for successful freezing, the density of the products should be higher than the solution density.

Aqueous solutions of NaCl, $CaCl_2$, or sucrose at low temperature (e.g., from −10 to −40 °C) can be used as immersion fluids. Immersion freezing has recognized advantages; it is one of the fastest freezing techniques, as the heat transfer coefficient is up to 20 times higher in liquid media than in air (Galetto et al. 2010). Moreover it is associated with higher quality of the final product. However, the main disadvantage that reduces its use is the uncontrolled solute uptake from the refrigerated solution into the product (Lucas and Raoult-Wack 1998). Immersion freezing systems have been commonly used for shell freezing of large particles due to the reducing ability of product dehydration when the outer layer is frozen quickly. A commonly observed problem in these freezing systems is the dilution of the solution with the product, which can change the concentration and process parameters. Immersion freezing is appropriate for irregularly shaped foods and may be used in continuous or batch operation.

Cryogenic freezers: In cryogenic freezing, the food is in direct contact with the refrigerant (liquid carbon dioxide at −80 °C or liquid nitrogen, −196 °C). The refrigerant evaporates or sublimates away removing the heat from the food and causing rapid freezing. The product has a temperature of −40 °C or lower in a few minutes. Liquid nitrogen and carbon dioxide refrigerants are colorless, odorless, and inert. In cryogenic freezing the investment costs are low and the freezing rate is very high, however, the cost of the refrigerant is high. The refrigerants used are liquefied in large industrial installations and shipped to the food-freezing factory in pressure vessels. Thus, the small size and mobility of cryogenic freezers allow for flexibility in design and efficiency of the freezing application (Miller and Butcher 2000).

Because the refrigerant is in direct contact with the food, heat transfer coefficients tend to be high, especially in the case of liquid nitrogen. One of the advantages of this technique is the flexibility in responding to changes in production rate. In contrast, in a blast freezer, the amount of heat given up to the evaporator coils is limited by the maximum air velocity inside the freezer that is fixed by the design. Common cryogenic tunnels have a single refrigerant spray zone, and one or more gas transfer fans to move the cold gas along the tunnel to the product inlet. Other equipments have multiple liquid nitrogen spray zones, providing better control and eliminating the fans.

Liquid nitrogen, with a boiling temperature of −196 °C at atmospheric pressure is sprayed into the freezer and evaporates both on leaving the spray nozzles and on contact with the products. The refrigerant passes in counter current to the movement of the products on the belt giving high transfer efficiency. The refrigerant consumption is in the range of 1.2-kg refrigerant per kg of the product. Typical food products used in this system are small pieces of fruits, berries.

Liquid carbon dioxide exists as either a solid or gas when stored at atmospheric pressure. When the pressurized gas is released to the atmosphere at −70 °C, half of the gas becomes dry-ice snow and the other half stays in the form of vapor. This unusual property of liquid carbon dioxide is used in a variety of freezing systems.

Cryomechanical freezing: In this system the product is first exposed to cryogenic freezing and then to mechanical refrigeration (ASHRAE 2010).

Surface heat transfer coefficients in the different freezing systems depend on the fluid dynamics of the cooling medium and the type of freezing equipment (Nesvabda 2008). Values of h range between 6 and 20 W/m^2 K for cold chambers with air in natural convection; 20–90 W/m^2 K for air under forced convection; 75–250 W/m^2 K for air in a fluidized bed; 100 (poor contact due to packaging) to 600 (good direct contact) W/m^2 K in plate heat exchanger; 500–1000 W/m^2 K for immersion in liquid nitrogen.

In commercial practice, freezing rates defined according to the International Institute of Refrigeration (IIR) vary between 0.2 and 100 cm/h; 0.2–0.5 cm/h corresponds to slow freezing (bulk freezing in cold chambers), 0.5–3 cm/h to quick freezing (air-blast and contact plate freezers), 5–10 cm/h to rapid freezing (individual quick freezing of small-sized products in fluidized beds), and 10–100 cm/h to ultra rapid freezing by spraying or immersion in cryogenic fluids (liquid nitrogen, carbon dioxide) (IIR 1986).

6.2.12 *Effect of Freezing and Frozen Storage on Quality Changes in Fruits*

Fruit tissues are very sensitive to freeze damage. Frozen fruits may undergo changes in color, flavor, taste, texture degradation, structural collapse, drip losses, and loss of turgor while thawing. These severe changes in fruit properties are related to the physical and chemical modifications induced by the formation of ice during freezing and subsequent storage of frozen fruits.

The attractive aroma, color, texture, and freshness are important characteristics that are difficult to dissociate from the raw product and hence the negative impact of freezing on the fruit quality is very high.

6.2.12.1 Physical Modifications

Physical modifications during freezing include changes in cell volume, water dislocation, mechanical damage, freeze-cracking, structural collapse, and exudate production during thawing (Zaritzky 2011). Fresh fruits contain a high proportion of water (80–95 %) therefore the formation of ice within the product causes damages in the cellular structure and quality losses. The amount of ice at a given subfreezing temperature depends on its initial content of soluble solids. Fruits with higher values of soluble solids have lower proportions of frozen water at a given subfreezing temperature. During freezing, changes in cell volume are produced because ice expands; however the volume change in the fruits is not uniform; zones containing ice crystals will expand and others will contract, leading to mechanical damage.

Slow freezing of tissues produces extracellular ice and leads to moisture movement through osmosis. When extracellular ice crystals are produced, dehydration and shrinkage of the cells may cause rupture or folding of cell membranes (Reid 1996; Fennema et al. 1973). Ice crystals continue to grow in size, and exert additional stress on fragile cellular structures.

Damage to fruit tissues during freezing and frozen storage can be attributed to microstructure alterations of the middle lamella, cell walls, and cell contents. The development of extracellular ice crystals may cause cell separation in the middle lamella region, causing cell-wall rupture and cell shrinkage. The cellular membranes lose their osmotic status and their semi-permeability (Tregunno and Goff 1996). Besides the metabolic system of the plant tissue is interrupted, dislocation of the enzymatic system occurs and biochemical deterioration reactions are also produced. Maintenance of structural integrity of plant membranes in highly vacuolated cells is important for retention of hydrostatic pressure within the cells, and for the prevention of drip and tissue softness (Fennema et al. 1973).

Texture is an important attribute contributing to the overall quality of fruits. The texture damage in frozen-thawed fruit tissues is attributed to the semirigid nature of the cells. Different from vegetables, fruits do not have a fibrous structure to resist freezing therefore fruits are less resistant to the freezing process than vegetables. Additionally, fruits to be frozen are harvested in a fully ripe state and are soft in texture. Formation of ice in fruit tissues results in undesirable changes in texture and loss of turgor (ability to retain water inside the cells); the tissue loses crispness and becomes soggy.

The cryodamage affects the textural quality of the thawed product not only due to a reduction of cell turgidity but also because of the production of thaw-exudate. Ice crystals that damage the integrity of the cellular compartments produce also high drip losses while thawing. The type and extent of cryodamage to plant tissue is related to the degree of tissue disruption and depends on ice crystal location and size, conditions that are determined by the rate of freezing and frozen storage. Factors that affect exudate production are size and location of ice crystals, rate of thawing, the status of the tissue before freezing, and the water-holding capacity of the tissue; exudate production during thawing leads to loss of nutrients.

The addition of sugar to plant tissue prior to freezing has been reported to decrease this type of cryodamage. Loss of membrane semi-permeability and disruption of cellular compartments in fruits can be minimized using rapid freezing rates.

Cell walls of peaches, strawberries, and raspberries were damaged by ice crystals during the freezing process (Fennema et al. 1973). Strawberries frozen slowly in air had more than 90 % of the cell walls in a fractured state, but with rapid freezing, cell-wall damage was only 10 %. Sterling (1968) subjected apple tissue to slow freezing conditions, noting cell separation, cell compression, and rupture of cell walls. When the rate of freezing of plant tissue was higher than 10 °C/min, intracellular ice crystals predominate (Mazur 1970).

At freezing rates higher than 10 °C/min, intercellular spaces in fruits did not become enlarged, cell walls did not separate or fracture, and the protoplasm remained close to the cell wall since water movement from the cells was restricted. Since intracellular ice should not damage the cell walls or protoplasm to any appreciable extent, rapid freezing should be advantageous in maintaining the textural quality of fruit tissue. Gutschmidt (1968) stated that strawberries frozen slowly in still air at −18 °C were inferior in texture to strawberries frozen in moving air at −40 °C, or in liquid nitrogen.

The loss in the characteristic crispness and turgidity in frozen fruits are to some extent reduced by selecting varieties which are resistant to these modifications (IIR 1986).

Many studies have focused on the effects of freezing on textural quality of fruits including cherries (Alonso and Canet 1994), raspberries and blackberries (Sousa et al. 2006, 2007), and strawberries (Van Buggenhout et al. 2006; Delgado and Rubiolo 2005) through mechanical and/or sensorial measurements of frozen/thawed fruits or through microscopic measurements.

As was remarked, most fruits and vegetables benefit from quick freezing, because the formation of small ice crystals contributes to a homogeneous structure which maintains the textural quality of the tissue.

However, some frozen fruits may suffer freeze-cracking when they are submitted to very high freezing rates, or very low temperatures, such as in cryogenic fluids.

Kim and Huang (1994) suggested that the crust formed on the surface of a product during freezing serves as a shell that prevents further volume expansion when the internal portion of the unfrozen material undergoes the phase transition. If the internal stress is higher than the strength of the frozen material, the product will crack during freezing. Systems with high void spaces show a higher probability that internal stress will dissipate, reducing the possibility of freeze-cracking. Precooling prevents freeze-cracking because it reduces the differences in temperature between the product and the freezing medium. Precooling also reduces the time delay between the freezing of the border and the center of the system; thus the center of the food expands during ice formation at an earlier stage. If the phase change of the core region occurs before the surface becomes brittle, food products can support the internal pressure and freeze-cracking is avoided. Rapid freezing coupled with low final temperatures will nearly always result in severe cracking of specimens containing large percentages of water. The cracking is probably the result of nonuniform contraction following solidification.

During frozen storage, several physical problems occur such as moisture migration, and ice recrystallization. Moisture migration and relocation of the water, both within and from the product is produced during frozen storage due to the existence of temperature gradients that create water-vapor pressure profiles. There is an overall tendency for moisture to move into the void spaces around the foodstuff and to accumulate on the product surface and on the internal package surface. In packaged frozen food, moisture migration leads to ice formation inside the packaging (Pham and Mawson 1997). This is a consequence of the temperature dependence of water vapor pressure; water vapor will tend to transfer from regions of high-vapor pressure to regions of low-vapor pressure. Temperature fluctuations (cooling–warming cycles) produce a net migration of moisture from the interior towards the surface of the foodstuff, or to the wrapping. The temperature of the packaging material responds to temperature fluctuations in the storage room faster than the product. As the surrounding temperature decreases, moisture inside the pores sublimes and diffuses to the packaging film; when ambient temperature increases, the ice on the wrapping tends to diffuse back to the surface of the food, however, reabsorption of water in the original location is impossible, and the process can be considered irreversible, producing undesirable weight losses. Moisture migration can be minimized by keeping temperature fluctuations and internal temperature gradients small, and by the inclusion of internal barriers within the product and within the packaging. Weight losses during freezing and frozen storage have economic consequences, unless the product is packaged in films of low-water-vapor permeability.

Recrystallization of ice is the process by which the average ice crystal size increases with time. Slow freezing results in the production of a small number of large ice crystals, whereas fast freezing leads to the formation of a large number of small ice crystals. However, during frozen storage, ice crystals undergo metamorphic changes. Small ice crystals are thermodynamically unstable, having a high surface/volume ratio and therefore a high excess of surface free energy. To minimize free energy, the number of crystals decreases at constant ice phase volume, but their mean size increases. Recrystallization basically involves small crystals disappearing, large crystals growing and crystals fusing together. It affects the quality of the products because small ice crystals help preserve quality, while large crystals often produce damage during freezing. Crystal size has an effect on its melting point; the melting point of a crystal is a function of its radius of curvature; for a given temperature there will be a critical radius which defines the minimum size that a crystal can have and still be stable. At the surface of the crystal, there is a constant interchange of water molecules between the solid and liquid phases. If the crystal surface is planar, then the number of molecules which leave the crystal will be equal to the number of molecules which will join it. If there is a corner on the crystal, with a given curvature, then the number of molecules leaving and joining will not be equal. The molecules that are part of the crystal at the corner will be less strongly joined to the crystal because they do not have as many neighbors to bond with, and so they are more easily removed from the crystal. At the same time, molecules from the liquid are less likely to join the crystal at the corner. This leads to a net loss of molecules from the corner, even though there is equilibrium at the planar crystal surface (Zaritzky 2000, 2008, 2011).

Recrystallization reduces the advantages of fast freezing and includes a number of changes in the number, size, shape, orientation, or perfection of crystals following initial solidification (Fennema et al. 1973).

As the temperature of an aqueous specimen increases within the subfreezing range, the rate of recrystallization also increases. Migratory recrystallization or grain growth refers to the tendency of large crystals to grow at the expense of the smaller ones. Melting-diffusion-refreezing and sublimation-diffusion-condensation are possible mechanisms leading to an increase in the average crystal size, a decrease in the number of crystals, and a decrease in surface energy of the entire crystalline phase. At constant temperature and pressure, migratory recrystallization is the result of differences in the surface energies of large and small crystals. Small crystals, with a very small radii of curvature, cannot bind the surface molecules as firmly as larger crystals, thus, small crystals exhibit lower melting points than large ones. Migratory recrystallization is enhanced by temperature fluctuations, which induce a melt-refreeze behavior due to ice content changes. Melt-refreeze behavior can lead to complete disappearance of smaller crystals during warming and growth of larger crystals during cooling, or to a decrease in the size of crystals during partial melting and re-growth of existing crystals during cooling. Melt-refreeze should occur to a greater extent at higher temperatures, and more rapidly for smaller crystals.

Recrystallization was studied in solutions and in different food systems (Donhowe and Hartel 1996; Sutton et al. 1996). Rates of ice recrystallization in frozen solutions and in frozen tissues were reported by Bevilacqua and Zaritzky (1982), and Martino and Zaritzky (1987, 1988, 1989). In these studies, it was proposed that the driving force for recrystallization of ice is the difference in surface energy of two adjacent crystals, with this energy being proportional to the crystal curvature. Ice crystal size distributions were measured from micrographs and a direct relationship between crystal size and the number of crystal sides was established. Small crystals with three or four sides show concave surfaces, and tend to disappear because the crystal boundaries move towards the center of curvature. Ice crystals with six sides have planar surfaces and are stable, and those with a higher number of sides tend to grow. Ice crystals in frozen plant tissues usually increase in size during frozen storage and in the early stages of thawing. Recrystallization can have a profound damaging effect on intact cells and can lead to changes in the microstructure and a loss of tissue firmness.

6.2.12.2 Chemical Modifications

The most common chemical changes related to quality deterioration in frozen fruits are reactions related to pigment degradation, enzymatic browning, production of off-odors and off-flavors, and the autoxidation of ascorbic acid. During freezing, the increasing concentration of solutes in the unfrozen matrix increases the ionic strength and can produce modifications that affect biopolymer structures. Besides, several properties of the unfrozen phase change; such properties include: pH, titratable acidity, and oxygen-reduction potential (Fennema et al. 1973).

Freezing contributes to cell rupture in the fruit tissue, so interactions between previously separated substances become possible. Unfrozen water is involved in deterioration reactions during freezing and frozen storage. The temperature and concentration of reactants in the unfrozen phase (freeze concentration effects) are the main factors responsible for changes in the reaction kinetics during freezing.

Chemical reactions slow down at low temperatures, but they continue during frozen storage conditions. In fruit tissues, the formation of ice crystals can release enzymes and chemical substances from cell organelles; enzymes come into contact with different substrates, leading to quality deterioration. Most enzymes exhibit substantial activity after freezing and thawing and many enzymes show significant activity in partially frozen systems. Some endogenous enzymes are responsible for undesirable changes such as off-flavors and off-odors, color and nutritive alterations during frozen storage of fruits.

Disruption of native cells by ice crystals can initiate enzymatic browning. In the presence of oxygen, certain frozen fruits undergo oxidative discoloration due to the action of *o*-diphenol oxidase (polyphenoloxidase, phenolase, tyrosinase) on phenolic compounds.

When raw peaches, pears, cherries or apples are frozen, stored, and thawed, undesirable brown pigments are formed. In pears and apples, chlorogenic acid and catechins are the major substrates of naturally occurring *o*-diphenol oxidase (polyphenoloxidase, phenolase, tyrosinase) (Fennema et al. 1973).

Browning of frozen plant tissue is usually most severe near the surface, because of the higher atmospheric oxygen concentration, than in the internal tissue. Internal as well as surface browning may occur in some frozen-thawed products such as the apple tissue since the numerous intracellular spaces generally contain some oxygen. Brown discoloration in frozen plant tissues can be minimized or prevented by heat inactivation of the enzymes (blanching), addition of browning inhibitors such as ascorbic acid, and exclusion of oxygen. Thermal inactivation of *o*-diphenol oxidase in pear tissue occurs slowly at temperatures between 50 and 80 °C and rapidly at temperatures of 90 °C and above (Fennema et al. 1973). Inactivation of *o*-diphenol oxidase in apple slices for pies can be achieved without a serious loss of quality by simply blanching prior to freezing. The loss of water-soluble minerals and vitamins during blanching should also be minimized by keeping blanching time and temperature at an optimum combination. Fruits such as peaches may be blanched and then cooled in cold water to make their skins loose and easy to slip off.

However blanching produces detrimental effects to fruit tissues, therefore, only a few types of fruits are blanched for inactivation of enzymes prior to freezing.

Most of the fruits cannot be subjected to a blanching treatment due to their tissue sensitivity, therefore, alternative pretreatments have to be used such as chemical treatments and the use of additives. Antioxidants such as ascorbic acid can be used to inhibit enzymatic reactions. At low concentrations, ascorbic acid does not inhibit the activity of *o*-diphenol oxidase but rather acts as a reducing agent to keep the oxidizable polyphenols in a reduced state. At high concentrations, the ascorbic acid maintains the polyphenols in the reduced state. In tissue with an active *o*-diphenol oxidase system, ascorbic acid is gradually oxidized and eventually the polyphenolic compounds are oxidized. Therefore under adverse conditions of frozen storage, ascorbic acid can be gradually lost and discoloration of tissue occurs.

Ascorbic acid has been used as an additive to sugar syrup of frozen cut fruits for restricting surface browning. Concentrations of 300–500 mg of ascorbic acid per kg of final product are usually adequate. The extent and intensity of browning of sliced frozen-thawed fruits such as peaches were inversely related to ascorbic acid content. Peach slices submerged first in ascorbic acid-containing syrup and then exposed to air at −6.7 °C, exhibited brown discoloration, within a few weeks, in contrast peach slices that remained completely submerged in syrup at the same temperature did not change in color during a 65-day storage period. Coverage of the slices with sugar syrup (50–55 %) restricted oxygen migration into the fruit tissue (Fennema et al. 1973; Guadagni 1969).

Several studies have demonstrated the influence of oxygen content in the head space on the rate of browning. Red sour cherries discolored when they were exposed to air in the head space during frozen storage for 50 days at −6.7 °C. The rate of browning of these fruits decreased when immersed in a 60 % sucrose syrup in hermetically sealed containers at the same temperature. However, if the storage temperature was maintained at −18 °C or lower, the type of container was not an important factor in the prevention of browning.

Enzymes can also affect the texture of fruits; exposing the cell wall to hydrolytic enzymes that attack pectins, hemicelluloses, and non-cellulose carbohydrate material constituents would dissipate the osmotic pressure (Sista et al. 1997). Hydrolytic enzymes, like chlorophylases and anthocynases present in plants, may catalyze the destruction of pigments in frozen tissues, which affects the color. The partial loss of anthocyanin pigments in frozen berries can be decreased by packaging the fruits with sugar or syrups; this method improves not only the color but also the flavor. Oxidative flavor deterioration can occur in non-blanched fruits during frozen storage. Pigment degradation and color quality deterioration are also related to lipid oxidation. In the case of frozen fruits and vegetables, chlorophyll can serve as a secondary substrate in lipid oxidation.

Apricots, peaches, cherries, and plums develop off-flavor during frozen storage; many of these changes are due to lipases and lipoxidases.

Several authors (Cano et al. 1990; Cano and Marin 1992; Marin et al. 1992; Skrede 1996; Moraga et al. 2006) have investigated the chemical modifications during freezing which are responsible for pH, soluble solid, water content, or color changes of frozen/thawed fruits.

Bunger et al. (2004) studied physical and chemical changes related to frozen fruit quality after freezing of apples; Botosoa et al. (2004) analyzed mangoes and González et al. (2002) worked on raspberries.

Botosoa et al. (2004) studied the effect of the freezing rate on mango quality (texture and color) and showed that rapid freezing (at −40 °C) yields a better preservation of the fruit than slow freezing (at −18 °C).

Chassagne-Berces et al. (2010) analyzed the effect of different freezing methods (at −20 °C in a cold chamber, at −80 °C in gas nitrogen convection chamber and after immersion in liquid nitrogen at −196 °C) on the quality of two fruit types (apple-*Malus domestica* Borkh- and mango *Mangifera indica* L. cv. 'KENT'-), two varieties of apples (*Golden Delicious* and *Granny Smith*) and two maturities of one variety (ripe and unripe *Granny Smith*). They measured texture, color, soluble solids, water activity, water content, pH, titrable acidity before and after the three different freezing protocols. Freezing induced significant changes mostly in texture, color, and soluble solids. Property variations due to freezing depended mainly on the type of fruit, to a lesser extent on the variety of apples studied and slightly on the maturity of *Granny Smith* apples. Among the freezing conditions tested, the authors reported that gas nitrogen convection at −80 °C was the best choice for limiting fruit quality degradation.

6.2.13 Nutritional Quality of Frozen Fruits

Freezing is considered to deliver a product comparable to the fresh product in nutritional quality. Available experimental data tend to show that freezing is less destructive than other preserving methods. The degradation of vitamins during the freezing process has generally a more significant impact on nutritional value. The main adverse effect of extended frozen storage on nutritive value may be the losses of the more labile vitamins, vitamin C (ascorbic acid), that are frequently used as indicators of the food processing effect (Jul 1984). Ascorbic acid losses have been studied in fruits and vegetables and are attributed to oxidative mechanisms during frozen storage. In the presence of dissolved oxygen, ascorbic acid in aqueous solution is oxidized to dehydroascorbic acid which is then oxidized irreversibly to 2,3-diketogulonic acid and its degradation products. Ascorbate oxidase exists naturally in many plant tissues, and if it is not inactivated, it catalyzes ascorbic acid oxidation during freezing.

When fruits are frozen in hermetically sealed metal containers with very little headspace oxygen, no appreciable amount of ascorbic acid is oxidized. However when fruits are packaged in containers having certain oxygen permeability, substantial deterioration of ascorbic acid during frozen storage is produced. The rate of ascorbic acid oxidation depends on the temperature of frozen storage and the pH of the fruits. As the pH is lowered, the stability of ascorbic acid in food generally improves (Fennema et al. 1973).

6.2.14 Microbial Stability of Frozen Foods

Microbial deterioration is not a problem in frozen foods because they are stored at temperatures below the lower limits of microbial growth (approximately −10 °C). However, with temperature fluctuations during storage and distribution, it may become significant. The major objective of freezing as a method for food preservation is to prolong storage life by retarding or inhibiting microbial growth. Freezing (and the subsequent frozen storage) can be lethal to some microorganisms, but this process is very slow and variable, depending on the type of foodstuff. Freezing cannot therefore be regarded as a method for reducing microbial contamination and for this reason, hygienic and sanitary conditions prior to processing are very important. Storage temperatures below −10 °C inhibit bacterial growth, whereas yeasts and moulds cannot multiply below −12 °C and −18 °C, respectively (Zaritzky 2000, 2008).

6.2.15 Preparatory Operations for Freezing

The characteristics of the raw materials (fruits) are of primary importance in determining the quality of the frozen product. The quality of the raw material depends on agronomic factors: production, sensitivity to disease, and suitability for mechanical harvesting. The quality deterioration of fruits can be minimized by cultivar selection suitable for freezing, determination of the optimum harvesting condition and with the application of pretreatments before freezing (Maestrelli 2000). Most fruits to be frozen are received directly from harvest. Freezing preservation of fruits can only help retain the inherent quality present initially in a product since the process does not improve the quality characteristics of raw materials. The quality level of the raw fruits is very important to obtain a good frozen product. Conventional preparatory operations include cleaning, washing, rinsing, sorting, peeling, slicing, and cutting depending on the type of fruit to be frozen and on the final use of the product. Berries and cherries are best frozen soon after harvest. Peaches, apricots, plums, apples, and pineapples may need to be held after harvest to fully ripen before freezing (Kendall 2008).

6.2.16 Pre-freeze Treatments

The use of pre-freeze treatments can help to reduce the physical and chemical changes that affect the quality of frozen fruits, either by inactivating the deterioration reactions, or by reducing the water content in the material (Torreggiani et al. 2000).

As it was mentioned, browning and off-flavors can be avoided for most products by blanching or application of chemical treatments, such as soaking in antioxidant solutions, before freezing. However, heat treatments such as blanching can be detrimental to food texture and nutritional quality in certain types of fruits. Many studies have focused on enhancing the quality of frozen/thawed fruit through pretreatments, such as the use of sugar syrups or the application of dehydrofreezing.

6.2.16.1 Pretreatments of Fruits Using Sugar Syrups

Addition of sugars is a traditional and important pretreatment for fruits prior to freezing since it has the effect of excluding oxygen from the fruit, which helps to retain color and appearance (Barbosa Canovas et al. 2005). Sugars when dissolved in solutions act by removing water from cells by osmosis, resulting in very concentrated solutions inside the cells. The high concentration of solutes depresses the freezing point reducing the ice content within the cells, and then the excessive structural damage. Sugar syrups in the range of 30–60 % sugar content are commonly used to cover the fruit completely, acting as a barrier to oxygen transmission and browning. Most fruits have a better color, flavor, and texture if packed with sugar or syrup. Some freeze well without sugar, especially those with a firm, waxy skin such as blueberries, huckleberries, cranberries, and grapes (Kendall 2008).

6.2.16.2 Dehydrofreezing

Dehydration can be applied as a pretreatment before freezing, removing a part of the water from the product in order to decrease the amount of crystals formed during the freezing process. Minor damage of the cellular membranes occurs and therefore fruit properties are better conserved. The process in which a food is dehydrated to a given moisture and then frozen is known as dehydrofreezing (Robbers et al. 1997; Spiazzi et al. 1998; Li and Sun 2002; Talens et al. 2003).

Dehydrofreezing has the following advantages over conventional freezing: (1) a reduction in moisture content would reduce the amount of water to be frozen, thus lowering refrigeration load during freezing (2) a reduction of distribution, storage and packaging costs; (3) better quality and stability (color, flavor) of the frozen product and (4) better thawing behavior (lower drip loss).

These techniques include partial air-drying, osmotic dehydration, and immersion chilling and freezing in concentrated aqueous solutions. The most commonly used techniques to reduce water content before freezing in fruits are partial air-drying and osmotic dehydration. Another method consisting in immersion chilling and freezing in concentrated aqueous solutions that makes it possible to combine osmotic dehydration with precooling was reported by Torreggiani et al. (2000).

Partial air-drying: When using this method, food ingredients of high water activity ($a_w > 0.96$) are generally obtained, since water removal is limited to 50–60 % of the original content. To avoid browning during air-drying, blanching, or other treatments such as dipping in antioxidant solutions (ascorbic or citric acid) can be used. Dehydrofrozen fruits may be used as fresh substitutes in frozen fruit salads, surface garniture or filling for tarts and pies. The wetting effect of the fruits is reduced, due to the lower water content. Such products are suitable for pastry foods where wetting has to be avoided and for yoghurt preparations where they can absorb moisture, avoiding the separation of whey. Partial water removal from the food prior to the freezing process leads to a higher concentration of cytoplasmatic components within the cells and to a decrease of the freezing point. Thus, a lower proportion of ice crystals to unfrozen phase, with a consequent reduction of structural and sensory modifications can be observed. Partial air-drying has proved to be effective for apple, pear and clingstone peach. However, the color of the fruit may be affected by heat during the pre-freeze air-drying. In such cases, air-drying can be replaced by osmotic dehydration. Other fruits such as kiwi and strawberry are susceptible to air-drying and other methods such as osmotic dehydration may be applied in these cases.

Osmotic dehydration: Conventional air-drying can be substituted by (or combined with) osmotic dehydration as a pre-freeze treatment. This process involves placing the fruit into solutions of high sugar concentration (0–70 g solute per 100 g solution) at a temperature range between 30 and 80 °C (Le Maguer 1988; Torreggiani et al. 2000). During soaking, the food undergoes simultaneous partial dehydration and solute penetration from the solution into the food. As compared to air-drying, osmotic dehydration presents many advantages such as processing away from oxygen, lower temperatures, shorter duration, etc. and the effect of added solutes of nutritional and functional properties.

Depending on the solutes used in the concentrated solutions, osmotic dehydration can lead to softer texture, improved pigment stability during storage and higher vitamin content, as compared to freezing alone or dehydrofreezing. Osmotic dehydration was reported as a practical pretreatment; during osmotic processing, water flows from the product into the osmotic solution, while osmotic solute is transferred from the solution into the product. Other mass transfer processes occur, such as leaching of small amounts of product-soluble compounds (sugars, acids, minerals, and vitamins), that may affect the sensory and nutritional characteristics of the product (Dixon and Jen 1977).

During immersion in these concentrated solutions an important outflow of water from the food is produced and a simultaneous transfer of solute from the concentrated solution into the food is observed under practical conditions, with an osmotic treatment of 1–2 h at ambient temperature, a solid gain of up to 5–10 % can be attained (expressed as gram of dry solids gained per 100 g of initial fresh product). This gain corresponds to a 50–100 % increase, if referred to an initial soluble solid content of 10 % (Talens et al. 2003). Osmotic dehydration has advantages over air-drying, such as adaptability to a wider variety of products and less-energy requirement. However, care should be taken when choosing the aqueous solution of high osmotic pressure since solute uptake often leads to substantial modification of the product composition. Sucrose, is used for osmotic dehydration of fruits; other osmotic agents include glucose, fructose, lactose, maltodextrin, corn syrup, etc. (Li and Sun 2002). It is possible to adapt the functional properties of the dehydrofrozen fruit, and to formulate new fruit products suitable for various industrial uses, by adjusting the physicochemical composition of the final product adding agents that decrease the water activity; incorporating additives with antioxidants, or other preservative properties (sugars, ascorbic acid, sulfur dioxide, etc.) into the food prior to freezing; adding solutes of nutritional or organoleptic interest (Torreggiani et al. 2000).

Owing to the soluble solid intake, the overall effect of osmosis is a decrease in water activity, with only a limited increase in consistency. Consistency can be associated with the plasticizing and swelling effect of water on the pectic and cellulosic matrix of the fruit tissues, however it depends mainly on the insoluble matter and water content (Erba et al. 1994; Torreggiani and Maestrelli 2006).

Compared to simple air dehydration, the combination of osmosis and air-drying can produce a softer product that is more acceptable as a snack or it can be incorporated into pastries, ice cream, etc. The selection of the solutes depends on taste, cost, and a_w lowering capacity. Sucrose, corn starch syrup at various fructose/glucose ratios, concentrated fruit juices, and other mono- and disaccharides have been used as osmotic solutions. The incorporation of different sugars into kiwi fruit slices significantly influenced chlorophyll stability during storage at −10 °C. This result was confirmed by studying both color and vitamin C retention in osmodehydrofrozen apricot cubes (Forni et al. 1997) and anthocyanin stability in osmodehydrofrozen strawberry.

Different studies have reported the beneficial effects of the osmotic dehydration as a pretreatment of fruits before freezing. Successful applications of dehydrofreezing on fruits such as strawberries and kiwi have been reported by Garrote and Bertone 1989; Robbers et al. 1997; Spiazzi et al. 1998. Garrote and Bertone (1989) found that strawberry halves osmotically treated in the presence of solutions of glycerol, glucose, and sucrose with different concentrations sustained a significantly smaller exudate loss, while untreated, fresh strawberry halves produced a larger amount of exudate (Li and Sun 2002).

Osmotic dehydration increases the cryoprotecting effect of solutes by means of impregnation (Tregunno and Goff 1996; Dermesonlouoglou et al. 2007). It decreases browning (Forni et al. 1997; Marani et al. 2007), structural collapse (Talens et al. 2002) as well as the exudate produced during thawing (Marani et al. 2007). Forni et al. (1997) reported higher ascorbic acid retention in pre-dehydrated and frozen apricots than in apricots frozen by the traditional procedure. Moraga et al. (2006) analyzed the effect of partial dehydration prior to the freezing of strawberries on the final quality of the product. Two processes were utilized to reduce the water content of fruit: osmotic dehydration in sucrose solution at 65 °Bx and air-drying at 45 °C. These authors found that air-drying markedly reduces the liquid exudate; similar results were reported by Maestrelli et al. (2001) in a study about partial water removal in melon before freezing.

Osmo-dehydrofreezing has been studied in several fruits: apricots (Forni et al. 1997), apples (Tregunno and Goff 1996; Marani et al. 2007; Blanda et al. 2008), kiwi fruits (Forni et al. 1990; Robbers et al. 1997; Talens et al. 2002; Chiralt et al. 2001; Marani et al. 2007), melons (Maestrelli et al. 2001), strawberries (Chiralt et al. 2001; Marani et al. 2007; Blanda et al. 2009), pears (Marani et al. 2007), pineapple (Ramallo and Mascheroni 2010).

Forni et al. (1997) studied the osmotic pretreatments of apricot cubes with different syrups (65 % sucrose, maltose, and sorbitol solutions added 1 % ascorbic acid and 0.1 % NaCl) combined with air-dried at 65 °C to improve the quality of frozen apricot. Marani et al. (2007) analyzed the quality of osmofrozen pears, kiwis, strawberries, and apples by measuring color, drip loss, and texture after osmotic dehydration for different periods of time and then after the freezing stage in a conventional air-blast tunnel at −40 °C. The effect of the osmotic solution on the whole process was also investigated using different osmotic solutions (sucrose, glucose, and corn syrup mixtures). The osmotic dehydration prior to freezing demonstrated to be useful for limiting drip loss and, in some cases, to decrease color change and improve texture. The authors indicated that the choice of the dehydrating agent and the usefulness of the pretreatment depend on the application intended for the final product.

Ramallo and Mascheroni (2010) analyzed the effect of: (1) osmotic dehydration in a sucrose solution of 60°Bx at 40°C and (2) hot air-drying, applied previous to the freezing process, on the quality of pineapple slices (mechanical properties, nutritional quality—based on the ascorbic acid content—and liquid loss by exudate). They reported that both, osmotic dehydration and air-drying, applied prior to freezing process, reduced the freezing time of pineapple samples. Ascorbic acid losses were higher during the osmotic dehydration than with air dehydration. Osmodehydrofreezing caused a loss of firmness as well as a pronounced diminution of breaking resistance. Hot air dehydration before freezing markedly decreased exudate volume during the product thawing.

Immersion chilling and freezing in concentrated aqueous solutions (*ICF*) was described by Torreggiani et al. (2000) as another pretreatment method. It is quite similar to osmotic dehydration, in that both involve direct contact between food pieces and a concentrated liquid solution but it is carried out at lower temperatures ranging from –20 °C to 0 °C whereas in the case of osmotic dehydration, operating temperatures range from 30 to 80 °C. Because of low operating temperatures and the freezing process occurring inside food during ICF, mass transfer rates are much lower than in osmotic dehydration. As a result, ICF should be considered as a quick precooling stage, associated with a slight surface formulation effect. Torreggiani et al. (2000) pointed out that this method offers numerous advantages that make it an interesting alternative to conventional freezing techniques such as air-blast freezing: rapid heat transfer, individual freezing, and lower operating and investment costs. Freezing time of small fruits (from 0 to −7 °C) can be reduced by a factor of 4–7 when using ICF instead of air-blast freezing. Consequently a better product quality can be achieved through ICF, as quick freezing preserves the texture of fruit more successfully and causes less dehydration during the freezing process. The addition of sugars and vitamins to the impregnation solution could improve food flavor and its nutritional or color qualities, and also reduce freezing and storage damage. Coating the frozen product with the remaining ICF solution can also help to improve the food color, mainly by increasing food brightness, and to retard food deterioration during frozen storage by creating a protective barrier to oxidation and water losses. In spite of the its potential, this process has not been developed on an industrial scale, mainly because of an inadequate control of mass transfer (water and solutes) between the product and the refrigerating solution (Torreggiani et al. 2000).

Incorporation of salts and enzymes in the pretreatments: Softening caused by freeze–thawing can sometimes be minimized by applying different pretreatments to the tissue such as with calcium chloride and/or sucrose. Immersion in sugar solutions containing hydrocolloids and calcium salts have been recommended as treatments that increase cellular structure resistance to freezing. A large part of these compounds act by interacting with the cell-wall components, maintaining the integrity of the microstructure after thawing. According to Sousa et al. (2007), the calcium influence on the texture of vegetables is due to the formation of calcium bridges among galacturonic acids.

The chelation of calcium ions as cross-links between carboxyl groups of adjacent polyuronide chains was reported as an important factor for the maintenance of fruit firmness. The formation of cation cross-bridges between uronic acids may make the cell wall less accessible to enzymes, that cause softening, or cell-wall-degrading enzymes produced by fungal pathogens (Suutarinen et al. 2002). Main et al. (1986) reported that pretreatments with calcium lactate increased the calcium content of the fruits and slightly enhanced their firmness. Suutarinen et al. (2002) used textural, as well as microscopic studies, to show that, in most cases, $CaCl_2$ pre-freezing treatments stabilized the structure of strawberry tissues. Alonso et al. (1995) reported that freezing resulted in an irreversible loss of turgor in thawed cherries but the treatments with calcium prevented the softness of the fruit produced by freezing and thawing.

Pectin methylesterase (PME) causes de-esterification of pectin molecules and the subsequent formation of calcium bridges between free carboxyl groups of adjacent pectin molecules (Chang et al. 1995). Therefore, addition of exogenous PME and calcium was reported as an useful technique for improving the textural properties of fruit and vegetables (Javeri et al. 1991). Suutarinen et al. (2002) reported that $CaCl_2$ and PME pretreatments, combined with rapid freezing in a liquid nitrogen freezer tunnel, presumably stabilized the original structure of the strawberries. Van Buggenhout et al. (2006, 2008), observed that the suitability of fruit tissue for freezing can be enhanced by vacuum infusion with PME and calcium, depending on the rate of freezing. Galetto et al. (2010) studied the immersion freezing of strawberries (IF) in $CaCl_2$ solutions, analyzing drip losses, pectin content, and the degree of esterification of the pectins, total and cell-wall-bonded calcium contents, the ratio-bonded calcium/total calcium, and textural parameters. The authors reported that the firmness of thawed fruit decreased by approximately 74 % with respect to fresh strawberries, and neither freezing by immersion in $CaCl_2$ solution nor immersion in pectin methylesterase solution followed by IF in $CaCl_2$ solution provided significant benefit in maintaining firmness when compared to slow freezing. They also reported that IF provided significant benefit in reducing drip loss of thawed strawberries in a 51 % compared to slow freezing, but pretreatment with PME did not provide any additional benefit to IF compared to slow freezing. IF in $CaCl_2$ solutions helped to retain the liquid and reduced the migration of moisture from the fruit to the environment during freezing and thawing; however, the retention of liquid did not prevent changes in texture after freezing.

6.2.17 Recommended Packaging and Industrial Freezing Methods for Fruits

Fruits exposed to oxygen are susceptible to oxidative degradation, resulting in browning and reduced storage life of products. Packaging of frozen fruits is based on excluding air from the fruit tissue. Some of the methods used for packaging frozen fruits are: the use of sugars, the replacement of oxygen by inert gas, consuming the oxygen by glucose-oxidase and/or the use of vacuum and oxygen-impermeable films (Barbosa Canovas et al. 2005; Kendall 2008). It is imperative that strict sanitary practices be imposed to minimize the presence of pathogenic organisms. The acidity of most fruits is a bacteriological barrier, but it is not a guarantee of bacteriological safety (ASHRAE 2010). Some fruits are packaged with sugar or syrup before freezing whereas others are frozen and then filled into polyethylene bags, cases, or bulk bins. There are several types of fruit packs suitable for freezing: syrup pack, sugar pack, unsweetened pack, tray pack, and sugar replacement pack. The type of pack is usually selected according to the use of the fruit. Syrup-packed fruits are generally used for cooking purposes, while dry-packed and tray-packed fruits are good for serving raw fruits (Barbosa Canovas et al. 2005).

Syrup pack: 40 % sugar syrup is recommended for most fruits. Lighter syrups are lower in calories and mostly desirable for mild-flavored fruits, while heavier syrups may be used for very sour fruits (Kendall 2008). Syrup is prepared by dissolving sucrose in warm water and cooling the solution before it is used; cooled syrup is used to cover the prepared fruit. The fruit is pressed down into the syrup before closing, then sealed and frozen (Beck 1996). Pectin can be used to reduce sugar content in syrups when freezing berries, cherries, and peaches. Pectin syrups are prepared by dissolving powdered pectin in water; the solution is stirred and boiled for 1 min; then sugar is added and dissolved; the solution is then cooled down with the addition of cold water. Pectin syrup is added to cover the fruit with a thin film and the pack is sealed and then frozen (Brady 2002).

Sugar packs: Sugar is first sprinkled over the fruit and then the container is agitated until the juice is drawn out and the sugar is dissolved. This type of pack is generally used for soft-sliced fruits such as peaches, strawberries, plums, and cherries, by using sufficient syrup to cover the fruit. Some whole fruits may also be coated with sugar prior to freezing (Beck 1996).

Unsweetened packs: These types of packs can be prepared in different forms: dry-packed, covered with water-containing ascorbic acid, or packed in unsweetened juice. When water or juice is used in syrup and sugar packs, the fruits are submerged in the liquid. Unsweetened packs yield a lower quality product when compared with sugar packs; however, some fruits such as blueberries, raspberries, scalded apples, gooseberries, currants, and cranberries maintain good quality without sugar (Beck 1996).

Tray packs: Unsweetened packs are prepared by using tray packs in which a single layer of prepared fruit is spread on trays, frozen, and packaged in freezer bags. This method allows the fruit sections to remain loose without clumping together, which offers the advantage of using individual frozen fruit pieces.

Sugar replacement packs: Artificial sweeteners can be used instead of sugar in the form of sugar substitutes. The sweet taste of sugar can be replaced by using these kinds of sweeteners, however, the beneficial effects of sugar-like color protection and thick syrup cannot be replaced. Fruits frozen with sugar substitutes will freeze harder and thaw more slowly than fruits preserved with sugar (Beck 1996) (Barbosa Canovas et al. 2005; Kendall 2008).

The commonly used freezing methods for whole fruits (small size) are air-blast, cryogenic, plate freezers (in packages) and fluidized bed (IQF). For large-size fruits (sliced) air-blast and plate freezers (in packages) are recommended. Fruits puree or pulp can be frozen in air-blast freezers and in plate freezers.

Another method is to pack the fruit into containers, with added sugar, and to freeze the product. The containers range from retail packages to large drums of 15 kg size. Fruit frozen in this manner is often destined for further processing. The freezing rate in the large packages can be quite slow, and there is a significant osmotic dehydration in the individual fruits.

Examples of products that are packaged in carton before freezing include sliced strawberries mixed with sugar in a 4–1 ratio and whole strawberries, other berries, mixed fruits, and melon balls in sugar syrup. The cartons and the containers must be liquid tight to prevent spills. These products are usually frozen in plate freezers (manual or automatic), stationary air-blast tunnels and push-through trolley freezers (ASHRAE 2010).

Small whole fruits such as blueberries, cranberries, strawberries, cherries, grapes, and pieces (slices, dices, or balls) of fruits such as apple, pineapple, mango, peach, and melon can be frozen individually in IQF freezers using trays, belt freezers, fluidized bed freezers, and cryomechanical freezers (North and Lovatt 2012). Many fruits and fruit pieces are sticky and fragile. Products from IQF freezers are packed in cartons or in polyethylene bags, into cases for bulk shipment. Other products such as fruit pies are usually frozen in automatic plate freezers or spiral belt freezers.

6.2.18 Shelf-Life of Frozen Fruits

The main factors affecting the quality of frozen fruits are: initial quality, prefreezing treatment, freezing process, packaging, and temperature and duration of storage. The period within which a fruit is safe to consume and has an acceptable quality that can be considered as its shelf-life. Microorganisms are not usually a problem in frozen foods, since they cannot grow at freezing temperatures (Zaritzky 2008).

The shelf-life of a frozen food is a complex concept that depends on the characteristics of the food product and the environmental conditions to which the food is exposed after being subjected to the freezing process. Packaging also plays an important role in maintaining the quality of foods.

Frozen fruits deteriorate mainly by slow chemical reactions, such as loss of nutritional value. For example, the vitamin C content may fall below the required standard before sensory quality becomes inadequate. The criteria for shelf-life may also vary depending on the sensitivity of the consumer. For consumers, taste, odor, and appearance are the most obvious criteria; in academia and in the industry, shelf-life assessments usually involve sensory evaluation and instrumental measurements based on a given quality index (e.g., vitamin C level).

Since 1950, different Time–Temperature–Tolerance (TTT) experiments have been performed by the USDA Western Regional Research Center in Albany, California, to analyze the stability of a great number of frozen foods stored at different temperatures for various periods of time (Van Arsdel et al. 1969; Jul 1984). Quality was measured by organoleptic testing carried out by taste panels, and by objective measurements such as ascorbic acid deterioration. Results were expressed as straight lines in a semi-logarithmic diagram of stability time (number of days taken to undergo a certain amount of quality deterioration) against storage temperature.

The practical shelf-life (PSL) of frozen foods depends not only on time and temperature of storage, but also to a large extent on PPP factors (product, process, and packaging) (Jul 1984). In general, quality changes during frozen storage at different temperatures are cumulative and irreversible; these changes are normally smaller at lower temperatures, thus storage temperature is the determining factor that governs quality and shelf-life. The International Institute of Refrigeration (1986) has recommended two definitions: Practical Storage Life (PSL) and High-quality life (HQL).

The *Practical storage life* (PSL) or acceptability time is the period of proper frozen storage after freezing of an initially high-quality product, during which the frozen food retains its quality characteristics and is suitable for consumption or for use in further processes.

High-quality life (HQL) is defined as the storage period from the time of freezing up to the point where 70 % of trained taste panel members detect a noticeable difference between the frozen food stored at different temperatures and the corresponding controls stored at −40 °C in a triangular sensory test. This parameter is also known as Just Noticeable Difference (JND).

Data on practical storage and high-quality life of different frozen foods have been reported in the literature (Van Arsdel et al. 1969; Jul 1984). IIR (1986) reported values of PSL of fruits at different storage temperatures. At −12 °C, −18 °C, and −24 °C, the values of PSL are 4 months, 18 months, and >24 months respectively in the case of raw peaches, apricots, and cherries; 5, 24, and >24 months for raw raspberries and strawberries, and 3, 24, and >24 months for raspberries and strawberries stored in sugar. The sensitivity of sugar-containing food products to temperature can be attributed to lower melting temperature and to the higher unfrozen water content.

As can be observed, PSL in these fruits is higher than 2 years at −24 °C decreasing at higher temperatures; the lower the storage temperature the greater the food stability. The ratios between PSL and HQL for frozen fruits range between 2.8 and 3.1 (Fu and Labuza 1997).

Frozen foods deteriorate during storage through different mechanisms and low, non-fluctuating temperatures are required to maintain food quality, thus storage and transport conditions have great influence on the quality of frozen foods. Constant and systematic control of handling temperatures is required throughout cold-chain frozen food distribution, from production to final consumption. According to EU Directive 89/108 (Quick Frozen Food Directive, QFF), after quick freezing, the product should be maintained at −18 °C or colder after thermal stabilization.

Fluctuations in storage temperature significantly affect the shelf-life of the product. The amplitude of ambient thermal fluctuation is reduced inside the package. It has been demonstrated mathematically that ambient temperature fluctuations with large oscillation periods cause higher amplitude fluctuations inside the product, this situation being more detrimental to the quality of frozen food than short oscillation periods (Zaritzky 1982).

6.2.19 New Trends in Freezing Technology

During decades, strategies such as lowering the refrigerating medium temperature, increasing heat transfer coefficients or reducing the size of the products to be frozen have been followed to increase the freezing rates. New technologies have been proposed and developed in order to accelerate and/or improve the freezing process such as ultrasound-assisted freezing and high-pressure shift freezing.

Ultrasound (with a frequency between 18 and 100 kHz and a high intensity >1 W/cm^2) has proved to accelerate the freezing rate through the effect of cavitation on ice crystallization. Cavitation ensures that small ice crystals are rapidly formed throughout the product increasing nucleation and crystal growth rates (Zheng and Sun 2005; Delgado and Sun 2011). Ultrasound can be used to assist immersion freezing (IF) in order to accelerate the freezing process. This method can be applied intermittently to increase freezing rates and also to improve sugar impregnation in fruits during IF. However research is still needed to establish the optimum operating parameters in order to evidence a positive effect on the quality of frozen fruits.

Another novel freezing technique is high-pressure shift freezing, HPSF (Li and Sun 2002; Fikiin 2008; Otero and Sanz 2011). This method takes advantage of the lower freezing point of water under pressure. At a pressure of 210 MPa the freezing point of ice I (normal ice at atmospheric pressure) drops to −22 °C.

The product to be frozen is introduced into the high pressure vessel and compressed up to the desired pressure level (for example 250 MPa); then it is cooled under pressure up to −21 °C, maintaining the unfrozen state condition. Once the desired temperature is reached in the whole product, pressure is released to the atmospheric value. Under such condition, the equilibrium freezing temperature of water is 0 °C, and the initial freezing point of the fruit is for example −1 °C, therefore this expansion induces a uniform and high supercooling in all the products (approximately 20 °C) due to the isostatic nature of the pressure. The extent of supercooling is the difference between the freezing point of the product and the minimum temperature reached after expansion. This high supercooling induces high nucleation rates and a large number of small ice crystals are formed in all the samples (Martino et al. 1998; Otero and Sanz 2011). This technology can be especially useful to freeze large pieces of foods where thermal gradients are pronounced when applying classical freezing methods.

Micrographs of peach and mango submitted to high pressure shift freezing showed a well-preserved fruit microstructure (Otero et al. 2000). Fruits such as strawberries in which texture depends more on the turgor pressure than on the integrity of the cell walls, are strongly affected by HPSF and the softness markedly increases. In this kind of products, pretreatments which enhance demethoxylation of pectin have been tested to improve the firmness of high-pressure shift-frozen strawberries (Otero and Sanz 2011). Besides, different studies show that HPSF up to 400 MPa was not able to inactivate polyphenoloxidase in different substrates causing color changes.

HPSF is a novel freezing technique that is still under research. Many experiments must be carried out to establish its effect on the quality of fruits, and to assess the economical viability of the process.

References

Allende, A., F.A. Tomás-Barberán, and M.I. Gil. 2006. Minimal processing for healthy traditional foods. *Trends in Food Science and Technology* 17: 513–519.

Alonso, J., and W. Canet. 1994. Mechanical assessment of texture of sweet cherries: Effects of freezing. *Journal of the Science of Food and Agriculture* 66(1): 1–7.

Alonso, J., T. Rodríguez, and W. Canet. 1995. Effect of calcium pre-treatments on the texture of frozen cherries. Role of pectinesterase in the changes in the pectic materials. *Journal of Agricultural and Food Chemistry* 43: 1011–1016.

Alothman, M., R. Bhat, and A. Karim. 2009. Effects of radiation processing on phytochemicals and antioxidants in plant produce. *Trends in Food Science & Technology* 20(5): 201–212.

Andrade Cuvi, M.J., A.R. Vicente, A. Concellón, and A.R. Chaves. 2011. Changes in red pepper antioxidants as affected by UV-C treatments and storage at chilling temperatures. *LWT - Food Science and Technology* 44: 1666–1671.

Artés, F., P. Gómez, E. Aguayo, V. Escalona, and F. Artés-Hernández. 2009. Sustainable sanitation techniques for keeping quality and safety of fresh-cut plant commodities. *Postharvest Biology and Technology* 51: 287–296.

ASHRAE. 2010. *Refrigeration. ASHRAE Handbook.* Atlanta: American Society of Heating, Refrigerating, and Air-conditioning Engineers.

Baka, M., J. Mercier, F. Corcuff, F. Castaigne, and J. Arul. 1999. Photochemical treatment to improve storability of fresh strawberries. *Journal of Food Science* 64: 1068–1072.

Barbosa Canovas, G., B. Altunakar, and D.J. Mejía-Lorío. 2005. *Freezing of fruits and vegetables. An agribusiness alternative for rural and semi-rural areas. FAO Agricultural Services Bulletin 158.* Italy: Food and Agriculture Organization of the United Nations.

Beck, P. 1996. *Freezing fruits.* Fargo: NDSU Extension Service, North Dakota State University of Agriculture and Applied Science.

Ben-Yehoshua, S., and J. Mercier. 2005. UV irradiation, biological agents, and natural compounds for controlling postharvest decay in fresh fruits and vegetables. In *Environmentally friendly technologies for agricultural produce quality*, ed. S. Ben-Yehoshua, 265–299. New York: CRC Press, Taylor and Francis Group.

Bevilacqua, A.E., and N.E. Zaritzky. 1980. Ice morphology in frozen beef. *Journal of Food Technology* 15: 589–597.

———. 1982. Ice recrystallization in frozen beef. *Journal of Food Science* 47: 1410–1414.

Bevilacqua, A., N. Zaritzky, and A. Calvelo. 1979. Histological measurements of ice in frozen beef. *Journal of Food Technology* 14: 237–251.

Bintsis, T., E. Litopailai-Tzanetaki, and R. Robinson. 2000. Existing and potential applications of ultraviolet light in the food industry – A critical review. *Journal of the Science of Food and Agriculture* 80: 637–647.

Blanda, G., L. Cerretani, A. Bendini, A. Cardinali, A. Scarpellini, and G. Lercker. 2008. Effect of vacuum impregnation on the phenolic content of Granny Smith and Stark Delicious frozen apple cvv. *European Food Research and Technology* 226: 1229–1237.

Blanda, G., L. Cerretani, A. Cardinali, S. Barbieri, A. Bendini, and G. Lercker. 2009. Osmotic dehydrofreezing of strawberries: Polyphenolic content, volatile profile and consumer acceptance. *Lebensmittel Wissenschaft und Technologie* 42: 30–36.

Botosoa, P.E., N. Chapleau, and A. Le-Bail. 2004. Influence de la congélation sur l'endommagement de tissus végétaux e Exemple de la mangue. *Revue Générale du Froid & du conditionnement d'air* 1048: 63–67.

Brady, P.L. 2002. *Freezing fruit*. Little Rock: UA Cooperative Extension Service, University of Arkansas.

Bunger, A., P.C. Moyano, R.E. Vega, P. Guerrero, and F. Osorio. 2004. Osmotic dehydration and freezing as combined processes on apple preservation. *Food Science and Technology International* 10(3): 163–170.

Canet, W., and M.D. Alvarez. 2006. Quality and safety of frozen vegetables. In *Handbook of frozen food processing and packaging*, ed. D.-W. Sun, 377–415. Boca Raton: CRC Press.

Cano, M.P., and M.A. Marin. 1992. Pigment composition and color of frozen and canned kiwi fruit slices. *Journal of Agricultural and Food Chemistry* 40: 2141.

Cano, P., M.A. Marin, and C. Fuster. 1990. Freezing of banana slices. Influence of maturity level and thermal treatment prior to freezing. *Journal of Food Science* 55(4): 1070.

Cantos, E., C. García-Viguera, S. de Pascual-Teresa, and F.A. Tomás-Barberán. 2000. *Journal of Agricultural and Food Chemistry* 48: 4606.

Cantwell, M., and T. Suslow. 2007. Cap. 36: Sistemas de manejo postcosecha:frutas y hortalizas precortadas (mínimamente procesadas). In *Tecnología poscosecha de cultivos hortofrutícolas*, ed. A.A. Kader, 497–518. Davis: Centro de Información e Investigación en Technología Postcosecha División de Agricultura y Recursos Naturales, Universidad de California Davis.

Chang, C.Y., L.R. Lai, and W.H. Chang. 1995. Relationships between textural changes and the changes in linkages of pectic substances of sweet pepper during cooking processes, and the applicability of the models of interactions between pectin molecules. *Food Chemistry* 53: 409–416.

Chassagne-Berces, S., F. Fonseca, M. Citeau, and M. Marin. 2010. Freezing protocol effect on quality properties of fruit tissue according to the fruit, the variety and the stage of maturity. *LWT - Food Science and Technology* 43: 1441–1449.

Chiralt, A., N. Martínez-Navarrete, J. Martínez-Monzó, P. Talens, G. Moraga, A. Ayala, and P. Fito. 2001. Changes in mechanical properties throughout osmotic processes: Cryoprotectant effect. *Journal of Food Engineering* 49(2–3): 129–135.

Cisneros-Zevallos, L. 2003. The use of controlled postharvest abiotic stresses as a tool for enhancing the nutraceutical content and adding-value of fresh fruits and vegetables. *Journal of Food Science* 68(5): 1560–1565.

Civello, P.M., and G.A. Martínez. 2008. Effect of heat treatments on gene expression and enzyme activities associated to cell wall degradation in strawberry fruit. *Postharvest Biology and Technology* 49: 38–45.

Civello, P.M., A.R. Vicente, and G.A. Martínez. 2006. UV-C technology to control postharvest diseases of fruits and vegetables. In *Recent advances in alternative postharvest technologies to control fungal diseases in fruits and vegetables*, ed. R. Troncoso-Rojas, M.E. Tiznado-Hernández, and A. González-León, 31–102. Kerakla, India: Transworld Research Network.

D'Hallewin, G., M. Schirra, E. Manueddu, A. Piga, and S. Ben-Yehoshua. 1999. Scoparone and scopoletin accumulation and ultraviolet-C-induced resistance to postharvest decay in oranges as influenced by harvest date. *Journal of the American Society for Horticultural Sciences* 124: 702–707.

D'hallewin, G., Schirra, M., Pala, M., Ben-Yehoshua, S., 2000. Ultraviolet C irradiation at 0.5 kJm–2 reduces decay without causing damage or affecting postharvest quality of star ruby grapefruit (*C. paradisi Macf.*). *Journal Agriculture Food Chemistry* 48: 4571–4575.

De Capdeville, G., C.L. Wilson, S.V. Beer, and J.R. Aist. 2002. Alternative disease control agents induce resistance to blue mold in harvested 'Red Delicious' apple fruit. *Phytopatology* 92: 900–908.

Delgado, A.E., and A.C. Rubiolo. 2005. Microstructural changes in strawberry after freezing and thawing processes. *LWT - Food Science and Technology* 38(2): 135–142.

Delgado, A., and D.W. Sun. 2011. Ultrasound accelerated freezing. In *Handbook of frozen food processing and packaging*, 2nd ed, ed. D.-W. Sun, 645–666. Boca Raton: CRC Press.

Dermesonlouoglou, E., M. Giannakourou, and P. Taoukis. 2007. Stability of dehydrofrozen tomatoes pretreated with alternative osmotic solutes. *Journal of Food Engineering* 78: 272–280.

Desrosier, W., and D.K. Tressler. 1977. *Fundamentals of food freezing*. Wesport: AVI Publishing Co. Inc.

Dixon, G.M., and J.J. Jen. 1977. Changes of sugars and acids of osmovacuum-dried apple slices. *Journal of Food Science* 42: 1126–1127.

Donhowe, D.P., and R. Hartel. 1996. Recrystallization of ice in ice cream during controlled accelerated storage. *International Dairy Journal* 6: 1191–1208.

El Ghaouth, A., C. Wilson, and A. Callahan. 2003. Induction of chitinase, beta-1,3-glucanase, and phenylalanine ammonia lyase in peach fruit by UV-C treatment. *Phytopathology* 93: 349–355.

El-Ghaouth, A., C.L. Wilson, and M. Wisniewski. 1998. Ultrastructural and cytochemical aspects of the biological control of Botrytis cinerea by Candida saitoana in apple fruit. *Phytopathology* 88(4): 282–291.

Erba, M.L., E. Forni, A. Colonnello, and R. Giangiacomo. 1994. Influence of sugar composition and air dehydration levels on the chemical-physical characteristics of osmodehydrofrozen fruit. *Food Chemistry* 50: 69–73.

Erkan, M., S.Y. Wang, and C.Y. Wang. 2008. Effect of UV treatment on antioxidant capacity, antioxidant enzyme activity and decay in strawberry fruit. *Postharvest Biology and Technology* 48(2): 163–171.

Falguera, V., J.P. Quintero, A. Jimenez, J.A. Muñoz, and A. Ibarza. 2011. Edible films and coatings: Structures, active functions and trends in their use. *Trends in Food Science & Technology* 22: 292–303.

Fallik, E. 2004. Pre-storage hot water treatments (immersion, rinsing and brushing). *Postharvest Biology and Technology* 32: 125–134.

Fennema, O. 1995. Metastable and non equilibrium states in frozen foods and their stabilization. In *Food preservation by moisture content: Fundamentals and applications*, ed. G.V. Barbosa Canovas and J. Welti-Chanes, 243. Lancaster: Technomic Publishing.

———. 1996. Water and ice. In *Food chemistry*, 3rd ed, ed. O. Fennema, 17–94. New York: Marcel Dekker.

Fennema, O.R., W.D. Powrie, and E.H. Marth. 1973. *Low temperature preservation of foods and living matter*. New York: Marcel Dekker.

Fikiin, K.A. 2008. Emerging and novel freezing processes. In *Frozen food science and technology*, ed. J.A. Evans, 101–123. Oxford: Blackwell Publishing.

Forni, E., D. Torreggiani, G. Crivelli, A. Maestrelli, G. Bertolo, and F. Santelli. 1990. Influence of osmosis time on the quality of dehydrofrozen kiwi fruit. *Acta Horticulturae* 282: 425–434.

Forni, E., A. Sormani, S. Scalise, and D. Torreggiani. 1997. The influence of sugar composition on the colour stability of osmodehydrofrozen intermediate moisture apricots. *Food Research International* 30: 87–94.

Franks, F. 1985. *Biophysics and biochemistry at low temperatures*, 39–52. Cambridge: Cambridge University Press.

Fu, B., and T. Labuza. 1997. Shelf life testing: Procedures and prediction methods. In *Quality in frozen food*, ed. M.C. Erickson and Y.C. Hung, 377–415. New York: Chapman & Hall.

Galetto, C.D., R.A. Verdini, S.E. Zorrilla, and A.C. Rubiolo. 2010. Freezing of strawberries by immersion in CaCl2 solutions. *Food Chemistry* 123: 243–248.

Garcia, M.A., M.N. Martino, and N.E. Zaritzky. 2001. Composite starch-based coatings applied to strawberries (*Fragaria ananassa*). *Nahrung/Food* 45: 267–272.

Gardner, D.W.M., and G. Shama. 2000. Modeling UV-induced inactivation of microorganisms on surfaces. *Journal of Food Protection* 63: 63–70.

Garrote, R.L., and R.A. Bertone. 1989. Osmotic concentration at low temperature of frozen strawberry halves. Effect of glycerol glucose and sucrose solution on exudate loss during thawing. *Food Science and Technology* 22: 264–267.

Goff, H.D. 1994. Measuring and interpreting the glass transition in frozen foods and model systems. *Food Research International* 27: 187–189.

———. 1997. Measurement and interpretation of the glass transition in frozen foods. In *Quality in frozen food*, ed. M.C. Erickson and Y.C. Hung, 29–50. New York: Chapman & Hall.

Gómez, P., S. Alzamora, M. Castro, and D. Salvatori. 2010. Effect of ultraviolet-C light dose on quality of cut-apple: Microorganism, color and compression behaviour. *Journal of Food Engineering* 98: 60–70.

González Aguilar, G., J.A. Villa Rodríguez, J.F. Ayala Zavala, and E.M. Yahia. 2010. Improvement of the antioxidant status of tropical fruits as a secondary response to some postharvest treatments. *Trends in Food Science and Technology* 21: 415–482.

González, E., B. de Ancos, and P. Cano. 2002. Preservation of raspberry fruits by freezing: Physical, physico-chemical and sensory aspects. *European Food Research and Technology* 215(6): 497–503.

González-Aguilar, G.A., C.Y. Wang, J.G. Buta, and D.T. Krizek. 2001. Use of UV-C irradiation to prevent decay and maintain postharvest quality of ripe 'Tommy Atkins' mangoes. *International Journal of Food Science and Technology* 36: 767–773.

Gonzalez-Aguilar, G.A., M.E. Tiznado-Hernandez, R. Zavaleta-Gatica, and M.A. Martinez-Tellez. 2004. Methyl jasmonate treatments reduce chilling injury and activate the defense response of guava fruits. *Biochemical and Biophysical Research Communications* 313(3): 694–701.

Gonzalez-Aguilar, G.A., M.A. Villegas-Ochoa, M.A. Martinez-Tellez, A.A. Gardea, and J.F. Ayala-Zavala. 2007a. Improving antioxidant capacity of fresh-cut mangoes treated with UV-C. *Journal of Food Science* 72(3): S197–S202.

Gonzalez-Aguilar, G.A., R. Zavaleta-Gatica, and M.E. Tiznado-Hernandez. 2007b. Improving postharvest quality of mango "Haden" by UV-C treatment. *Postharvest Biology and Technology* 45(1): 108–116.

Guadagni, D.G. 1969. Frozen fruits and juices. In *Quality and stability of frozen foods*, ed. W. Van Ardsel, 89–91. New York: Wiley-Interscience.

Guerrero-Beltrán, J.A., and G.V. Barbosa-Cánovas. 2004. Advantages and limitations on processing foods by uv light. *Food and Science Technology International* 10(3): 137–147.

Gutschmidt, J. 1968. Principles of freezing and low temperature storage, with particular reference to fruits and vegetables. In *Low temperature biology of foodstuff*, Recent advances on food science, vol. 4. London: Pergamon Press.

Hartel, R. 2001. *Crystallization in Foods*, 192–231. Frederick: Aspen Publishers Inc.

Heldman, D.R. 1974. Predicting the relationship between unfrozen water fraction and temperature during food freezing using freezing point depression. *ASHRAE Transactions* 91(Part 2B): 371–384.

International Institute of Refrigeration. 1986. *Recommendations of the Processing and Handling of Frozen Foods*. Paris: International Institute of Refrigeration.

Jacobi, K.K., E.A. MacRae, and S.E. Hetherington. 2001. Postharvest heat disinfestation treatments of mango fruit. *Scientia Horticulturae* 89(3): 171–193.

Javeri, H., R. Toledo, and L. Wicker. 1991. Vacuum infusion of citrus pectin methylesterase and calcium effects on firmness of peaches. *Journal of Food Science* 56: 739–742.

Jul, M. 1984. *The quality of frozen foods*. London: Academic Press Inc.

Kader, A.A. 2007a. Cap. 4. Biología y tecnología postcosecha: Un panorama. In *Tecnología poscosecha de cultivos hortofrutícolas*, ed. A.A. Kader, 157–168. Davis: Centro de Información e Investigación en Tecnología Postcosecha. División de Agricultura y Recursos Naturales, Universidad de California Davis.

———. 2007b. Cap. 14 Atmósferas modificadas en el transporte y el almacenamiento. In *Tecnología poscosecha de cultivos hortofrutícolas*, ed. A.A. Kader. Davis: Centro de Información e Investigación en Technología Postcosecha, División de Agricultura y Recursos Naturales, Universidad de California Davis.

Kader, A.A., and S. Ben-Yehoshua. 2000. Effects of superatmospheric oxygen levels on postharvest physiology and quality of fresh fruits and vegetables. *Postharvest Biology and Technology* 20(1): 1–13.

Kendall, P. 2008. Freezing vegetables. Colorado State University Cooperative Extension. Fact Sheet 9.331

Kennedy, C.J. 2000. Future trends in frozen foods. Chapter 13. In *Managing frozen foods*, ed. C.J. Kennedy. Cambridge: Woodhead Publishing Limited.

Kim, N.K., and Y.C. Huang. 1994. Freeze-cracking in foods as affected by physical properties. *Journal of Food Science* 59(3): 669–674.

Le Maguer, M. 1988. Osmotic dehydration: review and future directions. *Symposium on Progress in Food Reservation Processes*, Brussels, 1, 283–309.

Lamikanra, G. 2002. Enzymatic effects on flavor and texture of fresh-cut fruits and vegetables. In: G. Lamikanra (Ed.), *Fresh-cut Fruits and Vegetables* pp. 125–185. Science, Technology and Market. CRC Press, LLC., Boca Raton, Florida.

Le Meste, M., D. Champion, G. Roudaut, G. Blond, and D. Simatos. 2002. Glass transition and food technology. A critical appraisal. *Journal of Food Science* 67(7): 2444–2458.

Lee, J.Y., H.J. Park, C.Y. Lee, and W.Y. Choi. 2003. Extending shelf life of minimally processed apples with edible coatings and antibrowning agents. *Lebensmittel Wissenschaft und Technologie* 36: 323–329.

Levine, H., and L. Slade (eds.). 1991. *Water relationship in foods*, 251–273. New York: Plenum Press.

Li, B., and D.W. Sun. 2002. Novel methods for rapid freezing and thawing of foods—A review. *Journal of Food Engineering* 54(3): 175–182.

Lichtscheidl-Schultz, I. 1985. Effects of UV-C and UV-B on cytomorphology and water permeability of inner epidermal cells of *Allium* cepa. *Physiologia Plantarum* 63(3): 269–276.

Lim, Y.Y., T.T. Lim, and J.J. Tee. 2007. Antioxidant properties of several tropical fruits: A comparative study. *Food Chemistry* 103(3): 1003–1008.

Lucas, T., and A.L. Raoult-Wack. 1998. Immersion chilling and freezing in aqueous refrigerating media: Review and future trends. *International Journal of Refrigeration* 21: 419–429.

Lurie, S. 1998. Postharvest heat treatments. *Postharvest Biology and Technology* 14: 257–269.

Lurie, S., A. Handros, E. Fallik, and R. Shapira. 1996. Reversible inhibition of tomato fruit gene expression at high temperature. *Plant Physiology* 110: 1207–1214.

Maestrelli, A. 2000. Fruit and vegetables: The quality of raw material in relation to freezing. Chapter 3. In *Managing frozen foods*, ed. C.J. Kennedy. Cambridge: Woodhead Publishing Limited.

Main, G.L., J.R. Morris, and E.J. Wehunt. 1986. Effect of pre-processing treatments on the firmness and quality characteristics of whole and sliced strawberries after freezing and thermal processing. *Journal of Food Science* 51: 391–394.

Marani, C.M., M.E. Agnelli, and R.H. Mascheroni. 2007. Osmo-frozen fruits: Mass transfer and quality evaluation. *Journal of Food Engineering* 79: 1122–1130.

Marin, M.A., P. Cano, and C. Fuster. 1992. Freezing preservation of four Spanish mango cultivars (Mangifera indica L.): Chemical and biochemical aspects. *Zeitschrift für Lebensmittel untersuchung und -Forschung* 194(6): 566–569.

Martino, M.N., and N.E. Zaritzky. 1987. Effects of temperature on recrystallization of polycrystalline ice. *Sciences des Aliments* 7: 147–166.

———. 1988. Ice crystal size modifications during frozen beef storage. *Journal of Food Science* 53: 1631–1637. 1649.

———. 1989. Ice recrystallization in a model system and in frozen muscle tissue. *Cryobiology* 26: 138–148.

Martino, M.N., L. Otero, P.D. Sanz, and N. Zaritzky. 1998. Size and location of ice crystals in pork frozen by high pressure assisted freezing as compared to classical methods. *Meat Science* 50: 303–313.

Maverakis, E., Y. Miyamura, M.P. Bowen, G. Correa, Y. Ono, and H. Goodarzi. 2010. Light, including ultraviolet. *Journal of Autoimmunity* 34: 247–257.

Mazur, P. 1970. Cryobiology: The freezing of biological systems. *Science* 168: 939–949.

Miller, J.P., and C. Butcher. 2000. Freezer technology. Chapter 9. In *Managing frozen foods*, ed. C.J. Kennedy. Cambridge: Woodhead Publishing Limited.

Mittler, R., S. Vanderauwera, M. Gollery, and F. Van Breusegem. 2004. Reactive oxygen gene network of plants. *Trends in Plant Science* 9: 490–498.
Moraga, G., N. Martínez-Navarrete, and A. Chiralt. 2006. Compositional changes of strawberry due to dehydration, cold storage and freezing–thawing processes. *Journal of Food Processing and Preservation* 30: 458–474.
Nesvabda, P. 2008. Thermal properties and Ice crystal developments in frozen foods. In *Frozen foods science and technology*, ed. J. Evans, 1–25. Oxford: Blackwell Publishing.
North, M.F., and S.J. Lovatt. 2012. Freezing methods and equipment. Chapter 8. In *Handbook of frozen food processing and packaging*, 2nd ed, ed. D.-W. Sun. Boca Raton: CRC Press.
Oms-Oliu, G., R.M. Raybaudi-Massilia Martínez, R. Soliva-Fortuny, and O. Martín-Belloso. 2008. Effect of superatmospheric and low oxygen modified atmospheres on shelf-life extension of fresh-cut melon. *Food Control* 19: 191–199.
Otero, L., and P. Sanz. 2011. High pressure shift freezing. In *Handbook of Frozen Food Processing and Packaging*, 2nd ed, ed. Da-Wen Sun, 667–683. Boca Raton: CRC Press.
Otero, L., M. Martino, N. Zaritzky, M. Solas, and P.D. Sanz. 2000. Preservation of microstructure in peach and mango during high pressure-shift freezing. *Journal of Food Science* 65(3): 466–470.
Pan, J., A. Vicente, G. Martínez, A. Chaves, and M. Civello. 2004. Combined use of UV-C irradiation and heat treatment to improve postharvest life of strawberry fruit. *Journal of the Science of Food and Agriculture* 84: 1831–1838.
Paull, R.E., and N.J. Chen. 2000. Heat treatment and fruit ripening. *Postharvest Biology and Technology* 21: 21–37.
Pham, Q.T., and R.F. Mawson. 1997. Moisture migration and ice recrystallization in frozen foods. In *Quality in frozen foods*, ed. M.C. Erickson and Y. Hung. New York: Chapman and Hall.
Rahman, M.S. 2006. State diagram of foods: Its potential use in food processing and product stability. *Trends in Food Science and Technology* 17: 129–141.
———. 2007. Glass transition and other structural changes in foods. In *Handbook of food preservation*, 2nd ed, ed. M.S. Rahman, 75–93. New York: CRC Press.
———. 2009. Glass transitions data and models of foods. Chapter 8. In *Food properties handbook*, 2nd ed, ed. M.S. Rahman. Boca Raton: CRC Press.
Ramallo, L.A., and R.H. Mascheroni. 2010. Dehydrofreezing of pineapple. *Journal of Food Engineering* 99: 269–275.
Reid, D.S. 1996. Fruit freezing. In *Processing fruits: Science and technology*, vol. 1, 1st ed, ed. L.P. Somogyi, H.S. Ramaswamy, and Y.H. Hui, 275–286. Lancaster: Technomic Publishing Company.
———. 1997. Overview of physical/chemical aspects of freezing. In *Quality in frozen food*, ed. M.C. Erickson and Y.C. Hung, 10–28. New York: Chapman & Hall.
Robbers, M., R.P. Singh, and L.M. Cunha. 1997. Osmotic–convective dehydrofreezing process for drying kiwifruit. *Journal of Food Engineering* 62(5): 1039–1047.
Rodoni, L., N. Casadei, A. Concellón, A.R. Chaves, and A.R. Vicente. 2010. Effect of short-term ozone treatments on tomato (*Solanum lycopersicum* L.) fruit quality and cell wall degradation. *Journal of Agricultural and Food Chemistry* 58: 594–599.
Rodríguez-Félix, A., M. Rivera-Domínguez, and G. Gonzáles-Aguilar. 2005. Uso de atmósferas modificadas y controladas. In *Nuevas tecnologías de conservación de productos vegetales frescos cortados*, ed. G.A. Gonzáles-Aguilar, A.A. Gardea, and F. Cumea-Navarro, 447–474. Hermosillo: CIAD.
Rojas-Graü, M.A., R. Soliva-Fortuny, and O. Martin-Belloso. 2009. Edible coatings to incorporate active ingredients to fresh cut fruits: A review. *Trends in Food Science & Technology* 20: 438–447.
Roos, Y. 1995. *Phase transitions in foods*, Food science and technology international series. San Diego: Academic.
Roos, Y., and M. Karel. 1991a. Applying state diagrams to food processing and development. *Food Technology* 45: 66–71. 107.
———. 1991b. Non-equilibrium ice formation in carbohydrate solutions. *Cryo Letters* 12: 367–376.

Sandhya. 2010. Modified atmosphere packaging of fresh produce: Current status and future needs. *LWT - Food Science and Technology* 43: 381–392.

Shama, G. 2007. Process challenges in applying low doses of ultraviolet light to fresh produce for eliciting beneficial hormetic responses. *Postharvest Biology and Technology* 44(1): 1–8.

Shen, Y., Y. Sun, L. Qiao, J. Chen, D. Liu, and X. Ye. 2013. Effect of UV-C treatments on phenolic compounds and antioxidant capacity of minimally processed Satsuma mandarin during refrigerated storage. *Postharvest Biology and Technology* 76: 50–57.

Silveira, A.C., E. Aguayo, V.H. Escalona, and F. Artés. 2011. Hot water treatment and peracetic acid to maintain fresh-cut Galia melon quality. *Innovative Food Science and Emerging Technologies* 12: 569–576.

Sinha, R.P., and D.P. Häder. 2002. UV-induced DNA damage and repair: A review. *Photochemical and Photobiological Sciences* 1: 225–236.

Sista, R.V., M.C. Erickson, and R.L. Shewfelt. 1997. Quality deterioration in frozen foods associated with hydrolytic enzyme activities. In *Quality in frozen food*, ed. M.C. Erickson and Y.C. Hung, 101–110. New York: Chapman & Hall.

Skrede, G. 1996. Fruits. In *Freezing effects on food quality*, ed. L.E. Jeremiah, 183–245. New York: Marcel Dekker. Inc.

Slade, L., and H. Levine. 1995. Glass transitions and water-food structure interactions. In *Advances in food nutrition research*, vol. 38, ed. J.E. Kinsella and S.L. Taylor, 103–269. San Diego: Academic.

Sousa, M.B., W. Canet, M.D. Alvarez, and M.E. Tortosa. 2006. Effect of processing on the texture and structure of raspberry (cv. Heritage) and blackberry (cv. Thornfree). *European Food Research and Technology* 223: 517–532.

Sousa, M.B., W. Canet, M.D. Alvarez, and C. Fernandez. 2007. Effect of processing on the texture and sensory attributes of raspberry (cv. Heritage) and blackberry (cv. Thornfree). *Journal of Food Engineering* 78(1): 9–21.

Spiazzi, E. A., Raggio Z. I., Bignone K. A., Mascheroni, R. H. 1998. Experiments on dehydrofreezing of fruits and vegetables: Mass transfer and quality factors. Advances in the refrigeration systems, food technologies and cold chain, IIF/IIR, vol. 6, 401–408.

Sterling, C. 1968. Effect of low temperature on structure and firmness of apple tissue. *Journal of Food Science* 33: 577–580.

Stevens, C., C.L. Wilson, J.Y. Lu, V.A. Khan, E. Chalutz, S. Droby, M.K. Kabwe, Z. Haung, O. Adeyeye, L.P. Pusey, M.E. Wisniewski, and M. West. 1996. Plant hormesis induced by ultraviolet light-C for controlling postharvest diseases of tree fruits. *Crop Protection* 15: 129–134.

Sutton, R.L., A. Lips, G. Piccirillo, and A. Sztehlo. 1996. Kinetics of ice recrystallization in aqueous fructose solutions. *Journal of Food Science* 61(4): 741–745.

Suutarinen, J., K. Honkapää, R.L. Heiniö, K. Autio, A. Mustranta, and S. Kappinen. 2002. Effects of calcium chloride-based prefreezing treatments on the quality factors of strawberry jams. *Journal of Food Science* 67: 884–894.

Talens, P., N. Martínez-Navarrete, P. Fito, and A. Chiralt. 2002. Changes in optical and mechanical properties during osmodehydrofreezing of kiwi fruit. *Innovative Food Science and Emerging Technologies* 3: 191–199.

Talens, P., I. Escriche, N. Martínez-Navarrete, and A. Chiralt. 2003. Influence of osmotic dehydration and freezing on the volatile profile of kiwi fruit. *Food Research International* 36: 635–642.

Thompson, J.F., F.G. Mitchell, and R.F. Kasmire. 2007. Cap 11. Enfriamiento de productos hortofrutícolas. In *Tecnología poscosecha de cultivos hortofrutícolas*, ed. A.A. Kader, 157–168. Davis: Centro de Informaci6n e Investigaci6n en Technología Postcosecha. Divisi6n de Agricultura y Recursos Naturales, Universidad de California Davis.

Torreggiani, D., and A. Maestrelli. 2006. Quality and safety of frozen fruits. In *Handbook of frozen food processing and packaging*, ed. D.-W. Sun, 417–440. Boca Raton: CRC Press.

Torreggiani, D., T. Lucas, and A.-L. Raoult-Wack. 2000. The pre-treatment of fruits and vegetables. Chapter 4. In *Managing frozen foods*, ed. Christopher J. Kennedy. Cambridge: Woodhead Publishing Limited.
Tregunno, N., and H. Goff. 1996. Osmodehydrofreezing of apples: Structural and textural effects. *Food Research International* 29: 471–479.
Van Arsdel, W.B., M.J. Copley, and R.L. Olson. 1969. *Quality and stability of frozen foods: Time temperature-tolerance and its significance*. New York: Wiley Interscience.
Van Buggenhout, S., I. Messagie, V. Maes, T. Duvetter, A. Van Loey, and M. Hendrickx. 2006. Minimizing texture loss of frozen strawberries: Effect of infusion with pectin methylesterase and calcium combined with different freezing conditions and effect of subsequent storage/thawing conditions. *European Food Research and Technology* 223: 395–404.
Van Buggenhout, S., T. Grauwet, A. Van Loey, and M. Hendrickx. 2008. Structure/processing relation of vacuum infused strawberry tissue frozen under different conditions. *European Food Research and Technology* 226: 437–448.
Vicente, A.R., B. Repice, G.A. Martínez, A.R. Chaves, P.M. Civello, and G.O. Sozzi. 2004. Maintenance of fresh boysenberry fruit quality with UV-C light and heat treatments combined with low storage temperature. *The Journal of Horticultural Sciences and Biotechnology* 79: 246–251.
Vicente, A.R., C. Pineda, L. Lemoine, P.M. Civello, G.A. Martinez, and A.R. Chaves. 2005. UV-C treatments reduce decay, retain quality and alleviate chilling injury in pepper. *Postharvest Biology and Technology* 35: 69–78.
Vicente, A.R., G.A. Martínez, A.R. Chaves, and P.M. Civello. 2006. Effect of heat treatment on strawberry fruit damage and oxidative metabolism during storage. *Postharvest Biology and Technology* 40: 116–122.
Wang, J., L. Lite, and D. Yang. 2003. The correlation between freezing point and soluble solids of fruits. *Journal of Food Engineering* 60: 481–484.
Wang, H., H. Peng, and Y. Luo. 2007. Control of browning and microbial growth on fresh-cut apples by sequential treatment of sanitizers and calcium ascorbate. *Journal of Food Science* 72: M1–M7.
Wills, R., B. McGlasson, D. Graham, and D. Joyce. 1998. *Postharvest. An introduction to the physiology and handling of fruit, vegetable and ornamentals*. Wallingford: CAB International. 62 pp.
Wszelaky, A.L., and E.J. Mitchan. 2000. Effects of superatmospheric oxygen on strawberry fruit quality and decay. *Postharvest Biology and Technology* 20(2): 125–133.
Zaritzky, N.E. 2000. Factors affecting the stability of frozen foods. In *Managing frozen foods*, ed. C.J. Kennedy, 111–133. Cambridge: Woodhead Publishing Limited.
———. 2008. Frozen storage. In *Frozen food science and technology*, ed. J.A. Evans, 224–247. Oxford: Blackwell Publishing.
———. 2010. Chemical and physical deterioration of frozen foods. Chapter 20. In *Chemical deterioration and physical instability of food and beverages*, ed. Leif Skibsted, Jens Risbo, and Mogens Andersen, 561–607. Cambridge: Woodhead Publishing Limited.
———. 2011. Physical–chemical principles in freezing. Chapter 1. In *Handbook of frozen food processing and packaging*, 2nd ed, ed. Da-Wen Sun, 3–38. Boca Raton: CRC Press.
Zheng, L., and D.W. Sun. 2005. Innovative applications of power ultrasound. In *Emerging technologies for food processing*, ed. D.W. Sun, 603–626. San Diego: Elsevier Academic Press.
Zhu, S., and J. Zhou. 2007. Effect of nitric oxide on ethylene production in strawberry fruit during storage. *Food Chemistry* 100: 1517–1522.

Chapter 7
Thermal Drying of Foods

Henry T. Sabarez

7.1 Introduction

Drying is a process applied mainly for the purpose of extending the shelf-life of the products by reducing their water content to a level low enough to inhibit deteriorative reactions. The problems of drying are diverse as the intricacies and needs for various materials to be dried at different scales. Over 500 dryer types have been reported in the technical literature, and about 100 types are commercially available (Mujumdar and Law 2010). This large number of dryer designs is due to the differences in the physical attributes of the product, modes of heat input, operating temperatures and pressures, quality specifications on the dried product, and so on. According to Sablani and Rahman (Sablani and Rahman 2007), drying can be broadly classified based on the water-removing method applied such as, thermal drying, osmotic drying, and mechanical dewatering. Today, drying is employed in various industrial sectors (e.g. paper, wood, food, agriculture, waste management, etc.) utilising different techniques. As far as industrial sectors are concerned, food and agriculture remain the most dominant sectors with respect to the critical importance of drying to these industries (Mujumdar 2010a). This chapter mainly covers the development and application of thermal drying processes for food materials. Although this chapter emphasises thermal drying of food materials, the topics covered are applicable to other materials as well.

The preservation of food materials by drying dates back many centuries ago. Thermal drying (or dehydration) is the most common and cost-effective technique for preservation of foods and for the production of traditional as well as innovative processed products such as snacks with desired functionalities (Jangam 2012).

H.T. Sabarez (✉)
CSIRO, Werribee, VIC Australia
e-mail: henry.sabarez@csiro.au

A. Rosenthal et al. (eds.), *Fruit Preservation*, Food Engineering Series,
https://doi.org/10.1007/978-1-4939-3311-2_7

There are many different methods of drying food materials, each with their own advantages and disadvantages for particular applications. Numerous food products are routinely preserved by drying, which include grains, marine products, meat products, as well as fruits and vegetables (Jangam 2011). Most conventional dryers use hot air as the drying medium that applies the heat mainly by convection and carries away the water vapour, and are usually operated at atmospheric pressure under steady drying conditions (Mujumdar and Law 2010). To date, the convective air-drying continues to be commonly used at industrial scale because it is much simpler and easier to operate, in addition to its lower capital cost (for the time being) although it is inherently less energy-efficient.

The process of drying food materials in conventional hot air dryers is extremely complex involving coupled transient mechanisms of heat, mass and momentum transfer processes accompanied by physical, chemical and phase change transformations. This method requires large amounts of energy and usually imparts significant alterations in product quality and functionality attributes due to the exposure to longer drying times or higher temperatures. There is a significant interest in developing innovative drying approaches that will accelerate the drying process in order to increase production throughput and reduce the energy consumption (i.e. thereby decreasing both environmental and financial costs) with improved quality of the product. This is important due to the rising consumer demand for high-quality products coupled with need for eco-friendly and sustainable processes to maintain competitiveness with minimal impact on the environment. Today's consumer expectation for better quality, safety and high nutritional value drives research and improvement of drying technologies (Lewicki 2006).

The chapter starts with a brief overview of the drying mechanisms in conventional hot air-drying techniques, and the equipment and design used in food drying and their limitations. The latter part of this chapter focuses on the challenges and opportunities in drying R&D and a case study on modelling and optimisation.

7.2 Drying Mechanisms

A conceptual representation of the transport phenomena occurring during thermal drying of a solid food material is illustrated in Fig. 7.1. In convective air-drying of food materials, two distinct transport mechanisms occur simultaneously, involving heat transfer from the drying air to the food material and water transport (either as liquid or vapour) from the interior of the solid product to its surface and eventually to the air through evaporation. The heat transfer mechanism generally employs conduction, convection or radiation mode of energy transfer from a heat source (hot gas or heated metal surface) to the food material. On the other hand, the mass transfer mechanism during drying may be controlled by either the rate of moisture diffusion (liquid or vapour) within the food matrix (internal) or by the rate of moisture evaporation from the product surface to the drying medium (external). The liquid transport mechanisms include capillary flow, surface diffusion and liquid diffusion while the vapour

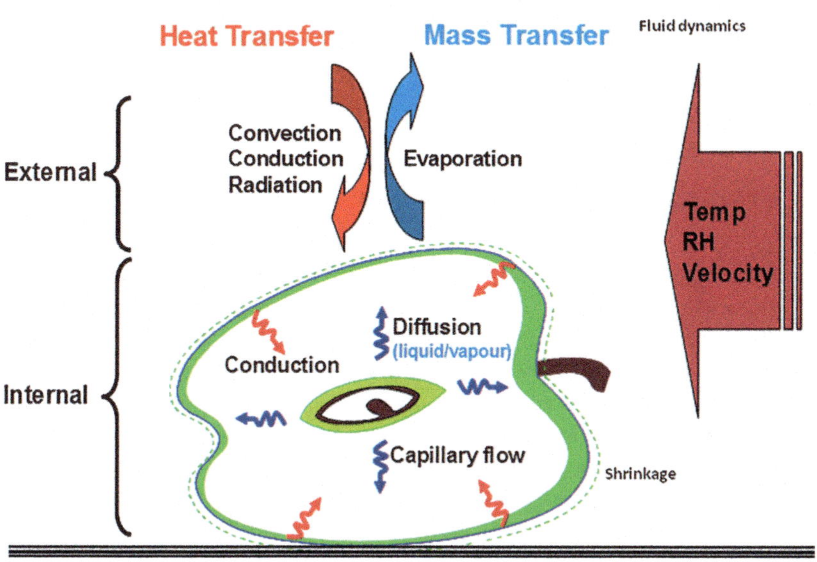

Fig. 7.1 A conceptual representation of the thermal drying process of a solid food material

transport mechanisms consist of Knudsen diffusion, mutual diffusion, Stefan diffusion, Poiseuille flow and condensation–evaporation (Sablani and Rahman 2007).

The heat and mass transfer mechanisms are usually influenced by both temperature and water concentration differences as well as the air velocity field, together with the properties of the materials itself. In general, the coupled transfer of heat, mass and momentum in at least two distinct sub-domains (air and food) simultaneously occurring both externally and internally to the food matrix during drying can be described as follows (Sabarez 2012): (1) convective and conductive heat transfer in the air, (2) convective and diffusive water transfer in the air, (3) heat transfer mainly by conduction within the solid interior, (4) mass transfer in the solid interior by diffusion (liquid or vapour), (5) moisture evaporation at the air–food interface, and (6) airflows (laminar or turbulent) around the food material.

The external transfer rates for both heat and mass are greatly influenced by air velocity field (fluid dynamics) and other drying air properties (i.e. temperature and relative humidity). The internally controlled drying process is mainly affected by temperature and predominates once the rate of replenishment of moisture from the interior to the surface of the product is slower than the external mass transfer rate. The internal mass transfer process often involves different mechanisms of moisture movement. According to Heldman and Hartel (1997), the moisture within the food product can migrate in several ways via a number of different mechanisms (liquid or vapour phase). In liquid diffusion, the rate at which moisture migrates depends on the nature of the food product, temperature and water concentration difference. Vaporisation in some cases may occur within the product and thus water diffuses in the form of vapour through the food matrix with the difference in vapour pressure as the driving force for moisture transfer. The differences in pressure between the drying medium and the internal food structure (pressure flow) and the differences in temperature between the surface and the interior of the product (thermal flow) may also influence the internal mobility of moisture.

7.3 Drying Equipment and Design

A large number of food materials are dehydrated in a variety of dryer designs with diverse processing conditions. The selection of the drying method for a particular food product is an important step as the drying technique and operating conditions affect the quality of the dried product as well as its cost. This depends on various factors, including the type of feed, amount of moisture, drying kinetics, heat sensitivity, physical structure of the material to be dried, quality requirements of a dried food, and many other factors (Jangam 2011). Also, the selection procedures of a drying system include cost estimation of various dryers including capital and operating costs.

Table 7.1 summarises a generalised classification of conventional dryers for drying food materials. It should be remembered that a dryer successfully used for one product can provide totally different performance for another product, and that a particular food material may need to be dried in different ways using a suitable dryer type. A vast number of dryer designs that can be found in the food industry alone exemplify the diversity of food materials that require different drying techniques for value-added processing into dried products and ingredients with the desired nutritional and functional specifications. The details of the various drying techniques for food materials can be found elsewhere (Bansal and Chung 2007). In particular, many articles have been published concerning different methods and conditions for drying fruit and fruit products (Table 7.2).

7.4 Drying Limitations of Conventional Dryers

The conventional thermal dryers currently used in the food industry have several limitations, including non-uniformity of product quality due to overdrying and/or underdrying caused by long, inadequate, or exposure of the product to the non-uniform drying conditions. The long drying times arise from the low thermal efficiency between the drying medium and materials being dried. In these dryers, there is also a possibility for thermal degradation of the nutrients and generation of off-flavours due to exposure to higher temperatures. In some cases, conventional dryers require large volumes at low production capacity. These limitations will give rise to poor drying performance and higher operating costs.

Table 7.1 A generalised classification of conventional thermal dryers for food materials

Classification	Types of dryers (general characteristics and applications)
Type of feed material	• Particles
	• Slurry/paste/sludge
	• Liquid suspension
Processing mode	• Batch
	• Continuous
Mode of heat transfer	• Convection
	• Conduction
	• Electromagnetic (RF, ohmic, infrared, microwave)
	• Combination (hybrid)
Energy sources	• Electricity
	• Gas (natural/LPG)
	• Solar/wind
	• Biomass
Mode of operation	• Cyclic
	• Intermittent
	• Continuous
Product temperature	• Above freezing point
	• Below freezing point
Operating pressure	• Atmospheric
	• Vacuum
	• High pressure

7.5 Challenges in Drying R&D

The main challenge in drying food materials is to get the most rapid removal of water from the food materials, with better retention on the product quality, without damaging the environment, at the lowest capital and operating costs of the process. Today, there are tens of thousands of different materials that need to be dried in hundreds of different types of dryers for different product specifications (Mujumdar 2010b). This gives rise to a massive challenge even to select the right dryer and optimal drying conditions. Moreover, a further complication arises from the fact that many food materials with very diverse physical/chemical properties need to be dried at different scales of production and with very different product quality specifications (Mujumdar and Wu 2010). In addition, new challenges are constantly emerging as new drying requirements appear for new products. Control of dryers based on quality measurement in real-time is also a challenge.

Scaling-up of most of the dryer types continues to be complex and empirical and is often equipment and product-specific because of the highly non-linear nature of the governing conservation equations of transport processes (Mujumdar and Huang

Table 7.2 A summary of relevant studies on the different drying methods applied to the drying of fruit and fruit products

Product	Drying method	Drying condition	Refs.
Apple	• Atmospheric freeze drying	• T_a: −5, −10 °C; RH_a: 30 %; ν_a: 1.5 m/s	Duan et al. (2013)
	• Drum drying	• T_a: 110 °C; drum rotation: 0.15 rpm	Henríquez et al. (2012)
	• Freeze drying	• T_a: −86 °C for 72 h	Noorbakhsh et al. (2013)
		• T_c: 52 °C; P: 5 mTorr	Henríquez et al. (2012)
	• Heat pump drying	• T_a: 35 °C; RH_a: 20 %; ν_a: 4.0 m/s	Chong et al. (2014)
		• T_a: 35 °C; RH_a: 20 %	Chong et al. (2013)
	• Hot air-drying	• T_a: 50 °C for 14 h	Schulze et al. (2014)
		• T_a: 70 °C; RH_a: 4.6 %; ν_a: 0.965 m/s	Chong et al. (2014)
		• T_a: 40, 60, 80 °C; ν_a: 0.9 m/s	Martynenko and Janaszek (2014)
		• T_a: 40 °C; ν_a: 1.8 m/min	Noorbakhsh et al. (2013)
		• T_a: 60 °C	Henríquez et al. (2012)
	• Microwave vacuum drying	• T_a: 50 °C; P: 20 hPa; MW power: 500–1000 W	Schulze et al. (2014)
		• T_a: 22 °C; ν_a: 1.0 m/s; P: 4–6 kPa; MW power: 240 W	Chong et al. (2014)
	• Radiant energy under vacuum	• T_a: 40 °C for 3 h; MW power: 1200 W for 9 min, 600 W for 3 min	Noorbakhsh et al. (2013)
	• Ultrasonic drying	• T_a: 30, 50, 70 °C; US power: 18.5, 30.8 kW/m^3	Rodríguez et al. (2014)
		• T_a: −10, −5, 0, 5, 10 °C; RH_a: 7 %; ν_a: 2 m/s; US power: 20.5 kW/m^3	Santacatalina et al. (2014)
		• T_a: 40 °C; ν_a: 1 m/s; US power: 0–31 kW/m^3	Ozuna et al. (2014)
		• T_a: 40, 60 °C; RH_a: 25 %; ν_a: 1.0 m/s; US power: 75, 90 W @ 20 kHz	Sabarez et al. (2012)
Apricot	• Hot air-drying	• T_a: 30, 40, 50, 60, 70 °C	Kholmanskiy et al. (2013)
		• T_a: 60 and 70 °C	Albanese et al. (2013)
		• T_a: 40 and 60 °C	Chayjan and Alaei (2013)
		• T_a: 50 °C; ν_a: 2 m/s	Contreras et al. (2012)
	• Microwave drying	• T_a: 60 and 70 °C; MW power: 2 kW @ 2.4 gHz	Albanese et al. (2013)
		• MW power: 90, 270, 450, 630, 900 W	Chayjan and Alaei (2013)
		• MW power: 100 W	Igual et al. (2012)
		• T_a: 50 °C; MW power: 0.4 W/g	Contreras et al. (2012)
	• Spray drying	• T_i: 155 °C; T_o: 75 °C	Chao et al. (2012)

(continued)

Table 7.2 (continued)

Product	Drying method	Drying condition	Refs.
Banana	• Freeze drying	• T_a: −40 to −45 °C; P: 100 Pa	Jiang et al. (2014)
	• Hot air-drying	• T_a: 40–70 °C; RH_a: 6.6–30.5 %; ν_a: 1.5–1.84 m/s	Pereira da Silva et al. (2014)
	• Microwave drying	• MW power: 180, 360 W @ 2450 MHz	Esehaghbeygi et al. (2014)
	• Microwave vacuum drying	• MW power: 2 W/g; P: 80 Pa	Jiang et al. (2014)
	• Rotary drying	• T_a: 70 °C; airflow: 5.3 kg/m^3; rotation: 0.3 s^{-1}	Cabrera-Padilla et al. (2014)
Grape	• Freeze drying	• T_a: −47 °C; P: <46 mBar	Gurak et al. (2013)
	• Hot air-drying	• T_a: 7 °C; RH_a: 35 %; ν_a: 12 m^3/s	Panceri et al. (2013)
		• T_a: 50, 60, 70, 80 °C	Gholami et al. (2013)
		• T_a: 40 °C; RH_a: 20 %	Marquez et al. (2013)
		• T_a: 60, 70, 80 °C; RH_a: 10,30, 50 %; ν_a: 2 m/s	Sabarez (2014)
		• T_a: 60 °C; ν_a: 1.2 m/s	Doymaz and Pala (2002)
		• T_a: 40–70 °C; ν_a: 1.0 m/s	Mohsen et al. (2007)
Mango	• Drum drying	• T_d: 152 °C for 54 s	Caparino et al. (2012)
	• Freeze drying	• T_a: −25 to 20 °C; P: 20–4 Pa	Caparino et al. (2012)
	• Heat pump drying	• T_a: 35 °C; RH: 20 %	Chong et al. (2013)
	• Hot air-drying	• T_a: 22 °C; ν_a: 0.52 m/s	Ochoa-Martínez et al. (2012)
		• T_a: 50–80 °C; ν_a: 1.76, 1.80, 1.91 m/s	Corso and Alvarez (2014)
		• T_a: 60, 70, 80 °C; ν_a: 3.5 m/s	Kabiru et al. (2013)
		• T_a: 58 °C for 24 h	Ismail and Nagy (2012)
	• Microwave vacuum drying	• MW power: 240 W	Chong et al. (2013)
	• Refractance Window drying	• T_w: 92 °C; RH: 75 % @ 24 °C	Ochoa-Martínez et al. (2012)
		• T_w: 95–97 °C; RH: 50–52 % @ 22 °C; ν_a: 0.7 m/s	Caparino et al. (2012)
	• Spray drying	• T_i: 190 °C; T_o: 90 °C for 3 s	Caparino et al. (2012)
	• Vacuum drying	• T_a: 80 °C for 4 h	Ismail and Nagy (2012)
Pear	• Heat pump drying	• T_a: 35 °C; RH_a: 20 %	Chong et al. (2013)
	• Hot air-drying	• T_a: 30, 40 °C; RH_a: 30, 35 %; ν_a: 1.2, 2.7 m/s	Silva et al. (2014)
		• T_a: 62, 76, 84, 92 °C	da Silva et al. (2013)
		• T_a: 50, 57, 64, 71 °C; ν_a: 2 m/s	Doymaz (2013)
		• T_a: 40, 60 °C	Santos et al. (2013)
		• T_a: 55, 65, 75 °C; ν_a: 2 m/s	Doymaz and Ismail (2012)
		• T_a: 30–70 °C; RH_a: 4–16 % @ 20 °C ν_a: 1.5 m/s	Mrad et al. (2012)
		• T_a: 60, 70 °C; RH_a: 40–45 %	Velescu et al. (2012)
		• T_a: 60 °C; RH_a: 30 %; ν_a: 1 m/s	Lozano et al. (1983)

(continued)

Table 7.2 (continued)

Product	Drying method	Drying condition	Refs.
	• Microwave drying	• MW power: 1000 W @ 2450 MHz	Arballo et al. (2010)
	• Microwave vacuum drying	• MW power: 240 W	Chong et al. (2013)
	• Spray drying	• T_i: 140–160 °C; airflow: 600 L/h	Saenz et al. (2009)
	• Ultrasonic-assisted drying	• T_a: 70 °C; US power: 400 W @ 24 kHz	Dujmic et al. (2013)
Pineapple	• Far Infrared (FIR) and hot air-drying	• T_a: 40–60 °C; ν_a: 0.5–1.5 m/s; FIR Intensity: 1–5 kW/m^2	Ponkhama et al. (2012)
	• Freeze drying	• T_a: −14 to −34 °C	Vieira et al. (2012)
	• Hot air-drying	• T_a: 40, 60, 75 °C; ν_a: 1.5 m/s	Ramallo and Mascheroni (2013)
		• T_a: 70 °C	Agarry et al. (2013)
		• T_a: 45, 60, 75 °C; ν_a: 1.5 m/s	Ramallo and Mascheroni (2012)
	• Spray drying	• T_i: 160 and 170 °C	Suhaimi et al. (2011)
Plum	• Hot air-drying	• T_a: 90 °C; ν_a: 1.0 m/s	Miletic et al. (2013)
		• T_a: 60 °C; ν_a: 1.5 m/s	Walkowiak-Tomczak (2012)
		• T_a: 50, 70, 75, 85 °C; ν_a: 0.6, 0.9, 1.2 m/s	Ioannou et al. (2011)
		• T_a: 55, 60, 75 °C; ν_a: 1.1 m/s	Zivkovic et al. (2011)
		• T_a: 60, 85 °C; RH$_a$: 40–30 %; ν_a: 1840 m^3/h	Madrau et al. (2010)
		• T_a: 85 °C; ν_a: 0.81 m/s	Jazini and Hatamipour (2010)
		• T_a: 70, 80 °C; RH$_a$: 15, 35 %; ν_a: 1.5, 2.9, 4.3, 5.7, 7.0 m/s	Martynenko and Janaszek (2014)
		• T_a: 70–100 °C; RH$_a$: 5–65 %; ν_a: 0.3 m/s	Sabarez (2012)
		• T_a: 50–70 °C; ν_a: 0.8 m/s	Sacilik et al. (2006)
		• T_a: 70, 75, 80 °C; RH$_a$: 3 %; ν_a: 0.25 m/s	Sabarez and Price (1999)
		• T_a: 40–90 °C; RH$_a$: 3 %; ν_a: 1.0 m/s	Price et al. (2000)
Strawberry	• Freeze drying	• T_a: −40 °C; P: 10^{-2} Pa for 48 h	Mosquera et al. (2012)
		• P: 6–150 Pa for 24 h	Oikonomopoulou and Krokida (2012)
	• Hot air-drying	• T_a: 50, 60, 70 °C	Aktas et al. (2013)
	• Ultrasonic drying	• T_a: 40–70 °C; US: 0, 30, 60 W	Gamboa-Santos et al. (2014a)
		• T_a: 40–70 °C; ν_a: 2 m/s; US power: 0, 30, 60 W	Gamboa-Santos et al. (2014b)
	• Vacuum drying	• T_a: 65 °C; P: 10 kPa	Orak et al. (2011)
		• T_a: 50, 60, 70 °C	Aktas et al. (2013)

Note: subscript: *a* air, *i* inlet, *o* outlet, *w* water, *d* drum surface

2007). Mathematical modelling has become a useful tool to simulate the drying process for scaling-up and optimisation, without excessive trial-and-error and the associated costs of physical experimentations. However, the processes of drying are very complex to model since they involve transient energy and mass and momentum transfer through a porous or nonporous medium with phase changes and with or without chemical reactions (Mujumdar and Huang 2007). Moreover, a further problem with modelling and scaling-up is the association of quality predictions that depend not only on the transport phenomena but also on the materials being dried. The development of a mathematical model considering all these phenomena is a major challenge. Modelling of the drying process at combined different scales of time (milliseconds to months) and length (nano to macro scales) has also remained a formidable task. In some cases, few experimental data are available for validation of the models. This is due to the complexity in measurements of the drying parameters (e.g. hostile environment of high temperature such as in spray drying) that would hamper the development of reliable drying models.

With advances in computing capabilities and analytical instrumentations, one can now look forward for a better understanding of the drying process at deeper scales than ever before. Such advances will intensify innovations and help elevate drying R&D to the next level of sophistication. These will facilitate the development of realistic drying theories and multi-scale models in the coming decade (Mujumdar and Huang 2007). A multidisciplinary approach to drying R&D will also be important for effective development of new and improved drying technologies.

7.6 R&D Opportunities in Drying

Increased competitiveness due to globalisation in addition to the growing consumer demand for better quality products will continue to provide the impetus for increased drying R&D seeking to further improve the energy efficiency and cost-effectiveness of the drying process with improved quality product. The environmental impact of large drying operations has also become a critical issue in recent years. Drying systems consume large amounts of energy and hence result in high emissions of greenhouse gases. This is concerning particularly in areas inevitably dependent on energy consumption from fossil fuels. The need for developments of new and improved drying technologies to improve efficiency, reduce environmental impact and enhance quality of the dried products will be crucial for a sustainable future of the food and other industries that heavily rely on drying processes.

Figure 7.2 illustrates an example of an approach for systematic development of new and improved drying processes for food materials. The process-product interaction requires a better understanding of the mechanisms not only the material drying kinetics as function of the drying operational variables, but also, the changes in the material properties during drying. It is therefore paramount that drying R&D efforts should be devoted to understanding the underlying drying fundamentals and the interplay between transport phenomena and the material properties to improve

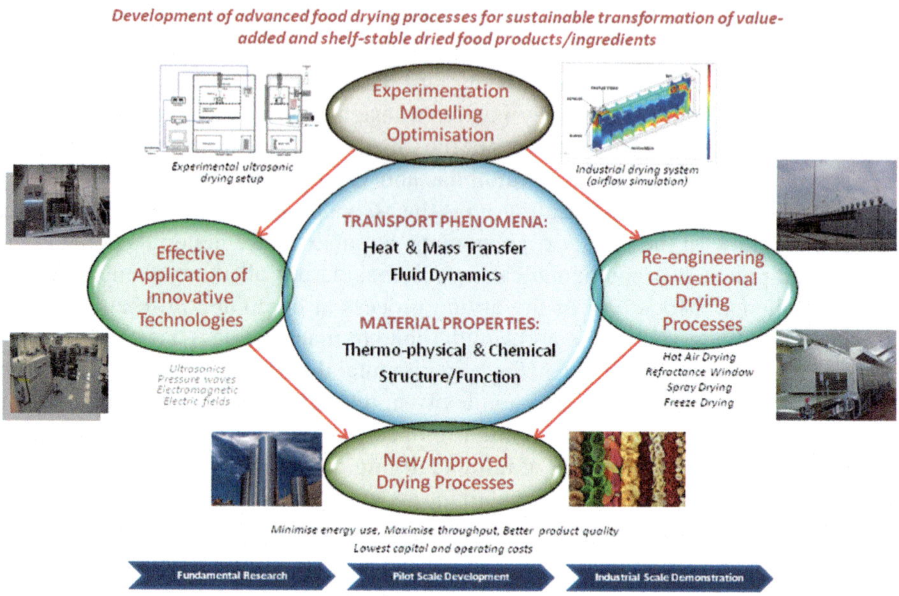

Fig. 7.2 Schematic diagram of an approach for development of new/improved drying processes for food materials

the performance of traditional drying techniques as well as for developing new ones. The new and improved drying technologies must be cost-effective (i.e. lowest capital and operating costs) to ensure market acceptance. In this approach, the research focus is categorised into two major areas of interest, including (1) development of innovative drying technologies, and (2) looking at ways of improving the performance of existing drying processes.

7.6.1 Development of Innovative Drying Technologies

In recent years, a number of innovative food processing technologies have been investigated and developed with the aim to improve or replace conventional processing technologies. These novel or emerging technologies take advantage of other physical phenomena such as sound waves, pressures, electric and electromagnetic fields. The limitations of the conventional drying processes may be overcome by the combined application of these innovative technologies for the development of new drying concepts for improving the quality of food products through gentle processing. In particular, the application of ultrasonic energy to assist drying of food materials has been explored for several decades. It is known for many years that the energy generated by sound pressure waves could enhance a wide range of processes due to a series of mechanisms activated by the ultrasonic energy such as heat, diffusion, mechanical rupture, chemical effects, etc. (Gallego-Juarez et al. 2007).

The ultrasonic systems proposed in literature for assisting the food drying processes may be grouped into two main types. The first type is made up of ultrasonic devices working directly coupled to food material during drying (Garcia-Perez et al. 2009). The direct contact system could promote an accelerated drying process as this system permits good transfer of ultrasonic energy from the vibrating element to the food material. Nevertheless, the main drawback of this technique may be its difficulty to adapt in conventional hot air-drying processes. In an attempt to make the ultrasonic systems easier to adapt to conventional drying processes, another type was developed based on the application of airborne ultrasound. The airborne ultrasonic system works without direct contact between the vibrating element and the food material. However, this results in a less efficient ultrasonic energy transfer. The main difficulties in this system come from the inefficient generation of ultrasonic energy in air and the transfer of such ultrasonic energy from air into the product due to the acoustic impedance mismatch and the energy absorption by the air at ultrasonic frequencies (Gallego-Juarez et al. 1999).

Several studies have reported the applications of ultrasonic technology in combination with convective drying processes. In particular, a number of investigations have shown the potential of power ultrasound to improve the drying process of various food materials. In these studies, the ultrasonic energy was transmitted as either airborne to the surface of food material (Garcia-Perez et al. 2007a, 2007b, 2009, 2010; Carcel et al. 2007, 2011a, 2011b; Khmelev et al. 2008, 2011; Ozuna et al. 2011; Soria and Villamiel 2010) or in direct contact between the product and the vibrating element (Gallego-Juarez et al. 2007; Gallego-Juarez 2010; Schossler et al. 2012). For instance, Gallego-Juarez et al. (1999) developed a series of high-power ultrasound prototypes for the application in forced-air-drying by airborne and in direct contact with the food sample. They observed a significant increase in drying rates for various food products tested when there was a direct contact between the vibrating element and the materials being dried which they attributed to the good acoustic impedance match (or coupling) between the transducer and the food materials. However, the effect of ultrasound on drying was reduced when carried out using an airborne technique.

In general, the studies that can be found in the literature have reported that ultrasound could enhance the drying process of food materials, and that the extent of benefit of ultrasound was dependent on process variables (i.e. product properties, drying conditions, method of ultrasound application, ultrasonic energy, etc.). A better preservation of the quality (internal food structure) due to the shortening of drying time has also been reported with the application of ultrasound during drying (Garcia-Perez et al. 2010). Despite all these works, no commercial-scale installation of the ultrasound drying technology has been developed to date (Soria and Villamiel 2010). This is due to some technical challenges to achieve an efficient transmission of acoustic energy and the practical difficulties to adapt the technology at industrial scale. In order to develop the ultrasound drying technology into fully industrial operations, it is important to gain a better understanding of the mechanisms and under what conditions ultrasound influences the drying kinetics and the associated changes in product quality attributes and in determining whether and how to apply such energy to the drying process effectively (Sabarez et al. 2012).

A promising approach for the application of ultrasound to assist in convective food drying of apple slices was developed and tested by Sabarez et al. (2012). This study was carried out to investigate the effect of ultrasound on the drying kinetics and product quality attributes using the alternative approach for the application of ultrasonic energy in the convective drying process. The approach is based on the transmission of ultrasonic energy using a vibrating stepped-plate ultrasound technique as a combination of airborne and through a series of solid contacts between the ultrasound element and the product tray as the ultrasonic vibration transmitting surface (Fig. 7.3). The results from this work indicate a significant reduction in drying time (up to 57 %) with the simultaneous application of ultrasound on convective drying of apple slices (Fig. 7.4). This corresponds to a reduction on energy consumption by up to 54 % with the ultrasound-assisted convective drying process. The processing variables (i.e. drying temperature, product thickness, ultrasonic power level) were observed to substantially influence the magnitude of the effect of ultrasonic energy in enhancing the drying process, indicating the necessity to establish the optimum drying conditions for specific product and ultrasonic applications. In particular, the ability of ultrasound to improve the efficiency of convective drying process seems to be maximised when using low temperature and high ultrasonic power level.

In a further study (Beck et al. 2014), the application of a specially designed ultrasonic horn for a completely airborne ultrasound transmission to assist in convective drying of a model food system was investigated. This work involved investigations of the impact of airborne ultrasound at various power levels and different levels of drying air conditions (i.e. temperature, relative humidity and velocity) using a Response

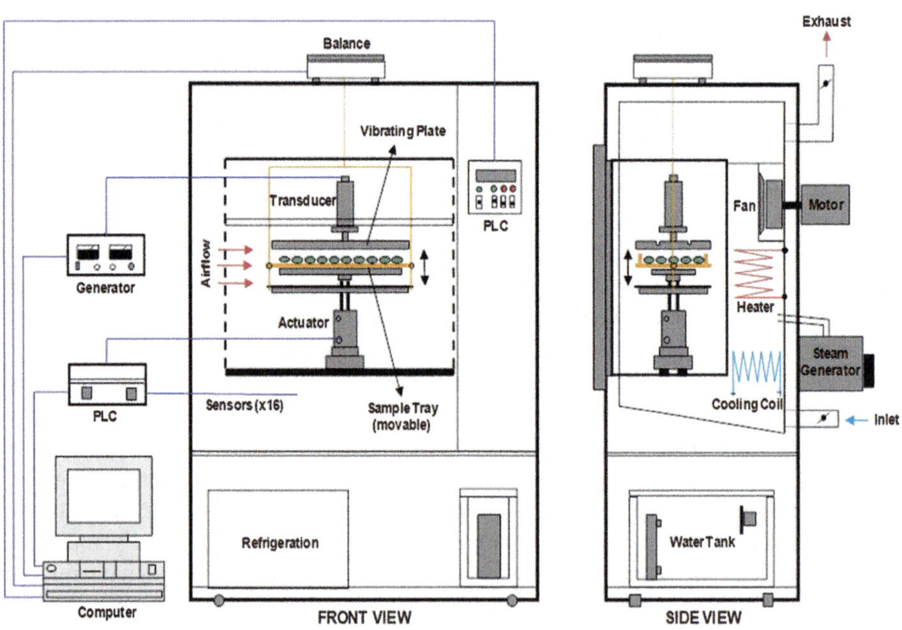

Fig. 7.3 Schematic diagram of a computerised ultrasound-assisted convective drying system

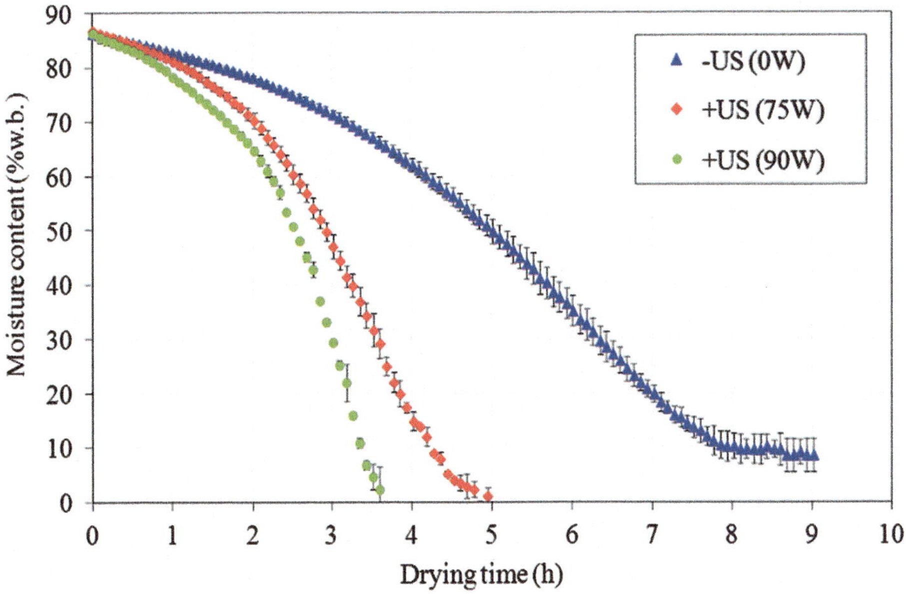

Fig. 7.4 Effect of ultrasound on the drying kinetics of apple slices at different ultrasonic power levels (T=40 °C; RH=25 %; ν=1.0 m/s; 5 mm thickness)

Surface Methodology (RSM) approach to examine the possible interactions between these parameters and to find the combination of these factors that yield the best response. The airborne ultrasound equipment tested in this work was found to enhance the conventional hot air-drying process by reducing the overall drying time significantly (i.e. by more than 60 %). The process parameters (temperature, air humidity, air speed and ultrasound power level) and their interactions affected the drying process substantially, with optimum conditions found using the RSM approach.

In general, the findings from these studies offer a promising alternative to facilitate the adaptability of the technology in industrial scale operations as there is no direct contact between the ultrasound element and the food sample to be dried. Further research efforts to optimise the technology for application in industrial food drying and the application to other drying techniques, together with future advancements in ultrasonic technology should provide the basis to build upon the development of the new ultrasonic drying technology for adoption in industrial drying practice.

7.6.2 *Process Improvements of Existing Drying Technologies*

The conventional drying processes that are currently used in the food industry will continue to play a significant role in food manufacturing while they are still viable and have not reached their limit of performance. The need for replacement with new drying technologies is typically limited due to the long life span (i.e. typically in the range of 20–40 years) for most dryers (Mujumdar 2010a). However, the environmental impact at large-scale operations apart from the increased global competitiveness has become a critical issue in recent years, requiring further improvements and optimisation of these existing drying technologies. These industrial dryers, most of which were designed during the era of cheap and abundant energy, typically operate at low thermal efficiencies in the range 20–60 % (Mujumdar and Wu 2010).

There is always a scope for improvements in almost all conventional dryers in practice to make these dryers more energy-efficient, environmentally friendly and safer to operate. Incremental improvements by optimisation of the design and operation of the conventional drying technologies are still readily embraced by the food industry due to reduced inherent risks than implementing new drying technologies. Various researchers have suggested different ways of reducing energy consumption, which included better control, proper insulation, recycling of a part of the drying air, and so on. This may also involve mechanical changes, drying media, gas distribution system, time cycles for drying, combining different heat transfer modes, multistage operation, etc. (Jangam 2011).

In industrial tunnel dehydrators, for example, the drying operation involves complex conditions interconnected with many factors associated in the design features and operational practices. These factors include the dynamic conditions of the drying air (i.e. airflow, temperature and humidity), tunnel operation (i.e. parallel-flow or counter-flow), tunnel design (e.g. dimensions, number of tunnels, insulation, tray spacing, etc.), and other parameters (e.g. tray loading, final moisture content, raw material size, etc.). It is difficult to provide specific assessment and optimisation considering all of these factors. Every dehydration system is unique and likely to have different energy requirement and drying characteristics. However, it is fundamental to identify critical factors that offer significant and measurable opportunities for improvements. The drying performance in a typical drying system of standard dimensions can be maximised through optimisation of operating conditions for improved design and operational practices. The conditions of the drying air (i.e. airflow, temperature and relative humidity) are often considered to be the main factors influencing the drying performance in a convective drying system.

7.6.2.1 Airflow

Air is required to transfer heat and remove moisture from the product during convective drying. The heat and mass transfer rates depend on both temperature and moisture concentration differences. The air velocity field greatly influences these transfer rates at the food–air interfaces. Therefore, the temperature and concentration of moisture in the product and the drying air are basically controlled by the level of air velocity and its distribution. The level of air velocity will play a significant role in achieving efficient drying of food products. Increased air movement over the product enhances the drying process as a consequence of improved heat and mass transfer rates. The movement of air is also important particularly during the early stages of drying when the external mass transfer mechanism (water evaporation process) predominates. It has little impact on drying when the diffusion rate (internal moisture transfer) is the limiting factor, mainly in the later stages of drying. It is this relationship that is extremely important and often overlooked. Many conventional dehydrators operate over a wide range of air velocities, implying significant opportunities for improvement for many of these dehydrators.

7.6.2.2 Temperature

It is well documented that the drying process is accelerated at higher temperatures. Elevated temperature would mean an increase in drying rate. Obviously, this shortens the drying time and consequently increases the production throughput. However, for most food materials caution is required because higher temperatures may alter their physical and chemical constituents, which affect the quality of the dried product. Hence, the drying temperature is limited to a point below which the undesirable characteristics are minimised. In prune drying, for example, the maximum allowable temperature was reported to be up to 85 °C (depending on fruit maturity, humidity, airflow, mode of tunnel operation, etc.), as drying above this temperature would result in excessive bleeding and splitting of the fruit and possibly caramelisation and off-flavour production (Sabarez and Price 1999; Gentry et al. 1965). Increase in temperature would also mean an increase in energy input requirement. It is therefore evident that establishing the optimal temperature level is important in obtaining efficient drying of food materials without compromising product quality.

7.6.2.3 Relative Humidity

The amount of moisture in the air is well known to affect the kinetics of moisture loss during air-drying of food materials. In a thermodynamic sense, decreasing the amount of moisture in the drying air increases the potential of the drying air to pickup and remove moisture from the product. This is due to the fact that the reduction in the relative humidity of the drying air increases the moisture concentration gradient between the product and the drying air. Consequently, this leads to an increase in the driving force for mass transfer from the product surface to the air stream. It is therefore possible to control the drying process by manipulating the relative humidity levels of the drying air.

In conventional hot air-drying processes, different levels of relative humidity of the drying air entering the system can be achieved by regulating a proportion of the make-up ambient air intake and the amount of moisture-laden exhaust air being recirculated. Lowering the relative humidity of the drying air is ideal to increase production throughput with reduced energy consumption. However, lower levels of relative humidity would have to be achieved by lower recirculation of humid exhaust air and therefore does have further energy consumption implications. A usual practice was to operate the dehydrators at a low relative humidity to obtain the maximum production throughput based on the fact that a large amount of food materials must be dried in a short time. However, a minimum relative humidity level of the drying air is required to avoid damage and burning. For example, Bertin and Blazquez (1986) reported a 15 % relative humidity as the minimum value for drying prunes. On the other hand, further energy savings with minimal effect on production throughput could be achieved if utilising the maximum acceptable relative humidity level. This is due to the fact that this will allow an increase in the recirculation of hot exhaust air to maximise heat recovery. Thus, better understanding and control of the optimal relative humidity level is also important in obtaining efficient drying operation.

7.6.2.4 Design and Operational Practices

There are other design parameters and operational practices that may provide further improvements to drying performance. In large-scale drying systems, there is a systematic variation in drying conditions (i.e. temperature and humidity) and consequently drying rate as the air moves from hot end to cold end of the drying chamber. As the air passes over successive wet products, the air progressively gains humidity and loses heat, resulting in humid conditions of the air (but still with considerable heat) exiting from the drying system. The temperature and amount of moisture in the exhaust air are dependent on many factors (i.e. dryer length, product loading, initial moisture content, air conditions entering the dryer, etc.). Usually, the temperatures of the exhaust air are still quite high, making recirculation of exhaust air inside the dehydrator ideal to recover its heat. However, the exhaust air also contains higher relative humidity, hence increased recirculation of exhaust air would increase the relative humidity of air entering the dryer.

An increase in relative humidity will decrease the drying rate consequently increasing the drying time (i.e. decrease in production throughput). This limits the amount of exhaust air that can be recirculated. It is therefore paramount that the degree of recirculation of exhaust air does not raise the humidity of the air entering the dryer to an unacceptable level. This underlines the importance of establishing the maximum permissible amount of the exhaust air that can be recirculated for energy conservation to improve energy efficiency whilst maintaining a high level of production throughput. The reason is obvious, the higher the recirculation of hot exhaust air the less the demand for cold fresh intake air and subsequently requiring less energy to raise the temperature of the drying air (mixture of recycle air and make-up ambient air) to the desired level. In this way, more energy is conserved and therefore the demand for fuel consumption is lower. It is clear that there is a considerable energy saving with the use of maximum air recirculation without affecting production throughput.

One of the major issues in drying operations is the non-uniformity in the moisture content of the final dried product, which is more apparent for large-scale industrial drying systems. This is inherently due to non-uniformity in drying conditions

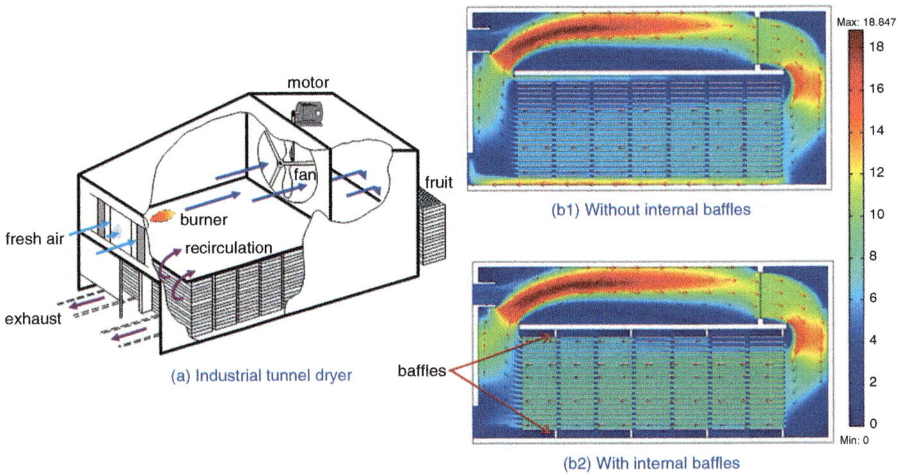

Fig. 7.5 A 2D computer simulation of the air velocity field (m/s) and its distribution across a typical (**a**) tunnel dryer, as affected by the internal baffle system (**b1**) without baffles and (**b2**) with baffles

(i.e. temperature and humidity) as a consequence of uneven airflow distribution. When drying high moisture foodstuffs, the problems of air velocity are not just too much or too little, they are often due to the uniformity of that air velocity. Also, if the velocity is too low, convection currents and other disturbances will cause wide variations in temperature and relative humidity, resulting in uneven drying (Sabarez 2012). Moisture non-uniformity will reduce drying effectiveness, increasing drying costs and reducing drying capacity. Non-uniformity in the moisture content of the final dried product could be minimised through proper guidance and distribution of air in the dryer. For instance, computer simulation studies show that airflow uniformity could be improved by adding suitable air baffles and guide vanes in a traditional tunnel drying system (Fig. 7.5). It should be noted that when designing a convective drying system, it is important to remember that air will always follow the path of least resistance. Thus blocking the air gap between the truck/trays and inside the dryer walls (sides, ceiling and floor) would force the hot air into the trays over the product. Hence, the dryer must be designed in such a way that the hot air is forced into the trays over the product.

Another key to achieving a cost-effective drying process is to dry the food materials as close as possible to the desired level of final moisture content. The maximum acceptable level of moisture content at which the food products should be dried is dependent on many factors (i.e. type of the product, ambient conditions during the subsequent storage, etc.). It should be noted that there is a very long tail in the last stages of the drying process, hence significant reduction in drying time can be achieved if food products could be dried close to the desired moisture content level. Obviously, drying down to lower moisture levels (i.e. overdrying) than the desired moisture level would increase the overall drying time, hence incurring additional fuel costs apart from reduced production throughput and possibly additional impact on product quality attributes (i.e. caramelisation, off-flavour, loss of nutrients, rehydration, etc.).

7.7 Modelling and Optimisation

Intensification of innovation can be accelerated through the use of a reliable tool to simulate and test new designs with minimal investment of time, manpower, and funds. Together with advances in computing capability, the development of computational models to accurately simulate complex processes with less time required is one of the great advances in process engineering research. This enables to predict outcomes for performance evaluation, optimisation and scaling-up of new and untested process designs, without the excessive need for expensive and labour intensive trial-and-error experimentations. Computational modelling can also be utilised to develop new conceptual designs and to optimise operating conditions as a cost-effective route to intensify improvements in existing conventional dryers. Intelligent utilisation of reliable mathematical models could intensify the search for disruptive drying technologies to supplant today's inefficient drying technologies.

There is a great deal of published literature regarding modelling of the drying process of food materials. Food materials are extremely complex in their structure and composition so there is no universally acceptable way to model their drying behaviour as exemplified from the vast volume of literature published on various modelling approaches (Kostoglou et al. 2013). This also reflects the extreme diversity of the drying mechanisms in food systems. Mujumdar and Huang (2007) stated that most models are applicable for specific product-equipment combinations. The most common ones are the empirical models resulting by simply fitting to the experimentally determined drying curves (Mujumdar 1987; Togrul 2005). These models are widespread due to their simplicity in the implementation while still adequately describing the drying process. Although empirical models would give good results for engineering applications in the food industry, they frequently do not allow the simulation of experiments carried out under conditions different to those used to identify the model parameters (Ah-Hen et al. 2013). In addition, these models are generally based on simplifying hypotheses which either considered only the internal or external resistance to mass transfer in an isothermal process where the product shrinkage and transport properties changes are often not taken into account (Sabarez 2012). During the drying process, variation in moisture content and temperature as a function of both time and space exists inside the material, but this is not included in empirical models, which may further limit their application to drying. The approaches are also based on simplifying assumptions which may not be applicable in some situations (e.g. complex food geometries) and dynamic operating conditions during the drying process. The dependence of thermo-physical and transport properties on product temperature and moisture content is not taken into consideration. Menges and Ertiken (2006) and Yaldiz et al. (2001) presented a comprehensive list of such empirical models.

A more comprehensive modelling approach involves numerical computation of a theoretical model describing the simultaneously coupled transfer of momentum (air only) heat and mass (both air and food) under transient conditions occurring during convective drying of food materials (Sabarez 2012; Curcio et al. 2008). These models are based on the fundamental physical principles of the drying process and take into account the variability of air flows (fluid dynamics) around the food material. For example, Sabarez (2012) successfully used similar approach to predict the moisture content and temperature distributions within the prunes during drying. Aversa et al. (2007) also employed a similar modelling approach to describe the transport phenomena occurring during drying of carrot slabs. It should be noted that at the air–food interface the external heat and mass transfer rates could be the controlling step in the drying process depending on the drying conditions, which are strongly dependent on the drying air velocity field. It is therefore important to further account for the momentum transport dynamics to improve the predictive precision of the model. However, a detailed analysis of the inherent complexity of the coupled transient phenomena (i.e. heat, mass, momentum and deformation) involved in the drying process is often regarded as time consuming for practical purposes. With the advent of increasing computing power, it is now possible to undertake simulations of these complex phenomena with less time required.

Case Study

A characteristic case study in convective drying of plums is presented to illustrate the application of the modelling approach for determining the optimal design and operating conditions at industrial scale tunnel drying systems. The mode of operation in tunnel dehydrators is one of the key factors that may influence the drying performance. The industrial tunnel dehydrators are currently operated in either counter-flow or parallel-flow mode of operation. The details of the mode of operations of this dehydrator can be found elsewhere (Sabarez 2012). The conditions of the drying air (i.e. airflow, temperature and relative humidity) are considered to be the main factors influencing the drying performance in tunnel dehydrators.

In industrial tunnel dehydrators (as in any large-scale drying systems), the materials being dried are typically exposed to the changing conditions of the drying air at any time and position. This requires modelling of the drying system that comprises both material and equipment models, in which the material model describes the drying kinetics and the equipment model determines the changes of the condition of the drying medium with time and space during drying. Together, these models constitute a complete modelling tool capable of predicting the dynamic behaviour of the drying system. Thus, the prediction of the drying air stream conditions flowing across the product surface which would affect the drying behaviour of the solid product at any time and position in the dryer is of particular importance in simulating the drying process of industrial drying systems where a systematic dynamic variation in drying conditions is typical (Sabarez 2012).

A two-dimensional axis-symmetric model was developed to describe the simultaneous transfer of momentum (air only), heat and mass (air and food) occurring in convective air-drying of fruits (e.g., plums). The governing partial differential equations (PDEs) describing the simultaneous transfer of heat, mass and momentum in two distinct sub-domains (air and food) during drying of plums were presented in previous studies (Sabarez 2010, 2012). The non-isothermal turbulent flow of air in the drying chamber is described according to the standard k-ε model (COMSOL Multiphysics 2007).

The resulting systems of highly coupled non-linear PDEs in the space–time domain together with the set of initial and boundary conditions were numerically solved by the finite element method (FEM) coupled to the Arbitrary Lagrange-Eulerian (ALE) procedure to account for the shrinkage phenomenon using a commercial software package (COMSOL Multiphysics 2007). The details of the numerical solution are presented in previous studies (Sabarez 2010, 2012). Also, the solution of the governing PDEs requires knowledge of the thermo-physical and transport properties of the product and air. The model parameters used in this work are given in previous studies (Sabarez et al. 1997; Sabarez 2012, 2014).

Figure 7.6 shows the measured surface and centre temperatures of the product together with the predicted values. It can be seen from this figure that there is a good agreement between the experimental data and predicted values. These results confirm the suitability of the model to describe the heat transfer process during drying of plums and demonstrate that the thermo-physical parameters used in the model are reasonable. Similar trends were also found for other drying conditions investigated (Sabarez 2010). This validates the dependency of the product's thermal properties on both temperature and moisture content.

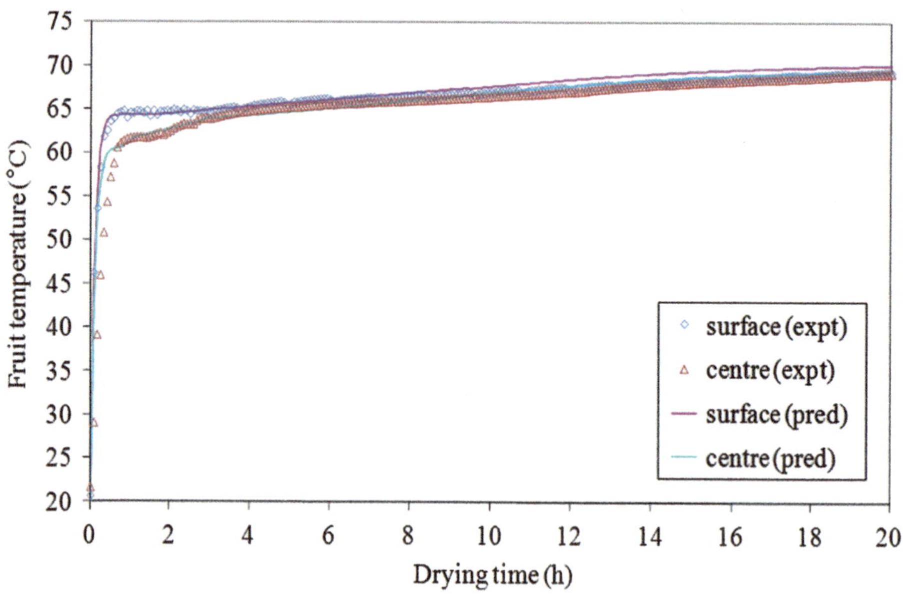

Fig. 7.6 Predicted versus experimental fruit temperature profile at different locations in the fruit (T=70 °C; RH=35 %; v=5.7 m/s)

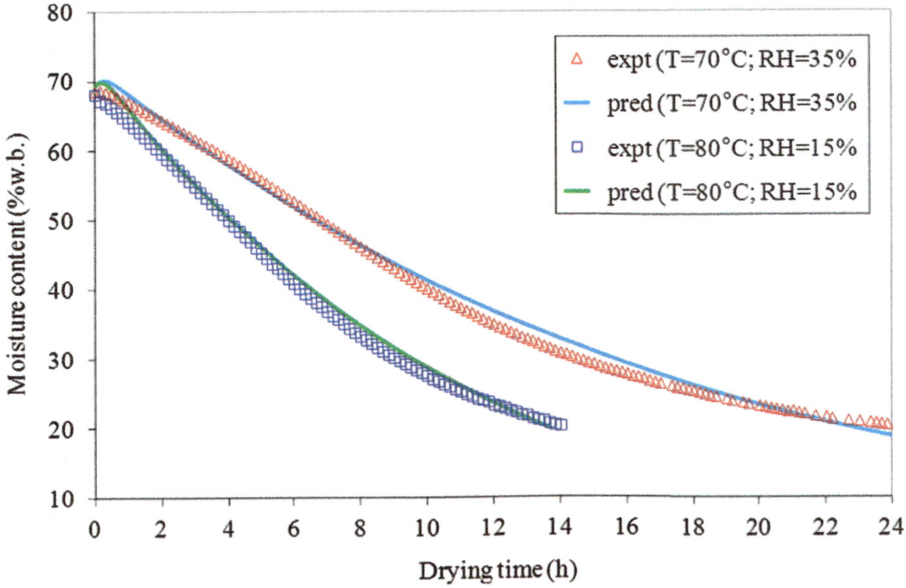

Fig. 7.7 Predicted versus experimental drying kinetics of plums (ν=5.7 m/s)

The figure indicates differences between the product surface temperature and the drying air temperature (about 5–10 °C) particularly in the early stages of drying. This can be explained by the evaporative cooling effect due to rapid moisture flux on the surface during this period. The rapid moisture flux on the surface of the product during this period requires more energy for moisture evaporation and hence less heat received by the product initially. Then as the moisture flux decreases in the later stages of drying, the product temperature gradually increased approaching near equilibrium with the drying air temperature.

Figure 7.7 shows the drying curves of plums predicted by the model for the two experimental drying tests performed at different air temperature and relative humidity levels under the same air velocity (5.7 m/s). In the moderate drying conditions, the air temperature was maintained at 70 °C with relative humidity of 35 % while in the more intense conditions the air temperature was 80 °C with a relative humidity of 15 %. As can be seen from this figure, the simulated results agree well with the experimental data. Also, Sabarez (2012) presented further validations to verify the predictive capability of the model over a range of conditions. The results confirm the validity of the model and demonstrate that the parameters used in the model are reasonable, indicating the suitability of the model to describe the drying process of plums under various conditions. The advantage of the proposed numerical model is that the temperature and moisture distributions across the solid food domain as well the changes of the condition of the drying air with location can be established at any time during drying (Fig. 7.8). This is important in simulating the drying process that will take into account the dynamic changes in the drying conditions.

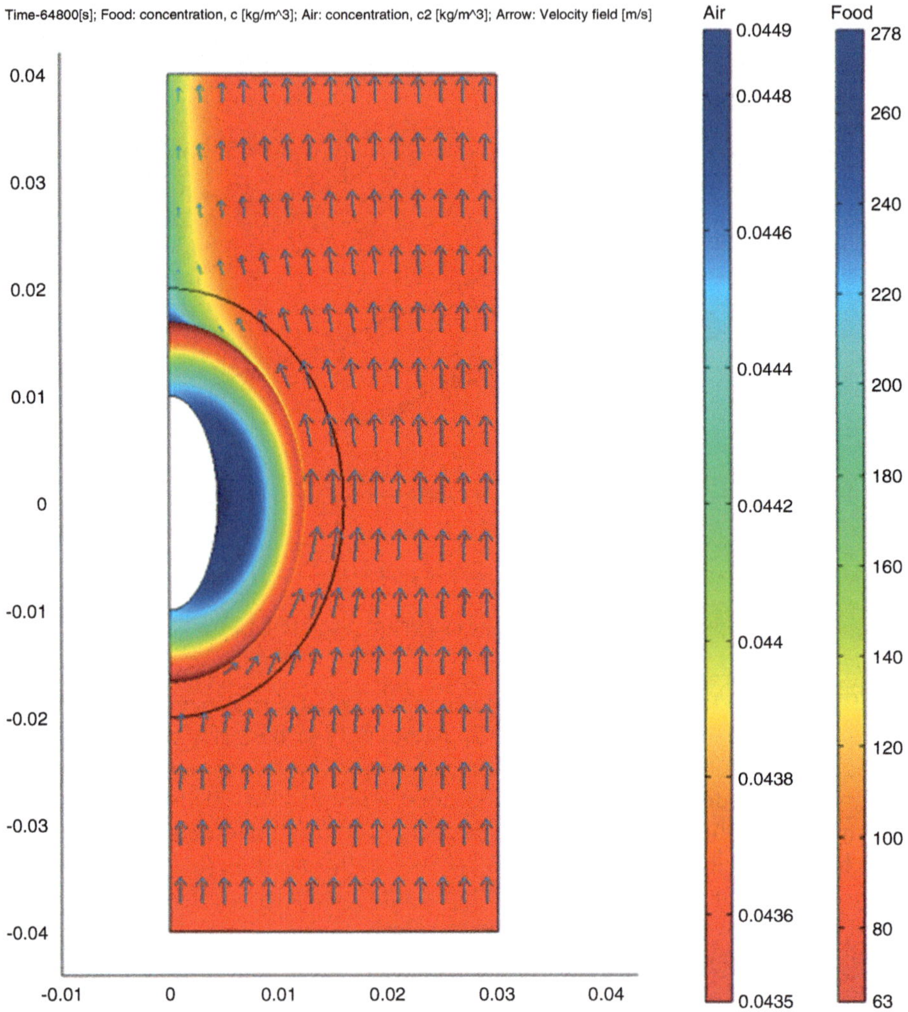

Fig. 7.8 Predicted product moisture concentration, moisture concentration and velocity profiles of the drying air during drying of plums ($T = 80$ °C; RH = 15 %; $\nu = 5.7$ m/s)

A number of computer simulations were undertaken to mimic the industrial scale tunnel drying of plums in a parallel-flow mode of operation. The selected conditions are representative for current commercial tunnel drying operation for plums (Sabarez 2010). The simulation runs were carried out for various inlet air velocity conditions. In this case, the effect of different air velocity levels was taken as an example to demonstrate the impact of this parameter on the drying kinetics and energy consumption. It is well known that the air velocity field greatly influences the heat and mass transfer rates at the food–air interfaces. Therefore, the temperature and concentration of moisture in the product and the drying air are basically

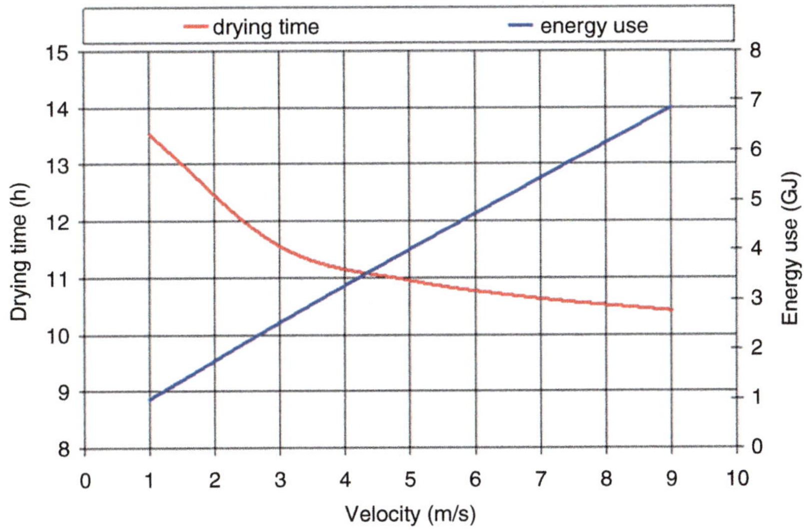

Fig. 7.9 Effect of air velocity on drying time and energy consumption (Inlet: $T=85$ °C, RH = 15 %; Recycle: $T=70$ °C, RH = 30 %, Ratio = 90 %; Ambient: $T=25$ °C, RH = 65 %)

controlled by the level of air velocity and its distribution. Figure 7.9 depicts the simulated effect of different levels of air velocity on both drying time and energy consumption. It shows that as the air velocity increases the energy consumption appears to increase. This is obvious as increases in air volume would result in increased energy requirement to heat the large volume of air to the desired temperature level. On the other hand, it appears that the drying time significantly decreases as the air velocity increases but only to certain point and then beyond this level the air velocity plays a proportionally decreasing role in reducing the drying time.

As can be observed from this plot, there appears to be an optimum level of air velocity to achieve better drying performance, which can be found at the intersection of the plots. Under these conditions, it can be seen that the optimum air velocity level is around 4–5 m/s. Hayashi (2007) suggested an air velocity of about 5 m/s as sufficient level in drying most products in tunnel dryers. The result indicates that further increase in air velocity beyond this level would significantly increase the energy consumption with minimal increase in throughput (i.e., slight reduction in drying time). For example, increasing the air velocity from 5 to 7 m/s would increase the energy consumption by as much as 26 % without gaining any significant reduction in drying time. Consequently, there is little to be gained by using a very high air velocity. If the air velocity is too high, the costs of energy required to heat the excessive air would tend to offset the benefits of slight reduction in drying time. On the other hand, it shows that large savings in energy could be achieved if operating at lower air velocities, however with the expense of longer drying times. Apart from reduced throughput, operating at longer drying times would also increase the labour costs associated in drying and possibly affect the product quality due to prolonged exposure times.

Also, the distributions of moisture content and temperature across the food materials are important to characterise the quality changes during drying. The development of mathematical models for improved understanding of the underpinning heat and mass transfer mechanisms controlling the drying process and the associated impact on product quality attributes is crucially important in achieving the optimum design and operating conditions of a drying system that maximises the retention of the desired quality attributes of the product. One of the important quality attributes that usually accompanies dehydration of food products, particularly for fruits (e.g. grapes) is the change of product colour due to browning reactions (i.e. enzymatic and non-enzymatic). The ability to predict the changes of the product colour during drying would be useful in optimising the drying process that gives the desired premium colour attributes. For example, a kinetic model was coupled to the heat and mass transfer calculations to describe the drying kinetics and the evolution of product colour during finish drying of trellis-dried sultanas (Sabarez 2014). This allows simultaneous predictions of moisture content, temperature and colour profiles of the product in a space–time domain during the drying process as a function of various operating conditions for establishing the optimal drying conditions that gives the desired premium colour attributes. It is envisaged that the approach can be extended to other food products and for incorporation of other product quality attributes.

7.8 Future Directions

Energy consumption is a major concern in industrial food drying operations, not only due to the increasing cost of fuel. Environmental concerns are also at the forefront of industry priorities, and that energy consumption needs to be addressed to reduce greenhouse gas emissions. The current concerns over potential energy shortages and global climate change will be likely to result in legislative actions to minimise fossil fuel usage. These, coupled with the increasing consumer demand for safe, healthy and high-quality shelf-stable processed foods continue to drive the need for research and innovation in drying technologies.

The limitations of the conventional drying processes may be overcome by the application of innovative or emerging technologies for gentle processing. However, there are still technical challenges to achieve efficient and cost-effective processes with the combined application of the innovative technologies and the practical difficulties to adapt these technologies at industrial scale. The use of eco-friendly sources of energy and the development of smart energy utilisation technologies will continue to become important research topics of interest in drying technology for the years to come. With advances in sensing, computing and visualisation capabilities, real-time model-based control of the drying process is becoming a reality to ensure efficient process, safe operation and better quality product. This will also place computational modelling in a good position to support and play a significant role to overcome these challenges with minimal investment of time, manpower, and funds.

The development of advanced and technically complex theoretical models that coupled the transport phenomena (heat and mass transfer, fluid dynamics), physical/structural changes, chemical reactions, phase changes, and complex food compositions will be crucial in understanding the length and time-scale interactions for optimisation and scaling-up. The additional physics (e.g. sound pressure waves) involved in the process with the application of innovative technologies further complicates the modelling task, aside from the association of quality predictions that synergistically depend not only on the physical phenomena but also on the materials being dried. This requires a multidisciplinary approach to obtain a better understanding of the underlying drying fundamentals and the interplay between transport phenomena and the material properties for the development of new efficient drying processes.

References

Agarry, S.E., A.O. Ajani, and M.O. Aremu. 2013. Thin layer drying kinetics of pineapple: Effect of blanching temperature-time combination. *Nigerian Journal of Basic and Applied Sciences* 21(1): 1–10.

Ah-Hen, K., C.E. Zambra, J.E. Aguero, A. Vega-Galvez, and R. Lemus-Mondaca. 2013. Moisture diffusivity coefficient and convective drying modelling of murta (*Ugni molinae* Turcz): Influence of temperature and vacuum on drying kinetics. *Food and Bioprocess Technology* 6: 919–930.

Aktas, T., H.H. Orak, F.H. Sahin, and N. Ekinci. 2013. Effects of different drying methods on drying kinetics and color parameters of strawberry tree (Arbutus unedo L.) fruit. *Journal of Tekirdag Agricultural Faculty* 10(2): 1–12.

Albanese, D., L. Cinquanta, G. Cuccurullo, and M. Di Matteo. 2013. Effects of microwave and hot-air drying methods on colour, β-carotene and radical scavenging activity of apricots. *International Journal of Food Science and Technology* 48: 1327–1333.

Arballo, J.R., L.A. Campañone, and R.H. Mascheroni. 2010. Modeling of microwave drying of fruits. *Drying Technology* 28: 1178–1184.

Aversa, M., S. Curcio, V. Calabro, and G. Iorio. 2007. An analysis of the transport phenomena occurring during food drying process. *Journal of Food Engineering* 78: 922–932.

Bansal, P.R., and K.Y. Chung. 2007. Food drying equipment and design. In *Food drying science and technology: Microbiology, chemistry, applications*, ed. Y.H. Hui, C. Clary, M.M. Farid, O.O. Fasina, A. Noomhorm, and J. Welti-Chenis, 359–402. Lancaster, PA: DESTech Publications. Chapter 15.

Beck, S.M., H.T. Sabarez, V. Gaukel, and K. Knoerzer. 2014. Enhancement of convective drying by application of airborne ultrasound: A response surface approach. *Ultrasonics Sonochemistry* 21(6): 2144–2150. https://doi.org/10.1016/j.ultsonch.2014.02.013.

Bertin, R., and M. Blazquez. 1986. Modeling and optimization of a dryer. *Drying Technology* 4(1): 45–66.

Cabrera-Padilla, R.Y., T.O. Oladeinde, J.R.D. Fincer, and J.R. Limaverde. 2014. Drying of banana paste in rotatory dryer with inert bed. *Brazilian Journal of Food Technology* 17(1): 41–50.

Caparino, O.A., J. Tang, C.I. Nindo, S.S. Sablani, J.R. Powers, and J.K. Fellman. 2012. Effect of drying methods on the physical properties and microstructures of mango (Philippine 'Carabao' var.) powder. *Journal of Food Engineering* 111: 135–148.

Carcel, J.A., J.V. Garcia-Perez, E. Riera, and A. Mulet. 2007. Influence of high intensity ultrasound on drying kinetics of persimmon. *Drying Technology* 25: 185–193.

———. 2011a. Improvement of convective drying of carrot by applying power ultrasound—Influence of mass load density. *Drying Technology* 29: 174–182.

Carcel, J.A., J.V. Garcia-Perez, J. Benedito, and A. Mulet. 2011b. Food process innovation through new technologies: Use of ultrasound. *Journal of Food Engineering* 110(2): 200–207.

Chao, L., Y. JiaJia, C. Kai, Z. ZhengLan, and L. HuanRong. 2012. Study of the technology of spray drying for apricot powder. *Xinjiang Agricultural Sciences* 49(6): 1110–1115.

Chayjan, R.A., and B. Alaei. 2013. Characteristics of thin layer microwave drying of apricot. *Electronic Journal of Polish Agricultural Universities* 16(4): 12–18.

Chong, C.H., C.L. Law, A. Figiel, A. Wojdyło, and M. Oziembłowski. 2013. Colour, phenolic content and antioxidant capacity of some fruits dehydrated by a combination of different methods. *Food Chemistry* 141: 3889–3896.

Chong, C.H., A. Figiel, C.L. Law, and A. Wojdyło. 2014. Combined drying of apple cubes by using of heat pump, vacuum-microwave, and intermittent techniques. *Food and Bioprocess Technology* 7: 975–989.

COMSOL Multiphysics. 2007. *Chemical Engineering Module model library*. Stockholm, Sweden: COMSOL AB.

Contreras, C., M.E. Martin-Esparza, and N. Martinez-Navarrete. 2012. Influence of drying method on the rehydration properties of apricot and apple. *Journal of Food Process Engineering* 35: 178–190.

Corso, O., and C. Alvarez. 2014. Color change kinetics of mango at different maturity stages during air drying. *Journal of Food Processing and Preservation* 38(1): 508–517.

Curcio, S., M. Aversa, V. Calabro, and G. Iorio. 2008. Simulation of food drying: FEM analysis and experimental validation. *Journal of Food Engineering* 87: 541–553.

da Silva, A.N., J.S.D. Coimbra, F.M. Botelho, N.M. de Moraes, J.T. de Faria, M.D.T. Bezerra, M.A. Martins, and A.M.D. Siqueira. 2013. Pear drying: Thermodynamics studies and coefficients of convective heat and mass transfer. *International Journal of Food Engineering* 9(4): 365–374.

Doymaz, I. 2013. Experimental study on drying of pear slices in a convective dryer. *International Journal of Food Science and Technology* 48: 1909–1915.

Doymaz, I., and O. Ismail. 2012. Experimental characterization and modelling of drying of pear slices. *Food Science and Biotechnology* 21(5): 1377–1381.

Doymaz, I., and M. Pala. 2002. The effects of dipping treatments on air-drying rates of the seedless grapes. *Journal of Food Engineering* 52(4): 413–417.

Duan, X., L. Ding, G. Ren, L. Liu, and Q. Kong. 2013. The drying strategy of atmospheric freeze drying apple cubes based on glass transition. *Food and Bioproducts Processing* 91: 534–538.

Dujmic, F., M. Brncic, S. Karlovic, T. Bosiljkov, D. Jezek, B. Tripalo, and I. Mofardin. 2013. Ultrasound-assisted infrared drying of pear slices: Textural issues. *Journal of Food Process Engineering* 36: 397–406.

Esehaghbeygi, A., K. Pirnazari, and M. Sadeghi. 2014. Quality assessment of electrohydrodynamic and microwave dehydrated banana slices. *LWT - Food Science and Technology* 55: 565–571.

Gallego-Juarez, J.A. 2010. High-power ultrasonic processing: Recent developments and prospective advances. *Physics Procedia* 2: 35–47.

Gallego-Juarez, J.A., G. Rodriguez-Corral, J.C. Galvez-Moraleda, and T.S. Yang. 1999. A new high intensity ultrasonic technology for food dehydration. *Drying Technology* 17(3): 597–608.

Gallego-Juarez, J.A., E. Riera, S. de la Fuente Balanco, G. Rodriguez-Corral, V.M. Acosta-Aparicio, and A. Blanco. 2007. Application of high-power ultrasound for dehydration of vegetables: Processes and devices. *Drying Technology* 25(11): 1893–1901.

Gamboa-Santos, J., A. Montilla, J.A. Carcel, A. Villamiel, and J.V. Garcia-Perez. 2014a. Airborne ultrasound application in the convective drying of strawberry. *Journal of Food Engineering* 128: 132–139.

Gamboa-Santos, J., A. Montilla, A.C. Soria, J.A. Cárcel, J.V. García-Pérez, and M. Villamiel. 2014b. Impact of power ultrasound on chemical and physicochemical quality indicators of strawberries dried by convection. *Food Chemistry* 161: 40–46.

Garcia-Perez, J.V., J.A. Carcel, J. Benedito, and A. Mulet. 2007a. Power ultrasound mass transfer enhancement in food drying. *Food and Bioproducts Processing* 85: 247–254.

Garcia-Perez, J.V., J.A. Carcel, J. Benedito, E. Riera, and A. Mulet. 2007b. Influence of process variables on hot air drying assisted by power ultrasound. In *Proceedings of the 19th International Congress on Acoustics, Madrid, Spain.*

Garcia-Perez, J.V., J.A. Carcel, E. Riera, and A. Mulet. 2009. Influence of the applied acoustic energy on the drying of carrots and lemon peel. *Drying Technology* 27(2): 281–287.

Garcia-Perez, J.V., A. Puig, I. Perez-Munuera, J.A. Carcel, and E. Riera. 2010. Kinetic and microstructural changes induced by power ultrasound application on convective drying of eggplant. In *Proceedings of the 20th International Congress on Acoustics (ICA 2010), Sydney, Australia.*

Gentry, J.P., L.L. Claypool, and M.W. Miller. 1965. Parallel-flow prune dehydration. California Agriculture.

Gholami, M., M. Mirmoradi, E. Abdolalizadeh, and M.S. Beni. 2013. Effect of two-step drying of black currant grape on the quality of raisin. *Journal of Innovation in Food Science and Technology* 5(1): 113–115.

Gurak, P.D., L.M.C. Cabral, and M.H. Rocha-Leão. 2013. Production of grape juice powder obtained by freeze drying after concentration by reverse osmosis. *Brazilian Archives of Biology and Technology* 56(6): 1011–1017.

Hayashi, E.M. 2007. Afghanistan fruit and vegetable dehydrator desk study. Report for the United States Agency for International Development (USAID) Accelerating Sustainable Agricultural Program (ASAP).

Heldman, D.R., and R.W. Hartel. 1997. *Principles of food processing*. New York: Chapman and Hall.

Henríquez, M., S. Almonacid, M. Lutz, R. Simpson, and M. Valdenegro. 2012. Comparison of three drying processes to obtain an apple peel food ingredient. *CyTA – Journal of Food* 11(2): 127–135.

Igual, M., E. García-Martínez, M.E. Martín-Esparza, and N. Martínez-Navarrete. 2012. Effect of processing on the drying kinetics and functional value of dried apricot. *Food Research International* 47: 284–290.

Ioannou, I., W. Guiga, C. Charbonnel, and M. Ghoul. 2011. Frozen mirabelle plum drying: Kinetics, modelling and impact on biochemical properties. *Food and Bioproducts Processing* 89: 438–448.

Ismail, O.M., and K.S.A. Nagy. 2012. Characteristics of dried mango slices as affected by pretreatments and drying type. *Australian Journal of Basic and Applied Sciences* 6(5): 230–235.

Jangam, S.V. 2011. An overview of recent developments and some R&D challenges related to drying of foods. *Drying Technology: An International Journal* 29(12): 1343–1357.

———. 2012. Guest editorial: Globalization of drying R&D. *Drying Technology: An International Journal* 30(7): 689–690.

Jazini, M.H., and M.S. Hatamipour. 2010. A new physical pretreatment of plum for drying. *Food and Bioproducts Processing* 88: 133–137.

Jiang, H., M. Zhang, A.S. Mujumdar, and R.X. Lim. 2014. Comparison of drying characteristic and uniformity of banana cubes dried by pulse-spouted microwave vacuum drying, freeze drying and microwave freeze drying. *Journal of the Science of Food and Agriculture* 94: 1827–1834.

Kabiru, A.A., A.A. Joshua, and A.O. Raji. 2013. Effect of slice thickness and temperature on the drying kinetics of mango (Mangifera indica). *International Journal of Research and Reviews in Applied Sciences* 15(1): 41–50.

Khmelev, V.N., R.V. Barsukov, B.S. Abramenko, and D.V. Genne. 2008. Research and development of ultrasonic device prototype for intensification of drying process. In *9th International Workshop and Tutorials EDM'2008.*

Khmelev, V.N., A.V. Shalunov, R.V. Barsukov, D.S. Abramenko, and A.N. Lebedev. 2011. Studies of ultrasonic dehydration efficiency. *Journal of Zhejiang University – Science A (Applied Physics and Engineering)* 12(4): 247–254.

Kholmanskiy, A.S., A.Z. Tilov, and E.Y. Sorokina. 2013. Drying kinetics of plant products: Dependence on chemical composition. *Journal of Food Engineering* 117: 378–382.

Kostoglou, M., N. Chrysafis, and N. Andritsos. 2013. Modelling tomato dehydration in a tunnel dryer using geothermal energy. *Drying Technology: An International Journal* 31: 5–15.

Lewicki, P.P. 2006. Design of hot air drying for better foods. *Trends in Food Science and Technology* 17(4): 153–163.
Lozano, J.E., E. Rotstein, and M.J. Urbicain. 1983. Shrinkage, porosity and bulk density of foodstuffs at changing moisture contents. *Journal of Food Science* 48: 1497–1502.
Madrau, M.A., A.N. Sanguinetti, A. Del Caro, C. Fada, and A. Piga. 2010. Contributions of melanoidins to the antioxidant activity of prunes. *Journal of Food Quality* 33: 155–170.
Marquez, A., M.P. Serratosa, and J. Merida. 2013. Colour and anthocyanin composition of sweet red wines made using different winemaking and grape drying techniques. *Enologos* 15(86): 30–32.
Martynenko, A., and M.A. Janaszek. 2014. Texture changes during drying of apple slices. *Drying Technology* 32: 567–577.
Menges, H.O., and C. Ertiken. 2006. Mathematical modelling of thin layer drying of golden apples. *Journal of Food Engineering* 77: 119–125.
Miletic, M., O. Mitrovic, B. Popovic, V. Nidovic, B. Zlatkovic, and M. Kandic. 2013. Polyphenolic content and antioxidant activity in fruits of plum (Prunus domestica L.) cultivars "Valjevka" and "Mildora" as influenced by air drying. *Journal of Food Quality* 36: 229–237.
Mohsen, E., S.G. Rahmat, and M.A.E. Mohammad. 2007. Influence of dipping on thin-layer drying characteristics of seedless grapes. *Biosystems Engineering* 98: 411–421.
Mosquera, L.H., G. Moraga, and N. Martínez-Navarrete. 2012. Critical water activity and critical water content of freeze-dried strawberry powder as affected by maltodextrin and Arabic gum. *Food Research International* 47: 201–206.
Mrad, N.D., N. Boudhrioua, N. Kechaou, F. Courtois, and C. Bonazzi. 2012. Influence of air drying temperature on kinetics, physicochemical properties, total phenolic content and ascorbic acid of pears. *Food and Bioproducts Processing* 90: 433–441.
Mujumdar, A.S. 1987. *Handbook of industrial drying*. New York: Marcel Dekker.
———. 2010a. Role of IDS in promoting innovation and global R&D effort in drying technologies. In *R&D needs, challenges and opportunities for innovation in drying technology*, ed. S.V. Jangam, and B.N. Thorat. e-Book.
———. 2010b. Drying R&D past, present and future. In *R&D needs, challenges and opportunities for innovation in drying technology*, ed. S.V. Jangam, and B.N. Thorat. e-Book.
Mujumdar, A.S., and L.X. Huang. 2007. Global R&D needs in drying. *Drying Technology* 25: 647–658.
Mujumdar, A.S., and C.L. Law. 2010. Drying technology: Trends and applications in postharvest technology. *Food and Bioprocess Technology* 3(6): 843–852.
Mujumdar, A.S., and Z.H. Wu. 2010. Thermal drying technologies: New developments and future R&D potential. In *R&D needs, challenges and opportunities for innovation in drying technology*, ed. S.V. Jangam, and B.N. Thorat. e-Book.
Noorbakhsh, R., P. Yaghmaee, and T. Durance. 2013. Radiant energy under vacuum (REV) technology: A novel approach for producing probiotic enriched apple snacks. *Journal of Functional Foods* 5: 1049–1056.
Ochoa-Martínez, C.I., P.T. Quintero, A.A. Ayala, and M.J. Ortiz. 2012. Drying characteristics of mango slices using the Refractance Window™ technique. *Journal of Food Engineering* 109: 69–75.
Oikonomopoulou, V.P., and M.K. Krokida. 2012. Structural properties of dried potatoes, mushrooms, and strawberries as a function of freeze-drying pressure. *Drying Technology* 30: 351–361.
Orak, H.H., T. Aktas, T.H. Yagar, S. Isbilir, N. Ekinci, and F.H. Sahin. 2011. Antioxidant activity, some nutritional and colour properties of vacuum dried strawberry tree (Arbutus unedo L.) fruit. *Acta Scientiarum Polonorum. Technologia Alimentaria* 10(3): 327–338.
Ozuna, C., J.A. Carcel, J.V. Garcia-Perez, and A. Mulet. 2011. Improvement of water transport mechanisms during potato drying by applying ultrasound. *Journal of the Science of Food and Agriculture* 91(14): 2511–2517.
Ozuna, C., T.G. Álvarez-Arenas, E. Riera, J.A. Cárcel, and J.V. Garcia-Perez. 2014. Influence of material structure on air-borne ultrasonic application in drying. *Ultrasonics Sonochemistry* 21: 1235–1243.

Panceri, C.P., T.M. Gomes, J.S. De Gois, D.L.G. Borges, and M.T. Bordignon-Luiz. 2013. Effect of dehydration process on mineral content, phenolic compounds and antioxidant activity of Cabernet Sauvignon and Merlot grapes. *Food Research International* 54: 1343–1350.

Pereira da Silva, W., I. Hamawand, and C.M.D.P.S. e Silva. 2014. A liquid diffusion model to describe drying of whole bananas using boundary-fitted coordinates. *Journal of Food Engineering* 137: 32–38.

Ponkhama, K., N. Meesoa, S. Soponronnaritb, and S. Siriamornpunc. 2012. Modeling of combined far-infrared radiation and air drying of a ring shaped-pineapple with/without shrinkage. *Food and Bioproducts Processing* 90: 155–164.

Price, W.E., H.T. Sabarez, R. Storey, and P.J. Back. 2000. Role of the waxy skin layer in moisture loss during dehydration of prunes. *Journal of Agricultural and Food Chemistry* 48: 4193–4198.

Ramallo, L.A., and R.H. Mascheroni. 2012. Quality evaluation of pineapple fruit during drying process. *Food and Bioproducts Processing* 90: 275–283.

———. 2013. Effect of shrinkage on prediction accuracy of water diffusion model for pineapple drying. *Journal of Food Process Engineering* 36: 66–76.

Rodríguez, O., J.V. Santacatalina, S. Simal, J.V. Garcia-Perez, A. Femenia, and C. Rosselló. 2014. Influence of power ultrasound application on drying kinetics of apple and its antioxidant and microstructural properties. *Journal of Food Engineering* 129: 21–29.

Sabarez, H.T. 2010. Improving prune dehydration cost efficiency. Unpublished report prepared by CSIRO Food and Nutritional Sciences for Horticulture Australia Limited (HAL), Sydney, Australia.

———. 2012. Computational modelling of the transport phenomena occurring during convective drying of prunes. *Journal of Food Engineering* 111(2): 279–288.

———. 2014. Mathematical modeling of the coupled transport phenomena and color development: finish drying of trellis-dried sultanas. *Drying Technology* 32: 578–589.

Sabarez, H.T., and W.E. Price. 1999. A diffusion model for prune dehydration. *Journal of Food Engineering* 42: 167–172.

Sabarez, H.T., W.E. Price, P.J. Back, and L.A. Woolf. 1997. Modelling the kinetics of drying of d'Agen plums (Prunus domestica). *Food Chemistry* 60(3): 371–382.

Sabarez, H.T., J.A. Gallego-Juarez, and E. Riera. 2012. Ultrasonic-assisted convective drying of apple slices. *Drying Technology: An International Journal* 30(9): 989–997.

Sablani, S.S., and M.S. Rahman. 2007. Fundamentals of food dehydration. In *Food drying science and technology: Microbiology, chemistry, applications*, ed. Y.H. Hui, C. Clary, M.M. Farid, O.O. Fasina, A. Noomhorm, and J. Welti-Chenis, 1–42. Lancaster, PA: DESTech Publications. Chapter 1.

Sacilik, K., A.K. Elicin, and G. Unal. 2006. Drying kinetics of Uryani plum in a convective hot-air dryer. *Journal of Food Engineering* 76: 362–368.

Saenz, C., S. Tapia, J. Chavez, and P. Robert. 2009. Microencapsulation by spray drying of bioactive compounds from cactus pear (Opuntia ficus-indica). *Food Chemistry* 114: 616–622.

Santacatalina, J.V., O. Rodríguez, S. Simal, J.A. Cárcel, A. Mulet, and J.V. García-Pérez. 2014. Ultrasonically enhanced low-temperature drying of apple: Influence on drying kinetics and antioxidant potential. *Journal of Food Engineering* 138: 35–44.

Santos, S.C.R.V.L., R.P.F. Guine, and A.I.A. Barros. 2013. Influence of drying on the properties of pears of the Rocha variety (Pyrus communis). *International Journal of Food Engineering* 9(2): 197–207.

Schossler, K., H. Jager, and D. Knorr. 2012. Effect of continuous and intermittent ultrasound on drying time and effective diffusivity during convective drying of apple and red bell pepper. *Journal of Food Engineering* 108: 103–110.

Schulze, B., E.M. Hubbermann, and K. Schwarz. 2014. Stability of quercetin derivatives in vacuum impregnated apple slices after drying (microwave vacuum drying, air drying, freeze drying) and storage. *LWT - Food Science and Technology* 57: 426–433.

Silva, V., A.R. Figueiredo, J.J. Costa, and R.P.F. Guiné. 2014. Experimental and mathematical study of the discontinuous drying kinetics of pears. *Journal of Food Engineering* 134: 30–36.

Soria, A.C., and M. Villamiel. 2010. Effect of ultrasound of the technological properties and bioactivity of food: A review. *Trends in Food Science and Technology* 21: 323–331.

Suhaimi, A.M., O.A. Malik, S. Asmui, and O.A. Shokri. 2011. Production and properties of spray dried pineapple juice. *Acta Horticulturae* 902: 525–528.
Togrul, H. 2005. Simple modelling of infrared drying of fresh apple slices. *Journal of Food Engineering* 71: 311–323.
Velescu, I., I. Tenu, P. Carlescu, and V. Dobre. 2012. Influence of temperature and geometry in the kinetics of drying pears. Lucrari Stiintifice, Universitatea de Stiinte Agricole Si Medicina Veterinara "Ion Ionescu de la Brad" Iasi. *Seria Agronomie* 55(2): 93–97.
Vieira, A.P., J.F. Nicoleti, and V.R.N. Telis. 2012. Freeze drying of pineapple slices: Evaluation of drying kinetics and product quality. *Brazilian Journal of Food Technology* 15(1): 50–58.
Walkowiak-Tomczak, D. 2012. Determination of chemical composition of plums during pre-treatment and drying. *Ecological Chemistry and Engineering* 19(1): 89–96.
Yaldiz, O., C. Ertekin, and H.I. Uzun. 2001. Mathematical modelling of thin layer solar drying of sultana grapes. *Energy* 26: 457–465.
Zivkovic, M., S. Rakic, R. Maletic, D. Povrenovic, M. Nikolic, and N. Kosanovic. 2011. Effect of temperature on physical changes and drying kinetics in plums (Prunus domestica L.) Pozegaca variety. *Chemical Industry & Chemical Engineering Quarterly* 17(3): 283–289.

Chapter 8
Membrane Technologies for Fruit Juice Processing

Manuel Dornier, Marie-Pierre Belleville, and Fabrice Vaillant

8.1 Introduction

For 40 years, membrane technologies have been developed to improve conventional liquid–solid separation processes or to envisage new fractionations of solutes from liquid mixtures. Nowadays, membrane technologies are largely used in the food industry mainly in the dairy industry for concentration or standardization of proteins or whey demineralization and in wine and beer processing for clarification and stabilization (Daufin et al. 2001; Peinemann et al. 2010). In the fruit juice industry, clarification of fruit juices using membrane technologies is usual. Nevertheless, over the last decades, interest for these technologies has increased in this field for other unit operations. Compared to conventional separation processes, the milder operating conditions generally used allow to better preserve the quality potential of the fruits.

After describing the basics of the main membrane processes and presenting their diversity, this review attempts to give a comprehensive overview of the applications of membrane technologies to fruit juice processing. It focuses more specifically on recent advances and potentialities and can therefore complete and up-date the reviews already reported in the literature on the subject (Decloux and Prothon 1998; Echavarria et al. 2011; Girard and Fukumoto 2000).

M. Dornier (✉)
Montpellier SupAgro, Montpellier, France
e-mail: manuel.dornier@supagro.fr

M.-P. Belleville
Université de Montpellier, Montpellier, France

F. Vaillant
CIRAD, Montpellier, France

A. Rosenthal et al. (eds.), *Fruit Preservation*, Food Engineering Series,
https://doi.org/10.1007/978-1-4939-3311-2_8

8.2 Membrane Processes

Developed over the years 1960–1970, membrane processes derive from the conventional filtration process, of which the objective is to separate a dispersed phase within a continuous phase but they are distinguished by the size of rejected solutes. Conventional filtration allows the retention of particles, while it is possible to separate macromolecules, small molecules, or ions from complex mixtures thanks to a membrane technique. Today, there is a large number of membrane processes based on different separation principles, but all of them involve a membrane that can be considered as the heart of the process. The membrane acts as a thin permselective barrier or interface between two phases and the mass transfer through the membrane takes place as the result of a driving force which can be gradients in pressure, concentration, electrical potential, or temperature (Fig. 8.1). An overview of different membranes processes which can be used in fruit juice processing is given hereafter.

8.2.1 Pressure-Driven Membrane Processes

Pressure-driven membrane processes (microfiltration (MF), ultrafiltration (UF), nanofiltration (NF) and reverse osmosis (RO)) are now well-established for concentration, fractionation, and purification of liquid solutions and water treatment (Mulder 1996). The first three technologies involve porous membranes which show different ranges of pore sizes (Fig. 8.2).

MF membranes are generally used for fine particles or bacteria removal or turbidity reduction processes (i.e., clarification) (El Rayess et al. 2011), whereas UF membranes which permit the rejection of species in the size range of 0.001–0.02 μm are used for macromolecule concentration and fractionation (van Reis and Zydney 2007). The NF membranes are characterized by a charged surface and mean pore

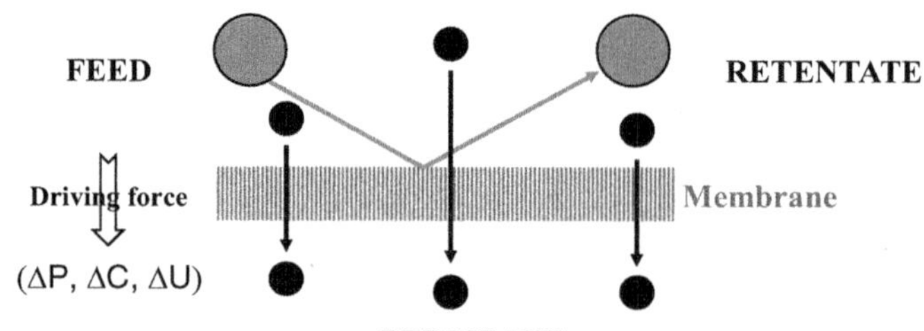

Fig. 8.1 Schematic representation of membrane processes

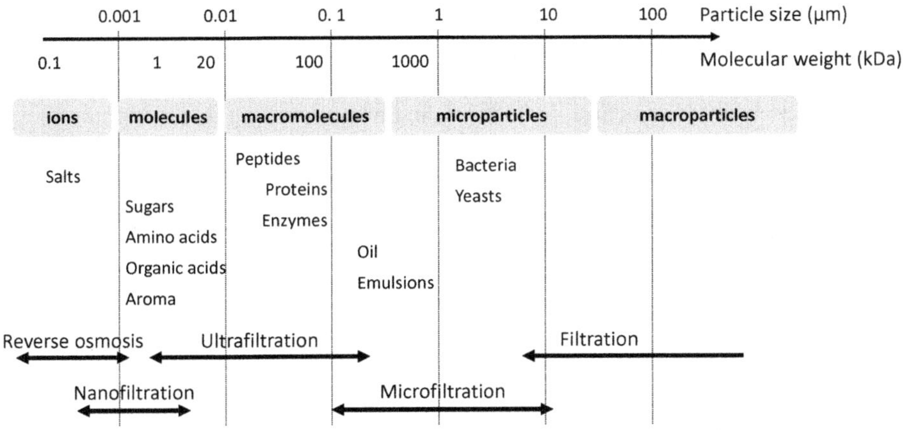

Fig. 8.2 Application size range of pressure-driven membrane processes

diameters are around 1–3 nm. Their main applications are water desalination (Mondal and Wickramasinghe 2008), whey demineralization (Román et al. 2009), and uncharged organic molecules concentration (Tylkowski et al. 2001). In all these processes, mass transfer occurs thanks to a pressure gradient maintained across the membrane and the rejection is achieved mainly by means of sieving mechanisms, although interactions between the membrane surface and the solution can influence the separation. In NF, differences in diffusivity and solubility of solutes and electrostatic interactions between the membrane surface groups and the ions play an important role in the separation efficiency (Nagy et al. 2011).

In contrast to the membrane techniques described above, RO involves nonporous membrane and the mechanisms of separation of species are based on the "solution-diffusion" model. The ionic species are generally totally rejected by the membrane, whereas small nonionic organic compounds which can be taken up by the membrane are rejected because they show a much lower mobility in the membrane than water molecules. Desalination is undoubtedly the largest application of RO (Peñate and García-Rodríguez 2012), but this technique is also widely used for the concentration of thermosensitive aqueous solutions as is reported hereafter (Alvarez et al. 1997, 1998, 2002; Gurak et al. 2010; Jesus et al. 2007; Matta et al. 2004; Petrotos et al. 1998; Rektor et al. 2004).

The major drawback of pressure-driven membrane processes is linked to the decrease of permeate flux observed throughout the process. In fact, as the membrane acts as a permselective barrier, an increase in the concentration of the rejected solutes can be observed on the feed side near the membrane surface. This boundary layer, also called the concentration polarization layer, results in an additional resistance to that of the membrane for overall liquid permeation. Furthermore, if the solute concentrations at the membrane surface become higher than the solubility value, salt precipitation or macromolecule gelation can occur. The thin gel layer formed at the membrane surface acts as a secondary membrane and results in changes in observed rejection. In addition, membrane fouling due to solute adsorption and/or accumulation inside the membrane porosity can occur reducing mass transfer across the membrane. Concentration polarization phenomena and membrane fouling which are major concerns for food applications have been widely studied over the past decades (Mulder 1996).

8.2.2 Electrical Membrane Processes

Electrodialysis is a membrane process which allows the selective separation of ions from a solution. In practice, an electrodialysis unit consists of many flat cationic and anionic membranes (cation and anion exchange membrane) arranged alternatively to form a membrane stack. This membrane arrangement forms distinct channels where the solution to be treated and an extracting solution flow (a saline solution) respectively. When this stack is placed between two electrodes as shown in Fig. 8.3, cations will try to move towards the cathode whereas anions will try to move towards the anode thanks to the applied electrical field.

However, their migration will be stopped either by anion or by cation exchange membrane. Indeed, ion exchange membranes are made of a cross-linked polymer matrix which has fixed charge groups to which mobile ions with opposite charge (counter ions)

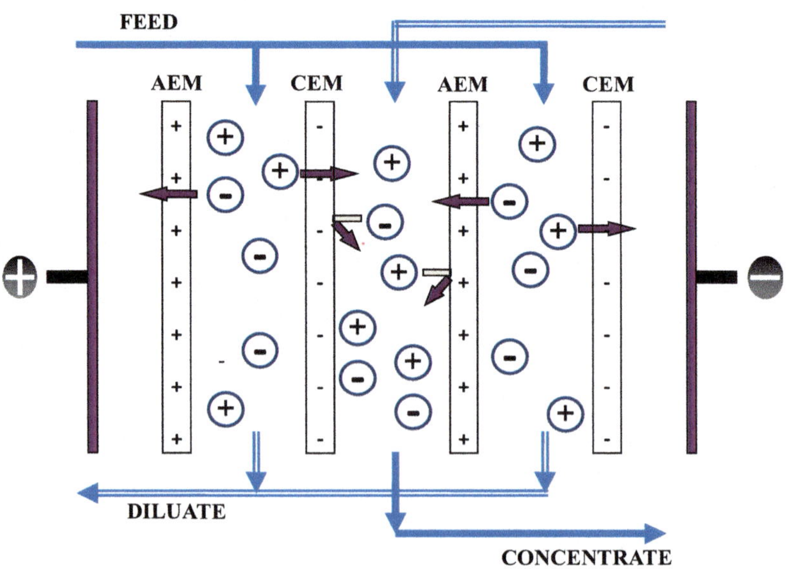

Fig. 8.3 Schematic diagram of electrodialysis separation. *AEM* anion exchange membrane, *CEM* cation exchange membrane

are attached. When an electrical current is applied, the ions with the same charge of the counter ions are thus free to move through the membrane, whereas the ions with the opposite charge are almost totally rejected. At the outlet of the ED unit, the salt concentration of the solution to be treated decreases, while in the other compartment the salt concentration increases. There is also another type of ionic exchange membrane called bipolar membranes. These membranes are composed of an anion exchange layer and a cation exchange layer separated by a hydrophilic layer. When a bipolar membrane is placed in the electrical field, the water present in the junction layer is split into H^+ and OH^- ions, which are transported in opposite directions to adjacent chambers. Conventional ED is mainly used to concentrate electrolyte solutions or to dilute or deionize solutions (i.e., demineralization of whey or grape must, etc.) (Strathmann 2010). The association of bipolar membranes to conventional ion exchange membranes allows the deacidification of fruit juices as presented hereafter (Vera et al. 2009).

8.2.3 Concentration-Driven Membrane Processes

Pervaporation (PV) is a membrane technique involving a non-porous permselective membrane which separates an upstream liquid mixture from a downstream side usually kept under vacuum (Fig. 8.4). The chemical potential gradient obtained by partial pressure reduction is thus responsible for the partial vaporization of the liquid mixture through the membrane.

The mass transport across the membranes involves three successive steps: (1) selective absorption into the membrane at the feed side of the membrane, (2) selective diffusion through the membrane, and (3) desorption into a vapor phase at the permeate side of the membrane. In PV processes, separation is more dependent on the solubility and the diffusivities of the components in the membrane

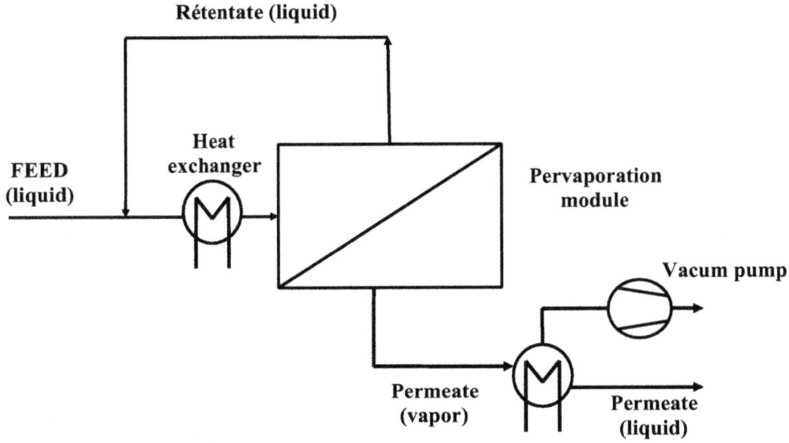

Fig. 8.4 Schematic diagram of pervaporation separation

rather than vapor/liquid equilibria. The permeate therefore has a significantly different composition from the vapor obtained from distillation processes. This latter point is the greatest advantage of PV; this separation technique is not limited by the formation of azeotropes. The selectivity of PV depends on the membrane properties. Hydrophilic membranes are used for dehydration applications, whereas hydrophobic ones allow organic compounds to be separated. Since equipment and operational costs of PV are relatively high, this technique is used for production of high value products. The main applications of PV are solvent dehydration (Jonquières et al. 2002), but PV can also be used for aroma recovery in food (Pereira et al. 2006).

In contrast to PV, membrane contactors involve a porous membrane in order to create an interface for mass transfer between two phases. The membranes generally used are made of classical hydrophobic polymers (polyethylene (PE), polypropylene (PP), polytetrafluoroethylene (PTFE), polyvinylidene fluoride (PVDF)) but ceramic membranes (Brodart et al. 2003; Romero et al. 2006) and metallic membranes (Hengl et al. 2007) have been successfully used as membrane contactors after being modified in order to present hydrophobic properties. The hydrophobic nature of the membrane prevents the penetration of the liquid phase and creates a liquid vapor interface at the entrance of each pore. Thus, evaporation occurs and the volatile compounds diffuse across the membrane and are condensed in a liquid phase: Direct Contact Membrane Distillation DCMD and Osmotic Distillation or Osmotic Evaporation (OD/OE) or on a cold wall (Air Gap Membrane Distillation AGMD), or removed by a gas phase located in the permeate side (Vacuum Membrane Distillation VMD and Sweep Gas Membrane Distillation SGMD) (El-Bourawi et al. 2006). It is important to note that in the last two processes, the condensation step does not occur in the membrane module, which is why these techniques are also referred to as membrane evaporation processes. DCMD and OD are the most studied processes; they are schematically represented in Fig. 8.5. In DCMD, the transfer occurs thanks to a thermal gradient applied across the membrane; the feed solution is heated up to 60–80 °C, whereas the temperature of the extractive phase is maintained as low as possible. The thermal gradient creates a partial vapor pressure gradient between the two solution–membrane interfaces and is responsible for the vapor transport throughout the immobile gas phase in the pores.

In OD, the driving force that generates the water flow is the difference of activity between the feed solution on one side and a hypertonic solution, typically a concentrated brine, on the opposite side. The OD flux depends on the nature of the osmotic agent; calcium chloride is the most effective (Celere and Gostoli 2005; Ravindra et al. 2006b). Compared to MD, this process can be carried out at lower temperatures avoiding any risk of thermal degradation of compounds.

The membrane contactors offer various advantages: emulsion problems are avoided, a large and constant exchanged area is involved, and independent fluid dynamics allow an easily controlled operation. MD is suited for both pure water production or for the concentration of nonvolatile components in aqueous solutions, whereas OD seems to be a valid alternative in the concentration of liquid food as is reported hereafter.

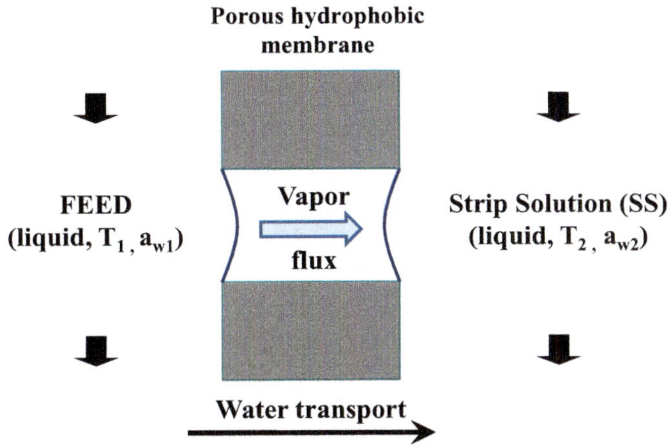

Fig. 8.5 Schematic representation of direct contact membrane distillation and osmotic distillation processes (T: temperature; a_w: water activity)—in the case of DCMC, SS: pure water, $T_1 > T_2$ and $a_{w1} \approx a_{w2}$—in the case of OD, SS: hypertonic solution, $T_1 \approx T_2$ and $a_{w1} < a_{w2}$

8.2.4 Interests and Limits

Membrane technologies have gained a huge importance in different application areas (water desalination and treatment, food processing industries, chemical industries, etc.) over the last decades because they have the potential to advantageously replace conventional techniques. All the different objectives of separation (i.e., concentration/clarification, purification, fractionation, extraction) can be achieved thanks to one or a combination of membrane processes. In addition, membrane techniques can be operated in continuous mode which limits start up and shut down procedures; the labor costs are thus reduced and consistent product quality is obtained. Nevertheless, the performances are sometimes lower than expected; the selectivity and stability for long-term application is not achieved. It may be due to the fact that the state of technological maturity of membrane processes is still far from the optimal one. Some efforts should be devoted to the development of new optimized membranes as well as to improvements in process engineering.

8.3 Applications of Membrane Processes to Fruit Juices

8.3.1 Potentialities of Membrane Processes Applied to Fruit Juices

As described above, membrane processes are a wide family of separation processes that can be used for concentration or removal of a wide range of constituents of food liquids. In order to evaluate their potentialities for fruit juice applications

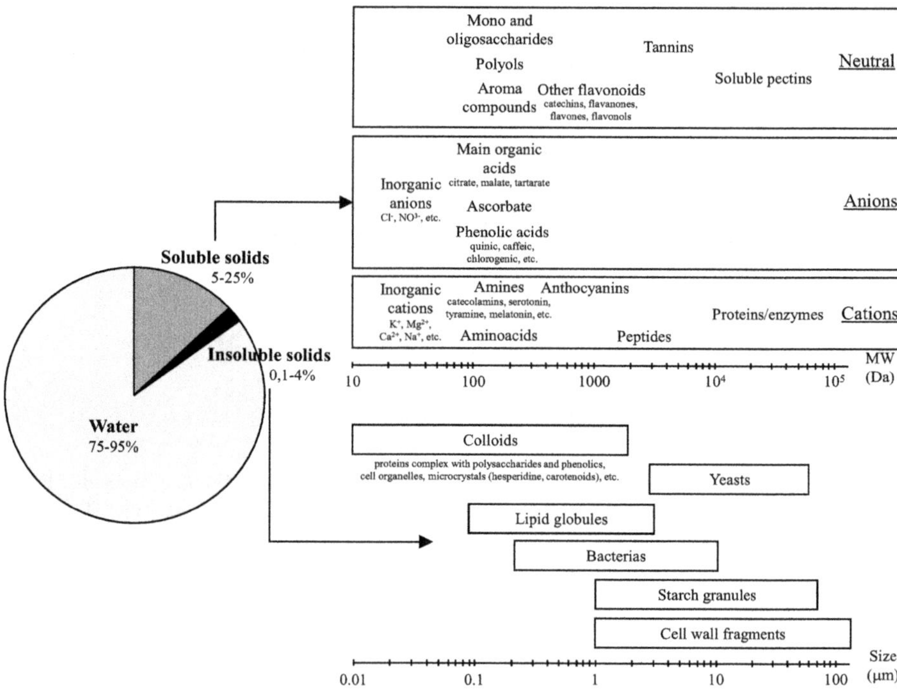

Fig. 8.6 Schematic representation of the different compounds or species present in fruit juices, classified according to their nature and size

specifically, we can do so by studying fruit juice composition. Of course, this composition greatly depends on many factors such as fruit species, variety, maturity, soil and climate conditions, cultural practices even, or juice extraction processing. Nevertheless, from a qualitative point of view, the composition of fruit juices is actually very similar. Whatever the product is, the same compounds are always present (Fig. 8.6).

First of all, fruit juices contain a very large quantity of water, from 75 to 95 % in weight. The concentration of the dry matter is therefore a first unit operation required by the fruit juice industry mainly for economic reasons. Water removal allows transport and storage costs to be greatly reduced. The main challenge is to concentrate the juice modifying the sensorial and nutritional quality of the raw materials as little as possible. Compared with vacuum evaporation, the reference technology used in the industry, some membrane processes could be of great interest. Because of the lower temperature used, they should damage the temperature-sensitive compounds less. Pressure-driven membrane processes with very low molecular weight cut-off, like reverse osmosis or nanofiltration, are able to separate water from the smaller solutes down to inorganic ions. Membrane contactors like osmotic evaporation can also be used to selectively extract water without heating the juice.

The dry matter of fruit juices mainly consists of a soluble fraction (5–25 % of the juice) and, in many cases, contains various suspended elements (0.1–4 %). These suspended insoluble solids (ISS) include pecto-cellulosic cell wall fragments, starch granules, microorganisms, and in certain cases lipid globules (phospholipids, sterols, glycolipids, carotenoids, triterpenoids, triacylglycerols, waxes), or colloid fraction. These elements are responsible for the turbid or opalescent appearance of raw juices. In the case where the clarification is required, for traditional clear juices for example (apple, grape, etc.) or for clarified juices used as ingredients in formulations (banana, citrus, etc.), micro or even ultrafiltration is well suited. Compared to classical techniques of clarification such as decantation, centrifugation, or dead-end filtration, membrane processes allow a better separation efficiency and can easily be implemented in continuous mode. Furthermore, by choosing the pore diameter below 0.2 μm, the product is sterilized at the same time (without heating) which could be an added advantage.

Glucose, fructose, sucrose, and fruit organic acids (citric, malic, and tartaric) are obviously the main soluble compounds of juices. A lot of other low molecular-weight compounds (MW < 1 kDa) are also present but in very lower amounts: minerals, especially potassium, N-containing compounds (amines, free amino acids), phenolic compounds (phenolic acids, flavonoid monomers), other organic acids, polyols, and aroma compounds. Soluble solids also include larger compounds up to 100 kDa: phenolic polymers (tannins), endogenous enzymes, or soluble pectic compounds. The contents of these compounds have a direct impact on the juice quality. In certain cases, it can be interesting to modulate fruit juice composition in order to increase the sensorial or nutritional quality (sugar/acid balance, bitterness, vitamin content, post-bottling haze, stability during storage, etc.) or to enhance a specific feature (antioxidant capacity, coloring power, etc.). Moreover, many of these solutes have interesting functional properties. They can be purified and concentrated thanks to separation processes after extraction from the juice or from the by-products of fruit juice processing (pectins, phenolics, carotenoids, etc.). In this context, pressure-driven membrane processes can be useful to separate them according to their molecular weight (ultra and nanofiltration). Electrically driven membrane processes like electrodialysis can be implemented to separate ionic forms with low molecular weight and modify ionic composition. In the particular case of aroma compounds that are very volatile in water, the membrane processes where a volatilization is necessary (pervaporation and some membrane contactors) can be considered in order to selectively extract them.

To sum up, the potential applications of membrane processes to fruit juice treatment are numerous and very diverse. Some of these processes like clarification by microfiltration are quite commonly used in the industry, especially for traditional clear juices like apple juice. The others are still developing and could be interesting for industry in the short-term because of the evolution of the fruit juice markets (high quality juices, product diversification, functional food) and thanks to technical improvements. In this work, we chose not to consider patents because they were often difficult to analyse from a scientific point of view.

8.3.2 Water Removal: Concentration

Because the production and consumption of fruit juices are generally far apart geographically, fruit juices are concentrated in order to reduce the storage and shipping costs and to achieve longer storage. The vacuum evaporation concentration technique, which allows high concentration efficiency to be reached, is the most widely used for fruit juice concentration. However, this thermal technique presents some drawbacks (i.e., loss of fresh juice flavor and vitamins, color degradation and "cooked" taste, etc.) and the final product does not satisfy the expectations of consumers who require more natural, tasty, and healthy food products. These last decades, many efforts have been devoted to develop improved methods for concentrated juice processing and cold process membrane technologies, especially pressure-driven membrane techniques and membrane contactors appear to represent an alternative to high-temperature treatments (Jiao et al. 2004).

8.3.2.1 Concentration by Pressure-Driven Membrane Processes

Among the pressure-driven membrane processes which can be used for fruit juice concentration, reverse osmosis (RO) is one of major interests. Firstly, RO can be operated at low temperature without any phase change. Compared with thermal concentration processes, energy costs are significantly reduced. Garcia-Castello et al. (2006), who compared energy consumption requirements by both RO and multiple effect evaporation, found that RO had a 7.7 times lower energy consumption. In addition, a limited thermal degradation of the product occurs during RO; the quality of the product is thus respected. From the first work published at the end of the 1960s (Merson and Morgan 1968), many studies (Table 8.1) are dedicated to the use of RO for the concentration of a large variety of fruit juices including apple, grape, orange, tomato, etc. These works generally focused on the retention of flavor compounds. It has been found that the retention of flavor components as well as the retention of other constituents depends on the membrane used. Aroma compound retention observed with polyamide membranes was greater than that measured with cellulose acetate membranes and spiral wound configuration was more convenient for concentration than the plate and frame configuration (Jiao et al. 2004). However, the transfer of aroma compounds through the membrane also depended on the type, the molecular structure, and the molecular weight of the volatile components (Pozderovic et al. 2007). Operating parameters such as transmembrane pressure, temperature, and crossflow velocity also affected the process efficiency (Jiao et al. 2004; Pozderovic et al. 2007). Pressure was found to be the most important parameter controlling the process. Alvarez et al. (1997) reported that increases in transmembrane pressure increased the permeate fluxes as long as the fluid velocity was kept lower than 2 m s^{-1}. High transmembrane pressures also favor the volatile retention (Alvarez et al. 1998; Pozderovic et al. 2007). As the mass transfer is favored when the temperature is increased, it is generally recommended to work under mild

Table 8.1 Examples of application of membrane technologies for concentration of fruit juices

Membrane processes	Type of juice	References
Reverse osmosis	Apple juice	Alvarez et al. (1997, 1998, 2002)
	Tomato juice	Petrotos et al. (1998)
	Orange juice	Jesus et al. (2007)
	Grape, grape must	Gurak et al. (2010), Rektor et al. (2004)
	Acerola juice	Matta et al. (2004)
Osmotic distillation	Apple juice	Bélafi-Bakó and Koroknai (2006), Onsekizoglu et al. (2010)
	Tomato juice	Durham and Nguyen (1994)
	Orange juice	Alves and Coelhoso (2006), Cisse et al. (2005)
	Melon juice	Vaillant et al. (2005)
	Grape, grape must	Bailey et al. (2000), Rektor et al. (2006)
	Kiwi	Cassano and Drioli (2007)
	Cactus pear juice	Cassano et al. (2007a)
	Citrus and carrot juices	Cassano et al. (2003)
	Passion fruit juice	Vaillant et al. (2001a)
	Pineapple juice	Hongvaleerat et al. (2008), Ravindra et al. (2008)
	Red fruit juices	Koroknai et al. (2008)
	Noni juice	Valdés et al. (2009)
Direct osmosis	Tomato juice	Petrotos et al. (1998, 1999, 2010)
	Pineapple	Nayak et al. (2011), Ravindra et al. (2006a)
	Beetroot, grape	Nayak et al. (2011)
	Orange	Herron et al. (1994)
	Raspberry	Herron et al. (1994), Wrolstad et al. (1993)

temperature conditions to avoid aroma loss (Jiao et al. 2004). However, Gurak et al. (2010) reported that concentration of grape can be carried out by RO at 50 °C without any decrease in its functional quality. The major drawback of RO is the decrease of permeate flux with an increase in the concentration. The increase of juice osmotic pressure, which reduces the driving force of the process, limits the capacity of RO to achieve concentration levels higher than 25–30 °Brix. At higher concentrations, the fouling problem and the high osmotic pressure of the juice sharply reduce water flux through the membrane. Therefore, RO must be considered as a pre-concentration process that can be coupled with other technologies to concentrate fruit juices up to 60 °Brix as required by the fruit juice industry. Nevertheless, this technique has in general high operational costs due to high pressure requirements (i.e., up to 6 MPa), so it is more appropriate for the treatment of high added value products.

To overcome the problem of high pressure values, some authors (Versari et al. 2003; Warczok et al. 2004) have proposed to replace the RO step by a NF step. However, this technique is not really suitable for fruit juice concentration because of solute leaks in the permeate. Indeed, Warczok et al. (2004), who tested several NF membranes in order to concentrate pear and apple juice, showed that it was possible to achieve high concentrations (i.e., up to 65–71 °Brix) at low transmembrane pressure (i.e., 12 bar). But a consequent fouling was observed and the rejections of total solutes were less than 80 %. Sugars and small organic compounds were found in the permeate.

8.3.2.2 Concentration Using Membrane Contactors

As osmotic distillation (OD) can be operated under mild conditions (room temperature and atmospheric pressure), it is a good candidate for the concentration of thermo-sensitive solutions such as fruit juices. Since the 1990s, the theoretical aspects of OD (Courel et al. 2000b; Nagaraj et al. 2006) and the application of this technique to the fruit juice concentration have been extensively studied (Table 8.1). Compared to RO, the main advantage of OD is the possibility to achieve high concentration levels. Fruit juices can be concentrated up to 65 °Brix (Cisse et al. 2005; Rodrigues et al. 2004; Vaillant et al. 2001a) in a single step process. In addition, the quality of the juices prepared from these concentrates (nutritional and organoleptic qualities) is generally better than those of juices prepared from concentrates obtained with thermal techniques (Cisse et al. 2011). Nevertheless, OD suffers from a weak evaporative capacity. The vapor flux values vary generally around 1 kg h^{-1} m^{-2} (Alves and Coelhoso 2006; Cassano et al. 2007a; Cisse et al. 2005; Vaillant et al. 2001a, 2005), but these values depend on the operating conditions. According to Courel et al. (2000a), the optimization of operating conditions in the OD module-like osmotic solution concentration, fluid velocities, and hydrodynamic conditions or temperature of the fluids can significantly improve the evaporating rate from 0.5 to 23 kg m^{-2} h^{-1}. Generally, OD fluxes decrease along the concentration process; the decrease is more drastic when the concentration is above 40 TSS. Vaillant et al. (2001a) reported that permeation flux continuously decreased from 0.7 to 0.6 kg h^{-1} m^{-2} when the total soluble solids content of orange juice increased from 15 to 40 °Brix and from 0.65 to 0.4 kg h^{-1} m^{-2} when it increased from 40 to 60 °Brix. The same evolution of evaporation fluxes was observed during kiwi juice concentration (Cassano and Drioli 2007). This phenomenon is mainly due to the decrease of the driving force concomitant with the increase of the juice concentration. But it also results in the drastic increase of juice viscosity, which significantly enhances the polarization phenomena at the feed side of the membrane (Courel et al. 2000a, 2000b; Vaillant et al. 2001a). In order to improve the productivity of the process, Vaillant et al. (2001a) suggested using a multistage procedure: a first step conducted at a constant total soluble solids content of around 40 °Brix during which a large part of the water was eliminated at a high vapor flux and a second step conducted at 60–65 °Brix in order to achieve the concentration process. It is important to note that low membrane fouling occurs during the OD process, and although it is possible to treat raw juices, best productivities are obtained when the juice is previously micro or ultrafiltrered

(Bailey et al. 2000; Cisse et al. 2005; Hongvaleerat et al. 2008). As the water vapor flow depends directly on the vapor pressure gradient, temperature that influences the vapor pressure values appears to be one of the most important factors in improving the vapor fluxes. Indeed, water flux were shown to exponentially increase with the temperature in some studies. Brodart et al. (2003), Courel et al. (2000a), and Ravindra et al. (2008) underlined the role of temperature polarization effect during pineapple juice concentration.

However, as the increase of temperature should be limited in order to preserve juice quality, the vapor pressure gradient can be significantly increased by thermostating both solutions separately at different temperatures (the osmotic solution was kept at low temperature while the aqueous solution to be concentrated was heated up to 40 °C). By adding a temperature difference of 15 °C to the transmembrane concentration difference (Bélafi-Bakó and Koroknai 2006), significantly enhanced water transport during apple juice concentration compared to single DCMD or OD operations was carried out in similar conditions. The membrane properties, in particular its thickness, also play a role in OD performance; the thinner the membrane, the higher the water flux. Courel et al. (2000a), Gryta (2005), and Brodart et al. (2003) showed that an increase in pressure on both membrane sides allows the liquid to penetrate deeper inside the porosity; the thickness of the gas layer was reduced and the water fluxes were enhanced. Nevertheless, the applied pressure should be lower than the penetration pressure value in order to avoid the membrane wetting and thus the mixing of the solutions. Finally, the design of the module seems to be very important; the highest flux values reported (i.e., about 10 kg h^{-1} m^{-2}) were obtained with the same flat experimental module and equipped with a PTFE membrane (Courel et al. 2000a; Hongvaleerat et al. 2008; Rodrigues et al. 2004).

As the quality of a fruit juice concentrate is related to its content of aroma compounds, some studies were focused on the transfer of volatiles during the OD process. Barbe et al. (1998), who compared the volatile loss during osmotic distillation with different types of membrane, suggested that the retention of aroma was correlated with membrane pore size; the higher the pore diameter, the higher the organic volatile retention. Ali et al. (2003) reported that the transfer of aroma compounds can be considerably slowed down by decreasing circulation velocity and temperature of the solution to be concentrated. Cisse et al. (2005) proposed to limit aroma losses during orange juice concentration by preconditioning the membrane with the microfiltrered juice and by avoiding thermal regeneration of brine during concentration; they suggested keeping the brine concentration near saturation (at 5.5 mol L^{-1}) by adding $CaCl_2$ crystals throughout the trials. Actually, compared with water transfer, transfer of aroma compounds is not significant and does not affect the juice quality, which was found much closer to that of the initial fresh juice than to the commercial thermal concentrate (Cassano and Drioli 2007; Cisse et al. 2005).

Even if OD appears to be an attractive alternative to thermal evaporation, it suffers from a serious drawback: the management of the stripping solution which drastically increases the cost of the process. Indeed, as brine often contains chlorides, it is very corrosive and its regeneration can be hardly accomplished by conventional evaporator. To reduce the additional costs linked to brine regeneration, Petrotos et al. (2010) proposed the use of electrodialysis as a more economical method for concentrating brines up to the saturation. To overcome the problem, Bui et al. (2007) proposed to replace OD by DCMD carried out at low temperature; the solution to be concentrated was heated at 40 °C while the stripping water was kept at 10 °C. They obtained fluxes of up to 2.9 kg m^{-2} h^{-1} for PVDF and 5.8 kg m^{-2} h^{-1} for Halar fibres when concentrating 30 % glucose solution at 40 °C. Gunko et al. (2006) and Kozák et al. (2009) reported that the same process can be used for concentrating apple and blackcurrant juices, respectively. Both studies emphasize the effect of transmembrane temperature gradient on process efficiency, but none of them studied the aroma transfer. Actually, Alves and Coelhoso (2006), who compared the OD and MD processes, concluded that OD has advantages over DCMD not only in terms of water flux, but also regarding the retention of aroma compounds.

The second bottleneck to the development of OD is linked to the membrane material. Hydrophobic membranes used in OD are much more expensive than hydrophilic ones. In addition, some juice components (i.e., peel oil and lipophilic compounds) as well as surfactants contained in cleaning solutions might change the membrane surface properties leading to the wetting of the membrane. The development of more hydrophobic, thinner, and more porous membrane with long-life stability will be necessary to favor the industrial development of OD. To face these material problems, some authors propose an alternative concentration process: direct osmosis (DO) also called forward osmosis (FO).

8.3.2.3 Concentration Using Direct Osmosis

Direct Osmosis (DO) is a membrane process which involves a dense hydrophilic membrane like in Reverse Osmosis (RO), but the driving force of water transfer across the membrane is a water activity gradient like in Osmotic Distillation (OD). As in a membrane contactor, the solution to be concentrated (a dilute aqueous solution) and the stripping solution (i.e., a brine or a high-concentrated sugar solution) flow tangentially along the membrane. Thanks to the osmotic gradient, water from the feed side is diffused into the membrane towards the stripping solution. The physical principles and applications of DO as well as their strengths and limitations have been recently reviewed by Cath et al. (2006) and Zhao et al. (2012). In particular, as DO is carried out at ambient pressure and low temperature, it is capable of concentrating liquid without product deterioration. Although DO represents another alternative to thermal evaporation processes in fruit juice processing, only few studies related to the concentration of fruit juices have been reported (Table 8.1). It is important to note that raw tomato juice can be treated by DO, but the use of ultrafiltered juices can markedly improve the process performance (Petrotos et al. 1999).

Like other membrane processes, the performance of the DO process depends on operating parameters such as final feed concentration, osmotic pressure difference between feed and osmotic agent, physical properties of feed as well as osmotic agent, hydrodynamic conditions, and membrane characteristics (i.e., thickness and water permeability).

Generally, concentrated solutions of sodium chloride, sucrose, fructose, or high fructose corn syrups are used as osmotic agents during direct osmosis process, but Petrotos et al. (1998), who compared the effect of different stripping solutions on water flux during DO concentration of tomato juice, reported that salt solutions were better as osmotic media than carbohydrate solutions. These authors pointed out the effect of the viscosity of stripping solutions which should be as low as possible. However, when concentrated salt solutions are used as osmotic agent, a migration of salt occurs and results in a salty taste in the fruit juice. To overcome this drawback, Ravindra et al. (2006a) used a mixed osmotic agent (sucrose 40 % and NaCl 12 % w/w) for the concentration of pineapple juice up to 60 °Brix. They showed that the presence of a low quantity of salt in sucrose solution significantly enhanced the water flux without leading to an adverse sensory quality of the juice. The presence of sucrose limited but did not stop the salt transfer; nevertheless, according to sensory analysis the quality of the concentrate was still acceptable.

It is worth noting that in the DO process, polarization phenomenon cannot be neglected. The adverse effect of polarization can be minimized by increasing flow velocity and turbulence at the membrane surface or by increasing the temperature (Petrotos et al. 1998; Ravindra et al. 2006a). In addition, the mass transfer is favored when thin membranes are used (Petrotos et al. 1998); it also depends on the membrane orientation. Nayak et al. (2011) observed higher transmembrane flux when feed solution was towards the active layer.

The advantages and drawbacks of DO as well as other membrane concentration processes are summarized in Table 8.2. Compared to reverse osmosis, direct osmosis allows fruit juices to be concentrated up to a high concentration level at low pressure, without membrane fouling but with the additional cost of the re-concentration of the stripping solution. Compared to osmotic distillation, the water fluxes are of the same order of magnitude (between 1 and 10 L h^{-1} m^{-2}) and they decrease with the feed concentration. In addition, DO shares the same drawback of brine management, but it offers a priori a better retention of volatile compounds, but no study focusing on aroma compound retention during DO process has been published. Nevertheless, due to the migration of osmotic agent (salt) into the product, the process seems to be more suitable for the concentration of vegetable juice.

8.3.3 Solid/Liquid Separation: Clarification and Microbial Stabilization

Solid/liquid separation in fruit juices using microfiltration or ultrafiltration has been studied for a long time. This is clearly the major utilization field of membrane technology in fruit juice processing. Industrial applications have already been developed in traditional clear juice industries especially apple juice for

Table 8.2 Comparison between the main membrane processes available for concentrating fruit juices

Membrane processes	Advantages	Drawbacks
Reverse osmosis	Broad industrial scale application	High pressure
	Low temperature	Fouling phenomena
	Low energy costs	Limited concentration level (<30 °Brix)
		Pulpy juices should be clarified
Osmotic distillation	Low temperature and pressure	New process: few experiments at industrial scale
	Low energy costs	Low evaporation capacity
	High concentration level (>60°Brix)	Management of the diluted brine
	No fouling problem	Production costs
Direct osmosis	Concentration of pulpy juices	New process: require an evaluation at industrial scale
		Low permeation flux
		Solute leaks from the strip solution
		Production costs

clarification. Scientific literature is abounding on the subject. To complete the numerous references mentioned by Girard and Fukumoto (2000) or Echavarria et al. (2011), Table 8.3 gives some recent application examples. More often than not, the membrane treatment is coupled with an enzymatic liquefaction that uses pectinolytic and cellulolytic enzymes in order to decrease viscosity and to improve juice filterability. Microfiltration is generally preferred because ultrafiltration often leads to higher solute retentions that greatly disturb the nutritional and organoleptic quality of the clarified juice. Actually, crossflow microfiltration can be applied to perform not only clarification, but also more variable unit operations as shown in Fig. 8.7.

Classically, partial or total destruction of a microorganism is achieved through severe thermal procedures, such as pasteurization or sterilization processes; nonetheless, these treatments seriously damage sensory, nutritional, and probably some functional properties of fruit juices. Therefore, the selective removal at ambient temperature of vegetative microorganisms and eventually their spores is of great interest to the fruit juice industry and has naturally attracted attention to membrane technology and more specifically to microfiltration. Actually, food-borne microorganisms and their spores are greater in size than 0.45 μm, and consequently, they can be retained by membranes with a lower pore diameter. Generally, commercial sterility of permeate is claimed to be achieved after filtration through a 0.2 μm pore diameter membrane. For larger pore sizes, the load of micro-organisms only can be reduced with magnitude of reduction decreasing when pore diameter

Table 8.3 Examples of recent applications of crossflow ultra or microfiltration for stabilization/clarification of fruit juices

Juice	Membrane	Operating conditions: Transmembrane pressure/crossflow velocity/temperature	Pretreatment	Permeate flux and volumetric reduction ratio (VRR)	Noticeable modification of juice composition	References
Acai	MF 0.2–0.8 μm, ceramic, 47 cm^2	1 bar/3 m s^{-1}/25 °C	Enzymatic treatment, dilution ¼	48–110 L h^{-1} m^{-2} at VRR 1.1–1.3	Losses of phenolics 43 %, of antioxydant activity 32 %	Machado et al. (2012)
Acerola	MF 0.14 μm, tubular, ceramic, 0.1 m^2	2 bar/10–40 °C	No	27–50 L h^{-1} m^{-2} at VRR 1.5–4		Wang et al. (2005)
	MF 0.3 μm, tubular, polyethersulfone, 500 cm^2	1 bar/30 °C	Enzymatic treatment	100–160 L h^{-1} m^{-2}		Matta et al. (2004)
Apple	UF 10 kDa, polyethersulfone, 200 cm^2	1–3 bar/20–40 °C	Enzymatic treatment, flocculation with gelatin and bentonite	4–41 L h^{-1} m^{-2} at VRR 5–6	Decrease of total phenolics	Bahceci (2012)
Cashew apple	MF 0.3 μm, tubular, polyethersulfone, 500 cm^2	2 bar/30 °C	Enzymatic treatment (tannase)	70 L h^{-1} m^{-2} at VRR 3	Losses of ascorbic acid 24 %	Campos et al. (2002)
	MF 0.1–1.4 μm, tubular, ceramic, 55 cm^2	0.8–2.3 bar/6–7 m s^{-1}/30 °C	Enzymatic treatment	36–79 L h^{-1} m^{-2} at VRR 1	Losses of tannins 98 %	Abreu et al. (2005)
Kiwi	UF 15 kDa, tubular, polyvinylidenefluoride, 0.23 m^2	0.9 bar/20–30 °C	Sulfites and enzymatic treatment	7–19 L h^{-1} m^{-2} at VRR 1–4	Losses of ascorbic acid 16 %, losses of antioxidant activity 8 %	Cassano et al. (2007b)

(continued)

Table 8.3 (continued)

Juice	Membrane	Operating conditions: Transmembrane pressure/crossflow velocity/temperature	Pretreatment	Permeate flux and volumetric reduction ratio (VRR)	Noticeable modification of juice composition	References
Melon	MF 0.2 µm, tubular, ceramic, 0.24 m^2	1.2–2.7 bar/7 m s^{-1}/35 °C	Enzymatic treatment	20–90 L h^{-1} m^{-2} at VRR 2–4	Losses of phenolics 19 %, of ascorbic acid 7 %, total retention of carotenoids	Vaillant et al. (2005)
Orange	UF 50 kDa, plate, polyamide, 35 cm^2	4.1 bar/30 °C	Enzymatic treatment, flocculation with gelatin and bentonite	5–20 L h^{-1} m^{-2}	Total retention of pectins	Rai et al. (2007)
	MF 0.2 µm, tubular, ceramic, 0.22 m^2	4 bar/7 m s^{-1}/20 °C	No	62 L h^{-1} m^{-2} at VRR 3.5	Modification of profile of aroma compounds, total retention of carotenoids	Cisse et al. (2005)
Passion fruit	MF 0.3 µm, tubular, ceramic, 50 cm^2	0.5–1 bar/25 °C	Centrifugation and enzymatic treatment	27–42 L h^{-1} m^{-2} at VRR 3	Losses of pectins 31–82 %, of ascorbic acid 3–52 %	De Oliveira et al. (2012)
	MF 0.3 mm, hollow fibers, polyamide, 158 cm^2	0.5–1 bar/25 °C	Centrifugation and enzymatic treatment	17–20 L h^{-1} m^{-2} at VRR 3	Losses of pectins 39–56 %, of ascorbic acid 2–16 %	De Oliveira et al. (2012)

Pineapple	MF 0.1 & 0.2 μm, hollow fibers, polysulfone, 110 cm^2	0.1–0.7 bar/1.5–3.5 m s^{-1}/20 °C	Enzymatic treatment	25 L h^{-1} m^{-2} at VRR 4	Losses of proteins 9 %, of vitamin C 9–22 %, of total phenolics 7–19 %, of antioxidant activity 1–7 %	Laorko et al. (2010)
	UF 30 & 100 kDa, hollow fibers, polysulfone, 110 cm^2	2 bar/1.2 m s^{-1}/20 °C	Enzymatic treatment	9–20 L h^{-1} m^{-2}	Losses of proteins 37–51 %, of vitamin C 18–31 %, of total phenolics 15–24 %, of antioxidant activity 5–24 %	Laorko et al. (2010)
	UF 0.01 μm tubular, ceramic 50 cm^2	2–6 bar/4.2 m s^{-1}/30–50 °C	Enzymatic treatment	68–124 L h^{-1} m^{-2} at VRR 1	Losses of pectins 80 %, of acidity 15 %	De Barros et al. (2003)
	UF 100 kDa hollow fibers polysulfone 0.12 m^2	0.2–2 bar/1.2 m s^{-1}/20–40 °C	Enzymatic treatment	34–46 L h^{-1} m^{-2} at VRR 1	Losses of pectins 62 %	De Barros et al. (2003)
Pomegranate	UF 15 kDa, tubular, ceramic, 75 cm^2	1–4 bar/1–3.8 m s^{-1}/20 °C	Enzymatic treatment	25–36 L h^{-1} m^{-2} at VRR 6	Losses of total phenolics 21–49 %	Baklouti et al. (2012)
Umbu	MF 0.2 μm, tubular, polypropylene, 380 cm^2	0.5–1.1 bar/4–6 m s^{-1}/35 °C	Enzymatic treatment, dilution 1/3	48–75 L h^{-1} m^{-2} at VRR 2.2	Acidity and total soluble solids decrease	Ushikubo et al. (2007)
Watermelon	MF 0.2 μm, plate, cellulose acetate, 35 cm^2	1.4–2.8 bar/30 °C	No	10–15 L h^{-1} m^{-2}	Losses of lycopen 65 %	Chhaya et al. (2008)

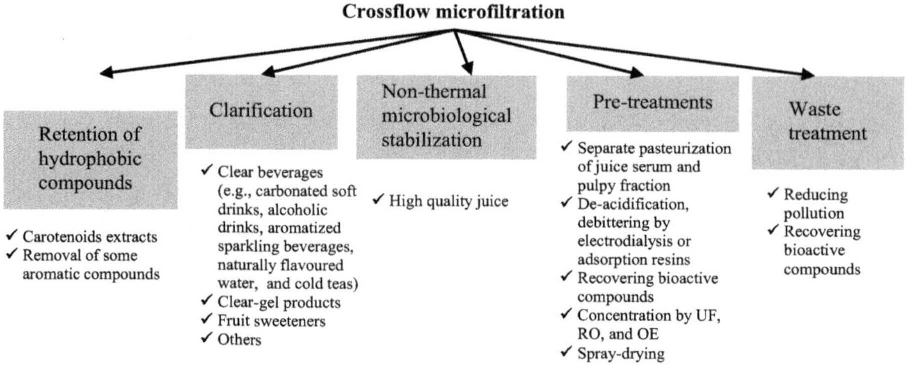

Fig. 8.7 Objectives of crossflow microfiltration in fruit juice industry

increases. Hence, the partial or complete retention of micro-organisms is not dependent on temperature as the process can be conducted at ambient, slightly lower, or higher temperatures. Crossflow filtration using membrane pore size below 0.2 μm can be equivalent to a sterilization process, but still require the same care in preserving sterility downstream, such as aseptic packaging. Membrane filtration process can be connected directly to sterile packaging equipment if membrane filter cartridge and housing can undergo steam-in place sterilization. Consequently, ceramic membranes and stainless steel housing are preferred in this case. If sterility cannot be preserved downstream but good manufacturing practices are implemented, levels of micro-organisms can be sufficiently reduced to provide an extended shelf-life.

8.3.3.1 Impact on Quality

A significant sensory quality attribute is the low turbidity of clear juice. Crossflow microfiltration with membrane having a pore size of 0.2 μm usually results in a clear juice with a turbidity value generally below 1 nephelometric turbidity unit (NTU). However, in some cases, the formation of a haze or precipitate can occur during storage, although the juice was found to be perfectly clear straight after microfiltration. Increase in turbidity may be due in some cases to protein-tannin haze formation. The color of clarified juice also depends on the properties of the biochemical compounds involved in color perception. If juice color is due to hydrophilic compounds such as anthocyanins or betalains, the microfiltered juice will preserve its original color and in most cases it is enhanced. However, when color is due to carotenoid compounds, the resulting permeate will be pale, as these hydrophobic compounds can be dispersed within the juice as emulsion droplets or they can even associate to form ISS, resulting in an important loss.

Quality of the retentate and that of permeate are also different as far as nutritional quality is concerned. Water-soluble compounds not associated with suspended

Table 8.4 Impact of crossflow microfiltration on the sensory, nutritional, and functional quality attributes of some tropical fruit juices

	Sensory quality		Nutritional and functional qualities	
Fruit juice	Permeate	Retentate	Permeate	Retentate
Passion fruit	Fresh notes	Color enhanced[a]		Carotenoids retained
Coconut water	Fresh notes		Minerals	
Pineapple	Fresh notes	Color enhanced		Carotenoids retained
Andean blackberry	Fresh notes	Color enhanced		
Noni	Undesirable aromas partially removed[b]		Phenolics and vitamin C	
Granadilla	Fresh notes			Carotenoids retained
Uvilla	Fresh notes	Color enhanced		Carotenoids retained
Orange	Poor aroma	Color enhanced	Preserved vitamin C	Carotenoids retained

[a]Color is enhanced because colored carotenoids concentrate in the retentate

[b]Fatty *octanoic acid* responsible for the rancid like aroma is partially retained

solids can pass easily through the membrane. In contrast, hydrophobic compounds associated with insoluble suspended solids or eventually arranged as emulsion droplets will be totally or partially retained. Consequently, water-soluble vitamins and most phenolic compounds will be found in the clarified juice at approximately the same concentrations as in the initial juice, whereas carotenoids will be found concentrated within the retentate. Consequently, the functional properties of clarified juice and pulpy fraction will depend on their respective initial composition in bioactive compounds. For example, if antioxidant capacity is due mainly to phenolic compounds and vitamin C, it will be preserved mostly in the clear juice, but if it is due to carotenoids, then it will be concentrated in the retentate. According to fruit type, either the permeate or retentate, or sometimes both, will be targeted as added-value products (Table 8.4).

Generally, impact of crossflow microfiltration on juice quality, except microbiological quality, is not dependant on membrane pore size within the normal range between 0.1 up to 5 μm. Within this range, only non-soluble compounds are retained corresponding mainly to cellular fragments; nonetheless, some compounds with much smaller size can be linked to cell walls and may be retained by the membrane. Nutritional compounds not linked in any way to the solid phase retained by the membrane will be found in approximately the same concentration levels in the microfiltered juice as in the initial juice. This is true among others for sugars, organic acids, phenolics, and some aromatic compounds. Nonetheless, for some aromatic compounds according to their chemical class, and affinity to the membrane and solid materials

Table 8.5 Retention efficiency observed for some compounds during crossflow microfiltration using a ceramic membrane with 0.2 μm pore diameter (Cisse et al. 2005; Courel et al. 2000a; Vaillant et al. 1999, 2005)

Characteristic	Retention efficiency[a] (%)
Suspended insoluble solids (SIS)	100
Microorganisms	100
Sugars	<1
Organic acids	<1
Acide L-ascorbique	<7
Micronutrients	
– Phenolic compounds	<1
– Anthocyanins	<1
– Elagitannins	<1
– Carotenoids	>90
– β-carotene	>90
– Lycopene	>90
– Betalains	<1
Aromatic compounds	
– Ester	<10
– Aldehydes	<10
– Aliphatic alcohols	<10
– Volatile acids	<10
– Terpenol	≈18
– Terpenic hydrocarbons	≈40
Functional quality	
Antioxidant capacity (H-ORAC)	<5

[a]Retention is calculated as $R_a = 1 - C_{permeate}/C_{edible\ juice}$

retained (Table 8.5), their distribution may vary between the permeate and retentate. For example, the more hydrophilic oxygenated compounds, e.g., esters, aliphatic alcohols and aldehydes, which are not linked to pulp or cloudy fractions (Brat et al. 2003), appear to pass easily through ceramic membranes, but they could also be partially retained by organic membranes. Therefore, selection of membrane material is important, ceramic membranes probably being the most neutral if compared to organic ones.

In contrast, to more hydrophilic oxygenated compounds, the more hydrophobic compounds, e.g., monoterpene and sesquiterpene hydrocarbons, are drastically retained and often found concentrated in the retentate, whatever the membrane material used. These compounds are strongly associated with cell-wall fragments (Brat et al. 2003), and they are consequently also separated from pulp, during strong centrifugation. Actually, in most cases, nutritional quality of juice permeate is totally comparable to the quality of juice supernatant obtained after performing a strong centrifugation (>4000×*g*). Therefore, after microfiltration or centrifugation, the quality of clear juice depends mainly on fruit juice composition.

It is interesting to note that the most hydrophilic oxygenated compounds usually determine a juice's freshly squeezed taste. These compounds will pass easily through the membrane and consequently permeate will preserve the characteristics of fresh fruit notes. On the other hand, terpene hydrocarbons are less volatile and often less thermo-sensitive. These compounds will concentrate in the retentate. Both observations have resulted in the development of the process that consists of thermally treating only the retentate and then mixing aseptically the permeate and retentate downstream, before homogenization and aseptic packaging (Fig. 8.8). The pulpy juice obtained is claimed to be of higher quality as it preserves the fresh aromatic notes coming from the permeate that has not suffered any thermal treatment. This process is particularly interesting for pulpy fruit juice such as citrus, of which the particular aroma is due mainly to the combination of highly volatile esters and aldehydes and terpenic compounds linked to the pulp.

On the other hand, most fruits including tropical ones are characterized mainly by their content in highly volatile aromatic compounds which are consequently also extremely thermo-sensitive. Most cannot be treated thermally without drastically degrading sensory properties. This is true for melon, papaya, and coconut water for example, as these fruit juices may almost totally lose their subtle aromas, which may not be recognized even by a trained sensorial panel. It has been shown that passion fruit juice, for example, loses as much as 90 % of aroma compounds during classical thermal processing; nonetheless, the juice is still attractive for its residual flavor. In these cases, crossflow microfiltration represents an alternative method for obtaining a juice fraction that is rich in highly volatile aromatic compounds and of high microbiological quality.

8.3.3.2 Engineering Aspects and Costs

As with all modular processes, crossflow microfiltration presents various advantages over other emerging technologies for application on a small or medium scale. Investment costs as well as running costs are almost proportional to the membrane surface, which also determines production capacity. Because crossflow microfiltration equipment may range in size from a few square centimetres to hundreds of square metres, this technology can be applied by small, medium, and large fruit juice industries. Two main factors must be taken into account when assessing the industrial feasibility of microfiltration, the average permeation flux (expressed in $L\ h^{-1}\ m^{-2}$) and the final yield of clarified juice. The average value of permeation flux depends on various factors, including fruit juice composition and characteristics, juice extraction procedure, final contents of ISS in the feed juice, pretreatments, and hydrodynamic conditions during microfiltration. It remains today a scientific challenge to understand why some juices with high content in suspended solids have higher average permeation flux than other juices with lower turbidity. Composition, microscopic shape of suspended solids, and specific affinity to membrane surface, among other nonidentified factors, may explain juice behavior during crossflow microfiltration. Therefore, up to now, experimental trials are required to obtain the average permeation flux under given hydrodynamic conditions, which in turn must be optimized to obtain the highest average permeation flux possible.

Schematic diagram of the fresh note process

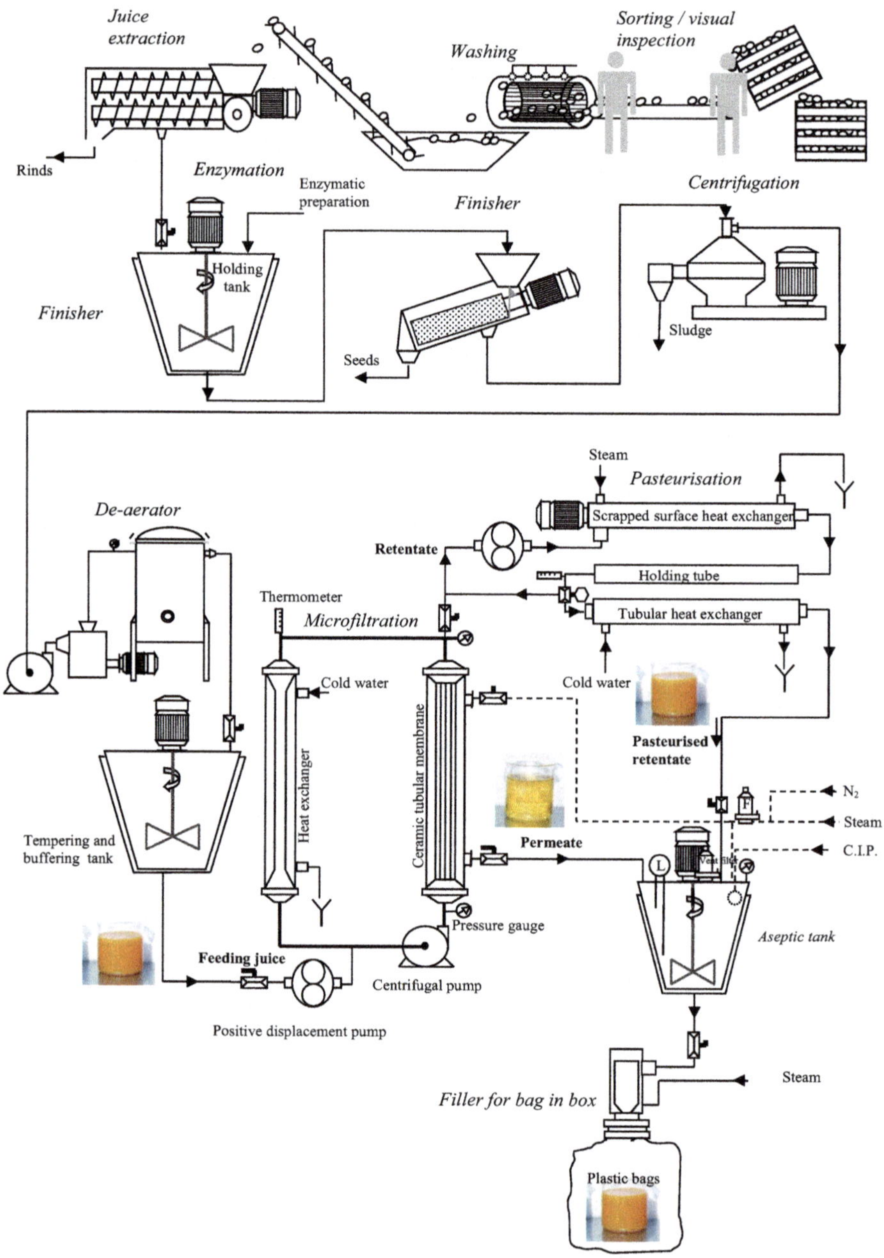

Fig. 8.8 Schematic lay out of the process coupling crossflow microfiltration and pasteurization

Depending on the added value of the final product, a critical average permeation flux exists, below which microfiltration is not sustainable. For example, if the technological objective is to clarify fruit juice, the critical average permeation flux is estimated to be around 50 L h^{-1} m^{-2}. For products with higher added value, this average permeation flux may be slightly lower. Permeation flux can be drastically improved by either mechanical or enzymatic pretreatment of the juice before filtration. The method followed during juice extraction also considerably influences filtration behavior. Often, pressing results in higher permeation fluxes than crushing and sieving, probably due to the generation of a higher content of cell fragments. Generally, the lower the initial contents of ISS, the higher the average permeation flux that can be achieved. Thus, mechanical procedures such as centrifugation that reduce ISS contents can potentially improve permeation fluxes.

Previous enzymatic treatment of fruit juices is also most often associated with crossflow microfiltration. Enzymatic treatments aim not only to reduce ISS content but also to reduce viscosity by hydrolyzing soluble macromolecules such as pectin, which is a critical compound in membrane fouling. Commercial enzyme preparations standardized for pectinase and cellulase activities are usually the most effective for pulpy juices, with optimal concentrations ranging between 10 and 150 mg L^{-1}. In all cases, the pretreatments before microfiltration should be assessed from the economic viewpoint, determining the additional costs of pretreatment and the advantage obtained in terms of the resulting increase in average permeation flux. Average permeation flux can be also improved by intermittent back-flow flushes cycle. The different pressure-driven systems of back-flow flushes generally allow an increase in permeation flux on average of 10–20 % (Wattananusorn 2008). Other methods such as ultrasounds have been also implemented on a commercial scale to destabilize the fouling layer and enhance microfiltration performance.

The other imperative parameter to be considered for the economic feasibility of a microfiltration unit is the final yield of clarified juice that can be obtained while maintaining a sufficiently high average permeation flux. The higher the yield of clarified juice, the richer the retentate becomes in retained particles and fouling of the membrane also increases drastically, gradually diminishing permeation flux. A critical yield can be reached above which permeation flux becomes unprofitable. For pulpy juices with a high content of suspended ISS, a maximum yield between 50 and 70 % of microfiltered juice can be obtained. For coconut water or cloudy apple juice previously centrifuged, yield of microfiltered juice may reach up to 90 %. If the yield of microfiltered juice is low, retentate is not discarded and must be used after pasteurization either in the "fresh note" process or in other fruit-based products (Vaillant et al. 2001b).

In the process presented in Fig. 8.8, the retained fraction of the juice is pasteurized and blended again with clarified juice. In this case, the fraction of the juice to be microfiltered is determined by considering the noticeable improvement of quality of the final pulpy juice. For passion fruit and orange juice, reaching a yield of microfiltered juice of 50–65 % is sufficient to significantly improve sensory and nutritional properties (Cisse et al. 2005).

Table 8.6 Average inflows and outflows obtained during trials with tropical fruit juices in a semi-industrial pilot plant using tubular ceramic membrane with a transmembrane pressure of 1.5 bar, a crossflow velocity of 7 m s^{-1} and at 30 °C (Vaillant et al. 2001b)

Fruit juice	Maximum yield clarified juice (%)	Average feed flux (L h^{-1} m^{-2})	Average permeation flux (L h^{-1} m^{-2})	Average extraction flux of retentate (L h^{-1} m^{-2})
Passion fruit	70	60	40	20
Andean blackberry	70	105	70	35
Mango	23	260	60	200
Pineapple	72	98	70	28
Mandarin	72	64	50	14
Orange	72	123	88	35
Coco water	97	155	150	5

These are common microfiltration yields for most pulpy fruit juices, which also allow relatively high average permeation flux to be achieved. Examples of experimental data for some tropical fruit juices are presented in Table 8.6, giving average inflows and outflows to and from the crossflow microfiltration unit and overall yield of clarified juice.

Another strategy can be implemented when maximum yield between 50 and 70 % of clear juice can be obtained and retained fraction of the juice must be discarded or is of insufficient value. The strategy consists in extracting just enough clear juice without significantly degrading the quality of the retentate. For example, from passion fruit juice that was previously treated with enzymes, at least 60 % of clarified juice can be extracted without noticeably degrading sensory quality of the retained juice. Also, because carotenoids and dietary fibers are mostly retained, the nutritional and functional quality is enhanced making this juice even more attractive to some consumers. Even the rheological properties of the retained juice may remain similar to those of the initial juice. Enzymic treatment before microfiltration thus implies a reduced viscosity, which can be offset by the concentration of insoluble compounds in the retentate and subsequent increase in viscosity.

Given the different operational modes presented, the implementation of a crossflow microfiltration system parallel to classical production of fruit juice has high potential for diversification and for addressing niche markets that demand the highest quality.

The design of a processing plant with a crossflow microfiltration system is relatively simple, as the process requires few inputs: energy for moving at least two pumps, cold water to remove heat produced by the pumps during filtration and cleaning-in-place system. The energy requirement of a crossflow microfiltration system is very low, compared with other processes such as pasteurization. The first pump is a positive feed pump that applies a static pressure to the circuit (generally between 1 and 4 bar) and the other, a circulation pump, that creates an appropriate crossflow velocity on the membrane surface (between 2 and 7 m s^{-1}). A crossflow microfiltration system can be operated in either batch or continuous mode and can be easily included in a classical processing line of fruit juice. Any pretreatments required, such as enzyme treatments and centrifugation, are traditional processes that already exist in many fruit-juice processing plants. The continuous operation of constant juice feed, permeate collection, and bleeding of retentate is the usual running mode in the fruit juice industry.

8.3.3.3 Membrane Cleaning

Membrane cleaning is an important limitation of crossflow microfiltration, as it must be performed periodically and the cost of detergents can be relatively important. For processing fruit juices, ceramic membranes are recommended, as they are more resistant to cleaning with standard chemical solutions (for instance nitric acid followed by sodium hypochlorite) used for cleaning-in-place operations. Depending on the process followed, usually one cleaning cycle (averaging 1.5 h) is performed for every 6–8 h of continuous processing. Tubular ceramic membranes are particularly suitable for pulpy juices if the channel diameter is equal to or higher than 4 mm. Although the cost of these membranes is much higher than organic spiral membranes, their significantly longer service life (>20 years) and relatively trouble-free operation mean that ceramic membranes are usually more cost-efficient over the long term. Nonetheless, costs of ceramic membranes have considerably decreased over recent years and they have become affordable by most agro-industries.

8.3.4 Modification of Solutes Composition

8.3.4.1 Acidity Modulation

In order to adjust the sugar/acid ratio in flawed fruit juices or to facilitate the use of very acidic juices, it could be of great interest to modulate their acid content. Because these carboxylic acids are always partially ionized in juices, ion-exchange processes are potentially able to remove these compounds. Electrodialysis was mentioned for the first time in the 1980s to correct acidity of apple, citrus, cherry, pineapple, and grape juices (Goloubev and Salem 1989; Nanjundaswamy and Chikkappaji 1989; Wucherpfennig and Keding 1982). Results indicated the technical feasibility of this membrane process using homopolar or bipolar membranes, but it led to weak pH variations and damaged organoleptic or nutritional quality of the juices. Focusing on passion fruit juice, Vera et al. (2003) evaluated new electrodialysis configurations at laboratory scale. They showed that electrodialysis with homopolar and bipolar membranes could allow the pH of clarified passion fruit juice to be increased from around 3–4. Compared with more conventional deacidification processes, i.e., calcium salts precipitation and ion-exchange chromatography, electrodialysis presented the advantages of avoiding the use of chemicals and of reducing damage to the flavor of the juice. Applied to different acidic tropical juices ($2.6 < pH < 3.2$), electrodialysis was then studied up to the pre-industrial scale (Vera et al. 2007a, 2007b). To reach a final pH of 4, the titratable acidity had to be decreased from 50 to 70 %. Better energetic performances were obtained using electrodialysis with homopolar membranes. Electrodialysis with bipolar membranes avoided soda consumption and allowed the organic acids extracted from the juice to be recovered. From a quality point of view, low differences were observed whatever the electrodialysis configurations and operating conditions used. Deacidification led to low changes of color, except for juices that contained anthocyanins. The organoleptic characteristics of juices were well-preserved, even if the aromatic intensity slightly decreased. Even if the interest of the process is clearly highlighted, its feasibility at an industrial scale requires a better optimization of performance and a limitation of membrane fouling that can occur in certain cases.

8.3.4.2 Phenolic Profile Modulation

In addition to having high antioxidant properties, phenolics from fruits have a lot of other specific biological actions. Modulation of the phenolic composition of fruit juice could present a great interest, especially in order to propose new natural products in accordance with the growing demands from the nutrition and health markets. Different membrane processes were cited in the literature for modifying phenolic composition of fruit juices.

Pressure-driven membrane processes such as ultrafiltration or nanofiltration could be employed to enrich juices with phenolic compounds or to fractionate them according to their molecular weight. For example, phenolic enrichment was studied in order to enhance antioxidant properties of apple juice (Saleh et al. 2006). Phenolics could be satisfactorily separated from sugars and concentrated up to four using a spiral-wound ultrafiltration membrane with 1 kDa molecular cut-off (MWCO) (1.2 m^2). In the same way, nanofiltration with plate and frame membrane was evaluated at laboratory scale to increase anthocyanins content of clarified açai juice (Simoes et al. 2011).

Phenolic fractionation was mentioned to modulate monomeric and polymeric anthocyanin fractions in Concord grape juice using 0.1 m^2 plate and frame ultrafiltration membrane with MWCO ranging from 10 to 1000 kDa (Kalbasi and Cisneros-Zevallos 2007). With a volumetric reduction ratio of 10 and a transmembrane pressure of 4 bar, permeate flux ranged between 3 and 15 $L\ h^{-1}\ m^{-2}$ at ambient temperature. By selecting the membrane or through sequential treatments, this process could be envisaged for tailoring the flavonoid profile, i.e., the color and the other functional properties of the product such as antioxidant power. In the same manner, flavonoids could be separated from chlorogenic acid using a 0.25 kDa membrane (Saleh et al. 2006). Permeate flux ranged between 5 and 25 $L\ h^{-1}\ m^{-2}$, depending on temperature (30–50 °C), transmembrane pressure (5–30 bar), and volumetric reduction chosen (up to 10).

Promising results were also obtained at laboratory scale with flat-sheet ultrafiltration membranes (MWCO from 1 to 150 kDa) in order to separate anthocyanins and ellagitannins from blackberry juice at 30 °C (Acosta et al. 2014). Except for the slackest membrane, ellagitannins were completely retained whereas anthocyanins were rejected by only 70–90 % (depending on membrane and transmembrane pressure). This difference of retention rates between ellagitannins and anthocyanins allowed the anthocyanin/ellagitannin balance to be modulated in the juice. Permeate flux obtained were encouraging for a future industrial application.

In order to overcome the limited selectivity of the pressure-driven membrane processes, electrodialysis associated with ultrafiltration has been suggested for the separation of ionized compounds with high molecular weight in aqueous extracts. This technology consists in intercalating an ultrafiltration membrane between two ion-exchange membranes in an electrodialysis stack. The property of molecular barrier of the ultrafiltration membrane is exploited thanks to an electric field as driving force. The process can be used to separate phenolics in fruit juices because these compounds are charged at low pH. It was evaluated for enrichment of cranberry juice with anthocyanins and proanthocyanidins using a 500 kDa MWCO ultrafiltration membrane (Bazinet et al. 2009, 2012). According to operating conditions, anthocyanin could be concentrated from 1.2- to 1.5-fold. Because of their higher molecular weight, migration of proanthocyanidins was severely slower and required higher treatment duration. The process also decreased the titratable acidity and so improved the taste of the juice. The applicability of the process at an industry level seems to be probable. Nevertheless, the main limitation would be the control of membrane fouling during the treatment.

8.3.5 Recovery of Functional Compounds from Juices or by-Products

Different membrane processes can be also envisaged to recover interesting compounds directly from juice or more commonly from by-products of fruit juice processing. Without guaranteeing the completeness, we propose here some examples found in the literature that are currently in development at laboratory or semi-industrial scale.

8.3.5.1 Carotenoids

Cashew apple processing into juice generates large quantities of solid waste that are usually discarded. Pinto de Abreu et al. (2013) proposed and evaluated a new process that provided an added value to this by-product, extracting carotenoids that were contained therein. The process comprised three successive steps: an extraction by pressing associated to an enzymatic maceration allowing carotenoids to be obtained in emulsion, a cold concentration of the emulsion by crossflow microfiltration, and a purification by diafiltration. Using inorganic tubular membranes of 0.2 μm average pore diameter, carotenoids could be concentrated 20-fold maintaining the permeate flux above 100 L h^{-1} m^{-2} (40 °C, 3.2 bar). Diafiltration allowed an increase in carotenoid purity by five in relation to dry matter. The final extract obtained presented a carotenoid content of 7 g kg^{-1}, which has a strong potential for use in the formulation of foods and beverages as a natural dye.

8.3.5.2 Phenolics

Because of its too bitter taste, bergamot juice is considered as a by-product of essential oil production. However, like in other citrus juices, it contains flavonoids of great interest for pharmaceutical and food industries. With a similar processing scheme as previously described, nanofiltration has been suggested for recovering these functional compounds from juice-limiting costs, without heating and without solvent (Conidi et al. 2011). Using a ceramic tubular nanofiltration membrane with 0.45 kDa MWCO, a differential retention between flavonoids (>91 %) and sugars (48 %) was obtained (24 °C, 33 bar). This selectivity allowed flavonoids to be separated from the other solutes.

Recovery of phenolics contained in olive mill wastewaters by using membrane processes has also been mentioned (Garcia-Castello et al. 2010). Wastewaters were first clarified by microfiltration and then treated by nanofiltration using a 1.6 m^2 spiral-wound membrane with 0.58 kDa MWCO (20 °C, 8 bar). In this case, sugars were better retained by the nanofiltration membrane compared to phenolic compounds that were therefore purified in the permeate. Finally, an extract enriched in polyphenolics could be obtained through an ultimate concentration step.

In order to recover phenolics contained in press liquors from blood orange peels, a by-product from citrus processing, nanofiltration has been evaluated (Conidi et al. 2012). The aim was to separate polyphenolic compounds from sugars and also to separate anthocyanins from the other flavonoids with low energy input and using mild conditions. Four spiral-wound nanofiltration membranes from 0.18 to 1 kDa MWCO were compared at semi-industrial scale (volumetric reduction ratio 3, around 2 m^2 of membrane area) at 20 °C. Logically, the higher the MWCO, the higher the retention of all the compounds. Whatever the membrane, anthocyanins retention was above 89 %. Depending on the MWCO of the membrane, retention varied from 22 to 93 % for sugars and from 70 to 85 % for the other flavonoids. Therefore, separation of these compounds was possible even if their molecular weights are similar. Electrostatic repulsion probably contributes to the high rejection of the membrane towards anthocyanins. The most promising results were obtained using the 1000 kDa membrane. Nevertheless, high fouling was highlighted and permeate flux were low (<4 L h^{-1} m^{-2}). The hydrophobicity of the membrane material could promote adsorptive fouling.

8.3.5.3 Aroma Compounds

Membrane processes can also be envisaged to extract volatiles from fruit juice without heating and without solvent. For that purpose, pervaporation is the most commonly cited technology. The potential of this process for recovering aroma compounds from fruit juices has been clearly demonstrated (Pereira et al. 2006). Nevertheless, many works found in literature were focused on model solutions and studies of its application with real juice are scarce. For example, Pereira et al. (2005) showed the potential of pervaporation in recovery aroma compounds from pineapple juice. Interesting results were obtained using flat and hollow fiber composite membranes (6–45 cm^2) at 25 °C, a permeate pressure of 400 Pa, and cooled trap with liquid nitrogen. Aroujalian and Raisi (2007) also tested the process with orange juice at laboratory scale using a composite flat membrane of 138 cm^2 varying temperature from 25 to 50 °C and permeate pressure from 133 to 5333 Pa. Even if the feasibility of the process is recognized, industrial scale-up needs further study to better characterize the quality of the aroma extracts obtained and to evaluate the economic aspects.

8.3.6 Endogenous Enzyme Inhibition

In many cases, enzymatic reactions constitute a real problem during processing or storage because they drastically alter the juice quality. The biochemical stabilization of the juice becomes therefore a crucial step for providing high quality products. In order to avoid thermal treatment (blanching) or the use of large amount of antioxidant such as ascorbic acid, alternative processes could be envisaged. In that field, two membrane processes are concerned: electrodialysis and ultrafiltration.

In one way, stabilization can be obtained by temporarily decreasing the pH of the juice in order to irreversibly inhibit the endogenous enzymes responsible of the quality damages. Because the direct addition of acid and then alkali drastically affects the product flavor, electrodialysis with bipolar membrane could be advantageously used to bring protons or hydroxyls from water dissociation. Tronc et al. (1997) first demonstrated that electrodialysis using a bipolar membrane surrounded by two ion-exchange membranes (anionic and cationic) allowed apple juice to be acidified from pH 3.8 to 2.7 without altering the flavor. Acidification partially inhibited polyphenoloxydase (PPO), and after 6 h storage, residual PPO activity in the juice was reduced by 80 %. Consequently, enzymatic browning evaluated through Lab color measurements was reduced in the treated juice. Nevertheless, potassium leaks limited the pH decrease. As a consequence, the process did not guarantee irreversible denaturation of PPO. The process was improved by Lam et al. (2000) using an anionic membrane surrounded by two bipolar membranes. This new configuration allowed pH to be decreased very rapidly up to two with reasonable energy consumption (14–18 kJ L^{-1}). After maintaining the juice for 1 h at pH 2 and then adjusting the pH to 3.35, total and irreversible denaturation of PPO was reached. Enzymatic browning was completely inhibited while preserving the composition and taste of the juice. Treatment duration can be drastically reduced combining electrodialysis with a mild heat treatment at 45 °C (Lam et al. 2006). In that case, PPO irreversible inactivation was obtained after only 4 min. The process also allowed pectin methylesterase to be inactivated that could destabilize the opalescence of cloudy apple juice during storage.

Ultrafiltration could also be implemented for enzymatic stabilization of clarified juices. Choosing a membrane with a 10–20 kDa molecular weight cut-off, this membrane process is able to physically retain some enzymes such as PPO or peroxydase limiting color change in the juice. This interest was mentioned, for example, for banana juice (Sims et al. 1994) or coconut water (Prades et al. 2012). However, because of the retention of many other solutes, the treatment often drastically modifies the sensory and nutritional quality of the juice.

8.4 Conclusion

Membrane technologies represent a wide family of very diverse separation processes. They allow many applications in fruit juice processing to be foreseen. For concentration and stabilization, they present high potential over conventional processes, especially because of their separation efficiency and the mild conditions used that better preserve the quality of raw fruits. Through modulating fruit juice composition or fractionation of functional compounds, membrane processes also allow the development of new products with low energy consumption and limited environmental impact.

Some membrane processes have already been implemented in the fruit juice industry like microfiltration for clarifying. However, most of them are still studied at laboratory or semi-industrial scales. On account of the technical progress in membrane material and process design and also of the evolving markets of fruit juices that tend more and more to high quality products and functional-health food, it is very likely that membrane processes will develop in the next few years. Of course, for some membrane processes technical barriers still exist and technology has to be improved in order to increase performance and better match the economic aspects.

In the future, the more probable trend could be a further integration of membrane technologies in the processing line. This integration will be definitely facilitated by using modelling and simulation tools. Associating or coupling membrane technologies with each other or with more conventional systems, the process should be designed in such a way that the whole product will be processed including by-products. Indeed, juice extraction residues such as press cake, peels, seeds, and so on contain many functional compounds of interest that could be recovered, concentrated, or purified using membrane separation processes. In a bio-refinery pattern, a deeper fractionation of fruit juices could be also envisaged in order to recover these functional compounds that could be used as natural ingredients in food, cosmetics, or pharmaceutical products.

The field of applications of membrane technologies is already very large in fruit juice processing and, without doubt, should continue to expand in the short term.

References

Abreu, F., A.M. Perez, M. Dornier, and M. Reynes. 2005. Potentialités de la microfiltration tangentielle sur membranes minérales pour la clarification du jus de pomme de cajou. *Fruits* 60: 33–40.

Acosta, O., F. Vaillant, A.M. Pérez, and M. Dornier. 2014. Potential of ultrafiltration for separation and purification of ellagitannins in blackberry (*Rubus adenotrichus* Schltdl.) juice. *Separation and Purification Technology* 125: 120–125.

Ali, F., M. Dornier, A. Duquenoy, and M. Reynes. 2003. Evaluating transfers of aroma compounds during the concentration of sucrose solutions by osmotic distillation in a batch-type pilot plant. *Journal of Food Engineering* 60: 1–8.

Alvarez, V., S. Alvarez, F.A. Riera, and R. Alvarez. 1997. Permeate flux prediction in apple juice concentration by reverse osmosis. *Journal of Membrane Science* 127: 25–34.

Alvarez, S., F.A. Riera, R. Alvarez, and J. Coca. 1998. Permeation of apple aroma compounds in reverse osmosis. *Separation and Purification Technology* 14: 209–220.

Alvarez, S., F.A. Riera, R. Alvarez, and J. Coca. 2002. Concentration of apple juice by reverse osmosis at laboratory and pilot-plant scales. *Industrial & Engineering Chemistry Research* 41: 6156–6164.

Alves, V.D., and I.M. Coelhoso. 2006. Orange juice concentration by osmotic evaporation and membrane distillation: A comparative study. *Journal of Food Engineering* 74: 125–133.

Aroujalian, A., and A. Raisi. 2007. Recovery of volatile aroma components from orange juice by pervaporation. *Journal of Membrane Science* 303: 154–161.

Bahceci, K.S. 2012. Effects of pretreatment and various operating parameters on permeate flux and quality during ultrafiltration of apple juice. *International Journal of Food Science and Technology* 47: 315–324.

Bailey, A.F.G., A.M. Barbe, P.A. Hogan, R.A. Johnson, and J.T. Sheng. 2000. The effect of ultrafiltration on the subsequent concentration of grape juice by osmotic distillation. *Journal of Membrane Science* 164: 195–204.

Baklouti, S., R. Ellouze-Ghorbel, A. Mokni, and S. Chaabouni. 2012. Clarification of pomegranate juice by ultrafiltration: Study of juice quality and of fouling mechanism. *Fruits* 67: 215–225.

Barbe, A.M., J.P. Bartley, A.L. Jacobs, and R.A. Johnson. 1998. Retention of volatile organic flavour/fragrance components in the concentration of liquid foods by osmotic distillation. *Journal of Membrane Science* 145: 67–75.

Bazinet, L., C. Cossec, H. Gaudreau, and Y. Desjardins. 2009. Production of a phenolic antioxidant enriched cranberry juice by electrodialysis with filtration membrane. *Journal of Agricultural and Food Chemistry* 57: 10245–10251.

Bazinet, L., S. Brianceau, P. Dubé, and Y. Desjardins. 2012. Evolution of cranberry juice physico-chemical parameters during phenolic antioxidant enrichment by electrodialysis with filtration membrane. *Separation and Purification Technology* 87: 31–39.

Bélafi-Bakó, K., and B. Koroknai. 2006. Enhanced water flux in fruit juice concentration: Coupled operation of osmotic evaporation and membrane distillation. *Journal of Membrane Science* 269: 187–193.

Brat, P., B. Rega, P. Alter, M. Reynes, and J.M. Brillouet. 2003. Distribution of volatile compounds in the pulp, cloud, and serum of freshly squeezed orange juice. *Journal of Agricultural and Food Chemistry* 51: 3442–3447.

Brodart, F., J. Romero, M.P. Belleville, J. Sanchez, C. Combe-James, M. Dornier, and G.M. Rios. 2003. New hydrophobic membranes for osmotic evaporation process. *Separation and Purification Technology* 32: 3–7.

Bui, V.A., M.H. Nguyen, and J. Muller. 2007. The energy challenge of direct contact membrane distillation in low temperature concentration. *Asia-Pacific Journal of Chemical Engineering* 2(5): 400–406.

Campos, D.C.P., A.S. Santos, D.B. Wolkoff, V.M. Matta, L.M.C. Cabral, and S. Couri. 2002. Cashew apple juice stabilization by microfiltration. *Desalination* 148: 61–65.

Cassano, A., and E. Drioli. 2007. Concentration of clarified kiwifruit juice by osmotic distillation. *Journal of Food Engineering* 79: 1397–1404.

Cassano, A., E. Drioli, G. Galaverna, R. Marchelli, G. Di Silvestro, and P. Cagnasso. 2003. Clarification and concentration of citrus and carrot juices by integrated membrane processes. *Journal of Food Engineering* 57: 153–163.

Cassano, A., C. Conidi, R. Timpone, M. D'Avella, and E. Drioli. 2007a. A membrane-based process for the clarification and the concentration of the cactus pear juice. *Journal of Food Engineering* 80: 914–921.

Cassano, A., L. Donato, and E. Drioli. 2007b. Ultrafiltration of kiwifruit juice: Operating parameters, juice quality and membrane fouling. *Journal of Food Engineering* 79: 613–621.

Cath, T., A.E. Childress, and M. Elimelech. 2006. Forward osmosis: Principles, applications, and recent developments. *Journal of Membrane Science* 281: 70–87.

Celere, M., and C. Gostoli. 2005. Heat and mass transfer in osmotic distillation with brines, glycerol and glycerol-salt mixtures. *Journal of Membrane Science* 257: 99–110.

Chhaya, C., P. Rai, G.C. Majumdar, S. Dasgupta, and S. De. 2008. Clarification of watermelon (*Citrullus lanatus*) juice by microfiltration. *Journal of Food Process Engineering* 31: 768–782.

Cisse, M., F. Vaillant, A. Perez, M. Dornier, and M. Reynes. 2005. The quality of orange juice processed by coupling crossflow microfiltration and osmotic evaporation. *International Journal of Food Science and Technology* 40: 105–116.

Cisse, M., F. Vaillant, S. Bouquet, D. Pallet, F. Lutin, M. Reynes, and M. Dornier. 2011. Athermal concentration by osmotic evaporation of roselle extract, apple and grape juices and impact on quality. *Innovative Food Science & Emerging Technologies* 12: 352–360.

Conidi, C., A. Cassano, and E. Drioli. 2011. A membrane-based study for the recovery of polyphenols from bergamot juice. *Journal of Membrane Science* 375: 182–190.

Conidi, C., A. Cassano, and E. Drioli. 2012. Recovery of phenolic compounds from orange press liquor by nanofiltration. *Food and Bioproducts Processing* 90: 867–874.

Courel, M., M. Dornier, J.M. Herry, G.M. Rios, and M. Reynes. 2000a. Effect of operating conditions on water transport during the concentration of sucrose solutions by osmotic distillation. *Journal of Membrane Science* 170: 281–289.

Courel, M., M. Dornier, G.M. Rios, and M. Reynes. 2000b. Modeling of water transport in osmotic distillation using asymmetric membrane. *Journal of Membrane Science* 173: 107–122.

Daufin, G., J.P. Escudier, H. Carrere, S. Berot, L. Fillaudeau, and M. Decloux. 2001. Recent and emerging applications of membrane processes in the food and dairy industry. *Food and Bioproducts Processing* 79: 89–102.

De Barros, S.D., C.M.G. Andrade, E.S. Mendes, and L. Peres. 2003. Study of fouling mechanism in pineapple juice clarification by ultrafiltration. *Journal of Membrane Science* 215: 213–224.

De Oliveira, R.C., R.C. Doce, and S.T.D. De Barros. 2012. Clarification of passion fruit juice by microfiltration: Analyses of operating parameters, study of membrane fouling and juice quality. *Journal of Food Engineering* 111: 432–439.

Decloux, M., and F. Prothon. 1998. Fruit juices, vegetable juices and sugar juices. In *Membrane separations in the process food industry*, ed. G. Daufin, F. René, and P. Aimar, 473–501. Paris: Lavoisier Tech&Doc.

Durham, R.J., and M.H. Nguyen. 1994. Hydrophobic membrane evaluation and cleaning for osmotic distillation of tomato puree. *Journal of Membrane Science* 87: 181–189.

Echavarria, A.P., C. Torras, J. Pagan, and A. Ibarz. 2011. Fruit juice processing and membrane technology application. *Food Engineering Reviews* 3: 136–158.

El Rayess, Y., C. Albasia, P. Bacchin, P. Taillandier, J. Raynala, M. Mietton-Peuchot, and A. Devatine. 2011. Cross-flow microfiltration applied to oenology: A review. *Journal of Membrane Science* 382: 1–19.

El-Bourawi, M.S., Z. Ding, R. Ma, and M. Khayet. 2006. A framework for better understanding membrane distillation separation process. *Journal of Membrane Science* 285: 4–29.

Garcia-Castello, E., J. Lora-Garcia, J. Garcia-Garrido, and A.D. Rodriguez-Lopez. 2006. Energetic comparison for leaching waste liquid from citric juice production using both reverse osmosis and multiple-effect evaporation. *Desalination* 191: 178–185.

Garcia-Castello, E., A. Cassano, C. Criscuoli, and E. Drioli. 2010. Recovery and concentration of polyphenols from olive mill wastewaters by integrated membrane system. *Water Research* 44: 3883–3892.

Girard, B., and L.R. Fukumoto. 2000. Membrane processing of fruit juices and beverages: A review. *Critical Reviews in Food Science and Nutrition* 40: 91–157.

Goloubev, V.N., and B. Salem. 1989. Traitement à l'électrodialyse du jus d'orange. *Industries Agricoles et alimentaires* 106: 175–177.

Gryta, M. 2005. Osmotic membrane distillation and other membrane distillation variants. *Journal of Membrane Science* 246: 145–156.

Gunko, S., S. Verbych, M. Bryk, and N. Hilal. 2006. Concentration of apple juice using direct contact membrane distillation. *Desalination* 190: 117–124.

Gurak, P.D., L.M.C. Cabral, M.H.M. Rocha-Leão, V.M. Matta, and S.P. Freitas. 2010. Quality evaluation of grape juice concentrated by reverse osmosis. *Journal of Food Engineering* 96: 421–426.

Hengl, N., A. Mourgues, E. Pomier, M.P. Belleville, D. Paolucci-Jeanjean, J. Sanchez, and G.M. Rios. 2007. Study of a new membrane evaporator for the concentration of heat sensitive aqueous solutions. *Journal of Membrane Science* 289: 169–177.

Herron J.R., E.G. Beaudry, C.E. Jochums, and L.E. Medina. 1994. *Osmotic concentration apparatus and method for direct osmosis concentration of fruit juice*. Patent US 5 281 430.

Hongvaleerat, C., L.M.C. Cabral, D. Dornier, M. Reynes, and S. Ningsanond. 2008. Concentration of pineapple juice by osmotic evaporation. *Journal of Food Engineering* 88: 548–552.

Jesus, D.F., M.F. Leite, L.F.M. Silva, R.D. Modesta, V.M. Matta, and L.M.C. Cabral. 2007. Orange (Citrus sinensis) juice concentration by reverse osmosis. *Journal of Food Engineering* 81: 287–291.

Jiao, B., A. Cassano, and E. Drioli. 2004. Recent advances on membrane processes for the concentration of fruit juices: A review. *Journal of Food Engineering* 63: 303–324.

Jonquières, A., R. Clément, P. Lochon, N. Néel, M. Dresch, and B. Chrétien. 2002. Industrial state-of-the-art of pervaporation and vapour permeation in the western countries. *Journal of Membrane Science* 206: 87–117.

Kalbasi, A., and L. Cisneros-Zevallos. 2007. Fractionation of monomeric and polymeric anthocyanins from concord grape (*Vitis labrusca* L.) juice by membrane ultrafiltration. *Journal of Agricultural and Food Chemistry* 55: 7036–7042.

Koroknai, B., Z. Csanádi, L. Gubicza, and K. Bélafi-Bakó. 2008. Preservation of antioxidant capacity and flux enhancement in concentration of red fruit juices by membrane processes. *Desalination* 228: 295–301.

Kozák, A., E. Békássy-Molnár, and G. Vatai. 2009. Production of black-currant juice concentrate by using membrane distillation. *Desalination* 241: 309–314.

Lam, Q.A., F. Lamarche, and J. Makhlouf. 2000. Acceleration of pH variation in cloudy apple juice using electrodialysis with bipolar membranes. *Journal of Agricultural and Food Chemistry* 48: 2160–2166.

Lam, Q.A., M. Mondor, F. Lamarche, D. Ippersiel, L. Bazinet, and J. Makhlouf. 2006. Effect of a combination of electrodialysis with bipolar membranes and mild heat treatment on the browning and opalescence stability of cloudy apple juice. *Food Research International* 39: 755–760.

Laorko, A., Z. Li, S. Tongchitpakdee, S. Chantachum, and W. Youravong. 2010. Effect of membrane property and operating conditions on phytochemical properties and permeate flux during clarification of pineapple juice. *Journal of Food Engineering* 100: 514–521.

Machado, R.M.D., R.N. Haneda, B.P. Trevisan, and S.R. Fontes. 2012. Effect of enzymatic treatment on the crossflow microfiltration of aci pulp: Analysis of the fouling and recovery of phytochemicals. *Journal of Food Engineering* 113: 442–452.

Matta, V.M., R.H. Moretti, and L.M.C. Cabral. 2004. Microfiltration and reverse osmosis for clarification and concentration of acerola juice. *Journal of Food Engineering* 61: 477–482.

Merson, R.L., and A.I. Morgan. 1968. Juice concentration by reverse osmosis. *Food Technology* 22: 631–634.

Mondal, S., and S.R. Wickramasinghe. 2008. Produced water treatment by nanofiltration and reverse osmosis membranes. *Journal of Membrane Science* 322: 162–170.

Mulder, M. 1996. *Basic principles of membrane technology*. Dordrecht, The Netherlands: Kluwer.

Nagaraj, N., G. Patil, B.B. Ravindra, U.H. Hebbar, K.S.M.S. Raghavarao, and S. Nene. 2006. Mass transfer in osmotic membrane distillation. *Journal of Membrane Science* 268: 48–56.

Nagy, E., E. Kulcsár, and A. Nagy. 2011. Membrane mass transport by nanofiltration: Coupled effect of the polarization and membrane layers. *Journal of Membrane Science* 368: 215–222.

Nanjundaswamy, A.M., and K.C. Chikkappaji. 1989. Deacidification of fruit juices by electrodialysis. In *Trends in food science and technology*, ed. M.R. Raghavendra Rao, 43–48. Mysore, India: Association of Food Scientists & Technologists.

Nayak, C.A., S.S. Valluri, and N.K. Rastogi. 2011. Effect of high or low molecular weight of components of feed on transmembrane flux during forward osmosis. *Journal of Food Engineering* 106: 48–52.

Onsekizoglu, P., K.S. Bahceci, and M.J. Acar. 2010. Clarification and the concentration of apple juice using membrane processes: A comparative quality assessment. *Journal of Membrane Science* 352: 160–165.

Peinemann, K.V., S. Pereira Nunes, and L. Giorno. 2010. *Membranes for food applications*, Membrane technology, vol. 3. Weinheim, Germany: Wiley.

Peñate, B., and L. García-Rodríguez. 2012. Current trends and future prospects in the design of seawater reverse osmosis desalination technology. *Desalination* 284: 1–8.

Pereira, C.C., J.R.M. Rufino, A.C. Habert, R. Nobrega, L.M.C. Cabral, and C.P. Borges. 2005. Aroma compounds recovery of tropical fruit juice by pervaporation: Membrane material selection and process evaluation. *Journal of Food Engineering* 66: 77–87.

Pereira, C.C., C.P. Ribeiro, R. Nobrega, and C.P. Borges. 2006. Pervaporative recovery of volatile aroma compounds from fruit juices. *Journal of Membrane Science* 274: 1–23.

Petrotos, K.B., P. Quantick and H. Petropakis. 1998. A study of the direct osmotic concentration of tomato juice in tubular membrane – module configuration. I. The effect of certain basic process parameters on the process performance." *Journal of Membrane Science* 150: 99–110.

Petrotos, K.B., P. Quantick, and H. Petropakis. 1999. Direct osmotic concentration of tomato juice in tubular membrane—module configuration. II. The effect of using clarified tomato juice on the process performance. *Journal of Membrane Science* 160: 171–177.

Petrotos, K.B., P. Quantick, and H. Petropakis. 1999. Direct osmotic concentration of tomato juice in tubular membrane—module configuration. II. The effect of using clarified tomato juice on the process performance. *Journal of Membrane Science* 160: 171–177.

Petrotos, K.B., A.V. Tsiadi, E. Poirazis, D. Papadopoulos, H. Petropakis, and P. Gkoutsidis. 2010. A description of a flat geometry direct osmotic concentrator to concentrate tomato juice at ambient temperature and low pressure. *Journal of Food Engineering* 97: 235–242.

Pinto de Abreu, F., M. Dornier, A.P. Dionisio, M. Carail, C. Caris-Veyrat, and C. Dhuique-Mayer. 2013. Cashew apple (*Anacardium occidentale* L.) extract from by-product of juice processing: a focus on carotenoids. *Food Chemistry* 138: 25–31.

Pozderovic, A., T. Moslavac, and A. Pichler. 2007. Influence of processing parameters and membrane type on permeate flux during solution concentration of different alcohols, esters, and aldehydes by reverse osmosis. *Journal of Food Engineering* 78: 1092–1102.

Prades, A., M. Dornier, N. Diop, and J.P. Pain. 2012. Coconut water preservation and processing: A review. *Fruits* 67: 157–171.

Rai, P., G.C. Majumdar, S. Dasgupta, and S. De. 2007. Effect of various pretreatment methods on permeate flux and quality during ultrafiltration of mosambi juice. *Journal of Food Engineering* 78: 561–568.

Ravindra, B.B., N.K. Rastogi, and K.S.M.S. Raghavarao. 2006a. Effect of process parameters on transmembrane flux during direct osmosis. *Journal of Membrane Science* 280: 185–194.

Ravindra, B.B., N.K. Rastogi, and K.S.M.S. Raghavarao. 2006b. Mass transfer in osmotic membrane distillation of phycocyanin colorant and sweet-lime juice. *Journal of Membrane Science* 272: 58–69

Ravindra, B.B., N.K. Rastogi, and K.S.M.S. Raghavarao. 2008. Concentration and temperature polarization effects during osmotic membrane distillation. *Journal of Membrane Science* 322: 146–153.

Rektor, A., N. Pap, Z. Kókai, R. Szabó, G. Vatai, and E. Békássy-Molnár. 2004. Application of membrane filtration methods for must processing and preservation. *Desalination* 162: 271–277.

Rektor, A., G. Vatai, and E. Békássy-Molnár. 2006. Multi-step membrane processes for the concentration of grape juice. *Desalination* 191: 446–453.

Rodrigues, R.B., H.C. Menezes, L.M.C. Cabral, M. Dornier, G.M. Rios, and M. Reynes. 2004. Evaluation of reverse osmosis and osmotic evaporation to concentrate camu-camu juice (*Myrciaria dubia*). *Journal of Food Engineering* 63: 97–102.

Román, A., J. Wang, J. Csanádi, C. Hodúr, and G. Vatai. 2009. Partial demineralization and concentration of acid whey by nanofiltration combined with diafiltration. *Desalination* 241: 288–295.

Romero, J., H. Draga, M.P. Belleville, J. Sanchez, C. Combe-James, M. Dornier, and G.M. Rios. 2006. New hydrophobic membranes for contactor processes—applications to isothermal concentration of solutions. *Desalination* 193: 280–285.

Saleh, Z.S., R. Stanley, and R. Wibisono. 2006. Separation and concentration of health compounds by membrane filtration. *International Journal of Food Engineering* 2: 4/1–4/14.

Simoes, C.D., M. Dornier, D. Pallet, M. Reynes, D. Dijoux, S. Freitas, and L. Cabral. 2011. Evaluation of nanofiltration membranes for the retention of anthocyanins of açai (*Euterpe oleracea* Mart.) juice. *Desalination and Water Treatment* 27: 108–113.

Sims, C.A., R.P. Bates, and A.G. Arreola. 1994. Color, polyphenoloxidase, and sensory changes in banana juice as affected by heat and ultrafiltration. *Journal of Food Quality* 17: 371–379.

Strathmann, H. 2010. Electrodialysis, a mature technology with a multitude of new applications. *Desalination* 264: 268–288.

Tronc, J.S., F. Lamarche, and J. Makhlouf. 1997. Enzymatic browning inhibition in cloudy apple juice by electrodialysis. *Journal of Food Science* 62(1): 75–78.

Tylkowski, B., I. Tsibranska, R. Kochanova, G. Peeva, and M. Giamberini. 2001. Concentration of biologically active compounds extracted from Sideritis ssp. L. by nanofiltration. *Food and Bioproducts Processing* 89: 307–314.

Ushikubo, F.Y., A.P. Watanabe, and L.A. Viotto. 2007. Microfiltration of umbu (*Spondias tuberosa* Arr. Cam.) juice. *Journal of Membrane Science* 288: 61–66.

Vaillant, F., A. Millan, G. O'Brien, M. Dornier, M. Decloux, and M. Reynes. 1999. Crossflow microfiltration of passion fruit juice after partial enzymatic liquefaction. *Journal of Food Engineering* 42: 215–224.

Vaillant, F., E. Jeanton, M. Dornier, G.M. O'Brien, M. Reynes, and M. Decloux. 2001a. Concentration of passion fruit juice on an industrial pilot scale using osmotic evaporation. *Journal of Food Engineering* 47: 195–202.

Vaillant, F., A. Millan, M. Dornier, M. Decloux, and M. Reynes. 2001b. Strategy for economical optimisation of the clarification of pulpy fruit juices using crossflow microfiltration. *Journal of Food Engineering* 48: 83–90.
Vaillant, F., M. Cissé, M. Chaverri, A. Perez, M. Dornier, F. Viquez, and C. Dhuique-Mayer. 2005. Clarification and concentration of melon juice using membrane processes. *Innovative Food Science & Emerging Technologies* 6: 213–220.
Valdés, H., J. Romero, A. Saavedra, A. Plaza, and V. Bubnovich. 2009. Concentration of noni juice by means of osmotic distillation. *Journal of Membrane Science* 330: 205–213.
van Reis, R., and A. Zydney. 2007. Bioprocess membrane technology. *Journal of Membrane Science* 297: 16–50.
Vera, E., J. Ruales, M. Dornier, J. Sandeaux, F. Persin, G. Pourcelly, F. Vaillant, and M. Reynes. 2003. Comparison of different methods for deacidification of clarified passion fruit juice. *Journal of Food Engineering* 59: 361–367.
Vera, E., J. Sandeaux, F. Persin, G. Pourcelly, M. Dornier, G. Piombo, and J. Ruales. 2007a. Deacidification of clarified tropical fruit juices by electrodialysis. Part II. Characteristics of the deacidified juices. *Journal of Food Engineering* 78: 1439–1445.
Vera, E., J. Sandeaux, F. Persin, G. Pourcelly, M. Dornier, and J. Ruales. 2007b. Deacidification of clarified tropical fruit juices by electrodialysis. Part I. Influence of operating conditions on the process performances. *Journal of Food Engineering* 78: 1427–1438.
Vera, E., J. Sandeaux, F. Persin, G. Pourcelly, M. Dornier, and J. Ruales. 2009. Deacidification of passion fruit juice by electrodialysis with bipolar membrane after different pretreatments. *J. Food Engineering* 90: 67–73.
Versari, A., R. Ferrarini, G.P. Parpinello, and S. Galassi. 2003. Concentration of grape must by nanofiltration membranes. *Food and Bioproducts Processing* 81: 275–278.
Wang, B.J., T.C. Wei, and Z.R. Yu. 2005. Effect of operating temperature on component distribution of West Indian cherry juice in a microfiltration system. *LWT - Food Science and Technology* 38: 683–689.
Warczok, J., M. Ferrando, F. López, and C. Güell. 2004. Concentration of apple and pear juices by nanofiltration at low pressures. *Journal of Food Engineering* 63: 63–70.
Wattananusorn, S. 2008. Determination of membrane resistance for cross-flow microfiltration in a rectangular module including back-flow. *Journal of Process Mechanical Engineering* 222: 75–77.
Wrolstad, R.E., M.R. McDaniel, R. Durst, N. Michaels, K.A. Lampi, and E.G. Beaudry. 1993. Composition and sensory characterization of red raspberry juice concentrated by direct-osmosis or evaporation. *Journal of Food Science* 58: 633–636.
Wucherpfennig, K., and K. Keding. 1982. Deacidification and acidification of stone, pome and berry juices and grape must by electrodialysis. *Flussiges Obst* 49: 590–601.
Zhao, S., L. Zou, C.Y. Tang, and D. Mulcahy. 2012. Recent developments in forward osmosis: Opportunities and challenges. *Journal of Membrane Science* 396: 1–21.

Chapter 9
Decision Aid Tools for the Preservation of Fruits by Modified Atmosphere Packaging

Carole Guillaume, Barbara Gouble, Valérie Guillard, Patrice Buche, and Nathalie Gontard

9.1 Introduction

Over the last decades, conventional "trial and error" approaches, time and cost consuming, were phased out in favour of integrated and requirement driven ones, based on modelling tools and reverse engineering, to develop decision-making tools in various fields of research and development (from oceanography to aerospace science) and used for different purposes: reduction of costs, increasing efficiency of a process, etc. However, this is not a generalized trend, and this powerful methodology still needs to be adapted and applied in some sectors as food packaging, for increasing the efficiency of packaging and consequently, for reducing food losses: in 2007, nearly half of the fresh fruits and vegetable production was lost before consumption and in western countries, up to 30% of the losses happened during their distribution where most of the produce are packed (Gustavsson et al. 2011). Such figures are not compatible with a future sustainable food chain and raise the need to abandon "pack and pray" procedures to properly design and dimension packaging of fresh commodities in a more rational way, taking into account needs of the produce as well as constraints and wishes of the stakeholders.

C. Guillaume (✉) • V. Guillard • N. Gontard
Université Montpellier 2, UMR1208 IATE, Bataillon, Montpellier, France
e-mail: c-guillaume@um2.fr

B. Gouble
INRA UMR408 SQPOV, Avignon, France

Université d'Avignon et des Pays de Vaucluse, UMR408, Avignon, France

P. Buche
INRA, UMR1208 IATE, Montpellier, France

A. Rosenthal et al. (eds.), *Fruit Preservation*, Food Engineering Series,
https://doi.org/10.1007/978-1-4939-3311-2_9

Fresh fruits are living and fragile products with short shelf life since they undergo physiological degradations that inevitably lower their quality. One way to improve their storage is to use chilled temperature whenever it is possible, as it is currently applied to fresh-cut produce (Manolopoulou and Papadopoulou 1998; Vargas et al. 2006); but it is costly, energy hungry, difficult to manage at selling points, and most of whole produce is distributed at ambient temperature. Another route, that can be combined to chilled temperature, is to replace normal air in the headspace of the packed produce by appropriate gas composition in order to slow down respiration and then delay senescence of the produce. To be efficient, gas concentrations should be as close as possible to the optimal atmosphere of storage, specific to each product since it depends on food requirements (Kader et al. 1989; Barron et al. 2002; Charles et al. 2008; Sandhya 2010). Such Modified Atmosphere Packaging (MAP) can be obtained either via a passive or an active way. Passive (or equilibrium) MAP is achieved by a natural interplay between the produce physiology (consumption or production of gases and vapours) and the mass transfer properties of the packaging material (gases and vapours permeation values). After a transient period, steady gas concentration (or equilibrium) is reached. To reduce the transient period, active technology can be used: same interplay occurs but nitrogen (or gas mixture) is previously flushed just before sealing. Note that other active systems can be used, as gas/vapour releaser or scavenger, but are not yet commonly applied in fruit packaging (except ethylene scavengers for long transports) and will not be tackled in this chapter. Then success of MAP relies on the selection of packaging material with suitable gas and vapours transfer properties in order to reach and/or maintain a steady optimal atmosphere. The coupling of mathematical models that describe produce physiology (e.g. respiration) with some describing mass transfer properties (e.g. gas permeation through material) has been recently developed and experienced to predict optimal packaging properties in passive (Mahajan et al. 2007) and active MAP (Charles et al. 2003, 2005).

This optimization procedure takes into account produce requirements and translates them into expected/optimal mass transfer properties according to a packaging geometry, but still cannot be considered as a powerful decision aid tool for MAP of fresh fruits as it stands. Firstly, because some physiological pathway as well as some mass transfer properties are not easy to model or not fully understood; and secondly, because such a tool should answer to a multi-criteria query by identifying and ranking a group of suitable materials on the basis of multiple characteristics requested by all the stakeholders about cost, transparency, compliance with food contact regulation, etc. Indeed the choice of a packaging material for fresh fruits relies on numerous other criteria than optimal transfer properties. For instance, packaging for strawberries must answer the following query "*I want a packaging material that preserve product shelf life (i.e. with optimal gas transfer properties), with a cost lower than 1.3 €/kg and that complies with the EU regulation on food contact material and if possible biodegradable and with mechanical characteristics suitable for the form-and-seal machine that I own…*". Criteria could be considered as wishes (something possible, e.g. "packaging material should be biodegradable") and constraints (something mandatory, e.g. cost of the material must be lower than 1.3 €/kg). In such a configuration MAP optimization procedure is part of the

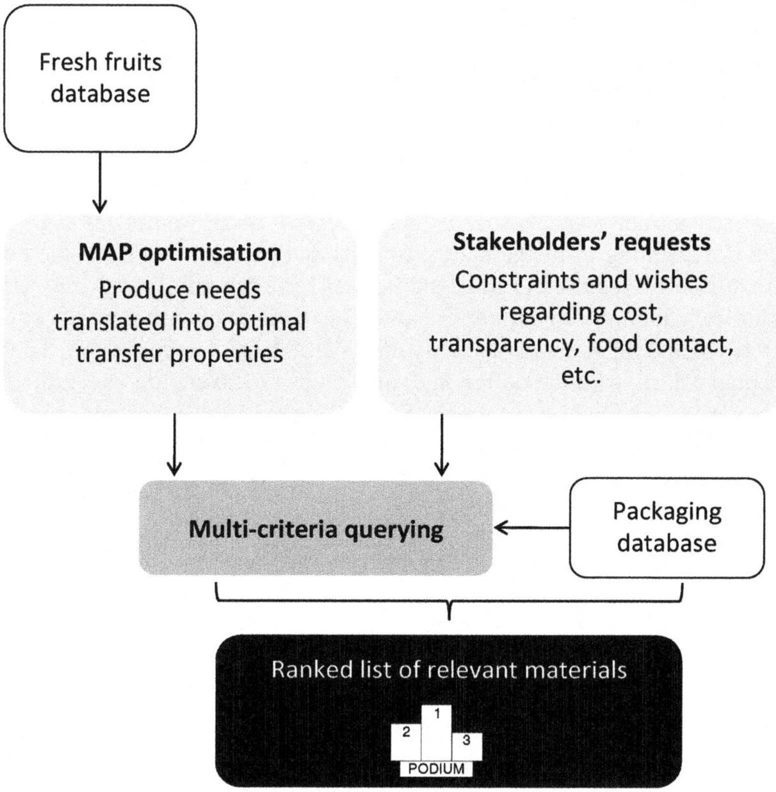

Fig. 9.1 Architecture of possible decision aid tool for MAP of fruits

decision-making tool and optimal transfer properties become constraints to guarantee produce quality during a certain period (i.e. expected shelf-life). As indicated in Fig. 9.1, this tool also relies on the creation of a database on fresh fruits, to feed mathematical models in MAP optimization step, and a database on packaging material to be consulted upon multi-criteria querying; then it should also manage missing as well as uncertain data (i.e. when one on the two O_2/CO_2 permeability values is unknown, for example).

The present chapter aims to bring insights on what could be a decision-making tool on modified atmosphere packaging for fruits as illustrated in Fig. 9.1. It provides current knowledge in biological, material, and computing science that could lead to the development of this tool and points out current bottlenecks that need to be overcome to succeed in it. Some reflections and examples are extracted from Tailorpack project (ANR-07-PNRA-029) funded by the ANR (French research agency) and EcoBioCAP project (FP7-265669) funded by EC (European Commission, seventh Framework Program). It is divided into three parts: the first one is devoted to the constitution of databases, the second one to MAP optimization procedure, and the third one to computing and statistical methods needed.

9.2 Constitution of Databases

Proper anticipation of data needed to be stored is a key factor to further develop powerful and relevant decision-making tools. In the foreseen application, two databases are required: one database on fresh fruits that gathers all input data used in MAP optimization procedure and one database on packaging materials to be queried about constraints and wishes of stakeholders. MAP optimization procedure relies on the coupling of mathematical models that permit to translate the needs of fruits in terms of required/optimal mass transfer properties. Such models used mathematical equations able to represent the physiology (respiration, transpiration) of fresh fruits and input data are mainly considered as numerical values. The multicriteria query deals with the preferences and needs of stakeholders regarding packaging materials, expressed as constraints and wishes with required/optimal mass transfer properties as mandatory constraints. Then part of data is numerical (e.g. mass transfer properties) and the other one may be qualitative (e.g. ability to biodegradation).

9.2.1 Database on Fresh Fruits

Data required for fresh fruits concern their optimal conditions of storage and their physiology, and more precisely parameters that permit to evaluate consumption and production of gases (i.e. oxygen and carbon dioxide) and vapours (i.e. moisture and ethylene) through respiration, transpiration and ripening pathways that can be described according to mathematical equations for some of them (presented hereinafter in Sect. 9.3). Once harvested, fresh fruits have to draw on their own reserves to maintain their cellular integrity. During respiration, stored carbohydrates are broken down into glucose, which is oxidized into CO_2, water and energy (adenosine triphosphate, ATP) according to several enzymatic steps but limited by the activity of the cytochrome oxidase. As soon as these substrates become unavailable, other carbonated resources that are essential (constitutive protein or membrane lipids) are consumed, leading to the death of the produce. Thus the potential shelf life of fruits is closely related to their carbohydrates content and their respiration rate, expressed as the quantity of O_2 consumed (or in a minor extend CO_2 released) per time and per mass of produce: the lower is the respiration rate, the longer is the potential shelf life (Marcellin 1975; Paull 1993). Water production during respiration, i.e. transpiration, can also affect the shelf life of the produce depending on the temperature and most of all the surrounding humidity. If it is too low—threshold relative humidity (RH) commonly admitted is of 85 % (Roy et al. 1996; Varoquaux and Ozdemir 2005)—dehydration quickly occurs leading to wilting symptoms until depression of the commercial value of the commodity as soon as water loss reaches 5–6 % (w/w) of fresh weight. Then assessment of weight loss in controlled condition of temperature and RH give information on the potential shelf life of the produce. Along respiration and transpiration, ripening also contributes to the evolution of fresh fruits.

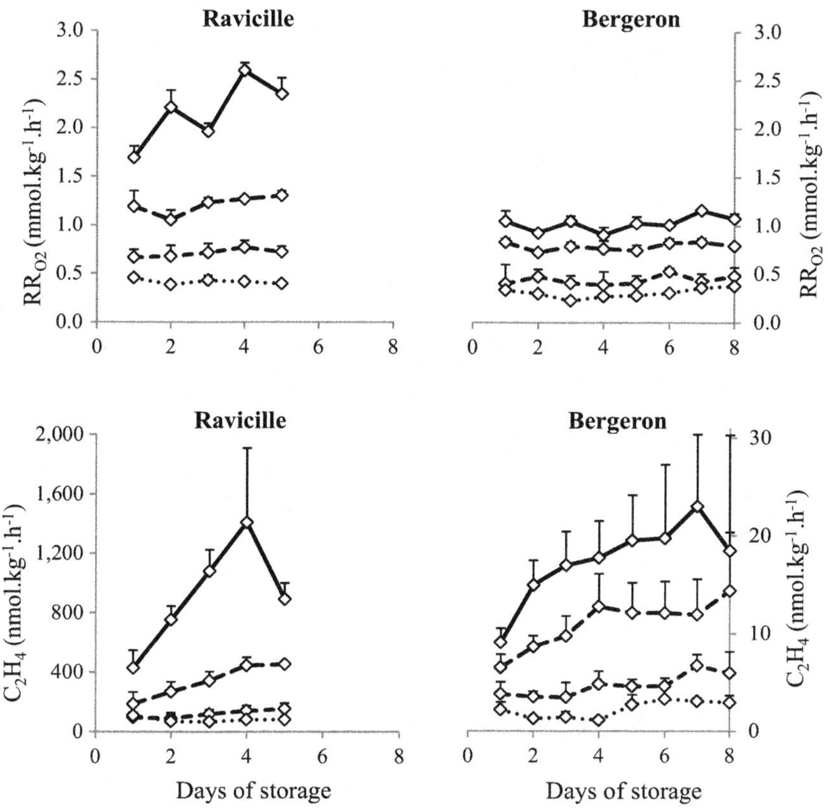

Fig. 9.2 Evolution of ethylene production (C_2H_4) and respiration rate (RR_{O_2}) in apricots (Ravilong)

Whether ripening is expected to bring optimal organoleptic qualities to the fruits, it is one of the major causes of their premature senescence and need to be delayed as much as possible during distribution and retailing. Since ripening pathway in climacteric fruits (based on the autocatalysis of ethylene) is now quite well understood compared to non-climacteric ones, it will be the only one discussed within this chapter. It can be characterized by ethylene production rate, expressed as the quantity of C_2H_4 produced per time and per mass of produce.

To develop powerful decision aid tools, database must be as large as possible; but acquisition of data that characterize those physiological pathways is a heavy work because of biological variability, possible contamination of fruits (e.g. microorganisms also contribute to O_2 consumption and CO_2 production and might affect the results on fruits respiration), absence of globally harmonized methodology of measurements (Fonseca et al. 2002), and dependence on temperature and RH. Some of these points are illustrated in Fig. 9.2 that presents oxygen consumption and ethylene production rates in two varieties of apricot during storage at different temperatures: Ravicille, a new variety with high blush, known for its fast ripening and high ethylene production, and Bergeron, main French variety, characterized by a slow ripening and low ethylene production. The former was overripened after 5 days of storage at the highest temperature, while storage of the latter could be extended to 8 days. Ravicille presented higher respiration rates than Bergeron whatever the temperature, but temperature effect was the same for the two varieties, it increases respiring activity. Respiration rates remained quite stable along the storage period for the two varieties at low temperature (i.e. 5 and 10 °C), whereas they were increasing with storage time at higher temperatures, which was clearly marked for Ravicille at 20 °C. Same patterns were observed for ethylene production (almost 100-fold higher in Ravicille than in Bergeron) and was attributed to their climacteric behaviour, with a respiration increase and an ethylene crisis associated with ripening changes. From these results, ethylene production, as respiration rate, is not only dependent on temperature and species, but also on variety (or cultivar) and postharvest age of the fruits (Gouble et al., 2006; Mc Glasson, 1984; Robertson, 2006). This points out the importance of proper fruit identification when assessing or collecting data on physiological pathways of fruits. An example of such identification is given in Table 9.1. With the increase of fresh-cut produce on the market, a set of data related to physiology of fresh fruits should ideally also be attributed to a given process, but this will not be discussed here. Minimal processing induces an increase in respiration activity from two- to eightfold (Brecht, 1995) and stimulates ethylene production from 5- to 20-fold (Pech et al. 1994).

9.2.1.1 Data on Optimal Storage Conditions

Lowering storage temperature is well known to reduce physiological reactions, but can lead to chilling injuries mainly in case of exotic fruits. More often, it is given as a range of optimal temperature of storage, T_{opt}, with a minimal and maximal level. Another factor affecting the produce physiology is the surrounding gas composition. Usually, lowering the O_2 level is really effective in reducing respiration, but anoxia (a switch to anaerobic catabolism and growth of anaerobic flora that produce undesirable off-flavours and off-odours) should be avoided (Nguyenthe and Carlin 1994; Varoquaux and Ozdemir 2005). High CO_2 levels (more than 10 %) might also reduce the respiration and ripening of several commodities (Mathooko 1996; Varoquaux and Ozdemir 2005) and can limit or inhibit the production of ethylene (Rothan and Nicolas 1994; Rothan et al. 1997). But such levels might lead to the development of anaerobic microorganisms. Moreover, injuries might occur when a fresh product is exposed to a level of CO_2 above its tolerance limit as, for example, formation of brown spots or yellowing that are common visual degradations caused by a high CO_2 content (Kader et al. 1989; Lopez Briones et al. 1992). Together with CO_2, 1-methylcyclopropene (1-MCP) is also considered as a competitive inhibitor of ethylene action and can be used to delay ethylene production and the respiration crisis (Fan et al. 2000; Hershkovitz et al. 2005) but when considering for MAP application, it would be part of active systems that are not discussed in this chapter. Since the optimal combination of O_2 and CO_2 greatly depends on the respiratory activity of the product (values range from less than 35 up to 300 mg $O_2\,kg^{-1}\,h^{-1}$) and its sensitivity to CO_2, there is no unique atmosphere composition that could be

Table 9.1 Example of data fields that should be required in the database on raw fresh fruits

Fruit identification					
Species	*Variety*				
Postharvest age					
Growing area					
Date of harvest					
Optimal storage conditions					
T_{opt}	*Min*	*Avg*	*Max*		
O_{2opt}	*Min*	*Avg*	*Max*		
CO_{2opt}	*Min*	*Avg*	*Max*		
RH_{opt}	*Min*	*Avg*	*Max*		
Respiration					
$RR_{O_2 \max}$	*Min*	*Avg*	*Max*	*Temp*	
Km_{app}	*Min*	*Avg*	*Max*	*Temp*	
Ki_{CO_2}	*Min*	*Avg*	*Max*	*Temp*	*Inhibition type*
$RR_{CO_2 \max}$	*Min*	*Avg*	*Max*	*Temp*	
RQ	*Min*	*Avg*	*Max*	*Temp*	
$^{Ea}RR_{O_2}$ or RR_{CO_2}	*Min*	*Avg*	*Max*	*Temp range*	
$^{Q_{10}}RR_{O_2}$ or RR_{CO_2}	*Min*	*Avg*	*Max*	*Temp range*	
Transpiration					
TR	*Min*	*Avg*	*Max*	*Temp*	*RH*
A_s	*Min*	*Avg*	*Max*	*Temp*	*RH*
a_{w_i}	*Min*	*Avg*	*Max*	*Temp*	*RH*
Ea TR	*Min*	*Avg*	*Max*	*Temp range*	
$Q_{10}TR$	*Min*	*Avg*	*Max*	*Temp range*	
Ripening					
$RR_{C_2H_4}$	*Min*	*Avg*	*Max*	*Temp*	
Constants	*Min*	*Avg*	*Max*	*Temp*	
$^{Ea}RR_{C_2H_4}$	*Min*	*Avg*	*Max*	*Temp range*	
$^{Q_{10}}RR_{C_2H_4}$	*Min*	*Avg*	*Max*	*Temp range*	
Other					
Density					

Note 1. *Min*, *Avg* and *Max* are the abbreviation for Minimal, Average and Maximal value, respectively
Note 2. Temp and RH are the abbreviation for temperature and relative humidity
Note 3. Meaning of each parameter is given in the text and found in Sect. 9.2.1

applied to all fresh commodities. Then critical concentrations of O_2 and CO_2 must be targeted for each fruit upon controlled atmosphere experiments in relation with the physiology and organoleptic qualities of the fruit. They can be referred to a window of optimal O_2, O_{2opt}, and optimal CO_2, CO_{2opt}, commonly expressed as a percentage or partial pressure (kPa). Related to dehydration phenomena, relative humidity (RH) is also essential for maintaining optimal fruits quality and should be filled in the database, whenever it is assessed, as optimal RH, RH_{opt}. Table 9.1 gives an overview of data needed to identify optimal storage conditions.

9.2.1.2 Data on Respiration

Fresh produce respiration models based on Michaelis–Menten type equations are the most common and the most fitting ones (Fonseca et al. 2002, Petracek et al. 2002, Varoquaux et al. 2002). From this basic model, first used by Chevillote (1973), two constants can be extracted and entered into the database as $RR_{O_2\,max}$, maximal O_2 respiration rate, and Km_{appO_2}, the Michaelis constant for O_2 consumption (apparent dissociation for enzyme/substrate complex). Since it has been improved by taking into account the potential inhibitory effect of CO_2 content on O_2 consumption through the four common types of Michaelis–Menten inhibitions type (competitive, uncompetitive, non-competitive and the combination of competitive and uncompetitive) as detailed in previous well-documented reviews on the subject (Peppelenbos and van't Leven 1996; Fonseca et al. 2002). The non-competitive inhibition model is preferred among the other due to its simplicity of use and good fit with most existing data concerning common produce (Fishman et al. 1995; Peppelenbos and van't Leven 1996), nevertheless uncompetitive inhibition might be more accurate in some case (Fonseca et al. 2002). Then another constant can be entered in the database, together with the inhibition type, Ki_{CO_2}, the CO_2 inhibition constant. While the scientific unit admitted for $RR_{O_2\,max}$ is mmol kg^{-1} h^{-1}, the quantity of O_2 can be found as a volume or mass unit as well as time can appear as 1 hour or 1 day. Then, it is very important to (1) indicate the units of each parameter and (2) consider conversion tool to properly use each of these parameter in MAP optimization procedure. Since there is no standardization neither on units nor measurements, maximal respiration rate can also be expressed as the amount of CO_2 released per time per mass of produce, $RR_{CO_2\,max}$. Considering the balance equation for respiration when sugars are catabolized, i.e. if 6 mol of O_2 are consumed then 6 mol of CO_2 are produced, maximal respiration rates should be equal whatever the gas considered, O_2 or CO_2 (if expressed as a volume or a number of moles). But, depending on the postharvest age of the fruit and the availability of substrates, this balance can change. Then, respiratory quotient, RQ, which is the ratio of CO_2 production to O_2 consumption is commonly determined and close to unit when sugars are catabolized (Peppelenbos and van't Leven 1996). As evidenced in Fig. 9.2, temperature greatly affects respiration and its effect can be evaluated through activation energy, Ea, or Q_{10} coefficient that derived from Arrhenius' or Gore's law, respectively. Q_{10} coefficient is commonly encountered in biological process and interpreted as the degree by which this process is accelerated by a rise of 10 °C and varies from two to four for fresh fruits respiration (Exama et al. 1993; Fonseca et al. 2002; Petracek et al. 2002). Determination of these coefficients can be done either on O_2 or CO_2 respiration rate.

9.2.1.3 Data on Transpiration and Ethylene Production

Several mathematical equations based on biophysical or thermodynamical models have been proposed to describe transpiration or vaporization rate. They are discussed further in 9.3 but their main drawback is that data required to solve them (e.g. thermal diffusivity, surface cellular structure, pores density, internal and specific heat) are quite difficult to assess because specific devices are required (e.g. isothermal or differential calorimeter, optical to electronic microscopes coupled with image analysis). Other mathematical equations have been recently proposed and validated to describe transpiration of fresh produce and relies on simple assessment of transpiration rate, TR, by weighing samples in controlled condition of temperature and RH and refers to weight loss (Mahajan et al. 2008; Sousa-Gallagher and Mahajan 2012). It is expressed as mass unit per time unit per mass of produce. For further mathematical purpose, surface area of the produce, A_s, and its initial water activity, a_w, are mandatory. Effect of temperature can also be considered and Q_{10} or Ea also assessed for transpiration.

When fruits are climacteric, ripening is related to ethylene production. Up to now, there is a lack of mathematical equation describing it. The only reference found in literature is the one of Grotte et al. (2006) who used regression equations to predict ethylene production during postharvest of Ravilong apricot. Neperian logarithm of ethylene production was expressed versus the sum of degree-days from blooming and parabolic functions were identified. They were characterized with parabola vertex constant (varying according to the postharvest age and the year of harvesting) and respective coefficient of determination. Then for the time being, input data related to ethylene production will be considered as ethylene production rate, $RR_{C_2H_4}$, the quantity of C_2H_4 produced per time and per mass of produce, and constants (Table 9.1) not precisely defined but that come from empirical model as the one from Grotte et al. (2006). As for other physiological mechanisms, temperature effect is indicated through Q_{10} or Ea.

All data discussed here regarding the fresh fruits database are gathered in Table 9.1. Specific apparent density of the produce might also be required. Indeed mathematical models used in the optimization procedure refer to gas or vapour exchanges in the headspace and it is more accurate to shift from the headspace to the free volume. The reader must be aware that Table 9.1 is just an example of data fields that are of interest for modelling purposes, and, by the way, these fields are tightly linked to mathematical models chosen (and displayed in 9.3). Some existing mathematical models have been deliberately discarded to not overload this chapter, whose aim is to give an approach on the construction of a decision aid tool on modified atmosphere packaging for fruits. Then mathematical models and associated data selected are just illustrations of what could be done and are the ones foreseen in the framework of Tailorpack (funded by ANR) and EcoBioCAP (funded by EC) projects. In addition, whether it does not appear in Table 9.1, each numerical data entered in the database needs to be associated with a reference, its units, and ideally with methodology used for assessment. This remark is also available for packaging materials database.

9.2.2 Database on Packaging Materials

Whereas the data entered into the database on fresh fruits are related to existing mathematical modelling, data entered in the database on packaging materials depends on wishes and constraints coming from stakeholder's requests that are not easy to anticipate (see Fig. 9.1). Guarantee the quality of the fresh fruit is considered obviously as the major constraint then, mass transfer properties imposed by fresh fruits requirements and determined thanks to the MAP optimization procedure will be search upon querying the packaging material database. So oxygen and carbon dioxide permeation (related to respiration of the produce), water vapour permeation (related to its transpiration) and ethylene permeation (related to its ripening, in case of climacteric fruit) must appear in the database on packaging materials. Permeation is a quantity of gas or vapour (volume, mass or moles) that pass through a known surface of material as a function of time, but depending on conditions of test, units differ:

- Transmission rate, given in volume, mass or moles to unit area and unit of time.
- Permeance, expressed as volume, mass or moles to unit area, unit of time and pressure unit (i.e. difference in partial pressure of the gas or vapour on both sides of the film).
- Permeability, given in volume, mass or moles multiplied by material thickness to unit area, unit of time and pressure unit.

For example, to express gas transfers, at least seven common different units are usually encountered: mL m^{-2} day^{-1}, barrer (10^{-10} cm^3 cm cm^{-2} s^{-1} cmHg), amol m^{-1} s^{-1} Pa^{-1}, cm^3 25 µm m^{-2} day^{-1} atm^{-1}, mL m^{-2} day^{-1} bar^{-1}, mL 100in^{-2} day^{-1}, mL mm m^{-2} atm^{-1} day^{-1}. In addition, gas and vapour transfers are dependent on temperature and RH, and this is particularly marked for hydrophilic materials (MujicaPaz and Gontard 1997; Mujica Paz et al. 2005). This points out the importance of introducing in identification of gas and vapour permeations other fields as (1) units, (2) error interval (or other statistic evaluation) and (3) conditions of test (including the method used when possible). Temperature effect is commonly considered using Arrhenius' law and then activation energy can be entered in the database, in a minor extend Gore's law is applied and Q_{10} extracted from it. However, these parameters are scarcely addressed in literature. Effect of relative humidity on gas and vapour permeability is till subject to study and there is still a need for theoretical approach describing it. Note that although O_2 and moisture transfer are usually given by material producers or studied, there is few attention paid to CO_2 transport and no to C_2H_4 transport in polymeric materials.

Numerical data related to mechanical properties of the material should be needed since they refer to a constraint or a whish from packaging manufacturer and fillers. Then several data fields related to material processing (screw or blow extrusion, injection moulding, etc.) must appear in the database, e.g. tensile stress and elongation at break, Young modulus, or Charpy impact, together with their associated units, errors, and conditions of test. Aside numerical data, qualitative ones are required and would be determined in association with all stakeholders: policy makers, safety authorities, waste management representatives, consumers and consumer group members, packers, fillers, importers, distributors, and of course food and packaging industries members. Some of them have been listed in the framework of EcoBioCAP project as sealability, transparency, direct printing, ovenable or microwaveable, fitness to selective waste collection (which one), cost, presence of additives (which

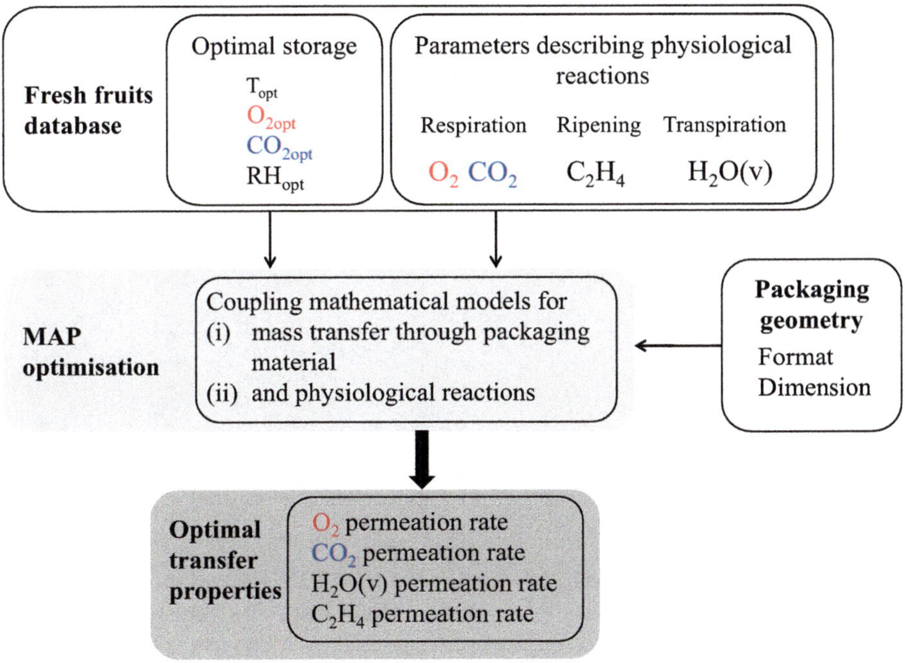

Fig. 9.3 Architecture of MAP optimization tool

one(s)), etc. Booming of researches and developments on bio-based materials for packaging leads to environmental considerations that should also appear in this database as the origin of the polymer (bio-sourced or from fossil resources), its elaboration (natural made, i.e. vegetal or microbial cells engineered, or chemically made, i.e. polymerization of monomers) and its end-life (biodegradable or not).

9.3 MAP Optimization Procedure

As represented in Fig. 9.3, MAP optimization procedure relies on the coupling of mathematical models for (1) physiological reactions and (2) mass transfers through packaging material; it can be found in literature as mass balance models that describe gas or vapour exchanges in food/packaging systems for a given food and a given packaging geometry. Then input data are related to parameters describing physiological reactions (from the fresh fruits database in Fig. 9.3) and to the format and dimension of the packaging. Focusing on the application to fresh fruits, they are able to predict changes in O_2 and CO_2 or water vapour partial pressures in the headspace of modified atmosphere packaging (note that whether they are mentioned in Fig. 9.3, ethylene exchanges are not approached due to the lack of mathematical model describing ethylene production). In this case permeation values should be

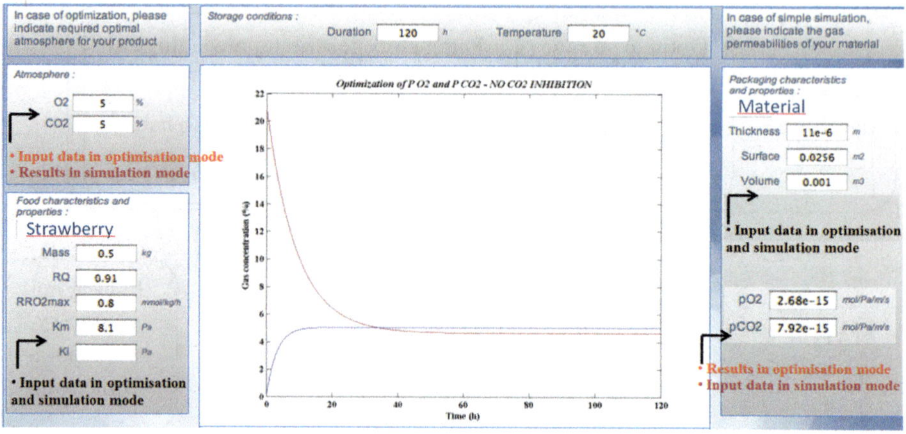

Fig. 9.4 Example of simulation or optimization carried on the web-application www.tailorpack.com for 500 g strawberries packed in a barrier tray sealed with a lid material

entered in the models as input data as illustrated in Fig. 9.4 and the mass balance model is used in a MAP simulation procedure. Note that some mass balance models even take into account the presence of scavengers to predict gas evolution in active MAP of fresh fruits and vegetable (Charles et al. 2003, 2005). However, goal of the MAP optimization procedure is to provide optimal mass transfer properties as output data of the mass balance models. This becomes possible if coupled models are used in a reverse manner in an identification procedure based on an optimization algorithm (e.g. Levenberg–Marquardt algorithm) that fit predicted partial pressures to optimal ones. Then, in addition to parameters describing physiological reactions and packaging geometry, optimal conditions of storage are required (see Fig. 9.3) and become the partial pressure targets (as in Fig. 9.4). Once optimal mass transfer properties are known, they can be used to query the packaging database (as illustrated in Fig. 9.1) or as targets to design packaging materials tailored to the application (Cagnon et al. 2012).

9.3.1 Mass Balance Model for Gas Exchanges

Mass balance models for oxygen and carbon dioxide exchanges in modified atmosphere packaging of fruits combine mathematical equations for fruit respiration and gas transfer though the packaging material. As previously indicated (see Sect. 9.2.1.2), respiration of most of the fresh produce is modelled according to Michaelis–Menten equation types but other approaches exist to model the respiration of fresh fruits (or vegetable). Linear, polynomial and exponential models have been tried along with Langmuir adsorption based ones, where the controlling mechanism was the adsorption of one molecule of O_2 at an active site of the cytochrome oxidase complex (Makino et al. 1996a; Makino et al. 1996b). Several combinations of these models (MM-exponential, exponential-polynomial, etc.) have also been considered and tested but did not give full satisfaction (Talasila 1992; Talasila et al. 1992). As they remain marginal in comparison to the model derived from Michaelis–Menten equation, they are cited but not considered for MAP optimization procedure. Transport of gases through material is ascribed to solution-diffusion mechanisms in dense polymer and obeys the first Fick's law, describing gradient driven fluxes (here a partial pressure difference) through a medium with a certain resistance (represented here by the coefficient of permeability or permeability of gas A, P_A):

$$J_A = P_A \times S \times \frac{p_2 - p_1}{e} \tag{9.1}$$

where J_A is the steady state flux of gas A through the film, S the film area, p_1 and p_2 gas partial pressures across the film, and e the film thickness.

Then by coupling the Michaelis–Menten type and Fick mathematical equations, gas evolution can be modelled as follows:

$$\dot{p}_{O_2}^{pkg} = \frac{\text{Pe}_{O_2} \times S}{e} \times \left(p_{O_2}^{ext} - p_{O_2}^{pkg}\right) - \text{RR}_{O_2} \times m = f_1 \tag{9.2}$$

$$\dot{p}_{CO_2}^{pkg} = \frac{\text{Pe}_{CO_2} \times S}{e} \times \left(p_{CO_2}^{ext} - p_{CO_2}^{pkg}\right) + \text{RR}_{O_2} \times m \times \text{RQ} = f_2 \tag{9.3}$$

With

$$\text{RR}_{O_2} = \frac{\text{RR}_{O_2\,\text{max}} \times p_{O_2}^{pkg}}{\text{Km}_{\text{app}O_2} + p_{O_2}^{pkg}} \tag{9.4}$$

or

$$\text{RR}_{O_2} = \frac{\text{RR}_{O_2\,\text{max}} \times p_{O_2}^{pkg}}{\text{Km}_{\text{app}O_2} + p_{O_2}^{pkg} \times \left(1 + \frac{p_{CO_2}^{pkg}}{Ki_{CO_2}^{n}}\right)} \tag{9.5}$$

or

$$\text{RR}_{O_2} = \frac{\text{RR}_{O_2\,\text{max}} \times p_{O_2}^{pkg}}{\text{Km}_{\text{app}O_2} + \left(1 + \frac{p_{CO_2}^{pkg}}{\text{Ki}_{CO_2}^{u}}\right) p_{O_2}^{pkg}} \tag{9.6}$$

where f_1 and f_2 are O_2 and CO_2 partial pressures, respectively (Eqs. 9.2 and 9.3). The first part of the right-hand side of f_1 and f_2 describes gas flux per time unit through the packaging material, while the second part describes oxygen consumption (i.e.

Table 9.2 Required input parameters in the mass balance model for gas exchanges

Parameter	Name	Units
Pe_{O_2}	O_2 permeability	mol m^{-1} s^{-1} Pa^{-1}
Pe_{CO_2}	CO_2 permeability	mol m^{-1} s^{-1} Pa^{-1}
S	Packaging surface	m^2
e	Packaging thickness	m
p_i^j	Partial pressure of j in i	Pa
RR_{O_2}	O_2 respiration rate	mmol kg^{-1} h^{-1}
$RR_{O_2\,max}$	Max O_2 respiration rate	mmol kg^{-1} h^{-1}
Km_{appO_2}	Mickaëlis–Menten constant	Pa
Ki_{CO_2}	CO_2 inhibition constant	Pa
m	Mass of food	kg
RQ	Respiration quotient	(–)

Eq. 9.2) and carbon dioxide emission (i.e. Eq. 9.3) by the fruit using a Michaelis–Menten type equation, considering no inhibition, non-competitive CO_2 inhibition or uncompetitive inhibition (Eqs. 9.4, 9.5 or 9.6, respectively). Parameters needed and their respective units are indicated in Table 9.2. Equations 9.2–9.5 have been programmed in Matlab software (The Mathworks Inc, USA) and using this program, simulation of the evolution with time of internal gas partial pressure $\dot{p}_{O_2}^{pkg}$ and $\dot{p}_{CO_2}^{pkg}$ can be performed. This Matlab program is implemented in an online free application: www.tailorpack.com which copy of the main window is visible on Fig. 9.4. As indicated in this figure, when running with an optimization procedure (here a Levenberg–Marquardt algorithm), $\dot{p}_{O_2}^{pkg}$ and $\dot{p}_{CO_2}^{pkg}$, partial pressures in O_2 and CO_2 at equilibrium, become the target and then are replaced by optimal values recommended for the targeted product. Therefore gas permeability values are the results of the optimization procedure. In the example chosen, 500 g of strawberries (var. Charlotte) are packed into a barrier tray (assumed to be barrier toward O_2 and CO_2) at 20 °C; therefore, gas exchange occurs only through the lid film. The objective was then to identify the optimal gas permeabilities of this lid film. Packaging geometry and data on strawberries came from Cagnon et al. 2012 (and appear on Fig. 9.4). The online application provided optimal O_2 and CO_2 permeability of 2680 and 7920 amol m^{-1} s^{-1} Pa^{-1}, respectively, for the lid material.

9.3.2 Mass Balance Model for Moisture Exchanges

Mass balance models for moisture exchanges in modified atmosphere packaging of fruits combine mathematical equations for fruit transpiration and water loss transfer though the packaging material. Basic models for fresh produce transpiration consider moisture transport through the skin as a function of biophysical (Fockens and Meffert 1972; Gaffney et al. 1985; Hayakawa and Succar 1982, Sastry and Buffington 1982; Veraverbeke et al. 2003) and thermophysical (Kang and Dong 1998; Song et al. 2002) parameters. Recent models considering the transpiration rate

Table 9.3 Required input parameters in the mass balance model for moisture exchanges

Parameter	Name	Units
P_{np}	Water vapour permeability	g m^{-1} h^{-1} Pa^{-1}
S	Packaging surface	m^2
e	Packaging thickness	m
$P_{kp}{}^{i}$	Water partial pressure in j	Pa
TR	Transpiration rate	g kg^{-1} h^{-1}
V_f	Headspace	m^3
m	Mass of food	kg
a_{w_i}	Water activity of the food	–
a_w	Water activity of the surrounding atmosphere	–
T	Temperature	K
K	Mass transfer coefficient	–
a	Constant coefficient	–

as a loss of water as a function of RH and temperature was reported and validated (Mahajan et al. 2008; Sousa-Gallagher and Mahajan 2012). They are considered hereinafter for coupling with Fick's law (that can also be used to describe moisture transfer through a polymeric film):

$$\mathrm{TR} = \mathrm{K}_i\left(a_{w_i} - a_w\right)\left(1 - e^{-aT}\right) \quad (9.7)$$

and

$$V_f \times \frac{d\left(y_{\mathrm{H_2O}}^{pkg}\right)}{dt} = \frac{P_{\mathrm{H_2O}}}{e} \times S \times \left(y_{\mathrm{H_2O}}^{ext} - y_{\mathrm{H_2O}}^{pkg}\right) + \mathrm{TR} \times m = f_3 \quad (9.8)$$

where f_3 represents water vapour exchange in a package containing a respiring produce. The first part of the right-hand side of f_3 describes moisture flux per time unit through the packaging material, while the second part describes transpiration of the produce. Meaning of the parameters needed and their respective units are indicated in Table 9.3. At the steady state, Eq. 9.8 is simplified to:

$$y_{\mathrm{H_2O}}^{\mathrm{target}} = y_{\mathrm{H_2O}}^{\mathrm{ext}} + \frac{\mathrm{TR} \times e \times m}{P_{\mathrm{H_2O}} \times S} \quad (9.9)$$

Equation 9.9 was tested to predict water vapour transmission rate of 300 g strawberries packed into a barrier tray (assumed to be barrier toward moisture) sealed with a lid by using Matlab software (The Mathworks Inc, USA). Optimal condition of storage was of 5 °C and 90 % RH, relative humidity outside the package set at 60 %, and all other required data came from Sousa-Gallagher and Mahajan (2012) (transpiration rate of 0.27 g kg^{-1} h^{-1}, lid thickness of 25.4 μm, surface area of 167 cm^2, volume of 1326 cm^3, mass of product of 300 g). To reach optimal storage condition, water vapour transmission rate should equal to 116 g m^{-2} day^{-1} at 5 °C. This water balance model has not been yet proposed online.

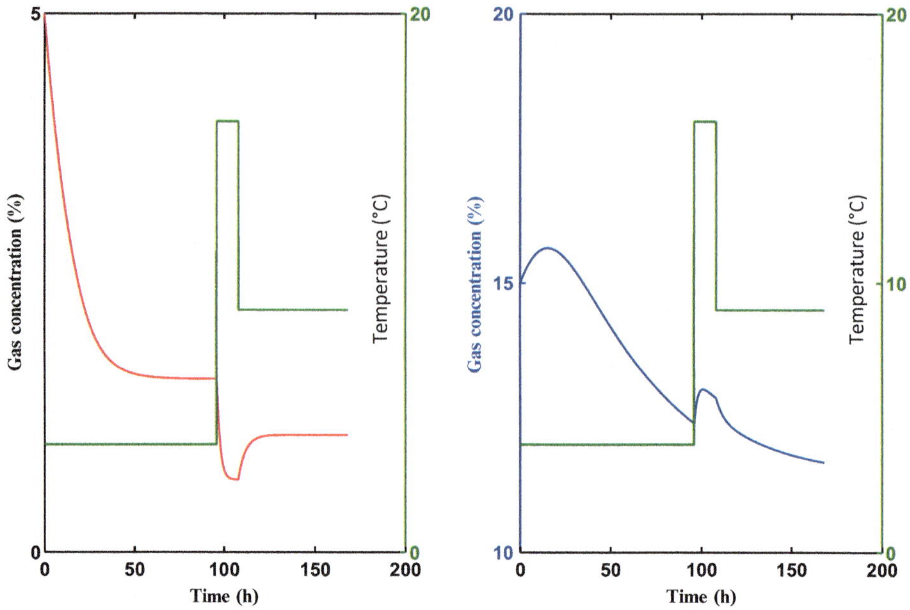

Fig. 9.5 Changes in O_2 (in red) and CO_2 (in blue) content in packed strawberries as a function of temperature of storage (in green)

9.3.3 Implementation of Temperature Variation

Temperature dependence of both physiological and permeation mechanisms can be modelled using Arrhenius' or Gore's laws referring to activation energy, Ea, and Q_{10}, respectively. An example of the use of Arrhenius' law is given in Eq. 9.7, to express change in transpiration rate as a function of temperature (in which *a*, the constant coefficient, takes Ea into account). An example of the use of the Gore's law is given below, to model the dependence of the permeability to temperature:

$$Q_{10} = \frac{\mathrm{Pe}(T+10)}{\mathrm{Pe}(T)} \tag{9.10}$$

where Pe is the permeability coefficient, and *T* the temperature (in °C).

This temperature dependence might be useful in MAP simulation procedure to predict changes in gas or water partial pressures within the packaging, when submitted to temperature variation mimicking realistic temperature cycle. Figure 9.5 gives an illustration of such a prediction for 500 g of packed strawberries (var. Ciflorette) in a LDPE (thickness of 50 μm) sachet with a total surface of 773 cm^2. The following scenario was assumed: strawberries were stored at 4 °C into the supermarket (for 4 days), then the core temperature of the product rose to 16 °C during home delivery (for 12 h) and finally, the product was stored 1 day in the fridge at 8 °C. Note that in this case initial content of O_2 and CO_2 was of 5 % and 15 %, respectively, due to gas flushing. Input parameters used for this simulation were Ea of 74.826 kJ mol^{-1} (Hertog et al. 1999); $RR_{O_2\,max}$ of 1.5 mmol kg^{-1} h^{-1}, RQ of 0.89, and Km of 2907 Pa all given at a reference temperature of 20 °C (from personal data); activation energies for O_2 and CO_2 permeabilities were taken from Charles et al. (2005): 35.97 and 32.16 kJ mol^{-1}, respectively, with a pre-exponential factor of 1 × 10–15 and 1.5 × 10–15^9 mol m^{-1} s^{-1} Pa^{-1}. During the first 4 days of storage at 4 °C, the O_2 is consumed by the strawberries until equilibrium around 2–3 %. Meanwhile, CO_2 produced permeates through the packaging material and goes out the package (due to high permeability of LDPE to gases) but no equilibrium is reached. Then, during home delivery (16 °C) and further storage at consumer's home (8 °C), temperature in the core of the product raises, leading to an increase in respiratory activity, and as expected, the atmosphere achieved at 4 °C is disrupted. Consequently, the product comes close to anoxia, which is unfavourable to the quality of strawberries. This points out mismatching between activation energy, Ea, or Q_{10} of the packaging material and fresh produces. Unfortunately, Ea (or Q_{10}) of synthetic materials currently available does not match Ea (or Q_{10}) of most of fresh produce; it is 2–3-fold lower (Varoquaux and Ozdemir, 2005).

9.3.4 *Implementation of Material Structures*

MAP optimization procedure might be part of a decision aid tool as illustrated in Fig. 9.1, but might also be used as the starting point for material design, i.e. tailored to the application. Mass balance models given in 9.3.1 and 9.3.2 have been used to determine optimal gas and moisture transfer properties of a single packaging material and dense polymeric film. However, although some fruits can be packed into sachet, i.e. the overall surface is available for mass transfer, they are commonly commercialized into trays sealed with lid, i.e. mass transfers are driven through the surface of the tray material and the surface of the lid material. In the examples given hereinabove, tray was considered as barrier and its respective surface area was not used in calculation, which was only based on the surface of the lid material. This is not realistic since gases and moisture transfer also occur through the material used for the tray. This material can be different from the one of the lid (but compatible for sealing) or at least exhibit a different thickness. Using optimization procedure in mass balance models, global optimal O_2 and CO_2 permeations are provided and are targets for the overall packaging. Then, if two different materials are foreseen, several combinations of permeation rates are possible depending on their respective area contribution to the overall O_2 and CO_2 permeation. So, it is easy to predict optimal combination of O_2 and CO_2 permeation by using a proportionality law. As an illustration, optimization procedure for gas exchanges was applied to 500 g of packed strawberries (var. Ciflorette) at 20 °C in a tray sealed with a lid exhibiting a total surface of 773 cm^2. Parameters were personally assessed and given in 9.3.3, and optimal O_2 and CO_2 considered at 10 % and 5 %, respectively. Global and

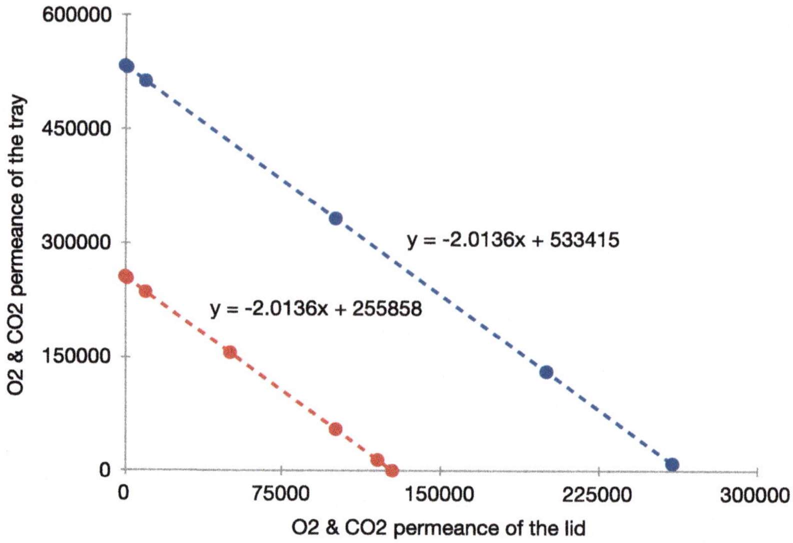

Fig. 9.6 Possible combinations in optimal O_2 (in red) and CO_2 (in blue) permeance (mL m^{-2} day^{-1} atm^{-1}) of the lid and the tray of a packaging for strawberries

optimal O_2 and CO_2 permeance predicted by the Tailorpack online application were 84,900 and 177,000 mL m^{-2} day^{-1} atm^{-1}. To reach these global and optimal permeances, Fig. 9.6 brings possible combinations of targeted O_2 and CO_2 permeance for the tray (surface of 516.5 cm2) and the lid film (surface of 256.5 cm^2). Of course, extreme of each lines presented in this figure are not feasible because the ratio of lid surface to tray surface varies always in a certain, realistic range. That means that, in practice, if an O_2 barrier material is chosen for the tray, a material with very high O_2 permeance must be chosen for the lid to compensate and in order to match the overall and target O_2 permeance. Same reasoning can be applied to moisture balance models in optimization procedures.

Most of plastic films available exhibit low to very low O_2 permeability values, which is a limitation for using them in MAP of fresh fruits. The material should be "breathable", with high to very high O_2 transfers therefore, plastic materials are usually micro- to macro-perforated to allow gas diffusion. It can be useful to predict in advance the density of perforation needed to apply to a given plastic film in order to reach the target optimal gas transfer properties of a material. Several mathematical model of gas transfer through perforated film have been reported and previously reviewed (Rodriguez-Aguilera and Oliveira 2009). The one presented below was already used and validated (Mahajan et al. 2007; Oliveira et al. 2012, Sousa-Gallagher and Mahajan 2013):

$$\mathrm{Pe}_{O_2,CO_2} = \mathrm{Pe}^{\mathrm{film}}_{O_2,CO_2} + \left(\frac{\pi R_P^2 \times D_{O_2,CO_2}}{e + R_P} \times N_P \right) \tag{9.11}$$

where R_P and N_P are the radius and the number of perforation, respectively, D, the diffusivity of each gas in air, Pe^{film} the permeability of the plastic film chosen, and Pe is the target permeability. It is part of a commercial online application (www.packinmap.com) in combination with a gas balance model and connected to databases. Therefore, this application can suggest to micro-perforate a particular film when appropriate.

Target transfer properties predicted by mass balance models might be used as the starting point for new material development taking into account multilayer, composite and nanocomposite technologies. To comply with it, it is necessary to implement current mass balance models with mathematical equations that describe gas transfer (such as permeability) as a function of composition or structural organization of the materials: from the addition of mass transfer resistances in multilayered structure to aspect ratio of pellet nanoparticles in nanocomposite materials, there is some existing models that have been previously reviewed (Guillaume et al. 2010).

9.4 Computing Advances to Develop Decision Aid Tool

As presented in the introduction section, the decision aid tool foreseen aims to select and rank a group of packaging materials that best suits user's needs to pack fruits under modified atmosphere as presented in Fig. 9.1. Then it relies on the interrogation of a packaging database upon a specific methodology names flexible querying. Stakeholder's queries can be bipolar, i.e. constituted of both negative and positive preferences. A negative preference is so-called a constraint and material that does not satisfy it must be rejected. A positive preference is so-called a wish and material that does not satisfy it is just consider as less desirable than others. In addition, preferences can be gradual or flexible, in a sense that different satisfaction levels are expected between fully satisfied and unsatisfied answers. Then flexible and bipolar multi-criteria querying technique can be applied in such a tool (Buche et al. 2012; Destercke et al. 2011). This technique copies human reasoning, facing a complex situation (multi-criteria) to make a choice (relevant packaging material). It allows a ranking of possible answers and the possibility to send back an answer to the user, even if this answer does not completely fulfil his preferences. Then for each criterion of the request, preferred (or guaranteed) answers are searching as well as possible ones (but not guaranteed) and it means that each answer interval is enlarged. For instance, the request on packaging material can be the following constraint "I want a packaging film that costs lower than 0.7 € m^{-2}". Guaranteed answers will be packaging materials not exceeding 0.7 € m^{-2}. Enlarging the answer interval by applying a possible variation of 40 % would lead to possible solutions where packaging material exhibit a price between 0.7 and 0.98 € m^{-2}. This can be represented in fuzzy set design, where a membership degree is applied to the requested criterion and varied from 0–1, i.e. from an element which is not part of the fuzzy set to an element which is preferred and part of the fuzzy set. In the example chosen, 0 is attributed to 0.98 € m^{-2} and 1–0.7 € m^{-2} and lower, as illustrated in Fig. 9.7. In a multi-criteria query,

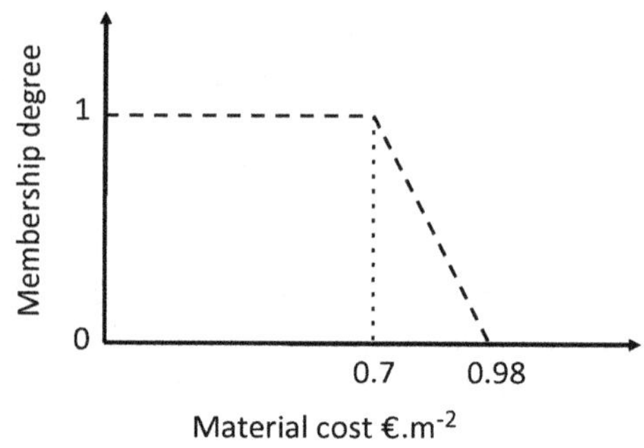

Fig. 9.7 Price preference represented by a membership function in fuzzy set

several fuzzy sets are generated considering each time preferred and possible answers and aggregated using a specific methodology described elsewhere (Destercke et al. 2011). Then algorithms of flexible querying are used for the ranking process (Destercke et al. 2011; Destercke and Guillard 2011). This points out the possibility of querying a database with missing data. When data related to a constraint query is missing, the decision support tool may consider that the material could potentially match this constraint and rank it, for instance, at the first position (because it fulfils all the other preferences); but the user must be informed that this data is required to ensure the material position.

Dealing with fruits means that the decision aid tool has to manage biological variability, and then uncertainties, to be as robust as possible (uncertainties might also come from packaging dimensions or data on material). It mainly concerns the output from the MAP optimization procedure that are considered as the major constraint in the foreseen application since it guarantee the fruit quality. Then optimal mass transfer properties should be given as a range of values instead of a single value, taking into account uncertainties through probabilistic or interval analysis. Probabilistic approaches as Monte Carlo simulations are commonly used to treat and model biological uncertainty (Baudrit et al. 2007, 2009; Hertog et al. 2007, 2009; Iqbal et al. 2009). It requires to specifying the distribution of and dependency between each variable. As it is often the case, this might be unknown and then assumptions might be chosen according to practical criteria (e.g. normal distribution and variables independent). But this procedure may provide misleading conclusions, leading to unguaranteed choices (Ferson and Ginzburg 1996). More robust models are obtained when considering conservative theories as interval analyses to perform uncertainty analysis (Jaulin et al. 2001). Only the bounds in which each input variable may vary are needed (i.e. minimal and maximal values). It might be interesting to combine these theories: starting with a quick and robust interval analysis and then Monte Carlo simulations to refine uncertainty prediction, depending on the statistical information available (Guillard et al. 2012). Imprecise probabilistic approaches (Baudrit et al. 2009) can also be considered with the advantages and drawbacks of the both previous theories: robustness and good statistical estimation against large uncertainty bounds and high computational cost. To not overload this chapter, only the use of interval analysis in propagation of uncertainties in dynamical systems as mass balance models is described hereinafter; but note that Monte Carlo simulations can be applied as it is already the case in the online PackinMap application that offers an option to define a percentage of incertitude and then send back a window of minimal and maximal P_{O_2} and P_{CO_2} optimal values based on Monte Carlo simulations. A comparison between those two approaches in gas balance models was also reported (Guillard et al. 2012).

In interval analysis, the precise value x of a variable is replaced by an interval whose bounds are the lowest and the highest possible values of the considered variable $[x]=[x-;x+]$ (i.e. taking into account its uncertainty). Consequently and if the variable is part of a function $f(x)$, it will also be replaced by the interval and bounds will be computed in order that $f([x])=\{f(x)|x\in[x]\}$. Therefore the solution to this equation will be a set of values obtained by selecting x in $[x]$. This theory has been applied to the gas balance model first in the simulation procedure and then in the optimization procedure. Equations 9.2 and 9.3 describe the evolution of O_2 (x_1) and CO_2 (x_2) as a function of time and are considered as ordinary differential equations and dynamically monotonic. At a given time, a set of x_1 and x_2 is possible depending on the interval chosen for each variable (e.g. $RR_{O_2\,max}$, Km). Results of interval analysis applied to gas balance model in simulation mode are the existence of a lower and upper envelope for each partial pressure (O_2 and CO_2) instead of a single one for each of them. An example is given in Fig. 9.8 where uncertainty propagation has been studied with interval analysis on change in O_2 and CO_2 partial pressures in LDPE packed blueberries; values of input parameters and their uncertainty comes from Song et al. (2002) and Guillard et al. (2012). Assuming that optimal O_2 partial pressure for blueberries should be higher than 4 kPa (Beaudry et al. 1992), the existence of such uncertainties does not guarantee this optimal condition since O_2 partial pressure may vary between 0.3 and 9.2 kPa. Then risk of anoxia exists if this combination of packaging geometry and material with blueberries is used. When going to the optimization procedure, same methodology can be applied (Destercke and Guillard 2011; Guillard et al. 2012) and result of it is a range of optimal O_2 and CO_2 permeability values. However, it may occur that no optimal solution is predicted. In such a configuration, fuzzy set can be used providing no guaranteed solution but possible ones, depending on the tolerated variation.

9.5 Concluding Remarks

As illustrated in Fig. 9.1, decision-making tools on modified atmosphere packaging for fruits must provide answers to the requests of stakeholders. The answer to such a multi-criteria query would be to identify a / or ranking a group of packaging materials fitting the wishes and constraints of stakeholders. Therefore, MAP optimization procedure takes a central place in the conception of such a tool, since it provides optimal mass transfer properties that guarantee the quality of the fresh fruit, which is obviously considered as the major constraint. This procedure relies on the

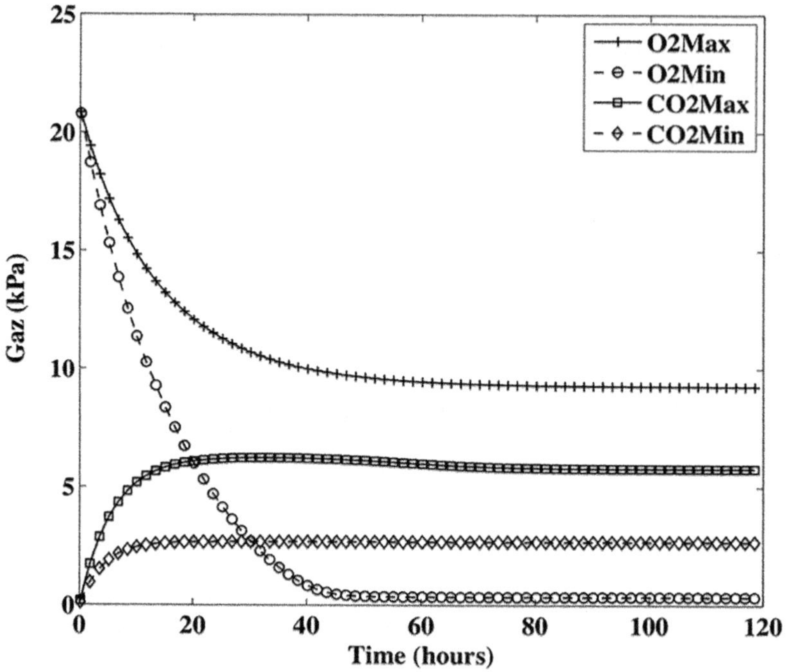

Fig. 9.8 Uncertainty propagation during the modelling of modified atmosphere packaging of blueberries packed in LDPE pouche

existence of mathematical equations describing physiological and mass transport phenomena and their coupling so-called mass balance models. As presented in this chapter, gas and moisture balance models already exist and are already used in some online applications. Combining with other equations they are very useful to anticipate the impact of packaging geometry (e.g. tray sealed with lid), packaging structure (e.g. perforations) or temperature (e.g. scenario of chill chain disruption) on the quality of fresh packed fruits. Actually it is more their ability to bring or maintain optimal conditions of storage than organoleptic quality. However, there is still some expectation in theoretical approaches to model for instance fruits ripening or RH dependence of gas transport in polymeric films. With the booming of fresh-cut produce, minimally fruit processing and active systems should also be considered in further mass balance models. Success of MAP optimization procedure is tightly related to a proper and large database on fresh fruits providing parameters for optimal storage conditions and physiological pathways. Each numerical data must be considered with its uncertainty that can be treated with probabilistic models or interval analysis in robust MAP optimization procedure, and with its unit that must be associated to unit converters for further modelling purpose. Collection of data can be eased with knowledge representation and reasoning (for automatic feeding) if data fields are well identified at the very beginning, but assessment of data still remains a heavy work. For a given species and variety, there is often a lack of complete studies and most of time only a handful of measurements are reported in several references. Regarding the packaging material database, there is plenty of data on oxygen and moisture transport whereas less interest is given to data on carbon dioxide and ethylene transport in materials. This constitutes a severe bottleneck to powerful decision aid tools that must manage missing data and may provide no or fair guarantee on fruits quality. That's why for the time being, each answer (or ranking) must be given in association with at least the percentage of known parameters.Acknowledgements The research leading to these results has received funding from the European Community's Seventh Framework Programme (FP7/2007–2013) under the grant agreement no FP7-265669-EcoBioCAP project and from the French Research Agency (ANR) through the Tailorpack project (ANR-07-PNRA-029).

References

Barron, C., P. Varoquaux, S. Guilbert, N. Gontard, and B. Gouble. 2002. Modified atmosphere packaging of cultivated mushroom (Agaricus bisporus L.) with hydrophilic films. *Journal of Food Science* 67: 251–255.

Baudrit, C., A. Helias, and N. Perrot. 2007. *Uncertainty analysis in food engineering involving imprecision and randomness*. Paper presented in the fifth International Symposium on Imprecise Probability: Theories and Applications, Prague, Czech Republic, July 16–19.

Baudrit, C., A. Helias, and N. Perrot. 2009. Joint treatment of imprecision and variability in food engineering: application to cheese mass loss during ripening. *Journal of Food Engineering* 3: 284–292.

Beaudry, R.M., A.C. Cameron, A. Shirazi, and D.L. Dostal-Lange. 1992. Modified atmosphere packaging of blueberry fruit: effect of temperature on package O_2 and CO_2. *Journal of the American Society for Horticultural Science* 117: 436–441.

Brecht, J.K. 1995. Physiology of lightly processed fruits and vegetables. *Hortscience* 30: 18–22.

Buche, P., S. Destercke, V. Guillard, O. Haemmerle, and R. Thomopoulos. 2012. Flexible bipolar querying of uncertain data using an ontology. In *Studies in Computational intelligence*, ed. J. Kacprzyk. London: Springer.

Cagnon, T., C. Guillaume, V. Guillard, and N. Gontard. *Nanostructuring and microstructuring of materials from a single agropolymer for sustainable MAP preservation of fresh food*. Paper presented at the 2012 IAPRI (International Association of Packaging Research Institutes) world packaging conference, San Luis Obispo, California, June 17–21.

Charles, F., C. Guillaume, and N. Gontard. 2008. Effect of passive and active modified atmosphere packaging on quality changes of fresh endives. *Postharvest Biology and Technology* 48: 22–29.

Charles, F., J. Sanchez, and N. Gontard. 2003. Active modified atmosphere packaging of fresh fruits and vegetables: Modeling with tomatoes and oxygen absorber. *Journal of Food Science* 68: 1736–1742.

Charles, F., J. Sanchez, and N. Gontard. 2005. Modelling of active modified atmosphere packaging of endives exposed to several postharvest temperatures. *Journal of Food Engineering* 72: 1–7.

Chevillote, P. 1973. Relation between reaction cytochrome oxidase-oxygen and oxygen uptake in cells in vivo. The role of diffusion. *Journal of Theoretical Biology* 39: 277–295.

Destercke, S., P. Buche, and V. Guillard. 2011. A flexible bipolar querying approach with imprecise data and guaranteed results. *Fuzzy Set and Systems* 169: 51–64.

Destercke, S., and V. Guillard. 2011. Interval analysis on non-linear monotonic systems as an efficient tool to optimise fresh food packaging. *Computers and Electronics in Agriculture* 79: 116–124.
Exama, A., J. Arul, R.W. Lencki, L.Z. Lee, and C. Toupin. 1993. Suitability of plastic films for modified atmosphere packaging of fruits and vegetables. *Journal of Food Science* 58: 1365–1370.
Fan, X., L. Argenta, and J.P. Mattheis. 2000. Inhibition of ethylene action by 1-methylcyclopropene prolongs storage life of apricots. *Postharvest Biology and Technology* 20: 135–142.
Ferson, S., and L.R. Ginzburg. 1996. Different methods are needed to propagate ignorance and variability. *Reliability Engineering and System Safety* 54: 133–144.
Fishman, S., V. Rodov, J. Peretz, and S. Benyehoshua. 1995. Model for gas exchange dynamics in modified atmosphere packages of fruits and vegetables. *Journal of Food Science* 60: 1078–1083.
Fockens, F.H., and H.F.T. Meffert. 1972. Biophysical properties of horticultural products as related to loss of moisture during cooling down. *Journal of the Food Science and Agriculture* 23: 285–298.
Fonseca, S.C., F.A.R. Oliveira, and J.K. Brecht. 2002. Modelling respiration rate of fresh fruits and vegetables for modified atmosphere packages: a review. *Journal of Food Engineering* 52: 99–119.
Gaffney, J.J., C.D. Baird, and K.V. Chau. 1985. Influence of airflow rate, respiration, evaporative cooling, and other factors affecting weight loss calculations for fruits and vegetables. *ASHRAE Transactions* 91: 690–707.
Gouble, B., M. Grotte, P. Reling, M. Reich, S. Bureau, J.M. Audergon, G. Clauzel, and G. Albagnac. 2006. Comparison between ripening on trees and post-harvest evolution for apricot cultivars contrasting in their rate of ethylene production. *Acta Horticulturae* 717: 333–336.
Grotte, M., B. Gouble, P. Reling, G. Clauzel, and J.M. Audergon. 2006. Kinetics of apricot fruit ethylene production in the orachard and post-harvest: A mathematical modelling approach. *Acta Horticulturae* 717: 343–346.
Guillard, V., C. Guillaume, and S. Destercke. 2012. Parameter uncertainties and error propagation in modified atmosphere packaging modelling. *Postharvest Biology and Technology* 67: 154–166.
Guillaume, C., V. Guillard, and N. Gontard. 2010. Modified atmosphere packaging of fruits and vegetables : modeling approach. In *Advances in Fresh-cut Fruits and Vegetables Processing*, ed. O. Martin-Belloso and R. Soliva-Fortuny, 255–284. Boca Raton: CRC Press.
Gustavsson, J., C. Cederberg, U. Sonesson, R. Van Otterdijk, and A. Meybeck. 2011. Global food losses and food wastes: extent, causes and prevention. Paper presented at the first Save Food congress, Dusseldorf, Germany, May 16–17.
Hayakawa, K., and J. Succar. 1982. Heat transfer and moisture loss of spherical fresh produce. *Journal of Food Science* 47: 596–605.
Hershkovitz, V., S.I. Saguy, and E. Pesis. 2005. Postharvest application of 1-MCP to improve the quality of various avocado cultivars. *Postharvest Biology and Technology* 37: 252–264.
Hertog, M.L.A.T.M., H.A.M. Boerriter, G.J.P.M. van den Boogaard, L.M.M. Tijskens, and A.C.R. van Schaik. 1999. Predicting keeping quality of strawberries (cv. 'Elsanta') packed under modified atmospheres: an integrated model approach. *Postharvest Biology and Technology* 15: 1–12.
Hertog, M.L.A.T.M., N. Scheerlinck, J. Lammertyn, and B.M. Nicolai. 2007. The impact of biological variation on postharvest behaviour of Belgian endive: the case of multiple stochastic variables. *Postharvest Biology and Technology* 43: 78–88.
Hertog, M.L.A.T.M., N. Scheerlinck, and B.M. Nicolai. 2009. Monte Carlo evaluation of biological variation: random generation of correlated non-Gaussian model parameters. *Journal of Computational and Applied Mathematics* 223: 1–14.
Iqbal, T., F.A.S. Rodrigues, P.V. Mahajan, and J.P. Kerry. 2009. Mathematical modelling of O2 consumption and CO2 production rates of whole mushrooms accounting for the effect of temperature and gas composition. *International Journal of Food Science and Technology* 44: 1408–1414.

Jaulin, L., M. Kieffer, O. Didrit, and E. Walter. 2001. *Applied Interval Analysis*. London: Springer.

Kader, A.A., D. Zagory, and E.L. Kerbel. 1989. Modified atmosphere packaging of fruits and vegetables. *Critical Reviews in Food Science and Nutrition* 28: 1–30.

Kang, J.S., and S.L. Dong. 1998. A kinetic model for transpiration of fresh produce in a controlled atmosphere. *Journal of Food Engineering* 35: 65–73.

Lopez Briones, G., P. Varoquaux, Y. Chambroy, J. Bouquant, G. Bureau, and B. Pascat. 1992. Storage of common mushroom under controlled atmospheres. *International Journal of Food Science and Technology* 27: 493–505.

Mahajan, P.V., F.A.R. Oliveira, and I. Macedo. 2008. Effect of temperature and humidity on the transpiration rate of whole mushrooms. *Journal of Food Engineering* 84: 281–288.

Mahajan, P.V., F.A.R. Oliveira, J.C. Montanez, and J. Frias. 2007. Development of user-friendly software for design of modified atmosphere packaging for fresh and fresh-cut produce. *Innovative Food Science and Emerging Technologies* 8: 84–92.

Makino, Y., K. Iwasaki, and T. Hirata. 1996a. Oxygen consumption model for fresh produce on the basis of adsorption theory. *Transactions of the American Society of Agricultural Engineers* 39: 1067–1073.

Makino, Y., K. Iwasaki, and T. Hirata. 1996b. A theoretical model for oxygen consumption in fresh produce under an atmosphere with carbon dioxide. *Journal of Agricultural Engineering Research* 65: 193–203.

Manolopoulou, H., and P. Papadopoulou. 1998. A study of respiratory and physico-chemical changes of four kiwi fruit cultivars during cool-storage. *Food Chemistry* 63: 529–534.

Marcellin, P. 1975. Conservation des fruits et légumes frais. *Cahiers de nutrition et de diététique* 10: 49–52.

Mathooko, F.M. 1996. Regulation of respiratory metabolism in fruits and vegetables by carbon dioxide. *Postharvest Biology and Technology* 9: 247–264.

Mc Glasson, W.B. 1984. Ethylene and fruit ripening. *Hortscience* 20: 51–54.

MujicaPaz, H., and N. Gontard. 1997. Oxygen and carbon dioxide permeability of wheat gluten film: Effect of relative humidity and temperature. *Journal of Agricultural and Food Chemistry* 45: 4101–4105.

Mujica Paz, H., V. Guillard, M. Reynes, and N. Gontard. 2005. Ethylene permeability of wheat gluten film as a function of temperature and relative humidity. *Journal of Membrane Science* 256: 108–115.

Nguyenthe, C., and F. Carlin. 1994. The microbiology of minimally processed fresh fruits and vegetables. *Critical Reviews in Food Science and Nutrition* 34: 371–401.

Oliveira, F., M.J. Sousa-Gallagher, P.V. Mahajan, and J.A. Teixeira. 2012. Evaluation of MAP engineering design parameters on quality of fresh-sliced mushrooms. *Journal of Food Engineering* 108: 507–514.

Paull R.E. 1993. Tropical fruit physiology and storage potential. Paper presented at the international conference on Postharvest Handling of Tropical Fruits, Chiang Mai, Thailand, July 19–23.

Pech, J.C., C. Balague, A. Latche, and M. Bouzayen. 1994. Postharvest physiology of climacteric fruits—Recent developments in the biosynthesis and action of ethylene. *Sciences des Aliments* 14: 3–15.

Peppelenbos, H.W., and J. van't Leven. 1996. Evaluation of four types of inhibition for modelling the influence of carbon dioxide on oxygen consumption of fruits and vegetables. *Postharvest Biology and Technology* 7: 27–40.

Petracek, P.D., D.W. Joles, A. Shirazi, and A.C. Cameron. 2002. Modified atmosphere packaging of sweet cherry (Prunus avium L., ev. 'Sams') fruit: metabolic responses to oxygen, carbon dioxide, and temperature. *Postharvest Biology and Technology* 24: 259–270.

Robertson, G.L. 2006. Packaging of horticultural products. In *Food Packaging Principles and Practice*, ed. H.A. Hughes, 470–506. New York: Marcel Dekker.

Rothan, C., S. Duret, C. Chevalier, and P. Raymond. 1997. Suppression of ripening-associated gene expression in tomato fruits subjected to a high CO_2 concentration. *Plant Physiology* 114: 255–263.

Rodriguez-Aguilera, R., and Oliveira, J.C. 2009. Gas exchange dynamics in modified atmosphere packaging of soft cheese. *Journal of Food Science* 74: E298–E303.
Rothan, C., and J. Nicolas. 1994. High CO_2 levels reduce ethylene production in kiwi fruits. *Plant Physiology* 92: 1–8.
Roy, S., R.C. Anantheswaran, and R.B. Beelman. 1996. Modified atmosphere and modified humidity packaging of fresh mushrooms. *Journal of Food Science* 61: 391–397.
Sandhya. 2010. Modified atmosphere packaging of fresh produce: Current status and future needs. *LWT—Food Science and Technology* 43: 381–392.
Sastry, S.K., and D.E. Buffington. 1982. Transpiration rates of stored perishable commodities. A mathematical model and experiments on tomatoes. *ASHRAE Transactions* 88: 391–397.
Song, Y., N. Vorsa, and K.L. Yam. 2002. Modeling respiration-transpiration in a modified atmosphere packaging system containing blueberry. *Journal o Food Engineering* 53: 103–109.
Sousa-Gallagher, M. J., and P.V. Mahajan. 2012. Quantification of mass transfer properties for engineering MAP design of fresh produce. Paper presented at the Environmental Microbiology and Biotechnology conference, Bologna, Italy, April 11–13.
Sousa-Gallagher, M.J., and P.V. Mahajan. 2013. Integrative mathematical modelling for MAP design of fresh-produce: Theoretical analysis and experimental validation. *Food control* 29: 444–450.
Talasila, P.C. 1992. *Modeling of heat and mass transfer in a modified atmosphere package*. Gainesville: University of Florida.
Talasila, P.C., K.V. Chau, and J.K. Brecht. 1992. Effects of gas concentrations and temperature on O2 consumption of strawberries. *Transactions of the American Society of Agricultural Engineers* 35: 221–224.
Vargas, M., A. Albors, A. Chiralt, and C. Gonzalez-Martinez. 2006. Quality of cold-stored strawberries as affected by chitosan-oleic acid edible coatings. *Postharvest Biology and Technology* 41: 164–171.
Varoquaux, P., B. Gouble, M.N. Ducamp, and G. Self. 2002. Procedure to optimize modified atmosphere packaging for fruit. *Fruits* 57: 313–322.
Varoquaux, P., and I. Ozdemir. 2005. Packaging and produce degradation. In *Produce Degradation: Pathways and Prevention*, ed. O. Lamikanra, 117–153. Boca Raton: CRC Press.
Veraverbeke, E.A., P. Verboven, V.O. Oostveldt, and B.M. Nicolai. 2003. Prediction of moisture loss across the cuticle of apple (Malus sylvestris subsp. Mitis (Wallr.)) during storage: Part 2. Model simulations and practical applications. *Postharvest Biology and Technology* 30: 89–97.

Chapter 10
Frying of Foods

Pedro Bouchon and Verónica Dueik

10.1 Introduction

According to the dictionary, a snack is "a small portion of food or drink, or a light meal, especially one eaten between regular meals". That old definition grossly underestimates the impact snacks have on the modern diet, because how we handle this small portion of our food makes a big statement about our health. In the modern world, snacks often undermine a healthy diet and contribute to the growing epidemic of obesity.

The increased consumption of energy-dense food with high levels of fat, sugar, and refined carbohydrates, combined with reduced physical activity, is seen as the main cause of the obesity epidemic (WHO 2000, 2003). This is why consumers are demanding snacks that taste good and are satiating and healthy; which encourages the food industry to develop new processing technologies and select new raw materials that allow such products to be obtained.

10.1.1 The Savory Snack Products Market and new Trends

The salty snacks category is one of the most diverse in the consumer packaged goods industry. Today, sales of potato chips, tortilla chips, snack nuts, other salted snacks and pretzels generate total US sales of $18.8 billion making it one of the most important categories for retailers and manufacturers alike (Mintel 2011). Potato chips and most savory snacks fall within the category of foods with high fat

P. Bouchon • V. Dueik (✉)
Pontificia Universidad Católica de Chile, Santiago, Chile
e-mail: vpdueik@uc.cl

A. Rosenthal et al. (eds.), *Fruit Preservation*, Food Engineering Series,
https://doi.org/10.1007/978-1-4939-3311-2_10

and salt contents. Despite this, snacking remains part of American culture, due to the unique characteristics of salty snacks.

Growing consumer awareness of the inherent unhealthy aspects of many of their favorite snack products has motivated manufacturers to invest and promote various innovations within these categories, given that a large number of the products are fried and are relatively high in fat and sodium. Some years ago, the negative point of potato chips resided in the fact that they were deep-fried in partially hydrogenated oil. Today, manufacturers have replaced the *trans* fat-rich hydrogenated oils with healthier vegetable oils, although for well-informed and demanding consumers this is not enough. Consumers are preferring snack products that offer a wide range of product alternatives, adhering to convenience, sensory and health trends. In relation to health, interest in salty snack products that are organic or all-natural, low-calorie, low-fat, low-carbohydrate, low-sodium, rich in fiber and vitamins, or offer some health-promoting benefit, are in great demand. Nevertheless, even health-conscious consumers are not willing to sacrifice organoleptic properties, and intense full-flavor snacks remain an important trend in the salty snack market (Mariscal and Bouchon 2008).

Understanding consumer trends can present new opportunities, which is the reason why food manufacturers are developing new products on a daily basis. Currently, aspects such as convenience, low-fat, nutrition and taste are a must in the production of snack products.

10.1.2 Traditional Processing Technologies of Savory Snacks

Savory snacks are snack products that are not sweet; and include potato chips, sheeted shaped dough chips, tortilla chips, corn chips, hard pretzels, popped popcorn, extruded snacks, processed seed snacks and roasted peanuts and other nuts, among others.

Traditional snack products considered as unhealthy, such as regular fried and salty extruded products, among others, are usually produced with simple industrial processes, and their production costs are indeed low. However, there are low-salt, low-sugar, low-fat, preservative-free products that in recent times have been labeled as healthy; the production of these entails higher costs.

A high proportion of the food we eat is produced by industrial processes: these combine a main ingredient with some additional ingredients to make the final product better looking and better tasting. Likewise, these procedures add value to the final product (Antoñanzas and Rodríguez-Ibeas 2011). Major salty snacks manufacturing technologies are baking, extrusion, and frying. All of them include food exposure to high temperatures (usually above 100 °C) and the presence of oxygen, processing conditions that may induce degradation of nutritional compounds, oxidation of beneficial compounds or even worse, formation of toxic compounds (Kubow 1992; Dueik and Bouchon 2011a). Apart from these aggressive processing conditions, they include the addition of other food ingredients such as oil, salt, fats, shortenings and preservatives, among others. In addition, snacks may contain important amounts of nitrates, preservatives, colorants, artificial flavors, *trans* and saturated fats, which undermine recent consumer trends towards healthy food and full-flavor snacks (Monteiro 2009).

Extruded snacks are manufactured at a high-temperature and short-time process in which moistened, expansive, starchy and/or proteinaceous food materials (biopolymers) are plasticized and cooked in a tube by a combination of moisture, pressure, temperature and mechanical shear, resulting in molecular transformation and chemical reactions (Castells et al. 2005). Despite the benefit of the low-cost process of this technology, some reactions during food extrusion, such as the Maillard reaction and loss of thermal labile vitamins may occur; reducing the nutritional value of extruded products. Changes in proteins and amino acid profile, dietary fiber, vitamins, mineral content, and some non-nutrient healthful components of food may be either beneficial or deleterious (Fellows 2000). When designing recipes, careful consideration must be given to the selection of cereal grains, starches, proteins, and other minor ingredients. For example, starch makes important contributions to the final product, including expansion, binding, caloric value and functionality, among others; while shortenings and fats not only contribute to the flavor and appearance of this type of snacks, but also increase their caloric and fat content.

Baked snacks are manufactured using tunnel ovens, where the food pieces are conveyed on a continuous baking support through a series of heated oven sections. Main changes occurring during the process are development of structure, dehydration, and surface coloring (Townsend 1990). Baked snacks have no oil added during the process and usually their flavor properties rely on the use of added natural or artificial flavors. However, the application of these flavors presents a challenge as these products are fragile and lack oil, which in fried products assists the adherence of powdered seasonings (Clark 2009).

Fried snacks are produced by immersing the raw material in edible oil at a temperature well above the boiling point of water, usually 160–180 °C, and therefore it may be classified as a dehydration process (Farkas 1994). The high temperatures and the presence of oxygen induce several changes in the product and the frying oil itself, which will be discussed in Sect. 10.2.5.

10.1.3 Regular Fried Snacks

One of the most important quality parameters of fried snacks is the amount of oil absorbed by the product, as excess consumption of fats and oils has been qualified as a key dietary contributor to coronary heart disease and possibly some types of cancer (Dueik and Bouchon 2011a).

Some of the most commonly found regular fried snacks are based on potatoes or potato flakes, which differ in their microstructure and final characteristics. Potato chips are a crunchy snack food, high in carbohydrates and deep-fried, in which almost half the calories come from fat. Potato chips are thin products composed just of crust in which the final moisture content is 2 % and 33–40 % is oil, both in terms of wet base. Thick products or French fries are composed by a core and a crust, which have final moisture content of around 60 % (w.b), whereas oil accounts for about 15 % of the final product. Other types of products include restructured potato products composed mainly of dehydrated potato flakes: these are thin products with a final moisture content of 2 % and oil content of 33–40 % (w.b).

Another kind of commonly consumed fried products are tortilla chips. A corn tortilla chip is a piece of tortilla fried in oil, and these are now widely available in supermarkets. This product is made from corn dough, and an important difference between tortilla and potato chips is that in this situation the food is firstly baked, prior to frying, and an important amount of starch gelatinizes during that stage. These products are generally high in salt and fat, reaching a final oil content of between 25 and 30 % of the weight of the final product.

As seen before, most common regular fried products have a similar denominator: the elevated amount of oil and salt present in the final product, without considering the high carbohydrate content of the materials from which they are manufactured. Increased consumer interest in healthy eating is causing the industry to rethink the production processes of existing products and to consider the development of new products that will respond to the interest in low-calorie, low-fat, low-cholesterol, and low-sodium snacks.

10.1.4 Novel Fried Snacks

There has been a clear trend towards more complex products, with intriguing shapes, attractive colors, and innovative tastes and textures, to provide differentiation from the existing products on the market, such as potato and potato flakes-based snacks, and tortilla chips, given that all of these are carbohydrate-based products. Niche products that offer novelty flavors, shapes, or unique ingredients have also been introduced. New raw materials such as parsnips, red beets, sweet potatoes, and carrots, as well as organic snack food products, are recent offerings on the market. Also, whole and multi-grain snacks are starting to appear on supermarket aisles.

10.2 Deep-Fat Frying of Food

10.2.1 History

Frying is one of the oldest and most widely food processes. Its popularity relates to the speed with which a food is cooked, the distinctive flavor and texture frying gives the food, and its contribution to increased shelf-life (Rossell 2001).

How far back in time frying with oil goes is difficult to tell (Morton 1998). Frying supposedly had its origins in China around 3000 BC, where foods were precooked prior to roasting, and Egyptian wall paintings show dough being fried in oil indicating that Europe and North Africa were using hot oil to cook foods before the time of Christ (Stier 2004). However, it is only in the past 50 years that frying has gained the attention it deserves from the scientific community.

Frying is an easy, delicious, and quick way to cook our foods, and in modern times it has become a wonderful solution to our hurried lifestyles and day-to-day living. One of the most popular deep-fat fried products are potato chips, which were inadvertently invented in 1853 by the George Crum, a chef at a New York resort. A customer ordered fries with his meal, but when the meal arrived, he pushed the fries away because they were too thick, sending them back to the chef. Crum made some thinner fries, but the customer was still not satisfied. Crum got angry and sliced the potatoes very thinly, and fried them in hot oil until they were curled to a chip. The customer loved the new product and they began to be sold in restaurants; soon after the first potato chip factories were built (Sleet 2011). Today, Americans eat about 17 lb of potato chips and French fries per capita, a quantity that is quickly increasing (USDA 2010). These statistics are worrying, as the amount of oil taken-up by the product during the frying process has increasingly become a public health concern, and there is a strong desire to reduce the oil content (Ni and Datta 1999). Oil consumption, particularly saturated fats, is recognized as one of the major factors playing a significant adverse role in health, such as coronary heart disease, cancer, diabetes, and hypertension (Saguy and Dana 2003).

10.2.2 The Frying Process

During atmospheric deep-fat frying the piece of food is immersed in edible oil at a temperature well above the boiling point of water, usually between 160 and 180 °C (Bouchon et al. 2003). This complex unit operation involves simultaneous heat and mass transfer. Heat transfer is by convection between the oil and the food surface, and by conduction within the food (Farkas et al. 1996). Mass transfer is characterized by the loss of water from the food as water vapor and the movement of oil into the food (Singh 1995). Figure 10.1 shows a schematic diagram of the process where the heat and mass transfer occurring during the frying process can be observed.

Farkas et al. (1996) observed that the temperature at any location in the core region is limited to values below the boiling point of the interstitial liquid (approximately 105 °C, due to the presence of solutes that increase the boiling point of water slightly above 100 °C). There is an initial rapid fall of the water content mainly due to the loss of surface and unbound inner water; as a consequence the surface becomes dry, so that the moving front propagates towards the interior and the surface temperature begins to rise, approaching the temperature of the cooking oil. Then the evaporation rate decreases because diffusion inside the chip is not enough and only a small amount of water is lost, although this takes longer. On the basis of visual observations and analysis of temperature profiles and moisture data, it has been suggested that the frying process is composed of four distinct stages: The first

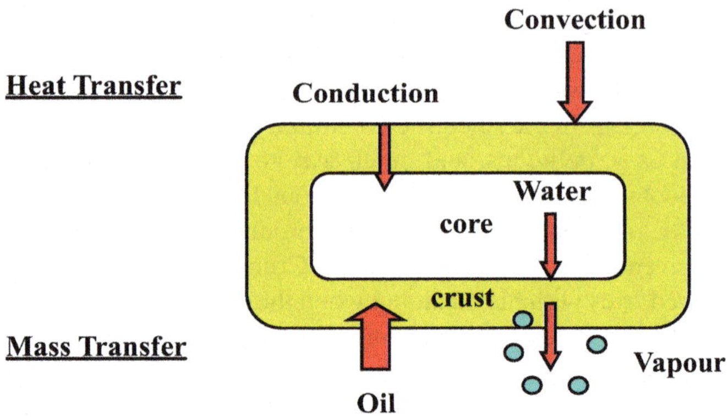

Fig. 10.1 Schematic diagram of simultaneous heat and mass transfer during frying (from Bouchon 2010)

stage, initial heating, is the period of time during which the surface of the product is heated from its initial temperature to the boiling point of water; this period is usually short and a negligible amount of water is lost from the food. In the second stage, surface boiling is characterized by the sudden loss of water and the beginning of the crust formation. The falling rate period, the third stage, represents the period of time during which the bulk of the moisture is lost. It is the longest of the stages and the temperature of the core region approaches that of the boiling point of water. The bubble-end point is the final stage, and describes the apparent end of moisture loss from the product (Singh 1995; Farinu and Baik 2005). Figure 10.2 presents a graph of the temperature history and moisture content monitored during atmospheric frying at 170 °C of a potato slice, in which it is possible to distinguish the different stages of the frying process (Dueik and Bouchon unpublished data).

The convective heat transfer coefficient for several oils has been previously determined by lumped capacity analysis, focusing on the non-boiling phase of frying, without considering the effect of moisture loss and bubbles leaving the surface, thus leading to convective heat transfer coefficients that may only apply to a limited period of time (absence of bubbling). Costa et al. (1999) showed that the h values determined in frying were up to two times greater than those obtained in the absence of vapor bubbling (between 250 and 280 W/(m^2 °C)) and varied with the rate of water loss, showing a maximum close to the period of maximum water loss rate. Hubbard and Farkas (1999) obtained a maximum convective heat transfer coefficient value of 1100 W/(m^2 °C) when frying potato cylinders in canola oil.

10.2.3 Frying Equipment

Frying technique has vastly improved over the years and has evolved from kitchen frying to large-scale industrial frying. Frying process can be divided into three major categories, such as (1) home frying, (2) restaurant or food service frying, and

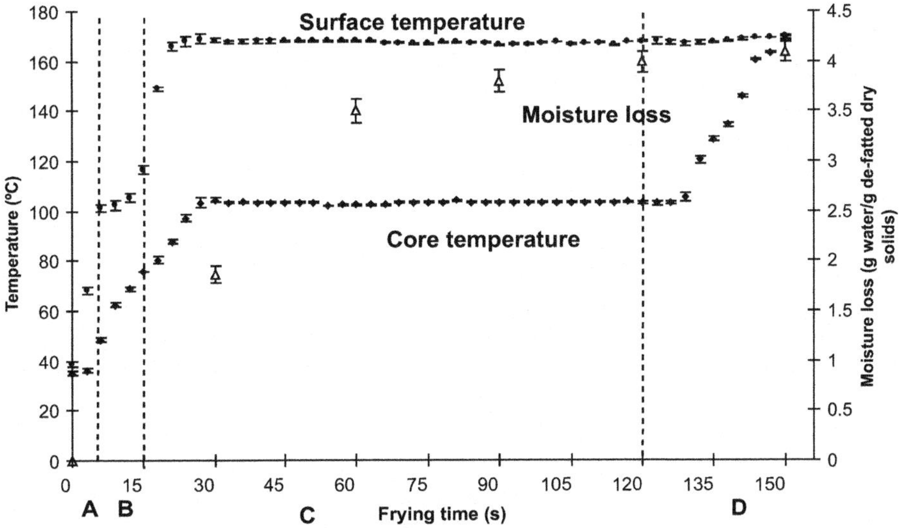

Fig. 10.2 Temperature history and moisture content during atmospheric frying at 170 °C of a potato slice (2 mm-thick), in which it is possible to distinguish the different stages of the frying process (**a**) initial heating; (**b**) surface boiling; (**c**) falling rate; (**d**) bubble end (Dueik and Bouchon unpublished data)

(3) industrial frying (Gupta 2005). This section will mainly focus on industrial frying, as it has become a major segment of the overall food industry, along with the innovations in packaging technologies that have helped industrial fried products maintain their quality over an extended period.

Industrial frying machines have evolved from the early days of frying potato chips to more sophisticated and automated frying machines. These facilities can be classified into batch and continuous fryers. The first ones are used for small volume productions, of about 5–20 kg of each frying unit, while continuous fryers allow processing of larger amounts of product with a higher production rate, with oil capacities ranging from 200 to 1000 kg and a throughput that varies from 250 to 25,000 kg product per hour (Moreira et al. 1999). Usually, this sort of equipment operates under atmospheric conditions, but recent innovations led to the development of fryers that can work under vacuum or high pressure conditions.

In batch fryers, the oil is placed in a large pan, where it is heated directly (for example by means of an electrical resistance) or direct gas flames, and when it reaches the frying temperature, the unfried product is added in specific amounts. When reaching the desired moisture content, the product is removed from the oil bath manually or using a takeout conveyor, to then be seasoned and packaged (Gupta 2005). Modern batch fryers are constructed using high-grade stainless steel, adequate thermostatic temperature controls, and auxiliary equipment for automatic frying (Kochhar 1998; Bouchon 2010). Normally in this type of unit operators immerse and remove the baskets manually from the oil; however, modern equipment may include an automatic basket lifter (Bouchon 2009). In order to produce a low-fat product, fried foods can be de-fatted using existing frying systems. De-fatting can be performed by injecting and forcing steam through the product to remove the excess of fat from the surface, or instead of adding steam, some units can be equipped with centrifuges to remove surface oil after frying. An example of such units is the one developed by Heat & Controls, Co. (United States), which includes a centrifugation step after removing the fried product from the oil bath in order to remove the surface oil and limit the amount of oil available for infiltration: this is able to reduce the final oil content of potato chips down to 22 %.

In continuous fryers, the oil can be heated either directly or indirectly in the frying vessel: the product is fed into the fryer at one end and the fried product is taken out from the opposite end by a takeout conveyor (Gupta 2005). Some of these continuous fryers are equipped with several heating zones to provide optimal temperature maintenance (Bouchon 2010). As the oil is absorbed by the fried product, the frying vessel has to be continuously topped up with fresh oil and an extractor has to be installed to eliminate the fumes, mainly consisting of moisture and a fine mist of fatty acids (Moreira et al. 1999; Dobraszczyk et al. 2006). Some of the main frying equipment manufacturers are Florigo B.V. in the Netherlands; along with Heat & Control, and Stein in the United States (Bouchon 2010).

As stated by Rossell (1998) the most important aspect of industrial frying is the frying oil and the factors that can affect its quality such as the interactions of the product to be fried with the oil; transport and storage of the oil; and frying equipment and conditions, among others. As explained by the author the quality of the frying oil affects significantly the quality of the fried product, which also outstand by Blumenthal (1991) and Kochhar (1998). This is because, inevitably, the elevated temperatures and the presence of oxygen reached in the fryer will promote chemical reactions in the frying oil (Dueik and Bouchon 2011a). In fact, in order to increase the shelf-life of the frying oil, there are some frying units that allow daily the periodic removal of any floating material and filtration of the oil, which if not removed can cause smoking, charring, darkening, and off-flavor development, affecting significantly the quality of the fried product (Kochhar 1998).

10.2.4 Oil Absorption Kinetics

Oil absorption is a complex phenomenon. It is not clearly understood when and how the oil penetrates into the structure; however, it has been shown that most of the oil is confined to the surface region of the fried product (Farkas et al. 1997; Saguy et al. 1997; Pedreschi et al. 1999; Bouchon et al. 2001), and it has been suggested that most of the oil is absorbed after removing the product from the oil bath during the cooling period (Bouchon et al. 2003; Ufheil and Escher 1996; Moreira et al. 1997). As a consequence, oil absorption is essentially a surface-related phenomenon, occurring mostly when the product is removed from the oil bath and begins to cool.

As stated previously, during frying, foods are exposed to high temperatures, so that the inner moisture is converted into steam, creating a positive pressure gradient. Water vapor escapes from the product through selective weaknesses, including cracks, defects, open capillaries, and channels in the cellular structure (Dana and Saguy 2006), so increasing surface porosity. Furthermore, some of this vapor may be trapped within the pores due to restrictive intercellular diffusion and expansion, becoming superheated, distorting the pore walls, and contributing to the porosity of the fried products (Ziaiifar et al. 2008). The overpressure inside the material is an obstacle to oil penetration inside the food structure, creating a barrier to oil absorption during frying. Then, when removing the product from the oil bath, surface oil can drain, becoming absorbed within the porous crust or remaining on the surface (Bouchon et al. 2003; Bouchon and Pyle 2005). In fact, Bouchon et al. (2003) showed that three different oil fractions can be identified as a consequence of the different absorption mechanisms in fried potato cylinders, that is: (1) Structural oil, which represents the oil absorbed during frying, (2) Penetrated surface oil, which represents the oil suctioned into the food during cooling after removal of the fryer, (3) and surface oil, which is the oil that remains on the surface. The authors determined that only a small amount of oil penetrated during frying as most of the oil was picked up at the end of the process, corresponding mainly to a surface phenomenon. Ufheil and Escher (1996) added a tracer dye to the frying oil at the end of frying and found that more than 80 % of the oil in potato chips was absorbed after the food was removed from the fat. Similarly, Moreira et al. (1997) found that the largest amount of oil penetrates into the structure of tortilla chips during the cooling period and not during frying. They observed that only 20 % of the total oil content is absorbed during frying and approximately 80 % remains on the product surface. In addition, they found that almost 64 % of the total oil content is absorbed during the cooling (post-frying) period. Thus, conditions after removal from the fat play a central role in fat uptake.

This phenomenon has been explained using several approaches, suggesting that when the product is removed from the oil, temperature drops and vapor inside the pores condensate and the overpressure will turn into underpressure. This pressure difference between the inner and outer surface of the pore creates a "vacuum effect", which results in surface oil penetration inside the pores left by water vaporization (Mellena 2003; Dana and Saguy 2006). Moreira and Barrufet (1998) outlined the mechanism of oil absorption during cooling solely in terms of capillary forces, as in a previous work Moreira et al. (1997) observed that most of the absorption took place when the temperature was still above the boiling point of water. Some studies explain that a critical element to initiate oil infiltration is the pressure barrier that needs to be overcome by the oil film, which is heavily reduced during cooling (Bouchon and Pyle 2005; Mellena 2003; Ziaiifar et al. 2008). This so-called condensation mechanism refers to the oil post-cooling suction due to water-vapor condensation, a mechanism that may well be mediated by capillary forces, as described by Bouchon and Pyle (2005).

Oil adherence and drainage also play an important role in oil absorption, given that they determine the available oil fraction that can be pulled into the pores. Rubnov and Saguy (1997) and Pedreschi et al. (2001) used fractal geometry analysis to study the surface topography of potato products, confirming the significant role of crust roughness in oil absorption. Using a similar approach, Moreno et al. (2010) examined the relationship between surface roughness and oil uptake in fried

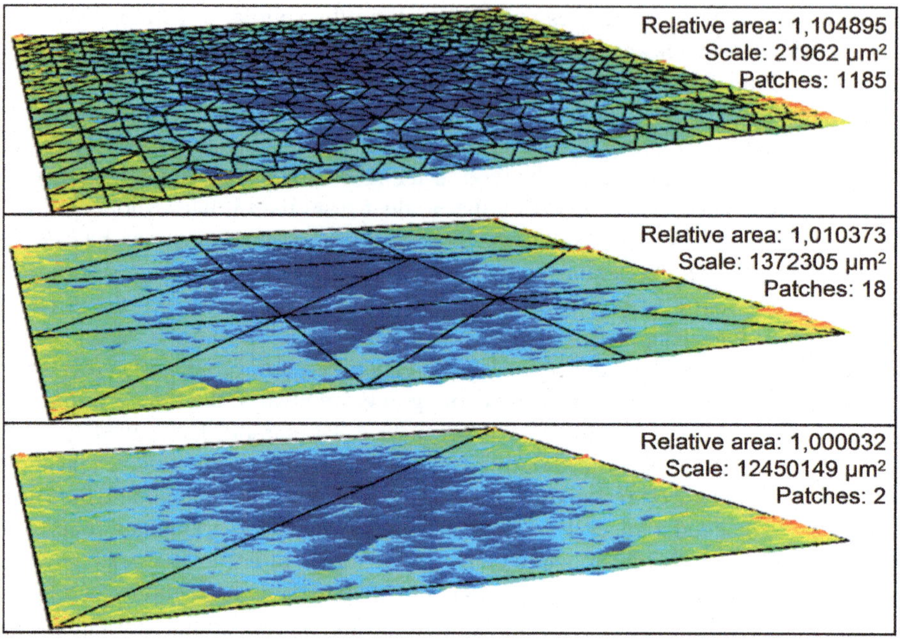

Fig. 10.3 Diagram showing a sequence of three tiling exercises on a gluten-based product measured at 10 μm sampling interval over a 5×5 mm surface region. From the bottom to the top, the size of the triangular patch, or scale, decreases progressively, and the relative area increases (Moreno et al. 2010)

formulated products, which were either based on potato flakes or wheat-gluten. The surface of fried products was measured using a scanning laser microscope and characterized by area-scale fractal analysis, as shown in Fig. 10.3. They determined that products with higher surface roughness absorbed more oil. However, this relationship was restricted to products of a similar nature (gluten or potato-flake-based product categories) and could not be extended when comparing different product categories, suggesting that other food-related properties may explain differences between product categories.

Oil adherence (and thereby oil absorption) to the surface of the product can also be influenced by frying oil degradation. The tendency of oil to adhere to the surface of the fried product and the thickness of the superficial oil layer are influenced by the presence of surface active agents. Based on this, Blumenthal (1991) developed the "surfactant theory of frying", that suggests that as oil degrades, surfactants are formed and that these act as wetting agents, reducing the interfacial tension between the food and the oil, and increasing the contact between them. This reduces the oil capability to drain from the product surface when it is removed from the frying oil. Also, as frying oil degrades, polymer formation increases its viscosity, thereby affecting its tendency to drain from the product surface. Hence, more oil is available to enter the product pores during the cooling period.

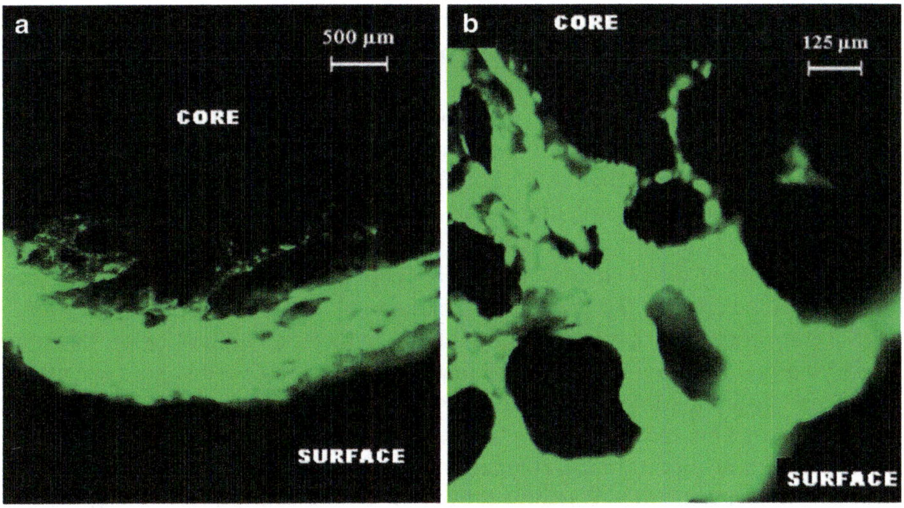

Fig. 10.4 CLSM photomicrographs of the cross Sect. of a potato cylinder fried at 185 °C for 5 min. Oil (in *green*) is located in a thin outer region or crust (**a**) and in fine pores or capillaries (**b**) (from Bouchon et al. 2003)

In relation to oil location, Keller et al. (1986) and Lamberg et al. (1990), when adding a heat-resistant and oil-soluble dye (Sudan Red B) to the frying medium, determined that oil uptake was restricted to a depth of a few cells. Farkas et al. (1992), used magnetic resonance imaging (MRI) to determine water location and oil penetration depth in immersion fried potato cylinders. They confirmed that oil was mainly located on the surface of the product and only slightly penetrated into the structure. This characteristic was also confirmed by Bouchon et al. (2001) using high spatial resolution infrared microspectroscopy to determine oil distribution profiles within fried potato cylinders. They observed that the triglyceride's ester-group stretch was confined to the outer region and that oil concentration ended sharply near the evaporation front. Figure 10.4 shows a confocal laser scanning microscopy (CLSM) photograph obtained by Bouchon et al. (2003) when analyzing oil location in potato cylinders fried at 185 °C for 5 min. They concluded that oil was mainly located in the crust region and that thermal degradation produced a porous network, which could later serve as a "sponge" for oil infiltration. Aguilera and Gloria (1997) observed using light microscopy and differential scanning calorimetry that the crust of frozen potato cylinders contained almost 6 times as much oil as the central core (23.6 % vs. 4 %, dry weight basis). In relation to potato chips, Pedreschi et al. (1999), using CLSM, showed that oil seems to surround intact dehydrated potato cells, without penetration, like an "egg-box". Observations made by Dueik et al. (2012a) using the same microscopy technique, corroborate these findings, where oil was located surrounding intact cells, which preserve their hexagonal shape.

Consumer trends are moving toward healthier and low-fat products, creating the need to reduce fat contents in industrial processing (Bouchon and Pyle 2005). A wide spectrum of factors affect oil uptake during frying. Some of these include temperature and time of frying, geometry and thickness of the sample (Krokida et al. 2001; Mehta and Swinburn 2001), product moisture content, pre-frying treatments, and frying oil quality and composition (Farinu and Baik 2005; Pinthus et al. 1993, 1995). The microstructure of the crust region has been considered to be the most important product-related determinant of the final oil uptake into the food (Saguy et al. 1997; Bouchon et al. 2001). Recently, Dueik et al. (2012a) demonstrated that product microstructure is one of the most important factors affecting oil absorption, even, more than moisture content. In fact, the authors observed that porosity and final oil content were linearly correlated, and that the size of those pores developed during the process is of paramount importance in oil absorption.

Several procedures have been proposed to reduce the amount of oil taken-up, pre-frying and post-frying treatments that attempt to achieve this goal are mainly based on two important research findings previously highlighted: (1) The marked effect that crust microstructure development has on oil absorption; (2) The fact that oil absorption results from the competition between drainage and suction into the porous crust once the food is removed from the oil bath and begins to cool.

Post-frying treatments, such as hot air-drying (Nonaka et al. 1977) and superheated steam drying (Kochhar 1999; Miranda and Aguilera 2006) have shown to reduce significantly oil absorption after frying, attempting to remove surface oil before post-cooling suction. These processes are based on surface oil removal before suction takes place and are able to reduce oil content in chips by 25 % (Kochhar 1999). Drainage of the superficial oil layer before oil absorption takes place can be critical in reducing oil absorption. Recently, Dueik et al. (2012a) reported oil absorption reductions of up to 70 % in atmospheric fried potato chips by including a post-frying centrifugation stage.

Among pretreatments, the reduction of surface permeability is a common practice, given the surface-related nature of oil absorption phenomenon. Partial moisture removal through drying (microwave, hot-air treatment, baking) prior to frying as shown to be an effective route to reduce oil uptake. It is important to understand that the effectiveness of such pretreatments is mainly due to the structural changes occurring at the surface of the food, which reduce surface permeability, and not really to a reduction of the moisture content on its own, as usually believed (Bouchon 2009). It is critical to select an adequate basis to carry out comparisons properly, when analyzing the effect of pretreatments. For instance, Moreno and Bouchon (2008) demonstrated that osmotic dehydrated samples showed a significant increase in the amount of oil absorbed after frying, contradicting previous findings reported in the literature. Actually, the decrease in oil absorption has shown to be real due to the increase in solids content occurring during the osmotic dehydration process rather than a reduction in the amount of oil taken-up (Moreno and Bouchon 2008). Several groups have studied the properties of different coatings to reduce oil migration (García et al. 2004; Albert and Mittal 2002; Balasubramaniam et al. 1997). Among different ingredients, cellulose and its derivates are of considerable interest due to their film-forming properties and unique thermal gelation ability (Balasubramaniam et al. 1997). These compounds can form a protective layer during the initial stages of frying due to thermally induced gelation above 60 °C, which inhibits the transfer of moisture and fat between the food and the frying medium (García et al. 2004). It must be remembered that frying is essentially a dehydration process and therefore, inhibition of water escape translates into a longer frying time, a fact that may well affect quality and nutritional attributes.

10.2.5 Effect of Processing Conditions on Food and Oil Quality

10.2.5.1 Composite Structure Development

As shown in Fig. 10.2, temperatures in the core and in the crust regions are quite different. Figure 10.5 presents the main changes occurring in both the food (crust and core) and the oil during frying, differentiating between changes that are derived from exposure to oxygen (left hand side) and those catalyzed by high temperatures (right hand side). Microstructural changes in the core are much milder and similar to those occurring during the cooking of potatoes (Bouchon and Aguilera 2001). In this inner section of the product (core), starch granules undergo gelatinization at around 60–70 °C, being swelled rapidly by intracellular water, occupying the whole interior of the cell. In a narrow temperature range (60–80 °C), the middle lamellae between cells disintegrates and cells separate giving the so-called mealy texture, and proteins may denature. The crust is the result of several alterations occurring in the outermost layers of the product, where the temperature exceeds 100 °C. These chemical and physical changes include: physical damage produced when the product is cut and a rough surface is formed with release of intracellular material; starch gelatinization and subsequent dehydration; protein denaturation; breakdown of the cellular adhesion; water evaporation and rapid dehydration of cells located in the forming crust, and oil uptake itself (Bouchon and Aguilera 2001). Since the first histological studies of deep-fat fried potatoes by Reeve and Neel (1960), using light microscopy, evidence has accumulated that except for the outermost layers damaged by cutting, the majority of the inner cells retain their individuality after frying and contain in their interior dehydrated but gelatinized starch granules. The microstructural aspect of the core tissue is similar to that of cooked potatoes (van Marle et al. 1992). In relation to the outer layers, cells seem to shrink during frying, with no extended rupture, while cell walls become wrinkled and convoluted around dehydrated gelled starch (Reeve and Neel 1960; Costa et al. 2000). It has been proposed that rapid dehydration reduces starch swelling, and therefore, cell walls do not break as sometimes occurs during ordinary cooking. Using hot-stage video microscopy, Bouchon and Aguilera (2001) assessed the geometrical changes in potato cells and starch granules during heating in oil, in real time, without observing major damage to the cell structure. Similar observations were conducted by Costa et al. (2000), when studying structural changes of potato during frying, and by McDonough et al. (1993) when evaluating the physical changes during deep-fat frying of tortilla chips. Aguilera and Gloria (1997) demonstrated using bright field

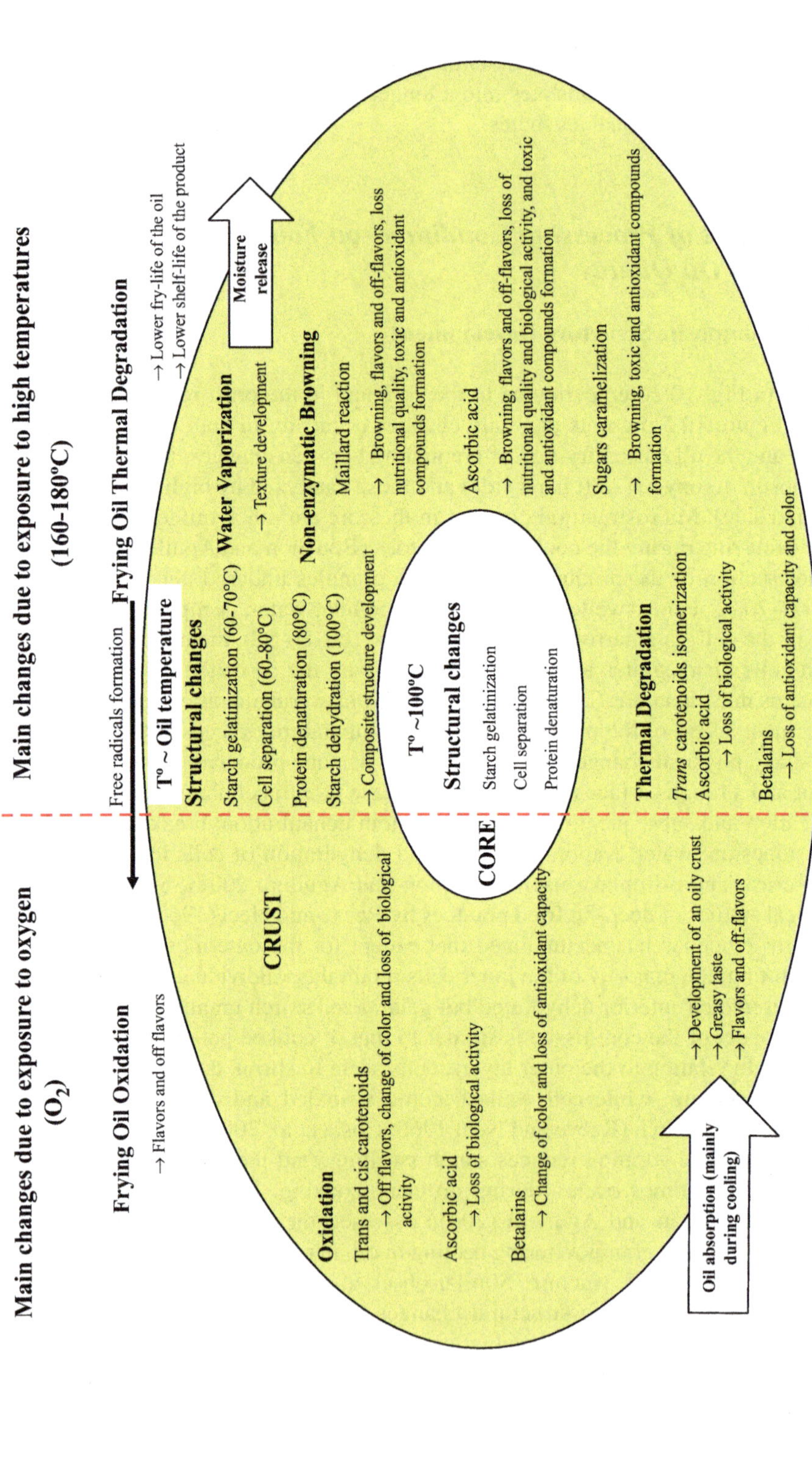

Fig. 10.5 Diagram showing the cross-section of a product during deep-fat frying, emphasizing the most important microstructural/quality changes occurring in the crust and the core regions, as well as in the oil, due to exposure to oxygen (*left hand side*) and high temperatures (*right hand side*). From Dueik and Bouchon (2011a)

microscopy that three distinct microstructures exist in frozen fried potatoes: (1) a thin outer layer (approx. 250 μm) formed by remnants of cell walls of broken or damaged cells by cutting; (2) an intermediate layer of shrunken intact cells which extends to the evaporation front, and; (3) the core with fully hydrated intact cells containing gelatinized starch.

10.2.5.2 Beneficial Compounds Degradation and Development of Toxic Compounds

Apart from the structural changes that occur during deep-fat frying, the components of the crust of the product can suffer a wide number of reactions that lead to the degradation of important nutritional compounds and, moreover, the formation of toxic compounds. Degradation of beneficial compounds depends on specific parameters during deep frying, such as temperature, oxygen, light, moisture content, pH, and obviously length of exposure (Lešková et al. 2006), among which the high temperatures and the presence of oxygen are the most detrimental factors for important nutritional compounds. Figure 10.5 shows the main reactions and their effect in the nutritional value and quality of the product that can occur when processing under atmospheric conditions. Examples in the Figure include important nutritional compounds, such as betalains, ascorbic acid, and carotenoids, which are particularly described in Figs. 10.6, 10.7, and 10.8. These compounds are degraded in the presence of atmospheric oxygen and heat, with consequent loss of the nutritive value of the fried product. Also, these beneficial compounds have an important role in the color of processed fruits and vegetables. Carotenoids are responsible for the orange color of carrot slices and undergo isomerization and oxidation during frying, causing loss of color, provitamin A activity, and off-flavors. Betalains impart the red violet color to red beets and their thermal degradation and oxidation produce loss of color and antioxidant properties. Furthermore, oxidized ascorbic acid (that is, dehydroascorbic acid), may participate in non-enzymatic browning reactions when exposed to elevated temperatures, yielding brown pigmented compounds, vitamin C loss, and the development of off-flavors.

10.2.5.3 Color and Acrylamide Formation

Maillard reaction, between superficial reducing sugars and amino acids, leads to the formation of brown pigments in potato chips and French fries (Márquez and Añon 1986). The surface color of the product changes gradually with temperature, to golden yellow and later to brown (Miranda and Aguilera 2006) as non-enzymatic browning reactions are a highly temperature-dependant group of reactions (Pedreschi et al. 2005). Certainly, these reactions are instrumental in the development of color, flavor, and aroma compounds in fried foods. However, despite its desired effects, the Maillard reaction is considered to be the principal mechanism in the generation of toxic compounds in fried food. In fact, Mottram et al. (2002) demonstrated that the

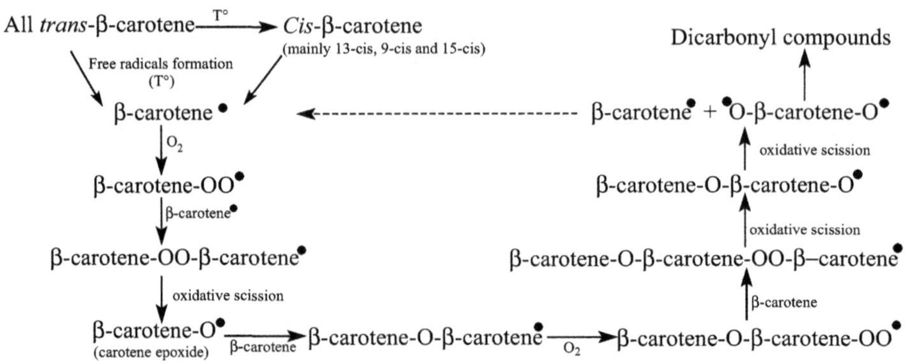

Fig. 10.6 Mechanism of all-trans-β-carotene degradation in the presence of air and elevated temperatures. Adapted from von Elbe and S.J. Schwartz (1996) and Choe and Min (2006)

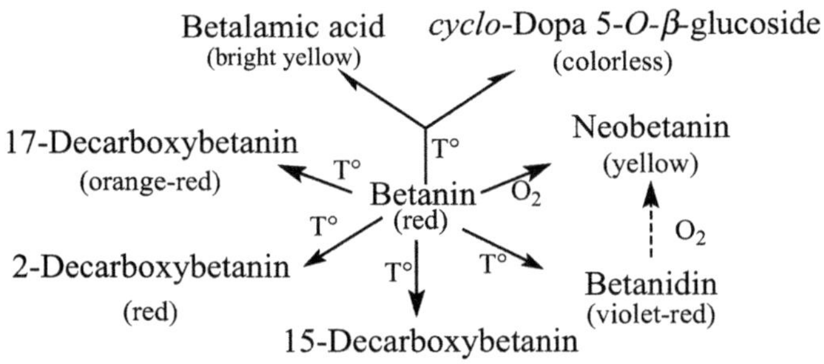

Fig. 10.7 Mechanism of betanin degradation in the presence of air and elevated temperatures. Adapted from Herbach et al. (2006)

reaction between amino acids (e.g., asparagine) and reducing sugars (e.g., fructose and glucose) at temperatures above 120 °C can be the main route for the formation of acrylamide (AA), a compound that has been classified as probably carcinogenic in humans. Non-enzymatic browning reactions may occur between reducing sugars and amino acids and between ascorbic acid, dehydroascorbic acid (that is, oxidized ascorbic acid) and other degradation products from ascorbic-acid oxidation, which enter into Maillard-type browning reactions. Furthermore, the dehydroascorbic-acid lactone ring can be irreversibly opened, giving rise to diketogulonic acid, which can be further degraded into several color compounds (e.g. furfural polymers) (Rojas and Gerschenson 1997; Belitz et al. 2004). Several studies have shown that increases in frying time and temperature during processing lead to higher concentrations of acrylamide (Gökmen et al. 2006).

In living organisms, acrylamide can be converted into its epoxide, glycidamide (GA), which is thought to be even more toxic than AA (Besaratinia and Pfeifer 2004; Koyama et al. 2006). In a recent study, Granvogl et al. (2008) proposed that lipid hydroperoxides can promote the conversion of AA into GA during the manu-

Fig. 10.8 Mechanism of ascorbic acid degradation in the presence of air and elevated temperatures. Adapted from Villamiel et al. (2006)

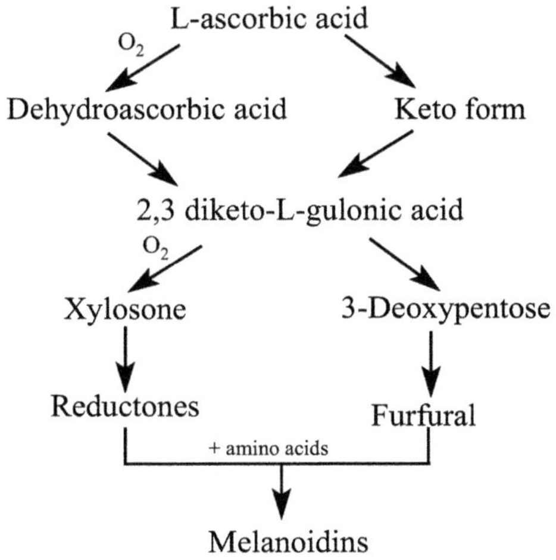

facture of potato chips and French fries. They prepared French fries by immersing potatoes either in coconut oil (mainly constituted of saturated fatty acids), or in sunflower oil (with a high percentage of unsaturated triglycerides), and found that French fries prepared using coconut oil had ~90 % less GA content than those fried in sunflower oil. Their results suggest that oils with a healthy fatty-acid profile might not be the best option for frying starchy products.

10.2.5.4 Frying Oil Degradation

The most important criteria used to select frying oils are long frying stability, fluidity, bland flavor, low tendency to foam or smoke formation, low tendency to gum (polymerize), the oxidative stability of the oil in the fried food during storage, good flavor stability of the product, and certainly price (Kochhar 1998; Rossell 1998). There is no ideal frying oil or fat that satisfies each and every frying application, as there are factors such as nutrition, cost, availability, and functionality that must be taken into account (Brinkmann 2000). Oils with a high content of saturated fatty acids have great stability in frying applications; however, these oils are undesirable from a nutritional point of view (Sanibal and Mancini-Filho 2004). On the other hand, oils high in polyunsaturated fatty acids show lower thermo-oxidative stability than rich monoenoic unsaturated fatty acids or saturated fatty-acid oils (Kita et al. 2005), although the latter are currently not often used due to health concerns. Commonly used fats are palm oil, tallow, and hydrogenated fats: although these are low-cost frying mediums and have a long frying life, they are also high in saturated and *trans* fatty acids (Mehta and Swinburn 2001). Today, the most popular oils used for frying are palm oil and its fractions, sunflower oil (especially high oleic sunflower oil), rapeseed (canola), and soybean oils. The last two are often partially hydrogenated to increase their thermo-oxidative stability (Kita et al. 2005).

Frying oil also suffers the effects of aggressive atmospheric conditions. During frying, oil undergoes thermal, oxidative, and hydrolytic degradation because of its exposure to elevated temperatures in the presence of air and moisture (Shyu et al. 1998; Kita et al. 2005). Oil degradation due to the aggressive atmospheric conditions is also shown in Fig. 10.2. Together with oil-breakdown and degradation, the leaching of food materials changes the oil from a medium that is almost pure triglyceride to a mixture of hundreds of compounds (Blumenthal 1991). Water released from the foodstuff hydrolyzes ester linkages of triacylglycerols, giving rise to di- and mono-acylglycerols, glycerols, and free-fatty acids; these free molecules are more susceptible to oxidative and thermal degradation than when esterified to the glycerol molecule (Choe and Min 2007; Mahungu et al. 1999). The oxidative degradation of free-fatty acids and triacylglycerols in vegetable oils is enhanced at frying temperature (185 °C). This reaction takes place by loss of hydrogen in the presence of trace metals, heat, and light (Mahungu et al. 1999). This alkyl radical can react easily with triplet oxygen producing a peroxy radical, which can then subtract hydrogen from another fatty acid, giving rise to a hydroperoxide and a further alkyl radical. Hydroperoxides are not stable under deep-fat frying conditions and may undergo fission to produce a wide variety of secondary lipid peroxidation products, including aldehydes, ketones, and other carbonyl-containing compounds. These compounds contribute to the volatile fraction of the degraded frying oil and thus determine the development of off-flavors in the fried product (Melton et al. 1994; Subramanian et al. 2000) some of them producing acrid, fishy, fruity, and plastic/waxy-flavored degraded products (Warner 2009).

Also, nonvolatile polar compounds as well as triacylglycerol dimers and polymers are major frying oil decomposition products. Hydroperoxides may remain in the triglyceride molecule as peroxy radicals, which can react with other radicals giving rise to olygomers by a combination of C–C, ether, and peroxide linkages (Choe and Min 2007). These products produce darkening, increase oil viscosity, and decrease the smoke point of the frying oil (Mahungu et al. 1999).

Moreover, apart from the effect of oil degradation in the sensory properties of frying oils, secondary lipid peroxidation products, such as HNE (4-hydroxy-2-*trans*-nonenal), derived from the oxidation of linoleic acid, have aroused particular interest as they have been shown to contain cytotoxic and mutagenic properties (Seppanen and Csallany 2004). Because fried foods are widely consumed, the extent of formation of toxic lipid degradation products in oils at frying temperature is important with regard to public health (Seppanen and Csallany 2006).

10.3 Vacuum Frying of Foods

10.3.1 History

Urban development and lack of time are the main responsible for modern dietary habits, making ready-to-eat foods, the mainstays of our diet. However, many of these foods are full of nitrates, preservatives, *trans* and saturated fats, and are also heavily processed, losing their nutritional quality or even worse, developing toxic compounds. This modern lifestyle is a challenge for the food industry, as many modern diseases such as obesity, diabetes, and cardiovascular diseases are linked to dietary lifestyles. As a consequence, consumers are looking for healthy snacks, but are unwilling to sacrifice full-flavor products and the unique characteristics of fried food (Mariscal and Bouchon 2008).

In the market we can find some ready-to-eat, low-fat fried snack products, but these products have not been as successful as was expected, as they are still unable to impart some of the main characteristic attributes of fried products, such as how they feel in the mouth, flavor, appearance, and texture. One alternative to develop low-fat healthy snacks is vacuum frying. The application of vacuum in food processing is a widely used technology to dehydrate heat-sensitive products. Products that decompose or undergo changes in structure, texture, appearance, and/or flavor as a consequence of high temperature can be dried under vacuum with minimum damage (Vega-Mercado et al. 2001). Vacuum application can lower the boiling point of moisture contained in raw foods thus leading to reduced drying temperature in an oxygen-deficient environment. Under these processing conditions it is possible to improve the quality and nutritive value of dried products. As regards these technologies, the ones that have been most studied are vacuum drying, freeze drying, microwave vacuum drying, and vacuum impregnation of fruits and vegetables.

Vacuum frying consists of a deep-fat frying process carried out in a closed system at an operating pressure well below atmospheric levels, which makes it possible to remove product moisture in a low-oxygen environment and at a low temperature, due to boiling-point depression (Garayo and Moreira 2002). In normal batch vacuum frying operation (see Fig. 10.9), the product is placed inside the frying basket once the oil reaches the target temperature, the lid is then closed and the chamber depressurized. Subsequently, the basket is immersed in the oil bath, where it remains for the required amount of time. It is then lifted and the vessel is pressurized. Oil absorption mainly occurs during this last stage, and the mechanism involved is still unclear (Dueik and Bouchon 2011a; Garayo and Moreira 2002). Most of the benefits of this type of frying technology are achieved by the lower oxygen exposure and low temperatures. They include natural color, flavor, and nutrient preservation (Dueik and Bouchon 2011b; Da Silva and Moreira 2008; Shyu and Hwang 2001), as well as oil quality protection (Shyu et al. 1998) and reduction in the generation of toxic compounds (Granda et al. 2004).

Today, vacuum frying technology is being used to maintain natural colors, flavors, and nutrients in high added-value products, such as vegetables and fruits (Dueik and Bouchon 2011b). There are some Chinese (Sunwell Shandong Green Food Co., Ltd; Xuzhou Kanata Food Co., Ltd.; Laiyang Hengrum Foodstuff Co., Ltd.), Japanese, and Thai companies that offer a wide range of healthy low-fat ready-to-eat vacuum-fried snacks. The acceptance of such foods into the market has been growing, as these snack products offer an intense flavor and color compared to regularly fried products, satisfying present nutritional trends for food that is low-fat and rich in antioxidants.

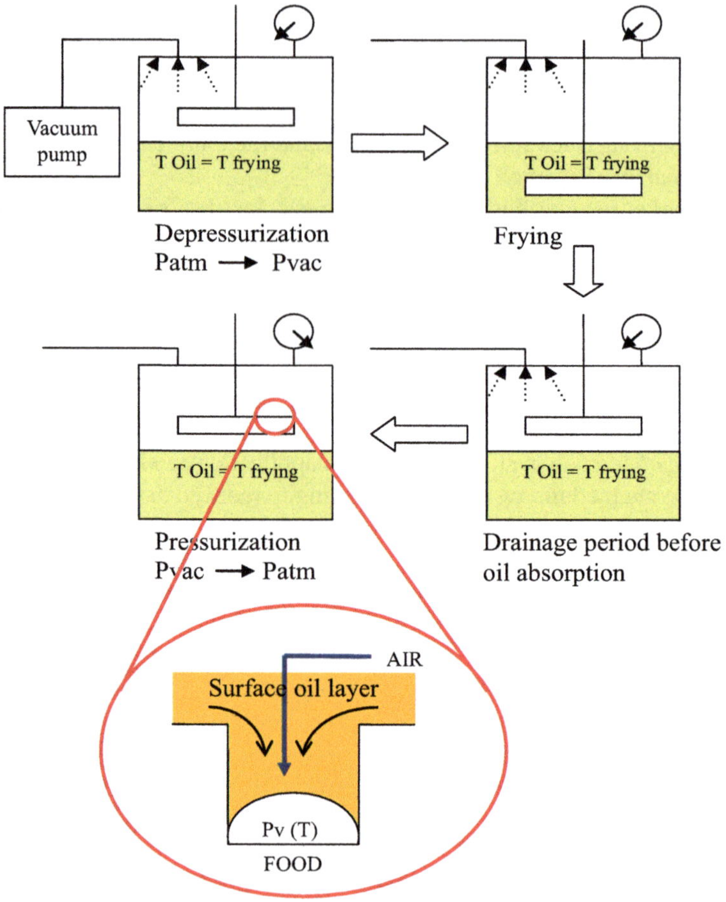

Fig. 10.9 Schematic representation of a normal batch vacuum frying operation

10.3.2 Industrial Equipment

In spite of the advantages mentioned regarding vacuum frying, it has not been widely used due to its high capital investment. The first vacuum frying equipment was developed by Florigo B.V. during the sixties to produce high-quality chips; however, due to the improvement in blanching technology and the quality of raw materials, the use of this technology became almost obsolete (Moreira et al. 1999). The recent advances in processing machines, such as the development of continuous industrial-size vacuum fryers (H&H Industry Systems B.V., The Netherlands; Qinhuangdao Tonghai Science & Technology Development Corporation, China) would make these products available at lower prices. Also, exports of continuous vacuum frying equipment have been growing considerably, as in 2004 to the European Union and South-East Asia, and in 2007 to the USA and India.

In South-East Asia (mainly the Philippines, Thailand, China, and Indonesia) batch-type vacuum fryers are mainly used for the relatively small production of fruit chips. For larger production quantities, continuously working vacuum fryers are available. In these installations, the vacuum is created by high vacuum pumps and the frying pan is installed in a stainless steel vacuum tube, where the in-feed of the raw product is carried out through a rotary air lock feeder and a conveyor belt takes the product along the vacuum tube. These units are equipped with centrifuges located in a vacuum chamber attached to the vacuum tube, which remove surface oil before it enters the product. Currently, there are a few other companies that have developed a working continuous vacuum fryer with a de-oiling system (Qinhuangdao Tonghai Science & Technology Development Corporation, China; Shandong Light M&E Co., China; patent given in 2002) or without it, but with a capacity of up to 1000 kg/h (Nissei Engineering Co., Ltd., Japan). Various patents, which include supplementary devices such as oil centrifuging systems, seasoning and packaging units, and oil recycling systems, will allow future differentiation (e.g. Van Der Doe 2001).

10.3.3 Vacuum and Atmospheric Frying Comparison

In order to adequately compare vacuum and atmospheric frying, Mariscal and Bouchon (2008) introduced the term "equivalent thermal driving force", which is the difference between the oil temperature and the boiling point of water at working pressure. This approach was designed assuming that dehydration during the frying process is mainly governed by heat transfer. However, we have observed a significant increase in the frying time (43 % higher) to reach bubble-end point, even when using equivalent thermal driving forces (Dueik et al. 2010). This can be a consequence of the lower diffusivity of liquid water and vapor at lower temperatures, the higher latent heat of vaporization of water in vacuum conditions and/or the higher specific volume of water vapor under vacuum conditions that might affect its escape within the product structure (Perry et al. 1999).

10.3.4 Oil Absorption During Vacuum Frying

In relation to oil absorption, the mechanism of its occurrence during vacuum frying is still not clear. Observations from Garayo and Moreira (2002) indicate that, as in atmospheric frying, post-frying cooling may play an important role, but in addition, the pressurization step should be critical. As explained earlier, in a normal vacuum-frying operation, the product is removed from the oil bath and the vessel is vented before cooling takes place. This translates into a sudden increase in the surrounding pressure at a constant temperature, which may force the vapor inside the pore to condense, decreasing P_{pore} and therefore initiating oil absorption before cooling begins (that is, $P_{surroundings} - P_{pore} > 0$). The authors explain that, because of the low pressure, air may diffuse much faster into the porous space, thus obstructing oil passage within the structure, before oil absorption while cooling occurs. This competition may drive a reduction in oil uptake of vacuum-fried snacks compared to those that are atmospheric fried. Accordingly, it was determined that the oil content of vacuum-fried potato chips was significantly lower compared to atmospheric fried potato chips, with a reduction of 44 %. Fan et al. (2005) when frying carrot slices under vacuum conditions found that oil content appears to be related to moisture content as explained by Gamble and Rice (1987) for atmospheric frying. This strong relationship between moisture loss and oil absorption was also observed by Mariscal and Bouchon (2008) when frying apple slices under vacuum conditions. In their study, when working with an equivalent thermal driving force of 60 °C, either at atmospheric conditions or under vacuum, they determined that vacuum-fried apple slices absorbed less oil compared to atmospheric fried slices. The authors attributed this behavior to the lower vapor-pressure of water during vacuum frying and to the higher temperatures reached during atmospheric frying, which induce added structural changes such as tissue/constituents degradation, leading to favorable conditions for oil infiltration. Similarly, Da Silva and Moreira (2008) observed that vacuum-fried sweet-potato slices absorbed about 24 % less oil than the atmospheric fried ones, but mango slices behaved differently, without showing any divergence in the amount of oil taken-up in both frying technologies. Dueik et al. (2010) found that vacuum-fried carrot chips absorbed about 50 % less oil when compared with their atmospheric counterparts. Similar results were obtained by Dueik and Bouchon (2011b) when frying carrot and potato slices; however when frying apple slices the reduction of oil content was only about 25 % and the final oil content was still extremely high (63.3 % w.b). Potatoes and carrots absorbed similar amounts of oil but their initial moisture content was rather different. Similarly, the initial moisture content of carrots and apples was similar, but, apples absorbed much more oil, concluding that there was no relationship between initial moisture content and final oil content, when analyzing products of different characteristics. This highlighted the importance of microstructure for oil absorption. Certainly, findings in other works, which relate initial moisture content to oil absorption (water replacement mechanisms), cannot be merely extrapolated when analyzing vegetable structures of a different nature, and other factors must be taken into account. Potatoes are a starchy and low-porous material, carrots are a non-starchy and low-porous material while apples are a non-starchy and porous material; therefore, their microstructure might play a key role. Based on the behavior of different food matrices, Dueik et al. (2012a) proposed an approach highlighting the importance of the different microstructures developed in vacuum and atmospheric fried products. In their work, they fried potatoes, apples, and carrots under vacuum and atmospheric conditions and determined key microstructural parameters (cumulative pore volume and pore size distribution) using gas adsorption at cryogenic temperatures. Their results showed that there was a strong linear relationship between porosity and final oil content in vacuum ($R^2=0.99$) and atmospheric ($R^2=0.99$) fried chips. However, this relationship could not be extended when analyzing the whole set of data, as oil absorption was radically different in atmospheric and vacuum-fried chips and was attributed to the development of pores of smaller diameter under atmospheric conditions supporting the role of capillary forces in the oil absorption phenomenon. The authors explained that the development of larger pores in vacuum-fried products can be a consequence of the higher specific volume of water vapor at low pressure.

Another important issue to be considered when talking about oil absorption during vacuum frying is the pressurization conditions. As determined by Dueik et al. (2012a) and Mir-Bel et al. (2009) drainage before pressurizing the vessel may significantly reduce the amount of oil absorbed. In fact, the first authors stated that a drainage period of 9 min may reduce the final oil content of vacuum-fried potato slices by 15 %. Other interesting efforts made in order to reduce the oil content of vacuum-fried products are from Moreira et al. (2009) that introduce a centrifugal stage (750 rpm, 40 s) before pressurizing the vessel. The system led them to reduce the total oil content of vacuum-fried potato chips by 77 %. A similar attempt was undertaken by Dueik et al. (2012a) when introducing a centrifugation post-frying step (400 rpm, 2 min) for decreasing the oil content in vacuum-fried potatoes and carrots (65 % less for both products) and apples (24 % less oil). The authors attributed the lower drainage capacity of vacuum-fried apples to their rougher surface, confirmed by area-scale fractal analysis of the products surface.

In relation to oil location, Dueik et al. (2012a) studied oil location using Nile Red- stained frying oil and CLSM in vacuum-fried potatoes, carrots, and apples. Figure 10.10 shows on its left hand the elevation map and on the right hand the oil location at different depths, where black zones mean absence of oil. In the Figure it can be observed that vacuum-fried potatoes appeared to preserve the hexagonal shape of potato cells, which look as if they are surrounded by oil, as observed for atmospheric fried potatoes. However, a difference between vacuum and atmospheric fried products was that the former revealed the presence of ungelatinized starch granules. Starch swelling and gelatinization requires the presence of water and temperature. It is believed that the depressurization step prior to frying may induce fast dehydration, limiting the amount of water required to initiate starch gelatinization.

10.3.5 Effect of Vacuum Frying Processing Conditions on the Quality of Food and Frying Oil

10.3.5.1 Composite Structure Development

Structural changes occurring during atmospheric frying are also desirable in vacuum-fried products, in order to obtain a similar product but with the described benefits of vacuum frying in relation to lower oil absorption, higher preservation of important nutritional compounds and color, and lower formation of toxic compounds.

As described earlier, to develop the desired composite structure, we need to achieve a minimum temperature, not only for moisture vaporization at the operating pressure, but also for starch gelatinization (60–70 °C) and cell separation (60–80 °C) to develop the so-called mealy texture. Subsequent moisture vaporization, at

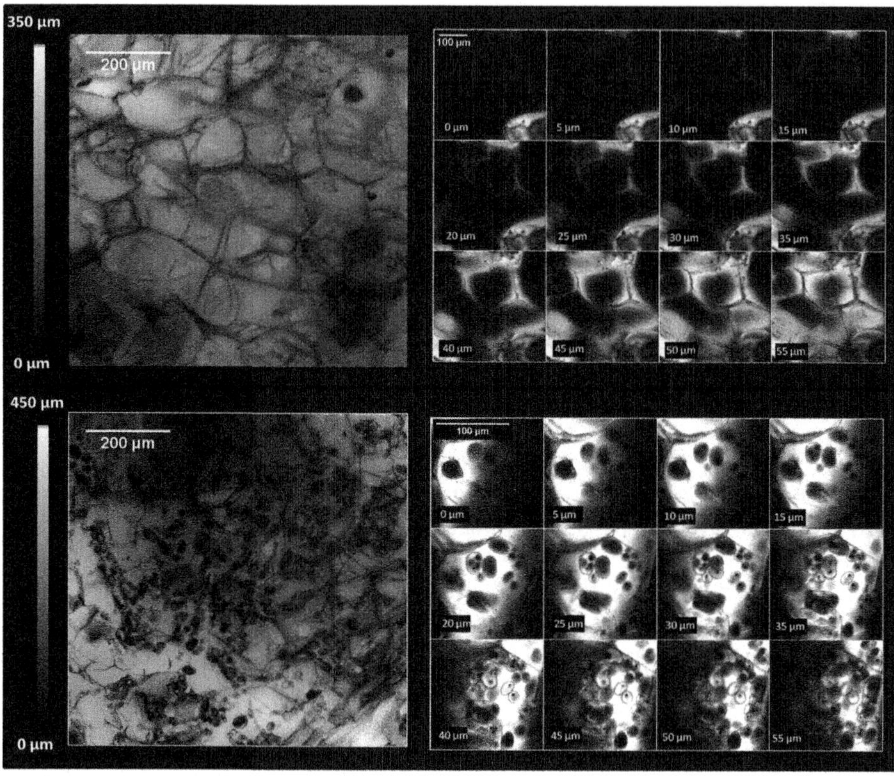

Fig. 10.10 Elevation map (*left*) and image gallery (*right*) of oil location in atmospheric (*up*) and vacuum-fried (*down*) potato slices using confocal microscopy. From Dueik et al. (2012a)

a temperature according to the operating pressure, led to surface starch dehydration, porosity development, oil uptake, and composite structure formation (Bouchon and Aguilera 2001). Gelatinization of starch, a process that is affected by adequate water and temperature, is a critical transformation that determines structure formation during frying of starchy products. Accordingly, the reactions that led to structural changes during frying are mainly dependent on the processing temperature, product composition, and moisture availability. For instance, Fukuoka et al. (2002) studied the effect of lower moisture content of the system on the gelatinization temperature of wheat starch, concluding that when decreasing the moisture content, a higher gelatinization temperature was achieved. Thus, the degree of gelatinization of a starchy product will depend on the temperature and water available in the system (Primo-Martin et al. 2007). In fact, in a work developed by Dueik et al. (2012a) the authors were able to observe the presence of ungelatinized starch in vacuum-fried potatoes, mainly due to the insufficient conditions of water and temperature.

Most studies on vacuum frying focus on agricultural products, such as potatoes, carrots, and apples, among others, but none of these studies have dealt with fabricated/formulated products. The advantages of such fabricated products, as pointed out by Gebhardt (1996) include reproducibility, uniformity, and lack of defects when compared with heterogeneous materials such as raw potato, which may instill important variations among products. So, a combination of use of controlled structures through product formulation and vacuum frying might be a choice for the desired expectations. In a work developed by Sobukola et al. (2012) the authors highlighted the importance of the boiling-point temperature of water (T_{bp}) and water availability for starch gelatinization in wheat-gluten and starch matrices fried under different vacuum levels. Their results showed that a higher T_{bp} favored the capacity of the matrix to form a continuous network, due to starch gelatinization in the presence of water and temperature. However, if the T_{bp} is not enough (for starch gelatinization), the crust region will have the required temperature, but will be dried out. On the other hand, the inner region will have enough liquid water, but its temperature will be restricted to the boiling point of water, and therefore, will be lower than the gelatinization temperature. Under these circumstances, most of the starch may not be gelatinized due to insufficient conditions, resulting in an unbound weaker structure, with a higher permeability to oil absorption. This study confirms the importance of food building blocks (in this case gluten and starch) and their changes during processing in the quality of fried snacks.

10.3.5.2 Important Nutritional Compounds and Natural Color Preservation

Vacuum deep-fat frying can be a promising technology to produce conventional and non-conventional healthy snacks from fruits and vegetables, preserving the favorable characteristics of raw products with the unique characteristics of fried products (Shyu and Hwang 2001; Da Silva and Moreira 2008; Mariscal and Bouchon 2008; Dueik and Bouchon 2011b). Vegetables, like carrots, apples, and beetroot slices are excellent raw materials to be exploited due to their high content of beneficial compounds. Healthy compounds present in the raw material, such as carotenoids in carrots, ascorbic acid in apples and potatoes, and betalains in red beets, may function as free radical scavengers, preventing human chronic diseases (Yen et al. 2002; Herbach et al. 2006); however, they are extremely sensitive to severe atmospheric frying conditions, undergoing oxidation and thermal degradation. Vacuum frying conditions are adequate for preserving these compounds, although there are only a few studies that deal with the preservation of beneficial compounds during the process.

The importance of carotenoids in food goes beyond their role as natural pigments, as biological functions have been increasingly attributed to these compounds. Carrots are the most important source of dietary carotenoids, however, release of carotenoids from foods occurs only when cells in the food matrix are disrupted, as is usually the case during food processing or mastication. Dietary lipids have been considered important cofactors in the bioavailability of carotenoids (Parada and Aguilera 2007). Also during food processing losses can occur due to the presence of oxygen and high temperatures as illustrated in Fig. 10.6, where all-trans-β-carotene may undergo isomerization at high temperatures; being its *cis* isomer less absorbed compared to its *trans*-isomers by humans (Schieber and Carle 2005). Also, both, *trans*- and *cis*-isomers of β-carotene can suffer subsequent epoxidation (at the end of the chain) and oxidative scission (mainly along the chain) resulting in fragmentations that can yield volatile compounds with a low molecular weight, similar to those produced during fatty-acids oxidation, thus, producing rancid odors and off-flavors; apart from loss of the attractive orange color. Kanasawud and Crouzet (1990) identified some of the main volatile compounds generated during the thermal degradation of β-carotene, concluding that it follows mainly an oxidative pathway as the exclusion of oxygen from the system reduced the formation of volatile compounds by around 90 %. Accordingly, vacuum frying of carrot slices can be an excellent alternative to produce healthy carrot snacks, as suggested by Dueik et al. (2010), in which work they used equivalent thermal driving forces during vacuum and atmospheric frying to compare all *trans*- α and β-carotene retention; concluding that minor degradation of *trans* α- and *trans* β-carotene occurs during vacuum frying (90 % retention), while in atmospheric fried carrots, retention was just about 30 %. Da Silva and Moreira (2008) studied retention of carotenoids during vacuum and atmospheric frying of several vegetables. They found that final total carotenoids (mg/g de-fatted dry solids) were 18, 19, and 51 % higher in vacuum-fried green beans, mango chips, and sweet-potato chips, respectively, compared with atmospheric fried products.

Natural sources of betalains are found in red beets and prickly pears, so giving them their characteristic color. The major betalain in red beet is betanin, containing a phenolic and a cyclic amine group, which are very good electron donors, acting as antioxidants (Kanner et al. 2001). Betalains are heat and oxygen-labile compounds (Fig. 10.7), but Huang and Von Elbe (1986) stated that the removal of oxygen greatly increases their stability towards heat. Thermal degradation of betanin produces red, yellow, and colorless compounds, affecting the antioxidant capacity and natural color. There are no studies available in the literature of the preservation of these compounds during vacuum frying, although there seems to be the potential technology to obtain red beet-based snack products with high antioxidant capacity and natural colors.

Some vegetable products, such as apples and potatoes, are excellent sources of ascorbic acid, which offers a wide spectrum of health benefits, such as protection from atherosclerosis and coronary disease and some types of cancer (Cameron et al. 1979). Ascorbic acid is highly sensitive to various modes of processing, being easily oxidized in a reversible reaction to dehydroascorbic acid, which is poorly active (10 % bioequivalence) and very unstable. Besides being reduced back to ascorbic acid, dehydroascorbic acid can also be rapidly hydrolyzed to products such as 2,3-diketogulonic acid, which breaks down to produce reductones and furfural. Ascorbic acid can also yield 2,3-diketogulonic acid anaerobically when exposed to high temperatures. These compounds can react with aminoacids to produce brown melanoidins (Choe and Min 2006), producing quality losses, and which have an important role in flavor, color, and nutritional quality (Rojas and Gerschenson 1997;

Mariscal and Bouchon 2008). Dueik and Bouchon (2011b) compared ascorbic acid retention in vacuum and atmospheric fried potato and apple slices, finding that retention was about 95 % in potato and apple slices when vacuum-fried at 98 °C; however, when atmospherically fried, retention was only about 50 % when frying at 160 °C. Interestingly, their results showed that the antioxidant capacity of chips may be related to both the presence of natural antioxidants and brown pigments developed at elevated temperatures, since the antioxidant capacity was significantly higher for atmospheric-fried samples.

10.3.5.3 Color and Acrylamide Formation

The aforementioned compounds are responsible for the color of fruits and vegetables, either directly (carotenoids and betalains) or because of the browning reactions in which they are involved (ascorbic acid). Color is a critical attribute, which may be heavily modified by aggressive deep-fat frying conditions; an effect that can be overcome using vacuum frying technology. Figure 10.11 shows digital pictures of vacuum (top row; 98 °C and 1.9 in. Hg) and atmospheric (bottom row; 160 °C) fried apple, red beet, carrot, and potato slices, using a thermal driving force of 60 °C (From Dueik and Bouchon 2011a). Dueik and Bouchon (2011b) reported that vacuum-fried potato chips (98 °C; 6.5 kPa) had higher L^* values, corresponding to lighter products than the atmospheric fried chips (160 °C), meanwhile, a^* values

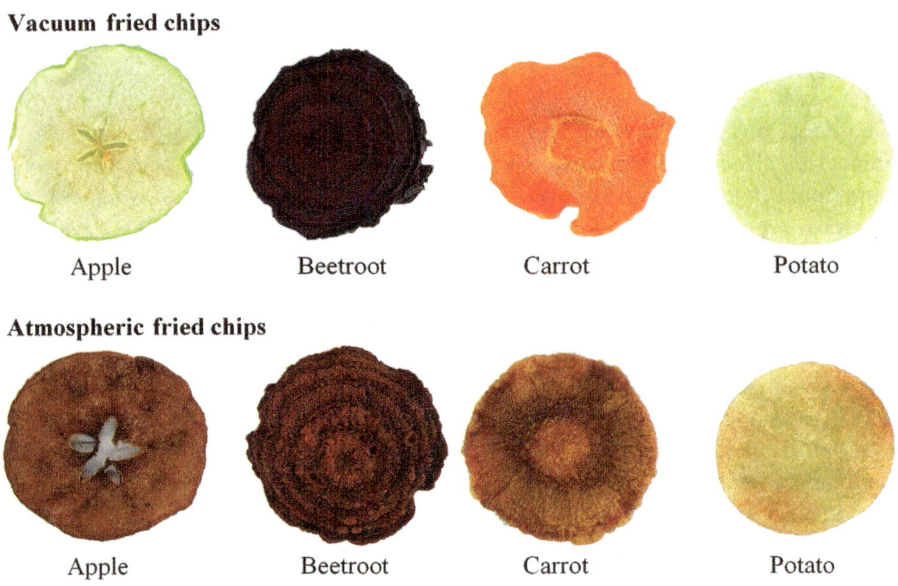

Fig. 10.11 Digital images of vacuum (*top row*; 98 °C and 1.9 in. Hg) and atmospheric (*bottom row*; 160 °C) fried apple, red beet, carrot, and potato slices, using a thermal driving force of 60 °C (From Dueik and Bouchon 2011a)

were significantly higher for atmospheric fried chips. The authors also observed that during atmospheric frying (160 °C) of potato, apple, and carrot slices, the formation of brown melanoidins (measured as the absorbance at 420 nm) was significantly higher than during vacuum frying (98 °C; 6.5 kPa), indicating the occurrence of non-enzymatic browning reactions.

Non-enzymatic browning reactions are associated with the formation of brown melanoidins, which can be desirable, as in the case of coffee and bread crusts, or undesirable, as in the case of potato chips that can suffer excessive browning due to the high temperatures involved and characteristics of the raw material (elevated amounts of reducing sugars). Vacuum frying can be an excellent processing alternative for the utilization of potato varieties that do not satisfy the requirement of low-reducing sugar content. Granda et al. (2004) studied the color development and acrylamide formation during vacuum and atmospheric frying using different potato varieties. The authors observed that in potato chips of different varieties fried under vacuum conditions (118 °C, 1.3 kPa, 8 min), acrylamide content was always significantly lower compared to atmospheric frying (165 °C, 4 min), with reductions ranging from between 85 and 99 %. Among vacuum-fried potatoes, the variety that presented the highest concentration of acrylamide was White-rose, which is unsuitable for frying, however it showed similar amounts of the carcinogenic compound as atmospheric fried Atlantic variety (recommended for frying). In relation to color, their study revealed that for all varieties *L*, *a* and *b* values were lower in the case of vacuum frying, concluding that vacuum-fried products had a low content of acrylamide and desirable color attributes.

10.3.5.4 Sensory Properties of Vacuum-Fried Products

The effect of main reactions and physical changes was discussed in terms of the impact on the quality of fried snacks. However, for consumers acceptance is of paramount importance as regards the sensory evaluation carried out when comparing vacuum and atmospheric fried chips. Da Silva and Moreira (2008) evaluated vacuum and atmospheric fried sweet potato, mango and blue potato slices, and green beans using a 30-member consumer panel and a 1–9 hedonic scale (9 = like it very much, 1 = dislike it very much, 5 = acceptable). The evaluated attributes were color, odor, texture, flavor, and overall quality, finding that the panel overwhelmingly preferred the vacuum-fried products for their color, texture, taste and overall quality. On a recent study developed by our research team (Dueik et al. 2012b), a flash profile test was applied to study major sensory differences between vacuum and atmospheric fried carrot chips. The trained panel decided that the discriminating attributes that allow for the differentiation of the samples were natural color, roasted aroma, carrot taste, bitter taste, and bitter aftertaste. Atmospheric fried samples showed a significant higher intensity in bitter taste, bitter aftertaste, and roasted aroma, and the lowest intensity in natural color and taste. These attributes were directly attributed to the exposure to high temperatures, which promotes the occurrence of non-enzymatic browning reactions, resulting in the formation of compounds responsible for flavors and off-flavors (Warner et al. 1996).

10.3.5.5 Frying Oil Degradation

As explained previously, frying oil degradation involves radical species for its propagation (formed due to the exposure to elevated temperatures). Thus, what if we can reduce their initial formation by means of substantially reducing the frying temperature, as in the case of vacuum frying? These free radicals can react with triplet oxygen to produce hydroperoxides and other carbonyl compounds and polymers with oxygenated functions. However, as oxygen is limited during vacuum frying, these free radicals can react only with other radical species present in the oil, so that the degradation products will be mainly polymers without oxygenated functions, but in much lower concentrations than in atmospheric frying.

Based on this, Shyu et al. (1998) studied the stability of soybean, palm oil, and lard during the vacuum frying of carrot slices, finding a significantly slower degradation rate for all the aforementioned frying mediums during vacuum frying, as determined by the lower kinetic constants of formation of hydroperoxides, carbonyl and polar compounds and viscosity increase, than observed in atmospheric frying. Aladedunye and Przybylski (2009) evaluated the performance of vacuum frying as an alternative to reduce the degradation of canola oil when frying French fries. The evaluated parameters were total polar compounds (TPC), anisidine value (AV), and tocopherols degradation. They observed a reduction of TPC formation of 76 % when frying under vacuum conditions instead of atmospheric ones, while the anisidine value (measurement of secondary oxidation products) was reduced in 93 %. Additionally, tocopherols degradation during vacuum frying was 12 times lower than in atmospheric frying. Our results suggest that vacuum frying is an excellent alternative to maintain frying oil quality, given that after the vacuum frying of 50 batches of potato chips, the TPC concentration was just 12.6 % of the total oil components. This represents an economical benefit as the TPC value was far lower than 25 %, which has been suggested as the limit beyond which a restaurant should discard its frying oil (Blumenthal 1991). In a normal frying operation this critical value is obtained at a much earlier stage (Moreira et al. 1997).

In terms of quality attributes, this will extend frying oil and fried product shelf-life, reduce off-flavors and rancid odor and inhibit the formation of toxic compounds such as HNE and GA (derived from the reaction of acrylamide with hydroperoxides).

The increasing concern about chronic diseases related to the consumption of saturated and trans fatty acids, creates the aim to select adequate frying oils, with a different fatty-acids profile than the ones used until now. At present, frying oils must have less than 3 % linolenic acid because it is readily degraded (Mehta and Swinburn 2001), however, processing conditions during vacuum frying leads to use these beneficial oils without affecting its quality.

10.3.6 New Trends

Manufacturers and ingredient companies are working together to help build the next generation of savory snacks. A next generation of processed foods will be designed for the brain–body axis. As such, they will be enjoyable to eat, nutritionally efficient, and will make us feel better. These new generations of snacks fall into the categories of low-fat, baked unfried, high fiber, made from rice or wheat bran, among others (Sajilata and Singhal 2005). Snack manufacturers are reformulating their product ranges in order to improve the nutritional profile, but this can be taken only so far before the most important characteristics of snacks—taste and texture—are seriously compromised. The race is on to develop new products that consumers will find satisfying, but that will also comfortably form part of a healthy, balanced diet.

Health-conscious consumers have already recognized the nutritional benefits of whole grain and multi-grain products in other sectors. Their use in snack products presents a clear market opportunity as manufacturers look for ways of responding to the need for products that are both more interesting and healthier.

Starches and starch derivatives have a long history of use in snack foods, especially as functional ingredients to help snacks achieve various textural attributes. Specialty starches have the potential for tremendous processing, textural, and feel-in-the-mouth advantages in snack development. With dozens of food starches available, it can be difficult to know which one will best meet a particular functional challenge (Huang 1995). Resistant starch availability opens new opportunities for snack manufacturers to develop high-quality, fiber-fortified snack products. In snack applications, resistant starch improves fiber-fortified product quality, increases expansion of fiber-fortified snacks, reduces oil pick-up in fried snacks, provides light, crispy texture, and can be used alone as a fiber source or as a functional complement to other types of fiber.

References

Aguilera, J.M., and H. Gloria. 1997. Determination of oil in fried potato products by differential scanning calorimetry. *Journal of Agricultural and Food Chemistry* 45: 781–785.

Aladedunye, F., and R. Przybylski. 2009. Protecting oil during frying: A comparative study. *European Journal of Lipid Science and Technology* 111: 893–901.

Albert, S., and G.S. Mittal. 2002. Comparative evaluation of edible coatings to reduce fat uptake in a deep-fried cereal product. *Food Research International* 35: 445–458.

Antoñanzas, F., and R. Rodríguez-Ibeas. 2011. Healthy vs. unhealthy food: A strategic choice for firms and consumers. *Health Economics Review* 1: 4–12.

Balasubramaniam, V., M. Chinnan, P. Mallikarjunan, and R. Phillips. 1997. The effect of edible film on oil uptake and moisture retention of a deep—Fat fried poultry product. *Journal of Food Process Engineering* 20: 17–29.

Belitz, H.D., W. Grosch, and P. Schieberle. 2004. *Food chemistry*. Berlin, Germany: Springer.

Besaratinia, A., and G. Pfeifer. 2004. Genotoxicity of acrylamide and glycidamide. *Journal of National Cancer Institute* 96: 1023–1029.

Blumenthal, M.A. 1991. New look at the chemistry and physics of deep-fat frying. *Food Technology* 45: 68–71.

Bouchon, P. 2009. Understanding oil absorption during deep-fat frying. In *Advances in food and nutrition research*, ed. S. Taylor. Amsterdam, NY: Academic.

Bouchon, P. 2010. Food deep-fat frying. In *Food processing handbook*, ed. J. Brennan and A. Grandison. Weinheim, Germany: Wiley-VCH.

Bouchon, P., and J.M. Aguilera. 2001. Microstructural analysis of frying potatoes. *International Journal of Food Science and Technology* 36: 669–676.

Bouchon, P., and D.L. Pyle. 2005. Modelling oil absorption during post-frying cooling I: Model development. *Food Bioproducts Processing* 83: 253–260.

Bouchon, P., P. Hollins, M. Pearson, D.L. Pyle, and M.J. Tobin. 2001. Oil distribution in fried potatoes monitored by infrared microspectroscopy. *Journal of Food Science* 66: 918–923.

Bouchon, P., J.M. Aguilera, and D.L. Pyle. 2003. Structure oil-absorption relationships during deep-fat frying. *Journal of Food Science* 68: 2711–2716.

Brinkmann, B. 2000. Quality criteria of industrial frying oils and fats. *European Journal of Lipid Science and Technology* 102: 539–541.

Cameron, E., L. Pauling, and B. Leibovitz. 1979. Ascorbic acid and cancer: A review. *Cancer Research* 39: 663–681.

Castells, M., S. Marin, V. Sanchis, and A.J. Ramos. 2005. Fate of mycotoxins in cereals during extrusion cooking: A review. *Food Additives and Contamination* 22: 150–157.

Choe, E., and D. Min. 2006. Chemistry and reactions of reactive oxygen species in food. *Critical Reviews in Food Science and Nutrition* 46: 1–22.

———. 2007. Chemistry of deep-fat frying oils. *Journal of Food Science* 72: R77–R86.

Clark, J.P. 2009. Snacks and baking. In *Case studies in food engineering: Learning from experience*, ed. J.P. Clark. New York: Springer.

Costa, R.M., F.A.R. Oliveira, O. Delaney, and V. Gekas. 1999. Analysis of the heat transfer coefficient during potato frying. *Journal of Food Engineering* 39: 293–299.

Costa, R.M., F.A.R. Oliveira, and G. Boutcheva. 2000. Structural changes and shrinkage of potato during frying. *International Journal of Food Science and Technology* 36: 11–24.

Da Silva, P., and R. Moreira. 2008. Vacuum frying of high-quality fruit and vegetable-based snacks. *Lebensmittel-Wissenschaft und Technologie* 41: 1758–1767.

Dana, D., and S. Saguy. 2006. Review: Mechanism of oil uptake during deep-fat frying and the surfactant effect-theory and myth. *Advances in Colloid and Interface Science* 128–130: 267–272.

Dobraszczyk, B.J., P. Ainsworth, S. Ibanoglu, and P. Bouchon. 2006. Baking, extrusion and frying. In *Food processing handbook*, ed. J.G. Brennan, 237–290. Weinheim, Germany: Wiley-VCH.

Dueik, V., and P. Bouchon. Understanding of moisture loss during vacuum frying of potato slices. Unpublished data.

———. Development of healthy low-fat snacks: Understanding the mechanisms of quality changes during atmospheric and vacuum frying. *Food Reviews International* 27: 408–432.

Dueik, V., and P. Bouchon. 2011. Vacuum frying as a route to produce novel snacks with desired quality attributes according to new health trends. *Journal of Food Science* 76: E188–E195.

Dueik, V., P. Robert, and P. Bouchon. 2010. Vacuum frying reduces oil uptake and improves the quality parameters of carrot crisps. *Food Chemistry* 119: 1143–1149.

Dueik, V., M.C. Moreno, and P. Bouchon. 2012a. Microstructural approach to understand oil absorption during vacuum and atmospheric frying. *Journal of Food Engineering* 111: 528–536.

Dueik, V., C.F. Marzullo, and P. Bouchon. 2012b. Effect of vacuum drying and vacuum frying on quality and sensory attributes of carrot snacks. *LWT--Food Science and Technology* 50(1): 361–365.

Fan, L., M. Zhang, G. Xiao, J. Sun, and Q. Tao. 2005. The optimization of vacuum frying to dehydrate carrot chips. *International Journal of Food Science and Technology* 40: 911–919.

Farinu, A., and O. Baik. 2005. Deep fat frying of foods-transport phenomena. *Food Reviews International* 21: 389–410.

Farkas, B.E. 1994. Modeling immersion frying as a moving boundary problem. Ph.D. dissertation. University of California, Davis.

Farkas, B.E., R.P. Singh, and M.J. McCarthy. 1992. Measurement of oil/water interface in foods during frying. In *Advances in food engineering*, ed. R.P. Singh and A. Wirakartakusumah, 237–245. Boca Raton, FL: CRC Press.

Farkas, B., R.P. Singh, and T. Rumsey. 1996. Modeling heat and mass transfer in immersion frying. I, model development. *Journal of Food Engineering* 29: 211–226.

Farkas, B.E., R.P. Singh, and M.J. McCarthy. 1997. Measurement of oil/water palmitic acid. *Journal of Agricultural and Food Chemistry* 45: 4286–4289.

Fellows, P. 2000. Extrusion. *Tecnología del procesado de los alimentos: Principios y práctica,* 355–372. Zaragoza: Ed. Acribia

Fukuoka, M., K. Ohta, and H. Watanabe. 2002. Determination of the terminal extent of starch gelatinization in a limited water system by DSC. *Journal of Food Engineering* 53(1): 39–42.

Gamble, M.H., and P. Rice. 1987. Effect of pre-fry drying on oil uptake and distribution in potato crisp manufacture. *Journal of Food Science and Technology* 22: 535–548.

Garayo, J., and R. Moreira. 2002. Vacuum frying of potato chips. *Journal of Food Engineering* 55: 181–191.

García, M.A., C. Ferrero, A. Campana, N. Bértola, M. Martino, and N. Zaritzky. 2004. Methylcellulose coatings applied to reduce oil uptake in fried products. *Food Science and Technology International* 10(5): 339–346.

Gebhardt, B. 1996. Oils and fats in snack foods. In *Bailey's industrial oil and fat products*, vol. 3, 5th ed, ed. Y.H. Hui. New York: Wiley.

Gökmen, V., T.K. Palazoglu, and H. Senyuva. 2006. Relation between the acrylamide formation and time-temperature history of surface and core regions of French fries. *Journal of Food Engineering* 77: 972–976.

Granda, C., R. Moreira, and S. Tichy. 2004. Reduction of acrylamide formation in potato chips by low-temperature vacuum frying. *Journal of Food Science* 69(8): E405–E411.

Granvogl, M., P. Koehler, L. Latzer, and P. Schieberle. 2008. Development of a stable isotope dilution assay for the quantification of glycidamide and its application to foods and model systems. *Journal of Agricultural and Food Chemistry* 56: 6087–6092.

Gupta, M.K. 2005. Frying of foods and snack food production. In *Bailey's industrial oil and fat products*, ed. F. Shahidi. New York: Wiley.

Herbach, K., F. Stintzing, and R. Carle. 2006. Betalain stability and degradation—Structural and chromatic aspects. *Journal of Food Science* 71: R41–R50.

Huang, D.P. 1995. New perspectives on starch and starch derivatives for snack applications. *Cereal Foods World* 40(8): 528–531. August, 1-5.

Huang, A.S., and J.H. Von Elbe. 1986. Stability comparison of two betacyanine pigments—Amaranthine and betanine. *Journal of Food Science* 51(3): 670–674.

Hubbard, L., and B. Farkas. 1999. A method for determining the convective heat transfer coefficient during immersion frying. *Journal of Food Process Engineering* 22: 201–214.

Kanasawud, P., and J. Crouzet. 1990. Mechanism of formation of volatile compounds by thermal degradation of carotenoids in aqueous medium. 1. β-carotene degradation. *Journal of Agricultural and Food Chemistry* 38: 237–243.

Kanner, J., S. Harel, and R. Granit. 2001. Betalains—A new class of dietary cationized antioxidants. *Journal of Agricultural and Food Chemistry* 49: 5178–5185.

Keller, C.H., F. Escher, and J. Solms. 1986. A method for localising fat distribution in deep-fat fried potato products. *Lebensmittel-Wissenschaft und Technologie* 19: 346–348.

Kita, A., G. Lisinska, and M. Powolny. 2005. The influence of frying medium degradation on fat uptake and texture of French fries. *Journal of the Science of Food and Agriculture* 85: 1113–1118.

Kochhar, S.P. 1998. Security in industrial frying processes. *Grasas y Aceites* 49: 282–302.

———. 1999. Safety and reliability during frying operations-effects of detrimental components and fryer design features. In *Frying of food*, ed. D. Boskou and I. Elmadfa, 253–269. Lancaster, PA: Technomic.

Koyama, N., H. Sakamoto, M. Sakuraba, T. Koizumi, Y. Takashima, M. Hayashi, H. Matsufuji, K. Yamagata, S. Masuda, N. Kinae, and M. Honma. 2006. Genotoxicity of acrylamide and glycidamide in human lymphoblastoid TK6 cells. *Mutation Research* 603: 151–158.

Krokida, M.K., V. Oreopoulou, Z.B. Maroulis, and D. Marinos-Kouris. 2001. Effect of osmotic dehydration pretreatment on quality of French fries. *Journal of Food Engineering* 49: 339–345.

Kubow, S. 1992. Routes of formation and toxic consequences of lipid oxidation products in foods. *Free Radical Biology and Medicine* 12: 63–81.

Lamberg, I., B. Hallstrom, and H. Olsson. 1990. Fat uptake in a potato drying/frying process. *Lebensmittel-Wissenschaft und Technologie* 23: 295–300.

Lešková, E., J. Kubíková, E. Kováčiková, M. Košická, J. Porubská, and K. Holčíková. 2006. Vitamin losses: Retention during heat treatment and continual changes expressed by mathematical models. *Journal of Food Composition and Analysis* 19: 252–276.

Mahungu, S.M., W. Artz, and E.G. Perkins. 1999. Oxidation products and metabolic processes. In *Frying of food*, ed. D. Boskou and I. Elmadfa, 25–45. Lancaster, PA: Technomic.

Mariscal, M., and P. Bouchon. 2008. Comparison between atmospheric and vacuum frying of apple slices. *Food Chemistry* 107: 1561–1569.

Márquez, G., and M.C. Añon. 1986. Influence of reducing sugars and aminoacids in the color development of fried potatoes. *Journal of Food Science* 51(1): 157–160.

McDonough, C., M.H. Gomez, J.K. Lee, R. Waniska, and L. Rooney. 1993. Environmental scanning electron microscopy evaluation of tortilla chip microstructure during deep-fat frying. *Journal of Food Science* 58: 199–203.

Mehta, U., and B. Swinburn. 2001. A review of factors affecting fat absorption in hot chips. *Critical Reviews in Food Science and Nutrition* 41(2): 133–154.

Mellena, M. 2003. Mechanism and reduction of fat uptake in deep-fat fried foods. *Trends in Food Science and Technology* 14: 364–373.

Melton, S.L., S. Jafar, D. Sykes, and M.K. Trigiano. 1994. Review of stability measurements for frying oils and fried food flavor. *Journal of the American Oils Chemical Society* 71(12): 1301–1308.

Mintel. 2011. Salty snacks: Chips, pretzels, snack nuts and seeds—US—March 2011. http://oxygen.mintel.com/sinatra/oxygen/display/id=393427/display/id=543039

Miranda, M., and J.M. Aguilera. 2006. Structure and texture properties of fried potato products. *Food Reviews International* 22: 173–201.

Mir-Bel, J., R. Oria, and M.L. Salvador. 2009. Influence of the vacuum break conditions on oil uptake during potato. *Journal of Food Engineering* 95: 416–422.

Monteiro, C. 2009. Nutrition and health. The issue is not food, nor nutrients, so much as processing. *Public Health Nutrition* 12: 729–731.

Moreira, R., and M.A. Barrufet. 1998. New approach to describe oil absorption in fried products: A simulation study. *Journal of Food Engineering* 35: 1–22.

Moreira, R., X. Sun, and Y. Chen. 1997. Factors affecting oil uptake in tortilla chips in deep-fat frying. *Journal of Food Engineering* 31: 485–498.

Moreira, R., M.E. Castell-Perez, and M.A. Barrufet. 1999. *Deep-fat frying: Fundamentals and applications*. Gaithersburg, MD: Aspen.

Moreira, R., P. Da Silva, and C. Gomes. 2009. The effect of a de-oiling mechanism on the production of high quality vacuum fried potato chips. *Journal of Food Engineering* 92: 297–304.

Moreno, M.C., and P. Bouchon. 2008. A different perspective to study the effect of freeze, Air, and osmotic drying on Oil absorption during potato frying. *Journal of Food Science* 73: E122–E128.

Moreno, M.C., C.A. Brown, and P. Bouchon. 2010. Effect of food surface roughness on oil uptake of deep-fat fried food. *Journal of Food Engineering* 101: 179–186.

Morton, I.D. 1998. Geography and history of the frying process. *Grasas y Aceites* 49: 247–249.

Mottram, D.S., B.L. Wedzicha, and A.T. Dodson. 2002. Acrylamide is formed in the Maillard reaction. *Nature* 419: 448–449.

Ni, H., and A.K. Datta. 1999. Moisture, oil and energy transport during deep-fat frying of food materials. *Trans IChemE, Part C (Food and Bioproducts Processing)* 77: 194–204.

Nonaka, M., R.N. Sayre, and M.L. Weaver. 1977. Oil content of French fries as affected by blanch temperatures, fry temperatures and melting point of frying oils. *American Potato Journal* 54: 151–159.

Parada, J., and J.M. Aguilera. 2007. Food microstructure affects the bioavailability of several nutrients. *Journal of Food Science* 72: R21–R32.

Pedreschi, F., J.M. Aguilera, and J.J. Arbildua. 1999. CSLM study of oil location in fried potato slices. *Microscopy and Analysis* 37: 21–22.

Pedreschi, F., J.M. Aguilera, and D.L. Pyle. 2001. Textural characterization and kinetics of potato strips during frying. *Journal of Food Science* 66: 314–318.

Pedreschi, F., P. Moyano, K. Kaack, and K. Granby. 2005. Color changes and acrylamide formation in fried potato slices. *Food Research International* 38: 1–9.

Perry, R.H., D.W. Green, and J.O. Maloney. 1999. *Perry's chemical engineers handbook*, 7th ed. New York: McGraw-Hill Handbooks.

Pinthus, E.J., P. Weinberg, and I.S. Saguy. 1993. Criterion for oil uptake during deep-fat frying. *Journal of Food Science* 58(204–205): 222.

———. 1995. Oil uptake in deep fat frying as affected by porosity. *Journal of Food Science* 60(4): 767–769.

Primo-Martin, C., N.H. van Nieuwenhuijzen, R.J. Hamer, and T. van Vliet. 2007. Crystallinity changes in wheat starch during the bread-making process: Starch crystallinity in the bread crust. *Journal of Cereal Science* 45(2): 219–226.

Reeve, R., and E. Neel. 1960. Microscopic structure of potato chips. *American Potato Journal* 37: 45–52.

Rojas, A., and L. Gerschenson. 1997. Influence of system composition on ascorbic acid destruction at processing temperatures. *Journal of the Science of Food and Agriculture* 74: 369–378.

Rossell, J.B. 1998. Industrial frying process. *Grasas y Aceites* 49: 282–295.

———.2001. Frying improving quality. Chapter 1

Rubnov, M., and I.S. Saguy. 1997. Fractal analysis and crust water diffusivity of a restructured potato product during deep-fat frying. *Journal of Food Science* 62: 135–137.

Saguy, S., and D. Dana. 2003. Integrated approach to deep fat frying: Engineering, nutrition, health and consumer aspects. *Journal of Food Engineering* 56: 143–152.

Saguy, I.S., E. Gremaud, H. Gloria, and R.J. Turesky. 1997. Distributions and quantification of oil uptake in French fries utilizing palmitic acid. *Journal of Agricultural and Food Chemistry* 45: 4286–4289.

Sajilata, M.G., and R.S. Singhal. 2005. Specialty starches for snack foods. *Carbohydrate Polymers* 59: 131–151.

Sanibal, E.A.A., and J. Mancini-Filho. 2004. Frying oil and fat quality measured by chemical, physical, and test kit analyses. *Journal of the American Oils Chemical Society* 81(9): 847–852.

Schieber, A., and R. Carle. 2005. Occurrence of *cis*-isomers in food: Technological, analytical, and nutritional implications. *Trends in Food Science & Technology* 16: 416–422.

Seppanen, C.M., and A. Csallany. 2004. Incorporation of the toxic aldehyde 4-hydroxy-2-trans-nonenal into food fried in thermally oxidized soybean oil. *Journal of the American Oils Chemical Society* 81: 1137–1141.

Seppanen, C.M., and A. Csallany. 2006. The effect of intermittent and continuous heating of soybean Oil at frying temperature on the formation of 4-hydroxy-2-trans-nonenal and other α-, β-unsaturated hydroxyaldehydes. *Journal of the American Oils Chemical Society* 83: 121–127.

Shyu, S.L., and L.S. Hwang. 2001. Effects of processing conditions on the quality of vacuum fried apple chips. *Food Research International* 34: 133–142.

Shyu, S., L. Hau, and L. Hwang. 1998. Effect of vacuum frying on the oxidative stability of oils. *Journal of the American Oils Chemical Society* 75: 1393–1398.

Singh, R.P. 1995. Heat and mass transfer in foods during deep-fat frying. *Food Technology* 49: 134–137.

Sleet, W. 2011. *Scaleless dieting: The essential survival kit for the overweight, obese and diabetics*. Bloomington, IN: AuthorHouse.

Sobukola, O., V. Dueik, and P. Bouchon. 2012. Understanding the effect of vacuum level in oil absorption and microstructure development of vacuum fried restructured wheat starch and gluten based snacks. *Food and Bioprocess Technology* 6(8): 2010–2017.

Stier, R.F. 2004. Frying as a science—An introduction. *European Journal of Lipid Science and Technology* 106: 715–721.

Subramanian, R., K.E. Nandini, P.M. Sheila, A.G. Gopalakrishna, K.S.M.S. Raghavarao, M. Nakajima, T. Kimura, and T. Maekawa. 2000. Membrane processing of used frying oils. *Journal of the American Oils Chemical Society* 77(3): 323–328.

Townsend, G.M. 1990. Cookies, crackers, and other flour confectionery. In *Snack food*, ed. R.G. Booth. New York: Springer.

Ufheil, G., and F. Escher. 1996. Dynamics of oil uptake during deep-fat frying of potato slices. *Lebensmittel-Wissenschaft und Technologie* 52: 640–644.

USDA. (2010). Fresh and processed potatoes, http://www.ers.usda.gov/briefing/potatoes/background.htm (accessed December 21)

Van Der Doe, M.H. 2001. Device and method for frying products. U.S. 6929812 (Cl. 426/438) 16 August 2005. Appl. 10/276299, 23 May 2001.

van Marle, J.T., A.C.M. Clerkx, and A. Boekstein. 1992. Cryo-scanning electron microscopy investigation of the texture of cooked potatoes. *American Potato Journal* 11: 209–216.

Vega-Mercado, H., M. Góngora-Nieto, and G. Barbosa-Cánovas. 2001. Advances in dehydration of foods. *Journal of Food Engineering* 49: 271–289.

Villamiel, M., M.D. Del Castillo, and N. Corzo. 2006. Browning Reactions. In *Food Biochemistry and Food Processing*, ed. Y.H. Hui, W. Nip, L. Nollet, G. Paliyath, and B. Simpson, 1st ed., 71–102. Iowa: Blackwell Publish.

von Elbe, J.H., and S.J. Schwartz. 1996. In *Colorants in food chemistry*, eds. Owen R., and Fennema E., 685–691. New York: Marcel Dekker.

Warner, K. 2009. Flavor changes during frying. In *Advances in deep-fat frying of foods*, ed. S. Sahin and S.G. Sumnu, 201–213. Boca Raton, FL: Contemporary Food Engineering Series, CRC Press.

Warner, K., P.S. Dimick, G.R. Ziegler, R.O. Mumma, and R. Hollender. 1996. 'Flavor-fade' and off-flavors in ground roasted peanuts as related to selected pyrazines and aldehydes. *Journal of Food Science* 61: 469–472.

World Health Organization (WHO). 2003. Diet, nutrition and the prevention of chronic diseases: Report of a joint WHO/FAO expert consultation. WHO technical report series 916. WHO: Geneva.

———. 2000. Obesity: Preventing and managing the global epidemic. Technical report series 894. WHO: Geneva.

Yen, G.C., P.D. Duh, and H.L. Tsai. 2002. Antioxidant and pro-oxidant properties of ascorbic acid and garlic acid. *Food Chemistry* 79: 307–313.

Ziaiifar, A.M., N. Achir, F. Courtois, I. Trezzani, and G. Trystram. 2008. Review of mechanisms, conditions, and factors involved in the oil uptake phenomenon during the deep-fat frying process. *International Journal of Food Science and Technology* 43: 1410–1423.

Chapter 11
Power Ultrasound Treatment of Fruits and Fruit Products

Hyoungill Lee, Bin Zhou, and Hao Feng

11.1 Introduction

Ultrasound refers to mechanical waves with frequencies in the range beyond human hearing (20–100 MHz). Ultrasound can be categorized into high-frequency ultrasound and power ultrasound (or high-intensity ultrasound) depending on the frequency and ultrasound intensity. The former has found applications in medical imaging (sonography), nondestructive inspection, and food composition determination, among others (McClements 1995; Coupland 2004). Power ultrasound has higher ultrasound intensity and acoustic power levels operating at a lower frequency (20–100 kHz) (Feng and Yang 2011). Many of the fruit-processing applications are conducted by submerging the fruits or fruit products in a fluid, either liquid or air utilizing power ultrasound to perform the operation. Therefore, ultrasound-assisted fruit processing is often a surface treatment for whole fruits, or a treatment with fruit juices, or pieces of fruits for the purpose of microbial and enzyme inactivation, extraction of certain components, or removal of moisture (dehydration). A relatively new application is to use ultrasound treatment to improve the postharvest quality of fruits. In this chapter, the focus will be placed on the applications of power ultrasound in fruit and fruit product treatments. Readers who are interested in the applications related to the use of high-frequency ultrasound in fruit processing can refer to the review articles of McClements (1995) and Coupland (2004).

H. Lee • B. Zhou • H. Feng (✉)
University of Illinois at Urbana—Champaign, Champaign, IL, USA
e-mail: haofeng@illinois.edu

A. Rosenthal et al. (eds.), *Fruit Preservation*, Food Engineering Series,
https://doi.org/10.1007/978-1-4939-3311-2_11

11.2 Principles of Power Ultrasound Treatment

11.2.1 Ultrasound Generation

The generation of ultrasound deals with two issues. The first is the generation of mechanical vibration at a right frequency and the second is to couple the vibration to a selected medium to form a propagating sound wave (Leighton 1994). The generation of oscillation is achieved by a transducer which transforms one form of energy into another. The transducers can be driven by gas, liquid, and electromechanical means (Mason and Lorimer 2002). The most widely used is piezoelectric transducers that utilize the reverse piezoelectric effect to convert electrical energy into mechanical wave energy at high efficiency. The frequency of the piezoelectric is determined by its geometry and its amplitude is the largest when it is driven near to resonance. The coupling of the vibration into a medium is a critical issue, especially for airborne ultrasound. Right impedance matching is required to minimize the losses and reduce the chance of damaging the transducer. The propagation of the sound wave in a medium can be in the form of longitudinal waves or transverse waves. The transverse waves refer to the waves where the displacement of the media particles is perpendicular to the direction of the wave propagation, while in a longitudinal wave, the particles are displaced parallel to the direction of the wave propagation. A special phenomenon associated with the longitudinal wave is that the particles in the medium will form rarefaction and compression regions. The longitudinal waves are often observed in a liquid, and if the "pulling-apart" force among the liquid molecules in the rarefaction regions is greater than the tensile stress limit between the molecules, cavities are formed in the liquid. The formation, growth, and collapse of the cavities are termed as cavitation. Cavitation plays a critical role in many power ultrasound applications.

11.2.2 Ultrasound Treatment Apparatus

Ultrasound-assisted fruit and fruit-product processing is usually performed in an ultrasonic cleaning bath, or with an ultrasonic probe (or horn) system when the operation is in a liquid, or with a specially designed airborne ultrasound unit. Cleaning baths or tanks are widely used in surface cleaning or other surface treatments. In a cleaning bath, a number of transducers are mounted at the bottom metallic wall of the bath (Fig. 11.1a) and the vibration of the transducer is transmitted into the tank via the metallic wall. Because of that, the volumetric acoustic power level is not high in a cleaning bath. In addition, the sound waves will form a standing wave pattern in the tank and thus the spatial ultrasound field distribution is not uniform. Probe systems (Fig. 11.1b) are useful when treating a liquid product or when solid–liquid extraction is used to extract valuable products from a fruit or fruit by-product. The ultrasound intensity distribution in a probe system is even more non-uniform than the cleaning baths. At the tip of the probe, the vigorous vibration of the metal probe forms a "reaction zone" as described by sonochemists, so strong that it will cause

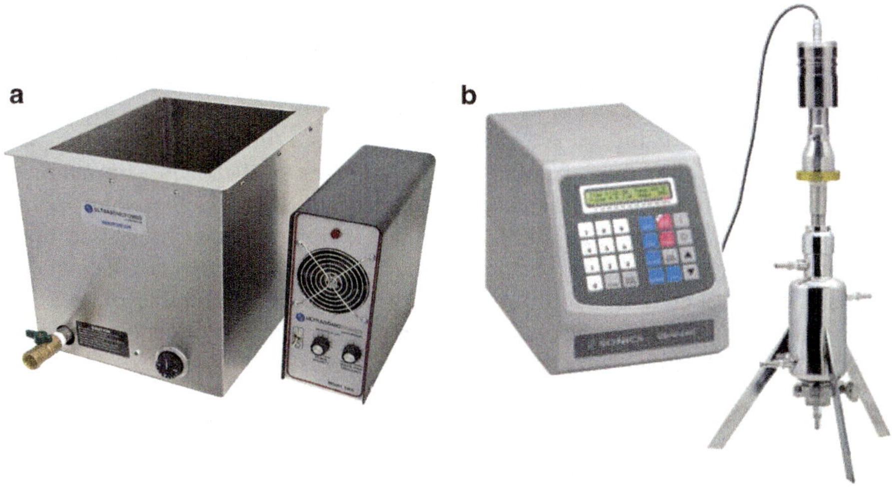

Fig. 11.1 (**a**) Ultrasonic cleaning bath, and (**b**) a probe unit

pitting or corrosion at the tip of the metallic probe. Because of the nature of wave propagation, airborne ultrasound also suffers from the problem of a non-uniform ultrasound field distribution on the product surface(s) to be treated. An important prerequisite for any ultrasound-assisted process is thus the uniform or near-uniform distribution of ultrasound field in the treatment chamber. Noticeably, this aspect has long been ignored in some previous ultrasound application studies. As a result, conflicting reports have been published in the literature causing confusions about the usefulness of this technology. Measures must be taken through careful engineering design to improve the acoustic field distribution in a treatment chamber, as reported by Zhou et al. (2012) in a recent study where a nearly uniform ultrasound distribution was realized in a pilot scale continuous-flow ultrasonic washing unit. In high-intensity treatments, such as those for extraction, and microbial and food enzyme inactivation, only probe type systems can provide enough acoustic power density (APD) to perform the task. The treatment can be further enhanced by combination of sonication with mild heat and static pressure in a process termed manothermosonication (MTS). However, the enhanced treatments are also accompanied by serious pitting problem. The metal powders produced from the probe or even treatment chamber walls will be released into the liquid product to be processed which should be avoided. Specially designed sonoreactors such as those proposed by Dion (2011) provide a good option for minimizing the pitting problem.

11.2.3 Mode of Action

Cavitation is regarded as the mode of action for sonication treatments conducted in a liquid medium. There are two types of cavitation, i.e., transient and stable cavitation, depending on the time taken from cavitation bubble formation to implosion.

Transient cavitation is produced by those bubbles that can stay in the liquid for a few acoustic cycles before collapse whereas the stable cavitation is formed by bubbles that will stay in the liquid for a few hundred acoustic cycles (Mason and Lorimer 2002). The implosion of transient cavitation bubbles produces extreme localized physical conditions, such as high temperature (up to 5000 K), high pressure (up to 1000 atm), high heating/cooling rate (up to 1010 K/s), formation of shock waves, and water jet having a speed of 156 km/h toward a solid surface (Leighton 1994; Suslick et al. 1999). Stable cavitation generates high shear force and microstreaming, which also help to enhance a process. An increase in cavitation intensity can thus enhance an ultrasound-assisted process. A number of measures can be taken to enhance a process, including utilizing multi-frequency technique, and conducting the treatment at elevated external pressure.

11.3 Surface Decontamination

Ultrasound has been used as an effective surface cleaning tool for years in a number of industrial applications, including cleaning of industrial parts, firearms, dental and medical devices, and semiconductor and disk drive, among others. The mechanism of ultrasound enhancement in cleaning is attributed to cavitation-related activities. It is postulated that the acoustic bubbles are oscillating at a distance of only a few tens of nanometers above the surface to be cleaned. The flow resulting from the bubble collapse can lead to important drag and shear forces on the surface, causing surface cleaning (Maisonhaute et al. 2001). Seymour et al. (2002) treated strawberries and seven selected vegetables with chlorine alone and chlorine in combination with ultrasound at three frequencies in an ultrasonic bath (25, 32, and 70 kHz, 15 W/L). While an additional reduction of 1.2 log CFU/g was achieved for iceberg lettuce when ultrasound was added to a chlorine wash, the reduction of *E. coli* NCIMB 12497 on strawberries was higher in the ultrasound+chlorine treatment than the chlorine-alone wash only at frequency of 32 kHz. The bacterial reduction of *Salmonella* and *E. coli* O157:H7-inoculated apples by ClO_2 (0–40 ppm) with and without ultrasonic treatment (170 kHz) for up to 10 min were studied by Huang et al. (2006). Over one log reduction was achieved for both *Salmonella* and *E. coli* inoculated on apples compared to a ClO_2 alone treatment when ultrasound was combined with ClO_2. São José and Vanetti (2012) evaluated the effectiveness of ultrasound treatment (45 kHz) in combination with selected sanitizers on decontamination of cherry tomatoes. A 10-min treatment in the presence of sodium dichloroisocyanurate (20–200 ppm), hydrogen peroxide (5 %), chlorine dioxide (10 ppm), or peracetic acid (40 ppm) resulted in 0.7–4.4 log CFU/g reduction in aerobic mesophilic bacteria count, and 1.1–3.4 log CFU/g reduction in the number of molds and yeasts. The combined treatment of ultrasound and 40 mg/L peracetic acid resulted in the highest reduction of the natural contaminant population and a reduction of *Salmonella* Typhimurium ATCC 14028 by 3.9 log CFU/g. The variations in the microbial count reduction reported in the literature may be caused by the non-uniform distribution of the ultrasound field in the unit used. The nonuniformity of the ultrasound field, and hence of cavitation, will contribute to spatial variation in microbial inactivation activities in a washing tank. As a result, during a washing operation, those fruit samples that have received adequate treatment and thus have a low microbial count are subject to cross-contamination by other samples that have received less ultrasound exposure (possibly because of the blockage of ultrasound propagation) (Zhou et al. 2012). It is thus important to understand how the treatment is affected by operating parameters, designing the wash systems capable of taking full advantage of ultrasound in fruits decontamination.

11.4 Postharvest Quality Enhancement

Using ultrasound treatment to maintain fruit postharvest quality is a relatively new research effort. Cao et al. (2010) immersed ripe strawberries in an ultrasonic bath (350 W) and treated them at 25, 28, 40, and 59 kHz for 10 min. The treated strawberries were stored at 5 °C for 8 days for quality evaluation. They reported a significantly slower decrease in firmness, less bacteria counts, and less decay for treatments at 40 and 59 kHz and thus concluded that an extended shelf-life could be achieved. After day 2 till the last day of the storage (day 8), the ultrasound treated strawberries retained significantly higher vitamin C compared to the control, again for samples treated at 40 and 59 kHz. Alexandre et al. (2011a) examined the effect of ultrasound and two other non-thermal processing methods on the quality of strawberries during storage. The samples treated with ozone, ultrasound (35 kHz and 120 W) and UV-C maintained higher ascorbic acid content than those washed with chemical sanitizers. Another work with strawberries was reported by Aday et al. (2012) who studied the effects of ultrasound power (30, 60, and 90 W) and time (5 and 10 min) on quality of strawberries. They found no significant differences in quality indexes for samples treated at two treatment times. Treatments with ultrasound all exhibited a marked decrease in the decay incidence compared to the control. Among the three ultrasound power levels, the quality attributes, such as pH, total soluble solids, and color were better in samples treated at 30 and 60 W. Fava et al. (2011) examined the effect of ultrasound at 20 kHz on the surface microstructure, color, and mechanical properties of grape berry after 5 min sonication at 30 °C with a probe system, with an aim to use it as a decontamination method. Under the tested conditions, the ultrasound treatment caused some ultrastructural and nanostructural changes on grape surface. They also reported a significantly lower L value but the changes in mechanical properties determined by puncture test were negligible. The combined effect of aqueous chlorine dioxide and sonication on the postharvest storage quality of plum was investigated by Chen and Zhu (2011). A combined treatment of ClO_2 (40 ppm) and sonication (40 kHz, 100 W, 10 min) extended the shelf-life of plum from 35 days for the control to 60 days. In addition, the total flavonoids, ascorbic acid, and titratable acid contents in ClO_2 + sonication treatment were significantly higher than that in the control. The ClO_2 + sonication-treated samples also exhibited a reduced respiration rate and less firmness loss compared to the control. The overall visual quality of the plums remained to be above 5 in a 9-point scale at day 60 for the ClO_2 + sonication-treated samples.

Postharvest blue mold decay caused by *Penicillium expansum* is one of the most economically significant postharvest diseases of pear. The use of synthetic chemical fungicides has long been the main method for controlling postharvest pathogens. However, concerns from the consumers about the pesticide residues and pathogen resistance to currently used pesticides have stimulated research activities looking into new strategies to control *P. expansum* and other plant disease-causing pathogens. The application of power ultrasound in combination with selected chemicals is among the new alternative methods tested by research scientists. Yang et al. (2011) treated peach fruit with water, salicylic acid (SA), and combination of ultrasound (40 kHz, 8.8 W/L) with SA for 10 min at 20 °C. They found that the combined treatment had a significantly lower decay incidence at day 6 than other treatments. In addition, the combined treatment resulted in higher activities of defense enzymes such as phenylalanine ammonia-lyase, polyphenol oxidase, and peroxidase. There were no significant changes in firmness, vitamin C, titratable acid, and total soluble solids among pear samples treated by different methods. Another work by Yang et al. (2012) examined the use of SA in combination of sonication (40 kHz, 10 min) for reducing chilling injury (CI) on cold-stored peach fruit. The combined action of SA and ultrasound resulted in greater CI inhibition than the SA alone wash. The SA + ultrasound samples exhibited significantly higher activities than the control in 7 out of 8 antioxidant enzymes which control the regulation of excess reactive oxygen (ROS) species in fruit tissues responsible for the symptoms of CI, thereby helping to inhibit CI in peach samples. It was also found that endogenous SA concentrations in peaches treated with SA + ultrasound were increased.

11.5 Ultrasound in Fruit Juice Processing

11.5.1 Microbial Inactivation

Inactivation of foodborne pathogens in juice products utilizes the lethal effects of cavitation-generated physical and chemical activities, such as shock waves, water jets, high shear, and formation of free radicals with a purpose to ensure microbial food safety while maintaining product quality. It is known that the resistance of different microorganisms to ultrasound treatment is different. Normally, Gram-positive organisms (i.e., *Listeria*) are more resistant than the Gram-negative cells (*Escherichia coli*, *Salmonella*, etc.). A comparison of the resistance of bacteria groups to ultrasonic inactivation is given in Fig. 11.2, expressed with D-values (Feng 2011). Since ultrasound treatment at sublethal temperatures often requires a relatively long time to achieve a 5 log reduction in the number of a target microorganism, efforts have been made to combine ultrasound with other lethal factors to enhance the killing and reduce the treatment time. Commonly used combinations include ultrasound with heat, pressure, and antimicrobials.

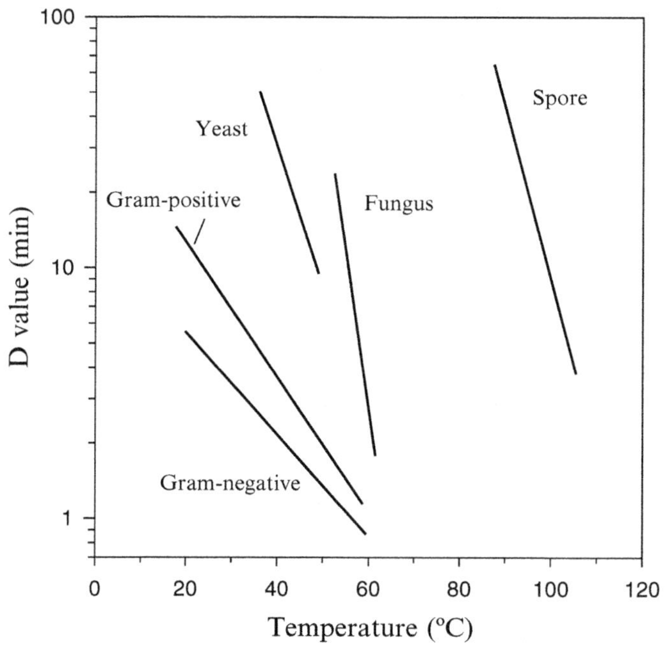

Fig. 11.2 D-values of bacteria in ultrasound inactivation with respect to temperature. Data for each bacterial group were collected from the literature. *Source*: Feng H. 2011. The thermodynamic and kinetic aspects of power ultrasound processes. In Ultrasound Technologies for Food and Bioprocessing, (eds.) Feng H, Barbosa GV, Weiss J, Springer, New York, NY

Apple juice and apple cider have been involved in a number of foodborne illness outbreaks in the United States and other countries. As a response, on January 19, 2001, the USFDA published a final rule in the Federal Register that requires juice producers to pasteurize their juice products, achieving a 5 log reduction in the population of the most resistant pathogen of public health significance. To mitigate the thermal damage caused by a thermal pasteurization, a number of alternative techniques, including ultrasound have been tested for inactivation of foodborne pathogens in apple juices. The reduction in microbial count is often reported with D-values, assuming a linear inactivation kinetic applies. The D-value of *E. coli* ATCC 25922 in a model apple juice by sonication at 20 kHz was 5.3 min (Patil et al. 2009). A non-linear inactivation of *E. coli* K12 was reported in apple cider by ultrasound (20 kHz, 0.46 W/mL) at 40 to and 60 °C (Ugarte-Romero et al. 2006). Compared to thermal treatment, sonication increased inactivation of *E. coli* by 5.3, 5.0, and 0.1-log cycles at 40 °C, 50 °C, and 60 °C, respectively (Fig. 11.3) (Ugarte-Romero et al. 2006). The treatment duration required to achieve a 5-log reduction of *E. coli* K12 in the cider were 3.6 and 17.7 min with sonication at 60 °C and 40 °C, respectively, compared to 4.2 min by heat at 60 °C and no lethal effect at 40 °C (Ugarte-Romero et al. 2006). D'amico et al. (2006) tested the inactivation of *E. coli* O157:H7 by ultrasound (20 kHz, 118 W/cm^2) in batch and continuous flow modes. The *E. coli* count reduction for the batch/continuous ultrasound treatments for

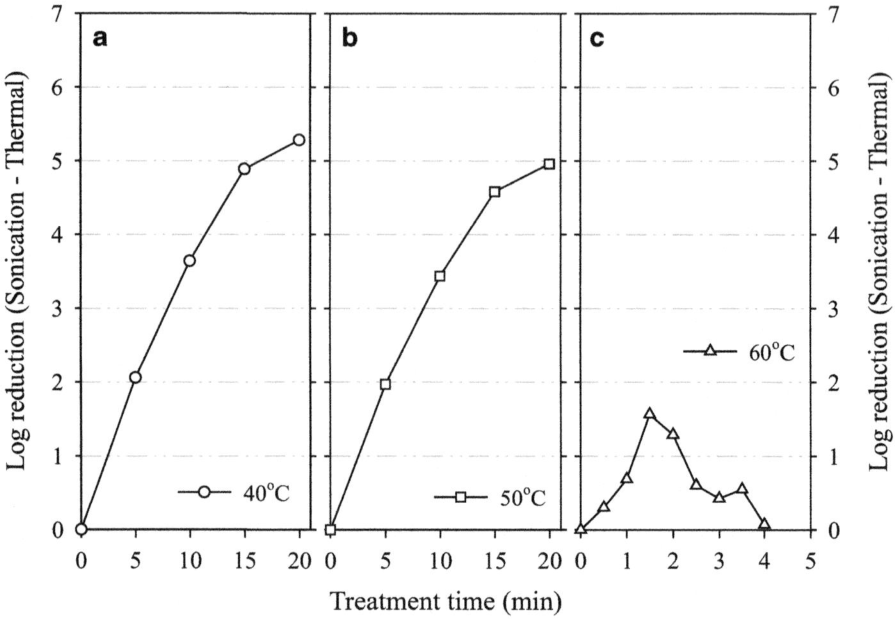

Fig. 11.3 Log reduction difference between sonication and thermal-alone treatment for *Escherichia coli* K12 in apple cider at 40 °C (**a**), 50 °C (**b**), and 60 °C (**c**). *Source*: Ugarte-Romero, E, Feng, H, Martin, SE, Cadwallader, KR, Robinson, SJ. 2006. Inactivation of *Escherichia coli* with Power Ultrasound in Apple Cider. Journal of Food Science 71: E102–108

6/18 min was 4.63/4.7 and 5.91/5.07 log reductions at 20 °C and 57 °C, respectively (D'amico et al. 2006). The D-values of *L. monocytogenes* by sonication (20 kHz, 457 mW/mL) at 20–60 °C in apple cider were between 1.0 and 12.3 min, while those by heat treatment at 20–60 °C were 1.6–117 min, respectively (Baumann et al. 2005). Gabriel (2012) treated cloudy apple juice with multi-frequency Dynashock ultrasound (28, 45, 100 kHz, alternately switched at a speed of 1 ms) for 30 min with initial temperature of 18 °C and final of 44 °C. The D-values of non-adapted/acid-adapted *E. coli* O157:H7, *Salmonella* spp., and *Listeria monocytogenes* were 27.49/32.45, 26.14/33.61, and 21.74/19.43 min, respectively (Gabriel 2012). Wang et al. (2010) applied ultrasound (25 kHz, 200–600 W, <50 °C, 30 min) to apple juice to inactivate *Alicyclobacillus acidiphilus* and *A. acidoterrestris*. A 2.69 and 3.71 log reduction of *A. acidiphilus*, and 4.21 and 4.56-log reduction of *A. acidoterrestris* at 200 and 600 W were observed after 30 min ultrasound. Palgan et al. (2011) reported around 4-log reduction of *Pichia fermentans* and a 5-log reduction of *E. coli* by manothermosonication (MTS, 20 kHz, 400 kPa) in an apple and cranberry juice blend. An additional log reduction of 2 and 1 log of *P. fermentans* and *E. coli* were achieved when combining ultrasound with high-intensity light pulses (Palgan et al. 2011). Arroyo et al. (2011) reported that D-values of *Cronobacter sakazakii* by manosonication (MS, 20 kHz, 35 °C, 200 kPa) in apple and orange juices were 0.95 and 0.75 min, respectively, which were higher than that in citrate-phosphate buffer (pH 7), 0.41 min. In apple cider, a 5-log reduction of *E. coli* K12 was achieved in 1.4 min by MTS (20 kHz, 59 °C, 400 kPa), 3.8 min by thermosonication (TS, 20 kHz, 59 °C), and 2.5 min by MS (20 kHz, 55 °C, 400 kPa) (Lee et al. 2012).

Orange juice is a high acid product with a typical pH of 3.5. Most orange juices are processed by a thermal treatment. The purpose of the thermal treatment is to inactivate two enzymes (pectinmethylesterase and polygalacturonase) that cause pulp separation in orange juice to prevent the juice from becoming transparent. The temperature (90 °C for 1 min) used to inactivate the enzymes is high enough to inactive vegetative foodborne pathogens that may present in the juice. Nevertheless, a number of reports have been published using ultrasound to inactivate microorganisms in orange juice. Patil et al. (2009) treated *E. coli* strains ATCC 25922 and ATCC 12900 in a model orange juice by ultrasound (20 kHz) at below 30 °C. The D-values of *E. coli* strains ATCC 25922/12900 by sonication at sublethal temperatures were 14.85/6.56, 2.92/6.14, and 2.45/5.4 min for amplitudes of 0.4, 7.5, and 37.5 μm, respectively (Patil et al. 2009). Zenker et al. (2003) reported that the D-values for *L. acidophilus* by ultrasound (20 kHz, 60 °C)/thermal (60 °C) treatments were 31.1/47.3 s in orange juice (pH 3.7). Gómez-López et al. (2010) reported inactivation of natural flora in calcium-added orange juice by 10 min ultrasound (20 kHz, 89.25 μm), achieving 1.38- and 0.56-log reduction in aerobic mesophilic count and yeast and molds count, respectively. Ultrasound application was extended to the inactivation of foodborne viruses in orange juice (Su et al. 2010). Two surrogates, feline calicivirus (FCV-F9) and murine norovirus (MNV-1), were inoculated into phosphate-buffered saline and orange juice followed by ultrasound (20 kHz, 2–3 min, 30 s on and 30 s off). In the buffer, a complete inactivation of feline calicivirus (FCV-F9) and murine norovirus (MNV-1) was achieved in 5- and 30-min treatment, while in orange juice a complete inactivation of feline calicivirus (FCV-F9) was achieved in 30 min and only 1.55 log PFU/mL reduction of norovirus (MNV-1) in the same treatment was reported (Su et al. 2010).

The inactivation of microorganisms with ultrasound in other juices has also been reported. The inactivation of yeast (*Pichia fermentans*) in tomato juice by ultrasound (20 kHz, final temperature of 24.4–39.9 °C) followed a non-linear kinetic (Adekunte et al. 2010a). The time required to achieve 5-log reduction of *P. fermentans* increase from 8 to 23 min with an increase of amplitude of 61.0–24.4 μm (Adekunte et al. 2010a). The D-values for *E. coli* K12 DH 5 α by ultrasound (20 kHz, 60 °C)/thermal (60 °C) treatments were 23.9/84.3 s in carrot juice (pH 5.9) (Zenker et al. 2003). Bermúdez-Aguirre and Barbosa-Cánovas (2012) tested pulsed and continuous TS (24 kHz, 40–60 °C) treatment to inactivate *Saccharomyces cerevisiae* in pineapple, grape, and cranberry juice. After 10-min continuous/pulse TS treatment, 6.4-/5.2-, 5.1-/5.7-, and no viable cell/5.9-log reductions were observed in pineapple, grape, and cranberry juice (Bermúdez-Aguirre and Barbosa-Cánovas 2012). Kasturi lime juice was treated by ultrasound (25 kHz, 20 °C, 30 and 60 min) (Bhat et al. 2011). The 1.9-reduction in total plate count and 0.5-log reduction in yeast and mold counts was observed in sonicated lime juice (Bhat et al. 2011).

It can be seen from Fig. 11.3 that, for TS, a sharp decrease in the net inactivation of sonication was observed compared to sonication at non-lethal temperature. At this temperature (60 °C, Fig. 11.3), a higher vapor pressure lowered the cavitation threshold but produced vapor-filled bubbles. The vapors inside of bubbles generate cushioning effect on the collapse of the vapor-filled bubbles, and then result in reduced effect on microbial inactivation (Mason 1999). For manosonication (MS), an increase in inactivation of microorganisms has been observed compared to sonication at same temperature. Enhanced inactivation by MS can be attributed to an increased intensity of bubble implosion. An increasing hydrostatic pressure surrounding a cavitating bubble in a liquid generally results in a decrease in the vapor pressure inside the bubble and hence an increase in the intensity of bubble implosion (Mason 1999). MTS has exhibited highest inactivation of microorganisms. Condón et al. (2005) proposed that heat at MTS caused changes in the envelopes in microorganisms and increased sensitivity to sonication under pressure. This phenomenon finally resulted in significant increases in inactivation of microorganisms by MTS (Condón et al. 2005).

11.5.2 Enzyme Inactivation

Food enzyme inactivation in fruit juices is aimed at preventing quality degradation caused by the enzymes. Pectinmethylesterase (PME) and polygalacturonases (PG) are two enzymes commonly found in citrus and tomato products. PME catalyzes the removal of the methyl groups from the polygalacturonic acid chain, and promotes cross-linking of the pectins.

The cross-linked pectins can aggregate and settle, leading to a loss of juice cloud. PG cleaves the polygalacturonic acid backbone of the pectin and reduces the average length of the pectin chains, contributing to reduce the viscosity of the juice. Therefore, the first step in the processing of orange juice and certain tomato products is to deactivate the two enzymes. This can be done with a thermal process at temperatures as high as 90 °C. Endeavors have been made to use less heat to perform the inactivation. Vercet et al. (2002) treated tomato juice to inactivate PG and PME by MTS (20 kHz, 200 kPa, 70 °C) for 1 min. MTS achieved a complete inactivation of PME and 62 % inactivation of the total PG, while thermal treatment (70 °C, 1 min) inactivated 38 % of the initial PME and unaffected PG (Vercet et al. 2002). Raviyan et al. (2005) applied sonication (20 kHz, 50 °C) and TS (61 and 72 °C) to inactivate PME in tomato juice at different cavitation intensities, 0.004–0.020 $mgL^{-1}min^{-1}$ (hydrogen peroxide yield rate). The D-values of PME by heat treatments at 50, 61, and 72 °C were 1571.4, 299.0, and 25.3 min, respectively. The D-values by sonication at 50 °C (min)/cavitation intensity ($mgL^{-1}min^{-1}$) were 240.6/0.007, 42.7/0.012, and 24/0.020, respectively (Fig. 11.4). The further decreases in D-values of PME were observed by TS at 61 °C, where D-values by TS at 61 °C (min)/cavitation intensity ($mgL^{-1}min^{-1}$) were 7.6/0.005, 1.5/0.007, and 0.8/0.012, respectively. TS at 72 °C resulted in lowest D-values, where D-values by

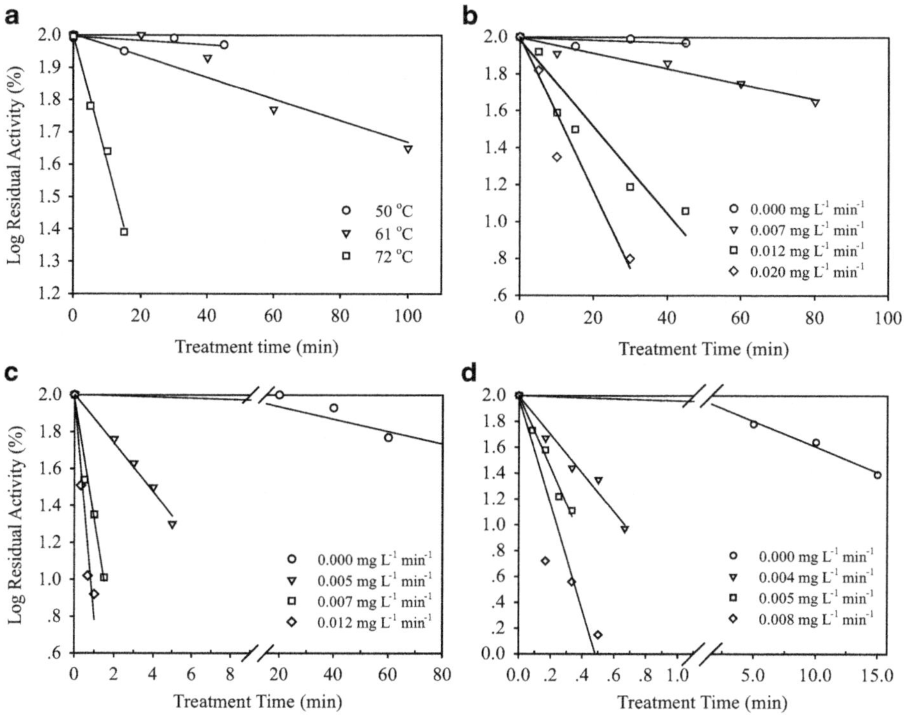

Fig. 11.4 Inactivation of tomato PME by thermal (**a**), sonication at 50 °C (**b**), 61 °C (**c**), and 72 °C (**d**) at different cavitation intensities. *Source*: Raviyan P, Zhang Z, Feng H. 2005. Ultrasonication for tomato pectinmethylesterase inactivation: effect of cavitation intensity and temperature on inactivation, Journal of Food Engineering 70: 189–196

TS at 72 °C (min)/cavitation intensity (mgL^{-1}min^{-1}) were 0.7/0.004, 0.4/0.005, and 0.3/0.008, respectively (Raviyan et al. 2005). Wu et al. (2008) also reported that TS (24 kHz) at 60 and 65 °C exhibited higher inactivation of PME in tomato juice compared to heat treatment at 60 and 65 °C. The times required to inactivate 90 % of initial PME were 41.8 and 11.7 min by TS at 60 and 65 °C, while those by heat treatments at 60 and 65 °C were 90.1 and 23.5 min (Wu et al. 2008). The D-values of PME in tomato juice by TS (20 kHz) at 50, 60, 70, and 75 °C were 86.5, 14.6, 4.1, and 2.8 min, respectively, while those by heat treatment at 60, 70, and 75 °C were 89, 16.2, and 4.04 min, respectively. There was no inactivation of PME by heat at 50 °C (Terefe et al. 2009). The inactivation of PG in tomato juice by TS and heat treatment followed non-linear inactivation kinetics, with inactivation by TS higher than the thermal treatment (Terefe et al. 2009). The thermal treatment at 50 °C did not show any inactivation of PG, and around 70 % inactivation of initial PG were achieved by 60, 70, and 75 °C for up to 60 min treatment. In contrast, TS at 50 °C for 60 min achieved around 60 % inactivation of PG, and over 80 % inactivation of PG were achieved by TS at 60, 70, and 75 °C within 30 min (Terefe et al. 2009).

Sonication inactivation of other food enzymes that cause quality degradation has also been investigated in recent years. Fonteles et al. (2012) treated cantaloupe melon juice for the inactivation of peroxidase (POD), polyphenol oxidase (PPO) and ascorbate peroxidase (APx) by sonication (19 kHz, 376 W/cm^2). After 10-min ultrasound treatment, the residual activity of POD, PPO, and APx were 23.3 %, 71.5 %, and 1.6 %, respectively. MTS (20 kHz, 400 kPa, initial temperature of 35 °C, final temperature of 63 °C) for 2.2 min was applied to inactivate PME in orange and carrot juice blend resulting in 78 % inactivation of PME (Caminiti et al. 2012). An enhanced PPO activity was reported in sonicated (35 kHz, 30 min) guava juice (Cheng et al. 2007). The authors postulated that ultrasound testament enhanced the disruption of biological cell wall followed by release of phenolic compounds and finally increased PPO activity.

Heat denatures the protein, which is the main mechanism of inactivation of enzyme by heating. The combination with heat and ultrasound has showed an increase in the inactivation of enzyme. O'Donnell et al. (2010) suggested that the free radicals or shear force generated by cavitation were mainly responsible of denaturing protein and enzyme inactivation. The further inactivation of enzyme has been observed by MTS. Pressure caused an increase in the intensity of bubble implosion and then an increase in formation of free radicals or shear forces (Mason 1999). More free radicals or shear forces produced by MTS can cause further increase in enzyme inactivation.

11.5.3 Effects of Ultrasound Treatment on Juice Quality

Ultrasound treatment of juice products may cause some quality alterations if the treatment time is long and cavitation intensity is high (Lee and Feng 2011). These changes include physical characteristics and chemical properties of the product due to the physical and chemical events taking place in the juice due to acoustic cavitation. Therefore, it is important to fine-tune the treatment conditions to minimize potential unfavorable quality changes. A number of reports have documented studies on the quality change during ultrasound treatments. Ugarte-Romero et al. (2006) and Lee et al. (2012) reported that sonication (40 °C for 17.7 min), TS (60 °C for 3.6 min, 59 °C for 5.1 min), MS (55 °C and 400 kPa for 2.2 min), and MTS (59 °C and 400 kPa for 1.4 min) at 20 kHz did not affect titratable acidity, °Brix, and pH, while all ultrasound treatments showed significantly lower turbidity compared to heat-treated and non-treated cider samples. The sonicated apple cider also exhibited less darkness than untreated sample. The reduced turbidity, and less darkness in sonicated apple cider might be explained by particle separation and reduction in particle sizes, as well as polyphenol oxidase (PPO) inactivation by ultrasound (Ugarte-Romero et al. 2006). PPO has been known to contribute to browning or darkening in apple cider (Zárate-Rodríguez et al. 2000). Aroma profiles in sonicated apple juice and cider were also reported. Apple juice treated by TS (20 kHz, 0.6 W/mL, 60 °C, 4.2 min) with an opened system resulted in 99 % reduction of

1-butanol-3-methyl acetate, the key aroma compounds in apple juice, compared to untreated apple juice (unpublished data). In contrast, in treatments of apple cider with MTS, TS and MS in a closed system filled with nitrogen, 9 key aroma compounds, including ethyl 2-methylbutanoate, butyl acetate, 1-butanol, ethyl hexanoate, 1-hexanol, butanoic acid, β-damascenone, hexanoic acid, and octanoic acid, were identified in all cider samples including untreated and thermally pasteurized samples (Lee et al. 2012). Right after the treatments and after 3-week storage at refrigerator, all sonicated samples showed more similar profiles of key aroma compounds to the raw apple cider than the thermally pasteurized sample.

The effects of ultrasound on orange juice qualities, such as color and nutrient, especially, ascorbic acid, have been investigated. An increased lightness of orange juice treated by TS (20 kHz, 75 °C, up to 30 s) and ultrasound alone (20 kHz, 10 °C, up to 10 min) was reported (Zenker et al. 2003; Gómez-López et al. 2010). The increase in lightness of sonicated orange juice was explained by partial precipitation of suspended, insoluble particles (Zenker et al. 2003). In terms of browning of orange juice, MTS-treated orange juice exhibited more browning pigments compared to thermally treated or raw orange juice. Vercet et al. (2001) reported an increase in browning pigments in MTS-treated (20 kHz, 69–111 °C, 200 kPa, 5 min) fruit juice model system compared to counterparts of thermally treated samples. Lee et al. (2005) reported that the browning index of continuous MTS (20 kHz, 70 °C, 400 kPa, 30 s)-treated orange juice was significantly higher than that of raw juice, but significantly lower than that treated by a commercial thermal method (Lee et al. 2005). Valero et al. (2007) treated orange juice by ultrasound at continuous and batch mode, and reported increased brown pigments only in sonicated sample at continuous mode, not in that at batch mode. It was reported that browning pigments could be formed by glucosyl radical and polymers in the presence of oxygen. Glucose in an aqueous phase could be converted to glucosyl radical by ultrasound treatment (Portenlänger and Heusinger 1994). The increase in browning pigments in sonicated sample at continuous mode might be explained by high exposure of juice to air at continuous mode (Valero et al. 2007). The ascorbic acid contents in sonicated orange juice were higher than those of untreated and thermally processed orange juice after 30-day storage at 10 °C (Tiwari et al. 2009a, 2009b). They treated orange juices with ultrasound (20 kHz, below 38 °C, 10 min) at three different APDs, 0.33, 0.74, and 0.88 W/mL. At Day 0, there were no significant differences in ascorbic acid contents among sonicated, untreated, and thermally treated (98 °C, 21 s) orange juices (Tiwari et al. 2009a). After 30-day storage at 10 °C, all sonicated orange juices exhibited higher ascorbic acid contents than untreated, and thermally treated orange juices (Tiwari et al. 2009a, 2009b). In the study by Gómez-López et al. (2010), the ascorbic acid content in calcium-added orange juice decreased with an increase in ultrasound (20 kHz, 10 °C, 10 min) treatment time up to 10 min. However, the retention of ascorbic acid in sonicated orange juice was higher than untreated juice through the storage at 4 and 10 °C for 10 days (Gómez-López et al. 2010). The TS-treated (20 kHz, 60 °C, 3.7 min) orange juice also exhibited better retention of ascorbic acid than thermally treated (60 °C, 6 min) after 35-day storage at 20 °C in darkness (Zenker et al. 2003). Zenker et al. (2003) postulated that degassing of juice by ultrasound might contribute to lower degradation of ascorbic acid in orange juice during storage. In another study, the MTS treatment (20 kHz, 62 °C, 200 kPa, 15, 30 s) resulted in around 10 % loss of ascorbic acid and carotenoid in orange juices while thermal treatment at the same temperature (62 °C, 15, 30 s) did not show any degradation (Vercet et al. 2001). The degradation of ascorbic acid and carotenoid was attributed to free radical produced by ultrasound (Vercet et al. 2001). It is noted that the temperature used in the thermal treatment by Vercet et al. (2001) was not high enough to deactivate PME and PG. For the storage tests of MTS-treated orange juice, Lee et al. (2005) reported that higher retention of ascorbic acid in continuous MTS-treated (20 kHz, 70 °C, 400 kPa, 30 s) orange juice than in thermally pasteurized one after storage at 4 °C for 63 days. More research is needed to determine exact mechanisms of ascorbic acid degradation in orange juice.

Other juice products, such as tomato, red grape, blackberry, strawberry, lime, and guava have also been treated by ultrasound and the quality changes after the treatment were reported. Tomato juice was sonicated at 20 kHz and 32–45 °C for 10 min and the quality attributes were reported by Adekunte et al. (2010b). Titratable acidity, °Brix, and pH did not show any differences between sonicated and untreated tomato juice while all color parameters (a, b, L values) and ascorbic acid content were lower in the sonicated juice than in the untreated one. An improvement of rheological properties of MTS-treated tomato juice was reported mainly caused by inactivation of detrimental enzymes, PG and PME in the juice (Vercet et al. 2002). Tiwari et al. (2010) treated red grape juice with ultrasound (20 kHz, 32–45 °C, 2–10 min), and reported an increase in lightness and significant retention of anthocyanin in the sonicated samples. Titratable acidity, °Brix, and pH did not exhibit any differences between untreated- and sonicated juice (Tiwari et al. 2010). Similar results in lightness and anthocyanin retention were also reported in sonicated (20 kHz, 9.24–22.79 W/cm^2, up to 10 min) blackberry juice, with an increase in lightness and significant retention of anthocyanin (Tiwari et al. 2009c). An increase in lightness in sonicated (20 kHz, 0.33–0.81 W/mL, below 40 °C, up to 10 min) strawberry juice has been reported, as well as reductions of anthocyanin and ascorbic acid by 3.2 and 11 % (Tiwari et al. 2008). For 10-day storage at 4 and 20 °C, higher degradations of anthocyanin and ascorbic acid in the strawberry juice sonicated above 0.47 W/mL than in untreated juice were observed (Tiwari et al. 2009d). Dubrović et al. (2011) reported that sonication (20 °C, 3–9 min) and TS (40, 60 °C, 3–6 min) treatment at 20 kHz on strawberry juice resulted in lower reduction of anthocyanin (0.7–4.4 %) than thermally treated (85 °C, 2 min) juice did (5.3–5.8 %) while only TS (55 °C, 9 min) treatment showed higher degradation of anthocyanin (5.8–7.1 %) than the thermal treatment. Cheng et al. (2007) reported better cloudiness, and higher ascorbic acid content in sonicated (35 kHz, 30 min) guava juice than in untreated juice, which might be caused by smaller particle size, and the degassing of dissolved oxygen in juice by ultrasound treatment. There was no significant difference in titratable acidity, °Brix and pH between sonicated- and untreated juice (Cheng et al. 2007). Kasturi lime juice was treated by ultrasound (25 kHz, 20 °C, 30 and 60 min) (Bhat et al. 2011). Titratable acidity, °Brix and pH of sonicated juice was not significantly different from those of untreated ones. For color measurements, L and a values decreased by ultrasound while b values increased compared to untreated juice. Ascorbic acid content, total phenolic, total flavonoids, total flavonols, and antioxidant activity in sonicated lime juice were significantly higher than those in untreated juice (Bhat et al. 2011).

11.6 Ultrasound-Assisted Drying of Fruits

Food dehydration is a traditional food preservation method. It is normally done by hot air drying at elevated temperatures and extended drying times. Because water molecules in a food matrix have a lower mobility than free water and thus a higher enthalpy of vaporization, the removal of water from a food with hot air to reach a water activity allowing no microbial activities is an energy-intensive process. On the other hand, exposing a food product at high temperatures and long times will cause inevitable quality degradation. Efforts have been made over the years to reduce drying time and minimize quality degradation in a drying process. Ultrasound, both airborne ultrasound and sonication has been investigated as a means to enhance a fruit-drying process. This is done in three different ways, depending on how the ultrasound is transmitted to a food product. Ultrasound can be emitted by a specially designed transducer and travels through air to reach the food to be dried. The food can be placed on a plate or in a chamber vibrated by a transducer so that the sound energy is directly coupled into the food product. Another arrangement is to place the food in a liquid and sonication is used to speed up the dewatering process (Gallego-Juárez and Riera 2011). The third approach uses ultrasound to enhance osmotic dehydration where dewatering is done in a hypertonic solution or to simply treat the samples in water, both are viewed as a pretreatment prior to a hot air or other drying process (Fernandes and Rodrigues 2007).

In airborne ultrasound-assisted drying, the application of high-intensity ultrasound to a food is performed by sirens and whistles, stepped vibrating plates, or vibrating cylinders (Mulet et al. 2011). Different systems have been used in airborne ultrasound generation because of the poor match between ultrasonic application systems and air, and the resulting high loss of power. Figure 11.5 shows a typical airborne ultrasound drying setup. The stepped surface design was reported to achieve high electroacoustic efficiency (75–80 %), and produce a maximum sound intensity level of 175 dB in the frequency range of 10–50 kHz. The vibrating cylinder design is aimed to direct application in an industrial drying unit. The idea is to vibrate the cylindrical drying chamber by a piezoelectric composite transducer at the resonant frequency of the selected vibration mode of the cylinder (Sanjuán et al. 2003). Computer simulation with finite element method was used to determine the resonant frequency of the system. Schössler et al. (2012) proposed a new method to conduct ultrasound-assisted drying. They mounted a screen into a ring sonotrode (24 kHz) and passed air through the screen to conduct hot air drying (70 °C), achieving a significant improvement in drying characteristics.

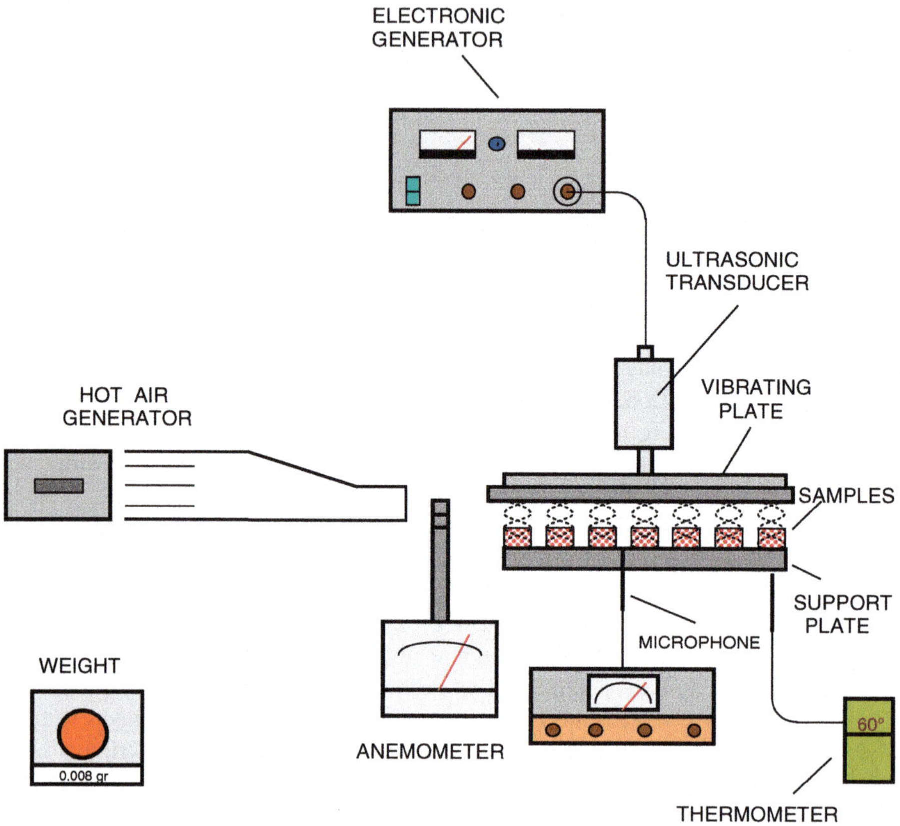

Fig. 11.5 Experimental setup for forced air drying assisted by airborne ultrasound. *Source*: Gallego-Juárez JA, Riera E. 2011. The thermodynamic and kinetic aspects of power ultrasound processes. In Ultrasound Technologies for Food and Bioprocessing, (eds.) Feng H, Barbosa GV, Weiss J, Springer, New York, NY

One of the early studies on the application of ultrasound in an osmotic solution to speed up osmotic dehydration process was reported by Simal et al. (1998) who treated apple cubes in a 70 °Brix sucrose solution at four temperatures (40, 50, 60, and 70 °C) in a ultrasonic bath (50 kHz, 150 W). They observed an increase in water losses and solute gain, and concluded that the solute gain for an ultrasound treatment at 40 °C was similar to an osmotic-alone treatment at 70 °C. Cárcel et al. (2007) conducted an experiment to determine the diffusivity of water and solute in apple cubes during an ultrasound-assisted osmotic dehydration process. For an ultrasound treatment with a probe unit at 30 °C, 20 kHz, and 11.5 W/cm^2 in a 30 °Brix sucrose syrup, the diffusivity of water increased by 117 % compared to a static osmotic treatment (without agitation). Deng and Zhao (2008) compared the effects of pulsed-vacuum (PV) and ultrasound (185 W) treatments on osmotic dehydration of Fuji apple cylinders in 60 % (w/w) high-fructose corn syrup (HFCS) solution.

Besides the significant increase in water losses and solute gain, they used SEM microimages to show the microstructure changes. The apple-cell deformation and cell structure collapse were the most severe in ultrasound-treated samples, but moderate in PV samples. SEM also revealed a larger amount of solute uptake in the cells of PV and ultrasound treated samples. The mass transfer enhancement in osmotic dehydration of fruits was also confirmed by Fernandes and Rodrigues (2010) in the drying of eight fruits (banana, genipap, jambo, melon, papaya, pineapple, pinha, and sapota), and by Garcia-Perez et al. (2012) in low-temperature drying (−14 °C) of apple cubes.

To reduce the sugar uptake in fruit drying, recent research has focused more on using ultrasound to pretreat a product in water, followed by a hot air drying or other drying processes. Fernandes et al. (2007) achieved a 16 % reduction in hot air drying time when pretreated papaya cubes by sonication in distilled water for 10–90 min at 100 kW/m^3 and 25 kHz. In addition, during ultrasound treatment, the papaya samples lost 13.8 % sugar in 30 min resulting in a low-sugar product. In another work of Fernandes et al. (2007), hot air drying (60 °C) of banana cylinders was conducted after treating the banana with ultrasound at 25 kH and 4870 W/cm^2 for 10–30 min. Oliveira et al. (2011) examined the influence of the ultrasonic pretreatment in water prior to air drying on dehydration of Malay apple and reported 28 % increase in water-effective diffusivity and 27.3 % reduction in total drying time in ultrasound-pretreated samples. The hot air drying time was reduced by 11 % after ultrasound treatment. The recent work of Nowacka et al. (2012) immersed apple cubes in an ultrasonic bath (35 kHz) for 10–30 min followed by hot air drying at 70 °C and 1.5 m/s air velocity. They reported a 31 % reduction in hot air drying time. At the same time, they noticed that the apples pretreated with ultrasound in water exhibited a 9–14 % higher porosity and 6–20 % lower density. Rithmanee and Intipunya (2012) used a probe ultrasound unit (20 kHz) to pretreat longan in water at 29–32.5 °C. The pretreated longan was then air dried at 60 °C. Quality attributes of the longan samples were evaluated. It was found that ultrasound-treated longan reduced polyphenol oxidase activity by 50–71 %, and the reduction for peroxidase was 47.6–94.1 %, compared to the non-pretreatment samples.

11.7 Extraction of Value-Added Chemicals from Fruit and Fruit Products

The US Council for Agricultural Science and Technology reported that 136–410 kg of waste is produced for every ton of input material in fruit- and vegetable-processing industries due to variability of input, high standards of production, and the amount of non-usable materials in fresh produce (The United Nations Environment Program [UNEP] 2002). The utilization of the waste and by-products in fruit processing will increase the value of the product and sustainability of the process. Many plant-based chemicals and bioactive molecules are rich in fruits processing by-products.

To recover the valuable compounds, different solid–liquid extraction or leaching methods have been used to extract the solutes that initially reside in the solid fruit or fruit by-products. Thus, from chemical engineering aspects, the extraction of bioactive and other value-added compounds from a fruit and fruit by-product is a two-step process. First, a molecular diffusion of the compound through the fruit product (the solid) into the solvent, and secondly, a convection and eddy diffusion of the molecules through a boundary layer into the bulk of the solvent. The use of power ultrasound to enhance a solid–liquid extraction process has been investigated to enhance the solid–liquid extraction process. The ultrasound-assisted extraction often shows an enhancement in extraction rate. It is attributed to the mechanical effect of cavitation, including increased interface mass transfer, alteration of the permeability of the cell membrane, and increased activity of the substances that are bound to cell structures, as well as the disruption of cell walls (Mason et al. 1996). Entezari et al. (2004) applied ultrasound in date syrup extraction and reported a higher extraction in a shorter time with improved physical quality of the date syrup extract. They also noticed significantly decreased microbial count in the syrup in comparison to the conventional method. The extraction of polyphenols from apples in water was increased by 6 % in an ultrasound-assisted extraction at 20–75 W s/mL and 80 °C (Vilkhu et al. 2011). Yue et al. (2012) utilized a Box–Behnken design to optimize an ultrasound-assisted extraction of polyphenols from unripe apples. They reported the optimum extraction conditions to be ultrasonic power of 519 W, extraction time of 30 min, extraction temperature 50 °C, and ethanol concentration of 50 % with a total polyphenols yield of 13.3 mg GAE/g. Annegowda et al. (2012) examined the effects of sonication treatments (0–60 min.) on phenolics and other antioxidant compounds in starfruits extracted in methanol and water. Methanolic extracts (total phenolics, total flavonoids, and total tannins) obtained after 30 min of sonication were significantly higher than that from water.

11.8 Other Applications

11.8.1 Pest Control

Different from the commercial pest control devices claimed to protect the house from pest attacks by generating (airborne) ultrasound waves to disturb the pests, the fruit pest control was done in a washing solution with an ultrasonic cleaning bath. Apples were infected by adult mites, or oviposit by codling moths, or infested with san jose scale before treating with ultrasound for 15–45 min for codling moth eggs, and 1–10 min for mites and san jose scales with addition of a detergent. Less than 60 % of the eggs were destroyed after 45 min treatment. The mortality of spider mite was directly related to sonication time, while no san jose scales were removed from apple surfaces (Hansen 2001).

11.8.2 Blanching

Blanching is a crucial step in fruits and vegetables processing prior to freezing, drying, or canning, with an aim to inactivate enzymes that would otherwise adversely affect the product quality. It is currently done by heating the fruits for a short period with either steam or hot water. Varying amounts of nutrients are lost in blanching, especially the water-soluble vitamins B-l, B-2, C, niacin, and folacin. Therefore, efforts have been made to develop alternative blanching methods to reduce quality loss during blanching. Ganjloo et al. (2008) investigated the use of ultrasound in blanching of seedless guava for inactivation of peroxidase (POD) at different temperatures and compared it with hot water blanching. A shorter time was recorded to achieve the same inactivation of POD when ultrasound was added to a hot water treatment. For instance, at 90 °C, 90 s were required to achieve a complete inactivation of POD with hot water. The time was reduced to 60 s when ultrasound (20 kHz, 25 % of the rating power) was applied. The quality changes of strawberries after heat plus ultrasound (thermosonication) blanching was examined by Alexandre et al. (2011b). The thermosonicated samples retained quality attributes better than heat blanched ones at the same temperatures ($p<0.05$).

11.9 Conclusion

There is an increasing interest in the application of power ultrasound in fruit and fruit products processing. This is partially due to the fact that acoustic energy has a "clean" image in the eyes of the public compared to some other new processing methods. The ultrasound generation is relatively simple and thus it is expected that with an increased usage and demand, the ultrasound generation technique will become more reliable and less costly. A reduced cost in the ultrasound generation technology will pave the road for more "real world" and large-scale applications. It can also be seen from this short summary that ultrasound-assisted processing deals with multiple unit operations and each has its own characteristics. The development of ultrasound processing technology needs a concerted effort between food processing engineers, ultrasound equipment manufacturers, processing equipment manufacturers, and food and agricultural industries. To explore the potential of ultrasound technology in fruit processing, a good understanding of the underlining physics of ultrasound and good engineering design of the system to achieve a uniform or near-uniform acoustic field distribution are indispensable.

References

Aday, M.S., R. Temizkan, M.B. Burak Büyükcan, and C. Caner. 2012. An innovative technique for extending shelf life of strawberry: Ultrasound. *LWT - Food Science and Technology* 52: 93–101. http://dx.doi.org/10.1016/j.lwt.2012.09.013.

Adekunte, A., B.K. Tiwari, A. Scannell, P.J. Cullen, and C.P. O'Donnell. 2010a. Modeling of yeast inactivation in sonicated tomato juice. *International Journal of Food Microbiology* 137: 116–120.

Adekunte, A.O., B.K. Tiwari, P.J. Cullen, A.G.M. Scannell, and C.P. O'Donnell. 2010b. Effect of sonication on color, ascorbic acid and yeast inactivation in tomato juice. *Food Chemistry* 122: 500–507.

Alexandre, E.M., T.R.S. Brandão, and C.L.M. Silva. 2011a. Efficacy of non-thermal technologies and sanitizer solutions on microbial load reduction and quality retention of strawberries. *Journal of Food Engineering* 108: 417–426.

Alexandre, E.M., D.M. Santos-Pedro, T.R.S. Brandão, and C.L.M. Silva. 2011b. Study on thermosonication and ultraviolet radiation processes as an alternative to blanching for some fruits and vegetables. *Food and Bioprocess Technology* 4: 1012–1019.

Annegowda, H.V., R. Bhat, L. Min-Tze, A.A. Karim, and S.M. Mansor. 2012. Influence of sonication treatments and extraction solvents on the phenolics and antioxidants in star fruits. *Journal of Food Science and Technology* 49: 510–514.

Arroyo, C., G. Cebrián, R. Pagán, and S. Condón. 2011. Inactivation of *Cronobacter sakazakii* by ultrasonic waves under pressure in buffer and foods. *International Journal of Food Microbiology* 144: 446–454.

Baumann, A., S.E. Martin, and H. Feng. 2005. Power ultrasound treatment of *Listeria monocytogenes* in apple cider. *Journal of Food Protection* 68: 2333–2340.

Bermúdez-Aguirre, D., and G.V. Barbosa-Cánovas. 2012. Inactivation of *Saccharomyces cerevisiae* in pineapple, grape and cranberry juices under pulsed and continuous thermo-sonication treatments. *Journal of Food Engineering* 108: 383–392.

Bhat, R., N.S.B.C. Kamaruddin, L. Min-Tze, and A.A. Karim. 2011. Sonication improves kasturi lime (*Citrus microcarpa*) juice quality. *Ultrasonics Sonochemistry* 18: 1295–1300.

Caminiti, I.M., F. Noci, D.J. Morgan, D.A. Cronin, and J.G. Lyng. 2012. The effect of pulsed electric fields, ultraviolet light or high intensity light pulses in combination with manothermosonication on selected physico-chemical and sensory attributes of an orange and carrot juice blend. *Food and Bioproducts Processing* 90: 442–448.

Cao, S., Z. Hu, B. Pang, H. Wang, H. Xie, and F. Wu. 2010. Effect of ultrasound treatment on fruit decay and quality maintenance in strawberry after harvest. *Food Control* 21: 529–532.

Cárcel, J.A., J. Benedito, C. Rosselló, and A. Mulet. 2007. Influence of ultrasound intensity on mass transfer in apple immersed in a sucrose solution. *Journal of Food Engineering* 78: 472–479.

Chen, Z., and C. Zhu. 2011. Combined effects of aqueous chlorine dioxide and ultrasonic treatments on postharvest storage quality of plum fruit (*Prunus salicina* L.). *Postharvest Biology and Technology* 61: 117–123.

Cheng, L.H., C.Y. Soh, S.C. Liew, and F.F. Teh. 2007. Effects of sonication and carbonation on guava juice quality. *Food Chemistry* 104: 1396–1401.

Condón, S., J. Raso, and R. Pagán. 2005. Microbial inactivation by ultrasound. In *Novel food processing techniques*, ed. G.V. Barbosa-Canovas, M.S. Tapia, and M.P. Can, 423–442. Boca Raton, FL: CRC Press LLC.

Coupland, J.N. 2004. Low intensity ultrasound. *Food Research International* 37: 567–577.

D'amico, D.J., T. Silk, J. Wu, and M. Guo. 2006. Inactivation of microorganisms in milk and apple cider treated with ultrasound. *Journal of Food Protection* 69: 556–563.

Deng, Y., and Y. Zhao. 2008. Effects of pulsed-vacuum and ultrasound on the osmodehydration kinetics and microstructure of apples (Fuji). *Journal of Food Engineering* 85: 84–93.

Dion, J. 2011. Contamination-free sonoreactor for the food industry. In *Ultrasound technologies for food and bioprocessing*, ed. H. Feng, G.V. Barbosa, and J. Weiss. New York: Springer.

Dubrović, I., Z. Herceg, A.R. Jambrak, M. Badanjak, and V. Dragović-Uzelac. 2011. Effect of high intensity and pasteurization on anthocyanin content in strawberry juice. *Food Technology and Biotechnology* 49: 196–204.

Entezari, H., H. Nazary, and H. Khodaparast. 2004. The direct effect of ultrasound on the extraction of date syrup and its micro-organisms. *Ultrasonics Sonochemistry* 11: 379–384.

Fava, J., K. Hodara, A. Nieto, S. Guerrero, S.M. Alzamora, and M.A. Castroa. 2011. Structure (micro, ultra, nano), color and mechanical properties of Vitis labrusca L. (grape berry) fruits treated by hydrogen peroxide, UV–C irradiation and ultrasound. *Food Research International* 44: 2938–2948.

Feng, H. 2011. The thermodynamic and kinetic aspects of power ultrasound processes. In *Ultrasound technologies for food and bioprocessing*, ed. H. Feng, G.V. Barbosa, and J. Weiss. New York: Springer.

Feng, H., and W. Yang. 2011. Ultrasonic process. In *Handbook on nonthermal processing technologies*, ed. F.F. Daniel, V.M. Balasubramaniam, and H. Zhang. Ames, IA: Wiley.

Fernandes, F.A.N., and S. Rodrigues. 2007. Ultrasound as pre-treatment for drying of fruits: Dehydration of banana. *Journal of Food Engineering* 82: 261–267.

———. 2010. Application of ultrasound and ultrasound-assisted osmotic dehydration in drying of fruits. *Drying Technology* 26: 1509–1516.

Fernandes, F.A.N., F.I.P. Oliveira, and S. Rodrigues. 2007. Use of ultrasound for dehydration of papayas. *Food and Bioprocess Technology* 1: 339–345.

Fonteles, T.V., M.G.M. Costa, A.L.T. de Jesus, M.R.A. de Miranda, F.A.N. Fernandes, and S. Rodrigues. 2012. Power ultrasound processing of cantaloupe melon juice: Effects on quality parameters. *Food Research International* 48: 41–48.

Gabriel, A.A. 2012. Microbial inactivation in cloudy apple juice by multi-frequency Dynashock power ultrasound. *Ultrasonics Sonochemistry* 19: 346–351.

Gallego-Juárez, J.A., and E. Riera. 2011. Technologies and applications of airborne power ultrasound in food processing. In *Ultrasound technologies for food and bioprocessing*, ed. H. Feng, G.V. Barbosa, and J. Weiss. New York: Springer.

Ganjloo, A., R.A. Rahman, J. Bakar, A. Osman, and M. Bimakr. 2008. Feasibility of high-intensity ultrasonic blanching combined with heating for peroxidase inactivation of seedless guava (*Psidium guajava* L.). In *19th National Congress on Food Technology, October 15–16, Moshhod, Iran.*

Garcia-Perez, J.V., J.A. Carcel, E. Riera, C. Rosselló, and A. Mulet. 2012. Intensification of low-temperature drying by using ultrasound. *Drying Technology* 30: 1199–1208.

Gómez-López, V.M., L. Orsolani, A. Martínez-Yépez, and M.S. Tapia. 2010. Microbiological and sensory quality of sonicated calcium-added orange juice. *LWT - Food Science and Technology* 43: 808–813.

Hansen, J.D. 2001. Ultrasound treatments to control surface pests of fruits. *HortTechnology* 11: 186–188.

Huang, T., C. Xu, K. Walker, P. West, S. Zhang, and J. Weese. 2006. Decontamination efficacy of combined chlorine dioxide with ultrasonication on apples and lettuce. *Journal of Food Science* 71(4): 134–139.

Lee, H., and H. Feng. 2011. Effect of power ultrasound on food quality. In *Ultrasound technologies for food and bioprocessing*, ed. H. Feng, G.V. Barbosa, and J. Weiss. New York: Springer.

Lee, J.W., H. Feng, and M.M. Kushad. 2005. Effect of manothermosonication on quality of orange juice. In *AIChE 2005 Annual Meeting, Cincinnati, OH.*

Lee, H., H. Kim, K.R. Cadwallader, H. Feng, and S.E. Martin. 2012. Inactivation of Escherichia coli K12 and quality assessment of apple cider treated with manosonication, thermosonication, and manothermosonication. *Ultrasonics Sonochemistry*. Submitted.

Leighton, T.G. 1994. *The acoustic bubbles*. London: Academic. 613p.

Maisonhaute, E., P.C. White, and R.G. Compton. 2001. Surface acoustic cavitation understood via nanosecond electrochemistry. *Journal of Physical Chemistry B* 105: 12087–12091.

Mason, T.J. 1999. An introduction to the uses of power ultrasound in chemistry. In *Sonochemistry*, 1–41. Oxford, England: Oxford Science Publication.

Mason, T.J., and J.P. Lorimer. 2002. *Applied sonochemistry: The uses of power ultrasound in chemistry and processing*. Weinheim, Germany: Wiley-VCH.

Mason, T.J., L. Paniwnyk, and J.P. Lorimer. 1996. The use of ultrasound in food technology. *Ultrasonics Sonochemistry* 3: 253–260.

McClements, D.J. 1995. Advances in the application of ultrasound in food analysis and processing. *Trends in Food Science and Technology* 6: 293–299.

Mulet, A., J.A. Cárcel, J.V. García-Pérez, and E. Riera. 2011. Ultrasound-assisted hot air drying of Foods. In *Ultrasound technologies for food and bioprocessing*, ed. H. Feng, G.V. Barbosa, and J. Weiss. New York: Springer.

Nowacka, M., A. Wiktor, M. Śledź, N. Jurek, and D. Witrowa-Rajchert. 2012. Drying of ultrasound pretreated apple and its selected physical properties. *Journal of Food Engineering* 113: 427–433.

O'Donnell, C.P., B.K. Tiwari, P. Bourke, and P.J. Cullen. 2010. Effect of ultrasonic processing on food enzymes of industrial importance. *Trends in Food Science and Technology* 21: 358–367.

Oliveira, F.I.P., M.I. Gallão, S. Rodrigues, and F.A.N. Fernandes. 2011. Dehydration of Malay apple (*Syzygium malaccense* L.) using ultrasound as pre-treatment. *Food and Bioprocess Technology* 4: 610–615.

Palgan, I., I.M. Caminiti, A. Muñoz, F. Noci, P. Whyte, D.J. Morgan, D.A. Cronin, and J.G. Lyng. 2011. Combined effect of selected non-thermal technologies on *Escherichia coli* and *Pichia fermentans* inactivation in an apple and cranberry juice blend and on product shelf life. *International Journal of Food Microbiology* 151: 1–6.

Patil, S., P. Bourke, B. Kelly, J.M. Frías, and P.J. Cullen. 2009. The effects of acid adaptation on *Escherichia coli* inactivation using power ultrasound. *Innovative Food Science & Emerging Technologies* 10: 486–490.

Portenlänger, G., and H. Heusinger. 1994. Polymer formation from aqueous solutions of α-D-glucose by ultrasound and γ-rays. *Ultrasonics Sonochemistry* 1: 125–129.

Raviyan, P., Z. Zhang, and H. Feng. 2005. Ultrasonication for tomato pectinmethylesterase inactivation: effect of cavitation intensity and temperature on inactivation. *Journal of Food Engineering* 70: 189–196.

Rithmanee, T., and P. Intipunya. 2012. Effects of high power ultrasonic pretreatment on physicochemical quality and enzymatic activities of dried longan. *Journal of Agricultural Sciences* 4: 299–306.

Sanjuán, N., M. Lozano, P. García-Pascual, and A. Mulet. 2003. Dehydration kinetics of red pepper (Capsicum annuum L var Jaranda). *Journal of the Science of Food and Agriculture* 83: 697–701.

São José, J.F.B., and M.C.D. Vanetti. 2012. Effect of ultrasound and commercial sanitizers in removing natural contaminants and *Salmonella enterica* Typhimurium on cherry tomatoes. *Food Control* 24: 95–99.

Schössler, K., H. Jäger, and D. Knorr. 2012. Effect of continuous and intermittent ultrasound on drying time and effective diffusivity during convective drying of apple and red bell pepper. *Journal of Food Engineering* 108: 103–110.

Seymour, I.J., D. Burfoot, R.L. Smith, L.A. Cox, and A. Lockwood. 2002. Ultrasound decontamination of minimally processed fruits and vegetables. *International Journal of Food Science and Technology* 37: 547–557.

Simal, S., J. Benedito, E.S. Sánchez, and C. Rosselló. 1998. Use of ultrasound to increase mass transport rates during osmotic dehydration. *Journal of Food Engineering* 36: 323–336.

Su, X.W., S. Zivanovic, and D.H. D'Souza. 2010. Inactivation of human enteric virus surrogate by high-intensity ultrasound. *Foodborne Pathogens and Disease* 7: 1055–1061.

Suslick, K.S., Y. Didenko, M.M. Fang, T. Hyeon, K.J. Kolbeck, W.B. McNamara, M.M. Mdleleni, and M. Wong. 1999. Acoustic cavitation and its chemical consequences. *Philosophical Transactions of the Royal Society of London A* 357: 335–353.

Terefe, N.S., M. Gamage, K. Vilkhu, L. Simons, R. Mawson, and C. Versteeg. 2009. The kinetics of inactivation of pectin methylesterase and polygalacturonase in tomato juice by thermosonication. *Food Chemistry* 117: 20–27.

The United Nations Environment Program [UNEP]. 2002. *Cleaner production fact sheet 3: Food manufacturing series*. Brisbane, QLD: University of Queensland and UNEP. www.geosp.uq.edu.au.

Tiwari, B.K., C.P. O'Donnell, A. Patras, and P.J. Cullen. 2008. Anthocyanin and ascorbic acid degradation in sonicated strawberry juice. *Journal of Agricultural and Food Chemistry* 56: 10071–10077.

Tiwari, B.K., C.P. O'Donnell, K. Muthukumarappan, and P.J. Cullen. 2009a. Ascorbic acid degradation kinetics of sonicated orange juice during storage and comparison with thermally pasteurized juice. *LWT - Food Science and Technology* 42: 700–704.

———. 2009b. Effect of sonication on orange juice quality parameters during storage. *International Journal of Food Science and Technology* 44: 586–595.

Tiwari, B.K., C.P. O'Donnell, and P.J. Cullen. 2009c. Effect of sonication on retention of anthocyanin in blackberry juice. *Journal of Food Engineering* 93: 166–171.

Tiwari, B.K., C.P. O'Donnell, A. Patras, N. Brunton, and P.J. Cullen. 2009d. Stability of anthocyanin and ascorbic acid in sonicated strawberry juice during storage. *European Food Research and Technology* 228: 717–724.

Tiwari, B.K., A. Patras, N. Brunton, P.J. Cullen, and C.P. O'Donnell. 2010. Effect of ultrasound processing on anthocyanin and color of red grape juice. *Ultrasonics Sonochemistry* 17: 598–604.

Ugarte-Romero, E., H. Feng, S.E. Martin, K.R. Cadwallader, and S.J. Robinson. 2006. Inactivation of *Escherichia coli* with power ultrasound in apple cider. *Journal of Food Science* 71: E102–E108.

Valero, M., N. Recrosio, D. Saura, N. Muñoz, N. Martí, and V. Lizama. 2007. Effects of ultrasonic treatments in orange juice processing. *Journal of Food Engineering* 80: 509–516.

Vercet, A., J. Burgos, and P. López-Buesa. 2001. Manothermosonication of foods and food-resembling system: Effect of nutrient content and nonenzymatic browning. *Journal of Agricultural and Food Chemistry* 49: 483–489.

Vercet, A., C. Sánchez, J. Burgos, L. Montañés, and P.L. Buesa. 2002. The effects of manothermosonication on tomato pectic enzymes and tomato paste rheological properties. *Journal of Food Engineering* 53: 273–278.

Vilkhu, K., R. Manasseh, R. Mawson, and M. Ashokkumar. 2011. Ultrasonic recovery and modification of food ingredients. In *Ultrasound technologies for food and bioprocessing*, ed. H. Feng, G.V. Barbosa, and J. Weiss, 345–368. New York: Springer.

Wang, J., X. Hu, and Z. Wang. 2010. Kinetics models for the inactivation of *Alicyclobacillus acidiphilus* DSM14558^{T} and *Alicyclobacillus acidoterrestris* DSM 3922^{T} in apple juice by ultrasound. *International Journal of Food Microbiology* 139: 177–181.

Wu, J., T.V. Gamage, K.S. Vilkhu, L.K. Simons, and R. Mawson. 2008. Effect of thermosonication on quality improvement of tomato juice. *Innovative Food Science & Emerging Technologies* 9: 186–195.

Yang, Z., S. Cao, Y. Cai, and Y. Zhang. 2011. Combination of salicylic acid and ultrasound to control postharvest blue mold caused by Penicillium expansum in peach fruit. *Innovative Food Science & Emerging Technologies* 12: 310–314.

Yang, Z., S. Cao, Y. Zheng, and Y. Jiang. 2012. Combined salicylic acid and ultrasound treatments for reducing the chilling injury on peach fruit. *Journal of Agricultural and Food Chemistry* 60: 1209–1212.

Yue, T., D. Shao, Y. Yuan, Z. Wang, and C. Qiang. 2012. Ultrasound-assisted extraction, HPLC analysis, and antioxidant activity of polyphenols from unripe apple. *Journal of Separation Science* 35: 2138–2145.

Zárate-Rodríguez, E., E. Ortega-Rivas, and G.V. Barbosa-Cánovas. 2000. Quality changes in apple cider as related to nonthermal processing. *Journal of Food Quality* 23: 337–349.

Zenker, M., V. Heinz, and D. Knorr. 2003. Application of ultrasound-assisted thermal processing for preservative and quality retention of liquid foods. *Journal of Food Protection* 66: 1642–1649.

Zhou, B., H. Feng, and A.J. Pearlstein. 2012. Continuous-flow ultrasonic washing system for fresh produce surface decontamination. *Innovative Food Science and Emerging Technologies* 16: 427–435.

Chapter 12
Fruit Preservation and Design of Functional Fruit Products by Vacuum Impregnation

Zamantha Escobedo-Avellaneda, Rébeca García-García, Aurora Valdez-Fragoso, Hugo Mújica-Paz, and Jorge Welti-Chanes

12.1 Introduction

Vacuum Impregnation (VI) is an emerging process used to obtain minimally processed fruits, intermediate moisture products, functional foods, and to improve different food preservation processes (Fito et al. 1996; Tapia de Daza et al. 1996). In this process, a porous solid (food) is immersed in a solution and then a pressure below atmospheric pressure is applied, and finally that pressure is restored to the atmospheric level; the pressure gradient causes the airflow out of the pores of the solid and the incorporation of the solution in the porous structure (Mújica-Paz et al. 2003a; Fito 1994). The process is called vacuum osmotic dehydration (VOD) when a hypertonic solution is used. VI causes solute and solvent exchange between the solution and the food, this exchange is favored by hydrodynamic mechanisms promoted by pressure differences between food and external fluid (Mújica-Paz et al. 2003b). Some deformation–relaxation phenomena also occur as a result of pressure changes originated by the expansion and compression of the gas occluded into the porous structure causing structure modifications. The expansion of the gas trapped in the pores promotes deformation of the solid matrix and when pressure becomes equal to the pressure of the system the hydrodynamic mechanism occurs, allowing the entry of external fluid in the pores due to capillary effect. Despite of this, VI allows obtaining products with better quality compared with those obtained with methods performed at atmospheric conditions such as osmotic dehydration (OD).

Z. Escobedo-Avellaneda • R. García-García • A. Valdez-Fragoso
• H. Mújica-Paz • J. Welti-Chanes (✉)
Tecnológico de Monterrey, Monterrey, Mexico
e-mail: jwelti@itesm.mx

A. Rosenthal et al. (eds.), *Fruit Preservation*, Food Engineering Series,
https://doi.org/10.1007/978-1-4939-3311-2_12

The possibility of incorporating different ingredients into the porous structure of the food allows generating compositional changes to increase quality and to achieve stability requirements. Microorganisms and compounds like minerals, vitamins, and phytochemicals can also be incorporated into the food matrix, which can lead to functional products with improved characteristics (Fito and Chiralt 1995; Fito et al. 2001; Salvatori et al. 1998; Zhao and Xie 2004).

12.2 Mass Transfer Mechanisms in Vacuum Impregnation Processes

Osmotic dehydration (OD) and VI are processes that cause water activity reduction in foods by immersing them usually in a concentrated aqueous solution, consisting regularly of sucrose or sodium chloride, although other water activity depressors such as lactose, maltodextrin, glucose, glycerin, and corn syrups have also been used. OD and VI allows obtaining foods of intermediate moisture content, and it is also used as a pretreatment before drying or freezing to increase the quality of final products. Osmotic pressure differences between the solution and the food or chemical potential gradient between the solution and the intracellular fluid act as the driving force for water removal (Shi and Le Maguer 2002).

During impregnation process, food acts as a semipermeable membrane in which water is mainly lost due to osmo-diffusion and capillary flow phenomena (osmotic dehydration), in addition, due to the complexity of foods, the not perfect permeability of membrane allows the transference of solutes (impregnation) from the osmotic solution into the free spaces and/or leaching out of soluble solids from the food to the solution by diffusion mechanisms and hydrodynamic flows inducing gain of external liquid which is especially important in porous foods treated under VI (Chiralt and Talens 2005; Khin et al. 2005; Shi and Le Maguer 2002).

Incorporation of liquid in the food due to VI is caused by the loss of gas entrapped within the pores of the food, once the pressure is restored to atmospheric pressure. VI takes place in two steps, during the first one, the food is placed in a closed recipient containing the impregnation or osmotic solution and a pressure below the atmospheric one is imposed causing that the gas trapped inside the pores of food expands and flows out. This step continues until external pressure (pe) in the container is equal to the pressure in the interior of the food (pi) allowing the transfer of a low amount of liquid from the impregnation solution to the food pores; in the second step the atmospheric conditions are restored inside the container causing the residual gas compresses and external liquid flows into the food pores by hydrodynamic mechanisms (Fig. 12.1) (Fito et al. 2002; Salvatori et al. 1998; Roa et al. 2001). This expansion of gas can cause structural changes whose intensity will depend on the mechanisms of mass transfer taking place in the food (Chiralt and Talens 2005). A unique transport mechanism does not occur during the process, the degree in which different mechanisms act determines the characteristics of the final product. Water transport can be performed by diffusion mechanism in foods with a nonporous

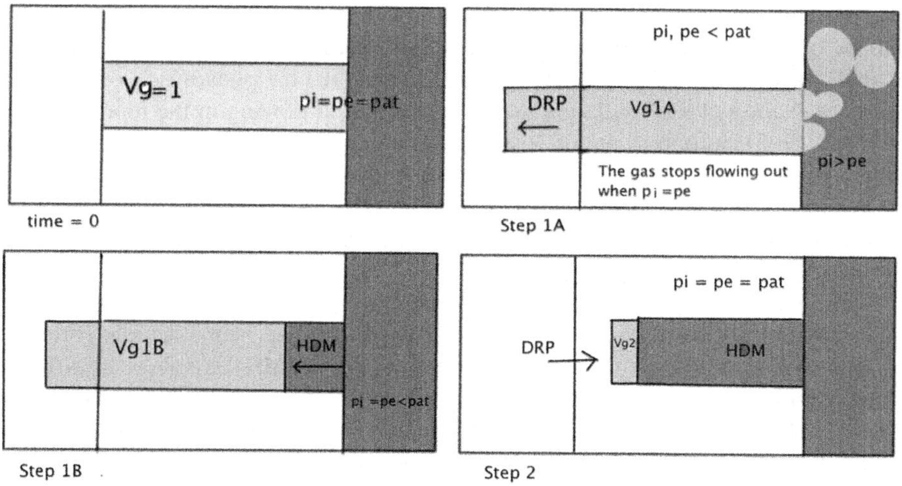

Fig. 12.1 Mechanisms of transport during vacuum impregnation. *DRP* deformation–relaxation phenomena, *HDM* hydrodynamic mechanism, *Vg* gas inside the pore volume, *pi* internal pressure, *pe* external pressure, *pat* atmospheric pressure. Adapted from Martínez-Navarrete et al. (1999)

structure, while in porous materials it can be transported by molecular diffusion, liquid diffusion, vapor diffusion (through gas flow), hydrodynamic flow, capillary transport, and surface diffusion (Shi and Le Maguer 2002).

Driving force for water and solute movement from the food to the osmotic solution and uptake of solutes from the solution to the food depends on the mechanisms implicated. Diffusion transport of water and solutes is driven by concentration gradients while capillary flow is due to pressure gradient between the system and external pressure, and water diffusion at pore surfaces is due to concentration gradients at the surfaces.

Fito and Chiralt (1995) point out that the fluid entering the porous of foods can occupy around 15 % of the fraction of pores located on the external surface of the product. Mass transfer can take place more rapidly in VI than at atmospheric conditions, and the solute uptake from impregnation solution occurs faster during the first 60 min of treatment decreasing slightly for longer times.

12.3 Factors Affecting Impregnation Process Effectiveness

Porosity, tissue structure, geometry, size, product ripeness, solution concentration, type of solutes, complexity of the impregnation solution composition (one solute, binary or ternary solution), pressure, temperature, time, agitation, ratio sample/solution, can affect the impregnation process and the final product properties (Chiralt and Talens 2005; Occhino et al. 2011; Shi and Le Maguer 2002; Roa et al. 2001).

Structural properties of food are fundamental in VI; pore distribution, size and shape of the pores determine mass transfer and deformation of the material. These properties affect: (a) the gas flowing out kinetics, (b) the phenomena of deformation–relaxation of the solid matrix and, (c) the liquid flowing in the food (Fito and Chiralt 1995). Although, liquid loss and solute gain in the food structure depends mainly on vacuum pressure and its application time, the diameter and the number of pores in the food structure are also determining factors, making necessary to know the effective porosity of food to predict the maximum solution that can be impregnated (Mújica-Paz et al. 2003b).

Effective porosity (ε_e), an important property in VI, is defined as the ratio between the volume of total gas in the sample and initial volume of the sample (Fito and Chiralt 1995). Mújica-Paz et al. (2003b) determinated the Ee values in different fruits: apple (0.33), mango (0.016), banana (0.048), papaya (0.042), melon (0.071), mamey (0.016) using different VI conditions. Other important parameter to consider in VI is the real porosity (ε_r), which constitutes a measure of the empty spaces in the food tissue it represents the maximum space that could be impregnated with an isotonic solution. Apple has large ε_r values (0.273) while mango and melon present intermediate values (0.152 and 0.133 respectively) and mamey, papaya, banana, and peach have the lower levels (0.016–0.058) (Mújica-Paz et al. 2003b). Vegetal tissues with wide intracellular spaces require short times to be impregnated, the time needed is of about 5 min when the external liquid is sugar syrup, however longer vacuum periods are required to reach mechanical equilibrium with smaller pores (Chiralt et al. 2001). Fruits with high porosity (20–30 %) will easily be impregnated, an example is apple with a highly porous tissue that allows its easy impregnation at vacuum and at atmospheric conditions (Degraeve et al. 2003; Salvatori et al. 1998; Mújica-Paz et al. 2003a, 2003b).

A study performed by Roa et al. (2001) in different fruits (melon, pineapple, papaya, apricot, and banana) showed the effect of food structure on the impregnation. By using a 65 °Bx sucrose syrup at 35 °C, five different vacuum pressures with 2 min pulses followed by 5 min at atmospheric conditions showed that melon had the highest impregnated volumetric fraction (13.8 %) when a vacuum pressure pulse at 3.07×10^4 Pa was used. On the other hand, pineapple and peach were the more difficult to impregnate (2.4 % at 1.74×104 Pa and 2.16 % at 5.29 Pa, respectively). These results reflect the complexity of the hydrodynamic penetration mechanism in a fruit, coupled with the relaxation–deformation phenomenon, which depends on many characteristics of the food and process conditions.

12.3.1 Effect of Pressure and Temperature on Vacuum Impregnation Effectiveness

As compared with atmospheric impregnation, the application of vaccum allows a more rapid and controlled incorporation of solutes in foods causing a greater a_w reduction as a result of higher solid gain and water loss. Water activity values obtained under VI are similar to those required on minimally processed fruits (Mújica-Paz et al. 2003a, 2003b; Tapia de Daza et al. 1996). Apple, melon, and mango slices, treated by VI using osmotic solutions (<50 °Brix) promoted a massive impregnation on the product by increasing the amount of the impregnation solution. However, at higher osmotic solution concentrations an increase in osmotic pressure hindering the entrance of solution to the pores in the food tissue favoring the water exit from the fruit. For each concentration of the osmotic solution, there is a pressure level of vacuum, which gets a minimum loss of water, in terms of solids gain, vacuum pressure increase favors the occupation of pores, and therefore the solid gain (Müjica-Paz et al. 2003b). The vacuum application time must be enough to achieve a mechanical balance within the product, with the subsequent internal gas outflow and the incorporation of external fluid. Müjica-Paz et al. (2003b) found a linear relationship between the VI time and the volume of papaya, mango, and mamey impregnated with an isotonic solution, and a quadratic effect for peach and melon.

Martínez-Monteagudo et al. (2006) pointed out that for obtaining the greatest incorporation of an isotonic solution (IS) in whole jalapeno pepper it is necessary to control the vacuum pressure, vacuum time, and relaxation time that affect the pattern of impregnation influenced by the phenomenon of deformation–relaxation of the pepper tissue. Whole jalapeño pepper was exposed to different vacuum pressure levels (133, 266, 400, 533, and 666 mbar) observing that the higher level of impregnation (0. 07 g IS per g pepper) occurs at 666 mbar since pressure gradient promotes the permeation and infiltration to a great extent; likewise, after 506 min of relaxation, impregnation rate decreases. With low vacuum pressure (133–400 mbar), the maximum level of impregnation achieved was 0.046 g IS/g pepper, and this value decreases as vacuum pressure decreases because of pressure difference between tissue and inner void which causes a migration of IS from the pepper tissue to the inner void (Fig. 12.2). As well as for 666 mbar, at low vacuum pressure (133–400 mbar) permeation level decreases after certain time of relaxation (339 min) showing that vacuum pressure and relaxation time influence impregnation rate of IS of whole pepper.

Temperature is the other important parameter to be considered in the VI process because it affects the impregnation effectiveness (water loss and solid gain); viscoelastic properties of the solid matrix change with temperature, so with higher temperatures there is a softening of the structure, and as a result, changes of pressure in soft matrices will affect deformation reducing the effectiveness of VI (Chiralt et al. 2001). Most of the studies with VI have shown that increasing the temperature favors mass transfer phenomena (water outlet and solids exchange); to describe these phenomena have been calculated diffusion coefficients and it has been shown that these parameters are related to the temperature using Arrhenius models like (Barat et al. 2001; Chiralt et al. 2001; Lombard et al. 2008; Zhao and Xie 2004). However, when using VI the goal is to reduce the temperature and time of the process and achieve the same a_w levels than would be achieved in osmotic processes performed at atmospheric pressure, these reductions of time and temperature reduces the quality loss of the impregnated food.

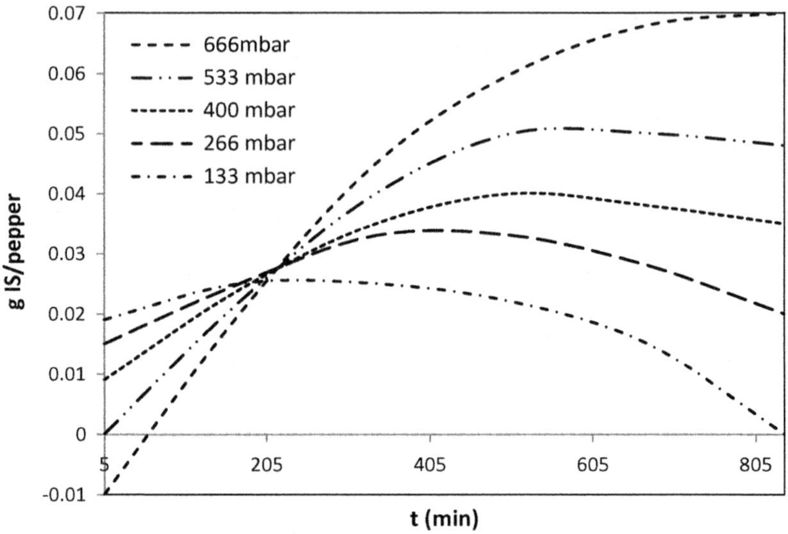

Fig. 12.2 Impregnation kinetics of jalapeño pepper with an isotonic solution at different vacuum levels and 5 min of vacuum application (Martínez-Monteagudo et al. 2006)

12.4 Solution Composition and the Impregnation Process

Main application of VI is the reduction of water activity of the food. The reduction achieved is affected by the type and concentration of solute used in the impregnation solution, but also the incorporation by impregnation of certain compounds like minerals, vitamins, phytochemicals, and microorganisms in the food matrix allows the development of functional foods with improved nutritional quality. Food impregnation is consequently a method of fortification and enrichment of foods (Alzamora et al. 2005; Watanabe et al. 2011; Zhao and Xie 2004). Functional foods have gained important niche markets due to their recognition as products with beneficial health effects beyond their role in normal growth or that can enhance physiological performance (Lockwood 2007; Wildman 2001a, 2001b).

12.4.1 Impregnation with Sugars and Salts

Sucrose and sodium chloride are the most common impregnation agents used for fruits and for vegetables, respectively. Glucose, fructose, lactose, dextrose, maltose, maltodextrins, corn starch syrup, and sorbitol, alone or in mixed are widely used to impregnate fruits, improving color, flavor, texture, and extending shelf-life due to water activity reduction. Products with better nutritional quality can also be obtained due to the solute layer formed in the food surface, layer that can reduce the loss by lixiviation of nutrients of the product (Monnerat et al. 2010).

Lombard et al. (2008) immersed pineapple cylinders in sucrose solutions of 45, 55, and 65 °Bx at 30, 40, and 50 °C for 20, 40, 60, 120, 180, and 240 min at atmospheric conditions and in a VI process at 200 mbar vacuum during the first 10 min, observing an increase in water loss and solid gain with temperature and solute concentration. The application of vacuum pulses facilitates water loss especially at higher concentrations and temperatures, in which temperature affects mostly the water loss while the solution concentration affects mostly the solid gain. Watanabe et al. (2011) studied the impregnation (4 kPa/5–60 min) of strawberry with sucrose (0.29–2.34 mol/L) to evaluate its effect on jam preparation and its anthocyanins content and color, obtaining that impregnation of strawberry stabilized anthocyanins when compared with only the addition of sucrose to the jam preparation. Mújica-Paz et al. (2003a) evaluated the impregnation of apple, mango, and melon slices at pressure of 135–674 mbar for 10 min followed by a 10-min relaxation period at atmospheric pressure in sucrose solutions of 41–60 °Brix (1:10 w/w, fruit: syrup ratio); additionally fruits were subjected at osmotic processes at atmospheric pressure for 20 min using solution with similar composition than the used in VI. These authors observed that the lowest final a_w levels were obtained in apple and mango treated at 50 °Brix syrup and vacuum pressure of 674 mbar and in melon with 57 °Brix and 593 mbar. It was observed that greater solids gain in apple than in melon and mango. Melon and mango presented weight losses of up to 8.9 % while the weight of apple increased.

Moreno et al. (2004) treated papaya slabs at atmospheric pressure and vacuum pulse at 50 mbar/10 min/30 °C in the beginning of the process, after that, atmospheric pressure was restored and samples allow standing in the impregnation sucrose solutions (55 and 65 % w/w) during 230 min more. Greater water loss was observed in samples treated with 65 °Brix solution than at 55 °Brix; likewise, solids gain was greater in those treatments when VI was applied. The application of vacuum pulses with less-concentrated osmotic solutions promotes hydrodynamic gain of osmotic solution into tissue pores, allowing to reach a higher solute concentration in the sample with less water loss than OD at atmospheric pressure. Samples subjected to atmospheric pressure exhibited initial water loss greater than samples treated with VI, and samples treated with a solution at 65 °Bx exhibited low level of weight loss as compared with the processed one at 55 °C.

The use of salts like sodium chloride, sodium, and potassium nitrite or nitrates in the impregnation process is named curing. Curing was originally developed to preserve meat and as salting process for fish and cheese by addition of sodium chloride. Salting of large food pieces is usually slow, and it takes some days because of the slow salt diffusivity. VI is a method that can be used to decrease the time of salting. The use of salts reduces the water content and a_w of foods and allows increasing the shelf-life and improving product characteristics in terms of color, flavor, and texture. Chiralt and Fito (1997), reported that VI reduced the salting process time of manchego-type cheese promoting a homogeneous distribution of salt in the product.

12.4.2 Impregnation with Minerals

Incorporation of minerals such as Ca, Fe, and Zn can be used to improve the nutritional quality of fruits and vegetables. Ostos et al. (2012) vacuum-impregnated mango cylinders with a solution of 6 % calcium lactate and 1 % calcium chloride, obtaining a final calcium concentration equivalent to 37.6 % of the recommended daily intake in 200 g of fresh mango without affecting its sensory characteristics. Fresh mushrooms cut into cylinders were vacuum impregnated with a calcium solution, promoting the incorporation of this mineral and achieving about 24–32 % of the recommended intake per each 100 g (Ortíz et al. 2003). In addition to increased calcium content, a study performed by Moraga et al. (2009) showed that the use of calcium to impregnate grapefruit by VI (50 mbar for 10 min with 180 min post treatment time) extend the shelf-life 3–6 days without altering the mechanical properties of the sample representing another advantage of this technology.

The impregnation with Ca has also been used to obtain foodstuffs of improved quality. It has been used in final products and during the drying process. González-Fésler et al. (2008) analyzed the effect of calcium impregnation treatments at atmospheric pressure or under vacuum (6666 Pa/15 min with subsequent atmospheric pressure restoration), with and without previous blanching, on the air drying kinetics of apple. The study was carried out in apple cylinders immersed in aqueous solutions containing 10.9 % (w/w) glucose and 5266 μg/g calcium salts consisting in a mixture of Ca^{2+} lactate and Ca^{2+} gluconate. VI samples were dried faster than the atmospheric impregnated ones at the beginning of the process, while samples blanched and VI impregnated were dried faster than the vacuum impregnated without blanching treatment. This study did not attempt to increase the calcium concentration in the product, but it shows the advantage of VI when drying products.

Gong et al. (2010) showed that some treatments applied before VI can improve the amount of calcium incorporated in foods; strawberries, carrots, corn, and blueberries air-dried at 65 °C, microwave-dried at 43.6 W, and freeze-dried followed by VI (with a solution of $CaCO_3$, 0.02–0.1 MPa for 30 min) and impregnation at atmospheric pressure for 90 min, showed a higher $CaCO_3$ concentration than those not submitted to VI.

Texture is an important quality parameter in fresh and processed vegetables, thus VI using solutions containing components that improve texture has been used. VI with $CaCl_2$ solutions reduces the loss of hardness in zucchini and even produces a hardening effect (Occhino et al. 2011). Impregnation with calcium combined with pectinmethylesterase (PME) has also been used to improve firmness of foods, the action of PME liberates free carboxylic acid, which could then interact with calcium, forming a gel, strengthening the cell-wall structure and increasing thus the firmness of fruits (Guillemin et al. 2006). VI process (50 mmHg/2 min) of previously pasteurized apples, strawberries, and raspberries, using $CaCl_2$ or $CaCl_2$ and PME showed that the firmness of fruits treated with PME and calcium was significantly superior than for the controls and that the use of $CaCl_2$–PME also improved firmness (Fig. 12.3) (Degraeve et al. 2003). Similar results were observed in apple pieces (Guillemin et al. 2008).

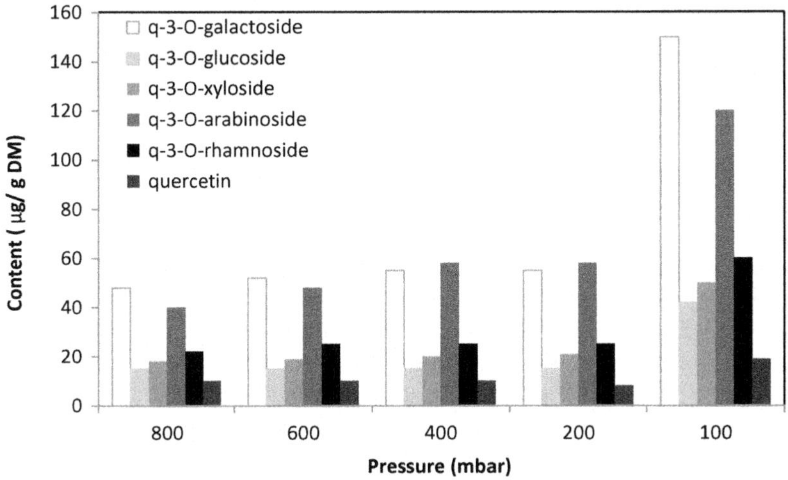

Fig. 12.3 Quercetin aglycone and quercetin glycoside content of apple slices at different pressures of the VI process (Schulze et al. 2012)

In addition to calcium, some studies with other minerals have been reported. Betoret et al. (2005) studied the VI of apple cylinders with aqueous solutions of sucrose containing calcium and iron gluconates (calcium and iron concentration up to 114.7 g/L and 2.98 g/L respectively) and also evaluated the location and distribution of the cations by Electron Dispersive X-ray Microanalysis (EDXMA), authors demonstrated an important effect of Ca concentration on the response of VI apple tissue, increasing matrix elasticity, and diminishing external liquid net fluxes, with no effect of Fe. Minerals were located mainly in the intercellular spaces, and their distribution was homogeneous.

VI (15 min vacuum at 50 mmHg followed by 30 min at atmospheric pressure) of strawberries with Ca and Zn before freezing with cryoprotectants (high fructose corn syrup or high merhoxyl pectin) showed that the VI pretreatment significantly increased the calcium and zinc content of frozen strawberries, and the cryoprotectant used improved the texture and reduced drip loss of frozen-thawed strawberries in comparison with the untreated product. Calcium increased the firmness, and zinc improved the color stability during the impregnation and freeze-thawing process. While fresh strawberries contribute with 2.3 % DRI of calcium and 1.2 % DRI of zinc, the vacuum-impregnated product contributed with 17.7 % DRI of calcium and 12.1 % DRI of zinc, respectively (Xie and Zhao 2004).

12.4.3 Impregnation with Phenolic Compounds

Phenolic compounds, widely distributed in nature, are recognized as functional molecules due to their health-promoting properties. They are classified as flavonoids (anthocyanins, flavonols, flavanols, flavones, flavanones, isoflavones, and proanthocyanidins) and nonflavonoids (phenolic acids, stilbenes, and gallotannins) with great influence in the taste, flavor, and appearance of foods (Cheynier 2005; Tapas et al. 2008; Tomás-Barberán and Espín 2001; Tripoli et al. 2007). Due to the different distribution of flavonoids in foods and the different beneficial properties that they exhibit, the impregnation of foods with these compounds represents a method to increase their concentration, or to incorporate specific compounds that are not normally presented in some foods.

Agar gel, apple, banana, and potato cubes impregnated at atmospheric conditions, showed increase of total phenolic content and antiradical scavenging capacity; moreover impregnation protected phenolic against the degradation that can occur during further convective air drying (Rózek et al. 2010). Apple slices were vacuum-impregnated (100–800 mbar/5 min, with atmospheric pressure restoration for 10 min) with quercetin glycosides obtained from commercial apple juice enriched with 0.3 % apple peel extract, and different concentrations of apple pectin and glucose to modify the viscosity and the soluble solid content (SS) of the product (Schulze et al. 2012). Vacuum improves the incorporation of quercetin in samples (Fig. 12.3) preferably in the inner apple sections and correlated with the firmness of the native apple and the increased apple weight (Schulze et al. 2012).

Gel cubes prepared with 4 % agar-agar, 9.6 % (w/w) sucrose, and distilled water were vacuum-impregnated during 0.5, 1, 2, 4, and 8 h using a multicomponent aqueous solution consisting of one or two osmo-active solutes (sucrose, sodium chloride, glycerol, and the mixture of sucrose/sodium chloride) and a commercial grape seed extract (6300 ± 45 mg GAE/kg) showed that total phenolic content increased with processing time. Treatment for 8 h using sodium chloride caused the highest increment in phenolics and antioxidant activity. When the sucrose–sodium chloride or glycerol and sucrose solutions were used as osmo-active solutes, the osmo-treated food showed a lower total phenolic content than product VI in the salt solution. This study demonstrates that the type of osmo-active agent is a key parameter in the design of intermediate moisture products with high phenolic content (Rózek et al. 2009).

12.4.4 Impregnation with Vitamins

Vitamins are a wide group of compounds with many specific functions in the body. Lin et al. (2006) evaluated the fortification (VI at 100 mmHg for 15 min followed by atmospheric restoration for 30 min) of pear slices with vitamin E using a 20 % diluted wildflower honey solution with 0.4–0.8 % α-Tocopherol from 3 different sources: α-Tocopherol-acetate (VE-acetate), free α-Tocopherol (V-OH), or water-soluble α-Tocopherol-acetate (VE-H_2O). Impregnation process increased vitamin content from 80 to 100 times, retaining from 65 to 80 % of vitamin E activity after 2 weeks storage when compared with the untreated control. Restrepo et al. (2010) vacuum-impregnated strawberries with vitamin E, obtaining 19.12 ± 3.01 mg vitamin E/100 g fruit. These results were consistent with the increase in antioxidant capacity in 14.7 % and 82.2 % evaluated by DPPH and FRAP respectively. Ursachi et al. (2009) studied the atmospheric impregnation and VI (500 mbar/10 min) of apples to incorporate ascorbic acid into the structure, and also evaluated its stability during refrigeration and freezing. Samples treated by VI showed 55 % more ascorbic acid than samples treated at atmospheric conditions, and they also presented lower degradation during storage. VI apples reduced their vitamin C content by 10.3 and 54.6 % while the impregnated ones at atmospheric pressure showed a reduction of 33.5 % and 83.5 after 3 and 9 days of refrigeration storage, respectively. Kikuchi et al. (2011) applied VI (70 cm Hg/0–60 min, followed by 3 h at atmospheric conditions) to whole potatoes for enriching them with ascorbic acid (10 % ascorbic acid solution), the effects of cooking and storage time were also evaluated. Results indicated that the ascorbic acid content of whole potatoes increased with vacuum time obtaining a maximum concentration of 150 mg/100 g.

Joshi et al. (2010) dried apple slices (2 mm-thick) that had been prepared by VI (6 in Hg/10 min, 22 min under atmospheric pressure) using a 1.6 % $CaCl_2$ (w/v), 0.05 % NaCl (w/v) and 0.1 % vitamin E (v/v) solution diluted in grape juice and an anti-browning treatment. Descriptive sensory analysis carried out by trained panelists revealed that the VI improved chip crispiness and crunchiness in comparison to the other treatments. Impregnated samples presented a calcium content of 760 mg/100 g and a vitamin E concentration of 168 mg/100 g meeting the daily requirement of these compounds in the consumer's diet.

12.4.5 Impregnation with Microorganisms

Probiotics are live microorganisms belonging to the lactic acid bacteria and bifidobacteria genus that improve host intestinal microbial balance causing improvement of gut health, lowering of blood cholesterol and improvement of the body's natural defense mechanisms. Betoret et al. (2003) combined the beneficial effects of probiotics with VI in fruits and vegetables, apple cylinders were vacuum-impregnated (50 mbar/10 min/10 min atmospheric pressure), either with commercial apple juice containing *Saccharomyces cerevisiae*, whole milk or apple juice containing *Lactobacillus casei* (spp. rhamnosus). After impregnation, apple samples contained around 10^7 CFU/g, and in order to increase stability and to assure fruit preservation, impregnated apple samples were air-dried at 40 °C and stored at room temperature for 2 months. *L. casei* viable cells population in dried and stored product was greater than 10^6 CFU/g. Marín et al. (2010a) vacuum-impregnated capegooseberry with *Lactobacillus plantarum* achieving a concentration of $1.5 \pm 0.6 \times 10^9$ CFU/g fresh fruit, which remains stable during 15 days at refrigeration conditions. In other study, Marín et al. (2010b) evaluated the effect of the impregnation of the same product with *Lactobacillus plantarum* and the commercial strain *Lactobacillus casei* ATCC 393 using a 14 % (p/p) glucose solution inoculated with the lactobacilli, on the sensorial acceptance of the product. Results showed that impregnated samples were sweeter, juicier, more translucent, more orange-colored, less acid, and less harsh than the untreated ones. Krasaekoopt and Suthanwong (2008) impregnated guava and papaya pieces under a vacuum pressure of 50 mBar with 15 and 30 °Bri x-extracted fruit juices containing 10^{10} CFU/mL of *Lactobacillus casei* by times of 1–15 min allowing the incorporation of 10^8–10^9 log CFU/g of probiotics. Samples impregnated for 10 and 5 min with fruit juice containing 15 °Brix were dried at 40 °C/36 h and storage at 4 °C for 4 weeks obtaining that the viable cell counts of *L. casei* 01 in both samples were approximately 10^7 log CFU/g, which is similar to the level found in dairy products.

Impregnation of foods with microorganism by combining VI and edible films has also been studied, Tapia and Carmona (2008) incorporate 10^7 log CFU of *Bifidobacterium lactis* Bb12/g in addition to other bioactive compounds in papaya cubes, demonstrating the advantages of VI to develop functional products.

12.5 Final Remarks

VI is a process designed to incorporate compounds in a food matrix and to exchange water from the food to the impregnation solution or both. Due to the water activity depression achieved during the process, this process allows the development of minimally processed or intermediate moisture foods with better organoleptic characteristics than products processed at atmospheric osmotic impregnation/dehydration or other more aggressive technologies. VI also enables the improvement of the nutritional quality and functionality of foods due to the incorporation of vitamins, minerals, bioactive compounds, and microorganisms, being an adequate process to develop functional foods. In the design of VI process, it is necessary to consider different parameters or characteristics of: (a) food (porosity, tissue structure, size and geometry), (b) impregnation solution (concentration and type of solute), (c) the process (vacuum pressure, exposure time, relaxation time at atmospheric pressure, temperature, product/solution relationship, and agitation). As these parameters are monitored, controlled, and applied, the nutritional and sensory characteristics of the foods will be less affected and inclusive, improved in achieving the goal to design new minimally processed or functional foods.

Acknowledgement Authors acknowledge the financial support from Tecnológico de Monterrey (Research Chair Funds CAT-200 and CDB081) and postdoctoral position CONACYT-Tecnológico de Monterrey.

References

Alzamora, S.M., D. Salvatori, M.S. Tapia, A. López-Malo, J. Welti-Chanes, and P. Fito. 2005. Novel functional foods from vegetable matrices impregnated with biologically active compounds. *Journal of Food Engineering* 67: 205–214.

Barat, J.M., A. Chiralt, and P. Fito. 2001. Effect of osmotic solution concentration, temperature and vacuum impregnation pretreatment on osmotic dehydration kinetics of apple slices. *Food Science and Technology International* 8(5): 451–456.

Betoret, N., L. Puente, M.J. Díaz, M.J. Pagán, M.J. García, M.L. Gras, J. Martínez-Monzo, and P. Fito. 2003. Development of probiotic-enriched dried fruits by vacuum impregnation. *Journal of Food Engineering* 56: 273–277.

Betoret, N., J. Martinez-Monzo, P.J. Fito, and P. Fito. 2005. Calcium and iron distribution in fortified vacuum-impregnated fruits determined by electron dispersion X-ray microanalysis. *Journal of Food Science* 70(1): 26–30.

Cheynier, V. 2005. Polyphenols in foods are more complex than often thought. *The American Journal of Clinical Nutrition* 81: 223S–229S.

Chiralt, A., and P. Fito. 1997. Salting of Manchego type cheese by vacuum impregnation. In *Food engineering 2000*, ed. P. Fito, E. Ortega, and G. Barbosa, 214–230. New York: Chapman & Hall.

Chiralt, A., and P. Talens. 2005. Physical and chemical changes induced by osmotic dehydration in plant tissues. *Journal of Food Engineering* 67: 167–177.

Chiralt, A., P. Fito, J. Barat, A. Andrés, C. Gónzalez-Martínez, I. Escriche, and M.M. Camacho. 2001. Use of vacuum impregnation in food salting process. *Journal of Food Engineering* 49: 141–151.

Degraeve, P., R. Saurel, and Y. Coutel. 2003. Vacuum impregnation pretreatment with pectinmethylesterase to improve firmness of pasteurized fruits. *Journal of Food Science* 68(2): 716–721.

Fito, P. 1994. Modeling of vacuum osmotic dehydration of food. *Journal of Food Engineering* 22: 313–328.

Fito, P., and A. Chiralt. 1995. An update on vacuum osmotic dehydration. In *Food preservation by moisture control*, ed. G. Barbosa-Cánovas and J. Welti-Chanes, 351–373. Lancaster, PA: Technomic Publishing.

Fito, P., A. Andrés, A. Chiralt, and P. Pardo. 1996. Coupling of hydrodynamic mechanism and deformation-relaxation phenomena during vacuum treatments in solid porous food liquid systems. *Journal of Food Engineering* 27: 229–240.

Fito, P., A. Chiralt, J. Barat, A. Andrés, J. Martíne-Monzó, and N. Martínez-Navarrete. 2001. Vacuum impregnation for development of new dehydrated products. *Journal of Food Engineering* 49: 297–302.

Fito, P., A. Chiralt, J. Barat, and J. Martínez-Monzó. 2002. Mass transfer and deformation relaxation phenomena in plant tissues. In *Engineering and food for the 21st century*, ed. J. Welti-Chanes, G. Barbosa-Cánovas, and J. Aguilera, 235–252. Boca Raton, FL: CRC Press.

Gong, Z., L. Gao, J. An, M. Zhang, A.S. Mujumdar, and J. Sun. 2010. Effects of predrying and vacuum impregnation with nano-calcium carbonate solution on strawberries, carrots, corn, and blueberries. *Drying Technology* 28: 36–41.

González-Fésler, M., D. Salvatori, P. Gómez, and S.M. Alzamora. 2008. Convective air drying of apples as affected by blanching and calcium impregnation. *Journal of Food Engineering* 87: 323–332.

Guillemin, A., P. Degraeve, F. Guillon, M. Lahaye, and R. Saurel. 2006. Incorporation of pectinmethylesterase in apple tissue either by soaking or by vacuum-impregnation. *Enzyme and Microbial Technology* 38: 610–616.

Guillemin, A., F. Guillon, P. Degraeve, C. Rondeau, M.C. Devaux, F. Huber, E. Badel, R. Saurel, and M. Lahaye. 2008. Firming of fruit tissues by vacuum-infusion of pectin methylesterase: Visualisation of enzyme action. *Food Chemistry* 109: 368–378.

Joshi, A.P.K., H.P.V. Rupasinghe, and N.L. Pitts. 2010. Sensory and nutritional quality of the apple snacks prepared by vacuum impregnation process. *Journal of Food Quality* 33: 758–767.

Khin, M.M., W. Zhou, and C. Perera. 2005. Development in the combined treatment of coating and osmotic dehydration of food—A review. *International Journal of Food Engineering* 1: 1–19.

Kikuchi, M., H. Koaze, T. Sato, M. Kojima, K. Yamamoto, K. Yasuda, M. Mori, and S. Tsuda. 2011. Ascorbic acid enrichment of whole potato tuber by vacuum-impregnation. *Food Chemistry* 127(3): 1114–1118.

Krasaekoopt, W., and B. Suthanwong. 2008. Vacuum impregnation of probiotics in fruit pieces and their survival during refrigerated storage. *Kasetsart Journal: Natural Science* 42(4): 723–731.

Lin, D.S., S.W. Leonard, C. Lederer, M.G. Traber, and Y. Zhao. 2006. Retention of fortified vitamin E and sensory quality of fresh-cut pears by vacuum impregnation with honey. *Journal of Food Science* 71(7): S553–S559.

Lockwood, B. 2007. *Nutraceuticals*, 1–18. London: Pharmaceutical Press.

Lombard, G.E., J.C. Oliveira, P. Fito, and A. Andrés. 2008. Osmotic dehydration of pineapple as a pre-treatment for further drying. *Journal of Food Engineering* 85: 277–284.

Marín, Z.T.A., M.R. Cortés, and O.I.C. Montoya. 2010a. Capegooseberry (*Physalis peruviana* L.) Colombian ecotype, minimally processed inoculated with native strain *Lactobacillus plantarum* LPBM10 by means of vacuum impregnation technique. *Revista Chilena de Nutricion* 37(4): 461–472.

———. 2010b. Sensory evaluation of cape gooseberry (*Physalis peruviana* L.) Colombian ecotype impregnated with native strain *Lactobacillus plantarum* LPBM10 and commercial strain *Lactobacillus casei* ATCC 393. *Vitae* 17(1): 21–28.

Martínez-Monteagudo, S., F. Salais-Fierro, J.R. Perez-Carrillo, A. Valdez-Fragoso, J. Welti-Chanes, and H. Mújica-Paz. 2006. Impregnation and infiltration kinetics of isotonic solution in whole jalapeño pepper using a vacuum pulse. *Journal of Food Science* 71(3): 125–131.

Martínez-Navarrete, N., A. Andrés-Grau, A. Chiralt-Boix, and P. Fito-Maupoey. 1999. *Termodinámica y Cinéticas de Sistemas Alimento Entorno*, 331. Valencia, Spain: Servicio de Publicaciones.

Monnerat, S.M., T.R.M. Pizzi, M.A. Mauro, and F.C. Menegalli. 2010. Osmotic dehydration of apples in sugar/salt solutions: Concentration profiles and effective diffusion coefficients. *Journal of Food Engineering* 100: 604–612.

Moraga, M.J., G. Moraga, P.J. Fito, and N. Martínez-Navarrete. 2009. Effect of vacuum impregnation with calcium lactate on the osmotic dehydration kinetics and quality of osmodehydrated grapefruit. *Journal of Food Engineering* 90: 372–379.

Moreno, J., G. Bugueño, V. Velasco, G. Petzold, and G. Tabilo-Munizaga. 2004. Osmotic dehydration and vacuum impregnation on physicochemical properties of Chilean papaya (*Carica candamarcensis*). *Journal of Food Science* 69(3): 102–106.

Mújica-Paz, H., A. Valdez-Fragoso, A. López-Malo, E. Palou, and J. Welti-Chanes. 2003a. Impregnation and osmotic dehydration of some fruits: Effect of the vacuum pressure and syrup concentration. *Journal of Food Engineering* 57: 305–314.

———. 2003b. Impregnation properties of some fruits at vacuum pressure. *Journal of Food Engineering* 56: 307–314.

Occhino, E., I. Hernando, E. Llorca, L. Neri, and P. Pittia. 2011. Effect of vacuum impregnation treatments to improve quality and texture of zucchini (*Cucurbita pepo*, L). *Procedia Food Science* 1: 829–835.

Ortíz, C.F., D.M. Salvatori, and S.M. Alzamora. 2003. Fortification of mushroom with calcium by vacuum impregnation. *Latin American Applied Research* 33: 281–287.

Ostos, A.S.L., M.A.C. Díaz, and M.H. Suarez. 2012. Evaluation process in different conditions of mango fortification (Tommy Atkins) with calcium by vacuum impregnation. *Revista Chilena de Nutricion* 39(2): 181–190.

Restrepo, A.M., M. Cortés, and B.A. Rojano. 2010. Enhancement of the antioxidant capacity of strawberries (*Fragaria ananassa* Duch.) by incorporation of vitamin E using the vacuum impregnation technique. *Vitae* 17(2): 135–140.

Roa, V., M.S. Tapia, and F. Millán. 2001. Mass balances in porous foods impregnation. *Journal of Food Science* 66(9): 1332–1336.

Rózek, A., I. Achaerandio, C. Güell, F. López, and M. Ferrando. 2009. Grape phenolic impregnation by osmotic treatment: Influence of osmotic agent on mass transfer and product characteristics. *Journal of Food Engineering* 94: 59–68.

Rózek, A., J.V. García-Pérez, F. López, C. Güell, and M. Ferrando. 2010. Infusion of grape phenolics into fruits and vegetables by osmotic treatment: Phenolic stability during air drying. *Journal of Food Engineering* 99: 142–150.

Salvatori, D., A. Andrés, A. Chiralt, and P. Fito. 1998. The response of some properties of fruits to vacuum impregnation. *Journal of Food Process Engineering* 21: 59–73.

Schulze, B., S. Peth, E.M. Hubbermann, and K. Schwarz. 2012. The influence of vacuum impregnation on the fortification of apple parenchyma with quercetin derivatives in combination with pore structures X-ray analysis. *Journal of Food Engineering* 109: 380–387.

Shi, J., and M. Le Maguer. 2002. Osmotic dehydration of foods: Mass transfer and modeling aspects. *Food Reviews International* 18(4): 305–335.

Tapas, A.R., D.M. Sakarkar, and R.B. Kakde. 2008. Flavonoids as nutraceuticals: A review. *Tropical Journal of Pharmaceutical Research* 7(3): 1089–1099.

Tapia, M.S., and A. Carmona. 2008. Uso de la impregnación a vacío y de películas comestibles en el desarrollo de un alimento funcional a base de papaya (*Carica papaya*) c.v Maradol. *Memorias del Instituto de Biología Experimental* 5: 110–114.

Tapia de Daza, M.S., S.M. Alzamora, and J. Welti-Chanes. 1996. Combination of preservation factors applied to minimal processing of foods. *Critical Reviews in Food Science and Nutrition* 36: 629–659.

Tomás-Barberán, F., and J.C. Espín. 2001. Phenolic compounds and related enzymes as determinants of quality in fruits and vegetables. *Journal of the Science of Food and Agriculture* 81: 853–876.

Tripoli, E., M. La Guadia, S. Giammanco, D. Di Majo, and M. Diammanco. 2007. Citrus flavonoids: Molecular structure, biological activity and nutritional properties: A review. *Food Chemistry* 104: 466–479.

Ursachi, C., R. Segal, and C. Muresan. 2009. Vacuum impregnation pretreatment of fresh cut vegetable. *Acta Technica Corviniensis – Bulletin of Engineering* 20: 17–20.

Watanabe, Y., K. Yoshimoto, Y. Okada, and M. Nomura. 2011. Effect of impregnation using sucrose solution on stability of anthocyanin in strawberry jam. *LWT – Food Science and Technology* 44: 891–895.

Wildman, R.E.C. 2001a. Classifying nutraceuticals. In *Handbook of nutraceuticals and functional foods*, ed. R.E.C. Wildman, 13–30. Boca Raton, FL: CRC Press.

———. 2001b. Nutraceuticals. In *Handbook of nutraceuticals and functional foods*, ed. R.E.C. Wildman, 1–12. Boca Raton FL: CRC Press.

Xie, J., and Y. Zhao. 2004. Use of vacuum impregnation to develop high quality and nutritionally fortified frozen strawberries. *Journal of Food Processing and Preservation* 28: 117–132.

Zhao, Y., and J. Xie. 2004. Practical applications of vacuum impregnation in fruit and vegetable processing. *Trends in Food Science & Technology* 15(9): 434–451.

Chapter 13
High Pressure Processing of Fruit Products

Amauri Rosenthal, Prashant Raj Pokhrel, Elisa Helena da Rocha Ferreira, Julia Hauck Tiburski, Gustavo V. Barbosa-Cánovas, and Jorge Welti-Chanes

13.1 Introduction

High pressure processing (HPP) is a nonthermal technology that has quite recently been adopted by the food industry. The potential of maintaining nutritional, sensory, and quality while inactivating pathogenic and deteriorative microorganisms and modifying enzymes activities makes this technology very appropriate for meeting increasing consumer demand for minimally processed, healthier and safer products. High pressure in many applications has little or no effect on nutrients and bioactive compounds or on other compounds responsible for the flavor and color of fruits. These characteristics of HPP are very appealing for fruit processing because, in thermal processing, high temperature greatly destroys heat-sensitive compounds and leads to cooked or altered flavors.

HPP is a post-packaging process in which a product kept in a flexible package is placed into a compression chamber. The chamber is sealed and filled with a pressure transmission fluid (commonly water and oil) and the system pressurized by a pumping and pressure intensification mechanism. HPP has been shown to provide efficient inactivation of vegetative microbial cells. Commonly, pressures between 400

A. Rosenthal (✉)
Embrapa Agroindústria de Alimentos, Rio de Janeiro, RJ, Brazil
e-mail: amauri.rosenthal@embrapa.br

P.R. Pokhrel • G.V. Barbosa-Cánovas
Washington State University, Pullman, WA, USA

E.H. da Rocha Ferreira • J.H. Tiburski
Federal Rural University of Rio de Janeiro, Seropédica, RJ, Brazil

J. Welti-Chanes
Tecnológico de Monterrey, Monterrey, Mexico

A. Rosenthal et al. (eds.), *Fruit Preservation*, Food Engineering Series,
https://doi.org/10.1007/978-1-4939-3311-2_13

and 800 MPa for 1 and 15 min have turned out to be effective for preserving quality and achieving microbial stability, thus providing prolonged shelf life under refrigerated conditions (Patterson et al. 2006; Rastogi et al. 2007; Penchalaraju and Shireesha 2013).

The application of HPP in food products was initially investigated by Hite in 1889, focusing on shelf-life extension of milk, and then later focused on other types of foods (Yaldagard et al. 2008). Although other studies were carried out in the twentieth century, only from 1980 onwards research on high pressure was intensified and disseminated, starting in Japan and followed by the United States and European countries. The first commercial product launched in Japan was a jam manufactured by Meidi-ya in the early 1990s. Most of the products initially processed by HPP were fruit based, such as jellies, jams, and juices. Fruit processing is still one of the main food sectors of high-pressure application, regardless of the fact that the application to other food products has significantly increased, such as on meat, seafood, dairy, and ready-to-eat meals. The use of HPP has been disseminated worldwide, being implemented in North America, Europe, Asia, Oceania, and more recently in Latin American countries (Tonello 2011). Guacamole was the first high pressure-processed product manufactured in Mexico and marketed in the United States by AVOMEX, now Fresherized Foods.

This technology has been mainly used for food preservation, although it has possibilities for other applications, such as texture modification, shift or assisted freezing, and extraction of food components (Knorr et al. 1998; Rastogi and Knorr 2013; Naik et al. 2013). The negligible effect of HPP on covalent bonds benefits the process of preserving nutritional and sensory attributes as compared to conventional thermal processes. On the other hand, high pressure can affect chemical bonds responsible for the conformational structure of complex molecules, such as polysaccharides and proteins, by affecting hydrophilic and hydrophobic bonds, and also to some extent by destabilizing hydrogen bonds (Cheftel 1992; Balasubramaniam et al. 2008). These effects lead to the possibility of application in gelation and gelatinization, which enable the production of different products typically prepared with the use of heating. Jacobo-Velázquez et al. (2017) have recently discussed the possibility that high hydrostatic pressure (HHP) treatment may act as an abiotic elicitor. The authors suggest that by appropriate HPP conditions, the accumulation of specific groups of bioactive compounds can be triggered during postharvest of horticultural crops, making it a promising application to be further investigated.

Despite being considered a nonthermal technology, during processing, temperature elevation occurs in all treatments due to compression heating. The increase in temperature depends on the pressure-transmitting medium (usually water) and on the food composition, and it is on average around 3 °C for every 100 MPa (Rasanayagam et al. 2003; Rastogi et al. 2007). Although HHP has proven to provide an efficient inactivation of vegetative cells of bacteria, yeasts, and molds, the process presents limitations regarding spores, which can significantly vary accordingly to the genus, specimens, and even strains of microorganisms. Also, the process is commonly considered to preserve bioactive compounds, but the retention of bioactive compounds depends on the food matrices and the applied processing conditions.

A review on the use of HPP to process various fruit products and some of the challenges in adopting this technology on the industrial scale are presented. The effect of HPP on various fruit products regarding microbial and enzymatic inactivation as well as on the retention of bioactive compounds and antioxidant activity is explored in this chapter.

13.2 Principles of High Pressure Processing (HPP)

Two basic principles govern high-hydrostatic-pressure processing: Le Chatelier's principle and the isostatic principle. According to Le Chatelier's principle, any phenomenon (phase transition, chemical reaction) that results in a decrease in total volume (negative activation volume) is enhanced by pressure. High pressure stimulates reactions that lead to a reduction in volume but opposes reactions that involve an increase in volume (Cheftel 1995; Farkas and Hoover 2000; Balasubramaniam et al. 2008; Mujica-Paz et al. 2011). According to the isostatic principle, high pressure processing is independent of volume. Pressure is transmitted instantaneously and uniformly throughout the product, whether the sample is in direct contact with the pressure medium or in indirect contact, i.e., in the package. HPP allows rapid and uniform pressure transmission throughout the food, independently of its size and geometry, which enables homogeneous products and facilitates scaling up the process (Patterson et al. 2006; Rastogi et al. 2007; Balasubramaniam et al. 2008).

13.3 High Pressure Processing Equipment and Related Costs

Some of the commercial scale manufacturers of high pressure processing systems are listed in Table 13.1. The design of HPP to process food is based on the same parameters used in nonfood industries such as ceramics and chemicals (Balda et al. 2012). There are basically two types of design: vertical and horizontal loading. The first horizontal HPP system was manufactured by GEL ALSTOM ACB in 1998 (Jung et al. 2010). HPP systems typically consist of a pressure vessel, two end closures, a device for restraining the end closures (yoke, thread, and pin), a pump, and

Table 13.1 Commercial scale high pressure processing manufacturers

Company	Country	Source
Avure Technologies Inc.	USA	http://www.avure-hpp-foods.com/
Engineered Pressure System Inc.	USA	http://epsi-highpressure.com/
Hiperbaric	Spain	http://www.hiperbaric.com/en/
MULTIVAC	Germany	http://us.multivac.com/
Kobe Steel Ltd.	Japan	www.kobelco.co.jp/english/

a control unit. The pressure vessel is the most important part of a high-pressure system, and it is made of low-alloy steel of high tensile strength (Elamin et al. 2015). Pilot-scale vessels have the capacity of 1–30 L, and commercial scale pressure vessels have capacities from 35 to 525 L. The design of the pressure vessel is very critical in the process of the manufacturing HPP units since it has to withstand very high levels of pressure and temperature (Elamin et al. 2015; USFDA 2014). The design of pressure vessel must follow the safety guidelines, as mentioned in ASME Section 8, Division 3 of the Boiler and Pressure Vessel Code, which requires the inner cylinder to crack and allow the release of pressure so as to have a leak before the break (USFDA 2014).

Commercial scale HPP systems (Figs. 13.1 and 13.2) use indirect pressurization by employing an electrohydraulic pump (Balasubramaniam et al. 2016), an intensifier, and water as the medium to transmit pressures from 100 to 800 MPa to foods. Food-grade oil is added to the water to prevent corrosion. Other fluids could be used as pressuring medium for better heat transfer efficiency and minimize corrosion of the inner part of the vessel. Control and monitoring of the processing pressure, time, and temperature are mastered by a computer connected to the HP system. HPP is used in the food industry as a preservation technique due to its ability to inactivate microorganisms and reduce enzymatic activity. In addition, it could be used, among other applications, to condition food ingredients, for texturization, and seafood deshelling.

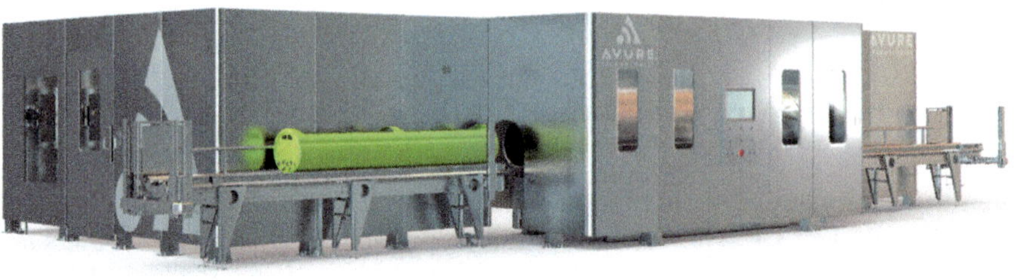

Fig. 13.1 Avure HPP unit (AV-70×) with a production capacity of 70 million pounds per year (picture courtesy of Avure Technologies, Inc.)

Fig. 13.2 Hiperbaric HPP unit (Hiperbaric 525) with a 525-L capacity and throughputs of 3000 kg of product per hour (picture courtesy of Hiperbaric)

The cost of an industrial HPP equipment can range from USD 500,000 to 4 million depending on the capacity and degree of automation (Balasubramaniam et al. 2008). The capital cost of a high pressure processing installation which includes several components such as high-pressure vessel, closures, yoke, pumping system, and process controls accounts for 75–80% for installing a HHP plant. Operating costs are in the following ranges: labor (5–10%), maintenance (5–10%), utilities (2–4%), and space (1–2%) (Balasubramaniam et al. 2016). As per Hiperbaric, processing costs in USD/lb. range from 0.064 (55 L unit) to 0.032 (525 L Unit) which clearly indicates how much it counts the size of the vessel at the time of making cost estimations (Balasubramaniam et al. 2016).

High pressure processing systems are also categorized based on the mode of application: batch or semicontinuous. In a batch process, the food product, packed in flexible containers, is placed in a basket, and then transferred to the pressure chamber where the pressure-transmitting medium is added after loading. Pressure increase to the desired level is achieved through compression of the pressure-transmitting fluid by the action of a pump or a piston connected to one or more intensifiers. The product is held for the desired time at the target pressure, and then, the pressure relief valve is opened which decompresses the vessel, and finally the product is removed from the vessel. (Balasubramaniam et al. 2008). During the pressure buildup, the temperature goes up due to compression heating of the pressure-transmitting fluid. HPP is characterized by three parameters: final pressure level (P), temperature (T), and pressure duration time (t). HPP, in principle, is independent of the size and geometry of the product, and it is a uniform and instantaneous process. Therefore, being mass/time independent gives great possibilities to control the processing conditions. An essential requirement for HPP is that water, or other pressurizing medium, is present to transmit the pressure throughout the food; thus, different types of food matrices, liquid or solid, can be high pressure processed.

In the semicontinuous high-pressure process the food product is treated before packaging. In these systems, only for pumpable products, the food is pumped into a pressure vessels until it is filled up, and then the supply valve is closed. The product inside the pressure vessels is pressurized by the high-pressure pump (Fig. 13.3). The processed product is finally transferred to the surge tank and then to the filling machine (Lelieveld and Hoogland 2016). The semicontinuous approach is not in use at this time by the food industry because of low throughputs and maintenance issues related to leaks. It is likely that in the future semicontinuous systems will be adopted by the food industry once these constraints are properly addressed.

13.4 High Pressure Homogenization (HPH)

High pressure homogenization (HPH) is an emerging nonthermal technologies that could be industrially used in liquid foods such as milk or juices. This technology has also shown promising results in microbial and enzyme inactivation in various other food products. HPH is suitable for pumpable products such as fruit juices and purées.

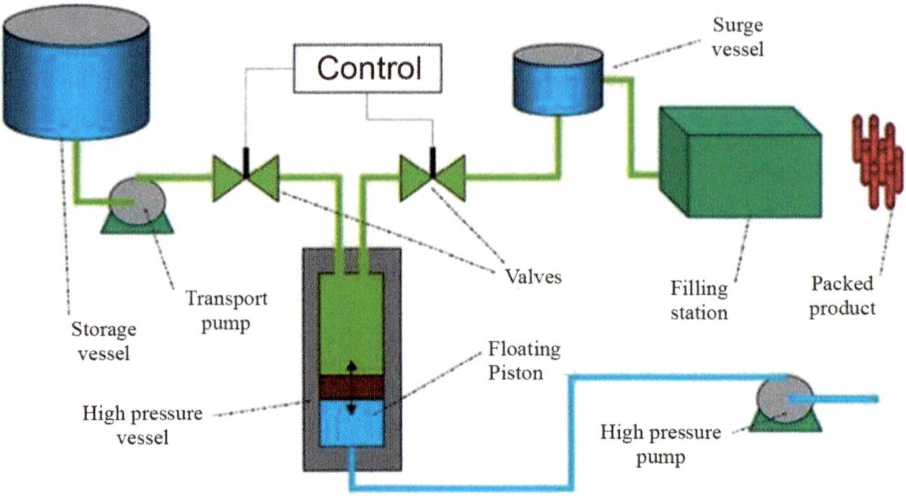

Fig. 13.3 Schemes of a semicontinuous high pressure processing system (source: Lelieveld and Hoogland 2016)

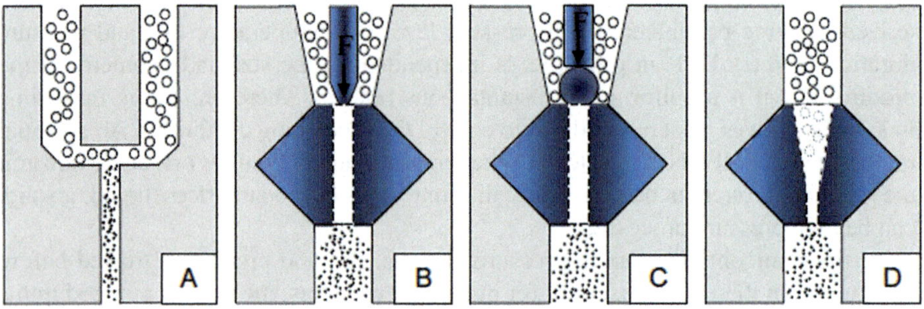

Fig. 13.4 High pressure homogenization valves [(**A**) microfluidics; (**B**) ceramic needle and seat; (**C**) ceramic ball and seat; (**D**) diamond, sapphire, or ruby nozzle; (*F*) force exerted on the needle] (source: Harte 2016)

HPH consists of a pressurized fluid being forced through a small orifice (valve, see Fig. 13.4), resulting in a great pressure gradient between the inlet and outlet of the orifice that results in intense shear forces, cavitation, turbulence, and short-life heating effects. The high pressure generated, in general, is in the range of 300 MPa (Stang et al. 2001) but recently higher pressures are used, up to 400 MPa. The product when passed through the narrow valve is subjected to high shear stress. This process involves the generation of significant heat which plays an important role during the pasteurization or other processes, as it is known that, in general, pressure and temperature act synergistically in microbial inactivation. When the food product is subjected to very high shear stress the formation of very fine emulsion droplets also takes place (Ferragut et al. 2015). Shear is caused due to the sudden flow restriction by forcing the fluid through the narrow valve (orifice) (Sanguansri and Augustin 2006). The shear force caused by the homogenizing valve as well as cavitation that takes place at the restriction due to pressure differences are responsible for the inactivation of microorganisms. The product usually undergoes a cooling step after passing through the valve. This technology can be used to pasteurize fruit juices and, without question, has great potential in the beverage processing industry (Welti-Chanes et al. 2009; Ruiz-Espinosa et al. 2013).

13.5 Fruit Products Processed by High Pressure

HPP is an industrial reality where a wide variety of foods are processed by this ever-growing technology. Cold cuts, seafood, vegetables, fruits, deeps, and dairy products are clear examples of the versatility of this technology which is revolutionizing the way a number of food products are processed. Preservation of fresh cuts, juices, smoothies, and pastes are some of the major uses in the fruit industry. Some of the applications of HPP in various fruit products are presented in this section.

13.5.1 Fresh-Cut and Dried Fruits

Preserving the quality of fresh-cut fruits is challenging as they can quickly undergo a series of reactions which may change their physical, chemical, and microbiological quality. Fresh-cut products are not very appropriate for conventional thermal processing as high temperature can cause thermal degradation of nutrients; in addition to that, there could be significant changes in their appearance and sensory attributes. By the use of HPP, fresh-cut fruits can be preserved without the use of high temperatures or chemical preservatives. HPP can preserve the quality of fresh-cut fruits in terms of color, texture, and nutritional content by inactivation of deteriorative enzymes and microorganisms (Hendrickx et al. 1998). For example, there was a three log reduction in total plate count in fresh-cut pineapples after a pressure treatment of 340 MPa for 15 min (Alemán et al. 1994). Likewise, browning in pre-cut mango was controlled after application of 800 MPa for 5 min and during storage at 3 °C (Boynton et al. 2002).

Besides being used for preservation of fresh-cut fruits, HPP has other applications such as a pretreatment before drying and dehydration. Rastogi et al. (2000) observed a higher diffusion coefficient in high pressure-pretreated pineapple cubes during osmotic dehydration compared to non-pressure-treated ones. Similarly, a study on textural changes and drying rates of high pressure-pretreated pineapple slices was conducted by Kingsly et al. (2009). These authors observed that hardness, springiness, and chewiness of pineapple slices were reduced by high pressure, whereas cohesiveness was not significantly affected. Studies show that high pressure pretreatment of fruit slices can minimize the time of drying and maintain the quality of the final products (Rastogi et al. 2000; Kingsly et al. 2009).

13.5.2 Fruit Juices, Blends, and Nectars

The increasing demand of healthy products with minimal processing is significantly promoting the use of HPP on fruit-based products such as juice, blends, nectars, and smoothies. It has been widely recognized that HPP is a sound alternative technology to thermal pasteurization of such products. HPP has successfully demonstrated at the industrial level its capability to preserve fruit juices and other fruit-based beverages with attractive shelf lives. The shelf life of orange juice was extended for more than 2 months under refrigeration conditions by the application of 350 MPa for 1 min at 30 °C (Donsi et al. 1996). Timmermans et al. (2011) observed reductions in the microbial population below detectable limit in orange juice immediately after processing at 600 MPa, 1 min at 17 °C, and during storage for up to 58 days at 4 °C. Bisconsin-Junior et al. (2014) conducted a study on orange juice (Pera Rio variety) with pH 4.18 and observed 3–4 log reduction of aerobic bacteria by the application of 550–600 MPa at 55–66 °C, with a pressure-holding time of 5.5–6 min. It has been found, as mentioned before, that high pressure has no effect on covalent bonds; this helps preserve the nutrients and sensory quality of fruit juices. Butz et al. (2003) observed that carotenoid content in orange juice and a blend of orange, carrot, and lemon juice were not significantly ($p > 0.05$) affected by high-pressure treatment at 600 MPa, 6 min. The effect of HPP on the volatile aromatic profile of strawberry juice was studied by Lambert et al. (1999). The authors observed that pressure treatments of 200 or 500 MPa at 20 °C with a holding time of 20 min did not affect the volatile aromatic profile. HPP also has the capability to inactivate enzymes present in the juice. Cano et al. (1997) observed a significant loss of PPO and POD in strawberry juice after the application of HPP in the range of 250–400 MPa. Cloud stabilization in juice is also an important aspect in its processing, application of 700 MPa for 1 min stabilized the cloud in freshly squeezed orange juice, and the shelf life of the processed juice was 90 days under refrigeration conditions (Goodner et al. 1999).

13.5.3 Fruit Pastes and Purées

Fruit-based pastes and purées such as guacamole, banana purée, avocado based salsas, among others, have a very short shelf life (5–10 days) due to the deteriorative action of various enzymes. Application of high temperature for pasteurization of such products adversely affects their organoleptic quality. High pressure is very appropriate to process those products since it has little or no effect on organoleptic quality but can effectively inactivate enzymes and deteriorative microorganisms.

Suthanthangjai et al. (2005) studied the stability of anthocyanin in raspberry puree; anthocyanin was more stable in the puree when it was treated under 200–700 MPa and stored at 4 °C. Krebbers et al. (2003) conducted a study on inactivation of *Bacillus stearothermophilus* spores in tomato puree and found that the application of 700 MPa at 90 °C for 30 s did not significantly affect lycopene content, but reduced the spores by 4.5 log. Patras et al. (2009) studied the effect of HPP on the bioactive compounds present in strawberry and blackberry purées; the study shows that there were no significant changes ($p > 0.05$) in ascorbic acid and anthocyanin content after pressure treatment of 400 MPa, 15 min. But in the same study the authors found significant ($p \leq 0.05$) changes (as compared to unprocessed) in those compounds after thermal processing ($P_{70} \geq 2$ min).

13.6 High-Pressure Inactivation of Microorganisms in Fruit Products

Several studies on the inactivation of pathogenic and spoilage microorganisms using high pressure have been published in recent years and have demonstrated that HPP is a preservation method that effectively inactivates microorganisms in fruit and vegetable products (Table 13.2). Processing pressure, pressure holding time, decompression time, and treatment temperature have an important role in the inactivation of microorganisms (Zimmermann et al. 2013).

13.6.1 Kinetics of Microbial Inactivation by High Pressure

Predictive kinetic models have been used to identify the resistance of target microorganisms and their inactivation behavior. Many studies on conventional thermal processing have used first-order kinetic models to predict the microbial inactivation. There is evidence that inactivation of microorganisms by high pressure processing as well as other nonthermal processes may not follow first-order kinetics (Peleg and Cole 1998). To describe a nonlinear survivor curve, several models, such as Weibull, modified Gompertz, log-logistic, and biphasic, have been proposed (Tola and Ramaswamy 2014; Zhao et al. 2014; Moody et al. 2014; Serment-Moreno et al. 2014, 2015). Table 13.3 presents some inactivation models and their mathematical expression to describe the linear and nonlinear survivor curves.

Weibull model was first introduced by Peleg and Cole (1998), and due to its simplicity, it has been used widely in predictive microbiology for a number of nonthermal processes. This model can be used to describe both linear and nonlinear

Table 13.2 High-pressure inactivation of microorganisms in fruit and vegetable products

Microorganism/virus	High-pressure treatment	Fruit/vegetable product	Log_{10} reduction	Reference
Aerobic microorganisms	350 MPa, 10 min, 20 °C	Cucumber juice (pH 6.4)	2.5–4.5	Zhao et al. (2014)
	400 MPa, 10 min, 20 °C			
	500 MPa, 2 min, 20 °C			
	400 MPa, 5 min, room temperature	Pomegranate juice	4.5	Chen et al. (2013)
	550 to 600 MPa, 5.5 to 6 min, 55 to 66 °C	Orange juice (Pera Rio variety) (pH 4.18)	3.0–4.0	Bisconsin-Junior et al. (2014)
	400 MPa, 15 min, 25 °C	Orange juice (pH 3.55)	7.0	Erkmen (2011)
E. coli O157:H7	400 MPa, 10 min	Mango juice	6.0	Hiremath and Ramaswamy (2012)
E. coli O157:H7	0.1–250 MPa, 20 min, 4 and 25 °C	Orange juice	6.5 (215 MPa, 4 °C)	Noma et al. (2004)
		Apple juice	7.0 (250 MPa, 25 °C)	
E. coli (ATCC 11775)	350 MPa, 5 min	Kiwifruit juice	5.0	Buzrul et al. (2008)
		Pineapple juice	2.5	
E. coli (CECT 515)	150–350 MPa, 5 min, 20, 40 and 60 °C	Apple juice	6.0 (203 MPa, 57 °C)	Muñoz et al. (2007)
		Orange juice	6.0 (248 MPa, 60 °C)	
E. coli (ATCC 29055)	150–400 MPa, 5 min, 25 °C	Apple juice	8.0 (400 MPa)	Ramaswamy et al. (2003)
E. coli	400 MPa, 10 min, 20 °C	Beetroot juice	6.2	Sokołowska et al. (2014)
	241 MPa, 3 min.	Orange juice	5.0	Guerrero-Beltrán et al. (2011)
E. coli (ATCC 11775)	300–600 MPa, 0.5–7 min, 21 °C	Apple juice	7.0	Moody et al. (2014)
S. typhimurium	400 MPa, 10 min	Orange juice	7.0	Erkmen (2011)

(continued)

Table 13.2 (continued)

Microorganism/virus	High-pressure treatment	Fruit/vegetable product	Log_{10} reduction	Reference
S. typhimurium	300–600 MPa, 4–369 s, 20 °C	Orange juice (Navel and Valencia variety)	5.0	Bull et al. (2005)
S. montevideo				
S. enteritidis				
L. monocytogenes	500 MPa, 1 to 5.5 min, 20 °C	Carrot juice	> 6.0	Patterson et al. (2012)
L. monocytogenes (NCTC 11994)	300 MPa, 5 min, 20 °C	Orange juice	3.0	Jordan et al. (2001)
		Apple juice	5.0	
L. innocua (ATCC 33090)	350 MPa, 5 min	Kiwifruit juice	5.0	Buzrul et al. (2008)
		Pineapple juice	3.5	
Norovirus	400 MPa, 2.5 min, room temperature	Strawberry purée	3.33	Kovač et al. (2012)
Hepatitis A	375 MPa, 5 min, room temperature	Strawberry purée	4.32	Kingsley et al. (2005)

survivors' curves where inactivation curve with shoulder and tails can be easily fitted. Yamamoto et al. (2005) used the logistic, Gompertz, and Weibull functions to model the inactivation kinetics of *E. coli* (ATCC 25922) by high pressure in phosphate-buffered saline. It was found that the Weibull function was the best to predict the inactivation of this microorganism.

The modified Gompertz model is used mainly to fit asymmetrical sigmoidal shape inactivation curves but this model does not predict the survivors very well when the inactivation curve is log-linear and when the data has a heterogeneous population.

The three-parameter biphasic model is widely used and assumes that the microbial population is divided into two subgroups, one treatment sensitive, and the other treatment resistant. The treatment-sensitive population is inactivated at the early stage, and the resistant portion at a later time forming two phases in the inactivation curve (Coroller et al. 2006).

13.6.2 Microorganisms of Concern in Fruit Products

13.6.2.1 *Salmonella*

Salmonella belongs to the family *Enterobacteriaceae* and comprises non-spore-forming gram-negative bacilli. Although most fruit juices have an acidic environment, some species of *Salmonella* can survive in these conditions. A study on high

Table 13.3 Kinetic models to describe linear and nonlinear microbial inactivation

Inactivation model	Mathematical expression	Notations
First order	$\text{Log}\left(\frac{N_t}{N_0}\right) = -\frac{t}{D}$	D: Decimal reduction time
Weibull	$\text{Log}\left(\frac{N_t}{N_0}\right) = -bt^n$	b: Scale factor
		n: Shape factor
Modified Gompertz	$\text{Log}\left(\frac{N_t}{N_0}\right) = a.\exp\left[-\exp(b + c.t)\right] - a.\text{Exp}\left[-\exp(b)\right]$	a, b, and c: Fitting parameters
Log-logistic	$\text{Log}\left(\frac{N_t}{N_0}\right) = \frac{A}{1+e^{4\sigma(\tau - logt)/A}} - \frac{A}{1+e^{4\sigma(\tau - \log t_0)/A}}$	A: Upper asymptote—lower asymptote
		σ: Maximum inactivation rate
		τ: Log time to reach maximum inactivation rate
Biphasic	$N_t = N_0\left(fe^{-k_1 t} + (1-f)e^{-k_2 t}\right)$	$(1-f)$: Fraction of treatment-sensitive cells
		f: Fraction of treatment-resistant cells
		k_1 and k_2: death rate constant

Note: N_t is the survivor at time t (min), and N_0 is the initial population of microorganisms

pressure processing of orange juice concluded that the application of 400 MPa for 10 min could result in 7 log reduction of *Salmonella* (Erkmen 2011). Likewise, Bull et al. (2005) observed 5 log reduction by the application of 300–600 MPa pressure for up to 6.08 min applied in Navel and Valencia orange juices. Erkmen (2011) also observed that *Salmonella typhimurium* is more baroresistant in milk than in peach and orange juices. This baroresistant difference may be related to the milk composition, i.e., proteins, carbohydrates, lipids, amino acids, and vitamins.

13.6.2.2 *Escherichia coli*

Escherichia coli not only provides information on hygienic conditions during product processing but also indicates possible occurrence of pathogens. This is explained by the fact that among the strains of this microorganism, there is a group capable of causing foodborne diseases (FBDs), called enteropathogenic *Escherichia coli* (EEC). There are currently six recognized classes of EEC, including the strains of *Escherichia coli* O157:H7, which belong to the group of enterohemorrhagic (EHEC) or verotoxin-producing (VTEC) *Escherichia coli*. To inactivate this strain using HPP, 215–400 MPa pressure treatments for 10–20 min were sufficient to achieve 6–7 log reductions in mango (Hiremath and Ramaswamy 2012), orange, and apple juices (Noma et al. 2004). Application of 200–600 MPa pressure for 5–10 min rendered 2.5 log reduction in pineapple juice (Buzrul et al. 2008) and 6 log in apple and orange, which required temperatures of 57 and 60 °C, respectively (Muñoz et al. 2007). 8 log reduction of *Escherichia coli* (ATCC 29055) was obtained in apple juice treated at 400 MPa for 5 min, as reported by Ramaswamy et al. (2003).

13.6.2.3 *Listeria monocytogenes*

Listeria monocytogenes is a serious foodborne pathogen of great concern to the fruit processing industry. *Listeria* can survive in the acidic environment (pH < 4.6) as well as in low temperature (4 °C). A study found that this microorganism survived in orange juice with pH 3.6 for 21 days at 4 °C and for 5 days at 30 °C (Parish and Higgins 1989).

Studies show that HPP is a possible alternative to a thermal process to inactivate *Listeria monocytogenes* in some fruit and vegetable products. Patterson et al. (2012) worked with pressures of 500 and 600 MPa for 1 min and 20 °C and obtained 6 log reduction of *Listeria monocytogenes* in carrot juice. Buzrul et al. (2008) observed 3.5 log reduction of *Listeria innocua* in pineapple juice and 5 log in kiwi treated at 350 MPa for 5 min. The psychotropic characteristic of *Listeria monocytogenes* makes the refrigerated environment of juices processed with HPP more favorable to its multiplication. However, after high-pressure treatment, the damaged cells generally become more sensitive to the acidity of fruit juices.

13.6.2.4 *Alicyclobacillus acidoterrestris*

Although not pathogenic, *Alicyclobacillus acidoterrestris* is an important microorganism of concern in fruit juices. This bacterium isolated from soil is gram positive, spore forming, and thermophilic, and grows at temperatures of 20–60 °C, and pH of 2.5–6.0. *Alicyclobacillus acidoterrestris* causes the formation of light sediment in fruit juices and beverages, teas, and herbal drinks (Splittstoesser et al. 1998). This microorganism is resistant to pasteurization and can deteriorate untreated or pasteurized fresh juices stored in environments where there is no temperature control. Table 13.4 shows studies on inactivation of *Alicyclobacillus acidoterrestris* combining high pressure and temperatures (200–621 MPa and 50–90 °C) for 10–45 min (Lee et al. 2002; Hartyáni et al. 2013; Sokołowska et al. 2013a). Synergistic inhibition of *Alicyclobacillus acidoterrestris* spores in apple

Table 13.4 Inactivation of *Alicyclobacillus acidoterrestris* in fruit juices by HHP

Type of bacteria	Juice	Treatment	Log_{10} reduction	Reference
Alicyclobacillus acidoterrestris spores (DMZ 2498)	Apple	600 MPa,10 min, 50 °C	3.0	Hartyáni et al. (2013)
Alicyclobacillus acidoterrestris spores (TO-29/4/02 and TO-117/02)	Apple	200 MPa, 45 min, 50 °C	4.0	Sokołowska et al. (2013a)
Alicyclobacillus acidoterrestris spores (ATCC 49025 and NFPA 1013)	Apple	414 MPa, 10 min, 71 °C	5.5	Lee et al. (2002)
Alicyclobacillus acidoterrestris (DSM 2492 vegetative cells)	Apple	300 MPa, 20 min, 50 °C	4.6	Alpas et al. (2003)
	Orange		4.4	

juice could be achieved by the combination of HHP and nisin. Various studies show reductions ranging from 4.2 to 4.7 log in the microbial population when nisin was applied, while the same process without nisin led to a reduction from 2.5 to 3.0 log (Sokołowska et al. 2012; Zhao et al. 2013). The synergistic inactivation of this microorganism was also observed when HPP was combined with other chemicals, such as citrate (Patrignani et al. 2013) and essential oils (Bevilacqua et al. 2012; Espina et al. 2013).

13.6.3 Baroprotective Effect on Some Fruit Products

The increase in resistance of microorganisms towards high pressure while increasing the solute concentration is commonly referred as a baroprotective effect (Palou et al. 1997). This effect is mainly due to the reduction in water activity due to enhancement in solute content. Reduction of water activity in fruit products protects microorganisms from being ruptured by the applied pressure. Sokołowska et al. (2013a) noticed that 200 MPa pressure treatments at 50 °C for 45 min in apple juice at a concentration of 71.1°Brix had no effect on *Alicyclobacillus acidoterrestris* spores. But in the juices of lower solid soluble concentration, 35.7, 23.6, and 11.2°Brix, reductions of 1.3–2.4 log, 2.6–3.3 log, and 2.8–4.0 log, respectively, were observed. The same effect was observed in *Saccharomyces cerevisiae* inoculated in the phosphate buffer and beet juice; 5 and 3.5 $\log_{10}$ reductions were found after a 300 MPa pressure treatment for 10 min at 20 °C, respectively (Sokołowska et al. 2013b). The baroprotective effect was also observed in hepatitis virus in samples with a concentration of salt between 1 and 6% (Kingsley and Chen 2009) and on *Byssochlamys nivea* processed by high pressure in pineapple nectar and juice (Ferreira et al. 2011).

13.7 High-Pressure Inactivation of Enzymes in Fruits

Enzymes are biological molecules that catalyze chemical reactions; their stability depends on environmental conditions like temperature, pH, solvent composition, and pressure. Therefore, changes in these conditions can disturb the delicate balance between stabilizing and destabilizing interactions leading to changes in overall enzyme conformation or changes at or near the active site, and thus resulting in loss of enzyme activity (Lopes et al. 2010; Tribst et al. 2011).

Fruits are rich in enzymes which can cause undesirable changes in color, flavor, and texture during preparation and processing. Blanching has been traditionally used in the fruit processing industry to lower the activity of the enzymes. High pressure processing is one of the very suitable alternatives to blanching to inactivate deteriorative enzymes present in fruit products.

The primary deteriorative enzymes in fruits are:

- Peroxidases (POD), related to the appearance of unpleasant odors, color changes, and loss of texture
- Polyphenol oxidases (PPO), responsible for enzymatic browning
- Pectin methylesterase (PME), which catalyzes the de-esterification of pectin and is responsible for the destabilization and changes in texture
- Lipoxygenase (LOX), associated with lipid oxidation and pigment degradation

13.7.1 Inactivation Kinetics

Currently, most of the literature regarding the kinetics of inactivation of enzymes by HPP is reported in isobaric and isothermal conditions as it is quite complex to separate the effects of the pressurization and depressurizations steps from those of the pressure-hold period (Chakraborty et al. 2014a, b). Furthermore, researchers do not usually take into account temperature changes and heat exchanges involving the product and the pressurization fluid (Serment-Moreno et al. 2014; Tejada-Ortigoza et al. 2015).

The standard kinetic models used to describe the inactivation of enzymes are presented in Table 13.5. Several empirical models have also been established to describe the kinetics of specific enzymes and microorganisms (Weemaes et al. 1998b; Van den Broeck et al. 2000; Polydera et al. 2004).

Table 13.5 Standard kinetic models to describe the inactivation of enzymes (Ludikhuyze et al. 2003; Serment-Moreno et al. 2014)

Inactivation model	Mathematical expression
First order	$A_t = A_0 e^{-kt}$
n^{th} order	$A_t = \left[A_0^{1-n} + (n-1)kt\right]^{\frac{1}{1-n}}$
Isozyme/biphasic	$A_t = A_l e^{-k_l t} + A_s e^{-k_s t}$
Fractional conversion	$A_t = A_\infty + (A_0 - A_\infty)e^{-kt}$

A enzyme activity, *n* reaction order, *k* inactivation rate (min^{-1}), *t* (min). Subscript: *t* after pressure-holding time of *t* min, *0* initial, *L* pressure-labile fraction, *S* pressure-stable fraction, ∞ after prolonged treatment

First-order model often describes the inactivation of enzymes by high pressure. The *n*th-order kinetic model is followed when a series of reactions are involved during the inactivation process (Buckow et al. 2009). A fractional conversion model is a particular case of a first-order model that assumes that the pressure treatment leaves a residual enzyme activity with much higher resistance (Rizvi and Tong 1997). A biphasic model is used when there are several isozymes present, with different temperature and pressure resistance, for example, a labile and a resistant fraction (Ramesh et al. 1998). The inactivation rate constant (k) can be estimated using a linear regression analysis of the natural logarithm of k versus inactivation time. To determine the temperature and pressure dependence of the inactivation rate constant, Arrhenius (Eq. 13.1) and Eyring (Eq. 13.2) equations are applied, generating the parameters activation energy (Ea) and activation volume (Va), respectively (Fachin et al. 2002):

$$\ln(k) = \ln(k_0) + \left[\frac{E_a}{R}.\left(\frac{1}{T_0} - \frac{1}{T}\right)\right] \quad (13.1)$$

$$\ln(k) = \ln(k_0) + \left[\frac{V_a}{RT}.(P - P_0)\right] \quad (13.2)$$

where k, k_0, R, E_a, and V_a are activation rate constant, reference rate constant at reference temperature (T_0) or pressure (P_0), universal gas constant, activation energy, and activation volume, respectively.

Pressure inactivation of enzymes from different sources has frequently been reported to follow first-order kinetics. Terefe et al. (2010) indicated that the inactivation kinetics of POD in strawberry purée by combining high pressure and thermal processing was well described by first-order kinetics, probably due to the inactivation of the labile fraction during the preheating and compression phases. This combination is known and more usually applied as high-pressure thermal processing (HPTP) or pressure-assisted thermal processing (PATS) where the food and the pressurizing media, in general water-based fluid, are preheated to the same temperature and then pressurized. Compression heating together with an external thermal energy source leads to the temperature required for the process to take place. Boulekou et al. (2010) proposed a first-order model that can accurately predict PME inactivation in peach pulp in a pressure range of 100–800 MPa and temperature of 30–70 °C. They found inactivation rate values ranging from 0.152 to 1.97 min^{-1} at 700 MPa, 30 and 70 °C, respectively. PME inactivation in orange juice was reported to follow first-order kinetics (Katsaros et al. 2010; Alexandrakis et al. 2014a). However, the results regarding inactivation rate varied greatly depending on the orange variety and purity of the enzyme.

Inactivation kinetics of orange PME in an orange juice–milk-based beverage was described using the biphasic model (Sampedro et al. 2008). The pressure-stable fraction was estimated to represent 7% of the initial activity, and based on the activation volume, the inactivation rate constant of the labile fraction was more pressure dependent than the inactivation rate constant of the stable portion. Additionally, until 550 MPa and 55 °C, only the labile fraction of PME was inactivated. Thermal and high-pressure inactivation of PME in grapefruit jam obtained in the range of 45–75 °C and 550–750 MPa was accurately described applying either the fractional or the biphasic model (Igual et al. 2013). The percentage of the pressure-stable fraction was estimated to be between 27 and 40%, and the k values increased with higher temperatures and pressures reaching a maximum of 1.2 min^{-1} at 700 MPa, 75 °C. However, at higher pressures, the inactivation rate constants of both fractions were less sensitive to temperature changes (greater Ea values).

The inactivation of purified white grapefruit PME was described by fractional conversion models with about 20% of the initial activity corresponding to the stable fraction (Guiavarc'h et al. 2005). An antagonistic effect of pressure and temperature on the inactivation rate of the heat-labile PME fraction was observed at temperatures higher than 58 °C and pressures up to 200 MPa.

13.7.2 Fruit-Based Enzymes

13.7.2.1 Peroxidase

A number of studies investigated the inactivation of peroxidase in fresh fruit by HPP (Table 13.6). In avocado slices, there was significant ($p \leq 0.05$) inactivation of POD with the application of either 400, 500, or 600 MPa for 6 min, but there was no reduction at lower pressures (200 or 300) for the same duration of treatment (Woolf et al. 2013). Yen and Lin (1996) evaluated POD inactivation in guava purée by the application of pressure of 400 and 600 MPa at 25 °C for 15 min and observed a residual activity of 90% and 74%, respectively, while thermal pasteurization

Table 13.6 Effect of high pressure processing on peroxidase activity in fruit products

Fruit	Product	Pressure, time, temperature[a]	Remaining enzyme activity (%)	Reference
Apricot	Nectar	300 MPa, 15 min, 25 °C	100	Huang et al. (2013)
		500 MPa, 15 min, 25 °C	140	
Avocado	Slices	300 MPa, 10 min, 20 °C	100	Woolf et al. (2013)
		600 MPa, 6 min, 20 °C	50	
Banana	Enzyme extract	110 MPa, 25 min, 70 °C	1.4	MacDonald and Schaschke (2000)
Kiwi	Juice	600 MPa, 15 min, 10 °C	80	Fang et al. (2008)
		600 MPa, 15 min, 50 °C	40	
Guava	Purée	400 MPa, 15 min, 25 °C	90	Yen and Lin (1996)
		600 MPa, 15 min, 25 °C	74	
Melon	Juice	500 MPa, 20 min, 22 °C	80	Ma et al. (2010)
Pineapple	Enzyme extract	600 MPa, 45 min, 60 °C	40	Rosenthal et al. (2002)
Strawberry	Fresh	300 MPa, 10 min, 60 °C	51	Terefe et al. (2009)
		600 MPa, 10 min, 20 °C	75	
	Purée	400 MPa, 5 min, 20 °C	113	Garcia-Palazon et al. (2004)
		600 MPa, 15 min, 20 °C	59	

[a]The temperatures indicated in the table refer to the temperature of the pressure vessel before pressure buildup

(88–90 °C, 24 s) completely inactivated guava's POD. The POD activity slightly decreased in melon juice treated at 500 MPa, 22 °C during the first 8 min and then had a reduction of 19% of the initial activity (i.e., maintained 81% of the original activity) when pressurization time was up to 10 min. Longer treatments did not yield higher inactivation, and the residual activity was still 78% after 20 min (Ma et al. 2010). Garcia-Palazon et al. (2004) reported that POD in strawberry was only significantly inactivated at room temperature when pressures exceeded 600 MPa; they observed a maximum inactivation of 35% after 15-min treatment.

As demonstrated by the studies above, the peroxidase is extremely pressure resistant; therefore, pressure treatments may be used in combination with other treatments to increase its inactivation. Mild heating is often considered the most appropriate method to use in conjunction with high pressure processing (Ludikhuyze et al. 2003). The application of high pressures in the range of 70–110 MPa in combination with mild temperatures (50–70 °C) in the treatment of bananas was successful on the inactivation of POD.

The inactivation of POD from fresh litchi was evaluated at different pressure levels and temperatures (Phunchaisri and Apichartsrangkoon 2005). At 200 MPa there was an increase in POD activity which was attributed to a higher availability of enzyme substrate due to cell disruption. With the application of higher pressures, at 400 and 600 MPa, a small reduction in POD activity was only observed when coupled with a mild heat treatment (60 °C). The same results were found in kiwifruit (Fang et al. 2008) as no POD inactivation was observed at pressure levels of 200 and 400 MPa, and at 600 MPa the inactivation was dependent on the temperature used. A reduction in enzyme activity of about 20%, 40%, and 60% was obtained when the treatment was performed at 10, 30, and 50 °C, respectively. Terefe et al. (2009) used response surface methodology to study the effect of pressure and mild temperature on the POD activity in fresh strawberries and observed that the residual activity ranged from 41.9 to 84.2% and that temperature had the greatest effect on POD inactivation. They predicted an optimum condition, at the studied range, at 600 MPa, 60 °C, and 10 min treatment time with a residual activity of 42%. Similarly, after treatment at 600 MPa and 60 °C, a POD residual activity of 45% was found in red grape juice by Rastogi et al. (1999).

In some cases, an increase in enzyme activity is observed after pressure treatments. Huang et al. (2013) reported 46.8% increase in POD activity in apricot nectars by the application of 500 MPa for 5 min. Similar trends on the increase of POD activity were also observed by various research groups (Cano et al. 1997; Garcia-Palazon et al. 2004; Terefe et al. 2009). The increase in enzyme activity might be attributed to the release of isoenzymes with higher enzyme activity due to their partial denaturation caused by HPP (Guerrero-Beltrán et al. 2005a).

13.7.2.2 Polyphenoloxidase

Polyphenoloxidase activity results in enzymatic browning of fruits and vegetables and also changes in organoleptic properties; therefore, its inactivation is highly desirable to maintain the quality of fruit products. Some significant results regarding PPO inactivation by high pressure are presented in Table 13.7. PPO is sensitive to pressure as compared to POD, so high pressure could be an alternative to high temperature for the irreversible inactivation of PPO (Seyderhelm et al. 1996; Hendrickx et al. 1998). However, its inactivation is primarily dependent on the type of processed fruit (Table 13.7). López-Malo et al. (1998) studied the effects of high hydrostatic pressure treatments of 345, 517, or 689 MPa, for 10, 20, or 30 min at initial pHs of 3.9, 4.1, or 4.3 on polyphenoloxidase (PPO) activity, color, and microbial inactivation in avocado puree during storage at 5, 15, or 25 °C and compared with untreated avocado puree. Significantly less ($p \leq 0.05$) residual PPO activity was obtained with increasing pressure and decreasing initial

Table 13.7 Effect of high pressure processing on polyphenoloxidase activity in fruit products

Fruit	Product	Pressure, time temperature[a]	Remaining enzyme activity (%)	Reference
Apple	Juice	600 MPa, 10 min, 30 °C	60	Buckow et al. (2009)
	Juice	750 MPa, 50 min, 50 °C	48	Valdramidis et al. (2009)
		750 MPa, 30 min, 30 °C	80	
Avocado	Purée (pH 4.1)	689 MPa, 20 min, 21 °C	20	López-Malo et al. (1998)
	Enzyme extract (pH 7.0)	900 MPa, 60 min, 25 °C	50	Weemaes et al. (1998a)
		600 MPa, 60 min, 60 °C	6	
	Guacamole	689 MPa, 20 min, 2 °C	23	Palou et al. (2000)
		4 × 689 MPa, 10 min, 2 °C	15	
	Paste (pH 6.5)	600 MPa, 3 min, 21 °C	50	Jacobo-Velázquez and Hernández-Brenes (2010)
	Slices (pH 6.5)	500 MPa, 6 min, 20 °C	100	Woolf et al. (2013)
Banana	Purée (pH 3.4)	517 MPa, 10 min, 21 °C	>100	Palou et al. (1999)
		689 MPa, 10 min, 21 °C	79	
	Enzyme extract	110 MPa, 25 min, 70 °C	25	MacDonald and Schaschke (2000)
Guava	Purée	400 MPa, 15 min, 25 °C	86	Yen and Lin (1996)
		600 MPa, 15 min, 25 °C	63	
Longan	Juice	300 MPa, 30 min, 25 °C	135	Chaikham and Apichartsrangkoon (2012)
		500 MPa, 30 min, 25 °C	95	
Litchi	Fresh	600 MPa, 10 min, 60 °C	15	Phunchaisri and Apichartsrangkoon (2005)
	Syrup	400 MPa, 20 min, 60 °C	>100	
	Syrup	600 MPa, 20 min, 30 °C	35	Dajanta et al. (2012)

(continued)

Table 13.7 (continued)

Fruit	Product	Pressure, time temperature[a]	Remaining enzyme activity (%)	Reference
Mango	Purée (pH 4.5)	379 MPa, 15 min, 25 °C	74	Guerrero-Beltrán et al. (2005b)
		379 MPa, 15 min, 25 °C + 500 ppm ascorbic acid	4.5	
Melon	Juice	500 MPa, 20 min, 22 °C	9	Ma et al. (2010)
Mulberry		365 MPa, 3.5 min, 90 °C	12	Engmann et al. (2014)
Nectarine	Purée	600 MPa, 5 min, 10 °C	60	García-Parra et al. (2014)
Peach	Purée	517 MPa, 5 min, 20 °C+ 250 ppm cysteine	5.5	Guerrero-Beltrán et al. (2005c)
Strawberry	Pulp	600 MPa, 25 min, 25 °C	48.5	Cao et al. (2011)
		500 MPa, 25 min, 25 °C	58	
	Fresh fruit	600 MPa, 15 min, 25 °C	0	Garcia-Palazon et al. (2004)
		800 MPa, 10 min, 25 °C	0	
	Fresh fruit	600 MPa, 10 min, 60 °C	71.8	Terefe et al. (2009)
Strawberry, apple, banana, orange	Smoothie	450 MPa, 5 min, 20 °C	65	Keenan et al. (2012)
		600 MPa, 10 min, 20 °C	40	

[a]The temperatures indicated in the table refer to the temperature of the pressure vessel before pressure buildup

pH. Browning was related mainly with changes in the *a* color component. Weemaes et al. (1998a, 1998b) reported that avocado PPO exhibits high pressure stability at room temperature and neutral pH and inactivation are only achieved upon pressurizing at very high pressures (700–900 MPa) or with the combination of a mild heat treatment (60–70 °C). However, they concluded that to inhibit browning during storage is necessary to combine pressure with low pH and low storage temperatures. Nevertheless, decreasing the pH to inactivate enzymes may increase the perception of sour flavor, which is not always wanted depending on the product (Jacobo-Velázquez and Hernández-Brenes 2010). Palou et al. (2000) observed that another way to increase PPO inactivation in guacamole was the use of multiple pressurization-decompression cycles. The lowest residual PPO activity value (15%) was obtained after four HP cycles at 689 MPa with a 5-min holding time each. Jacobo-Velázquez and Hernández-Brenes (2010) reported a PPO residual activity of approximately 50% for HHP processing of avocado paste (pH 6.5) at 600 MPa for 3 min which increased in the subsequent days following pressurization, peaked at a maximum value (10 days), and then started a declining phase until the end of the storage period (45 days). On the other hand, Woolf et al. (2013) reported no inactivation of PPO whatsoever in avocado slices submitted to pressures from 200 to 600 MPa; on the contrary, in pressures up to 400 MPa they observed a higher PPO activity compared to the untreated sample. They suggested that this phenomenon might be related to cell membrane breakdown due to pressurization facilitating enhanced enzyme release/extractability, which indirectly results in an increase in the measured enzyme activity.

Palou et al. (1999) reported an increase in PPO activity after HHP treatment at 517 MPa for 10 min in banana purée and 79% residual activity when 689 MPa pressure was applied for 10 min. Further inactivation was only achieved when the banana purée was submitted to blanching before HPP treatment. Likewise, MacDonald and Schaschke (2000) reported that the inactivation of PPO from banana was maximized when mild temperatures were associated with the pressure treatment. Phunchaisri and Apichartsrangkoon (2005) combined mild temperatures (up to 60 °C) with pressure (600 MPa, 60 °C, 10 min) to treat litchis and obtained a PPO residual activity of around 15%. Another study with litchis in syrup treated by high pressure (600 MPa, 30 °C or 50 °C, 20 min) reported a reduction in PPO activity by 33–51%, whereas pasteurization markedly reduced that activity by 90% (Dajanta et al. 2012). PPO from guava was shown to be very resistant to pressure treatments; the application of 400 and 600 MPa (15 min, 25 °C) resulted into a residual activity of 86% and 63%, respectively (Yen and Lin 1996). PPO activity in longan juices pressurized at 300 MPa significantly increased (135%) compared to fresh juice and juices pressurized at 500 MPa, the latter of which had PPO activity of about 95% of fresh juice (Chaikham and Apichartsrangkoon 2012). PPO from apple is also high pressure resitant since its inactivation becomes noticeable at ambient temperature only at pressures greater than 600 MPa (Weemaes et al. 1998a; Valdramidis et al. 2009).

The impact of high pressure and temperature on PPO activity of cloudy apple juice was investigated by Buckow et al. (2009). They reported that the application of pressures in the 200–500 MPa range associated with mild temperatures (20–55 °C) caused PPO activation of up to 65%, which they suggested can occur due to protein dissociation, interactions with other constituents in the extract, or release of membrane-bound enzymes. They also observed a clear antagonistic effect of pressure and temperature on PPO inactivation in the low-pressure region (≤300 MPa), and synergistic with pressures above 300 MPa. In that sense, they suggested that a viable high pressure treatment to inactivate apple PPO at low temperatures is only feasible at 600 MPa or greater, or instead with the use of initial temperatures of 60–70 °C. Valdramidis et al. (2009) confirmed the pressure resistance of apple PPO at ambient temperature and concluded that to achieve high PPO inactivation it is recommended to use either temperature greater than 50 °C or pressures greater than 750 MPa. Falguera et al. (2013) studied the effect of apple variety in PPO inactivation using six different varieties (Braeburn, Fuji, Gala, Golden Delicious, Granny Smith, and Red Delicious) and reported that the variety affected not only the inactivation pattern but also the final inactivation. At 80 °C Granny Smith PPO was completely inactivated after 8 min at 600 MPa, while Fuji PPO activity was only reduced to 27.7% after 16 min under the same conditions. PPO activity from apricot nectar also showed an increase after pressurization at 300–500 MPa for 5–20 min, the increase varying between 30 and 45% (Huang et al. 2013).

The different matrixes of the same fruit (i.e., fresh fruit, purée, pulp, juice) can result in differences in enzyme inactivation. PPO activity in strawberry purée gradually decreased with increasing pressure levels (400–600 MPa) and longer treatment times (5–25 min) and the highest reduction of PPO activity resulted in 51.5% at 600 MPa for 25 min (Cao et al. 2011). In fresh strawberries there are contradicting results; Garcia-Palazon et al. (2004) reported complete inactivation of PPO after treatment at either 600 MPa for 15 min or 800 MPa for 10 min. In contrast, Terefe et al. (2009) observed a high resistance of fresh strawberry PPO. Even in the most extreme condition studied (600 MPa, 60 °C, 10 min) there was still a residual activity of 71.8%, whereas in strawberry PPO extracts the enzyme remained stable until 400 MPa at 40 °C and until 500 MPa at 25 °C (15 min treatment) and then a sharp decrease was observed. PPO was completely inactivated at 600 MPa at 40 °C, but at ambient temperature, there was still 5% residual activity observed at 800 MPa.

Keenan et al. (2012) pressurized fruit smoothies at 450 MPa, 20 °C, 5 min and 600 MPa, 20 °C, 10 min and observed that PPO activities were reduced by 35% and 60%, respectively, while PPO in pasteurized samples was completely inactivated. PPO from melon was less resistant to pressure than the PPOs from other fruits cited in this review, as a treatment at 500 MPa for 20 min and 22 °C was able to inactivate 91% of PPO activity (Ma et al. 2010).

The application of high pressure to some fruit purées at industrial level is limited due to the high resistance of browning-related enzymes, such as PPO. Therefore, it is common to use antioxidants to prevent these undesirable reactions (García-Parra et al. 2014). Guerrero-Beltrán et al. (2005b) studied the synergistic effect of anti-browning agents (ascorbic acid and cysteine) and HPP treatments on PPO activity in mango purées at pH 4.5. The residual activities were 74% and 4.5%, with HPP treatment alone and HPP combined with ascorbic acid (500 ppm), respectively. PPO activity of nectarine purées treated by 400 MPa or 600 MPa for 5 min was evaluated; in both cases, the activity decreased by 30% and 60%, respectively (García-Parra et al. 2014). Guerrero-Beltrán et al. (2005c) observed that the addition of ascorbic acid in nectarine purée caused an increase in PPO activity in the untreated and the pressurized samples, as was reported for peach purée.

Considerable differences in PPO pressure resistance are dependent on the plant source and even the plant variety; however, it is usually recognized that pressures higher than 700 MPa are needed to inactivate PPO completely without the use of mild temperatures (Palou et al. 1999).

13.7.2.3 Lipoxygenase

Lipoxygenase (linoleate: oxygen oxidoreductase) catalyzes the oxidation of unsaturated fatty acids and can affect sensory and nutritional properties of fruits (Akyol et al. 2006). LOX is less stable to pressure than PPO and POD (Seyderhelm et al.

1996). Only a few studies have been conducted about the inactivation of LOX in fruits, as shown in Table 13.8.

Palou et al. (2000) studied the pressure stability of LOX in guacamole, observing that processing conditions of 689 MPa for 15 min at 21 °C were necessary to achieve complete enzyme inactivation, whereas four cycles were needed to reduce PPO activity to 14%. Pressurization of avocado paste at 600 MPa for 3 min at 23 °C resulted in average residual LOX activity of 55%, which increased during storage, reaching a maximum at 14 days, and then decreased until the end of storage time (Jacobo-Velázquez and Hernández-Brenes 2010). The reactivation of enzymes after high pressure processing has been mainly related to the unfolding and refolding of enzymes back to the initial or other active structures (Gomes and Ledward 1996; Sun et al. 2002). In melon juice, LOX was readily inactivated after 3 min of treatment at 500 MPa and 22 °C (Ma et al. 2010).

13.7.2.4 Pectic Enzymes

Pectate lyase is associated with the degradation of the pectic fraction and thus with changes of the rheological properties of fruit juices (Huang et al. 2013). Yen and Lin (1996) studied the pressure inactivation of pectinesterase from guava purée at two pressure levels, 400 and 600 MPa for 15 min, obtaining a residual activity of 90% and 76%, respectively. Calligaris et al. (2012) studied the inactivation of pectate lyase in banana juice by high pressure homogenization. Pectate lyase was completely inactivated by the lower pressure level used, 150 MPa, with a flow rate of 200 L/h. There was no recovery of activity after 1 day of storage, and the inactivation was attributed to the mechanical stress caused by the process, which caused irreversible conformational changes of pectate lyase leading to its complete inactivation.

High hydrostatic pressure processing was suggested as an alternative process to stabilize freshly squeezed orange juice and to extend its shelf life (Nienaber and Shellhammer 2001). However, results from different studies show that the same processing conditions will generate distinct degrees of PME inactivation depending on the source, pH, ionic strength, composition of food, and purification level of the enzyme (Alexandrakis et al. 2014a).

Polydera et al. (2004) investigated the effect of high hydrostatic pressure (100–800 MPa) combined with moderate temperature (30–60 °C) on PME inactivation in freshly squeezed orange juice and observed that inactivation increases with increasing processing pressure at all temperature levels tested. However, a synergistic effect of pressure and temperature was reported at low temperature (up to 60 °C), whereas at high temperatures an antagonistic effect was noted.

Bisconsin-Junior et al. (2014) determined the optimum processing conditions to obtain an orange juice with low PME residual activity and low microorganism count. They found that treatments at 550–600 MPa, 55–60 °C, and 330–360 s were

Table 13.8 Effect of high pressure processing of lipoxygenase and pectic enzymes in fruit products

Fruit	Product	Enzyme	Pressure, time, temperature[a]	Remaining enzyme activity (%)	Reference
Avocado	Guacamole	LOX	689 MPa, 15 min, 21 °C	0	Palou et al. (2000)
	Avocado paste	LOX	600 MPa, 3 min, 23 °C	55	Jacobo-Velázquez and Hernández-Brenes (2010)
Melon	Juice	LOX	500 MPa, 3 min, 22 °C	7	Ma et al. (2010)
Feijoa	Purée	PME	450 MPa, 5 min, 25 °C	50	Ortuño et al. (2013)
Grapefruit	Jam	PME	700 MPa, 15 min, 65 °C	20	Igual et al. (2013)
			400 MPa, 10 min, 45 °C	30	
Orange	Juice	PME	400 MPa, 15 min, 30 °C	20	Polydera et al. (2004)
			600 MPa, 5 min, 50 °C	5	
	Juice with pulp	PE	600 MPa, 1 s, 10–15 °C	90	Goodner et al. (1998)
			800 MPa, 1 s, 10–15 °C	18	
	Juice	PME	600 MPa, 3.2 min, 45 °C	15	Bisconsin-Junior et al. (2014)
	Juice	PME	200 MPa, 15 min, 30 °C	25	Cano et al. (1997)
	Juice	PME	400 MPa, 12 min, 50 °C	50	Nienaber and Shellhammer (2001)
	Juice	PME	300 MPa, 5 min, 30 °C	20	Katsaros et al. (2010)
Peach	Pulp	PME	700 MPa, 2 min, 70 °C	0	Boulekou et al. (2010)
			700 MPa, 5 min, 30 °C	30	
Persimmon	Pulp	PME	700 MPa, 20 min, 40 °C	80	Katsaros et al. (2006)
			800 MPa, 15 min, 60 °C	60	
Sea buckthorn	Juice	PME	400 MPa, 15 min, 25 °C	60	Alexandrakis et al. (2014b)
			600 MPa, 10 min, 30 °C	20	

[a]The temperatures indicated in the table refer to the temperature of the pressure vessel before pressure buildup

able to inactivate more than 80% of initial orange juice PME. Katsaros et al. (2010) combined high pressure with mild temperatures to inactivate orange PME and observed that a treatment of 300 MPa at 40 °C was able to inactivate more than 90% of PME in less than 5 min.

Following the results of PME inactivation presented in Table 13.6, it is clear that high inactivation rate can be achieved by combining high pressures (>600 MPa) with mild temperatures. According to Alexandrakis et al. (2014a), an increase in both pressure and temperature generally enhances PME inactivation, and frequently a synergistic effect of these variables is observed (i.e., pressurization at high temperatures leads to greater inactivation than just a heat treatment at the same temperature). However, in some cases, antagonistic effect of pressure and temperature was observed, for example, Polydera et al. (2004) found an antagonistic effect of pressure and temperature when low pressure (100–250 MPa) was combined with high temperature (60 °C).

Recently, HPP-induced configurational changes on PME molecules were evaluated using circular dichroism (CD) spectroscopy. The results evidenced that pressurization may lead to a structurally molten globule-like state. The PME molecule kept the secondary structure intact while the tertiary structure was significantly changed, leading to modifications on the substrate–enzyme-binding interactions and thus reducing its activity (Alexandrakis et al. 2014a).

13.8 HPP Effect on Bioactive Compounds and Antioxidant Activity

Fruits are an excellent source of nutrients and functional compounds with antioxidant capacity, including carotenoids, ascorbic acid, tocopherols, flavonoids, and phenolic acids. A diet rich in fruits and vegetables, and consequently in antioxidants, is related to decreased risk of chronic diseases, including cardiovascular disease, diabetes, dementia, and some cancers (Scalbert et al. 2005). One of the advantages of high pressure processing is the retention of these compounds as compared to conventional thermal processing. Nevertheless, the effect of pressure on antioxidant capacity is not the same for all fruit products (Oey et al. 2008b).

13.8.1 Some Important Bioactive Compounds in Fruits

13.8.1.1 Carotenoids

Carotenoids are quite abundant in fruits and vegetables and are responsible for their red, orange, and yellow hues, which is probably the primary attribute to determine fruit quality for consumers. Furthermore, carotenoids have various biological functions, such as provitamin A activity, antioxidant capacity, and protection against macular degeneration (Stahl and Sies 2003). High pressure processing generally has little effect on the stability of carotenoids in fruits, the degree of which depends on the type of fruit and product, and may increase in some cases the rate of carotenoids extraction.

Jacobo-Velázquez and Hernández-Brenes (2012) treated avocado paste at 600 MPa for 3 min and observed that high pressure induced a significant increase (~56%) in carotenoid concentrations. The highest increases for individual carotenoids were found for neoxanthin-β (513%), followed by α-cryptoxanthin (312%), α-carotene (284%), β-cryptoxanthin (220%), β-carotene (107%), and lutein (40%). The levels declined during storage, but not to values lower than those initially present in the unprocessed avocado paste. Other authors also observed an increase in carotenoid content for three commercial melon varieties after HPP, being in particular β-carotene levels significantly increased in all evaluated varieties (Wolbang et al. 2008). de Ancos et al. (2000) processed persimmon pulp at pressures ranging from 50 to 400 MPa at room temperature for 15 min. There was an increase in the extraction of total carotenoids at 50 MPa (19%) and 400 MPa (16%) in comparison with the control pulp, which was attributed to a better extraction of the β-carotene, violaxanthin, lutein, anteroxantin, and β-cryptoxanthin. No change in the concentration of zeaxanthin was observed after treatment. Effects of high pressure processing (HPP) (200–400 MPa, 25 °C, 1–6 min) on the carotenoid content of astringent and non-astringent persimmon fruits of two maturity levels were studied by Plaza et al. (2012). Overall, treatment at 200 MPa produced a significant increase in extracted carotenoid content of astringent samples that were up to 86% and 45% in maturity stages III and V, respectively. Accordingly, orange juice processed at 400 MPa, 1 min, 40 °C exhibited an increase of 50% in total carotenoid content immediately after treatment, and the same pattern was observed for all individual carotenoids. Furthermore, after 40 days of refrigerated storage, the pressurized orange juice still presented 137% of the initial carotenoid content, whereas the non-pressurized sample presented only 87% (Plaza et al. 2011).

In apricot nectar, total carotenoids and β-carotene exhibited no significant changes after HPP treatments, except at 500 MPa, 20 min, where there was 17% increase in total carotenoids and 19% increase in β-carotene as compared to untreated samples. However, some individual carotenoids, β-cryptoxanthin, zeaxanthin, and lutein showed significant decreases in some treatments (Huang et al. 2013). High-pressure treatment of 600 MPa, 5 min, preserved the carotenoid content of nectarine purées, whereas losses were observed in the thermally processed purée. After 30 days of refrigerated storage, the carotenoid content of the pressurized sample decreased only 20%, whereas a reduction of 42% was observed in the non-treated sample (García-Parra et al. 2014). Liu et al. (2014) treated mango nectar at 600 MPa for 1 min and observed no changes in total carotenoid content after treatment, in agreement with Hernández-Brenes et al. (2013) who reported no changes in carotenoid content in mango pulp after treatment at 600 MPa for 3 min.

Watermelon juice treated at different combinations of pressure and temperature showed no change in the concentration of all-trans-lycopene, total cis-lycopene, and total lycopene when high-pressure treatment (up to 600 MPa) was applied at low temperatures (Liu et al. 2012). However, when pressures greater than 900 MPa or temperatures over 60 °C were used there was a reduction in lycopene level (Liu et al. 2012). Therefore, HPP could be a useful treatment to improve the extraction of health-related compounds and, in consequence, to modify their bioaccessibility (Hernández-Brenes et al. 2013; Lemmens et al. 2013).

13.8.1.2 Anthocyanins

It has been reported that anthocyanins present in fruit juices and purées are generally quite stable to high-pressure treatment (up to 600 MPa) at ambient temperature. Kinetic studies have demonstrated that anthocyanins are more sensitive to temperature increase, and even though they are more rapidly degraded as the pressure increases, the pressure effect is smaller (Verbeyst et al. 2010; Verbeyst et al. 2011).

Engmann et al. (2014) studied the effects of four independent variables (temperature, heating time, pressure, and pressurizing time) on the content of anthocyanin in mulberry juice treated with high pressure homogenization. The temperature had the greatest effect on the decrease of anthocyanins, followed by the heating time. When mulberry juice was treated with high pressure homogenization at 200 MPa for 1–3 cycles significant reductions in the content of anthocyanins were observed. Furthermore, the addition of ascorbic acid (400 mg/L) completely inhibited anthocyanin degradation during homogenization.

Cao et al. (2011) reported that total monomeric anthocyanins in strawberries exhibited no significant changes after high pressure treatments regardless of the pressure (400–600 MPa) or treatment time (5–25 min). Likewise, Bodelón et al. (2013) observed that the concentration of anthocyanins remained similar after all pressure treatments (100–400 MPa) at both temperatures (20 and 50 °C). Combining high pressure with mild temperatures did not significantly affect the total anthocyanin content of strawberries despite the lowest value corresponded to the samples processed at the most intense processing condition (600 MPa, 60 °C, 10 min) (Terefe et al. 2009). The same preservation effect was observed with blood orange juice; treatments at pressure levels of 400, 500, and 600 MPa for 15 min did not alter the initial anthocyanin content (Torres et al. 2011). Similarly, a retention of more than 98% of anthocyanin content was obtained in bayberry juice processed at 400, 500, and 600 MPa at room temperature for 10 min (Yu et al. 2013).

Dajanta et al. (2012) observed that pressurized lychee (600 MPa, 30 °C or 50 °C, 20 min) displayed significantly lower anthocyanin content than those pasteurized samples. Chen et al. (2013) studied the effect of high pressure on anthocyanin content of cloudy pomegranate juice and verified that there was a significant reduction in its content after processing, around 11%. However, the impact was smaller than that of HTST (high temperature short time) thermal treatment. A decrease in anthocyanin content was also reported in plum purée treated at 300–500 MPa, and the authors suggested that this degradation could be caused by the residual PPO activity (not inactivated by HPP treatment) along with residual oxygen in the sample (González-Cebrino et al. 2013). Ferrari et al. (2010) also investigated the effects of HPP treatment (400–600 MPa) at 25, 45, and 50 °C for 5 or 10 min on anthocyanin content of pomegranate juice. They found that at room temperature, the concentration of these molecules decreases with the increase in the treatment duration and pressure level. However, when higher temperatures were used there was an increase in anthocyanin content, indicating that there was a range of processing conditions where the mechanisms of anthocyanins degradation were modified, probably by affecting the molecules involved in the kinetics of the reaction, such as enzymes. The same mechanisms were suggested by Carbonell-Capella et al. (2013), who observed a positive influence of high pressure treatment on total anthocyanin content of a fruit juice mixture of papaya (32.5%, v/v), mango (10%, v/v), and orange (7.5%, v/v). The highest anthocyanin content was observed at the highest levels of both pressure and time (500 MPa, 15 min). The anthocyanin level of fruit smoothies was evaluated after HPP treatments (450 MPa, 20 °C, 5 min or 600 MPa, 20 °C, 10 min) and it was observed that samples processed at 450 MPa presented an increased concentration of anthocyanins. However, at 600 MPa their level was reduced (Keenan et al. 2010b). Similarly, the application of high pressure had a positive effect on the anthocyanin content of blueberry juice, with a maximum increase of 16% at 400 MPa for 15 min (Barba et al. 2013).

13.8.1.3 Ascorbic Acid

Factors such as oxygen concentration and the food matrix are known to affect ascorbic acid stability during high pressure treatments. In addition, higher losses are usually found in fruit juices compared to buffer systems because of the presence of endogenous pro-oxidants such as metal ions and enzymes (Oey et al. 2008b). Recent literature regarding the effect of HPP on ascorbic acid of fruit products is presented in Table 13.9.

Queiroz et al. (2010) evaluated the impact of HPP on the ascorbic acid content of cashew apple juice and observed that ascorbic acid content did not change significantly in all samples treated at 250 MPa. However, samples pressurized at 400 MPa for 7 min presented a significant decrease in AA content compared to the control. Although statistically significant, the maximum reduction was only 0.9%. Kiwi purée was subjected to HPP at 500 MPa for 3 min at room temperature, and the remaining level of ascorbic acid was around 6% lower. The authors suggested a correlation between this initial loss in vitamin C and the parameter of color a* (green color) that is associated with nonenzymatic browning (Fernández-Sestelo et al. 2013). Fresh litchi fruits were HP treated (100, 200, and 300 MPa for 5, 10, and 15 min at 27 °C) and a decrease in ascorbic acid of around 17% was observed after processing. During storage, ascorbic acid content of litchi fruits decreased, but processed samples showed higher retention (69% after 42 days—300 MPa for 15 min) than the untreated sample (48% after 17 days) (Kaushik et al. 2014a).

Yen and Lin (1996) investigated the effect of high pressure on the ascorbic acid content of guava purée during storage at 4 °C, and no change was observed, as compared with fresh samples, after treatment at a pressure of 400 or 600 MPa and 25 °C for 15 min. Furthermore, the authors reported that the guava purée stored at 4 °C for 40 days retained a high quality, similar to freshly extracted purée.

Table 13.9 Effect of high pressure processing on the ascorbic acid content of fruit products

Fruit	Product	Pressure, time, temperature[a]	Remaining ascorbic acid content (%)	Reference
Apple	Purée	400 MPa, 5 min, 20 °C	57	Landl et al. (2010)
		600 MPa, 5 min, 20 °C	14	
Blood orange	Juice	400 MPa, 15 min, 20 °C	95	Torres et al. (2011)
		600 MPa, 15 min, 20 °C	92	
Cashew apple	Juice	250 MPa, 7 min, 25 °C	99.5	Queiroz et al. (2010)
		400 MPa, 7 min, 25 °C	98.8	
Guava	Purée	400 or 600 MPa, 15 min, 25 °C	100	Yen and Lin (1996)
Kiwi	Purée	500 MPa, 3 min, 20 °C	94	Fernández-Sestelo et al. (2013)
Litchi	Fresh	200 MPa, 10 min, 27 °C	92.5	Kaushik et al. (2014a)
		300 MPa, 15 min, 27 °C	83	
Longan	Juice	300 MPa, 30 min	90	Chaikham and Apichartsrangkoon (2012)
		500 MPa, 30 min	79	
Mango	Nectar	600 MPa, 1 min, 25 °C	100	Liu et al. (2014)
	Nectar	Homogenization—200 MPa, 60–85 °C	~50	Tribst et al. (2011)
	Pulp	600 MPa, 5 min, 27 °C	126	Kaushik et al. (2014b)
Melon vs. *Northern Sky*	Cubes	600 MPa, 10 min, 25 °C	14	Wolbang et al. (2010)
Melon vs. *Chantele*	Cubes	600 MPa, 10 min, 25 °C	57	Wolbang et al. (2008)
Orange	Juice	Homogenization—250 MPa, 22 °C	100	Welti-Chanes et al. (2009)
	Juice	Homogenization—200 MPa, T_{max} 70 °C	95	Velázquez-Estrada et al. (2013)
	Juice	400 MPa, 1 min, 40 °C	92	Sánchez-Moreno et al. (2005)
Straw-berry	Purée	100–400 MPa, 15 min, 20 °C	100	Bodelón et al. (2013)
		400 MPa, 15 min, 50 °C	84	

[a]The temperatures indicated in the table refer to the temperature of the pressure vessel before pressure buildup

Tribst et al. (2011) investigated the effect of high pressure homogenization on the ascorbic acid content of mango nectar. Results showed that all treatments (200 MPa and 60–85 °C) resulted in about 50% ascorbic acid degradation. The authors suggested that the high degradation was due to the presence of oxygen, high shear, and mild temperature. These results are in disagreement with other studies with orange juice, where no significant ascorbic acid loss was observed after homogenization at 250 MPa, even after multiple passes (Welti-Chanes et al. 2009).

HPP was also used to treat orange and blood orange juice and was shown to efficiently preserve the AA content in these fruit juices (Torres et al. 2011; Sánchez-Moreno et al. 2005). Another study with mango nectar also reported no significant changes in L-ascorbic acid after processing at 600 MPa, 1 min. After 16 weeks of storage at either 4 or 25 °C, the authors observed a L-ascorbic acid decrease of 16% and 30%, respectively (Liu et al. 2014). Working with mango pulp, Kaushik et al. (2014b) observed an increase in ascorbic acid content after a single brief pulse or a 5-min treatment at pressures higher than 200 MPa, reaching a maximum of 126% at 600 MPa. At longer treatment times (10–20 min) there was no difference between pressure treated and untreated regarding ascorbic acid content. The observed increase could be related to a better extraction under high pressure, which causes cells to rupture under compression and the release of cytosol content into extracellular space (Prasad et al. 2009).

Wolbang et al. (2008) investigated the effect of high pressure processing on the vitamin C retention in three different melon cultivars and found different patterns for each cultivar (retention of 14–57%), which confirms the importance of the food matrix in bioactive compound retention. Another important factor that can affect the vitamin C retention is the use of mild temperatures coupled with pressure, which are usually used to inactivate enzymes and microorganisms. However, in some foods, high temperatures can impair nutritional losses. In the case of strawberry purée, treatments at 100–400 MPa at 20 °C did not affect ascorbic acid content, but an increase to 50 °C caused a small decrease (15%) in comparison to the untreated sample (Bodelón et al. 2013).

13.8.1.4 Phenolic Compounds

Phenols appeared to be relatively resistant to the effect of pressure, as presented in Table 13.10. Pomegranate juice is an important beverage from the nutritional point of view, and it is frequently consumed due to its high content of phenolic compounds (Gil et al. 2000). Ferrari et al. (2010) reported that the total phenolic content of pomegranate juice decreased between 7 and 42% after pressure treatments at 25 and 45 °C at all studied pressures (400–600 MPa). However, at 400 MPa and 50 °C there was an increase of 8% and 41% after pressurization for 5 and 10 min, respectively. Similar results were also found by Varela-Santos et al. (2012) in pomegranate juice; total phenolic content increased after all pressure treatments, ranging from 1 to 16%. Chen et al. (2013) compared the effect of thermal treatment and HPP on total phenolic content of pomegranate juice and confirmed that HPP preserved these compounds, while a significant decrease was found in the heat-treated sample.

Queiroz et al. (2010) reported a significant increase (17–28%) in the polyphenol content of cashew apple juice after treatment at 250 MPa, 5 min, and 400 MPa, 5 min, respectively. Similarly, García-Parra et al. (2014) observed an increase in total phenolic content in nectarine purée pressurized at 600 MPa for 5 min. On the other hand, total polyphenolic compounds in granny smith apple purée were signifi-

Table 13.10 Effect of high pressure processing on phenolic compounds (PC) content of fruit products

Fruit	Product	Pressure, time, temperature[a]	Remaining PC content (%)	Reference
Apple	Purée	400 MPa, 5 min, 20 °C	100	Landl et al. (2010)
		600 MPa, 5 min, 20 °C	75	
Apricot	Nectar			Huang et al. (2013)
Blackberry	Purée	600 MPa, 15 min, 20 °C	105	Patras et al. (2009)
Blueberry	Juice	200 MPa, 5 min, 25 °C	127	Barba et al. (2013)
		600 MPa, 15 min, 25 °C	106	
Cashew apple	Juice	400 MPa, 5 min, 25 °C	128	Queiroz et al. (2010)
Longan	Juice	300 MPa, 30 min, 25 °C	100	Chaikham and Apichartsrangkoon (2012)
Mango	Nectar	600 MPa, 1 min, 25 °C	100	Liu et al. (2014)
	Pulp	600 MPa, 1 pulse, 27 °C	119	Kaushik et al. (2014b)
Nectarine	Purée	600 MPa, 5 min, 25 °C	130	García-Parra et al. (2014)
Orange	Juice	Homogenization—200 MPa, T_{max} 70 °C	98	Velázquez-Estrada et al. (2013)
Strawberry	Purée	600 MPa, 15 min, 20 °C	109	Patras et al. (2009)
	Pulp	500 MPa, 25 min, 25 °C	115	Cao et al. (2011)
		400 MPa, 10 min, 25 °C	75	
Plum	Purée	600 MPa, 150 s, 20 °C	103	González-Cebrino et al. (2013)
Pomegranate	Juice	400 MPa, 10 min, 50 °C	141	Ferrari et al. (2010)
	Juice	450 MPa, 150 s, 20 °C	116	Varela-Santos et al. (2012)
	Juice	400 MPa, 5 min, 20 °C	103	Chen et al. (2013)

[a]The temperatures indicated in the table refer to the temperature of the pressure vessel before pressure buildup

cantly reduced by pressure treatment at 600 MPa and remained unchanged during processing at 400 MPa (Landl et al. 2010). This increase in total phenolic content may be linked to an increased extractability of some of the antioxidant components following high pressure processing.

Several studies have reported the increase of polyphenol concentration in red fruits treated by high hydrostatic pressure. Patras et al. (2009) observed an 8.3–9.8% increase in total phenolics in strawberry purées treated at 500 and 600 MPa and up to 5% increase in blackberry purée treated at 600 MPa. Cao et al. (2011) processed strawberry purée and also found that higher pressures (500–600 MPa) caused an increase in phenolic content, whereas at 400 MPa a slight decrease was observed. This reduction may be due to the higher residual activity of PPO and POD that are not completely inactivated at 400 MPa. Total phenolic content was also significantly increased (13–27%) after 200 MPa, 5–15 min, in blueberry juice (Barba et al. 2013).

Apricot nectar treated at 500 MPa presented a significant increase (6%) in total phenolics as compared to untreated apricot nectar, and the samples pressurized at 300 MPa, 5–15 min, and 400 MPa, 15 min, also exhibited a significant increase. Seven individual phenolics were identified, including catechin, caffeic acid, epicatechin, neochlorogenic acid, ferulic acid, *p*-coumaric acid, and chlorogenic acid, and a significant increase of these compounds was observed except for *p*-coumaric acid that exhibited no significant change (Huang et al. 2013).

Kaushik et al. (2014b) demonstrated the effect of pressure-induced extraction on total phenolic content in mango pulp. After a single pressure pulse, they observed an increase varying from 9 to 19% at 100 MPa and 500 MPa, respectively. The authors proposed that this could be the result of plant cell disruption caused by pressure, leading to a higher extractability. On the other hand, longer treatments (≥10 min) either decreased (at 100 and 200 MPa) or preserved (300–600 MPa) the total phenolic content of mango pulp.

Total phenolic content of smoothies (made of strawberry, apple, banana, and orange) treated at 450 MPa for 5 min was slightly higher (~11%) compared to thermally processed samples (Keenan et al. 2010b). In a subsequent study Keenan et al. (2010a) investigated the content of individual phenolic compounds (procyanidin B1, hesperidin, chlorogenic acid, and p-coumaric acid) of pressure-treated fruit smoothies and observed that the effect differed depending on the type of compound. Procyanidin B1 content remained stable after being treated at 450 MPa for 1 or 3 min, whereas the levels of hesperidin and p-coumaric acid increased and chlorogenic acid content decreased with the same treatments. Carbonell-Capella et al. (2013) reported that phenolic compounds of a fruit juice mixture (orange, mango, and papaya) sweetened with stevia were resistant to HPP, and in some treatments even increased, 22% after 300 MPa, 10 min, with 1.25% of stevia, and 18% after 300 MPa, 5 min, with 2.5% of stevia.

Some studies have investigated the content of total phenolic of pressure-treated fruit products during storage. According to Liu et al. (2014), the content of total phenols in mango nectar showed no significant changes after HPP treatments. However, they reported that after 16 weeks of storage at 4 and 25 °C there was a gradual loss of 19% and 28%, respectively. Cao et al. (2012) observed a reduction of 16% and 28% of total phenols in HPP-treated (600 MPa, 4 min) in cloudy strawberry juice, and 14% and 29% in HPP-treated (600 MPa, 4 min) in clear strawberry juice after 6 months' storage at 4 and 25 °C, respectively. However, Varela-Santos et al. (2012) found a much higher decrease (39.6%) of total phenolics in HHP-treated (350 MPa, 150 s) pomegranate juice after 35 days of storage at 4 °C.

13.8.2 Effect on Antioxidant Activity

The effect of HPP on antioxidant content of several fruit products is presented in Table 13.11. The effect of high pressure on the total antioxidant activity of avocado purée (sum of the oxygen radical absorbance capacity, ORAC values of the hydrophilic (H-ORAC), and lipophilic (L-ORAC) fractions) was studied by

Table 13.11 Effect of high pressure processing on the antioxidant capacity of fruit products

Fruit	Product	Pressure, time, temperature[a]	Method	Remaining antioxidant capacity (%)	Reference
Avocado	Paste	600 MPa, 3 min	ORAC	64	Jacobo-Velázquez and Hernández-Brenes (2012)
Blackberry	Purée	600 MPa, 15 min, 20 °C	DPPH	167	Patras et al. (2009)
Blueberry	Juice	200 MPa, 5–9 min	TEAC	100	Barba et al. (2013)
		600 MPa, 15 min		83	
Cashew apple	Juice	250 MPa, 3 min	FRAP	100	Queiroz et al. (2010)
			DPPH	140	
Grapefruit	Jam	550 MPa, 5 min, 45 °C	ORAC	100	Igual et al. (2013)
			DPPH	100	
Mango	Nectar	600 MPa, 1 min	FRAP	100	Liu et al. (2014)
			DPPH	83	
Mango	Pulp	200 MPa, 5 min, 30 °C	DPPH	100	Kaushik et al. (2014b)
Melon	Cubes	600 MPa, 10 min, 25 °C	FRAP	45	Wolbang et al. (2008)
Mulberry	Juice	Homogenization—200 MPa	ORAC	42	Yu et al. (2014)
Nectarine	Purée	600 MPa, 5 min, 20 °C	TEAC	152	García-Parra et al. (2014)
Orange	Juice	Homogenization—200 MPa	TEAC	105	Velázquez-Estrada et al. (2013)
			FRAP	89	
	Juice	400 MPa, 40 °C, 1 min	DPPH	105	Sánchez-Moreno et al. (2005)
Persimmon	Fresh	200 MPa, 6 min, 25 °C	FRAP	100	Hernández-Carrión et al. (2014)
Plum	Purée	600 MPa, 5 min, 25 °C	TEAC	93	González-Cebrino et al. (2013)
Pomegranate	Juice	400 MPa, 5 min	FRAP	96	Chen et al. (2013)
			DPPH	94	
	Juice	450 MPa, 30 s	DPPH	120	Varela-Santos et al. (2012)
		550 MPa, 90 s		93	
Strawberry	Purée	500 MPa, 15 min, 20 °C	DPPH	84	Patras et al. (2009)

ORAC oxygen radical absorbance capacity, *FRAP* ferric-reducing antioxidant power, *DPPH* 2,2-diphenyl-1-picryl-hydrazyl-hydrate, *TEAC* Trolox equivalent antioxidant capacity

[a]The temperatures indicated in the table refer to the temperature of the pressure vessel before pressure buildup

Jacobo-Velázquez and Hernández-Brenes (2012). They observed a 36% decrease in total ORAC values and a 10% and 40% reduction in H-ORAC and L-ORAC, respectively. The decrease in H-ORAC values may be attributed to the degradation of phenolic compounds or vitamin C, and that of L-ORAC values is possibly related to the deterioration of some carotenoids. Chen et al. (2013) studied the effect of HP on the antioxidant activity of pomegranate juice and observed that HPP better preserved nutritional content than thermal treatments, with losses that varied from 4% to 6% and 6% to 11%, respectively. During storage at 4 °C, a decrease in both radical scavenging assay (DPPH; 2,2-diphenyl-1-picryl-hydrazyl-hydrate) antioxidant activity and ferric-reducing antioxidant power (FRAP) antioxidant activity was observed, and this phenomenon was associated with the loss of total anthocyanins and total phenols. Fresh melon cut into pieces and packaged was treated with HP and presented a decrease in antioxidant activity after treatment (55%). However, a significant reduction (18%) was already observed after the process of cutting and packing, i.e., prior to HP treatment (Wolbang et al. 2008). Barba et al. (2013) observed that pressure treatments at 200 MPa for 5–15 min obtained similar TEAC values as fresh juice. Nevertheless, higher pressures as 400 MPa, 15 min, and 600 MPa at all times gave the lowest values for Trolox equivalent antioxidant capacity, TEAC. They attributed these losses to ascorbic acid degradation by residual PPO activity.

Several researchers have documented the preservation of antioxidant activity after pressure treatment. HPP did not alter FRAP-measured antioxidant capacity in cashew apple juice, while with the DPPH method there was an enhancement of 40% on antioxidant capacity after 3 min at 250 MPa (Queiroz et al. 2010). Fresh persimmon treated at 200 MPa for 6 min at 25 °C showed no difference in antioxidant activity in comparison with the untreated sample (Hernández-Carrión et al. 2014).

Another study by Varela-Santos et al. (2012) evaluated the effect of HP (350–550 MPa for 30, 90, and 150 s) on bioactive compounds in pomegranate juice during 35 days of storage at 4 °C. The sample treated at 450 MPa and 550 MPa exhibited the strongest antioxidant capacity, followed by the control sample at 350 MPa. The values obtained slightly increased during the first 5 days of storage and then decreased. This increase in antioxidant activity value could be related to the extraction of some of the hydrolyzable tannins present in the fruit rind, and related to the increase in ellagic acid, ellagic structures polymerized into ellagitannins and/or anthocyanin polymers formed during the storage period. Liu et al. (2014) studied the effect of HP (600 MPa,1 min) on antioxidant activity during 16 weeks of storage at 4 and 25 °C in mango nectar. HHP treatments caused no significant change in antioxidant capacity of mango nectars using FRAP assay, whereas a significant decrease using DPPH assay was observed. After storage for 16 weeks, the antioxidant capacity using DPPH assay reduced by 31.76% and 32.63% at 4 and 25 °C, respectively, and the antioxidant capacity using FRAP assay decreased by 20.76% and 21.08%.

13.9 HPP Effect on Color Characteristics of Fruit Products

The color of fruits and vegetables is mainly due to three families of pigments, chlorophylls, carotenoids, and anthocyanins, which are responsible for green, red-yellow, and red to blue-purple colors, respectively. The major role of these pigments is to attract insects and animals to disseminate their seeds, but these same pigments are also responsible for the visual attractiveness of fruits to potential consumers. Therefore, the preservation of the color of fruit juices, pulps, and even fresh fruits is one of the key advantages of high pressure processing since HPP has a limited effect on natural pigments. However, the color of HP-treated fruit products can change during storage due to residual enzymatic activity.

Fernández-Sestelo et al. (2013) treated kiwi purée at 500 MPa for 3 min at room temperature and evaluated its color during 40 days of storage at 4 °C. The control sample showed a significant loss in greenness (parameter a^*) after 1 day of storage, while in the pressurized samples this loss started after 5 days of storage. They also observed that the browning index was lower in the pressurized sample than in the control over the entire storage period.

Blueberry juice showed no significant changes in a^* values (0.36) after different pressure treatments (200–600 MPa, 12–15 min), and beverage luminosity (L^*) significantly decreased in relation to the unprocessed blueberry juice when the treatment time was longer. The color difference between the treated and untreated samples was less than 1.5 for all treatments; therefore, it is considered only "slightly changed" (Barba et al. 2013). A study with bayberry juice showed that HPP treatment caused a minor decrease in L^* value and higher pressures resulted in a more significant reduction of a^* and b^* values, meaning that the juice appeared less red and yellow after HPP (Yu et al. 2013). The same was observed for strawberry purée, in which the a^* and L^* values decreased after processing, and the latter was associated with enzymatic browning reaction (Cao et al. 2012). However, other researchers observed increased L^* values, indicating that the sample presented lighter color or was brighter after HPP in plum purée (González-Cebrino et al. 2013), banana purée (Calligaris et al. 2012), and apple purée (Landl et al. 2010).

The color of mango nectar treated by high pressure homogenization (HPH) (300 MPa) or HPH combined with a thermal process (200 MPa, 61.5 °C, 5 min) differed from the control, but the differences were lower than with a heat treatment only (100 °C, 10 min). Changes caused by 300 MPa treatment or HPH combined with heat affected the nectar by primarily reducing the red intensity. Moreover, the pasteurization process caused greater change in the total color difference of mango nectar (5.9) and the contrary, 200 MPa, 61.5 °C, 5 min, processing presented the best nectar color retention (2.5) (Tribst et al. 2011).

According to Woolf et al. (2013), pressure level and not treatment duration had a greater effect on the color of avocado slices. Green and yellow intensity were reduced, and avocado turned slightly darker when treated at pressures higher than 300 MPa. However, those differences were not visually detected. Similarly, avocado paste did not present significant color variations as an effect of the application of pressure, and the only parameter that increased was the a^* value (reduction of green) (Jacobo-Velázquez and Hernández-Brenes 2010).

13.10 HPP of Fruits: Advantages and Limitations

High pressure processing has numerous advantages over conventional thermal processing. The foremost benefit is the ability of this technology to retain most of the nutritional and sensory attributes of fresh products, which will satisfy the consumer who is expecting more fresh-like products in the market. Various studies have shown a higher retention of nutrients and bioactive compounds in food when processed by HPP as compared with other preservation technologies such as thermal processing (Oey et al. 2008a; Barrett and Lloyd 2012). This technology has great potential in the treatment of fresh and raw cleansing juices rich in nutrients, as HP-processed juices have been shown to have a longer shelf life as compared to unpasteurized fresh juices and blends. Since the products have a longer shelf life, it enables a broader and longer commercial distribution, thus minimizing the logistic costs and increasing the potential market. Today's consumers desire to move towards "clean label" (related to free of additives), healthy, and convenient products. HPP is one of the novel technologies that can satisfy these consumer expectations.

From the processing point of view, this technology is independent of the shape and size of the product, as it is based on the isostatic principle. Therefore, a product of any shape and size in a proper flexible package can be processed very easily, and the whole product undergoes equal treatment, unlike the outcome of conventional thermal processing and even novel microwave technologies, which have limitations in terms of a uniform treatment. Moreover, HPP is a post-packaging process; therefore there is no risk of post-processing contamination. Both pasteurization and sterilization of fruit products are possible if this technology is combined with appropriate hurdles such as temperature, pH reduction, and antimicrobial addition. Furthermore, this technology is environmentally friendly, so there is no generation of by-products or residues from the process itself; the only required medium is a pressurizing liquid (usually water) which can be easily recycled during the process. All of these relevant features have made this technology superior in the processing of fruit products over many others, either conventional or novel.

However, like other processing methods, HPP also has some limitations for processing fruit products. This technology runs either in a batch or in a semicontinuous mode, unless high pressure homogenization is considered. The batch mode of operation in HPP has a significant impact on the throughput of the industry. Moreover, high pressure processed products are currently slightly expensive if compared with other products in the same category. Reasons for this are the capital costs and the throughput of the system, although the operational costs cannot be considered of great impact, as previously discussed, considering the potential of adding meaningful value to the processed products. In fact some studies showed that the consumers are willing to pay more for pressurized products ready for consumption than those conventionally treated (Hicks et al. 2009; Romano et al. 2016) and thus it may withstand the higher costs to allow a certain margin (between price and total costs) to make a profit.

Processing also depends on the nature of the food. Products with highly porous structures may be modified with the processing; for example, the texture of whole or sliced strawberry may be altered if it is HP treated. HPP is also not suitable for a product with low moisture content, and several studies have shown that some deteriorative food enzymes and spores are highly resistant to pressure, as previously mentioned. This technology is also limited by the packaging materials. Glass or tin cannot be used with this system; only flexible bottles or pouches that can withstand the applied pressure are suitable for the process.

13.11 Final Remarks

The application of high pressure processing for the preservation of fruit products such as juices, smoothies, jams, and fresh cuts has immense potential in addressing the present customer demands for maintaining nutritional, functional, and sensory quality while assuring food safety. Perspectives of enhancing the use of this technology for new fruit-based products are encouraging. In addition to pasteurization, this technology has also been demonstrated pretreatment benefits including drying time reduction, and an increase in diffusion coefficient in some fresh-cut fruits, which can be of potential application for processes such as nutrient impregnation or extraction of functional compounds. Low-energy consumption and retention of bioactive compounds have made this technology attractive to the food industry. Likewise, retention of sensory attributes and by offering more fresh-like fruit products have made this technology appealing to consumers. This technology may continue to grow in the fruit processing industry, mainly in the sector of minimally processed fruit juices and fruit cuts. Proper optimization of processing parameters to reduce the pressure level and holding time is a current need for this technology, which could significantly reduce capital and operating costs. The use of multiple hurdles such as pH, mild temperature, and antimicrobials along with HPP could play a major role in meeting these current requirements and offer even better quality products that those just treated by HPP.

References

Akyol, Ç., H. Alpas, and A. Bayındırlı. 2006. Inactivation of peroxidase and lipoxygenase in carrots, green beans, and green peas by combination of high hydrostatic pressure and mild heat treatment. *European Food Research and Technology* 224 (2): 171–176.

Alemán, G., D.F. Farkas, J.A. Torres, E. Wilhelmsen, and S. McIntyre. 1994. Ultra-high pressure pasteurization of fresh cut pineapple. *Journal of Food Protection* 57 (10): 931–934.

Alexandrakis, Z., G. Katsaros, P. Stavros, P. Katapodis, G. Nounesis, and P. Taoukis. 2014a. Comparative structural changes and inactivation kinetics of pectin methylesterases from different orange cultivars processed by high pressure. *Food and Bioprocess Technology* 7 (3): 853–867.

Alexandrakis, Z., K. Kyriakopoulou, G. Katsaros, M. Krokida, and P. Taoukis. 2014b. Selection of process conditions for high pressure pasteurization of sea buckthorn juice retaining high antioxidant activity. *Food and Bioprocess Technology* 7 (11): 3226–3234.

Alpas, H., L. Alma, and F. Bozoglu. 2003. Inactivation of Alicyclobacillus acidoterrestris vegetative cells in model system, apple, orange and tomato juices by high hydrostatic pressure. *World Journal of Microbiology and Biotechnology* 19 (6): 619–623.

Balasubramaniam, V.M., G.V. Barbosa-Cánovas, and H.L. Lelieveld. 2016. High-pressure processing equipment for the food industry. In *High Pressure Processing of Food*, 39–65. New York: Springer.

Balasubramaniam, V.M., D. Farkas, and E.J. Turek. 2008. Preserving foods through high-pressure processing. *Food Technology* 11: 32–38.

Balda, F.P., B.V. Aparicio, and C.T. Samson. 2012. Industrial high pressure processing of foods: Review of evolution and emerging trends. *Journal of Food Science and Engineering* 2 (10): 543.

Barba, F.J., M.J. Esteve, and A. Frígola. 2013. Physicochemical and nutritional characteristics of blueberry juice after high pressure processing. *Food Research International* 50 (2): 545–549.

Barrett, D.M., and B. Lloyd. 2012. Advanced preservation methods and nutrient retention in fruits and vegetables. *Journal of the Science of Food and Agriculture* 92 (1): 7–22.

Bevilacqua, A., M.R. Corbo, and M. Sinigaglia. 2012. Use of natural antimicrobials and high pressure homogenization to control the growth of Saccharomyces bayanus in apple juice. *Food Control* 24 (1): 109–115.

Bisconsin-Junior, A., A. Rosenthal, and M. Monteiro. 2014. Optimisation of high hydrostatic pressure processing of Pêra Rio orange juice. *Food and Bioprocess Technology* 7 (6): 1670–1677.

Bodelón, O.G., J.M. Avizcuri, P. Fernández-Zurbano, M. Dizy, and G. Préstamo. 2013. Pressurization and cold storage of strawberry purée: Colour, anthocyanins, ascorbic acid and pectin methylesterase. *Food Science and Technology* 52 (2): 123–130.

Boulekou, S.S., G.J. Katsaros, and P.S. Taoukis. 2010. Inactivation kinetics of peach pulp pectin methylesterase as a function of high hydrostatic pressure and temperature process conditions. *Food and Bioprocess Technology* 3 (5): 699–706.

Boynton, B.B., C.A. Sims, S. Sargent, M.O. Balaban, and M.R. Marshall. 2002. Quality and stability of precut mangos and carambolas subjected to high-pressure processing. *Journal of Food Science* 67 (1): 409–415.

Buckow, R., U. Weiss, and D. Knorr. 2009. Inactivation kinetics of apple polyphenol oxidase in different pressure–temperature domains. *Innovative Food Science & Emerging Technologies* 10 (4): 441–448.

Bull, M.K., E.A. Szabo, M.B. Cole, and C.M. Stewart. 2005. Toward validation of process criteria for high-pressure processing of orange juice with predictive models. *Journal of Food Protection* 68 (5): 949–954.

Butz, P., A.F. García, R. Lindauer, S. Dieterich, A. Bognar, and B. Tauscher. 2003. Influence of ultra-high pressure processing on fruit and vegetable products. *Journal of Food Engineering* 56 (2): 233–236.

Buzrul, S., H. Alpas, A. Largeteau, and G. Demazeau. 2008. Inactivation of *Escherichia coli* and *Listeria innocua* in kiwifruit and pineapple juices by high hydrostatic pressure. *International Journal of Food Microbiology* 124 (3): 275–278.

Calligaris, S., M. Foschia, I. Bartolomeoli, M. Maifreni, and L. Manzocco. 2012. Study on the applicability of high-pressure homogenization for the production of banana juices. *Food Science and Technology* 45 (1): 117–121.

Cano, M.P., A. Hernandez, and B.D. de Ancos. 1997. High pressure and temperature effects on enzyme inactivation in strawberry and orange products. *Journal of Food Science* 62 (1): 85–88.

Cao, X., X. Bi, W. Huang, J. Wu, X. Hu, and X. Liao. 2012. Changes of quality of high hydrostatic pressure processed cloudy and clear strawberry juices during storage. *Innovative Food Science & Emerging Technologies* 16: 181–190.

Cao, X., Y. Zhang, F. Zhang, Y. Wang, J. Yi, and X. Liao. 2011. Effects of high hydrostatic pressure on enzymes, phenolic compounds, anthocyanins, polymeric color and color of strawberry pulps. *Journal of the Science of Food and Agriculture* 91 (5): 877–885.

Carbonell-Capella, J.M., F.J. Barba, M.J. Esteve, and A. Frígola. 2013. High pressure processing of fruit juice mixture sweetened with Stevia rebaudiana Bertoni: Optimal retention of physical and nutritional quality. *Innovative Food Science & Emerging Technologies* 18: 48–56.

Chaikham, P., and A. Apichartsrangkoon. 2012. Comparison of dynamic viscoelastic and physicochemical properties of pressurised and pasteurised longan juices with xanthan addition. *Food Chemistry* 134 (4): 2194–2200.

Chakraborty, S., N. Kaushik, P.S. Rao, and H.N. Mishra. 2014a. High-pressure inactivation of enzymes: A review on its recent applications on fruit purees and juices. *Comprehensive Reviews in Food Science and Food Safety* 13 (4): 578–596.

Chakraborty, S., P.S. Rao, and H.N. Mishra. 2014b. Effect of pH on enzyme inactivation kinetics in high-pressure processed pineapple (*Ananas comosus L.*) purée using response surface methodology. *Food and Bioprocess Technology* 7 (12): 3629–3645.

Cheftel, J.C. 1992. Effects of high hydrostatic pressure on food constituents: An overview. In *High Pressure and Biotechnology*, ed. C. Balny, R. Hayashi, K. Heremans, and P. Masson, vol. 224, 195–209. Paris: INSERM and John Libbey.

———. 1995. Review: High-pressure, microbial inactivation and food preservation. *Food Science and Technology International* 1 (2–3): 75–90.

Chen, D., H. Xi, X. Guo, Z. Qin, X. Pang, X. Hu, et al. 2013. Comparative study of quality of cloudy pomegranate juice treated by high hydrostatic pressure and high temperature short time. *Innovative Food Science & Emerging Technologies* 19: 85–94.

Coroller, L., I. Leguérinel, E. Mettler, N. Savy, and P. Mafart. 2006. General model, based on two mixed Weibull distributions of bacterial resistance, for describing various shapes of inactivation curves. *Applied and Environmental Microbiology* 72 (10): 6493–6502.

Dajanta, K., A. Apichartsrangkoon, and S. Somsang. 2012. Comparison of physical and chemical properties of high pressure-and heat-treated Lychee (*Litchi chinensis Sonn.*) in syrup. *High Pressure Research* 32 (1): 114–118.

de Ancos, B., E. Gonzalez, and M.P. Cano. 2000. Effect of high-pressure treatment on the carotenoid composition and the radical scavenging activity of persimmon fruit purees. *Journal of Agricultural and Food Chemistry* 48 (8): 3542–3548.

Donsi, G., G. Ferrari, and M. Di Matteo. 1996. High pressure stabilization of orange juice: evaluation of the effects of process conditions. *Italian Journal of Food Science* 8 (2): 99–106.

Elamin, W.M., J.B. Endan, Y.A. Yosuf, R. Shamsudin, and A. Ahmedov. 2015. High pressure processing technology and equipment evolution: A review. *Journal of Engineering Science and Technology Review* 8 (5): 75–83.

Engmann, F.N., Y. Ma, H. Zhang, L. Yu, and N. Deng. 2014. The application of response surface methodology in studying the effect of heat and high hydrostatic pressure on anthocyanins, polyphenol oxidase, and peroxidase of mulberry (*Morus ni*gra) juice. *Journal of the Science of Food and Agriculture* 94 (11): 2345–2356.

Erkmen, O. 2011. Effects of high hydrostatic pressure on Salmonella typhimurium and aerobic bacteria in milk and fruit juices. *Romanian Biotechnological Letters* 16 (5): 6540–6547.

Espina, L., D. García-Gonzalo, A. Laglaoui, B.M. Mackey, and R. Pagán. 2013. Synergistic combinations of high hydrostatic pressure and essential oils or their constituents and their use in preservation of fruit juices. *International Journal of Food Microbiology* 161 (1): 23–30.

Fachin, D., A.V. Loey, O. Indrawati, L. Ludikhuyze, and M. Hendrickx. 2002. Thermal and high-pressure inactivation of tomato polygalacturonase: A kinetic study. *Journal of Food Science* 67 (5): 1610–1615.

Falguera, V., F. Gatius, A. Ibarz, and G.V. Barbosa-Cánovas. 2013. Kinetic and multivariate analysis of polyphenol oxidase inactivation by high pressure and temperature processing in apple juices made from six different varieties. *Food and Bioprocess Technology* 6 (9): 2342–2352.

Fang, L., B. Jiang, and T. Zhang. 2008. Effect of combined high pressure and thermal treatment on kiwifruit peroxidase. *Food Chemistry* 109 (4): 802–807.

Farkas, D.F., and D.G. Hoover. 2000. High pressure processing. *Journal of Food Science* 65 (s8): 47–64.

Fernández-Sestelo, A., R.S. de Saá, C. Pérez-Lamela, A. Torrado-Agrasar, M.L. Rúa, and L. Pastrana-Castro. 2013. Overall quality properties in pressurized kiwi purée: Microbial, physicochemical, nutritive and sensory tests during refrigerated storage. *Innovative Food Science & Emerging Technologies* 20: 64–72.

Ferragut, V., M. Hernández-Herrero, M.T. Veciana-Nogués, M. Borras-Suarez, J. González-Linares, M.C. Vidal-Carou, and B. Guamis. 2015. Ultra-high-pressure homogenization (UHPH) system for producing high-quality vegetable-based beverages: physicochemical, microbiological, nutritional and toxicological characteristics. *Journal of the Science of Food and Agriculture* 95 (5): 953–961.

Ferrari, G., P. Maresca, and R. Ciccarone. 2010. The application of high hydrostatic pressure for the stabilization of functional foods: Pomegranate juice. *Journal of Food Engineering* 100 (2): 245–253.

Ferreira, E., L.M.P. Masson, A. Rosenthal, M. Souza, L. Tashima, and P.R. Massaguer. 2011. Termorresistência de fungos filamentosos isolados de néctares de frutas envasados assepticamente. *Brazilian Journal of Food Technology* 14 (3): 164–171.

Garcia-Palazon, A., W. Suthanthangjai, P. Kajda, and I. Zabetakis. 2004. The effects of high hydrostatic pressure on β-glucosidase, peroxidase and polyphenoloxidase in red raspberry (Rubus idaeus) and strawberry (Fragaria× ananassa). *Food Chemistry* 88 (1): 7–10.

García-Parra, J., R. Contador, J. Delgado-Adámez, F. González-Cebrino, and R. Ramírez. 2014. The applied pretreatment (blanching, ascorbic acid) at the manufacture process affects the quality of nectarine purée processed by hydrostatic high pressure. *International Journal of Food Science and Technology* 49 (4): 1203–1214.

Gil, M.I., F.A. Tomás-Barberán, B. Hess-Pierce, D.M. Holcroft, and A.A. Kader. 2000. Antioxidant activity of pomegranate juice and its relationship with phenolic composition and processing. *Journal of Agricultural and Food Chemistry* 48 (10): 4581–4589.

Gomes, M.R.A., and D.A. Ledward. 1996. Effect of high-pressure treatment on the activity of some polyphenoloxidases. *Food Chemistry* 56 (1): 1–5.

González-Cebrino, F., R. Durán, J. Delgado-Adámez, R. Contador, and R. Ramírez. 2013. Changes after high-pressure processing on physicochemical parameters, bioactive compounds, and polyphenol oxidase activity of red flesh and peel plum purée. *Innovative Food Science & Emerging Technologies* 20: 34–41.

Goodner, J.K., R.J. Braddock, and M.E. Parish. 1998. Inactivation of pectinesterase in orange and grapefruit juices by high pressure. *Journal of Agricultural and Food Chemistry* 46 (5): 1997–2000.

Goodner, J.K., R.J. Braddock, M.E. Parish, and C.A. Sims. 1999. Cloud stabilization of orange juice by high pressure processing. *Journal of Food Science* 64 (4): 699–700.

Guerrero-Beltrán, J.A., G.V. Barbosa-Cánovas, and B.G. Swanson. 2005a. High hydrostatic pressure processing of fruit and vegetable products. *Food Reviews International* 21 (4): 411–425.

Guerrero-Beltrán, J.A., G.V. Barbosa-Cánovas, and J. Welti-Chanes. 2011. High hydrostatic pressure effect on natural microflora, *Saccharomyces cerevisiae*, *Escherichia coli*, and *Listeria Innocua* in navel orange juice. *International Journal of Food Engineering* 7 (1).

Guerrero-Beltrán, J.A., B.G. Swanson, and G.V. Barbosa-Canovas. 2005b. High hydrostatic pressure processing of mango puree containing antibrowning agents. *Food Science and Technology International* 11 (4): 261–267.

Guerrero-Beltrán, J.A., B.G. Swanson, and G.V. Barbosa-Cánovas. 2005c. Shelf life of HHP-processed peach puree with antibrowning agents. *Journal of Food Quality* 28 (5–6): 479–491.

Guiavarc'h, Y., O. Segovia, M. Hendrickx, and A. Van Loey. 2005. Purification, characterization, thermal and high-pressure inactivation of a pectin methylesterase from white grapefruit (*Citrus paradisi*). *Innovative Food Science & Emerging Technologies* 6 (4): 363–371.

Harte, F. 2016. Food processing by high-pressure homogenization. In *High Pressure Processing of Food*, 123–141. New York: Springer.

Hartyáni, P., I. Dalmadi, and D. Knorr. 2013. Electronic nose investigation of Alicyclobacillus acidoterrestris inoculated apple and orange juice treated by high hydrostatic pressure. *Food Control* 32 (1): 262–269.

Hendrickx, M., L. Ludikhuyze, I. Van den Broeck, and C. Weemaes. 1998. Effects of high pressure on enzymes related to food quality. *Trends in Food Science & Technology* 9 (5): 197–203.

Hernández-Brenes, C., P.A. Ramos-Parra, D.A. Jacobo-Velázquez, R. Villarreal-Lara, and R.I. Díaz-De la Garza. 2013. High hydrostatic pressure processing as a strategy to increase carotenoid contents of tropical fruits. *Matrix* 5: 11.

Hernández-Carrión, M., J.L. Vázquez-Gutiérrez, I. Hernando, and A. Quiles. 2014. Impact of high hydrostatic pressure and pasteurization on the structure and the extractability of bioactive compounds of persimmon "Rojo Brillante". *Journal of Food Science* 79 (1): C32–C38.

Hicks, D.T., L.F. Pivarnik, R. McDermott, N. Richard, D.G. Hoover, and K.E. Kniel. 2009. Consumer awareness and willingness to pay for high-pressure processing of ready-to-eat food. *Journal of Food Science Education* 8 (2): 32–38.

Hiremath, N.D., and H.S. Ramaswamy. 2012. High-pressure destruction kinetics of spoilage and pathogenic microorganisms in mango juice. *Journal of Food Processing and Preservation* 36 (2): 113–125.

Huang, W., X. Bi, X. Zhang, X. Liao, X. Hu, and J. Wu. 2013. Comparative study of enzymes, phenolics, carotenoids and color of apricot nectars treated by high hydrostatic pressure and high temperature short time. *Innovative Food Science & Emerging Technologies* 18: 74–82.

Igual, M., F. Sampedro, N. Martinez-Navarrete, and X. Fan. 2013. Combined osmodehydration and high pressure processing on the enzyme stability and antioxidant capacity of a grapefruit jam. *Journal of Food Engineering* 114 (4): 514–521.

Jacobo-Velázquez, D.A., and C. Hernández-Brenes. 2010. Biochemical changes during the storage of high hydrostatic pressure processed avocado paste. *Journal of Food Science* 75 (6): S264–S270.

———. 2012. Stability of avocado paste carotenoids as affected by high hydrostatic pressure processing and storage. *Innovative Food Science & Emerging Technologies* 16: 121–128.

Jacobo-Velázquez, D.A., M.R. Cuéllar-Villarreal, J. Welti-Chanes, J.L. Cisneros-Zevallos, P.A. Ramos-Parra, and C. Hernández-Brenes. 2017. Nonthermal processing technologies as elicitors to induce the biosynthesis and accumulation of nutraceuticals in plant foods. *Trends in Food Science and Technology.* 60: 80–87.

Jordan, S.L., C. Pascual, E. Bracey, and B.M. Mackey. 2001. Inactivation and injury of pressure-resistant strains of *Escherichia co*li O157 and *Listeria monocytogenes* in fruit juices. *Journal of Applied Microbiology* 91 (3): 463–469.

Jung, S., Samson, C. T., & de Lamballerie, M. (2010). High hydrostatic pressure food processing. In *Alternatives to Conventional Food Processing* (pp. 254–306).

Katsaros, G.I., M. Tsevdou, T. Panagiotou, and P.S. Taoukis. 2010. Kinetic study of high pressure microbial and enzyme inactivation and selection of pasteurization conditions for Valencia orange juice. *International Journal of Food Science and Technology* 45 (6): 1119–1129.

Katsaros, G., Apseridis, J., & Taoukis, P. (2006). Modelling of High Hydrostatic Pressure inactivation of pectinmethylesterase from persimmon fruit. In *13th World Congress of Food Science & Technology,* (pp 753–753) IUFoST Edpsciences.

Kaushik, N., B.P. Kaur, and P.S. Rao. 2014a. Application of high pressure processing for shelf life extension of litchi fruits (*Litchi chinensis* cv. Bombai) during refrigerated storage. *Food Science and Technology International* 20 (7): 527–541.

Kaushik, N., B.P. Kaur, P.S. Rao, and H.N. Mishra. 2014b. Effect of high pressure processing on color, biochemical and microbiological characteristics of mango pulp (*Mangifera indica* cv. *Amrapali*). *Innovative Food Science & Emerging Technologies* 22: 40–50.

Keenan, D.F., N.P. Brunton, T.R. Gormley, F. Butler, B.K. Tiwari, and A. Patras. 2010b. Effect of thermal and high hydrostatic pressure processing on antioxidant activity and colour of fruit smoothies. *Innovative Food Science & Emerging Technologies* 11 (4): 551–556.

Keenan, D.F., N. Brunton, R. Gormley, and F. Butler. 2010a. Effects of thermal and high hydrostatic pressure processing and storage on the content of polyphenols and some quality attributes of fruit smoothies. *Journal of Agricultural and Food Chemistry* 59 (2): 601–607.

Keenan, D.F., C. Rößle, R. Gormley, F. Butler, and N.P. Brunton. 2012. Effect of high hydrostatic pressure and thermal processing on the nutritional quality and enzyme activity of fruit smoothies. *Food Science and Technology* 45 (1): 50–57.

Kingsley, D.H., and H. Chen. 2009. Influence of pH, salt, and temperature on pressure inactivation of hepatitis A virus. *International Journal of Food Microbiology* 130 (1): 61–64.

Kingsley, D.H., D. Guan, and D.G. Hoover. 2005. Pressure inactivation of hepatitis A virus in strawberry puree and sliced green onions. *Journal of Food Protection®* 68 (8): 1748–1751.

Kingsly, A.R.P., V.M. Balasubramaniam, and N.K. Rastogi. 2009. Effect of high-pressure processing on texture and drying behavior of pineapple. *Journal of Food Process Engineering* 32 (3): 369–381.

Knorr, D., O. Schlüter, and V. Heinz. 1998. Impact of high hydrostatic pressure on phase transitions of foods. *Food Technology (USA)* 52 (9): 42–45.

Kovač, K., M. Diez-Valcarce, P. Raspor, M. Hernández, and D. Rodríguez-Lázaro. 2012. Effect of high hydrostatic pressure processing on norovirus infectivity and genome stability in strawberry puree and mineral water. *International Journal of Food Microbiology* 152 (1): 35–39.

Krebbers, B., A.M. Matser, S.W. Hoogerwerf, R. Moezelaar, M.M. Tomassen, and R.W. van den Berg. 2003. Combined high-pressure and thermal treatments for processing of tomato puree: evaluation of microbial inactivation and quality parameters. *Innovative Food Science & Emerging Technologies* 4 (4): 377–385.

Lambert, Y., G. Demazeau, A. Largeteau, and J.M. Bouvier. 1999. Changes in aromatic volatile composition of strawberry after high pressure treatment. *Food Chemistry* 67 (1): 7–16.

Landl, A., M. Abadias, C. Sárraga, I. Viñas, and P.A. Picouet. 2010. Effect of high pressure processing on the quality of acidified Granny Smith apple purée product. *Innovative Food Science & Emerging Technologies* 11 (4): 557–564.

Lee, S.Y., R.H. Dougherty, and D.H. Kang. 2002. Inhibitory effects of high pressure and heat on Alicyclobacillus acidoterrestris spores in apple juice. *Applied and Environmental Microbiology* 68 (8): 4158–4161.

Lelieveld, H.L., and H. Hoogland. 2016. Continuous high-pressure processing to extend product shelf life. In *High Pressure Processing of Food*, ed. V.M. Balasubramaniam, G.V. Barbosa-Cánovas, and H.L.M. Lelieveld, 67–72. New York: Springer.

Lemmens, L., E.S. Tchuenche, A.M. Van Loey, and M.E. Hendrickx. 2013. Beta-carotene isomerisation in mango puree as influenced by thermal processing and high-pressure homogenisation. *European Food Research and Technology* 236 (1): 155–163.

Liu, F., Y. Wang, R. Li, X. Bi, and X. Liao. 2014. Effects of high hydrostatic pressure and high temperature short time on antioxidant activity, antioxidant compounds and color of mango nectars. *Innovative Food Science & Emerging Technologies* 21: 35–43.

Liu, Y., X. Hu, X. Zhao, and H. Song. 2012. Combined effect of high pressure carbon dioxide and mild heat treatment on overall quality parameters of watermelon juice. *Innovative Food Science & Emerging Technologies* 13: 112–119.

Lopes, M.L.M., V.L. Valente Mesquita, A.C.N. Chiaradia, A.A.R. Fernandes, and P. Fernandes. 2010. High hydrostatic pressure processing of tropical fruits. *Annals of the New York Academy of Sciences* 1189 (1): 6–15.

López-Malo, A., E. Palou, G.V. Barbosa-Cánovas, J. Welti-Chanes, and B.G. Swanson. 1998. Polyphenoloxidase activity and color changes during storage of high hydrostatic pressure treated avocado puree. *Food Research International* 31 (8): 549–556.

Ludikhuyze, L., A. Van Loey, O. Indrawati, C. Smout, and M. Hendrickx. 2003. Effects of combined pressure and temperature on enzymes related to quality of fruits and vegetables: from kinetic information to process engineering aspects. *Critical Reviews in Food Science and Nutrition* 43 (5): 527–586.

Ma, Y., X. Hu, J. Chen, G. Zhao, X. Liao, F. Chen, et al. 2010. Effect of UHP on enzyme, microorganism and flavor in cantaloupe (*Cucumis melo L.*) juice. *Journal of Food Process Engineering* 33 (3): 540–553.

MacDonald, L., and C.J. Schaschke. 2000. Combined effect of high pressure, temperature and holding time on polyphenoloxidase and peroxidase activity in banana (Musa acuminata). *Journal of the Science of Food and Agriculture* 80 (6): 719–724.

Moody, A., G. Marx, B.G. Swanson, and D. Bermúdez-Aguirre. 2014. A comprehensive study on the inactivation of *Escherichia coli* under nonthermal technologies: High hydrostatic pressure, pulsed electric fields and ultrasound. *Food Control* 37: 305–314.

Mujica-Paz, H., A. Valdez-Fragoso, C.T. Samson, J. Welti-Chanes, and J.A. Torres. 2011. High-pressure processing technologies for the pasteurization and sterilization of foods. *Food and Bioprocess Technology* 4 (6): 969.

Muñoz, M., B. de Ancos, C. Sánchez-Moreno, and M.P. Cano. 2007. Effects of high pressure and mild heat on endogenous microflora and on the inactivation and sublethal injury of *Escherichia coli* inoculated into fruit juices and vegetable soup. *Journal of Food Protection* 70 (7): 1587–1593.

Naik, L., R. Sharma, Y.S. Rajput, and G. Manju. 2013. Application of high pressure processing technology for dairy food preservation-future perspective: A review. *Journal of Animal Production Advances* 3 (8): 232–241.

Nienaber, U., and T.H. Shellhammer. 2001. High-pressure processing of orange juice: Kinetics of pectinmethylesterase inactivation. *Journal of Food Science* 66 (2): 328–331.

Noma, S., C. Tomita, M. Shimoda, and I. Hayakawa. 2004. Response of Escherichia coli O157: H7 in apple and orange juices by hydrostatic pressure treatment with rapid decompression. *Food Microbiology* 21 (4): 469–473.

Oey, I., M. Lille, A. Van Loey, and M. Hendrickx. 2008a. Effect of high-pressure processing on colour, texture and flavour of fruit-and vegetable-based food products: A review. *Trends in Food Science & Technology* 19 (6): 320–328.

Oey, I., I. Van der Plancken, A. Van Loey, and M. Hendrickx. 2008b. Does high pressure processing influence nutritional aspects of plant based food systems? *Trends in Food Science & Technology* 19 (6): 300–308.

Ortuño, C., T. Duong, M. Balaban, and J. Benedito. 2013. Combined high hydrostatic pressure and carbon dioxide inactivation of pectin methylesterase, polyphenol oxidase and peroxidase in feijoa purée. *The Journal of Supercritical Fluids* 82: 56–62.

Palou, E., C. Hernández-Salgado, A. López-Malo, G.V. Barbosa-Cánovas, B.G. Swanson, and J. Welti-Chanes. 2000. High pressure-processed guacamole. *Innovative Food Science & Emerging Technologies* 1 (1): 69–75.

Palou, E., A. López-Malo, G.V. Barbosa-Cánovas, J. Welti-Chanes, and B.G. Swanson. 1997. Effect of water activity on high hydrostatic pressure inhibition of *Zygosaccharomyces bailii*. *Letters in Applied Microbiology* 24: 417–420.

———. 1999. Polyphenoloxidase activity and color of blanched and high hydrostatic pressure treated banana puree. *Journal of Food Science* 64 (1): 42–45.

Parish, M.E., and D.P. Higgins. 1989. Survival of Listeria monocytogenes in low pH model broth systems. *Journal of Food Protection* 52 (3): 144–147.

Patras, A., N.P. Brunton, S. Da Pieve, and F. Butler. 2009. Impact of high pressure processing on total antioxidant activity, phenolic, ascorbic acid, anthocyanin content and colour of strawberry and blackberry purées. *Innovative Food Science & Emerging Technologies* 10 (3): 308–313.

Patrignani, F., G. Tabanelli, L. Siroli, F. Gardini, and R. Lanciotti. 2013. Combined effects of high pressure homogenization treatment and citral on microbiological quality of apricot juice. *International Journal of Food Microbiology* 160 (3): 273–281.

Patterson, M.F., A.M. McKay, M. Connolly, and M. Linton. 2012. The effect of high hydrostatic pressure on the microbiological quality and safety of carrot juice during refrigerated storage. *Food Microbiology* 30 (1): 205–212.

Patterson, M.F., D.A. Ledward, and N. Rogers. 2006. High pressure processing. In *Food Processing Handbook*, ed. J.C. Brenan, 173–197. Weinheim, Germany: Wiley-VCH Publishing.
Peleg, M., and M.B. Cole. 1998. Reinterpretation of microbial survival curves. *Critical Reviews in Food Science* 38 (5): 353–380.
Penchalaraju, M., and B. Shireesha. 2013. Preservation of foods by high-pressure processing: A review. *Indian Journal of Scientific Research and Technology* 1 (3): 30–38.
Phunchaisri, C., and A. Apichartsrangkoon. 2005. Effects of ultra-high pressure on biochemical and physical modification of lychee (*Litchi chinensis Sonn.*). *Food Chemistry* 93 (1): 57–64.
Plaza, L., C. Colina, B. de Ancos, C. Sánchez-Moreno, and M.P. Cano. 2012. Influence of ripening and astringency on carotenoid content of high-pressure treated persimmon fruit (Diospyros kaki L.). *Food Chemistry* 130 (3): 591–597.
Plaza, L., C. Sánchez-Moreno, B. de Ancos, P. Elez-Martínez, O. Martín-Belloso, and M.P. Cano. 2011. Carotenoid and flavanone content during refrigerated storage of orange juice processed by high-pressure, pulsed electric fields and low pasteurization. *Food Science and Technology* 44 (4): 834–839.
Polydera, A.C., E. Galanou, N.G. Stoforos, and P.S. Taoukis. 2004. Inactivation kinetics of pectin methylesterase of Greek Navel orange juice as a function of high hydrostatic pressure and temperature process conditions. *Journal of Food Engineering* 62 (3): 291–298.
Prasad, K.N., E. Yang, C. Yi, M. Zhao, and Y. Jiang. 2009. Effects of high pressure extraction on the extraction yield, total phenolic content and antioxidant activity of longan fruit pericarp. *Innovative Food Science & Emerging Technologies* 10 (2): 155–159.
Queiroz, C., C.F.F. Moreira, F.C. Lavinas, M.L.M. Lopes, E. Fialho, and V.L. Valente-Mesquita. 2010. Effect of high hydrostatic pressure on phenolic compounds, ascorbic acid and antioxidant activity in cashew apple juice. *High Pressure Research* 30 (4): 507–513.
Ramaswamy, H.S., E. Riahi, and E. Idziak. 2003. High-pressure destruction kinetics of E. coli (29055) in Apple Juice. *Journal of Food Science* 68 (5): 1750–1756.
Ramesh, M.N., K. Sathyanarayana, and A.B. Girish. 1998. Biphasic model for the kinetics of vegetable cooking at 100 °C. *Journal of Food Engineering* 35 (1): 127–133.
Rasanayagam, V., V.M. Balasubramaniam, E. Ting, C.E. Sizer, C. Bush, and C. Anderson. 2003. Compression heating of selected fatty food materials during high-pressure processing. *Journal of Food Science* 68 (1): 254–259.
Rastogi, N.K., and D. Knorr. 2013. *Recent Developments in High Pressure Processing of Foods*. New York: Springer.
Rastogi, N.K., A. Angersbach, and D. Knorr. 2000. Synergistic effect of high hydrostatic pressure pretreatment and osmotic stress on mass transfer during osmotic dehydration. *Journal of Food Engineering* 45 (1): 25–31.
Rastogi, N.K., M.N. Eshtiaghi, and D. Knorr. 1999. Effects of combined high pressure and heat treatment on the reduction of peroxidase and polyphenoloxidase activity in red grapes. *Food Biotechnology* 13 (2): 195–208.
Rastogi, N.K., K.S.M.S. Raghavarao, V.M. Balasubramaniam, K. Niranjan, and D. Knorr. 2007. Opportunities and challenges in high pressure processing of foods. *Critical Reviews in Food Science and Nutrition* 47 (1): 69–112.
Rizvi, A.F., and C.H. Tong. 1997. Fractional conversion for determining texture degradation kinetics of vegetables. *Journal of Food Science* 62 (1): 1–7.
Romano, K.R., F.D.B.A. Finco, A. Rosenthal, M.V.A. Finco, and R. Deliza. 2016. Willingness to pay more for value-added pomegranate juice (Punica granatum L.): An open-ended contingent valuation. *Food Research International* 89: 359–364.
Rosenthal, A., D. Ledward, A. Defaye, S. Gilmour, and L. Trinca. 2002. Effect of pressure, temperature, time and storage on peroxidase and polyphenol oxidase from pineapple. *Progress in Biotechnology* 19: 525–532.
Ruiz-Espinosa, H., G.G. Amador-Espejo, M.E. Barcenas-Pozos, J.O. Angulo-Guerrero, H.S. Garcia, and J. Welti-Chanes. 2013. Multiple-pass high-pressure homogenization of milk

for the development of pasteurization-like processing conditions. *Letters in Applied Microbiology* 56 (2): 142–148.

Sampedro, F., D. Rodrigo, and M. Hendrickx. 2008. Inactivation kinetics of pectin methyl esterase under combined thermal–high pressure treatment in an orange juice–milk beverage. *Journal of Food Engineering* 86 (1): 133–139.

Sánchez-Moreno, C., L. Plaza, P. Elez-Martínez, B. de Ancos, O. Martín-Belloso, and M.P. Cano. 2005. Impact of high pressure and pulsed electric fields on bioactive compounds and antioxidant activity of orange juice in comparison with traditional thermal processing. *Journal of Agricultural and Food Chemistry* 53 (11): 4403–4409.

Sanguansri, P., and M.A. Augustin. 2006. Nanoscale materials development–a food industry perspective. *Trends in Food Science & Technology* 17 (10): 547–556.

Scalbert, A., C. Manach, C. Morand, C. Rémésy, and L. Jiménez. 2005. Dietary polyphenols and the prevention of diseases. *Critical Reviews in Food Science and Nutrition* 45 (4): 287–306.

Serment-Moreno, V., G. Barbosa-Cánovas, J.A. Torres, and J. Welti-Chanes. 2014. High-pressure processing: Kinetic models for microbial and enzyme inactivation. *Food Engineering Reviews* 6 (3): 56–88.

Serment-Moreno, V., C. Fuentes, G. Barbosa-Cánovas, J.A. Torres, and J. Welti-Chanes. 2015. Evaluation of high pressure processing kinetic models for microbial inactivation using standard statistical tools and information theory criteria, and the development of generic time-pressure functions for process design. *Food and Bioprocess Technology* 8 (6): 1244–1257.

Seyderhelm, I., S. Boguslawski, G. Michaelis, and D. Knorr. 1996. Pressure induced inactivation of selected food enzymes. *Journal of Food Science* 61 (2): 308–310.

Sokołowska, B., S. Skąpska, M. Fonberg-Broczek, J. Niezgoda, M. Chotkiewicz, A. Dekowska, and S. Rzoska. 2012. The combined effect of high pressure and nisin or lysozyme on the inactivation of *Alicyclobacillus acidoterrestris* spores in apple juice. *High Pressure Research* 32 (1): 119–127.

Sokołowska, B., S. Skąpska, M. Fonberg-Broczek, J. Niezgoda, M. Chotkiewicz, A. Dekowska, and S.J. Rzoska. 2013a. Factors influencing the inactivation of *Alicyclobacillus acidoterrestris* spores exposed to high hydrostatic pressure in apple juice. *High Pressure Research* 33 (1): 73–82.

Sokołowska, B., S. Skąpska, M. Fonberg-Broczek, J. Niezgoda, M. Rutkowska, M. Chotkiewicz, et al. 2013b. The effect of high hydrostatic pressure on the survival of *Saccharomyces cerevisiae* in model suspensions and beetroot juice. *High Pressure Research* 33 (1): 165–171.

Sokołowska, B., S. Skapska, J. Niezgoda, M. Rutkowska, A. Dekowska, and S.J. Rzoska. 2014. Inactivation and sublethal injury of *Escherichia coli* and *Listeria innocua* by high hydrostatic pressure in model suspensions and beetroot juice. *High Pressure Research* 34 (1): 147–155.

Splittstoesser, D.F., C.Y. Lee, and J.J. Churey. 1998. Control of Alicyclobacillus in the juice industry. *Dairy, Food and Environmental Sanitation* 18: 585–587.

Stahl, W., and H. Sies. 2003. Antioxidant activity of carotenoids. *Molecular Aspects of Medicine* 24 (6): 345–351.

Stang, M., H. Schuchmann, and H. Schubert. 2001. Emulsification in high-pressure homogenizers. *Engineering in Life Sciences* 1 (4): 151–157.

Sun, N., S. Lee, and K.B. Song. 2002. Effect of high-pressure treatment on the molecular properties of mushroom polyphenoloxidase. *Food Science and Technology* 35 (4): 315–318.

Suthanthangjai, W., P. Kajda, and I. Zabetakis. 2005. The effect of high hydrostatic pressure on the anthocyanins of raspberry (*Rubus idaeus*). *Food Chemistry* 90 (1): 193–197.

Tejada-Ortigoza, V., Z. Escobedo-Avellaneda, A. Valdez-Fragoso, H. Mújica-Paz, and J. Welti-Chanes. 2015. Combined effect of high hydrostatic pressure and mild heat treatments on pectin methylesterase (PME) inactivation in comminuted orange. *Journal of the Science of Food and Agriculture* 95 (12): 2438–2444.

Terefe, N.S., K. Matthies, L. Simons, and C. Versteeg. 2009. Combined high pressure-mild temperature processing for optimal retention of physical and nutritional quality of strawberries (Fragaria× ananassa). *Innovative Food Science & Emerging Technologies* 10 (3): 297–307.

Terefe, N.S., Y.H. Yang, K. Knoerzer, R. Buckow, and C. Versteeg. 2010. High pressure and thermal inactivation kinetics of polyphenol oxidase and peroxidase in strawberry puree. *Innovative Food Science & Emerging Technologies* 11 (1): 52–60.

Timmermans, R.A.H., H.C. Mastwijk, J.J. Knol, M.C.J. Quataert, L. Vervoort, I. Van der Plancken, et al. 2011. Comparing equivalent thermal, high pressure and pulsed electric field processes for mild pasteurization of orange juice. Part I: Impact on overall quality attributes. *Innovative Food Science & Emerging Technologies* 12 (3): 235–243.

Tola, Y.B., and H.S. Ramaswamy. 2014. Combined effects of high pressure, moderate heat and pH on the inactivation kinetics of *Bacillus licheniformis* spores in carrot juice. *Food Research International* 62: 50–58.

Tonello, C. 2011. Case studies on high-pressure processing of foods. In *Nonthermal Processing Technologies for Food*, ed. H.Q. Zhang, G.V. Barbosa-Cánovas, V.M.B. Balasubramaniam, C.P. Dunne, D.F. Farkas, and J.T.C. Yuan, 36–50. New Delhi: Wiley-Blackwell/IFT Press.

Torres, B., B.K. Tiwari, A. Patras, P.J. Cullen, N. Brunton, and C.P. O'Donnell. 2011. Stability of anthocyanins and ascorbic acid of high pressure processed blood orange juice during storage. *Innovative Food Science & Emerging Technologies* 12 (2): 93–97.

Tribst, A.A.L., M.A. Franchi, P.R. de Massaguer, and M. Cristianini. 2011. Quality of mango nectar processed by high-pressure homogenization with optimized heat treatment. *Journal of Food Science* 76 (2): M106–M110.

USFDA. (2014). *Kinetics of Microbial Inactivation for Alternative Food Processing Technologies—High Pressure Processing*. Retrieved September 28, 2016, from http://www.fda.gov/Food/FoodScienceResearch/SafePracticesforFoodProcesses/ucm101456.htm.

Valdramidis, V.P., W.D. Graham, A. Beattie, M. Linton, A. McKay, A.M. Fearon, and M.F. Patterson. 2009. Defining the stability interfaces of apple juice: implications on the optimisation and design of high hydrostatic pressure treatment. *Innovative Food Science & Emerging Technologies* 10 (4): 396–404.

Van den Broeck, I., L.R. Ludikhuyze, A.M. Van Loey, and M.E. Hendrickx. 2000. Inactivation of orange pectinesterase by combined high pressure and temperature treatments: a kinetic study. *Journal of Agricultural and Food Chemistry* 48 (5): 1960–1970.

Varela-Santos, E., A. Ochoa-Martinez, G. Tabilo-Munizaga, J.E. Reyes, M. Pérez-Won, V. Briones-Labarca, and J. Morales-Castro. 2012. Effect of high hydrostatic pressure (HHP) processing on physicochemical properties, bioactive compounds and shelf-life of pomegranate juice. *Innovative Food Science & Emerging Technologies* 13: 13–22.

Velázquez-Estrada, R.M., M.M. Hernández-Herrero, C.E. Rüfer, B. Guamis-López, and A.X. Roig-Sagués. 2013. Influence of ultra high pressure homogenization processing on bioactive compounds and antioxidant activity of orange juice. *Innovative Food Science & Emerging Technologies* 18: 89–94.

Verbeyst, L., I. Oey, I. Van der Plancken, M. Hendrickx, and A. Van Loey. 2010. Kinetic study on the thermal and pressure degradation of anthocyanins in strawberries. *Food Chemistry* 123 (2): 269–274.

Verbeyst, L., K. Van Crombruggen, I. Van der Plancken, M. Hendrickx, and A. Van Loey. 2011. Anthocyanin degradation kinetics during thermal and high pressure treatments of raspberries. *Journal of Food Engineering* 105 (3): 513–521.

Weemaes, C.A., L.R. Ludikhuyze, I. Van den Broeck, and M.E. Hendrickx. 1998b. Kinetics of combined pressure-temperature inactivation of avocado polyphenoloxidase. *Biotechnology and Bioengineering* 60 (3): 292–300.

Weemaes, C., L. Ludikhuyze, I. Broeck, and M. Hendrickx. 1998a. High pressure inactivation of polyphenoloxidases. *Journal of Food Science* 63 (5): 873–877.

Welti-Chanes, J., C.E. Ochoa-Velasco, and J.A. Guerrero-Beltrán. 2009. High-pressure homogenization of orange juice to inactivate pectinmethylesterase. *Innovative Food Science & Emerging Technologies* 10 (4): 457–462.

Wolbang, C.M., J.L. Fitos, and M.T. Treeby. 2008. The effect of high pressure processing on nutritional value and quality attributes of *Cucumis melo L. Innovative Food Science & Emerging Technologies* 9 (2): 196–200.

Wolbang, C.M., D.P. Singh, S.R. Sykes, J.K. McInerney, A.R. Bird, and M.T. Treeby. 2010. Influence of pre-and postharvest factors on β-carotene content, its in vitro bioaccessibility, and antioxidant capacity in melons. *Journal of Agricultural and Food Chemistry* 58 (3): 1732–1740.

Woolf, A.B., R. Wibisono, J. Farr, I. Hallett, L. Richter, I. Oey, et al. 2013. Effect of high pressure processing on avocado slices. *Innovative Food Science & Emerging Technologies* 18: 65–73.

Yaldagard, M., S.A. Mortazavi, and F. Tabatabaie. 2008. The principles of ultra high pressure technology and its application in food processing/preservation: A review of microbiological and quality aspects. *African Journal of Biotechnology* 7 (16): 2739–2767.

Yamamoto, K., M. Matsubara, S. Kawasaki, M.L. Bari, and S. Kawamoto. 2005. Modeling the pressure inactivation dynamics of *Escherichia coli*. *Brazilian Journal of Medical and Biological Research* 38 (8): 1253–1257.

Yen, G.C., and H.T. Lin. 1996. Comparison of high pressure treatment and thermal pasteurization effects on the quality and shelf life of guava puree. *International Journal of Food Science and Technology* 31 (2): 205–213.

Yu, Y., Y. Lin, Y. Zhan, J. He, and S. Zhu. 2013. Effect of high pressure processing on the stability of anthocyanin, ascorbic acid and color of Chinese bayberry juice during storage. *Journal of Food Engineering* 119 (3): 701–706.

Yu, Y., Y. Xu, J. Wu, G. Xiao, M. Fu, and Y. Zhang. 2014. Effect of ultra-high pressure homogenisation processing on phenolic compounds, antioxidant capacity and anti-glucosidase of mulberry juice. *Food Chemistry* 153: 114–120.

Zhao, L., S. Wang, F. Liu, P. Dong, W. Huang, L. Xiong, and X. Liao. 2013. Comparing the effects of high hydrostatic pressure and thermal pasteurization combined with nisin on the quality of cucumber juice drinks. *Innovative Food Science & Emerging Technologies* 17: 27–36.

Zhao, L., Y. Wang, S. Wang, H. Li, W. Huang, and X. Liao. 2014. Inactivation of naturally occurring microbiota in cucumber juice by pressure treatment. *International Journal of Food Microbiology* 174: 12–18.

Zimmermann, M., D.W. Schaffner, and G.M. Aragão. 2013. Modeling the inactivation kinetics of *Bacillus coagulans* spores in tomato pulp from the combined effect of high pressure and moderate temperature. *LWT-Food Science and Technology* 53 (1): 107–112.

Chapter 14
Safety and Quality of Irradiated Fruits and Vegetables

Brendan A. Niemira

14.1 Introduction

In the US, per capita consumption of fresh produce is increasing, with fresh and fresh-cut fruits and vegetables among the fastest growing market segments (Calvin 2007; ERS 2008; CNPP 2010). However, as produce consumption has increased in the US, so have produce-related outbreaks. Responsible for only 0.7 % of all reported foodborne outbreaks in the 1970s, contaminated produce was the causative agent in 6 % of all outbreaks in the 1990s (FDA 2004; Sivapalasingam et al. 2004). The Centers for Disease Control and Prevention reported that foodborne outbreaks associated with fresh produce doubled between the period 1973–1987 and 1988–1992 (Buck et al. 2003; CDC 2011). The plant commodity group accounted for 66 % of viral, 32 % of bacterial, 25 % of chemical, and 30 % of parasitic illnesses from 1998 to 2008, with leafy greens being the single food most commonly associated with foodborne illness (Painter et al. 2013). From 1990 to 2005, produce accounted for 22 % of the most common outbreaks, 713 of total 3204 (CSPI 2007). Irrespective of the human cost of illness and death, medical costs and lost productivity due to foodborne illness has been estimated at $6.5–34.9B (IFT 2004). Scharff (2010) estimated that illnesses, produce recalls, and loss of consumer confidence cost approximately $39B annually.

Fresh and fresh-cut produce support the growth of commensal bacteria and bacterial human pathogens on their surfaces (Annous et al. 2005; Gómez-López et al. 2009). *Salmonella* spp., *Escherichia coli* O157:H7, *Listeria monocytogenes*, and *Shigella* spp. are regular contaminants of intact and minimally processed fruits and vegetables (Sivapalasingam et al. 2004; Mandrell 2009). In a recent comparison among antimicrobial sanitation methods, irradiation was identified as one of the

B.A. Niemira (✉)
USDA-ARS Eastern Regional Research Center, Wyndmoor, PA, USA
e-mail: Brendan.Niemira@ars.usda.gov

A. Rosenthal et al. (eds.), *Fruit Preservation*, Food Engineering Series,
https://doi.org/10.1007/978-1-4939-3311-2_14

most reliably effective interventions, surpassing chemical sanitizers and newly developed nonthermal methods (Goodburn and Wallace 2013). This chapter will present the scientific context and the latest research on irradiation as a method to inactivate human pathogens on fresh produce. The irradiation technologies available will be discussed, along with the relevant regulatory framework within which irradiation may be employed.

14.2 Irradiation Technologies

When irradiation is used as a food processing treatment, the dose is defined by the energy absorbed during exposure, measured in grays (Gy). One gray represents one joule of energy absorbed per kilogram of irradiated product, the equivalent of 100 rad, or 0.1 Mrad. Food irradiation research typically reports doses in kilogray (kGy). Energy absorbed depends on the product mass, bulk density, and thickness. An excessive dose may render the food unacceptable for consumption, while inadequate doses fail to achieve desired effects. The qualitative terminology used to describe doses are generally characterized as low (<3 kGy), medium (>3 and <10 kGy), or high (>10 kGy).

Commercial food processing uses e-beam (high energy electrons), X-rays, or gamma rays (high energy photons). An overview of the advantages and disadvantages of each is presented in Table 14.1. Each of the technologies induce ionization of molecules in the food target, leading to the generation of radicals, breakage of DNA, and other radiochemical effects. The irradiation equipment used in food irradiation is not energetic enough to induce radioactivity in the target. This process is discussed in greater detail below, under Sect. "General Mode of Action. Because of the ionization of the air within the treatment equipment, ozone and heat are generated during the irradiation process. For each of the technologies, proper ventilation and temperature control are key. This is especially important for the irradiation of fruits and other ozone-sensitive products. Failure to manage these secondary effects can potentially result in product degradation (Maxie and Abdel-Kader 1966).

14.2.1 Electron Beam

E-beams are produced by linear or cyclotron accelerators. These impart a high velocity, and therefore a high kinetic energy, to a stream of electrons. Commercial e-beam irradiators produce a focused beam of electrons with an energy of up to 10 MeV. The full processing dose is delivered quickly, taking typically <5 s. The electrons are aimed at the target, typically with a cone-shaped guide. As irradiation doses are additive in nature, higher doses may be delivered by repeated exposure. Short exposure time prevents any significant rise in temperature during processing.

Table 14.1 Major irradiation technologies—advantages and disadvantages

Factors	Electron beam	X-ray	Gamma
Source	*Electric power*: Electrons are generated using electronics and accelerated to high energy using magnetic fields, 5–10 MeV[a]. When accelerator is powered off, no radiation is emitted.	*Electric power*: Created when high energy electrons (up to 5 MeV) strike a metal plate (e.g., tungsten or tantalum alloys), typical conversion efficiency is 5–10 %. When accelerator is powered off, no radiation is emitted.	*Radioisotopes*: Radioactive decay of ^{60}cobalt (2.5 MeV) or ^{137}cesium (0.51 MeV). Radioisotope source is always emitting radiation—shielding of source must be the default position.
Mechanism	*High-energy electrons* cleave water molecules, creating oxygen and hydroxyl radicals that damage DNA, membranes. Direct cleavage of DNA also occurs.	*High-energy photons* stimulate atoms within target to release high-energy electrons, which cleave water molecules into radicals. Direct cleavage of DNA also occurs.	*High-energy photons* stimulate atoms within target to release high-energy electrons, which cleave water molecules into radicals. Direct cleavage of DNA also occurs.
Infrastructure required	*Shielding*: >2 m concrete or <1 m steel/iron/lead.	*Shielding*: >2 m concrete or <1 m steel/iron/lead.	*Shielding*: Depending on design, >5 m water or >2 m concrete or <1 m steel/iron/lead.
	Cooling: extensive for high-voltage electronics and accelerator.	*Cooling*: extensive for high-voltage electronics and accelerator. Additional cooling systems required for plate target.	*Cooling*: moderate for control equipment
	Ventilation: for ozone removal while unit is operating.	*Ventilation*: for ozone removal while unit is operating.	*Ventilation*: at all times for ozone removal when source is exposed to air.
Speed[b]	*Seconds*	*Seconds*	*Minutes* (depending on source strength)
Penetrability[c]	*6–8 cm*, suitable for relatively thin or low density products.	*30–40 cm*, suitable for all products.	*30–40 cm*, suitable for all products.

[a]*MeV* million electron volts
[b]Speed of dose delivery. The desired dose will vary depending on the target organism and commodity irradiated
[c]Penetrability in food, avg. density approximately 1 g/cm^3. This figure will vary for individual commodities due to localized variation in density associated with bone, voids, fibrous matter, etc.

The penetrability of the electron beam is lower than that of gamma rays or X-rays (Table 14.1); product must therefore be packaged accordingly. In commercial practice, many products are treated from multiple sides to enhance dose uniformity. A recent study by Moreira et al. (2012) demonstrated that the absorbed dose in a typical vegetable product (i.e., a tomato slice) can vary from 1.04 to 0.86 kGy over a distance of less than 4 cm. This nonuniformity is a function of the electron beam; material handling during irradiation is therefore crucial. Product can typically be stacked no thicker than 6–8 cm before absorption results in unacceptable dose distributions.

An advantage of e-beam irradiators over gamma irradiators is the electronic nature of the radiation source, which can be completely deactivated. As high-power, industrial scale electronic devices, electron accelerators require specialized training and maintenance for their operation, and the shielding required for an operating e-beam unit is comparable to that required for a gamma irradiator. Furthermore, the technical advantage of operation without radioisotopes is, to an extent, offset by the greater complexity of the electron accelerator.

14.2.2 X-ray

Familiar to the general public due to their widespread medical applications, X-rays are high-energy photons with much higher penetrability than the electrons of e-beams. Because photons will penetrate food products with very good uniformity, there is much less dose variation as compared with e-beam. Product can typically be stacked as thick as 30–40 cm without significant loss of penetration. The high-energy X-rays used in food irradiation require the same type of accelerator systems used in an e-beam irradiator. The beam of high-energy electrons is directed at a dense metal target. Electrons are absorbed by the metal atoms and the high-energy photons (X-rays) are emitted. The efficiency of this process is typically 5–10 %. Alloys of tungsten, tantalum, and various types of stainless steel have been used in the target plate; these vary in durability, cost, and efficiency.

The energy of the X-ray output depends on the energy of the originating e-beam and the conversion efficiency. Typically, 90–95 % of the energy that is directed at the metal plate is lost as heat; cooling subsystems function to stabilize operation and the temperature of the food processing chamber. Shielding for an X-ray irradiator is comparable to that required for the other two types of units. X-rays have a similar penetrability to gamma rays (Table 14.1) and can be used for bulk packages or higher density foodstuffs. Unlike gamma rays, X-rays are produced electronically. As they do not use radioisotopes, they can be completely inactivated in a power-off state. Some commercial irradiators can convert from e-beam to X-ray operation as needed by installation and removal of the metal target plate. While X-rays would seem to be an attractive middle ground between gamma and e-beam irradiators, the low energy efficiency, heat buildup in the metal plate and the low, but measurable, neutron scattering are drawbacks.

14.2.3 Gamma Ray

Gamma rays are high-energy photons produced by the disintegration of radioactive isotopes. These have the same excellent penetration as the photons of X-rays. In food irradiators, this is typically cobalt-60 and, less commonly, cesium-137. When cobalt-60 (half-life of 5.27 years) disintegrates into nickel-60 (a stable isotope), it emits two gamma rays of average energies of 1.17 and 1.33 MeV, respectively. Disintegration of cesium-137 (half-life of 30.17 years) to barium-137 produces gamma rays with energy of 0.66 MeV. Cobalt-60 has historically been the preferred radioisotope for food irradiation. Gamma rays from cobalt-60 are more energetic than those from cesium-137; higher dose rates mean shorter processing times. Also, cobalt-60 is the less expensive of the two isotopes. Finally, the CsCl form of cesium-137 in irradiators is water-soluble, a point of environmental concern. Cobalt-60 is a pure metal and is not water-soluble. It should be noted that these advantages of cobalt-60 are offset by its much shorter half-life. Cobalt-60 irradiators must be replenished with fresh radioisotope ("recharged") much more frequently than cesium-137 irradiators. After 10.5 years of operation, a cobalt-60 irradiator will have approximately 25 % of its initial strength, while a cesium-137 irradiator will have retained approximately 80 % of its initial strength. A lower dose rate will lengthen treatment times.

Gamma rays have excellent penetrability, suitable for irradiating pallet- or crate-sized packages of product. The shielding required for a gamma irradiator is roughly comparable to that required for an e-beam or X-ray irradiator (Table 14.1). Gamma irradiators are more energy-efficient than e-beam or X-ray, as the photons are produced by radioactive decay, not electron acceleration. However, this also means that the source cannot be "turned off"; radioactive material must always be shielded, even during maintenance and downtimes. The time required for processing depends on the target dose and source strength. Longer processing times may prompt some form of temperature control for temperature-sensitive products such as fruits.

14.2.4 General Mode of Action

An e-beam interacts with the atoms directly, while high energy-photons (gamma rays or X-rays) energize electrons within the food target. These electrons may raise the energy of the electrons to a higher level within the atom (excitation) or leave the atom completely (ionization). Both processes yield free radicals, i.e., atoms with unpaired electrons on their outer shell. Free radicals are very reactive because their unpaired electrons readily pair with outer shell electrons of surrounding atoms within the food target. Water makes up the bulk of fresh fruits and vegetables; therefore, the majority of the absorbed irradiation energy hits water molecules, creating hydrogen and hydroxyl radicals (Diehl 1995). Interaction of these water-derived free radicals with the organic molecules that make up the food is the primary mode of action. Under conditions of limited free water, such as in dried or frozen products, this process is less efficient and fewer radicals are produced. Also, without free water, the radicals have reduced mobility and are more likely to recombine with each other than to interact with target microorganisms. In these products, higher doses become necessary for microbial control (Thayer and Boyd 1995; Nieto-Sandoval et al. 2000).

14.3 Food Safety Treatments

Low doses of irradiation effectively inactivate pathogens on produce (Fan et al. 2008; Gomes et al. 2008). In the US, regulations have long allowed for irradiation as a treatment to preserve freshness and control insect pests. Since 2008, processors have been permitted to irradiate spinach and iceberg lettuce to control human pathogens (FDA 2008). Commercialization has recently expanded with the establishment of an irradiation facility designed for treating leafy greens (Beach 2013).

14.3.1 Control of Human Pathogens

Radiation resistance of microorganisms is stated as D_{10} values, which are radiation doses (in kGy) required to inactivate 90 % of specific pathogens. Table 14.2 summarizes some of the typical D_{10} values obtained for three common foodborne pathogens (*E. coli, Salmonella* spp., and *L. monocytogenes*) on fresh and fresh-cut produce. The D_{10} values of the three pathogens range from 0.04 to 0.54 kGy. Radiation resistance of a pathogen can be influenced by many factors such as temperature and the gas atmosphere at which fresh-cut produce is irradiated, type of produce, strains of pathogens, and location of pathogen in/on fresh-cut produce.

For *Salmonella*-inoculated coriander, the efficacy of irradiation was maintained by cold storage. Reduced pathogen levels were seen on produce stored at 5 °C after irradiation, but storage at 22 °C led to pathogen regrowth after 2 days (Villagomez et al. 2010). During post-irradiation refrigerated storage, a 2-log reduction dose (0.42 kGy) led to regrowth of *L. monocytogenes* on endive, but no regrowth was seen after a 4-log equivalent dose (0.84 kGy) throughout the entire 19-days storage period (Niemira et al. 2003). Similarly, Romaine lettuce inoculated with *L. monocytogenes* gave D_{10} values of 0.16–0.25 kGy; the pathogen showed a persistent suppression during refrigerated storage after a 5-log dose (Mintier and Foley 2006).

E. coli O157: H7, *L. monocytogenes, Salmonella enterica,* and *Shigella flexneri* inoculated onto shredded Iceberg lettuce were significantly reduced by X-ray treatment (Mahmoud 2009). A 1-kGy dose gave reductions of 4.1–4.8 log10 cfu/g, while 2 kGy resulted in >5 log10 cfu/g reductions. The same four pathogens (*E. coli* O157: H7, *L. monocytogenes, Salmonella enterica,* and *Shigella flexneri*) were inoculated onto spinach leaves and irradiated (Mahmoud et al. 2009). Reductions obtained after 1 kGy ranged from 4 to 6 log10 cfu/g; a 2.0 kGy dose completely eliminated the contamination.

Incorporating irradiation into conventional produce processing is a key element in research. Interactions of human pathogens with produce surfaces, and the implications for dose optimization, are a subject of ongoing investigation. Post-irradiation re-growth of human pathogens can be influenced by when in the product handling chain the produce is irradiated.

Table 14.2 Radiation sensitivity of notable pathogens and indicator organisms associated with fresh-cut fruits and vegetables (modified from Niemira and Fan 2013)

	Escherichia coli D_{10}[a]	*Salmonella* D_{10}	*Listeria* spp. D_{10}
Apple			0.24 (Lm) *(1)*
Broccoli			0.22 (Lm) *(2)*
Cabbage	0.17 *(3)*	0.29 (Sp) *(3)*	0.19 (Lm) *(2)*
Carrots	0.12–0.26 *(4)*	0.16 (St) *(5)*	0.31 (Lm) *(5)*; 0.3,0.5 (Lm) *(4)*; 0.36 (Lm) *(6)*
Celery	0.20 *(7)*; 0.14 *(8)*		0.23 (Lm) *(8)*
Cilantro	0.16 (H7) *(9)*		
Cucumber	0.19 *(3)*; 0.47 *(10)*	0.25 (Sp) *(3)*; 0.43 (St) *(10)*; 0.18 (St) *(5)*	0.30 (Li) *(10)*; 0.35 (Lm) *(5)*
Green onions	0.26, 0.28 *(11)*		
Lettuce, iceberg	0.12 (H7) *(12)*; 0.10 (H7) *(14)*; 0.11–0.12 (H7) *(16)*; 0.04, 0.08 *(17)*; 0.23 (H7) *(18)*	0.25 *(15)*; 0.16–0.23 *(16)*; 0.21 *(18)*	0.20 (Li) *(13)*; 0.20 (Lm) *(15)*; 0.24 (Li) *(18)*
Lettuce, Red leaf	0.14 (H7) *(12)*	0.23 *(15)*	0.19 (Lm) *(15)*
Lettuce, Green leaf	0.12 (H7) *(12)*	0.31 *(15)*	0.19 (Lm) *(15)*
Lettuce, Boston	0.14 (H7) *(12)*	0.24 *(15)*	0.19 (Lm) *(15)*
Lettuce, Romaine	0.39 (H7, internalized) *(19)*		0.17–0.19 (Lm) *(20)*; 0.39 (Lm, int) *(20)*
Mint	0.17 (H7) *(21)*	0.17 *(21)*	
Pineapple		0.24 *(22)*	
Spinach	0.24 (H7) *(11)*; 0.11 (H7) *(23)*; 0.29 *(24)*	0.12 *(23)*; 0.29 *(24)*	0.18 *(24)*
Sprouts, alfalfa	0.30 (H7) *(25)*	0.46 *(25)*	
Sprouts, broccoli	0.46 (H7) *(26)*	0.13 *(26)*	0.16 (Lm) *(26)*
Sprouts, mung bean	0.20 (H7) *(27)*		0.20 (Lm) *(2)*
Sprouts, radish	0.41 (H7) *(26)*	0.16 *(27)*	0.22 (Lm) *(26)*
Tomato		0.25–0.39 *(28)*	0.24 *(2)*
Tomato cubes		0.29–0.54 *(29)*	

Li = *Listeria ivanovii*, Lm = *Listeria monocytogenes*, St = *Salmonella* Typhimurium, Sp = *Salmonella* Paratyphi, H7 = *E. coli* O157:H7, Int = internalized

[a]D_{10} values (in kGy) are followed by the citation reference (*in italics*)

Citations: (*1*) Fan (2005), (*2*) Bari et al. (2005), (*3*) Khattak et al. (2005), (*4*) Kamat et al. (2005), (*5*) Dhokane et al. (2006), (*6*) Caillet et al. (2006), (*7*) López et al. (2005), (*8*) Prakash et al. (2000), (*9*) Foley et al. (2004), (*10*) Lee et al. (2006), (*11*) Fan et al. (2008), (*12*) Niemira et al. (2002), (*13*) Kim et al. (2006), (*14*) Foley et al. (2002), (*15*) Niemira (2003), (*16*) Goularte et al. (2004), (*17*) Jeong et al. (2010), (*18*) Mahmoud (2009), (*19*) Niemira (2007a, 2007b), (*20*) Mintier and Foley (2006), (*21*) Hsu et al. (2010), (*22*) Shashidhar et al. (2007), (*23*) Neal et al. (2008), (*24*) Mahmoud et al. (2009), (*25*) Rajkowski and Thayer (2000), (*26*) Waje et al. (2009), (*27*) Bari et al. (2004), (*28*) Prakash et al. (2007), (*29*) Schmidt et al. (2006)

As the response to irradiation is specific to individual commodities, specific handling practices, and combinations with other treatments, significant research gaps remain as to how best to use this technology.

14.3.2 Biofilm-Associated Pathogens

Time in storage prior to irradiation allows internalization of pathogens in stomata and biofilm formation on the surfaces of leaves (Niemira and Cooke 2010). Commensal bacteria and pathogens typically live as a biofilm, a complex community bound to inert and biological surfaces in a durable exopolysaccharide matrix (Korber et al. 1997; Fett 2000). Biofilms effectively reduce the efficacy of ozone, chlorine, hydrogen peroxide, and other antimicrobial processes, frequently by orders of magnitude (Stewart et al. 2004). Research suggests that although this attenuation is less dramatic for irradiation than for other antimicrobial processes, it remains a significant factor.

There is evidence that the irradiation response of biofilm-associated pathogens is dependent on strains and growth conditions. Biofilm-associated cells of *Salmonella* Stanley and *Salmonella* Enteritidis were significantly more sensitive than planktonic cells to irradiation (Niemira and Solomon 2005). In that study, the biofilm habitat did not change the radiation sensitivity of *S. anatum*. In a later study, biofilms of *E. coli* O157:H7 ATCC 43894 were grown at 37 °C for 24, 48, or 72 h. These biofilm-associated cells were significantly more sensitive to irradiation than respective planktonic cells (Niemira 2007a). *E. coli* O157:H7 ATCC 35150 biofilms were less sensitive to irradiation after 24 h of growth, but *E. coli* O157:H7 C9490 biofilms of the same age were more sensitive. Longer cultivation times (48 and 72 h) did not result in differential sensitivity for these isolates (Niemira 2007a). In that study, D_{10} values varied as much as 27 % above or below the D_{10} values obtained for planktonic cells. Temperature of biofilm cultivation engendered a complex response in *L. monocytogenes* and *L. innocua*. These biofilms were either equally sensitive or more sensitive to irradiation as compared with planktonic cells grown under the same conditions. (Niemira 2008).

Leaf-internalized *E. coli* O157:H7 cells had D_{10} values of 0.30–0.45 kGy, higher than those for surface-associated cells which had D_{10} of 0.12–0.14 kGy (Niemira 2007b). Biofilm-associated *E. coli* O157:H7 cells on the leaf surface were also less sensitive to irradiation (Niemira and Cooke 2010). While biofilm attenuation of chemical sanitizers is typically by orders of magnitude, irradiation was typically reduced in efficacy by ~50 %. Thus, the penetrating nature of ionizing radiation may make it uniquely suited to treating internalized, biofilm-associated, or otherwise protected pathogens on or in produce.

14.4 Phytosanitary Applications

14.4.1 Insect Pests and Import Controls

Irradiation inactivates insect pests on fruits and vegetables by: (1) sterilization (0.03–0.2 kGy) that prevents egg hatching and/or reproduction of the insects during or after storage; (2) killing insects outright (1–3 kGy). Irradiation is used as an alternative to fumigation, cold treatment, heat treatment, and other techniques. Given the environmental impact of chemical treatments, phytosanitary fumigation with ethylene dibromide and similar chemicals is increasingly restricted by legislation. Irradiation is an effective quarantine treatment for control of fruit flies and other insects in a number of commodities and is increasingly adopted worldwide. Australia and New Zealand recently approved the use of irradiation doses of up to 1 kGy to control insect pests on tomatoes and peppers (FSANZ 2013). The US Animal Plant Health Inspection Service has approved irradiation for protection against all insect pests including fruit flies, the mango seed weevil, and others. APHIS is currently seeking to facilitate the expansion of irradiation used in this way by streamlining the administrative oversight and data collection associated with its use (APHIS 2013).

14.5 Shelf-Life Extension

Irradiation with doses of 0.2–0.5 kGy can delay ripening of fruits by interference with biochemical processes (Dubery et al. 1984; Niemira et al. 2005). This extends shelf-life and marketability. A dose of 0.44 kGy reduced ethylene production as well as a variety of volatile esters and alcohols in Gala apples (Fan et al. 2001). That study used gamma rays, as the penetration of photons is appropriate for the density profiles of fruits and vegetables. The limited penetration of e-beam requires careful processing conditions to ensure dose uniformity; applications for changing ripening metabolism and increasing shelf-life has been demonstrated with fruits such as avocado (Grandison 1993). Climacteric fruits (i.e., fruits that show a transient increase in their rate of respiration and ethylene production at the onset of ripening) require irradiation before the onset of the climacteric increase in ethylene production (Maxie and Abdel-Kader 1966; Dubery et al. 1984). The necessity of treatment at an early stage of maturity can lead to abnormal ripening and uneven coloring; as in the case of tomato, an irradiation treatment applied too early may actually initiate a wound-response increase in ethylene production, accelerating ripening (Larrigaudiere et al. 1991).

The majority of tropical fruits are climacteric, with a characteristically short shelf-life. These also tend to be sensitive to low temperature. Irradiation causes a shift from glycolysis toward the pentose phosphate shunt in bananas and toward the glyoxylate cycle in bananas and mangoes (Thomas 1986). Thus, irradiation interfered with the onset of senescence rather than delaying ripening of mango per se (Dubery et al.

1984). Respiration was enhanced in Gala apples treated with 1-methylcyclopropene, an ethylene action inhibitor (Fan and Mattheis 2001). Therefore, irradiation slows ripening in climacteric fruits by reducing (1) ethylene sensitivity and, (2) ethylene biosynthesis (Maxie and Abdel-Kader 1966; Larrigaudiere et al. 1991; Strydom et al. 1991). Irradiated Bartlett pears showed reducing ripening rates, even when subsequently exposed to ethylene (Bramlage and Couey 1965).

14.5.1 Sensory Impacts

In general, irradiation tends to cause hydrolysis of pectins and concomitant release of sugars following depolymerization of carbohydrate polymers (Amour et al. 1993; Yu et al. 1996). The delay of ripening notwithstanding, this pectinase-like action renders irradiated fruits softer and sweeter. Treated fruits are somewhat more susceptible to damage during handling and shipping. X-ray irradiation of 0.75 kGy reduced the firmness of papaya, rambutan, and Kau orange (Boylston et al. 2002). After 1.0 or 2.0 kGy, strawberries were softer, with increased water-soluble pectin and decreased oxalate-soluble pectin (Yu et al. 1996). In separate studies, irradiated strawberries and apples showed reduced titratable acidity and increased sweetness compared to unirradiated controls (Lovell and Flick 1966; Fan and Mattheis 2001). Grapefruit were able to tolerate 0.3 kGy with little loss of quality of the fruit, pulp, or juice if they were also treated with gibberellic acid (Miller and McDonald 1966). However, a higher dose of 0.6 kGy led to unacceptable levels of skin pitting, softening, and loss of juice quality.

In vegetables, overall sugar content is less than that for intact fruits. Fan et al. (2012) showed that irradiation (0.5 and 1.0 kGy) induced symptoms similar to russet spotting and other discolorations (pink ribs, rusty brown, and vein browning) in heads of Iceberg lettuce. This was true for both external and internal leaves. Neither pretreatment with 1 ppm 1-methylcyclopropene (an anti-browning agent) nor irradiation in the absence of oxygen had significant effect. The researchers determined that maintaining low oxygen atmosphere during the entire 14-day storage almost eliminated the disorders. Irradiation induced higher respiration, compared with non-treated control, as indicated by significantly lower headspace O_2 and higher CO_2 levels. The authors conclude that modified atmosphere packaging could mitigate irradiation-induced disorders in Iceberg lettuce.

14.6 Regulations

14.6.1 US and International

A joint FAO/IAEA/WHO study (1997) examined the wholesomeness of food irradiated above 10 kGy. In that study, it was concluded that food may be safely irradiated to any dose sufficient to achieve the desired physiological or microbial outcome,

Table 14.3 United States Code of Federal Regulations 21CFR179.26: applications and dose limits for irradiated food

Commodity and purpose	Dose limits
Control of *Trichinella* in pork	0.3–1.0 kGy
Suppression of growth and maturation in fresh food	Maximum dose 1.0 kGy
Disinfestation of insect pests	Max. 1.0 kGy
Antimicrobial treatment of dry enzymes	Max. 10.0 kGy
Antimicrobial treatment of dry herbs and spices	Max. 30.0 kGy
Control of pathogens in fresh and frozen raw poultry	Max. 3.0 kGy
Sterilization of food intended for use by NASA	Minimum dose 44.0 kGy
Control of pathogens and extension of shelf-life of refrigerated and frozen meats	Max. 4.5 kGy (refrigerated), Max. 7.0 kGy (frozen)
Control of *Salmonella* in fresh shell eggs	Max. 3.0 kGy
Control of pathogens in seeds used to produce sprouts	Max. 8.0 kGy
Control of *Vibrio* species and other foodborne pathogens in fresh or frozen molluscan shellfish	Max. 5.5 kGy
Control of foodborne pathogens and extension of shelf-life in fresh Iceberg lettuce and fresh spinach	Max. 4.0 kGy

without appreciable loss of nutritional adequacy. Regulations governing commodities to be irradiated, the maximum (and in some cases, the minimum) dose to be applied, and the purposes for which a product may be irradiated are determined by individual countries. Approval and inspection of food irradiation plants are also nation-specific. However, international trade has trended toward increased harmonization of regulations in order to facilitate import and export.

The applicable regulations for food irradiation in the United States are presented in Table 14.3. Many European nations (e.g. Austria, Germany, Ireland) have approved only dry herbs, spices, and vegetable seasonings for irradiation, with a maximum dose of 10 kGy, a lower limit than that allowed for such food in the US (Table 14.3). A number of facilities have been licensed throughout the European Union for the purpose of irradiating herbs, spices, and seasonings, as well as onions and potatoes (for sprout suppression) (OJEU 2011). In the UK, commodities allowed to be irradiated include fruit (max. 2 kGy), vegetables (1 kGy), cereals (1 kGy), bulbs and tubers (0.2 kGy), dried aromatic herbs, spices, and vegetable seasonings (10 kGy), fish and shellfish (3 kGy), and poultry (7 kGy) (FSA 2012).

Chinese regulations encompass six general classes of food approved nationally for irradiation (Niemira and Gao 2011). These are (1) dried nuts and preserved fruits, (2) dried spices, (3) fresh fruits and vegetables, (4) frozen packaged meat of livestock and poultry, (5) beans, grains, and their products, and (6) cooked meat food of livestock and poultry. In addition to these general classes of food, three specific food items have been approved for irradiation: pollen, hog carcasses, and sweet potato wine. It can therefore be understood that regulatory approvals may be general or quite specific, based on the approach taken by the controlling legal authorities. As of 2009, the total amount of irradiated food in China is about 200,000 tons. Irradiation-induced extension of shelf-life serves to broaden the domestic market that Chinese manufacturers can economically reach.

Fig. 14.1 The "radura" logo required for labeling on irradiated foods. The logo must be accompanied by the text "*Treated with radiation*" or "*Treated by irradiation*"

In the United States, products which have been irradiated must be labeled with an indicative symbol called a "radura" (Fig. 14.1). Regulations for labelling vary internationally, but product labelling must adhere to the regulations in effect in the point-of-sale nation, rather than those in effect for the county of origin.

14.6.2 Packaging

Like other food treated by irradiation, the use of this process with fruits involves constraints on the types of packaging materials available. Irradiation may affect the physical performance characteristics of packaging as well as undergo unwanted physical changes as a result of the irradiation itself. Traditional packaging materials such as metal, glass, and paper are relatively resistant to ionizing radiation. Although irradiation has minimal effects below 10 kGy, it is possible that volatiles may be generated from unsuitable plastics, producing unpleasant aromas or flavors. Irradiation acts on plastics primarily by cross-linking and/or cleavage of the polymers. The specific type of polymer used will determine the relative importance of each (Wilson 1974). Guidelines for the packaging materials permitted for irradiation in the United States are presented in Tables 14.4 and 14.5.

Plastics have historically been basic resin polymers, augmented with stabilizers, antistatic compounds, antislip agents, plasticizers, or other adjuvants. Plastic film laminates are increasingly used to provide the desired mixture of strength, durability, gas permeability, ability to take inks, etc. Each plastic component of the laminate has a unique response to irradiation including specialty laminates which may incorporate a metal foil barrier layer.

Table 14.4 United States Code of Federal Regulations 21CFR179.45: Packaging materials approved for irradiated food

Material	Maximum dose (kGy)
Nitrocellulose-coated or vinylidene chloride copolymer-coated cellophane	10
Glassine paper	10
Wax-coated paperboard	10
Films of polyolefin or polyethylene terephthalate. These may contain:	10
1. Sodium citrate, sodium lauryl sulfate, polyvinyl chloride[a]	
2. Coatings comprising a vinylidene chloride copolymer containing a minimum of 85 % vinylidene chloride with one or more of the following comonomers: Acrylic acid, acrylonitrile, itaconic acid, methyl acrylate, and methyl methacrylate	
Kraft paper (only as a container for flour)	0.5
Polystyrene film	10
Rubber hydrochloride film	10
Vinylidene chloride-vinyl chloride copolymer film	10
Nylon 11	10
Ethylene-vinyl acetate copolymers	30
Vegetable parchments	60
Polyethylene film[a]	60
Polyethylene terephthalate film[a]	60
Nylon 6 films[a]	60
Vinyl chloride-vinyl acetate copolymer film[a]	60
Acrylonitrile copolymers[a]	60

[a]This material may be amended with additional materials, listed in Table 14.5

Table 14.5 United States Code of Federal Regulations 21CFR179.45: Adjuvants and amendments approved for incorporation into certain packaging materials approved for irradiated food

Adjuvant/amendment	Limit (by wt. of polymer) (%)
Amides of erucic, linoleic, oleic, palmitic, and stearic acid.	1
BHA (butylated hydroxyanisole)	1
BHT (butylated hydroxytoluene)	1
Calcium and sodium propionates	1
Petroleum wax	1
Mineral oil	1
Stearates of aluminum, calcium, magnesium, potassium, and sodium	1
Triethylene glycol	1
Polypropylene, noncrystalline	2

Active packaging, biodegradable plastics, and materials which release preservatives and antimicrobial compounds are an important area of innovation within packaging technology (Jin and Niemira 2011). These advanced materials can effectively suppress the growth of spoilage organisms or pathogens, enhance flavor, aroma, and color, or serve other important functions. The Codex General Standard for Irradiated Foods (Codex 1984) established general guidelines for packaging of irradiated food. For these materials to be used with irradiated food products to be sold in the US, the packaging materials compounds must be cleared by the U.S. Food and Drug Administration (Table 14.4). Some of these plastics may be amended with various adjuvants, including preservatives, etc. (Table 14.4).

14.7 Conclusions

Irradiation can effectively preserve and improve the safety, sensory properties, and shelf-life of a wide variety of fresh and fresh-cut produce. Preparation methods, storage protocols, shipment and transshipment conditions, and market forces will be key influencing factors in the application of this technology. Used singly or in combination with other treatments, irradiation is a processing intervention that can help domestic and international fresh and fresh-cut produce growers and suppliers provide consumers with the safest, highest quality produce possible.

Acknowledgments The author extends thanks to David Geveke, Ph.D., and Joseph Sites, PE, for their thoughtful reviews of this manuscript. Mention of trade names or commercial products in this publication is solely for the purpose of providing specific information and does not imply recommendation or endorsement by the United States Department of Agriculture (USDA), which is an equal opportunity provider and employer.

References

Amour, J., C. Gosselin, J. Arul, F. Castaigne, and C. Willemot. 1993. Gamma-radiation affects cell wall composition of strawberries. *Journal of Food Science* 58: 182–185.

Annous, B.A., E.B. Solomon, P.H. Cooke, and A. Burke. 2005. Biofilm formation by *Salmonella* spp. on cantaloupe melons. *Journal of Food Safety* 25: 276–287.

APHIS (Animal Plant Health Inspection Service). 2013. Notice of request for extension of approval of an information collection; irradiation phytosanitary treatment of imported fruits and vegetables, http://www.gpo.gov/fdsys/pkg/FR-2012-10-31/html/2012-26727.htm (accessed Feb 22, 2103).

Bari, M.L., M.I. Al-Haq, T. Kawasaki, et al. 2004. Irradiation to kill *Escherichia coli* O157:H7 and *Salmonella* on ready-to-eat radish and mung bean sprouts. *Journal of Food Protection* 67: 2263–2268.

Bari, M.L., M. Nakauma, S. Todoriki, et al. 2005. Effectiveness of irradiation treatments in inactivating *Listeria monocytogenes* on fresh vegetables at refrigeration temperature. *Journal of Food Protection* 68(2): 318–323.

Beach, C. 2013. Canadian firm prepares to irradiate leafy greens. *The Packer*, p. B4. Jan 21, 2013.

Boylston, T.D., C.A. Reitmeier, J.H. Moy, G.A. Mosher, and L. Taladriz. 2002. Sensory quality and nutrient composition of three Hawaiian fruits treated by x-irradiation. *Journal of Food Quality* 25: 419–433.

Bramlage, W.J., and H.M. Couey. 1965. *Gamma radiation of fruits to extend market life Marketing Res Rep No 717*. Washington, DC: ARS-USDA.

Buck, J.W., R.R. Walcott, and L.R. Beuchat. 2003. Recent trends in microbiological safety of fruits and vegetables. Online. *Plant Health Progress*. doi:10.1094/PHP-2003-0121-01-RV.

Caillet, S., M. Millette, M. Turgis, S. Salmieri, and M. Lacroix. 2006. Influence of antimicrobial compounds and modified atmosphere packaging on radiation sensitivity of *Listeria* monocytogenes present in ready-to-use carrots (*Daucus carota*). *Journal of Food Protection* 69: 221–227.

Calvin, L. 2007. Outbreak linked to spinach forces reassessment of food safety practices. Amber Waves, USDA Economic Research Service, http://ideas.repec.org/a/ags/uersaw/125239.html (accessed Feb 26, 2103).

CDC (Centers for Disease Control and Prevention). 2011. CDC estimates of foodborne illness in the united states, http://www.cdc.gov/foodborneburden/PDFs/FACTSHEET_A_FINDINGS_updated4-13.pdf (accessed Feb 26, 2013).

Codex Alimentarius Commission. 1984. *Codex general standard for irradiated foods and recommended international code of practice for operation of radiation facilities used for the treatment of foods*, vol. XV. Geneva, Switzerland: WHO.

CNPP (Center for Nutrition Policy and Promotion). 2010. U.S. Dept. Agriculture—Dietary Guidelines for Americans, http://www.cnpp.usda.gov/dietaryguidelines.htm (accessed Feb 26, 2013).

CSPI (Center for Science in the Public Interest). 2007. Outbreak alert: Closing the gaps in our federal food-safety net, http://www.cspinet.org/foodsafety/outbreak_alert.pdf (accessed Feb 26, 2013).

Dhokane, V.S., S. Hajare, R. Shashidhar, A. Sharma, and J.R. Bandekar. 2006. Radiation processing to ensure safety of minimally processed carrot (*Daucus carota*) and cucumber (*Cucumis sativus*): Optimization of dose for the elimination of *Salmonella* Typhimurium and *Listeria monocytogenes*. *Journal of Food Protection* 69(2): 444–448.

Diehl, J.F. 1995. *Safety of irradiated foods*, 2nd ed. New York: Marcel Dekker.

Dubery, I.A., L.J. van Rensburg, and J.C. Schabort. 1984. Malic enzyme activity and related biochemical aspects during ripening of irradiated mango fruit. *Phytochemistry* 23: 1383–1386.

ERS (Economic Research Service), USDA. 2008. Food availability (per capita) data system, http://www.ers.usda.gov/Data/FoodConsumption (accessed Feb 26, 2013)

FAO/IAEA/WHO. 1997. High-dose irradiation: Wholesomeness of food irradiated with doses above 10 kGy, in joint FAO/IAEA/WHO study group on high-dose irradiation. WHO Technical Report Series 890.

Fan, X., and J.P. Mattheis. 2001. 1-Methylcyclopropene and storage temperature influence responses of Gala apple fruit to gamma irradiation. *Postharvest Biology and Technology* 23: 143–151.

Fan, X., L. Argenta, and J. Mattheis. 2001. Impacts of ionizing radiation on volatile production by ripening Gala apple fruit. *Journal of Agricultural and Food Chemistry* 49: 254–262.

Fan, X. 2005. Antioxidant capacity of fresh-cut vegetables exposed to ionizing radiation. *Journal of the Science of Food and Agriculture* 85: 995–1000.

Fan, X., B.A. Niemira, and A. Prakash. 2008. Irradiation of fresh and fresh-cut fruits and vegetables. *Food Technology* 3: 36–43.

Fan, X., K. Sokorai, B.A. Niemira, R.S. Mills, and M.Y. Zhen. 2012. Quality of gamma ray-irradiated iceberg lettuce and treatments to minimize irradiation-induced disorders. *Hortscience* 47: 1108–1112.

FDA (U.S. Food and Drug Administration). 2004. Produce safety from production to consumption: 2004 action plan to minimize foodborne illness associated with fresh produce consumption, http://www.fda.gov/Food/FoodSafety/Product-SpecificInformation/FruitsVegetablesJuices/FDAProduceSafetyActivities/ProduceSafetyActionPlan/ucm129487.htm (accessed Feb 26, 2013).

———. 2008. Final Rule (73 FR 49593), Irradiation in the production, processing and handling of food. 21 CFR Part 179, http://www.fda.gov/ForConsumers/ConsumerUpdates/ucm093651.htm (accessed Feb 26, 2013).

Fett, W.F. 2000. Naturally occurring biofilms on alfalfa and other types of sprouts. *Journal of Food Protection* 63(5): 625–632.

Foley, D.M., A. Dufour, L. Rodriguez, F. Caporaso, and A. Prakash. 2002. Reduction of *Escherichia coli* 0157:H7 in shredded Iceberg lettuce by chlorination and gamma irradiation. *Radiation Physics and Chemistry* 63: 391–396.

Foley, D.M., M. Euper, F. Caporaso, and A. Prakash. 2004. Irradiation and chlorination effectively reduces *Escherichia coli* O157:H7 inoculated on cilantro (*Coriandrum sativum*) without negatively affecting quality. *Journal of Food Protection* 67(10): 2092–2098.

FSANZ (Food Standards Australia and New Zealand). 2013. Application A1069—Irradiation of Tomatoes & Capsicums, http://www.foodstandards.gov.au/foodstandards/applications/applicationa1069irra5511.cfm (accessed Feb 22, 2013).

FSA (Food Standards Agency). 2012. Irradiated food, http://www.food.gov.uk/policy-advice/irradfoodqa/#.USfLRvLBmM0 (accessed Feb 22, 2013)

Gomes, C.R., G. Moreira, M.E. Castell-Perez, et al. 2008. E-Beam irradiation of bagged, ready-to-eat spinach leaves (Spinacea oleracea): An engineering approach. *Journal of Food Science* 73: E95–E102.

Gómez-López, V.M., A. Rajkovic, P. Ragaert, N. Smigic, and F. Devlieghere. 2009. Chlorine dioxide for minimally processed produce preservation: A review. *Trends in Food Science & Technology* 20(1): 17–26.

Goodburn, C., and C.A. Wallace. 2013. The microbiological efficacy of decontamination methodologies for fresh produce: A review. *Food Control* 32(2): 418–427.

Goularte, L., C.G. Martins, I.C. Morales-Aizpurua, et al. 2004. Combination of minimal processing and irradiation to improve the microbiological safety of lettuce (*Lactuca sativa* L.). *Radiation Physics and Chemistry* 71: 155–159.

Grandison, A.S. 1993. Combination treatments including electron beam irradiation for extending the shelf-life of fresh fruits. *Communication at the 2nd FAO/IAEA Coordination Meeting on Irradiation in Combination with Other Processes for Improving Food Quality*. FRDC, Saint-Hyacinthe, Quebec.

Hsu, W.Y., A. Simonne, P. Jitareerat, and M.R. Marshall Jr. 2010. Low-dose irradiation improves microbial quality and shelf life of fresh mint (*Mentha piperita* L.) without compromising visual quality. *Journal of Food Science* 75: M222–M230.

Jeong, S., B.P. Marks, E.T. Ryser, and S.R. Moosekian. 2010. Inactivation of *Escherichia coli* O157:H7 on lettuce, using low-energy X-ray irradiation. *Journal of Food Protection* 73: 547–551.

Jin, T., and B.A. Niemira. 2011. Application of antimicrobial polylactic acid coating in reduction of *Escherichia coli* O157:H7 and *Salmonella* Stanley on apples. *Journal of Food Science* 76(3): M184–M188.

Kamat, A.S., N. Ghadge, M.S. Ramamurthy, and M.D. Alur. 2005. Effect of low-dose irradiation on shelf life and microbiological safety of sliced carrot. *Journal of the Science of Food and Agriculture* 85(13): 2213–2219.

Khattak, A.B., N. Bibi, M.A. Chaudry, et al. 2005. Shelf life extension of minimally processed cabbage and cucumber through gamma irradiation. *Journal of Food Protection* 69(11): 2648–2663.

Kim, J., J. Lee, J. Kim, et al. 2006. Effect of gamma irradiation on *Listeria ivanovii* inoculated to Iceberg lettuce stored at cold temperature. *Food Control* 17: 397–401.

Korber, D.R., A. Choi, G.M. Wolfaardt, S.C. Ingham, and D.E. Caldwell. 1997. Substratum topography influences susceptibility of *Salmonella enteritidis* biofilms to trisodium phosphate. *Applied and Environmental Microbiology* 63: 3352–3358.

Larrigaudiere, C., A. Latche, J.C. Pech, and C. Triantaphylides. 1991. Relationship between stress ethylene production induced by gamma irradiation and ripening of cherry tomatoes. *Journal of the American Society for Horticultural Science* 116: 1000–1003.

Lee, N.Y., C. Jo, D.H. Shin, W.G. Kim, and M.W. Byun. 2006. Effect of gamma-irradiation on pathogens inoculated into ready-to-use vegetables. *Food Microbiology* 23(7): 649–656.

López, L., S. Avendaño, J. Romero, S. Garrido, J. Espinoza, and M. Vargas. 2005. Effect of gamma irradiation on the microbiological quality of minimally processed vegetables. *Archivos Latinoamericanos de Nutrición* 55(3): 287–292.

Lovell, R.T., and G.J. Flick. 1966. Irradiation of Gulf Coast area strawberries. *Food Technology* 29: 99–102.

Mahmoud, B.S.M. 2009. Effects of X-ray radiation on *Escherichia coli* O157:H7, *Listeria monocytogenes*, *Salmonella enterica* and *Shigella flexneri* inoculated on shredded iceberg lettuce. *Food Microbiology* 27(1): 109–114.

Mahmoud, B.S.M., G. Bachman, and R.H. Linton. 2009. Inactivation of *Escherichia coli* O157:H7, *Listeria monocytogenes*, *Salmonella enterica* and *Shigella flexneri* on spinach leaves by X-ray. *Food Microbiology* 27(1): 24–28.

Mandrell, R.E. 2009. Enteric human pathogens associated with fresh produce: Sources, transport and ecology. In *Microbial Safety of Fresh Produce*, ed. X. Fan, B.A. Niemira, C.J. Doona, F.E. Feeherry, and R.B. Gravani, 5–41. Ames, IA: Blackwell.

Maxie, E.C., and A.S. Abdel-Kader. 1966. Food irradiation-physiology of fruits as related to the feasibility of the technology. *Advances in Food Research* 15: 105–145.

Miller, W.R., and R.E. McDonald. 1966. Postharvest quality of GA-treated Florida grapefruit after gamma irradiation with TBZ and storage. *Postharvest Biology and Technology* 7: 253–260.

Mintier, A.M., and D.M. Foley. 2006. Electron beam and gamma irradiation effectively reduce *Listeria monocytogenes* populations on chopped romaine lettuce. *Journal of Food Protection* 9(3): 570–574.

Moreira, R.G., A.F. Puerta-Gomez, J.S. Kim, and M.E. Castell-Perez. 2012. Factors affecting radiation D-values (D10) of an *Escherichia coli* cocktail and *Salmonella* Typhimurium inoculated in fresh produce. *Journal of Food Science* 77: E104–E111.

Neal, J.A., E. Cabrera-Diaz, M. Márquez-Gonzaález, J.E. Maxim, and A. Castillo. 2008. Reduction of *Escherichia coli* O157:H7 and *Salmonella* on baby spinach, using electron beam radiation. *Journal of Food Protection* 71: 2415–2420.

Niemira, B.A., C.H. Sommers, and X. Fan. 2002. Suspending lettuce type influences recoverability and radiation sensitivity of *Escherichia coli* O157:H7. *Journal of Food Protection* 65(9): 1388–1393.

Niemira, B.A. 2003. Radiation sensitivity and recoverability of *Listeria monocytogenes* and *Salmonella* on 4 lettuce types. *Journal of Food Science* 68(9): 2784–2787.

Niemira, B.A., X. Fan, K.J.B. Sokorai, and C.H. Sommers. 2003. Ionizing radiation sensitivity of *Listeria monocytogenes* and *L. innocua* inoculated on endive (*Cichorium endiva*). *Journal of Food Protection* 66: 993–998.

Niemira, B.A., and E.B. Solomon. 2005. Sensitivity of planktonic and biofilm-associated *Salmonella* to ionizing radiation. *Applied and Environmental Microbiology* 71(5): 2732–2736.

Niemira, B.A., C.H. Sommers, and D. Ukuku. 2005. Mechanisms of microbial spoilage of fruits and vegetables. In *Produce Degradation: Reaction Pathways and their Prevention*, ed. O. Lamikanra, S.H. Imam, and D.O. Ukuku, 463–482. New York, NY: Taylor & Francis.

Niemira, B.A. 2007a. Irradiation sensitivity of planktonic and biofilm-associated *Escherichia coli* O157:H7 isolates is influenced by culture conditions. *Applied and Environmental Microbiology* 73(10): 3239–3244.

———. 2007b. Relative efficacy of sodium hypochlorite wash vs. irradiation to inactivate *Escherichia coli* o157:h7 internalized in leaves of romaine lettuce and baby spinach. *Journal of Food Protection* 70(11): 2526–2532.

———. 2008. Influence of growth temperature on irradiation sensitivity of planktonic and biofilm-associated *Listeria monocytogenes* and *L. innocua*. *Food and Bioprocess Technology*. doi:10.1007/s11947-008-0079-5.

Niemira, B.A., and P. Cooke. 2010. *Escherichia coli* O157:H7 biofilm formation on lettuce and spinach leaf surfaces reduces efficacy of irradiation and sodium hypochlorite washes. *Journal of Food Science* 75(5): M270–M277.

Niemira, B.A., and M. Gao. 2011. Irradiation as a non-thermal process for beverages, juices and fluid foods (Ch. 7). In *Novel thermal and non-thermal technologies for fluid foods*, ed. P.J. Cullen, B. Tiwari, and V. Valdramidis. Amsterdam, Netherlands: Elsevier.

Niemira, B.A., and X. Fan. 2013. *Advances in processing technologies to preserve and enhance the safety of fresh and fresh-cut fruits and vegetables*, Encyclopedia of food microbiology, 2nd ed. Ames, IA: Wiley-Blackwell.

Nieto-Sandoval, J.M., L. Almela, J.A. Fernandez-Lopez, and J.A. Muñoz. 2000. Effect of electron beam irradiation on color and microbial burden of red paprika. *Journal of Food Protection* 63: 633–637.

OJEU (Official Journal of the European Union). 2011. List of approved facilities for the treatment of foods and food ingredients with ionising radiation in the Member States, http://eur-lex.europa.eu/LexUriServ/LexUriServ.do?uri=OJ:C:2011:077:0014:0017:EN:PDF (accessed Feb 22, 2013).

Painter, J.A., R.M. Hoekstra, T. Ayers, R.V. Tauxe, C.R. Braden, F.J. Angulo, et al. 2013. Attribution of foodborne illnesses, hospitalizations, and deaths to food commodities by using outbreak data, United States, 1998–2008. *Emerging Infectious Disease*, http://dx.doi.org/10.3201/eid1903.111866 (accessed Feb 26, 2013).

Prakash, A., P. Inthajak, H. Huibregtse, F. Caporaso, and D.M. Foley. 2000. Effects of low-dose gamma irradiation and conventional treatments on shelf life and quality characteristics of diced celery. *Journal of Food Science* 65: 1070–1075.

Prakash, A., N. Johnson, and D. Foley. 2007. Irradiation D values of *Salmonella* spp. in diced tomatoes dipped in 1% calcium chloride. *Foodborne Pathogens and Disease* 4: 84–88.

Rajkowski, K.T., and D.W. Thayer. 2000. Reduction of *Salmonella* spp. and strains of *Escherichia coli* O157:H7 by gamma radiation of inoculated sprouts. *Journal of Food Protection* 63: 871–875.

Scharff, R.L. 2010. Health-related costs from foodborne illness in the United States. Produce Safety Project at Georgetown University, http://www.producesafetyproject.org/admin/assets/files/Health-Related-Foodborne-Illness-Costs-Report.pdf-1.pdf (accessed Feb 26, 2013).

Schmidt, H.M., M.P. Palekar, J.E. Maxim, and A. Castillo. 2006. Improving the microbiological quality and safety of fresh-cut tomatoes by low-dose electron beam irradiation. *Journal of Food Protection* 69: 575–581.

Shashidhar, R., V.S. Dhokane, S.N. Hajare, A. Sharma, and J.R. Bandekar. 2007. Effectiveness of radiation processing for elimination of *Salmonella* Typhimurium from minimally processed pineapple (Ananas comosus Merr.). *Journal of Food Science* 72: M98–M101.

Sivapalasingam, S., C.R. Friedman, L. Cohen, and R.V. Tauxe. 2004. Fresh produce: A growing cause of outbreaks of foodborne illness in the United States, 1973 through 1997. *Journal of Food Protection* 67: 2342–2353.

Stewart, P.S., P.K. Mukherjee, and M.A. Ghannoum. 2004. Biofilm antimicrobial resistance. In *Microbial biofilms*, ed. M.A. Ghannoum and G.A. O'Toole, 250–268. Washington, DC: ASM Press.

Strydom, G.J., J. Van Staden, and M.T. Smith. 1991. The effect of gamma radiation on the ultrastructure of the peel of banana fruits. *Environmental and Experimental Botany* 31: 43–49.

Thayer, D.W., and G. Boyd. 1995. Radiation sensitivity of *Listeria monocytogenes* on beef as affected by temperature. *Journal of Food Science* 60: 237–240.

Thomas, P. 1986. Radiation preservation of foods of plant origin. III. Tropical fruits: Bananas, mangoes, and papayas. *CRC Critical Reviews in Food Science and Nutrition* 23: 147–205.

Villagomez, N.S., E.A.N. Herrera, L.R. Orozco, G. Wild-Padua, and M.H. Iturriaga. 2010. Effect of storage time on the efficacy of chemical and irradiation treatments to inactivate *Salmonella* on cilantro (*Coriandrum sativum* L.). *Journal of Food Safety* 30: 584–598.

Waje, C.K., S.Y. Jun, Y.K. Lee, et al. 2009. Microbial quality assessment and pathogen inactivation by electron beam and gamma irradiation of commercial seed sprouts. *Food Control* 20: 200–204.

Wilson, J.E. 1974. Irradiation of polymers: Crosslinking versus scission. Ch 7. In *Radiation chemistry of monomers, polymers and plastics*. New York: Marcel Dekker.

Chapter 15
Microwave Processing of Fruits

Katia Nicolau Matsui, Cynthia Ditchfield, and Carmen Cecilia Tadini

15.1 Introduction

Although the development of microwave technology was driven during the Second World War, its domestic use for food heating begun in the decade of 50. Since then, the use of microwaves for cooking is widespread, but the industrial application of this technology for processing materials is still relatively new.

Heating by microwave energy, also known as dielectric heating, is considered a direct heating method. In conventional thermal processing, energy is transferred to the material through convection, conduction, and/or radiation of heat. However, microwave energy is transferred across the materials through molecular interaction with the electromagnetic field. This mechanism of heat transfer promotes greater penetration depth and faster heating rates that would potentially improve retention of thermolabile constituents in the food.

Microwaves are a form of lower energy electromagnetic radiation, with frequency range from 300 to 3000 MHz, corresponding to wavelength range from 100 μm up to 1 m. The microwaves are generated by a *magnetron* from an electrical power source with very short time of start up and shut down. As soon as the *magnetron* is activated, microwaves are generated travelling through the product without heating the oven cavity.

At the customary microwaves frequencies, the magnetron used in conventional microwave ovens generates power at the order of a 1 kW in the 2–3 GHz range,

K.N. Matsui (✉)
Federal University of Rio Grande do Norte, Natal, Brazil

C. Ditchfield
University of São Paulo, Pirassununga, SP, Brazil

C.C. Tadini
University of São Paulo, São Paulo, SP, Brazil

A. Rosenthal et al. (eds.), *Fruit Preservation*, Food Engineering Series,
https://doi.org/10.1007/978-1-4939-3311-2_15

and in industrial ovens with output to 1 MW. Industrial microwave processing is usually accomplished at the frequencies set aside for industrial use, 915 MHz, 2.45, 5.8, and 24.124 GHz. For food, the frequencies commonly applied are 915 and 2450 MHz.

Detailed descriptions of component parts of microwave ovens are given by Buffler (1993).

An important element of microwave process development and system design is the capability to model electromagnetic interactions. An understanding of the variation of dielectric properties with temperature and processing state is crucial to obtain the desired processing conditions.

15.1.1 Mechanisms of Microwave Heating for Food

The common experience of using microwaves to heat food is based primarily on the dipole behavior of the water molecule in the food and the dipole's interaction with microwaves. Because microwaves generate rapidly changing electric fields, these dipoles rapidly change their orientations in response to the changing fields. If the field change is occurring near the natural frequency at which reorientation happens, energy consumed is maximized and heating is optimal. In the terminology of microwave processing, when this happens it is said the material is well "coupled."

When microwaves are directed towards a food, part of the energy is reflected, part is transmitted through the surface, and part is absorbed. The proportions of energy which fall into these three categories have been defined in terms of the dielectric properties.

Dielectric materials are heated when exposed to microwave's energy and the quick modification of electromagnetic field tries to polarize its molecules (ions, dipoles). However, molecules can't follow the quick polarization, so part of this energy is lost and converted to heat. It's usually admitted that water and ions (polar molecules) are responsible for ohmic loss of microwave's energy inside the food, as expressed in Eq. (15.1):

$$\frac{\dot{q}}{V} = 2\pi f \varepsilon_0 \varepsilon'' F_{CE}^2 \tag{15.1}$$

wherein $\frac{\dot{q}}{V}$ is the volumetric rate of heat transferring $[\mathrm{W} \cdot \mathrm{m}^{-3}]$, f is the frequency $[\mathrm{s}^{-1}]$, ε_0 is the permittivity of the free space $[8.854 \times 10^{-12}\ \mathrm{F} \cdot \mathrm{m}^{-1}]$, ε'' is the dielectric loss factor [dimensionless], which describes the ability of a material to dissipate energy in response to an applied electric field, and F_{CE} is the electric field strength inside the load $[\mathrm{V} \cdot \mathrm{m}^{-1}$ or $\mathrm{kg \cdot s^{-3} \cdot A^{-1}}]$.

15.1.2 Dielectric Properties

Knowledge of the dielectric properties of foods is essential in research, modeling, and development of thermal treatment based on radio frequency (RF) and microwaves (MW) energy. The dielectric properties of materials that are of interest in most applications can be defined in terms of their relative permittivity. Permittivity is a complex quantity generally used to describe the dielectric properties that influence reflection of electromagnetic waves at interfaces and attenuation of the wave energy within materials (Sosa-Morales et al. 2010). The relative complex permittivity ($\varepsilon*$) describes permittivity related to free space and it is mathematically expressed as:

$$\varepsilon* = \varepsilon^{'} - j\varepsilon^{''} \tag{15.2}$$

wherein ε' is the dielectric constant [dimensionless], ε'' is the loss factor [dimensionless], and $j = \sqrt{-1}$.

The real part, ε', describes the ability of a food to store energy in response to an applied electric field and influences the electric field distribution and the phase of waves travelling through the material. The imaginary part, ε'', influences both energy absorption and attenuation, and describes the ability of a food to dissipate energy in response to an applied electric field, which typically results in heat generation. The amount of thermal energy in the food is proportional to the value of the loss factor (Tang 2005).

The characterization of dielectric properties is vital for understanding the response of a food when submitted to microwaves, since most useful quantities needed in the design of microwave thermal processes.

The heating effect in food materials (loss factor, ε'') is the result of two mechanisms: dipolar rotation and ionic conduction. When the electric field is applied, the response of dipoles to an oscillating field is an increase in rotational and vibrational energies, depending on the degree of symmetry of the molecule, with a resulting frictional generation of heat. This mechanism is responsible for water heating, a main component of most foods.

The second mechanism is ionic conduction. The electrical field causes the migration of positively and negatively charged ions towards oppositely charged regions. This results in multiple collisions and disruption of hydrogen bonds, both of which result in the generation of heat. Ionic conduction has a larger influence on ε'' than ε', and therefore decreases the penetration depth.

The increase of temperature in a food related to dielectric heating can be expressed by the Eq. (15.3):

$$\frac{\dot{q}}{V} = \rho C_P \frac{dT}{dt} = 2\pi f \varepsilon_0 \varepsilon^{''} F_{CE}^2 \tag{15.3}$$

wherein ρ is the density of the food [$kg \cdot m^{-3}$], C_P is specific heat [$J \cdot kg^{-1} \cdot K^{-1}$], and dT/dt is the time rate of temperature increase [$K \cdot s^{-1}$].

Another property related to dielectric parameters is penetration depth (d_p), which is usually defined as the depth into a sample where the microwave has dropped to $1/e$ ($e = 2.718$) or 36.8 % of its transmitted value. The penetration depth is a function of dielectric properties and can be described by Eq. (15.4):

$$d_p = \frac{\lambda_0 \sqrt{\varepsilon'}}{2\pi\varepsilon''} \tag{15.4}$$

wherein λ_0 is the free space microwave wavelength (for 2450 MHz, $\lambda_0 = 12.2$ cm). Another expression to calculate the d_p is:

$$d_p = \frac{c}{2\pi f \sqrt{2\varepsilon' \left[\sqrt{1 + \left(\frac{\varepsilon''}{\varepsilon'} \right)^2} - 1 \right]}} \tag{15.5}$$

wherein c is the speed of light in free space [3×10^8 m·s^{-1}] and f is the frequency [Hz].

Common food products have $\varepsilon'' < 25$, which implies a d_p of (0.6–1.0) cm. According to Tang (2005) and Wang et al. (2003a, 2003b) the penetration of microwaves at (915 and 2450) MHz in foods with high moisture content, at room temperature, is typically between 0.3 and 0.7 cm, depending on the salt content and frequency.

The power dissipated inside a material is proportional to ε''. The ratio $\varepsilon''/\varepsilon'$ called the loss tangent or dissipation factor, a descriptive dielectric parameter, is also used as an index of the material's ability to generate heat:

$$\tan\delta = \frac{\varepsilon''}{\varepsilon'} \tag{15.6}$$

wherein $\tan\delta$ is the loss tangent [dimensionless].

Over the last years, many potential applications of electromagnetic heating for foods have emerged and been published in the literature; however, new uses or research in food products to be treated with microwaves may be limited due to lack of dielectric properties data. Table 15.1 shows data for dielectric constants and loss factors of fresh fruits based in literature's data.

Table 15.1 Dielectric constant (ε') and loss factor (ε'') of fruits as function of moisture content (MC), temperature and microwave's frequency (f)

Fruit	MC [g/100 g w.b.[a]]	f [MHz]	T [°C]	ε'	ε''	References
Apple	88	915	23	57.0	8.0	[b]
		2450	23	54.0	10.0	
Avocado	71	915	23	47.0	16.0	[b]
		2450	23	45.0	12.0	
Banana	78	915	23	64.0	19.0	[b]
		2450	23	60.0	18.0	
Cantaloupe	92	915	23	68.0	14.0	[b]
		2450	23	66.0	13.0	
Cherry	–	915	20	73.7	16.4	[c]
		1800	20	70.9	16.0	
Grape	82	915	23	69.0	15.0	[b]
		2450	23	65.0	17.0	
Grapefruit	91	915	23	75.0	14.0	[b]
		1800	20	72.1	12.6	[c]
		2450	23	73.0	15.0	[b]
Kiwifruit	87	915	23	70.0	18.0	[b]
		2450	23	66.0	17.0	
Lemon	91	915	23	73.0	15.0	[b]
		2450	23	71.0	14.0	
Lime	90	915	23	72.0	18.0	[b]
		2450	23	70.0	15.0	
Mango	86	915	23	64.0	13.0	[b]
		2450	23	61.0	14.0	
Mango "Tommy Atkins"	86	915	20	74.0	13.8	[e]
Orange	87	915	23	73.0	14.0	[b]
		1800	20	72.5	14.8	[c]
		2450	23	69.0	16.0	[b]
Papaya	88	915	23	69.0	10.0	[b]
		2450	23	67.0	14.0	
Passion fruit	–	915	20	59.7	15.0	[d]
Peach	90	915	23	70.0	12.0	[b]
		2450	23	67.0	14.0	
Pear	84	915	23	67.0	11.0	[b]
		2450	23	64.0	13.0	
Strawberry	92	915	23	73.0	14.0	[b]
		2450	23	71.0	14.0	
White sapote	–	915	20	62.6	24.0	[d]

[a] w.b.: wet basis
[b]Venkatesh and Raghavan (2004)
[c]Wang et al. (2003a)
[d]Wang et al. (2005)
[e]Sosa-Morales et al. (2009)

15.1.3 Factors Influencing Dielectric Properties and Microwaves Heating

Several important factors influence dielectric properties determined for a given material. Some of these factors are related to the nature of the material while others are associated with the conditions when electro-heating is applied.

The correct prediction of food's dielectric heating behavior can only result from rigorous control of some parameters, especially temperature and moisture content.

To set temperature profiles and heating rates during exposure to electromagnetic radiation, some factors might be known, such as (1) variables related with equipment (power, frequency, geometry of system), (2) relationship between dielectric properties and variables of equipment, (3) relationship between dielectric properties and composition of food (moisture content, salts content, thermophysical properties as thermal conductivity, specific heat, and thermal diffusivity), and (4) influence of product parameters (mass, size, structure, geometry) on dielectric properties.

In parallel, to obtain temperature profiles, more and more accurate techniques are being applied, these include thermocouples, optical fiber probes, magnetic resonance, bioindicators, time/temperature integration devices, models, and temperature distribution in 3D. Thermocouples, which are used in conventional thermal processes due to their lower cost, have their application restricted to areas with no microwave contact because their metallic structure reflects the waves and could damage the equipment. On the other hand, optical fiber sensors have huge potential to monitor temperature in areas inside the microwave oven without interference of electromagnetic field and can be used in continuous and batch processes.

15.1.3.1 Frequency, Temperature, and Moisture Content

The dielectric properties of most materials vary considerably with the frequency of the applied electric field. Thus, an important phenomenon contributing to the frequency dependence of the dielectric properties is the polarization of molecules arising from the orientation with the imposed electric field, which have permanent dipole moments, like water (Venkatesh and Raghavan 2004). Ionic conductivity plays a major role at lower frequencies (e.g., <200 MHz), whereas both ionic conductivity and the dipole rotation of free water are important at microwave frequencies. For example, ionic conduction was the dominant mechanism for dielectric dispersion at frequencies until 300 MHz in mangoes (Sosa-Morales et al. 2009). For pure liquids with polar molecules, like alcohols or water, polar dispersion dominates the frequency characteristics of dielectric properties. The combined effect of temperature and frequency can be observed in Fig. 15.1.

The influence of temperature on the dielectric properties of foods depends on many factors, including food composition, especially moisture and salt contents and the involved frequencies. Thus, the temperature dependence of the dielectric constant is quite complex, and it may increase or decrease with temperature depending on the material. Generally ε'' increases as temperature increases at low frequencies

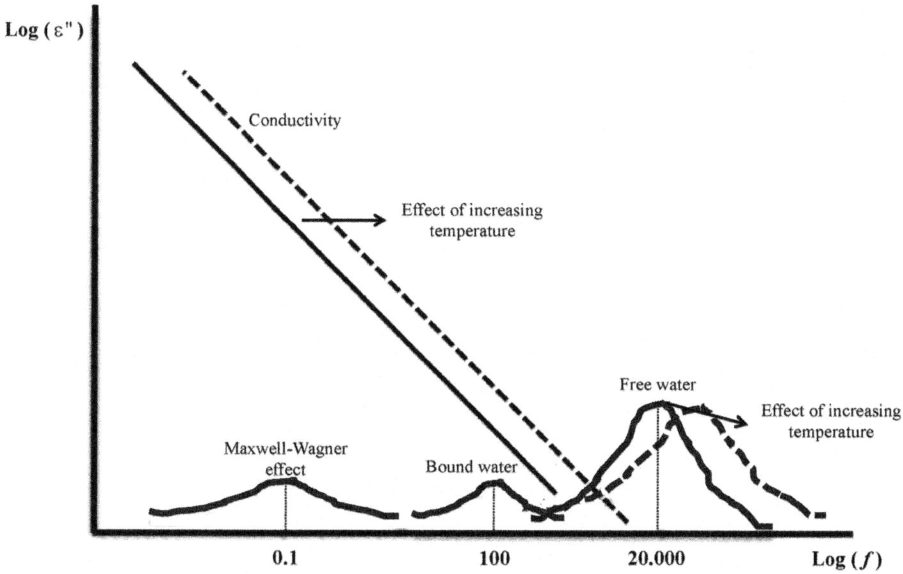

Fig. 15.1 Contribution of various mechanisms to the loss factor of high moisture materials as functions of frequency and temperature (Adapted from Wang et al. 2003b, with permission)

due to ionic conductance, while it decreases with temperature increase at high frequencies, due to the free water dispersion (Sosa-Morales et al. 2010).

Water is the greatest microwave energy absorber in food: the larger the water volume the more effective is the heating. That is why water is considered the principal component in the determination of the dielectric properties of hygroscopic foods. The temperature increase causes evaporation of part of this water and reduces the moisture content, so dielectric constant (ε') and dielectric loss factor (ε'') are directly affected.

15.1.3.2 Density, Mass, Geometry, and Thermophysical Properties of Food

Physical structure also affects the dielectric properties of materials. The relationship between the amount of mass per unit of volume (density) of the food product and its influence on heating is an interesting one. Food materials with low density (as coffee grains) present low permittivity values. The presence of air within foods also affects the dielectric constant; this is especially notable with particulates, such as, pulverized or granular materials.

The size, shape, and homogeneity of products have large influence on dielectric properties: foods with regular forms present a more uniform heating. However, if the size of food is much larger than wavelength, the effect of waves is compromised; in this case, the center of food is heated by conduction mechanism, and therefore the thermal conductivity is another parameter to be considered in process design.

15.1.4 Dielectric Properties of Fruits Reported in Literature

Characterization of dielectric properties as mentioned above is very important to understand the response of a material when subjected to microwave fields in heating or drying process of fruits. Some recent data for fresh fruits, at common frequencies, are shown in Table 15.1. For all fruits measured, the dielectric constant decreases as frequency increases.

As can be observed in Table 15.1, different conditions have been applied to verify the influence of radio frequency and microwaves energy on fruit treatments. An interesting approach was conducted by Wang et al. (2003a, 2003b) using these treatments to control postharvest inscet pests present in fruits and nuts. The authors found that the dielectric loss factor of fresh fruits and insects decreased with increasing frequency at constant temperatures. They also observed that the loss factors at RF frequencies of insects and fruits were clearly greater than that of nuts, suggesting possible differential and faster heating of fruits and insects than nuts, when treated simultaneously.

15.2 Advantages and Disadvantages

Microwaves energy for heating materials can potentially be employed in any application that requires heating/thermal processing of foods. The mechanisms of heating due to microwaves energy are completely different from those of conventional heating techniques (steam, hot air, hot water, etc.) and therein lay the advantages and disadvantages of its use.

Heating is applied to foods for many reasons like inactivation of enzymes and microorganisms, removal of water by evaporation, cooking, and browning. For each intended application the differences between microwaves and conventional methods must be examined to determine whether they are desirable or not and how they can be tailored for food products to ensure consistent results and high quality. Successful industrial applications exploit either the volumetric heating effect of microwaves or the selective nature of their interaction with food (Summu and Sahin 2005). Microwave equipments tend to be smaller than other alternatives which results in space savings, on the other hand they usually have a higher initial cost.

The heat transfer mode is through interaction of the microwaves with the product, generating heat by ionic polarization and dipole rotation within the material structure. The penetration depth of microwaves varies according to wave frequency and food composition (which affects its dielectric properties). Therefore heat is generated volumetrically, inside the product and at a very fast rate, especially if the product has high water content (as is usually the case with fruits), thus thermal degradation of essential nutrients is substantially reduced and retention of food quality factors is increased (Salazar-González et al. 2012).

The heat generated in this manner is transferred by conduction and convection and is distributed throughout the product.

Another advantage is that the heating systems can be turned on or off instantly. Moreover, heating with microwaves is more efficient than conventional heating: due to the efficiency of high-power magnetrons (85 % at 900 MHz, 80 % at 2450 MHz), the overall efficiency of microwave system is very high (Meredith 1998).

While the fast heating rates are an interesting feature of microwave heating, they also present problems.

Some problems that are commonly associated with microwave heated products are non-uniform temperature distributions, soggy surfaces, firmer texture, unacceptable color and flavor. Non-uniform temperature distribution occurs for a variety of reasons among them that the microwave field is not completely even, that the composition of the food varies at different points (heterogeneous food materials), the geometry can influence microwave absorption, etc. (Brewer 2005).

The rapid increase in temperature causes an elevation of pressure by formation of steam and this affects product texture (by expansion of the food matrix) and also drives out moisture very quickly. The elevated internal pressure can cause material explosion and puffing. The water tends to condense on the surface because the surrounding atmosphere is cool and the product becomes soggy. The depletion of moisture inside the product can create a firmer texture.

The formation of color and flavor associated with cooking depends on extended time at high temperatures mainly on the product surface. Microwave heating involves a short time that is insufficient for the occurrence of browning and also usually the temperatures on the product surface are lower because of water condensation.

The high temperature and short processing time does however favor retention of nutrients and lower color and aroma degradation. Fruit products that are processed by microwaves tend to present a higher perceived quality than those submitted to conventional techniques. The short processing time provides lower energy consumption, thus presenting higher energy efficiency and less environmental impact. These are the main advantages of microwave processing.

Several factors influence the outcome of microwave heating such as position of the product inside the cavity, variation of the mains supply voltage, microwave generator, and applicator design features. Some of these factors can be controlled while others are difficult to address. Extensive training of operators is required to minimize repeatability issues.

As research provides greater understanding of microwave processing and its interaction with food materials it is expected that effective industrial application will develop further. More efficient equipment designs and process modeling and control should favor use of microwaves, alone or in combination with conventional technologies. In fruit technology, microwave processing has been applied to drying, blanching, tempering and thawing of frozen fruit, thermal treatment of fruit juices and purées, and juice extraction. Further discussion of these applications is presented.

15.3 Examples of Use in Fruit Technology

Applications of microwave heating are found in many heat treatment operations in the food-processing reported in the literature. Some of these concerning fruits are minimal processing of apple puree (Picouet et al. 2009), avocado puree (Guzmán-Geronimo et al. 2008), and pasteurization of orange juice (Tajchakavit and Ramaswamy 1997a, 1997b; Cinquanta et al. 2010); drying of pistachios (Kouchalzadeth and Shafeei 2010), macadamias (Silva et al. 2006), banana slices (Drouzas and Schubert 1996), orange slices (Díaz et al. 2003), strawberries, apples (Erle and Schubert 2001), grapes (Kassem et al. 2011), and pumpkin slices (Alibas 2007).

This chapter is focused on some operations with fruit as blanching, drying, and thermal processing.

15.3.1 Blanching

Blanching is a processing step that is required for many fruit products before frying, freezing, canning, and drying. The main objective is to inactivate enzymes (peroxidase, polyphnenol oxidase, pectic enzymes) and also to cleanse and decrease the initial microbial load, exhaust gas from the plant tissue, increase the bioavailability of certain nutrients, and preheat the fruit. This is usually carried out by immersion in hot water (88–99 °C) or steam (Dorantes-Alvarez and Parada-Dorantes 2005; Summu and Sahin 2005).

Conventional blanching methods can cause leaching of water soluble components like vitamins and other nutrients. The wastewater generated is an environmental concern. Other effects of blanching include texture alterations as well as degradation of color and flavors.

Microwave blanching can be accomplished by direct heating of the fruit, with no need for addition of water thus minimizing leaching and wastewater generation. Enzyme inactivation occurs rapidly so there is a decrease in processing times, resulting in economy of energy and greater nutrient retention. Texture of vegetable tissues is less affected by microwave blanching; studies have reported a better preservation of cellular structure (Moreno et al. 2000). The sensory acceptance of microwave blanched fruits when compared to conventional methods is usually better. Color and flavor of microwave blanched fruits is preserved to a higher extent. These are the main reasons that indicate that microwave blanching has a promising future.

Despite these advantages, little industrial application is seen and studies indicate that results may be controversial. A successful application of microwave blanching depends on the product and the manner of processing. Non-uniform temperature distribution is a major issue and research has concentrated in improving oven designs to reach more uniform temperature within the product. Combination with other heating methods is also an alternative. Initial blanching with water or steam can provide energy economy since these methods are cheaper and will raise the product temperature before application of microwaves to blanch the product internally. On the other hand, blanching in hot water after microwave blanching evens temperature distribution and diminishes possible excessive surface desiccation.

Wrolstad et al. (1980) applied microwave blanching to strawberries used to produce juice and concentrated juice. The strawberries were heated up to an internal temperature of (82–88) °C for (3–4) min. The anthocyanin pigments, color, phenolics, and ascorbic acid were determined and microwave blanched strawberries presented better retention of all parameters studied when compared to unblanched fruits.

Cano et al. (1990) blanched banana slices in boiling water (11 min) and in a microwave oven (650 W for 2 min). Under these conditions, blanching in boiling water was more effective in enzyme inactivation and produced a higher sensory quality.

De Ancos et al. (1999) applied microwave blanching to strawberry, papaya, and kiwi. Microwave treatments occurred at 285, 570, and 850 W for 30 s and at 475 W for 15, 30, 45 and 60 s. Enzyme activities (polyphenol oxidase and peroxidase) were reduced for all treatments applied, though different resistances were determined for different fruits and treatment results were satisfactory. Color of purees was affected by the treatment with increasing difference according to higher treatment exposure. Degradation of anthocyanins was very small for strawberries treated with microwaves. For papaya samples, total carotenoid content decreased with microwave treatment though the qualitative carotenoid composition remained unchanged. Chlorophyll content of kiwis was severely affected by microwave treatment, causing a significant loss of green color. These results demonstrate that different fruits and processing conditions produce variable results and that to effectively apply this technology, it is important to conduct experiments.

Moreno et al. (2000) tested microwave blanching and steam blanching of strawberries prior to osmotic dehydration. Microwave-blanched strawberries presented a firmer texture, less structural damage, and less color change than steam blanched strawberries. However sucrose uptake in osmotic dehydration was higher for steam blanched strawberries resulting in a lower water activity and greater stability of the product. Also steam blanching was more effective in reducing microbial load. This study demonstrates that it is important to consider the effects of the blanching technology employed on the posterior processing and thus determine the most adequate technology.

With the improvement of oven designs and knowledge gained in microwave heat transfer modeling, microwave blanching of fruits should increase due to higher nutrient retention, better sensory quality, and energy efficiency when compared to other processing options.

15.3.2 Drying

The physical principles of microwave drying provide a distinct advantage when compared to conventional air drying. Air drying consists of three main phases: constant rate period, first falling rate period, and second falling rate period. In air drying, water has to diffuse to the surface in order to be transported by the gas phase.

At the beginning of the drying process, the surface is wet due to migration of water from within the particle by capillary driven flow. As the amount of water present decreases, heat has to penetrate inside the particle to evaporate water that then has to move to the surface to be transported by the gas phase. If the temperature gradient is increased in an attempt to increase rate of drying, overheating of the surface can occur and if the surface dries out, there is the possibility of "case hardening" which further diminishes the drying rate and significantly impairs product sensory quality (Erle 2005; Summu and Sahin 2005).

Microwave drying is an alternative to increase drying rate particularly during the first falling rate period. Mass transfer is accelerated because microwaves penetrate the food and the center becomes warmer than the surroundings, generating steam inside the product and pumping moisture to the surface because of the increased vapor pressure. With microwave drying, there is no risk of "case hardening" because the surface is always wet. Air drying involves very long process times while microwave drying is much faster, favoring product nutritional and sensory quality. Larger particles can be dried more efficiently because of microwave penetration. There are however important drawbacks and care must be taken to ensure an economical and efficient process.

The generation of steam inside the particle can cause high pressure that can result in product puffing, which may or may not be desirable. Water absorbs microwave energy very well but it also has a very high specific heat capacity, thus a large amount of microwave energy is required to raise the temperature of a high moisture content product, rendering the process expensive. Since water is usually the main constituent of foods to absorb microwave energy, as drying progresses, microwave absorption will decrease. The drying rate in microwave processing is directly proportional to microwave power but increasing microwave power towards the end of drying can create potential problems.

As water content decreases, its evaporation from the surface may not be enough to lower product temperature and as drying progresses molecules within the product are able to move more freely and absorb more energy which can lead to overheating and burning of the product. The electrical strength of the microwave field becomes higher when less energy is absorbed by the product. So at lower moisture contents, there is a risk of creating arcing or plasma with a discharge of the electrical field. This will damage the equipment and the product. Increasing microwave power also causes more pronounced differences due to uneven absorption of microwaves, creating regions of hotter and colder material within the product and generating quality issues. Another concern is if the heated particle is much smaller than the wavelength of the electromagnetic energy because absorption is then very poor.

Sensory quality of microwave-dried products is usually better when compared to air-dried products, though freeze drying is still the best. Rehydration of microwave-dried products is better because the product softens rapidly in water and shrinkage can be lower. The best results have been obtained by a combination of drying techniques. Microwave drying has been applied in combination with air drying, vacuum drying, osmotic dehydration, and spouted bed drying. Also use of intermittent microwave power deliverance has been found to improve temperature distribution and product quality.

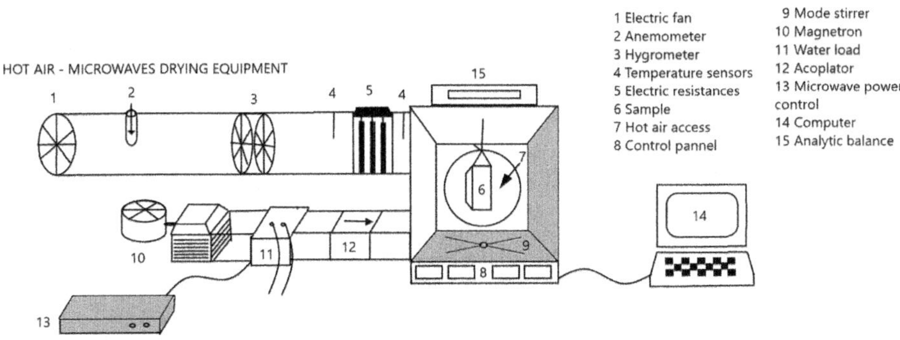

Fig. 15.2 Schematic illustration of the combined hot air-microwave drying equipment (From Andrés et al. 2004, with permission)

Several studies of the application of microwaves to drying of fruits have been conducted. Apple dices were dried by a spouted bed microwave combined process proposed by Feng and Tang (1998). The combined process resulted in more uniform temperature distribution, less discoloration, higher rehydration rate, and up to 80 % less drying time. Sham et al. (2001) applied air drying followed by microwave vacuum drying to apple chips. The authors concluded that a higher vacuum resulted in greater crispness and lower density.

Andrés et al. (2004) and Bilbao-Sáinz et al. (2006) applied combined microwave and air drying to apple cylinders. These authors performed the drying experiments in a specially designed hot air-microwave oven equipped with continuous output-power microwave energy (Fig. 15.2). Each time the oven temperature and the incident microwave power were prefixed, the microwave power levels were set at 0, 3, 5, 7 and 10 $W \cdot g^{-1}$ combined with air at 25, 30, 40 and 50 °C. Air velocity was set at 1.0 ± 0.1 $m \cdot s^{-1}$ in all cases and the relative humidity of the ambient was 62 ± 8 %. Sample was suspended from a balance and the drying kinetics was studied by continuously weighing the sample during drying. In the drying curves, four drying periods were observed and the microwave power effect was higher than air temperature, decreasing significantly the drying time. The initial moisture content of fresh apples was determined obtaining less shrinkage and higher porosity (puffing) of the microwave-dried products.

Askari et al. (2006) observed similar results for coated (pectin, CMC, starch, $CaCl_2$) apple slices subjected to hot air and microwave drying. Higher rehydration rates were also registered in this work.

Li et al. (2010) developed quite interesting volatiles monitoring and control system designed for microwave drying. The system consisted of a volatiles detection unit, a microwave drying unit, a temperature and power control unit, and a PC-based data acquisition unit. The volatiles detection unit included an electronic nose to maximize retention of volatiles in apple samples dried by microwave vacuum technology (Fig. 15.3). Apple color and appearance were unchanged by employing this control strategy and less time and energy were consumed.

Cranberries have also been dried by employing microwave technology and results are favorable. Yongsawatdigul and Gunasekaran (1996) osmotically dehydrated cran-

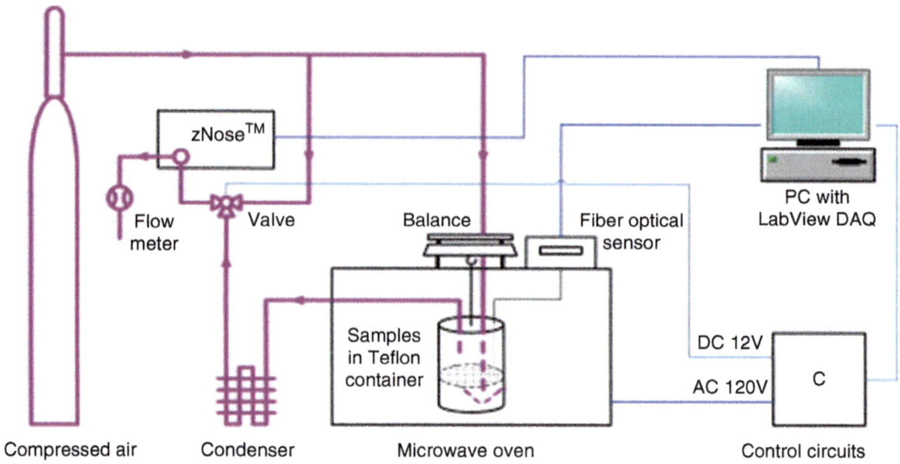

Fig. 15.3 Schematic diagram of the microwave drying system (From Li et al. 2010, with permission)

berries then applied microwave drying in continuous and pulsed modes. Pulsed treatments and lower vacuum pressure resulted in better quality products. Beaudry et al. (2004) compared different drying technologies: hot air, microwave-assisted hot air, vacuum and freeze drying. Freeze-dried cranberries presented the highest rehydration rates, but microwave-assisted drying resulted in better characteristics than the other options tested. Leusink et al. (2010) compared microwave vacuum-dried cranberries to freeze-dried and air-dried cranberries. Vacuum microwave-dried product had a higher porosity when compared to air-dried and freeze-dried ones and the antioxidants retention was similar for both freeze-dried and microwave vacuum-dried cranberries.

Pineapple slices were microwave-assisted air dried after osmotic dehydration by Botha et al. (2012). Variable microwave power was employed to minimize charring of slices. Significant reduction of drying times was obtained with good quality and minimum losses due to charring.

Díaz et al. (2003) dried orange slices by a combined microwave air drying method. Rehydration was not affected by the drying technique but processing times were greatly reduced by microwaves application.

Drouzas and Schubert (1996) studied microwave vacuum drying of banana slices obtaining better taste, aroma, and rehydration rates. Maskan (2000) compared microwave-assisted air drying to air drying followed by microwave finish drying. Microwave-finish drying produced the best results with lighter product color, greater rehydration rates, and reduced drying times (approximately 64 % less). Mousa and Farid (2002) determined that thermal and drying effects were enhanced with greater vacuum in the microwave vacuum drying of banana slices. Mui et al. (2002) compared air drying, vacuum microwave drying, combined air and microwave drying to freeze drying of banana chips. Banana chips subjected to 90 % air drying and 10 % vacuum microwave drying presented the highest volatile levels, sensory ratings, and were crispier than the others tested.

Erle and Schubert (2001) applied osmotic dehydration followed by microwave vacuum drying of apples and strawberries. Osmotic dehydration prior to microwave vacuum drying ensured a better texture. Vitamin C retention was of approximately 60 % for both apples and strawberries. Raghavan and Silveira (2001) studied shrinkage of strawberries subjected to osmotic dehydration followed by microwave vacuum drying and concluded that higher power levels and osmotic dehydration result in greater volume variation. Different power levels of microwave vacuum drying were applied to strawberries by Wojdylo et al. (2009). Strawberries treated at 240 W produced the highest quality and the reduced processing time is an advantage when compared to other techniques.

Maskan (2001) tested combinations of air drying and microwave drying applied to kiwi slices. Independent of the method employed the color change observed was significant, with considerable darkening of the samples.

The studies reviewed here clearly demonstrate that microwave drying presents some interesting features, especially if combined to other methods like vacuum drying. The results depend on the fruits and the processing parameters employed. Industrially a powder made from orange juice concentrate is produced by microwave vacuum drying and a combination of air drying and microwave puffing is employed in Germany and Poland. Comparison of energy demands and investment costs shows that microwave vacuum drying investment cost is one of the most expensive options (though less than freeze drying) while the energy demands are lower than spray drying and air drying (Erle 2005). As food quality becomes more important for consumers, microwave drying/microwave-assisted drying should see further industrial employment. Energy efficiency and better design of microwave dryers should also stimulate this.

15.3.3 Thermal Processing

Thermal processing is employed to reduce microbial counts and inactivate enzymes. Microwaves processing is capable of both. Batch processing of fluid foods in vessels has been studied but due to non-uniform temperature distribution and difficulties in predicting the heating pattern continuous systems have been proposed. Continuous microwave systems are composed of a tube through which the fluid food is pumped and subjected to microwaves energy. Focalized microwave applicator designs have improved heating profiles and reduced processing times. Combination of microwave pasteurization systems with other technologies such as UV light, gamma irradiation, and H_2O_2 have been tested (Summu and Sahin 2005).

This section presents investigations of applications of microwave heating with pasteurization or sterilization purposes, mainly against inactivation of target microorganisms and the inactivation of enzymes or retention of either important bio-compounds or sensory attributes has been studied, as summarized in Table 15.2.

One of the main reasons for pasteurizing citrus juice is the inactivation of pectin methylesterase (PME), the enzyme responsible for the loss of cloudiness, which is a very important quality attribute for the consumers. Since the first reported study on the use of microwave energy to pasteurize orange juice some authors have investigated the effect of this treatment on PME of orange and other citrus juices.

Tajchakavit and Ramaswamy (1997a, 1997b) compared PME inactivation in orange juice pasteurized by employing a continuous flow microwave system and by a conventional system. PME inactivation was much more efficient in the microwave system. Vikram et al. (2005) compared ohmic heating, microwave batch processing, infrared heating, and conventional pasteurization (water bath) of orange juice. Vitamin C retention and color of treated samples were compared. Microwave pasteurization resulted in least color difference but the vitamin C retention was low. The authors however pointed out that the microwave system employed did not allow precise temperature control, which might have influenced the results.

Peroxidase (POD) catalyzes reactions that cause undesirable changes including off-flavor, aroma and color, as well as loss of some nutrients. The main enzyme responsible for the browning reaction in light colored fruits and vegetables is polyphenoloxidase (PPO). Since the enzymes PPO and POD are very resistant to heat, they are usually considered as biological indicators of thermal processing.

Direct application of microwave energy to whole pieces of fruits or vegetables is limited by temperature gradients that can vaporize internal water and damage food texture. Combining microwave and hot-water bath treatment can completely inactivate PPO in a short time with minimal antioxidant loss, weight loss, and shrinkage (Devece et al. 1999). Microwave inactivation of polyphenol oxidase in bananas has been shown to be comparable to traditional blanching methods (Prem Akumar and Khurdiya 2002).

The inactivation kinetics of PPO and POD at 2450 MHz was studied in the batch microwave treatment of green coconut water to temperatures between 52.5 and 92.9 °C by Matsui et al. (2008). The objective of the treatment was to pasteurize this water while avoiding the loss of thermolabile components and sensory changes using a rapid heating method. The thermal inactivation of PPO and POD during microwave processing of green coconut water was significantly faster in comparison with data reported for green coconut water using conventional thermal treatments.

Apple juice pasteurization by microwaves has been tested in both batch and continuous systems. Cañumir et al. (2002) pasteurized apple juice in a batch microwave system and compared inactivation of *E. coli* to juice pasteurized in a conventional system. The results indicated that similar reductions could be achieved by microwaves. Tajchakavit et al. (1998) employed a continuous flow microwave heating system to pasteurize apple juice. Destruction of *S. cerevisiae* and *L. plantarum* was found to be much greater in microwave processing than for conventional pasteurization under similar time-temperature conditions.

Table 15.2 Studies on microwave heating for fruits and fluid foods

Objectives	Results	References
Apple		
Evaluate destruction of *S. cerevisiae* and *L. plantarum* in apple juice with a continuous flow microwave heating system and compare with batch conventional pasteurization under similar time-temperature conditions	Continuous microwave heating:	Tajchakavit et al. (1998)
	S. cerevisae:	
	$D_{52.5°C}=4.8$ s; $D_{55°C}=2.1$ s; $D_{57.5°C}=1.1$ s; z=7 °C	
	L. plantarum:	
	$D_{57.5°C}=14$ s; $D_{60°C}=3.8$ s; $D_{62.5°C}=0.79$ s; z=4.5 °C	
	Batch conventional heating:	
	S. cerevisae:	
	$D_{50°C}=58$ s; $D_{55°C}=25$ s; $D_{60°C}=10$ s; $D_{70°C}=1.9$ s; z=13.4 °C	
	L. plantarum:	
	$D_{55°C}=52$ s; $D_{60°C}=22$ s; $D_{70°C}=8.4$ s; $D_{80°C}=1.2$ s; z=15.9 °C	
Evaluate the effect of pasteurization at different microwave power levels (270–900) W on the microbiological quality (reduction of *E. coli*) of apple juice using a home 2450 MHz microwave. Data obtained were compared with conventional pasteurization	*E. coli* populations were lower at (720–900) W for (60–90) s resulted in (2–4) logs reduction of *E. coli*	Cãnumir et al. (2002)
	Data obtained were similar when applied conventional pasteurization at 83 °C for 30 s	
	E. coli: D-values ranged from (0.42 ± 0.03) min at 900 W and (0.48 ± 0.10) min at 720 W with temperatures of (70.3 ± 2.1) °C and (76.2 ± 1.9) °C, respectively. z = 58.5 ± 0.4 °C	
Design a lab-scale continuous flow microwave pasteurization system for apple cider and characterize the process parameters. Pasteurization of this system was evaluated by the inactivation of a *E. coli* strain (25922) inoculated in apple cider	In conditions of continuous flow microwave pasteurization:	Gentry and Roberts (2005)
	73 °C, 130 s, 0.231 L•min^{-1}	
	900 W; 73 °C, 216 s, 0.381 Ltmin^{-1}	
	2000 W resulted in a 5 $\log_{10}$ reduction of *E. coli*	
Avocado		

(continued)

Table 15.2 (continued)

Objectives	Results	References
Study the behavior of PPO and POD treated with microwave energy (1200 and 2450 MHz; range of time: 0–70 s)	At the 20 s of treatment with microwaves a clear decrease in activities of both enzymes is presented	Jiménez-Vieyra et al. (2004)
	Without major variations at 40 s	
	The time of treatment applied do not damage the original appearance of fruit	
Grapefruit juice		
Evaluate the effect of batch conventional pasteurization (the juice took 80 s to reach 80 ± 2.5 °C and it remained at this temperature for 11 s) and microwave pasteurization (30 s and 900 W) on the main bioactive compounds of grapefruit juice and PME activity	Conventional treatment led to a significant decrease in citric acid from (1538 to 1478) mg/100 g and ascorbic acid from (36 to 34.3) mg/100 g, while microwave pasteurization preserved these compounds	Igual et al. 2010
	PME residual activity detected after thermal treatments was 12.04 ± 3.86 % and 10.07 ± 0.6 3 % in traditional pasteurization and microwave pasteurization, respectively	
Green coconut water		
The inactivation kinetics of enzymes PPO and POD was studied for the batch microwave treatment (52.5–92.9 °C, 0.30–33.25 s) and the results were compared with conventional pasteurization	The thermal inactivation of PPO and POD during microwave processing of green coconut water was significantly faster in comparison with conventional processes reported in the literature	Matsui et al. (2008)
	PPO: $D_{92.2°C} = 52$ s. The z-value was 17.6 °C	
	POD: $D_{92.9°C} = 16$ s. The z-value was 11.5 °C	
Orange juice		
Compare the efficiency of continuous microwave system and a conventional plate heat exchanger (PHE) in the inactivation of PME	Microwave system: reduced 99 % of PME activity at 85 °C and 10 s	Nikdel et al. (1993)
	PHE system: to the same reduction needed 90.5 °C and 15 s	

Estimate D and z-values for PME in continuous microwave system and to compare with conventional plate heat exchanger (PHE)	Microwave system: $D_{60°C}=7.4$ s while PHE system: $D_{60°C}=152$ s	Tajchakavit and Ramaswamy (1997a, 1997b)
Study the effect of microwaves energy in fresh orange peels and to investigate pectin yield	Microwave treatment led to destructive changes in the plant tissue that resulted in an increase in the capillary-porous characteristics and the water absorption capacity of the fruit	Kratchanova et al. (2004)
	These changes gave an opportunity for the considerable increase in the yield of extractable pectin and the heating inactivated the PME activity	

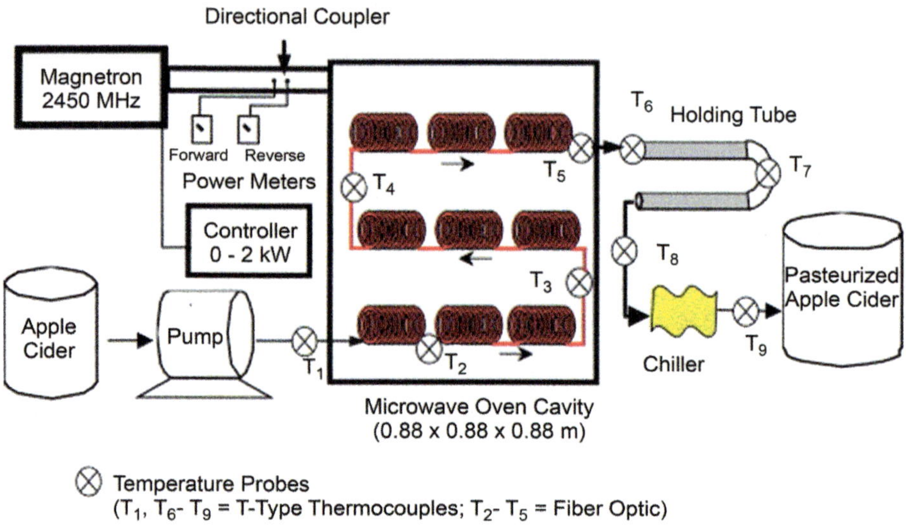

Fig. 15.4 Schematic of the continuous flow microwave pasteurization system (From Gentry and Roberts 2004, with permission)

Gentry and Roberts (2004) proposed employing a continuous flow microwave pasteurization system for apple cider at 2450 MHz (Fig. 15.4). The process parameters evaluated in the project were volume load size 0.5 and 1.38 L, input power 900 and 2000 W, and inlet temperature 3, 21 and 40 °C. The microwave pasteurization system consisted of helical coils located throughout a large cavity oven through which the product was pumped. The lethality of the process was evaluated by the inactivation of *E. coli* inoculated to the sample. The results showed that the larger capacity system was more efficient (1.38 L) and improved the power absorption within the microwave cavity.

Guzmán et al. (2002) treated avocado purée in a domestic microwave at 633 W. Samples treated for 30 s with addition of copper or zinc chloride preserved color better during 7 days refrigerated storage.

Gerard and Roberts (2004) heated apple mash by microwaves to improve juice yield and quality. Juice yield increased with microwave processing and sensory characteristics were similar to those of juice produced with mashes that were not treated.

Maskan (2006) employed microwaves for concentration of pomegranate juice and compared the results to rotary vacuum and atmospheric heating concentration. Microwave concentrated juice presented a similar color to atmospheric heating, but better than rotary vacuum concentrated juice

Industrial application of microwave thermal processing is still incipient but research results indicate that juice and purée quality can be improved by application of this technology. The most desirable scenario would be that, through the knowledge of the dielectric properties, one could determine the appropriate conditions to apply microwave energy and obtain desired process lethality. More adoption by the food industry is expected in the near future in order to provide better foods for consumers.

15.4 Other Potential Applications

Many food manufacturing plants employ frozen fruit pulps, purées, and juice concentrates as raw materials. These need to be thawed or tempered before use and microwave systems are successfully employed to thaw these materials (Swain and James 2005).

Cendres et al. (2011) have proposed an innovative process for extraction of fruit juices employing microwave heating. Plums, grapes, and apricots were treated in an extraction vessel placed inside the cavity of a microwave oven. Fresh and frozen fruits were tested. An opening in the lower surface of the oven allowed the juice to flow into graduated cylinders by the action of gravity and pressure generated by steam. Extraction occurred very rapidly with greatest yields from frozen fruit at low power. The juices obtained had a bright color and a fresh fruit flavor. Apricot and plum juices were highly acidic. The yields did not reach commercial levels for grapes and plums so further studies are needed.

References

Alibas, I. 2007. Microwave, air and combined microwave-air-drying parameters of pumpkin slices. *LWT - Food Science and Technology* 48: 1445–1451.

Andrés, A., C. Bilbao, and P. Fito. 2004. Drying kinetics of apple cylinders under combined hot air–microwave dehydration. *Journal of Food Engineering* 63: 71–78.

Askari, G.S., Z. Emam-Djomeh, and S.M. Mousavi. 2006. Effects of combined coating and microwave assisted hot-air drying on the texture, microstructure and rehydration characteristics of apple slices. *Food Science and Technology International* 12: 39–46.

Beaudry, C., G.S.V. Raghavan, C. Ratti, and T.J. Rennie. 2004. Effect of four drying methods on the quality of osmotically dehydrated cranberries. *Drying Technology* 22(3): 521–539.

Bilbao-Sáinz, C., A. Andrés, A. Chiralt, and P. Fito. 2006. Microwaves phenomena during drying of apple cylinders. *Journal of Food Engineering* 74: 160–167.

Botha, G.E., J.C. Oliveira, and L. Ahrné. 2012. Microwave assisted air drying of osmotically treated pineapple with variable power programmes. *Journal of Food Engineering* 108: 304–311.

Brewer, M. 2005. Microwave processing, nutritional and sensory quality. In *The microwave processing of foods*, ed. H. Schubert and M. Regier, 76–101. Cambridge, England: Woodhead.

Buffler, C.R. 1993. Introduction to microwaves. In *Microwave cooking and processing: Engineering fundamentals for the food scientist*, ed. C.R. Buffler, 1–13. New York: Van Nostrand Reinhold.

Cano, M.P., M.A. Marin, and C. Fúster. 1990. Effects of some thermal treatments on polyphenoloxidase and peroxidase activities of banana *(Musa cavendishii, var enana). Journal of Science Food and Agriculture* 51: 223–231.

Cañumir, J.A., J.E. Celis, J. de Bruijn, and L.V. Vidal. 2002. Pasteurisation of apple juice by using microwaves. *LWT - Food Science and Technology* 35: 389–392.

Cendres, A., C. Farid, J.F. Maingonnat, and C.M.G.C. Renard. 2011. An innovative process for extraction of fruit juice using microwave heating. *LWT - Food Science and Technology* 44: 1035–1041.

Cinquanta, L., D. Albanese, G. Cuccurullo, and M. Di Matteo. 2010. Effect on orange juice of batch pasteurization in an improved pilot-scale microwave oven. *Journal of Food Science* 75: E46–E50.

De Ancos, B., M.P. Cano, A. Hernandez, and M. Monreal. 1999. Effects of microwave heating on pigment composition and colour of fruit purees. *Journal of the Science of Food and Agriculture* 79: 663–670.

Devece, C., J.N. Rodriguez-Lopez, L.G. Fenoll, J. Tudela, J.M. Catala, E. de Los Reyes, and F. Garcia-Canovas. 1999. Enzyme inactivation analysis for industrial blanching applications: comparison of microwave, conventional, and combination heat treatments on mushroom polyphenoloxidase activity. *Journal Agriculture and Food Chemistry* 47(11):4506–4511.

Díaz, G.R., J. Martínez-Monzó, P. Fito, and A. Chiralt. 2003. Modelling of dehydration-rehydration of orange slices in combined microwave/air drying. *Innovative Food Science & Emerging Technologies* 4: 203–209.

Dorantes-Alvarez, L., and L. Parada-Dorantes. 2005. Blanching using microwave processing. In *The microwave processing of foods*, ed. H. Schubert and M. Regier, 153–173. Cambridge, England: Woodhead.

Drouzas, A.E., and H. Schubert. 1996. Microwave application in vacuum drying of fruits. *Journal of Food Engineering* 28: 203–209.

Erle, U. 2005. Drying using microwave processing. In *The microwave processing of foods*, ed. H. Schubert and M. Regier, 142–152. Cambridge, England: Woodhead.

Erle, U., and H. Schubert. 2001. Combined osmotic and microwave-vacuum dehydration of apples and strawberries. *Journal of Food Engineering* 49: 193–199.

Feng, H., and J. Tang. 1998. Microwave finish drying of diced apples in a spouted bed. *Journal of Food Science* 63(4): 679–683.

Gentry, T.S., and J.S. Roberts. 2004. Formation kinetics and application of 5-hydroxymethylfurfural as a time–temperature indicator of lethality for continuous pasteurization of apple cider. *Innovative Food Science & Emerging Technologies* 5: 327–333.

———. 2005. Design and evaluation of a continuous flow microwave pasteurization system for apple cider. *LWT - Food Science and Technology* 38(3): 227–238

Gerard, K.A., and J.S. Roberts. 2004. Microwave heating of apple mash to improve juice yield and quality. *LWT - Food Science and Technology* 37: 551–557.

Guzmán, G.R., A.L. Dorantes, U.H. Hernández, S.H. Hernández, A. Ortiz, and E.R. Mora. 2002. Effect of zinc and copper chloride on the color of avocado puree heated with microwaves. *Innovative Food Science & Emerging Technologies* 3: 47–53.

Guzmán-Geronimo, R.I., M.G. López, and L. Dorantes-Alvarez. 2008. Microwave processing of avocado: Volatile flavor profiling and olfactometry. *Innovative Food Science & Emerging Technologies* 9: 501–506.

Igual, M., M.M. García-Martínez, and N. Martínez-Navarrete. 2010. Effect of thermal treatment and storage on the stability of organic acids and the functional value of grapefruit juice. *Food Chemistry* 118: 291–299.

Jiménez-Vieyra, M.E., M.L. Zambrano-Zaragoza, M.R. Aguillar-Rodríguez. 2004, Mar. *Effect of treatment with microwave energy on polyphenoloxidase and peroxidase activity in fruits*. In: International Congress on Engineering and Food, ICEF 9, Montpellier, France.

Kassem, A.S., A.Z. Shokr, A.R. El-Mahdy, A.M. Aboukarina, and E.Y. Hamed. 2011. Comparison of drying characteristics of Thompson seedless grapes using combined microwave oven and hot air drying. *Journal of the Saudi Society of Agricultural Sciences* 10: 33–40.

Kouchalzadeth, A., and S. Shafeei. 2010. Modeling of microwave-convective drying of pistachios. *Energy Conversion and Management* 51: 2012–2015.

Kratchanova, M., E. Pavlova, and I. Panchev. 2004. The effect of microwave heating of fresh orange peels on the fruit tissue and quality of extracted pectin. *Carbohydrate Polymers* 56: 181–185.

Leusink, G.J., D.D. Kitts, P. Yaghmaee, and T. Durance. 2010. Retention of antioxidant capacity of vacuum microwave dried cranberry. *Journal of Food Science* 75(3): C311–C316.

Li, Z., G.S.V. Raghavan, and N. Wang. 2010. Apple volatiles monitoring and control in microwave drying. *LWT - Food Science and Technology* 43: 684–689.

Maskan, M. 2000. Microwave/air and microwave finish drying of banana. *Journal of Food Engineering* 44: 71–78.

———. 2001. Kinetics of colour change of kiwifruits during hot air and microwave drying. *Journal of Food Engineering* 48: 169–175.

———. 2006. Production of pomegranate (*Punica granatum* L.) juice concentrate by various heating methods: Colour degradation and kinetics. *Journal of Food Engineering* 72: 218–224.

Matsui, K.N., J.A.W. Gut, P.V. de Oliveira, and C.C. Tadini. 2008. Inactivation kinetics of polyphenol oxidase and peroxidase in green coconut water by microwave processing. *Journal of Food Engineering* 88(2): 169–176.

Meredith, R. 1998. *Engineer's handbook of industrial microwave heating*. London: The Institution of Electrical Engineers.

Moreno, J., A. Chiralt, I. Escriche, and J.A. Serra. 2000. Effect of blanching/osmotic dehydration combined methods on quality and stability of minimally processed strawberries. *Food Research International* 33: 609–616.

Mousa, M., and M. Farid. 2002. Microwave vacuum drying of banana slices. *Drying Technology* 20(10): 2055–2066.

Mui, W.W.Y., T.D. Durance, and C.H. Scaman. 2002. Flavor and texture of banana chips dried by combinations of hot air, vacuum, and microwave processing. *Journal of Agricultural and Food Chemistry* 50: 1883–1889.

Nikdel, S., C.S. Chen, M.E. Parish, D.G. Mackellar, and L.M. Friedrich. 1993. Pasteurization of citrus juice with microwave energy in continuous flow unit. *Journal of Agricultural and Food Chemistry* 41: 2116–2119.

Picouet, P.A., A. Landl, M. Abadias, M. Castellari, and I. Viñas. 2009. Minimal processing of a Granny Smith apple puree by microwave heating. *Innovative Food Science & Emerging Technologies* 10: 545–550.

Premakumar, K., and D.S. Khurdiya. 2002. Effect of microwave blanching on the nutritional qualities of banana puree. *Journal of Food Science and Technology* 39(3): 258–260.

Raghavan, G.S.V., and A.M. Silveira. 2001. Shrinkage characteristics of strawberries osmotically dehydrated in combination with microwave drying. *Drying Technology* 19(2): 405–414.

Salazar-González, C., M.F. San Martín-González, A. López-Malo, and M.E. Sosa-Morales. 2012. Recent studies related to microwave processing of fluid foods. *Food and Bioprocess Technology* 5: 31–46.

Sham, P.W.Y., C.H. Scaman, and T.D. Durance. 2001. Texture of vacuum microwave dehydrated apple chips as affected by calcium pretreatment, vacuum level, and apple variety. *Journal of Food Science* 66(9): 1341–1347.

Silva, F.A., A. Marsaioli Jr., G.J. Maximo, M.A.A.P. Silva, and L.A. Gonçalves. 2006. Microwave assisted drying of macadamia nuts. *Journal of Food Engineering* 77: 550–558.

Sosa-Morales, M.E., G. Tiwari, S. Wang, J. Tang, H.S. Garcia, and A. Lopez-Malo. 2009. Dielectric heating as a potential post-harvest treatment of disinfesting mangoes. Part II: Development of RF-based protocols and quality evaluation of treated fruits. *Biosystems Engineering* 103(3): 287–296.

Sosa-Morales, M.E., L. Valerio-Junco, A. López-Malo, and H.S. García. 2010. Dielectric properties of foods: Reported data in the 21st century and their potential applications. *LWT - Food Science and Technology* 43: 1169–1179.

Summu, G., and S. Sahin. 2005. Recent developments in microwave heating. In *Emerging technologies for food processing*, ed. D.-W. Sun, 419–444. San Diego, CA: Elsevier.

Swain, M., and S. James. 2005. Thawing and tempering using microwave processing. In *The microwave processing of foods*, ed. H. Schubert and M. Regier, 175–191. Cambridge, England: Woodhead.

Tajchakavit, S., and H.S. Ramaswamy. 1997a. Continuous-flow microwave inactivation kinetics of pectin methyl esterase in orange juice. *Journal of Food Processing and Preservation* 21: 365–378.

———. 1997b. Thermal vs. microwave inactivation kinetics of pectin methylesterase in orange juice under batch mode heating conditions. *LWT - Food Science and Technology* 30: 85–93.

Tajchakavit, S., H.S. Ramaswamy, and P. Fustier. 1998. Enhanced destruction of spoilage microorganisms in apple juice during continuous flow microwave heating. *Food Research International* 31(10): 713–722.

Tang, J. 2005. Dielectric properties of food. In *The microwave processing of foods*, ed. H. Schubert and M. Regier, 22–38. Cambridge, England: Woodhead.

Venkatesh, M.S., and G.S.V. Raghavan. 2004. An overview of microwave processing and dielectric properties of agri-food materials. *Biosystems Engineering* 88(1): 1–18.

Vikram, V.B., M.N. Ramesh, and S.G. Prapulla. 2005. Thermal degradation kinetics of nutrients in orange juice heated by electromagnetic and conventional methods. *Journal of Food Engineering* 69: 31–40.

Wang, Y., et al. 2003a. Dielectric properties of fruits and insect pests as related to radio frequency and microwave treatments. *Biosystems Engineering* 85: 201–212.

Wang, Y., T. Wig, J. Tang, and L. Hallberg. 2003b. Dielectric properties of food relevant to RF and microwave pasteurization and sterilization. *Journal of Food Engineering* 57: 257–268.

Wang, S., M. Monzon, Y. Gazit, J. Tang, E. J. Mitcham, and J. W. Armstrong. 2005. Temperature-dependent dielectric properties of selected subtropical and tropical fruits and associated insect pests. *Transactions of the ASAE* 48(5):1873–1881.

Wojdylo, A., A. Figiel, and J. Oszmiański. 2009. Effect of drying methods with the application of vacuum microwaves on the bioactive compounds, color, and antioxidant activity of strawberry fruits. *Journal of Agricultural and Food Chemistry* 57: 1337–1343.

Wrolstad, R.E., D.D. Lee, and M.S. Poei. 1980. Effect of microwave blanching on the color and composition of strawberry concentrate. *Journal of Food Science* 45: 1573–1577.

Yongsawatdigul, J., and S. Gunasekaran. 1996. Microwave-vacuum drying of cranberries: Part I. Energy use and efficiency. *Journal of Food Processing and Preservation* 20: 121–143.

Chapter 16
Fruit Preservation by Ohmic Heating and Pulsed Electric Fields

Olga Martín-Belloso and Mariana Morales-de la Peña

16.1 Introduction

Modern lifestyle has caused considerable changes to the traditional eating habits of most population worldwide. At present, people claim for food with fresh-like characteristics and showing antioxidant properties. Since fruits represent a rich source of nutrients and phytochemicals, their consumption has significantly increased over the past few decades. However, due to their composition and physical–chemical characteristics, they represent ideal media for microbial growth. Therefore, during the last years, scientists and technologists have been prompted to develop technologies capable of preserving food with minimal effects on functional properties and quality attributes. At the same time, the use of low-energy or energy-efficient methods is also a target for food processors.

Among the different preservation processes existing today, thermal pasteurization and sterilization have been predominantly used in the food industry to preserve fruit products due to their efficacy and product safety record (Lado and Yousef 2002). However, high temperature reached during these processes causes undesirable changes to the overall product quality, affecting its nutritional properties and organoleptical characteristics. Fortunately, intensive research in technology has allowed the improvement of fruit preservation processes. In this sense, Cohn and Mendelsohn (1879) proved the possibility of microorganism inactivation with electricity. They reported a research on sterilization by passage of a DC-current through a nutrient solution obtaining interesting and positive results. Beyond this date, sev-

O. Martín-Belloso (✉)
Department of Food Technology, University of Lleida,
Rovira Roure 191, Lleida 25198, Spain
e-mail: omartin@tecal.udl.es

M. Morales-de la Peña
Tecnológico de Monterrey, Monterrey, Mexico

A. Rosenthal et al. (eds.), *Fruit Preservation*, Food Engineering Series,
https://doi.org/10.1007/978-1-4939-3311-2_16

eral studies focused on food sterilization by using electricity have been reported (de Haan et al. 2002).

Within electrical methods, ohmic heating (OH) and high-intensity pulsed electric fields (HIPEF) have shown high efficacy to inactivate a great variety of microorganisms at different levels. Moreover, due to processing characteristics, OH and HIPEF usually reduce or avoid the common detrimental effects caused by heat in treated products. Hence, they have gained increased attention in the food industry and, nowadays, their development is becoming one of the most interesting topics in food research.

Generally, fruits and vegetables exhibit sufficient conductivity for being treated by OH or HIPEF processes. In this sense, the efficacy of these techniques to preserve fruit and vegetable products has been evaluated throughout the last decades. Most of the results indicate that microorganism inactivation by these novel technologies could reach levels as high as those achieved by conventional treatments. Moreover, authors report that microbial inactivation rates are faster than those for conventional treatments. Regarding OH, it has been observed that fruit products reach the required temperatures in less than one minute at relatively low electric field strengths (Palanniapan and Sastry 1991; Sarang et al. 2008). Otherwise, electroporation effects in microorganisms caused by the application of electric fields set at high intensities have been observed during HIPEF treatment. Therefore, these electrical technologies represent a potential alternative to heat processes in order to obtain stable fruit products with high nutritional values, good quality characteristics, and suitable sensorial properties.

This chapter mainly reviews the basic principles of OH and HIPEF as food preservation methods, their general effects over different microorganisms, and other important aspects of fruit products. Furthermore, future trends and requirements have been considered in order to achieve their commercial implementation.

16.2 Basic Principles

The principle of OH and HIPEF treatments is the passage of electric current, at an appropriate voltage, through a food placed or flowing between two electrodes in batch or continuous flow systems (Ghnimi et al. 2008). The main differences between both processes are the way and intensity of the electric field application. While electric fields are lower than 100 V/cm in OH, very short pulses (1–10 μs) of high-intensity electric fields (20–80 kV/cm) are applied in HIPEF.

OH process is considered a conduction-heating technique for liquids and pumpable particles with sufficient conductivity, such as fruit and vegetables (Butz and Tauscher 2002; Palanniapan and Sastry 1991; Sarang et al. 2008). During the process, an important movement of ions occurs into the medium which results in an internal energy transformation (from electric to thermal) that warm up the product (Butz and Tauscher 2002; Sastry and Barach 2000). In comparison with conventional thermal process, OH enables to heat materials at extremely rapid rates (from a few seconds to a few minutes) leading to safe products with high quality and added value

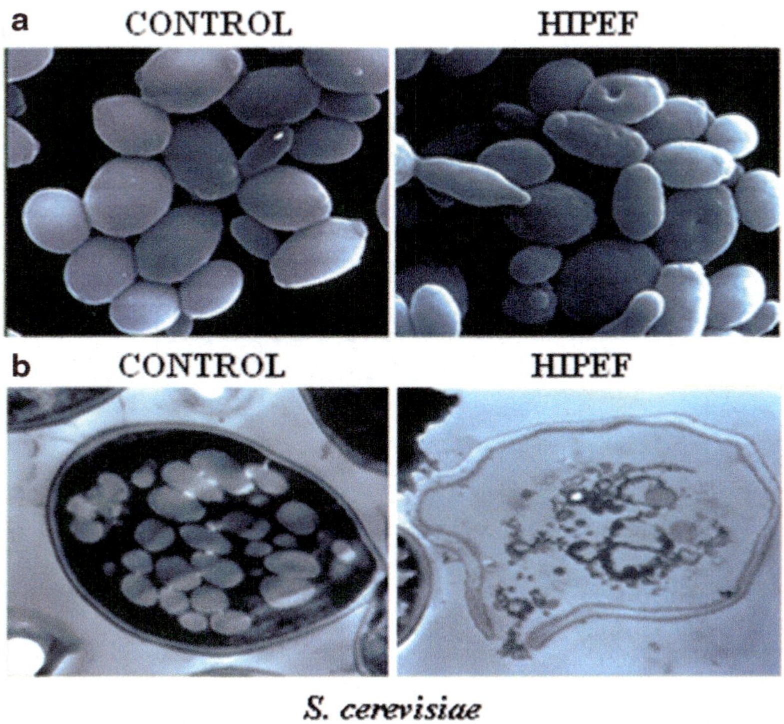

Fig. 16.1 Untreated and PEF-treated cells of *Saccharomyces cerevisiae* in orange juice. (**a**) Scanning electron micrographs and (**b**) transmission electron micrographs of untreated and PEF-treated cells. From Elez-Martínez et al. (2004)

(Sastry 2005; Parrott 1992). In this way, it can be considered an advanced thermal process where food acts as an electrical resistor (Castro et al. 2003). Moreover, since the conductivity of most food increases with temperature, OH becomes more effective as temperature increases.

Initially, it was thought that the effects of pasteurization or sterilization by OH were purely thermal, and that the applied electric field had no effects on biological cells. However, today, many studies have demonstrated that non-thermal effects of electricity also contribute to microbial inactivation by the effects of cell electroporation (Cho et al. 1999; Sun et al. 2008; Sastry 2008). Electroporation is defined as the formation of pores in cell membranes (Fig. 16.1) due to the presence of an electric field. As a consequence, the permeability of the membrane is enhanced and material diffusion throughout the membrane is achieved by electro-osmosis (An and King 2007). This observation resulted in the development of a nonthermal technology termed HIPEF (Sastry 2008).

HIPEF process involves the application of pulses of high voltage (20–80 kV/cm) to food placed between two electrodes for less than 1 s (Butz and Tauscher 2002; Odriozola-Serrano et al. 2008a). Nonetheless, due to the characteristics of this technology, its application is restricted for fluid food that can withstand high electric fields, have low electrical conductivity, and do not contain or form bubbles. Moreover,

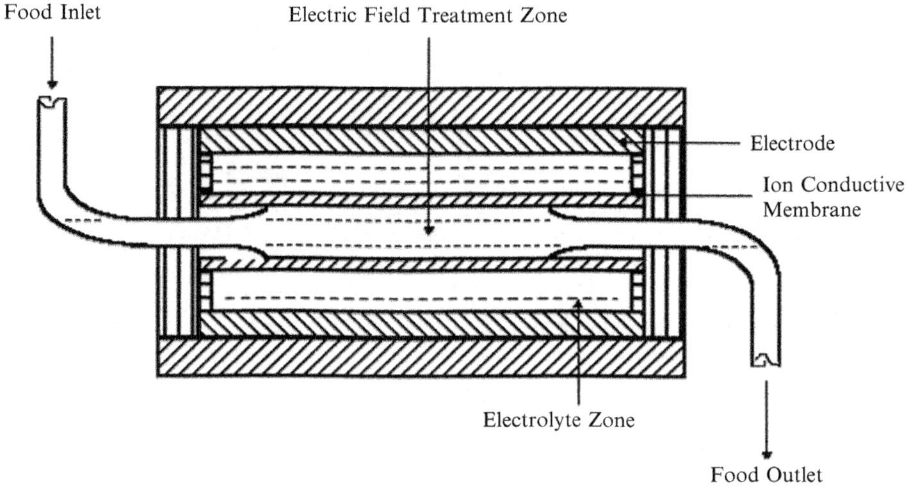

Fig. 16.2 Basic treatment chamber for food preservation by OH or HIPEF

in order to maintain proper processing operation, they must have a maximum particle size smaller than the gap of the treatment region (Ortegas-Rivas 2007; Toepfl et al. 2005; Heinz et al. 2002).

It could be said that the effectiveness of both process, OH and HIPEF, is mainly due to the cell membrane permeabilization (Kulsshrestha and Sastry 2003), which takes place in an extremely short time (Dimitrov 1984). Several authors have observed that vegetative microorganisms can be inactivated at sufficient field-strength in liquid food. Nonetheless, the inactivation effects under electric field strengths associated with OH are typically far lower than those associated with HIPEF (Sastry 2008).

Basically, OH or HIPEF equipments consist of a treatment chamber composed by two electrodes that are in direct contact with the food (Fig. 16.2), whereby electricity is passed using a variety of voltage and current combinations. The electrical process occurred inside the treatment chamber causes electrochemical interactions (Sobrino-López and Martín-Belloso 2009). Depending on the magnitude of the process, electrode corrosion can be produced resulting in the migration of some materials into the treated food (Roodenburg et al. 2005). Furthermore, some non-desired reactions, such as electrical breakdown, sparking, bubble creation, and microbial stagnant zones, could occur. Lima et al. (1999) reported that gas production and dissolution appeared to take place with stainless steel electrodes and it was not evident with specially coated titanium electrodes. Fortunately, those reactions did not affect the ascorbic acid content or the flavor of treated products. Furthermore, authors also observed that there was a noticeable rust tinge in the solution when stainless steel electrodes were used and no tinge with titanium electrodes.

Therefore, a proper design of the treatment chambers is an important aspect in order to impart safe and uniform treatments. Moreover, the components of treatment chambers must be food grade, washable, and autoclavable (Huang and Wang 2009). As well, the use of electrode material with particular characteristics cannot be disregarded (Butz and Tauscher 2002).

16.3 Potential Applications

The increasing interest for electrical processes has suggested the need for deeper and detailed understanding of OH and HIPEF as preservation treatments. It is well known that choosing an adequate technology to achieve high product stability requires determining its capability to inactivate pathogenic and spoilage microorganisms. Hence, with the aim of evaluating the feasibility of these emerging technologies to get this objective, different applications have been tried during the last years. First, investigations were implemented using model and buffer solutions. Nonetheless, today, different products such as fruit and vegetable juices are being considered as treatment medium in order to obtain reliable results for commercial implementation.

Since OH can be applied in fluids with solid particles, investigations have been done in whole fruits, fruit purees, and fruit juices, while HIPEF studies, due to the process characteristics, have been mainly carried out in fruit juices. The obtained results demonstrate that both treatments are able to inactivate microorganisms at levels as high as those achieved by conventional thermal process. On the one hand, temperature is considered as the principal critical factor in OH (Sastry and Barach 2000). Additionally, electric field strength also influences the effectiveness of the process (Palanniapan and Sastry 1991). On the other hand, electric field strength (E), treatment time (t), pulse width (τ), number of pulses (n), pulse frequency (f), pulse shape, and pulse polarity (p) are the main variables affecting a HIPEF process (Fox 2007; Elez-Martínez et al. 2007a, 2007b). Moreover, the extent of microbial inactivation achieved by OH or HIPEF also depends on other factors such as microbial characteristics, food electrical conductivity, thermo-physical properties, composition, and the design of processing device (Liezerson and Shimoni 2005a, 2005b).

At present, industrial applications of OH are numerous and include blanching, drying, evaporation, dehydratation, fermentation, pasteurization, and sterilization. Currently, this emerging thermal process is being used to preserve whole fruits in Japan and the United Kingdom (Sastry and Barach 2000) and for the production of syruped fruit-salad and fruit juices (Anonymous 2002). Furthermore, a number of continuous ohmic heating plants are currently operating in North America, Europe, and Asia. Although OH technology appears to be promising and highly effective, there is little information in open literature concerning its effects on specific products (Liezerson and Shimoni 2005a, 2005b).

Sastry and Barach (2000) mentioned that there are no particular microorganism strains with a unique resistance to OH technology. Hence, the most resistant pathogens would likely be the same as those for thermal process (USA-FDA 2000). In this sense, various studies have been focused on the comparison of OH and conventional treatments effects over different microorganisms and spores in fruit products. Baysal and Icier (2010) observed that OH had higher lethality effects on *Alyciclobacillus acidoterrestris* spores in orange juice than conventional pasteurization. Their results showed that conventional heating was ineffective for pasteurizing the juice, whereas the maximum OH treatment applied (30 V/cm) was sufficient to inactivate 5 log units of the spores. According to them, as temperature increased, the number of spores sharply decreased and a moderate increase in the voltage gradient seemed to enhance the spore inactivation effect of OH (Fig. 16.3). Similarly, Pereira et al. (2007) evaluated the influence of OH on the heat resistance of *Bacillus licheniformis* spores in cloudberry jam and compared the results with those obtained after conventional heating. Spore inactivation during OH was more efficient than that obtained with conventional treatment. According to the authors, this was probably due to the thermal effect with an additional killing effect caused by the electric current. In a different study, the same authors (Pereira et al. 2007) reported lower D and z values for the inactivation of *E. coli* and *B. licheniformis* when submitted to OH. They concluded that electric current may have affected the microbial cell death rate due to the electroporation effect.

Regarding HIPEF process research, there are only a limited number of technical scale prototypes and commercial application. The evolution of successful results from laboratory to an industrial scale has been shown to be a difficult task. According to Toepfl et al. (2005), before an industrial exploitation of this nonthermal technique to preserve food, it has to be demonstrated that the process is economically interesting in comparison to the existing pasteurization methods, in terms of costs of operation and investment as well as product quality and, in particular, consumer acceptance. In this sense, a huge effort has been done by scientist through intensive investigations.

Namely, microbial inactivation effects of HIPEF process have been evaluated and compared with conventional treatments applied in fruit juices. Results have indicated that HIPEF is a technology capable of achieving optimal pasteurization levels in this kind of products. Nevertheless, it mainly depends on treatment conditions, the type of microorganism under study, and the intrinsic properties of the product. According to several authors, E and t are the variables with major influence during HIPEF processing (Toepfl et al. 2005; Rodrigo et al. 2003a; Yang et al. 2004; Giner et al. 2005). Generally, the higher the E or t, the greater the microorganism inactivation (Fox 2007; Esplugas et al. 2001; Álvarez et al. 2003; Monfort et al. 2010). Based on scientific research, there is a threshold E after which cell inactivation occurs. An irreversible membrane damage leads to cell break down and intracellular leakage and, consequently, microbial death (El-Hag et al. 2008). Though microbial inactivation also is enhanced when t is prolonged, authors have observed that there is a critical point along t where few microbial log reductions are obtained, indicating that reduction of microorganisms at long t became gradual or long tailing (Odriozola-Serrano et al. 2008a).

Generally, vegetative cells of bacteria and yeasts can be effectively inactivated by HIPEF (Álvarez et al. 2006); however, bacteria spores present a higher resistance (Pol et al. 2001). Therefore, at present, HIPEF process can be applied for pasteurization but not for sterilization purposes. Among vegetative cells, gram-positive bacteria, which have a thick mucopeptide backbone layer (Wan et al. 2009), are more resistant to HIPEF than gram-negative microorganisms (García et al. 2007). Since the induced voltage across the cell membrane is proportional to its geometric size (Toepfl et al. 2007), small microorganisms, such as *Listeria*, are more resistant to HIPEF (Heinz et al. 2003). Additionally, the growth stage and initial microbial concentration in food could influence the levels of inactivation after HIPEF processing.

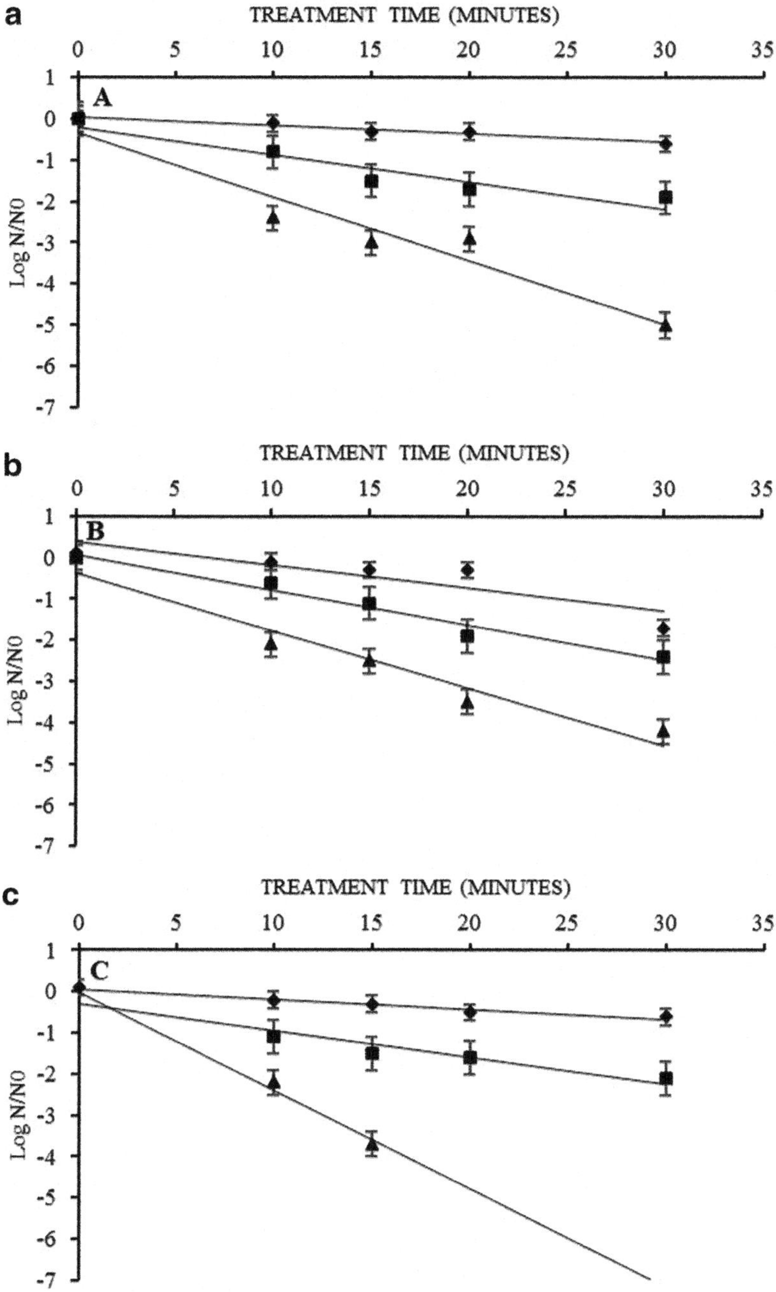

Fig 16.3 Survival curves for *A. acidoterrestris* DSM 3922 spores in orange juice (pH 3.637) treated with ohmic heating at 30 V/cm (**a**), 40 V/cm (**b**), and 50 V/cm (**c**) and at 70 °C (*filled diamond*), 80 °C (*filled square*), and 90 °C (*filled triangle*). From Baysal and Icier (2010)

On the one hand, cells in stationary phase have shown less sensitivity to HIPEF processing than those in logarithmic growth stage (Álvarez et al. 2000; Rodrigo et al. 2003b). On the other hand, according to Donsì et al. (2007), microbial inactivation level increases by decreasing the initial concentration of microorganisms.

Rivas et al. (2006), Akin and Evrendilek (2009), Sampedro et al. (2009), and Fernández-Molina et al. (2006) reported that HIPEF process set at high *E* or *t* values allows a significant reduction of different microorganisms, such as *Escherichia coli, Lactobacillus plantarum, Listeria innocua,* or *Pseudomonas fluorescens,* inoculated in different food. Likewise, Marsellés-Fontanet et al. (2009) found that the optimal microbial inactivation of different microorganisms usually present in grape juice (*Kloeckera apiculata, Saccharomyces cerevisiae, Lactobacillys plantarum, Lactobacillus hilgardii,* and *Gluconobacter oxydans*) could be obtained by applying a HIPEF treatment of 35 kV/cm during 1000 μs with *f* of 303 Hz.

Nowadays, after the evaluation of OH and HIPEF capability to produce safe food, researchers and technologists are dedicating their efforts to evaluate other interesting aspects of fruit preservation by those technologies. Namely, enzymatic activity, concentration of bioactive compounds, sensorial properties, physical–chemical attributes, and product stability during storage as affected by electrical technologies are being studied. Although information related to OH with reference to these aspects is limited in literature, some interesting studies have been reported.

Regarding enzyme inactivation, a study measuring the changes on polyphenol oxidase (PPO) activity in fresh grape juice ohmically heated (20, 30 and 40 V/cm, 20–90 °C) was carried out by Içier et al. (2008). They found that the PPO critical deactivation temperatures were 60 °C at 40 V/cm and 70 °C at 20 and 30 V/cm. Otherwise, Liezerson and Shimoni (2005a, 2005b) observed that OH reduced pectin metylesterase (PE) activity in orange juice by 98 %. On the other hand, it has been observed that a higher intensity of HIPEF treatment is required to inactivate enzymes in fruit products in comparison to the intensity required for microbial inactivation. Different enzymes such as peroxidase (POD, Fig. 16.4), PPO, or lipoxygenase (LOX) have been significantly inactivated when *E* or *t* were set at high values (up to 40 kV/cm and 2000 μs) (Zhong et al. 2005; Marselles-Fontanet and Martín-Belloso 2007; Li et al. 2008; Aguiló-Aguayo et al. 2008, 2010). The increase on *E* and *t* during HIPEF processing might cause alterations on the enzyme structure and, therefore, the substrate cannot fit the active site, preventing its conversion into products and resulting in a reduction of the enzymatic activity (Elez-Martínez et al. 2007a, 2007b).

In order to determine whether the application of OH or HIPEF can lead to a longer shelf-life of fruit products with minimum quality loss, some works have been completed recently. Pataro et al. (2011) investigated the shelf-stability of apricots in syrup ohmically heated. They also evaluated the changes on quality attributes (pH, soluble solid content, and color) as well as ascorbic acid retention of the apricots immediately after processing and during storage. Their results demonstrated that the product kept its microbial stability during 52 weeks and the quality characteristics of the apricots were not adversely affected by OH, except for ascorbic acid, which was slightly reduced over time. Similarly, Liezerson and Shimoni 2005a, 2005b compared the effects of OH and conventional heat treatments on orange juice stability during storage at 4 °C. Although both thermal technologies prevented the growth of

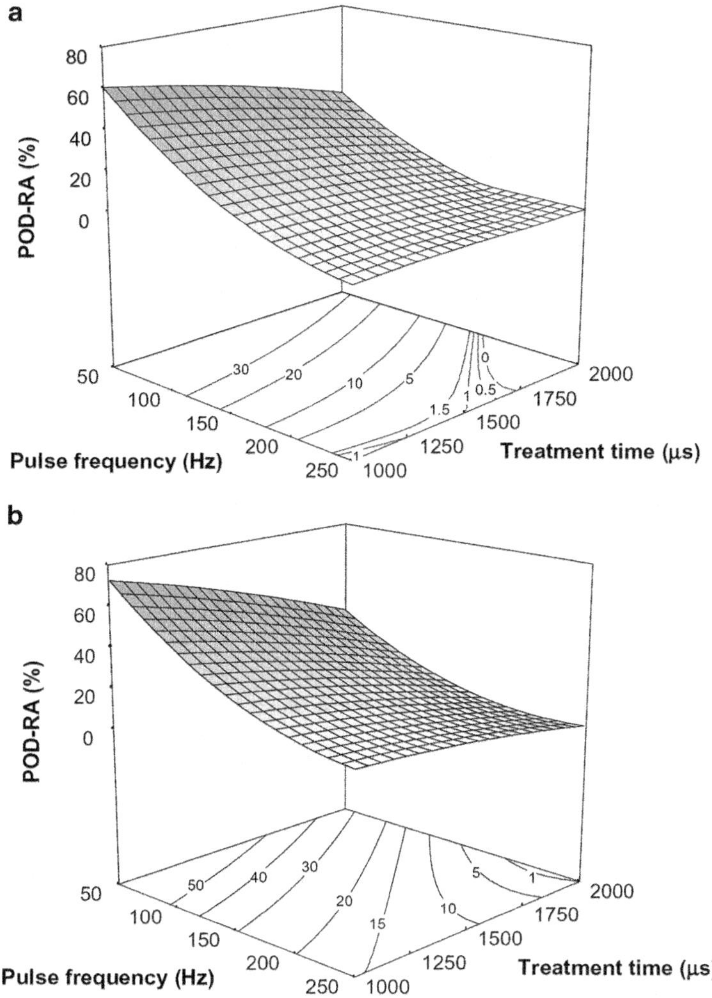

Fig. 16.4 Effect of the pulse frequency and total treatment time on residual POD activity when pulsed electric fields (PEF) treatment was set at 35 kV/cm, 7 μs of pulse width in bipolar (**a**) or monopolar (**b**) mode. From Aguiló-Aguayo et al. (2008a)

microorganisms for 105 days, the sensory shelf-life of the OH-treated orange juice was higher than 100 days, almost two times longer than that of conventionally pasteurized juice. Moreover, authors reported that degradation curves of ascorbic acid followed a linear decrease pattern. Lima et al. (1999) studied the degradation of vitamin C in orange juice subjected to OH or conventional pasteurization. Results from this research indicated that the electrical treatment had no significant effect on ascorbic acid degradation.

With regard to microbial stability achieved by HIEPF processing, some authors have stated that this nonthermal technology is able to prolong the shelf-life of different

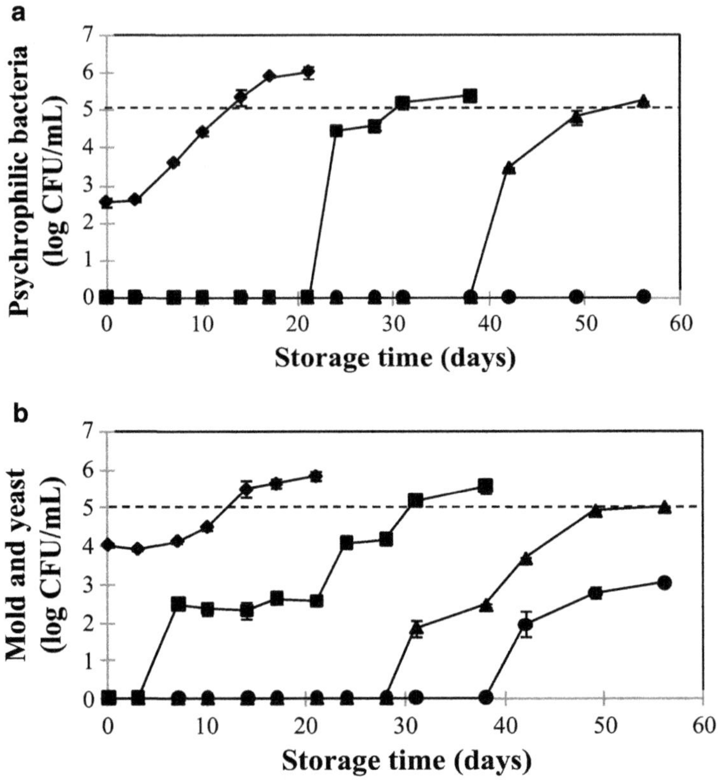

Fig. 16.5 Microbial stability of a fruit juice-soymilk beverage throughout storage at 4 °C. Psychrophilic bacteria: (**a**) mold and yeast: (**b**). Fruit juice-soymilk beverages: (*filled diamond*) untreated, (*filled square*) 800 µs-PEF-treated, (*filled triangle*) 1400 µs-PEF-treated, (*filled circle*) thermally treated (*dashed line*). From Morales-de la Peña et al. (2010)

fruit juices in comparison to the fresh ones. Nevertheless, microbial stability of HIPEF-treated fruit juices is not as long as that of thermally treated products (Fig. 16.5). Through this limitation, it has been observed that other important aspects such as bioactive compounds content, quality aspects, and sensorial attributes can be better preserved in HIPEF-processed juices than in those thermally treated (Morales-de la Peña et al. 2010; Odriozola-Serrano et al. 2008a, 2008b; Mosqueda-Melgar et al. 2008; Elez-Martínez et al. 2007a, 2007b). Many authors agreed that vitamin C content in fruit juices treated by HIPEF decreases as the storage time increases; however, degradation rate is slower compared to that observed in heated juices (Min et al. 2003; Elez-Martínez et al. 2006; Plaza et al. 2006; Torregrosa et al. 2006; Odriozola-Serrano et al. 2008b; Zulueta et al. 2007; Morales-de la Peña et al. 2010). Likewise, Quitao-Texeira et al. (2009) observed that total carotenoid concentration of fresh, HIPEF- or heat-treated carrot juices diminished with storage time. Nevertheless, HIPEF-treated juices showed smaller degradation rate than those processed by heat.

16.4 Commercial Implementation

Since it has been mentioned before, OH implementation at industrial levels has been relatively easier than the implantation of HIPEF treatment. Since the mid 1990s, the number of commercial installations of OH processes has grown and a number of new manufacturers have entered the market (Sastry 2008). Namely, Emmepieme SRL in Piacenza, Italy, uses OH to process many varieties of food including fruit juices (Anderson 2003). According to Sastry (2008), OH systems are now better-engineered, more sophisticated, and far less expensive than their predecessors. Otherwise, at present, a lot of research in the field of HIPEF engineering has been done; however, existing information focused on the development of HIPEF units at industrial levels and economical analysis of its implementation is scarce.

In order to scale up HIPEF technology, accurate data about process conditions for microbiologically safe products, food characteristics information, and available facilities for plant installation are required. Hoogland and de Haan (2007), Barbosa-Cánovas and Altunakar (2006), and Toepfl et al. (2006a, 2006b) concluded that one of the main limitations for HIPEF processing of food at commercial levels is the generation of high voltages pulses with sufficient peak power to obtain safe products. Therefore, this is an important challenge for researchers in order to make commercial processing available in the near future.

According to literature data, the cost of commercial OH systems, including installation, can be in excess of USD9,000,000, which is a costly investment for a manufacturing facility (Anderson 2008). A commercial pasteurizer used to produce juice products will cost a company from $0.11 to $0.17 per gallon to generate a finished product (Stark 2008). Regarding HIPEF, since industrial systems are not available, costs can only be estimated based on data accessible from other pulse power applications, or scaled up from pilot-scale equipments (Toepfl et al. 2006a, 2006b). Hoogland and de Haan (2007) stated that the installation of a HIPEF system to process 5000 L/h would be roughly 1 million Euros, which is more expensive than a standard thermal processing plant. Similarly, in a previous study, Braakman (2003) reported an investment cost of around 2 million Euros for a flow capacity of 5 ton/h or 4 million Euros for 10 ton/h capacity. Barbosa-Cánovas and Altunakar (2006) affirmed that the major concern for commercialization of HIPEF technology is the initial investment cost. Nonetheless, if there were more companies dedicated to the design of HIPEF equipments and a high number of industrial equipments installed, the cost of each system would dramatically decrease, making its implementation more accessible.

16.5 Environmental Considerations

According to Pereira and Vicente (2010), the emergence of novel thermal and non-thermal technologies such as OH and HIPEF allows producing high-quality products with improvements in terms of heating efficiency, and consequently, in energy savings. Both technologies are considered locally clean process, resulting in a lower environmental impact than that caused by traditional process. OH and HIPEF can provide not only energy and water savings, but also increased reliability, reduced emissions, higher product quality, and improved productivity (Masanet et al. 2008). In this sense, these technologies appear to be more environmentally friendly.

Lung et al. (2006) evaluated the energy saving of HIPEF in comparison to the energy of conventional treatment for orange juice preservation. Authors reported that electricity savings of HIPEF were rounding 18 %, based on the assumed electricity consumption range of the thermal processing. In this sense, HIPEF pasteurization seems to be less energy-intensive than heat, resulting in annual savings of 791.2–1055 TJ per year of fossil-fuel equivalents. Furthermore, HIPEF also contribute to the reduction of CO_2 emissions (Lelieved 2005).

Food processing systems powered only by electricity such as OH or HIPEF are considered environmental-friendly, once they may eliminate completely or at least reduce significantly the local use of boilers or steam generation systems. As a result, these processes diminish wastewater and increase water and energy savings. Moreover, if the electricity is generated by an environmentally clean and renewable energy source such as hydroelectric power, they will effectively contribute to reduce the pollution load, helping to preserve the environment.

16.6 Conclusions and Future Trends

Electrical treatments have demonstrated to be feasible technologies to provide safe fruit products. In comparison with traditional thermal process, OH and HIPEF preserve fruits with higher content of bioactive compounds and better quality attributes. Furthermore, OH and HIPEF have demonstrated to be environmentally friendly techniques, while reducing processing costs and improving the added value of the products. These facts represent a great advantage in order to satisfy current consumers' claims for tasty, healthy, and easy-to-handle food with acceptable shelf-life. Therefore, electrical treatments are increasingly attracting the attention of scientists and technologists. However, more research is essential to completely understand and exploit both processes, since they may have significant economic consequences to the industry. Whereas OH has been successfully implemented at industrial levels, HIPEF process is still under scientific evaluation due to specific gaps that need to be covered before commercial applications. Current limitations, related to high investment costs, full control of processing parameters, and lack of regulatory approval, have been delaying a wider implementation of HIPEF at industrial scale. In this sense, further studies need to be done with the aim to optimize HIPEF process.

References

Aguiló-Aguayo, I., I. Odriozola-Serrano, L.J. Quintao-Texeira, and O. Martín-Belloso. 2008. Inactivation of tomato juice peroxidase by high-intensity pulsed electric fields as affected by process conditions. *Food Chemistry* 107: 949–955.

Aguiló-Aguayo, I., R. Soliva-Fortuny, and O. Martín-Belloso. 2010. Optimizing critical high-intensity pulsed electric fields treatments for reducing pectolytic activity and viscosity changes in watermelon juice. *Innovative Food Science and Emerging Technologies*. 11: 299–305.

Akin, E., and G.A. Evrendilek. 2009. Effect of pulsed electric fields on physical, chemical, and microbiological properties of formulated carrot juice. *Food Science and Technology International* 15: 275–282.

Álvarez, I., J. Raso, A. Palop, and F. Sala. 2000. Influence of different factor son the inactivation of Salmonella senftenberg by pulsed electric fields. *International Journal of Food Microbiology* 55: 143–146.

Álvarez, I., R. Pagán, S. Condón, and J. Raso. 2003. The influence of process parameters for the inactivation of Listeria monocytogenes by pulsed electric fields. *International Journal of Food Microbiology* 87: 87–95.

Álvarez, I., S. Condón, and J. Raso. 2006. Pulsed electric fields technology for the food industry. In *Microbial inactivation by pulsed electric fields*, ed. J. Raso and H. Volker, 97–130. New York: Springer Science + Bussines Media, LLC.

An, H.J., and J.M. King. 2007. Thermal characteristics of ohmically heated rice starch and rice flours. *Journal of Food Science* 72: C84–C88.

Anderson, D. 2008. *Ohmic heating as an alternative food processing technology*. Master in Food Science diss., Kansas State University.

Anonymous. 2000. Ohmic heating system for sterilization and aseptic filling. *Italian Food and Beverage Technology*. 28–29.

Barbosa-Cánovas, G.V., and B. Altunakar. 2006. Pulsed electric field processing of foods: An overview. In *Pulsed electric field technology for the food industry*, ed. J. Raso and V. Heinz. New York: Springer.

Baysal, A.H., and F. Icier. 2010. Inactivation kinetics of Alicyclobacillus acidoterrestris Spores in orange juice by ohmic heating: Effects of voltage gradient and temperature on inactivation. *Journal of Food Protection* 73: 299–304.

Butz, P., and B. Tauscher. 2002. Emerging technologies: Chemical aspects. *Food Research International* 35: 279–284.

Castro, I., J.A. Teixera, S. Salengke, S.K. Sastry, and A.A. Vicente. 2003. Ohmic heating of strawberry products: Electrical conductivity, measurements and ascorbic acid degradation kinetics. *Innovative Food Science and Emerging Technologies* 5: 27–36.

Cho, H.Y., A.E. Yousef, and S.K. Sastry. 1999. Kinetics of inactivation of Bacillus subtillis spores by continuous or intermittent ohmic or conventional heating. *Biotechnology & Bioengineering* 63: 368–372.

Cohn, F., and B. Mendelsohn. 1879. *Beitrage Biologie der Pflauzen* 3: 141.

Dimitrov, D.S. 1984. Electric field-induced breakdown of lipid bilayers and cell membranes: Athin viscoelastic film model. *Journal of Membrane Biology* 78: 53–60.

Donsì, G., G. Ferrari, and G. Pataro. 2007. Inactivation kinetics of Saccharomyces cerevisiae by pulsed electric fields in batch treatment chamber: The effect of electric field unevenness and initial cell concentration. *Journal of Food Engineering* 78: 784–792.

El-Hag, A., A. Otunola, S.H. Jayaram, and W.A. Anderson. 2008. Reduction of microbial growth in milk by pulsed electric fields. *IEEE International Conference on Dielectric Liquids*, ICDL, 4622478.

Elez-Martínez, P., R. Soliva-Fortuny, and O. Martín-Belloso. 2006. Comparative study on shelf-life of orange juice processed by high intensity pulsed electric fields or heat treatment. *European Food Research and Technology* 222: 321–329.

Elez-Martínez, P., and O. Martín-Belloso. 2007a. Impact of pulsed electric fields on food enzymes and shelf-life. In *Food preservation by pulsed electric fields*, ed. H.L.M. Lelieveld, S. Notermans, and S.W.H. de Haan. Boca Raton, FL: CRC Press.

———. 2007b. Effects of high intensity pulsed electric field processing conditions on vitamin C and antioxidant capacity of orange juice and gazpacho, a cold vegetable soup. *Food Chemistry*, 102: 201–209.

Esplugas, S., R. Pagan, G.V. Barbosa-Cánovas, and B.G. Swanson. 2001. Engineering aspects of the continuous treatment of fluid foods by pulsed electric fields. In *Pulsed electric fields in food processing*, ed. S. Esplugas, R. Pagan, G.V. Barbosa-Cánovas, and B.G. Swanson. Lancaster, PA: Technomic.

Fernández-Molina, J.J., D. Bermúdez-Aguirre, B. Altunakar, G.G. Swanson, and G.V. Barbosa-Cánovas. 2006. Inactivation of Listeria innocua and Pseudomonas fluorescens by pulsed electric fields in skim milk: Energy requirements. *Journal of Food Engineering* 29: 561–573.

Fox, M.B. 2007. Microbial inactivation kinetics of pulsed electric field treatment. In *Food Preservation by Pulsed Electric Field*, ed. H.L.M. Lelieved, S. Notermans, and S.W.H. de Haan, 127–134. New York: CRC Press.

García, D., N. Gómez, P. Mañas, J. Raso, and R. Pagán. 2007. Pulsed electric fields cause bacterial envelopes permeabilization depending on the treatment intensity, the treatment medium pH and the microorganism investigated. *International Journal of Food Microbiology* 133: 219–227.

Ghnimi, S., N. Flach-Malaspina, M. Dresch, G. Delaplace, and J.F. Maingonnat. 2008. Design and performance evaluation of an ohmic heating unit for thermal processing of highly viscous liquids. *Chemical Engineering Research and Design* 86: 626–632.

Giner, J., P. Grouberman, V. Gimeno, and O. Martín. 2005. Reduction of pectinesterase activity in a commercial enzyme preparation by pulsed electric fields: Comparison of inactivation kinetics models. *Journal of Food Science and Agricultural* 85: 1613–1621.

de Haan, S.W.H., B. Roodenburg, J. Morren, and H. Prins. 2002. Technology for preservation of food with pulsed electric fields (PEF). *IEEE AFRICON Conference* 2: 791–796.

Heinz, V., I. Alvarez, A. Angersbach, and D. Knorr. 2002. Preservation of liquid foods by high intensity pulsed electric fields—Basic concepts for process design. *Trends in Food Science and Technology* 12: 103–111.

Heinz, V., S. Toepfl, and D. Knorr. 2003. Impact of temperature on lethality and energy efficiency of apple juice pasteurization by pulsed electric fields treatment. *Innovative Food Science and Emerging Technologies*. 12: 103–111.

Hoogland, H., and W. de Haan. 2007. Economic aspects of pulsed electric field treatment of food. In *Food preservation by pulsed electric fields*, ed. S. Lelieve, S. Nontermans, and S.W.H. de Haan. Cambridge, MA: CRC Press.

Huang, K., and J. Wang. 2009. Designs of pulsed electric fields treatment chambers for liquid foods pasteurization process: A review. *Journal of Food Engineering* 95: 227–239.

Içier, F., H. Yildiz, and T. Baysal. 2008. Polyphenoloxidase deactivation kinetics during ohmic heating of grape juice. *Journal of Food Engineering* 85: 410–417.

Kulsshrestha, S.A., and S.K. Sastry. 2003. Frequency and voltage effects on enhanced diffusion during moderate electric field (MET) treatment. *Innovative Food Science* 4: 189–184.

Lado, B., and A.E. Yousef. 2002. Alternative food-preservation technologies: Efficacy and mechanisms. *Microbes and Infection* 4: 433–40.

Lelieved, H. 2005. PEF—A food industry's view. In *Novel food processing technologies*, ed. G.V. Barbosa-Cánovas, M.S. Tapia, P.M. Cano, O. Martin-Belloso, and A. Martinez. Boca Raton, FL: CRC Press.

Li, Y.Q., Q. Chen, X.H. Liu, and Z.X. Chen. 2008. Inactivation of soybean lipoxygenase in soymilk by pulsed electric fields. *Food Chemistry* 109: 408–414.

Liezerson, S., and E. Shimoni. 2005. Effect of ultrahigh-temperature continuous ohmic heating treatment of fresh orange juice. *Journal of Agriculutre and Food Chemistry* 53: 3519–3524.

———. 2005b. Stability and sensory shelf-life of orange juice pasteurized by continuous ohmic heating. *Journal of Agricultural and Food Chemistry* 53: 4012–4018.

Lima, M., B.F. Heskitt, L.L. Burianek, S.E. Nokes, and S.K. Sastry. 1999. Ascorbic acid degradation kinetics during conventional and ohmic heating. *Journal of Food Processing Preservation* 23: 421–434.

Lung, R., E. Masanet, and A. McKane. 2006. The role of emerging technologies in improving energy efficiency: Examples from the food processing industry. In *Proceedings of the Industrial Energy Technologies Conference*. New Orleans, LO.

Masanet, E., E. Worrell, W. , Graus, and C. Galistiky. 2008. Energy efficiency improvement and cost saving opportunities for the fruit and vegetable processing industry, and energy star guide for energy and plant managers. *Technical Report: Environmental Energy Technologies Division.*

Marselles-Fontanet, A.R., and O. Martín-Belloso. 2007. Optimization and validation of PEF processing conditions to inactivate oxidative enzymes of grape juice. *Journal of Food Engineering* 83: 452–462.

Marsellés-Fontanet, A.R., A. Puig, P. Olmos, S. Mínguez-Sanz, and O. Martín-Belloso. 2009. Optimising the inactivation of grape juice spoilage organisms by pulse electric fields. *International Journal of Food Microbiology* 130: 159–165.

Min, S., S.K. Min, and Q.H. Zhang. 2003. Inactivation kinetics of tomato juice lipoxygenase by pulsed electric fields. *Journal of Food Science* 68: 1995–2001.

Monfort, S., E. Gayán, G. Saldaña, E. Puértolas, S. Condón, J. Raso, and I. Álvarez. 2010. Inactivation of Salmonella typhimurium and Staphylococcus aureus by pulsed electric fields in liquid whole egg. *Innovative Food Science and Emerging Technologies.* 11: 306–313.

Morales-de la Peña, M., L. Salvia-Trujillo, M.A. Rojas-Graü, and O. Martín-Belloso. 2010. Impact of high intensity pulsed electric field on antioxidant properties and quality parameters of a fruit juice-soymilk beverage in chilled storage. *LWT – Food Science and Technology* 43: 872–881.

Mosqueda-Melgar, J., R.M. Raybaudi-Massilia, and O. Martín-Belloso. 2008. Non-thermal pasteurization of fruit juices by combining high-intensity pulsed electric fields with natural antimicrobials. *Innovative Food Science and Emerging Technologies* 9: 328–340.

Odriozola-Serrano, I., R. Soliva-Fortuny, and O. Martín-Belloso. 2008a. Changes of health-related compounds throughout cold storage of tomato juice stabilized by thermal or high intensity pulsed electric field treatments. *Innovative Food Science and Emerging Technologies* 9: 272–279.

———. 2008b. Modeling changes in health-related compounds of tomato juice treated by high-intensity pulsed electric fields. *Journal of Food Engineering* 89: 210–216.

Ortegas-Rivas, E. 2007. Processing effects for safety and quality in some non-predominant food technologies. *Critical Reviews in Food Science and Nutrition* 47: 161–173.

Palanniapan, S., and S. Sastry. 1991. Electrical conductivities of selected solid foods during ohmic heating. *Journal of Food Process and Engineering* 14: 36–221.

Parrott, D.L. 1992. Use of ohmic heating for aseptic processing of food particulates. *Food Technology* 12: 68–72.

Pataro, G., G. Donsì, and G. Ferrari. 2011. Aseptic processing of apricots in syrup by means of a continuous scale ohmic unit. *Food Science and Technology* 44: 1546–1554.

Pereira, R.N., and A.A. Vicente. 2010. Environmental impact of novel thermal and non-thermal technologies in food processing. *Food Research International* 43: 1936–1943.

Pereira, R., J. Martins, C. Mateus, J.A. Teixeira, and A.A. Vicente. 2007. Death kinetics of Escherichia coli in goat milk and Bacillus licheniformis in cloudberry jam treated by ohmic heating. *Chemical Papers* 61: 121–126.

Plaza, L., C. Sánchez-Moreno, P. Elez-Martínez, B. De Ancos, O. Martín-Belloso, and M.P. Cano. 2006. Effect of refrigerated storage on vitamin C and antioxidant activity of orange juice processed by high-pressure or pulsed electric fields with regard to low pasteurization. *European Food Research and Technology* 223: 487–493.

Pol, I.E., W.G.C. van Arendonk, H.C. Mastwijk, J. Krommer, E.J. Smid, and R. Moezelaar. 2001. Sensitivities of germinating spores and carvacrol-adapted vegetative cells and spores of Bacillus cereus to nisin and pulsed-electric-field treatment. *Applied and Environmental Microbiology* 67: 1693–1669.

Quitao-Texeira, L.J., I. Odriozola-Serrano, R. Soliva-Fortuny, A. Mota-Ramos, and O. Martín-Belloso. 2009. Comparative study on antioxidant properties of carrot juice stabilised by high-intensity pulsed electric fields or heat treatments. *Journal of Food Science and Agricultural* 89: 2636–2642.

Rivas, A., D. Rodrigo, A. Martínez, G.V. Barbosa-Cánovas, and M. Rodrigo. 2006. Effect of PEF and heat pasteurization on the physical-chemical characteristics of blended orange and carrot juice. *LWT – Food Science and Technology* 39: 1163–1170.

Rodrigo, D., G.V. Barbosa-Cánovas, A. Martínez, and M. Rodrigo. 2003a. Pectin methyl esterase and natural microflora of fresh mixed orange juice and carrot juice treated with high intensity pulsed electric fields. *Journal of Food Protection* 66: 2336–2342.

———. Kinetic model for the inactivation of Lactobacillus plantarum by pulsed electric fields. *International Journal of Food Microbiology* 81: 223–229.
Roodenburg, B., J. Morren, H.E. Berg, and S.W.H. de Haan. 2005. Metal release in a stainless steel Pulsed Electric Field (PEF) system. Part I. Effect of different pulse shapes; theory and experimental method. *Innovative Food Science and Emerging Technologies* 6: 327–336.
Sampedro, F., D.J. Geveke, X. Fan, and H.Q. Zhang. 2009. Effect of PEF, HHP and thermal treatment on PME inactivation and volatile compounds concentration of an orange juice-milk based beverage. *Innovative Food Science and Emerging Technology* 10: 463–469.
Sarang, S., S. Sastry, and L. Kinipe. 2008. Electrical conductivity of fruits and meats during ohmic heating. *Journal of Food Engineering* 87: 6–351.
Sastry, S.K. 2005. Advances in ohmic heating and moderate electric field (MEF) processing. In *Novel food processing technologies*, ed. G.V. Barbosa-Cánovas, M.S. Tapia, and M.P. Cano. Boca-Raton, FL: CRC.
———. 2008. Ohmic heating and moderate and moderate electric field processing. *Food Science and Technology International* 14: 419–422.
Sastry, S.K., and J.T. Barach. 2000. Ohmic and inductive heating. *Journal of Food Science* 65: 42–46.
Sobrino-López, A., and O. Martín-Belloso. 2009. Review: Potential of high-intensity pulsed electric fields technology for milk processing. *Food Engineering Reviews* 2(1): 17–27. doi:10.1007/s12393-009-9011-7.
Stark, C. 2008. Personal communication. *Country Pure Foods* (Ellington, CT).
Sun, H.X., S. Kawamura, J.I. Himoto, K. Itoh, T. Wada, and T. Kimura. 2008. Effects of ohmic heating on microbial counts and denaturation of proteins in milk. *Food Science and Technology Research* 14: 117–123.
Toepfl, S., V. Heinz, and D. Knorr. 2005. Overview of pulsed electric field processing for food. In *Emerging technologies for food processing*, ed. Da-Wen Sun. San Diego, CA: Elsevier Academic Press.
———. Applications of pulsed electric field technology for the food industry. In *Pulsed electric field technology for the food industry*, ed. J. Raso and V. Heinz, 197–221. Berlin, Germany: Springer.
Toepfl, S., A. Mathys, V. Heinz, and D. Knorr. 2006. Review: Potential of high hydrostatic pressure and pulsed electric fields for energy efficient and environmentally friendly food processsing. *Food Reviews International* 22: 405–423.
Toepfl, S., V. Heinz, and D. Knorr. 2007. History of pulsed electric field treatment. In *Food preservation by pulsed electric fields*, ed. H.L.M. Lelieves, S. Notermans, and S.W.H. Haan. Boca Raton, FL: CRC Press.
Torregrosa, F., M.J. Esteve, A. Frigola, and C. Cortés. 2006. Ascorbic acid stability during refrigerated storage of orange-carrot juice treated by high pulsed electric field and comparison with pasteurized juice. *Journal of Food Engineering* 73: 339–345.
USA-FDA. 2000. *Kinetics of microbial inactivation for alternative food processing technologies: Ohmic and inductive heating*. Silver Spring, MD: United States of America, Food and Drug Administration, Center of Food Safety and Applied Nutrition.
Wan, J., J. Converntry, P. Swiergon, P. Sanguansri, and C. Versteeg. 2009. Advances in innovative processing technologies form microbial inactivation and enhancement of food safety—Pulsed electric field and low-temperature plasma. *Trends Food Science and Technology* 20: 414–424.
Yang, R.J., S.Q. Li, and Q.H. Zhang. 2004. Effects of pulsed electric fields on the activity of enzymes in aqueous solutions. *Journal of Food Science* 69: FCT241–FCT248.
Zhong, K., X. Hu, G. Zhao, F. Chen, and X. Liao. 2005. Inactivation and conformational change of horseradish peroxidase induced by pulsed electric field. *Food Chemistry* 92: 473–479.
Zulueta, A., M.J. Esteve, I. Frasquet, and A. Frígola. 2007. Fatty acid profile changes during orange juice-milk beverage processing by high-pulsed electric field. *European Journal of Lipid Science and Technology* 109: 25–31.

Chapter 17
Fruits and Fruit Products Treated by UV Light

Tatiana Koutchma, Marta Orlowska, and Yan Zhu

Abbreviations

A	Absorption
AIN	Aluminum nitride
DF	Divergence factor
EL	Excimer lamp
EPA	US Environmental Protection Agency
EVA	Ethylene vinyl acetate
EVOH	Ethyl vinyl alcohol copolymer
FDA	US Food and Drug Administration
GaN	Gallium nitride
LED	Light emitting diodes
LPHO	Low-pressure high-output lamp
LPM	Low-pressure mercury lamp
MCL	Maximum contaminant level
MPM	Medium pressure mercury lamp
PBS	Phosphate-buffered saline
PET	Polyethylene terephthalate
PF	Petri factor
PL	Pulsed lamp
PLT	Polyethylene
PME	Pectin methylesterase
PPO	Polyphenol oxidase
R	Reflection
RDA	Recommended Daily Allowance
RF	Reflection factor
T or UVT	Transmittance or transmittance of material in the ultraviolet range
TiO_2	Titanium dioxide
UV	Ultraviolet

T. Koutchma (✉) • M. Orlowska • Y. Zhu
Agriculture and Agri-Food Canada, Guelph, ON, Canada
e-mail: Tatiana.Koutchma@agr.gc.ca

A. Rosenthal et al. (eds.), *Fruit Preservation*, Food Engineering Series,
https://doi.org/10.1007/978-1-4939-3311-2_17

UV-A	Ultraviolet light range: 315–400 nm
UV-B	Ultraviolet light range: 280–315 nm
UV-C	Ultraviolet light range: 200–280 nm
cUV	Continuous ultraviolet mode
VUV	Vacuum ultraviolet radiation (100–200 nm)
WF	Water factor

17.1 Introduction

During the last decade an increase of fresh fruit and fruit products production constantly grew due to fruits health properties. A large number of studies have associated the consumption of fruits and their products with decreased risks of development of diseases such as cancer and coronary heart disease (Hansen et al. 2003). This may be due to the presence of health promoting phytochemicals such as carotenoids, flavonoids, phenolic compounds, and vitamins (Gardner et al. 2000) which have in some cases been shown to have disease preventing properties.

Fruit products are consumed in raw, minimally processed or processed ready-to-eat or ready-to-drink forms as whole fresh fruits, fresh-cut fruits, and fruits as ingredients, beverages, juices, and jams. Processing of fruits starts after harvesting and four activities can be distinguished: stabilization or preservation, transformation, production of ingredients, and production of fabricated foods. The role of processing technology in each activity implies the control of microbiological, chemical, and biochemical changes occurred as a result of microbial and enzymatic activities, oxidation reactions that can lead to safety, color, flavor, taste, and texture problems. Processing technologies that do not significantly alter the organoleptic or nutritional qualities of the fruits and do not form any undesirable chemical compounds in the product would have obvious advantages in modern food production. The interest in so-called minimal processing technologies led to the broad development of nonthermal or mild heat high tech methods that have a potential to replace traditional thermal preservation techniques and also result not only in better quality and longer shelf life but potentially in higher nutritional value or products with health benefits. In this respect, it is of paramount importance to develop processing methods which preserve not only the safety of fruits but also sensorial and nutritional quality and bioactivity of the constituents present in fruits and their products.

UV light treatment of foods is a nonthermal physical method of processing that is cost effective, free of chemicals and waste effluents, which makes it an ecologically friendly and sustainable technology. It does not produce by-products and it is safe to use, although precautions must be taken to avoid human exposure to UV light and to evacuate ozone generated by vacuum and far UV wavelengths.

The discovery of UV inactivation of the chlorine-resistant parasites *Cryptosporidium parvum* and *Giardia* sp. has catalyzed the use of UV light in the drinking water industry (Hijnen et al. 2006) and treatment of waste and processing water. UV has been utilized similarly in the disinfection of air, nonfood contact and

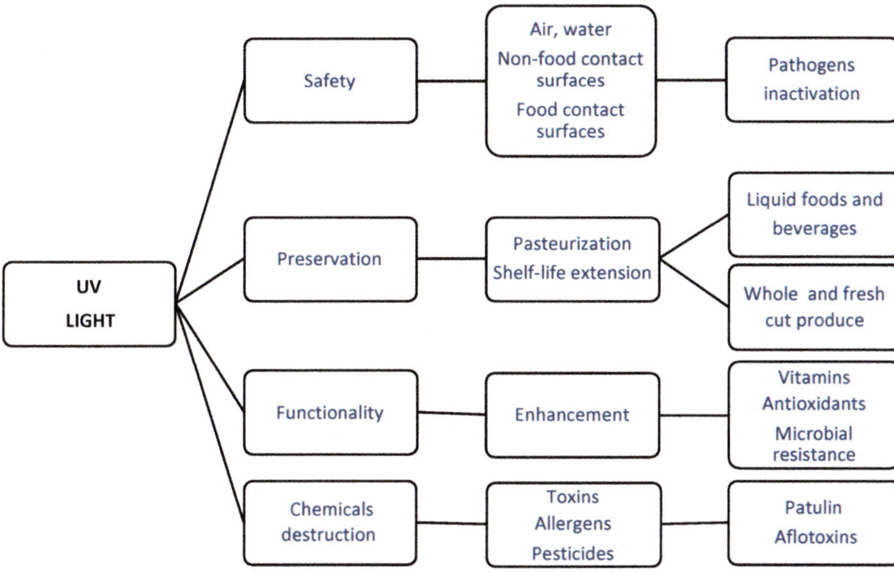

Fig. 17.1 Potential application of UV light in fruit production

food contact surfaces, and recently was used for treatments of surfaces of solid foods, liquid foods, beverages and their ingredients. Based on engineering advances and new scientific data, ultraviolet (UV) light technology in continuous and pulsed modes (cUV and PL) offers promise of improved microbiological and chemical safety and enhanced functionality of whole fresh fruits, fresh-cut fruits, and juice products. Applications of UV treatments demonstrated better quality preservation of fruit products that have a freshness of flavor, color, texture, and nutritional value closer to non-treated products. Additionally, UV light not only minimally affects quality attributes but also has beneficial effects on foods functional properties such as content of bioactive compounds and has a potential for obtaining premium quality products that can lead to the faster commercialization. Reports are available that application of UV light can also improve toxicological safety of foods of plant origin through its ability to reduce levels of toxins such as patulin mycotoxin in fresh apple cider (Dong et al. 2010), and possibly to control browning through its effects on enzymes (Manzocco et al. 2009). The schematic diagram of potential areas of applications of UV light technology in fruit processing is shown in Fig. 17.1.

This chapter aims to review the latest applications of continuous and pulsed UV light for processing fresh fruits and fruits products. The fundamental principles and features of UV light generation, propagation, and evaluation of UV light parameters will be briefly reviewed. Prevention control measures where UV light can be utilized to improve safety during fruit production will be analyzed. A particular focus will be given to the effects of UV light on survival of pathogenic and spoilage microorganisms typical for fruits and fruit plants environment and essential for the establishment of UV preservation processes followed by the discussion of recent research of effects of UV light on quality and enhancement of bioactive compounds. The effects of UV light on the destruction of mycotoxins will be presented.

17.2 UV Light Technology Fundamentals

17.2.1 Basic Principles

The wavelength range for UV light for food processing varies from 100 to 400 nm. This range may be further subdivided into UV-A (315–400 nm) normally responsible for tanning in human skin; UV-B (280–315 nm) that causes skin burning and can lead to skin cancer; UV-C (200–280 nm) called the germicidal range since it effectively inactivates bacteria and viruses. Vacuum UV range (100–200 nm) can be absorbed by almost all substances and thus can be transmitted only in a vacuum. Radiation from UV light and the adjacent visible spectral range as well as other less energetic types are termed nonionizing radiation. In contrast, ionizing radiation which includes X-rays, gamma-rays, and ionizing particles (beta-rays, alpha-rays, protons) is capable of ionizing many atoms and molecules. The absorption of nonionizing radiation, however, leads to electronic excitation of atoms and molecules. Light is emitted from the gas discharge at wavelengths dependent upon its elemental composition and the excitation, ionization, and kinetic energy of those elements. The gas discharges are responsible for the light emitted from UV lamps.

17.2.2 UV Light Sources

Light is emitted from the gas discharge at wavelengths dependent upon its elemental composition and the excitation, ionization, and kinetic energy of those elements. The gas discharges are responsible for the light emitted from UV lamps. UV light transfer phenomenon is defined by the emission characteristics of the UV source along considering long-term lamp aging and absorbance/scattering of the product. Consequently, performance of UV system depends on the correct matching of the UV source parameters to the demands of the UV application. The commercially available UV sources include low and medium pressure mercury lamps (LPM and MPM), excimer (EL), pulsed lamps (PL), and light emitting diodes (LED). The LPM and excimer lamps are monochromatic sources whereas emission of MPM and PL is polychromatic. There are no reports on the application of EL in fruit processing so this UV source won't be discussed in this chapter.

17.2.2.1 Mercury Lamps

The mercury vapor UV lamp sources have been successfully used in water treatment for nearly 50 years and well understood as reliable sources for other disinfection treatments that benefit from their performance, low cost, and quality. Typically three general types of mercury UV lamps are used: low-pressure (LPM); low-pressure high-output (LPHO); and medium-pressure (MPM). These terms are based on the vapor pressure of mercury when the lamps are operating. LPM lamps are operated at nominal total gas pressures of 10^2–10^3 Pa that corresponds to the vapor pressure of mercury at temperature of 40 °C. The emission spectrum of LPM is concentrated at the resonance lines at 253.7 nm (85 % of total intensity) and 185 nm. The wavelength of 253.7 nm is most efficient in terms of germicidal effect since photons are absorbed most by the DNA of microorganisms at this specific wavelength. Light with a wavelength below 230 nm is most effective for the dissociation of chemical compounds. The photons with the wavelength of 185 nm are responsible for ozone production and the combination of both wavelengths is a very effective means for photochemical air treatment. The US FDA regulations approved the use of a LPM lamps for juice processing and they have already been successfully commercialized (US FDA 2000a).

MPM lamps are operated at a total gas pressure of 10^4–10^6 Pa. Compared to the LPM lamps, the coolest possible temperature of the MPM is about 400 °C, whereas it goes up to 600 and even 800 °C in a stable operation. The emission spectrum of MPM covers wavelengths from about 250 nm to almost 600 nm, which results from a series of emissions in the UV and in the visible ranges. MPM lamps are not considered to be useful for targeted germicidal treatment. However, their strong UV radiation flux results in high penetration depth. By varying the gas filling, doping, and the quartz material, the spectrum as well as the radiation flux of the UV lamps can be varied and matched to suit specific food processing applications, especially for oxidation or photo degradation.

Recently, LPHO amalgam lamps that contain a mercury amalgam were developed and incorporated into disinfection applications; however, LPM and MPM are the dominant sources for UV disinfection treatment.

17.2.2.2 Pulsed Lamps

The efficacy of pulsed flash lamps (PL) is potentially greater than continuous sources due to high intensity, broader spectrum, instant start, and robust packaging with no mercury in the lamp. In this technology, alternating current is stored in a capacitor and energy is discharged through a high-speed switch to form a pulse of intense emission of light within about 100 ms. The emission is similar in wavelength composition to the solar light. The UV pulsed devices can deliver high intensity UV which can both penetrate opaque fluids better than mercury lamps and provide enhanced treatment rates. More research is needed to establish them for fruit treatments applications.

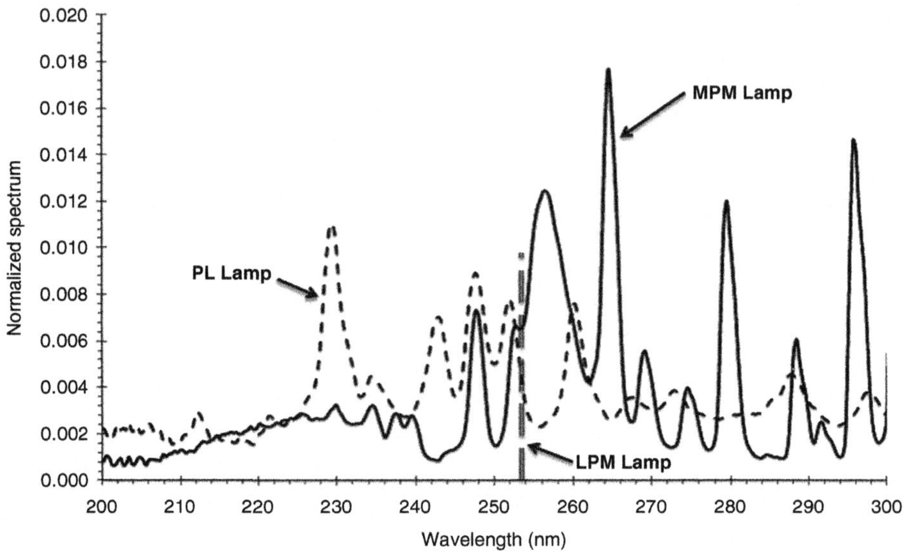

Fig. 17.2 Comparison of spectrums of continuous (LPM and MPM) lamps and PL UV sources

Figure 17.2 shows the normalized spectra of continuous UV (cUV) sources such as LPM, MPM, and PL. Individual spectra are not comparable on a UV intensity basis but are comparable on a spectral basis regarding which wavelengths dominate the respective wavelength outputs.

17.2.2.3 Light Emitting Diodes

In recent years, UV-LEDs have been developed with the following advantages: low cost, energy-efficient, long life, easy control of emission, and no production of mercury waste. The wavelength of the commercial UV-LED is in the range 240–400 nm and enables new applications in existing markets as well as in new research areas. A LED is a semiconductor device that emits light when carriers of different polarities (electron and holes) combine generating a photon. The wavelength of the photon depends on the energy difference the carriers overcome in order to combine. The example of UV-LED system that operates between 210 and 365 nm is the one formed by aluminum nitride (AlN), gallium nitride (GaN), and intermediate alloys. Currently, UV-LEDs are commercially available at research grade in limited quantities and their lifetime reach on the order of 200 h. It is very likely that in the near future, many applications that today make use of mercury lamps will be carried out by UV-LEDs.

Table 17.1 provides a summary of some of the basic characteristics of common UV sources in commercial use and under development and can be used for compari-

Table 17.1 Comparison of efficiency characteristics of continuous pulsed UV lamps and LEDs

UV source	Electrical efficiency, %	UV efficiency, %	UV intensity, W•cm−2	Lamp surface, T, °C	Lifetime, hours	Output spectrum
LPM	50	38	0.001–1	40	2000	Monochromatic 253.7 nm
MPM	15–30	12	12	400–1000	400	Polychromatic 200–400 nm
Flash Xenon	45–50	9	600	1000–10,000	800	Polychromatic 100–1000 nm
Surface Discharge	15–20	17	30,000	NA	NA	Polychromatic 200–800 nm
LED	1–4 %	NA	700	50–60	10,000	Monochromatic 200–400 nm Selectable

son purposes. It is evident that no single lamp technology will represent the best source for all food applications. However, situation-specific requirements may dictate a clear advantage for a given process technology. For UV reactors containing LPM or LPHO mercury lamps, UV absorbance and transmittance at 253.7 nm are important design parameters. However, for broadband UV lamps, such as MPM or PL, it is important to measure the full scan of absorbance or transmittance in the germicidal region from 200 to 400 nm. Special technologies lamps as PL UV, LEDs are promising due to different spectral bands or specific wavelength that they can provide considering effects on quality attributes. More research is needed to establish their suitability for fruit processing applications.

17.2.3 UV Light Propagation

UV light emitted from the atoms and ions within the gas discharge of a UV source will propagate away from those atoms and ions. As UV light propagates, it interacts with the materials it encounters through absorption, reflection, refraction, and scattering. Each of these phenomenon influences the intensity and wavelength of the UV light reaching the bacteria or chemical compound on the surface or in the liquid.

Absorption (*A*) of light is the transformation of energy of light photons to other forms of energy as it travels through a substance. *Reflection* (*R*) is the change in the direction of propagation experienced by light deflected by an interface. *Scattering* is the phenomenon that includes any process that deflects electromagnetic radiation from a straight path through an absorber when photons interact with a particle. The scattering phenomenon plays an important role in disinfecting food liquids containing particles. Experimental measurements are usually made in terms of *transmittance* of a substance (*T*) or (UVT), which is defined as the ratio of the transmitted to the incident light irradiance. A convenient way of presenting information about UVT of materials is to give the values of their absorption coefficient at various wavelengths, over a given depth (e.g., 1 cm). Knowing this, the transmittance for any particular depth and the depth of the liquid which will absorb 90 % of the energy at 253.7 nm can be calculated.

Photochemical reactions proceed as a direct result of radiation energy (photons) being introduced to a system. In view of the wavelengths used in most UV-light treatments, the molecules (*A*) are primarily affected by energy absorption that results in photochemical reactions. In the general case, the process may be viewed as

$$A + h\nu ® A_{+}^{+} ® \text{Products} \quad (17.1)$$

The first step in this reaction is the absorbance of a photon by a reactant molecule (*A*), leading to the production of an electronically excited intermediate. The excited state can be for period of 10^{-10} to 10^{-8} s in which the energy of the electrons is increased by the amount of photon energy. Under some conditions, the intermediate state may undergo a chemical change to yield products that are relatively stable. For a photochemical reaction to proceed, photons must have sufficient energy to promote reactions to break or form a bond and photon energy must be absorbed to promote reactions. The extent of chemical reaction depends upon the quantum yield and fluence of incident photons. A quantum yield is the ratio of absorbed photons that cause a chemical change to the total absorbed photons. UV light at 253.7 nm has a radiant energy of 472.27 kJ/Einstein or 112.8 kcal/Einstein (1 Einstein represents 1 mole of photons). It is theoretically possible for 253.7 nm light to affect the O–H, C–C, C–H, C–N, H–N, and S–S bonds if it's absorbed.

17.2.4 UV Fluence and Dose Definition and Determination

Fluence rate, fluence, and dose are other important terms to characterize UV light treatments in fruit processing. Fluence rate is the total radiant power incident from all directions onto an infinitesimally small sphere of cross-sectional area dA, divided by dA (Bolton and Linden 2003). Fluence is defined as the fluence rate multiplied by the exposure time. The term UV dose should be avoided as synonym of fluence because dose refers in other contexts to absorbed energy, but only a small fraction of all incident UV light is absorbed by microorganisms (Bolton and Linden 2003). In the case of PL, fluence is determined as energy per pulse multiplied by the number of pulses. The absorbed fluence indicates radiant energy is available for driving the solution reaction. However, when UV light is absorbed by the solution, it is no longer available for inactivating the microorganisms. The remaining interactions including reflection, refraction, and scattering change the direction of UV light but the light is still available for inactivation. The radiant energy delivered to the

Table 17.2 Controlling pulse parameters in timer IC-based pulse generator and microcontroller-based pulse generator

Symbol	Definition	Unit
α	Absorption coefficient of total sample	cm^{-1}
ε	Extinction coefficient	$L\bullet mol^{-1}\bullet cm^{-1}$
λ	Wavelength	m
τ	Residence time	s
Φ	Quantum yield	$mol\bullet Einstein^{-1}$
Ω	Solid angle	Sr
c	Concentration of an absorber	$mol\bullet L^{-1}$
d	Path length of light	cm
D_{eff}	Effective (delivered) UV dose	$mJ\bullet cm^{-3}$
H_{abs}	Absorbed UV fluence	$mJ\bullet cm^{-2}$
H_{app}	Applied UV fluence	$mJ\bullet cm^{-2}$
H_{trans}	Transmitted UV fluence	$mJ\bullet cm^{-2}$
I_0	Incident UV fluence rate	$mW\bullet cm^{-2}$
$I_{\lambda,\Omega}(x,t)$	Specific intensity for monochromatic radiation (λ) and for a particular direction (Ω)	$mW\bullet cm^{-2}\bullet sr^{-1}$
k_1	First order rate constant	s^{-1}
l	UV path length of sample	cm
L	Distance between UV source and sample surface	cm
N	Chemical concentration	$mol\bullet L^{-1}$
N_0	Initial chemical concentration (before UV exposure)	$mol\bullet L^{-1}$
$q_{n,p}$	Photon flux	$Einstein\bullet s^{-1}$
t	UV exposure time	s
U_λ	Energy per Einstein of photons	$mJ\bullet Einstein^{-1}$
V	Volume of sample	L

molecule or microorganism is called the effective or delivered germicidal UV dose. Microbial inactivation depends primarily on the effective dose.

UV fluence and consequently UV dose depends on the nature of media, the manner of radiation exposure, the target material to be irradiated, and the purpose of study. A general expression of UV fluence was given by Labas et al. (2006):

$$H = \frac{1}{V} \int_{V} \int_{\lambda_1}^{\lambda_2} \int_{©} I_{\lambda,©}(\underline{x},t) \cdot d© \cdot d\lambda \cdot dV \cdot \tau \tag{17.2}$$

where $I_{\lambda,\Omega}(x,t)$ is the specific intensity for monochromatic radiation (λ) and for a particular direction (Ω). V is reaction volume. τ is residence time. Table 17.2 summarizes nomenclature used in Sect. 17.2. In order to apply the equation for specific calculation, many other equations were derived for various UV reactor and wavelength.

Bolton and Linden (2003) established a standard method of UV fluence determination in bench-scale collimated beam UV experiments for microbial inactivation. For a LPM lamp the UV fluence is calculated by Eq. (17.3) considering corrections of petri factor (PF), reflection factor (RF), divergence factor (DF), and water factor (WF). As only free photons transmitted through the media can be used to inactive the microbes, this UV fluence is also called as transmitted UV fluence.

$$H_{\text{trans}} = I_0 \cdot (\text{PF}) \cdot (\text{RF}) \cdot (\text{DF}) \cdot (\text{WF}) \cdot t \tag{17.3}$$

where I_0 is radiometer reading at the center of the dish and t is exposure time. The unit of transmitted UV fluence is $\text{mJ} \cdot \text{cm}^{-2}$.

The PF is defined as the ratio of the average of the incident irradiance over the area of the Petri dish to the irradiance at the center of the dish. The RF represents the decrease of a small fraction of beam due to the reflection between two different media. For finite distances of the cell suspension from the UV lamp, the beam is not perfectly collimated and diverges significantly, so the DF should be considered (Eq. 17.3a).

$$\text{DF} = \frac{L}{L + l} \tag{17.3a}$$

where l is UV path length of sample, L is a distance between UV source and sample surface.

If the water or other tested liquid absorbs UV at the wavelength of interest, then it is necessary to account for the decrease in irradiance arising from absorption as the beam passes through the sample. The WF is defined as Eq. (17.3b).

$$\text{WF} = \frac{1 - 10^{-\alpha l}}{\alpha l \cdot \ln 10} \tag{17.3b}$$

where α is absorption coefficient of total sample at 253.7 nm.

Equation (17.3) provides a method to calculate UV dose but it must be limited to collimated LPM UV lamp and microbial inactivation application. Other UV fluence and dose calculations may apply under different conditions and for various purposes.

Applied UV fluence is generated by an applied incident UV intensity modified by petri factor on the surface of sample in a certain exposure time. For a collimated beam UV lamp, it can be calculated based on Eq. (17.4) with unit of $\text{mJ} \cdot \text{cm}^{-2}$.

$$H_{\text{app}} = I_0 \cdot (\text{PF}) \cdot t \tag{17.4}$$

Applied fluence reflects the energy emission from the UV source and it is independent to the material to be irradiated. Knowledge of the applied fluence is important to select a correct power and type of UV source by taking into the account their UV efficiency as shown in Table 17.1 in order to achieve a targeted degradation or inactivation of material.

Absorbed UV fluence is the energy absorbed by the media and may result in the photochemical reaction (Eq. 17.1). For a collimated beam UV lamp, it can be calculated based on Eq. (17.5) with unit of $mJ \cdot cm^{-2}$.

$$H_{abs} = I_0 \cdot (PF) \cdot (RF) \cdot (DF) \cdot \int_0^t \left(1 - 10^{-\alpha i}\right) \cdot dt \tag{17.5}$$

If the absorption coefficient is constant, Eq. (17.5) can be rewritten as:

$$H_{abs} = I_0 \cdot (PF) \cdot (RF) \cdot (DF) \cdot \left(1 - 10^{-\alpha l}\right) t \tag{17.5a}$$

Absorbed UV fluence can be used to measure the degradation of chemicals in the liquid media. Totally absorbed energy may destroy the target chemical when liquid media itself does not absorb UV radiation. However, absorbed fluence is not suitable to measure the inactivation of microorganisms because the UV light is no longer available for the inactivation when it is absorbed by media.

Effective or delivered UV dose is the energy delivered and absorbed by the targeted component in the sample and result in the photochemical reaction, which can be calculated through chemical actinometry using Eq. (17.6)

$$D_{eff} = \int_0^t \frac{-dN / dt \cdot U_\lambda}{!} dt \tag{17.6}$$

where Φ is quantum yield of chemical compound, N is concentration of chemical compound, U_λ is energy per Einstein of photons, and t is UV exposure time. The unit of effective dose is $mJ \cdot cm^{-3}$. If the degradation reaction compliance with the first order reaction, Eq. (17.6) can be rewritten as following Eq. (17.6a).

$$D_{eff} = \frac{N_0 \cdot U_\lambda \cdot \left(1 - e^{-k_1 t}\right)}{\Phi} \tag{17.6a}$$

where N_0 is initial concentration of chemical compound, k_1 is a first order reaction rate constant of photoreaction of chemical.

17.3 UV Light Based Control Measures in Fruits Processing Facilities

During manufacturing process, fruits can be exposed to microbiological cross contamination from the air, water, and surfaces. The traditional approach to controlling such contamination has been to target specific sites within the manufacturing environment with cleaning and disinfection regimes. UV light is an economical step towards improved hygiene control measures in the food industry. Sanitation, disinfection, and oxidation with UV light is a versatile, environmental-friendly technology, which can be used in the fruits processing and storage facilities to reduce microbial contamination and consequently to improve safety of fruits.

17.3.1 Air Treatment

Clean, fresh air is the basis in the industrial production of fruits. Microorganisms in the air, such as viruses, bacteria, yeasts, and fungi, can contaminate raw materials and intermediate products and spoil finished products during their processing and packaging. LPM sources are used very successfully in these applications, for disinfection in air intake ducting and store rooms and to ensure air of very low germ content in production areas. Short wave VUV radiation at 185 nm produces ozone from the oxygen in the ambient air so that this is activated for the oxidation process. UV oxidation breaks down pollutants in the exhaust air. For providing clean air in sensitive manufacturing food facilities, a combination of filters and UV light has been recommended. Basically two applications of UV are becoming common. In one, the moving air stream is disinfected in much the same manner as with a water system. In the other application, stationary components of the system such as air conditioning coils, drain pans, and filter surfaces are exposed to help prevent mold and bacteria growth or to disinfect the filter to aid in handling. The UVT in air is higher than in water and, therefore, the number of lamps required in a large duct is quite reasonable. Common airborne virus and bacteria are readily deactivated with UV. Fungi (molds and spores) require much higher doses. In the moving air stream, high wattage lamps are used, usually without a quartz sleeve. UV lamp fixtures are placed in such a manner as to completely irradiate surfaces where bacteria and mold might collect and grow. Mathematical modeling software and bioassay testing have been developed to allow efficient design and validation of these systems. Low operating costs and reasonable equipment costs can make UV very cost effective.

17.3.2 Water Treatment

Control of microorganisms in industrial process waters is often necessary to maintain the quality of the product or process. The fruit industry is a large volume consumer of water, and the potential for reuse or recycling of fruit processing water represents an attractive economic and sustainable benefit to the industry. A combination of UV light and ozone is a powerful oxidizing action to reduce microbial load and the organic content of water to very low levels.

17.3.3 Disinfection of Nonfood and Food Contact Surfaces

Mold and biofilms can develop on nonfood surfaces (ceilings, walls, floors) and equipment including tanks and vats, cooling coils, and food contact surfaces of equipment such as cutting equipment and conveyor belts (Kowalski 2006). In general, standard cleaning and disinfection procedures are adequate to contain these problems but alternatives are available, including antimicrobial coatings like copper and TiO_2. UV irradiation of food processing equipment and surfaces, cooling coils disinfection systems, whole area UV disinfection, and after-hours irradiation of rooms when personnel are not present are all viable control options for maintaining high levels of sanitation and disinfection in fruit processing facilities (Kowalski and Dunn 2002). UV light kills up to 99.9 % of total germs on conveyor belts used for transporting fruits and vegetables.

17.3.4 Packaging

The packaging technologies play important role in extending the shelf life of fruits. UV light might be applied as pre- or post-packaging technology to reduce the microbial spoilage. As a pre-packaging control measure UV treatment of packaging in fruit filling plant, e.g., for lids, cups, sealing and packaging foils for drinks and beverages help to extend fruits shelf life. When using cUV and PL as post-packaging treatment for packaged fruits, the considerations about transparency are referred to the packaging materials. For example, materials such as glass, polystyrene, and PET, which allow visible light to penetrate through the container, are not transparent to the UV wavelengths that are essential for microbial inactivation and therefore they are not suitable for cUV and PL treatments. On the other hand, polymers such as polyethylene, polypropylene, polybutylene, EVA, nylon, Aclar, and EVOH transmit UV light and hence meet the requirements for PLT very well (Anonymous 2000). In addition, ink printed labels or drawings could interfere with the light absorption of the treated item and should be avoided on the surface of packaging materials. Besides the intrinsic transparency of the material, it is critical that the "condition" of the item to be treated is suitable for the penetration of the light. This means that the product surface should be smooth, clear and without roughness, pores and grooves which could "shadow" the microbial cells from the light, causing less complete light diffusion and thus reducing process effectiveness; for the same reason, the item to be treated should be clean and free of contaminating particulates. In addition, items having a complex geometry could have areas hidden from the light and could require a more accurate design of the treatment chamber in order for the light pulses to reach each point of the product surface.

17.4 UV Treatment of Whole Fresh Fruits to Enhance Functionality and Safety

17.4.1 Functional Foods and UV Hormesis

In the recent years, there has been an increasing interest by the consumers in functional food products that may help to maintain optimal health condition, performance, and well-being. Functional foods can be defined as foods that are clinically proven to provide health benefits and/or reduce the risk of chronic diseases beyond their basic nutritional value due to presence of physiologically bioactive compounds. Functional foods include natural foods (fruits, vegetables) and processed foods that have been enriched or fortified with nutrients, phytochemicals, or botanicals. The nutraceutical potential of plant foods can be also naturally enhanced through special growing conditions or postharvest exposure to abiotic stresses, such as UV light (Shama and Alderson 2005; Shama 2007). The latter treatment is known as "hormesis." According to Shama (2007) "hormesis" involves the use of small levels of potentially harmful stressors directed against a living organism or living tissue in order to induce a beneficial or protective response. Recent studies on a variety of different fruits, such as berries (Baka et al. 1999; Allende et al. 2007; Pombo et al. 2011), apples (Ubi et al. 2006; Hagen et al. 2007), tropical fruits (Gonzalez-Aguilar et al. 2010; Srilaong et al. 2011), and mushrooms (Mau et al. 1998; Jasinghe and Perera 2006) proved that UV light can be successfully applied as a hormetic agent. In addition to enhanced levels of bioactive compounds, prolonged storability, delayed senescence, and microbial deterioration were observed in UV treated fruits.

17.4.2 UV Effects on Fruits Functionality

Fruits hormetic response is a sophisticated process, not fully understood yet. It has been shown that UV light stimulates cellular protective mechanisms that include changes in the metabolic activity with the activation of particular genes and enzymes. This includes: (1) the enzymes peroxidase and reductase that are responsible for the oxidative burst and formation of lignin polymers generating structural barriers against invading pathogens; (2) glucanases and chitinases that exhibit lytic activities towards major fungal cell wall components; and (3) l-phenylalanine ammonia lyase (PAL)—involved in biosynthesis of phenolics which are characterized by strong UV absorptive properties (Gonzalez-Aguilar et al. 2010). The exemplary UV absorbing plant phytochemicals, i.e., chlorogenic acid, gallic acid, epicatechin, and quercetin, are presented in Fig. 17.3.

Through the synthesis of phenolic compounds, plants primarily protect the DNA and also activate their antioxidant and antimicrobial defense system (El Ghaouth et al. 2003; Erkan et al. 2008; Interdonato et al. 2011; Pombo et al. 2011;

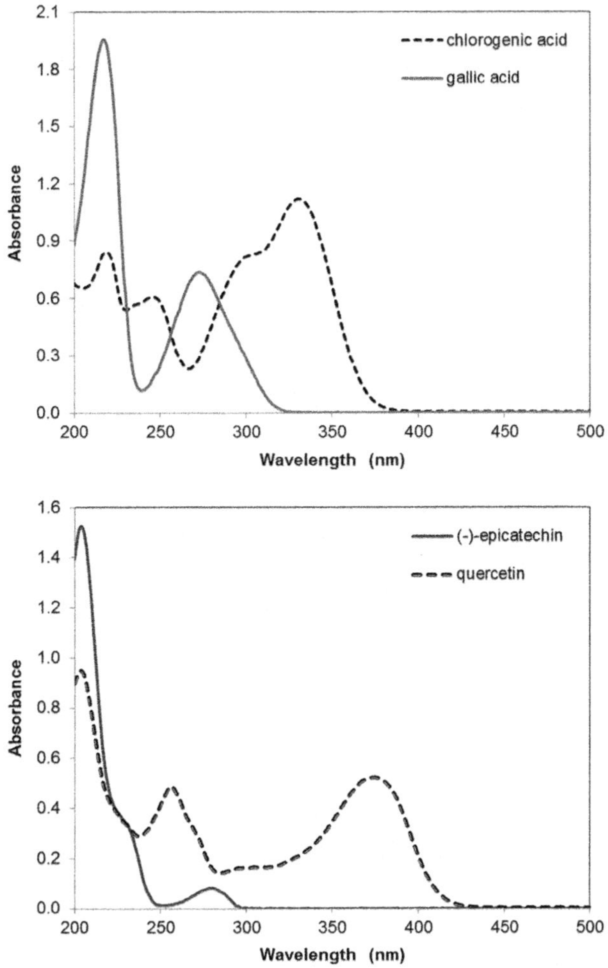

Fig. 17.3 Absorption spectra of 0.005 M chlorogenic and gallic acids, 0.001 M (−)-epicatechin and quercetin, measured in quartz demountable cuvettes with path lengths of 0.1 and 0.2 mm, respectively (Agriculture and Agri-Food Canada, unpublished data)

Zhang et al. 2012). Bioactive compounds are formed mainly in the peel of treated fruits (Hagen et al. 2007). However, Bakhshi and Arakawa (2006) reported that fruit flesh has also the ability to accumulate phytochemicals. In post UV-B/visible treated apples, authors observed increased levels of phenolic acids, anthocyanin, and flavonols. Flavanols, procyanidins, and dihydrochalcones were not affected by the applied treatment.

Accumulation of antioxidants within plant tissues enhances nutritional quality of UV treated commodities. Phenolics, stilbenes, vitamins C and D, carotenoids, anthocyanins, and polyamines are essential ingredients in human diet due to health promoting activities, such as anticancer, anti-inflammatory, and antihistaminic. Table 17.3 summarizes data on the UV effects on functional fruit properties. In general, under optimal treatment conditions an increase in the levels of physiologically active compounds was observed.

Table 17.3 Microcontroller specifications

Commodity	UV treatment, L/#/P/F	Enhanced nutraceuticals (relative change, %)	Health benefits	References
Grapes	UV-C/3/NA/3.6 kJ•m^{-2}	Trans-resveratrol (980-2500)	Enhance longevity, cardioprotective, neuroprotective, anti-cancerogenic	Li et al. (2008), Guerrero et al. (2010)
Pears	Vis/2 + UV-B/3/36 W + 20 W/PFD = 4.56 μmol•m^{-2}s	Anthocyanins: 12.5 mg/100 g after 240 h of irradiation at 27 °C; non detectable in control fruits	*Anthocyanins*—protect liver; reduce blood pressure; improve eyesight; anti-inflammatory and antimicrobial activities; *Vit. C and polyphenols*—antioxidants; prevent age-related diseases, such as heart disease, immune system decline, and brain dysfunction; anti-inflammatory, antihistaminic, and antitumor activities	Zhang et al. (2012)
Apples	Vis/1 + UV-B/2//400 W + 20 W/0.20 W•m^{-2}	Anthocyanins (56) Quercetin glycosides (12-15) Chlorogenic acid (142) Ascorbic acid (6.5)		Hagen et al. (2007), Konczak and Zhang (2004)
Blueberries	UV-C/15/8 W/4.30 kJ•m^{-2}	Anthocyanins (54) Quercetin glycosides (30-85) Chlorogenic acid (11) Resveratrol (33.5)		Wang et al. (2009)
Strawberries	UV-C/3/8 W/2.15 kJ•m^{-2}	Antioxidant capacity (18.5) Total phenolic content (30)		Erkan et al. (2008)
Pepper fruits	UV-C/4/30 W/7 kJ•m^{-2}	Antioxidant capacity (10.5)		Vicente et al. (2005)
Mature green-tomatoes fruits	UV-B/2/NA/40 kJ•m^{-2}	Total phenolic content (7) Total flavonoid content (12)		Liu et al. (2011a)
		Lycopene (11)	Antioxidant; prevents cardiovascular disease and cancers (prostate and gastrointestinal tract)	
	UV-C 3.47 kJ•m^{-2}	Putrescine, agmatine, tyramine	Polyamines acts as anti-inflammatory agents, prevent cardiovascular and age associated diseases, have radical scavenging properties	Maharaj (1995)
Peaches	UV-C/NA/15 W/2.47 kJ•m^{-2}	Putrescine (35) Spermidine (44) Spermine (40)		Gonzalez-Aguilar et al. (2004), Soda (2011)
Mangoes	UV-C/NA/15 W/4.93 kJ•m^{-2}	Putrescine (160) Spermine (16.5)		González-Aguilar et al. (2001, 2007)

Bitter orange	UV-C/1/NA/0.72 kJ•m^{-2}	Naringin (7) Tangeretin (55)	Flavonoids have antioxidant, anticancer, and blood lipid lowering activities	Arcas et al. (2000)
Oranges	UV-C/4/3.8 W/3.0 kJ•m^{-2}	*Scoparone and scopoletin*—levels of both compounds was not detectable in non-UV treated fruits	Phytoalexins posses antioxidant activity, anti-inflammation activity, and cholesterol-lowering ability	Rodov et al. (1992), D'hallewin et al. (1999), Boue et al. (2009)
Kumquat				
Grapefruits	UV-C/4/3.8 W/0.5 kJ•m^{-2}			D'hallewin et al. (2000)
Common mushrooms	UV-C/1/NA/6.06 kJ•m^{-2}	Vitamin D_2 (173)	Plays crucial role in bone health; aids in the functioning of the pancreas, fetal development, neural function, and immunity; anticancerogenic; cardioprotective	Mau et al. (1998)

L—band of UV light, #—number of UV sources, P—power of UV source, F—UV fluence; relative change = ((S − C)/C) × 100 %, S—UV treated fruit, C—control without UV exposure; PFD—photon flux density

17.4.3 Factors Affecting the Formation of Nutraceuticals

The overall effect of postharvest UV irradiation on the bioactive compounds depend on growing conditions, crop commodity and cultivar, temperature at which UV treatment is performed, applied UV bandwidth, and dose. Knowledge of these parameters allows optimizing the process in order to yield satisfactory nutritional, quality, and safety levels.

Growing conditions. The variable levels of sun light exposition during fruit growth can result in different postharvest fruits characteristics. Hagen et al. (2007) reported that apples grown in the shady side of the tree were characterized by ~40–50 % lower initial content of phytochemicals in comparison to those grown in sun-exposed canopies. The postharvest UV treatments of apples grown in shade resulted in higher yields of bioactive compounds than in apples grown in sun. Plant functional properties can be also modified by special growing conditions. Tsormpatsidis et al. (2011) cultivated "Elsanta" strawberry plants under UV opaque (blocked UV radiation up to 380 nm) and UV transparent film. UV radiation increased the rate of color development and resulted in higher levels of anthocyanins (14–31 %), flavonoids (9–21 %), and phenolics (9–20 %) content at strawberry harvesting. Moreover fruits ripened under UV transparent film were firmer, smaller but greater in number than fruit ripened under a UV opaque film. Authors also observed an increase in flavonoid (16 %) and phenolic (8 %) concentrations in plant leaves exposed to UV radiation.

Crop commodity and cultivar. In general, UV exposure results in enhanced antioxidant properties of treated fruits. However, different compounds that are characteristic for a given fruit will contribute to the antioxidant capacity of irradiated commodity. For example, resveratrol is characteristic to grapes and its remarkable accumulation was reported after UV-C exposure (Li et al. 2008). Citrus fruits are rich sources of flavonoids (naringin, tangeretin) and increased levels of those compounds were observed by Arcas et al. (2000) in UV-C treated bitter oranges. It is also necessary to mention that within a given specie the response to UV treatment may differ amongst cultivars. Ubi et al. (2006) noted different levels of anthocyanins induced by UV-B treatment at 17 °C in several tested apple cultivars. The highest levels of nutraceuticals were found in Tsugaru, whereas the lowest in Sansa apple cultivar (Tsugaru > Akane > Iwai > Sansa).

Temperature. Several studies were performed on the photo-stimulation of anthocyanins production in the fruits exposed to UV-B/visible light treatment at different temperature conditions. In the case of several apples cultivars (Iwai, Sansa, Tsugaru, and Akane) Ubi et al. (2006) found the treatment at 17 °C more effective, in comparison to that performed at 27 °C. On the contrary, Arakawa (1991) and Reay and Lancaster (2001) observed a higher yield of anthocyanins in "Jonathan," "Gala," and "Royal Gala" apples irradiated at higher temperatures (20–25 °C) than at lower temperature (10–15 °C) conditions. Similarly Zhang et al. (2012) reported UV-B/visible irradiation of Red Chinese sand pears to be more effective at 27 °C than at 17 °C. Postharvest exposure to UV-B/visible light at −0.5/−0.5 °C (day/night), 20/20, and 20/6 °C resulted in higher levels of anthocyanins in apples but not in

European pears (Marais et al. 2001). Therefore, the choice of the optimal temperature conditions for postharvest UV treatment has to be experimentally defined for a given commodity and cultivar.

UV bandwidth. Effects of different UV bands, i.e., UV-C (200–280 nm), UV-B (280–315 nm), and UV-A (315–400 nm), alone and in combination with visible light on accumulation of physiologically active compounds in treated fruits were studied. Beneficial effects on the plant functional properties were observed in the case of combined UV-B–visible light treatments for apples (Arakawa 1991; Ubi et al. 2006; Hagen et al. 2007) and pears (Zhang et al. 2012). Ubi et al. (2006) and Hagen et al. (2007) found UV-B/visible light treatment to be more effective in the accumulation of apple phytochemicals in comparison to the application of UV-B (Ubi et al. 2006) or visible light (Hagen et al. 2007) treatment alone. Mau et al. (1998) studied the effects of UV-B and UV-C on the transformation of ergosterol to vitamin D_2 in common (*Agaricus bisporus*) mushrooms. Both tested treatments yielded in vitamin D_2 formation, however UV-B light was found to be more effective. The UV-B exposure (4.93 kJ•m^{-2}) resulted in the increase of vitamin D_2 by 387 %, whereas UV-C (6.06 kJ•m^{-2}) by 173 %. In another studies, Jasinghe and Perera (2006) compared the effects of UV-C (23.0 kJ•m^{-2}) with the UV-A (25.2 kJ•m^{-2}) on the formation of vitamin D_2 in edible mushrooms. The UV-C exposure resulted in higher levels of vitamin D_2 in all tested mushrooms: Shiitake, Oyster, Abalone, and Button. UV-C light was also successfully applied to a variety of other fruits. As a result of UV-C irradiation, an increase and/or better maintenance of the phenolic compounds during storage was observed in the case of mangoes (González-Aguilar et al. 2001, 2007), blueberries (Wang et al. 2009), pepper fruits (Vicente et al. 2005), and green tomatoes fruits (Liu et al. 2011a).

UV dose and optimal treatment conditions. González-Aguilar et al. (2001) observed the highest accumulation of phytochemicals in mangoes exposed to 4.93 kJ•m^{-2} whereas treatments at 2.46 or 9.86 kJ•m^{-2} resulted in lower yield of phenols and polyamine compounds. Lammertyn et al. (2004) and Allende et al. (2007) recommended 1.0 kJ•m^{-2} as optimal fluence for the UV-C processing of strawberries since at higher treatments browning and dehydration of the sepals occurred. Moreover, overdosing can result in accelerated ripening and senescence processes as well as lower resistance to microbial and/or fungal decay, leading to reduced fruit storability and economical losses (Nigro et al. 1998). Therefore in order to obtain the most satisfactory levels of nutraceuticals without affecting adversely appearance and shelf life of a given fruit commodity, the optimal UV treatment conditions must be applied.

17.4.4 Synergistic Antimicrobial Effects of UV Light and Hormetic Plant Response

The germicidal effects of UV light against naturally occurring pathogenic and non-pathogenic microflora on the surface of fresh produce can be synergistically enhanced by the hormetic response of irradiated fruits. For instance Li et al. (2010)

reported a higher inhibition of *Monilinia fructicola* growth in the pears inoculated with the pathogen before the UV-C treatment than in those being inoculated after UV-C exposure. Similarly Pombo et al. (2011) observed a reduction in growth of *Botrytis cinerea* inoculated on the strawberries 8 h after UV-C treatment (4.1 kJ•m^{-2}). In other studies Obande et al. (2011) studied the shelf life of tomatoes that were first exposed to UV-C light at 8 kJ•m^{-2} and then were inoculated with *Penicillum digitatum*. After 10 days of storage at 20 °C, the UV treated fruits were firmer and the diameter of fungal lesions was considerably smaller in comparison to controls. Therefore higher resistance to postharvest diseases of UV treated commodities can be partially attributed to the physiological changes stimulated by UV light. These include accumulation of phytochemicals, known to have antimicrobial and antifungal activities, and increased activities of lignifying enzymes that strengthen structural barriers against invading pathogens. Enhanced levels of phytoalexins (scoparone) and flavonoids (naringin, tangeretin) were associated with reduced fungal decay caused by *P. digitatum* in UV treated lemons (Ben-Yehosua et al. 1992), grapefruits (Lers et al. 1998), and oranges (Arcas et al. 2000). Lower susceptibility to grey mold rot (*B. cinerea*) was attributed to accumulation of rishitin in tomatoes (Charles et al. 2008) and resveratrol in grapes (Nigro et al. 1998) exposed to UV-C fluences of 3.7 kJ•m^{-2} and 0.5 kJ•m^{-2}, respectively.

Besides the molds, pathogenic bacteria can be present on the surface of fresh produce, such as *Salmonella* spp., O157:H7 and non-O157 shiga toxin producing *Escherichia coli* that constitute a threat to human health and safety. It was presented by several authors that either UV-C or pulsed light (PL) treatments have the ability to reduce the population of these pathogens. For instance Yaun et al. (2004) reported a reduction of *E. coli* O157:H7 by approximately 3.3 log on apples exposed to UV-C light at 240 W•m^{-2}. The same UV irradiation conditions resulted in slightly lower log reduction of *Salmonella* spp. on tomatoes (2.19 log). Pulsed light (Xenon Corp.) with the emission spectrum in the UV/Visible range (100–1100 nm) was applied for 5, 10, 30, 45, and 60 s to raspberries inoculated with *E. coli* O157:H7 and *Salmonella* spp. Bialka et al. (2008) reported reductions between 0.7 and 3.0 $\log_{10}$ CFU/g of *E. coli* O157:H7 and 1.2 and 3.4 $\log_{10}$ CFU/g of *Salmonella* on treated berries. However, fruit processing with PL light was accompanied by temperature increase and therefore microbial reduction might result from the combined light-heat effects.

These examples demonstrated that the postharvest UV processing of variety of fresh produce can be effective against both pathogenic and nonpathogenic microflora. More cases of successful UV applications are presented in Table 17.4.

Fresh produce have tender skin that can be easily injured during harvesting and handling stages. The positive effects of UV treatments were also observed in the case of damaged fruits, which are normally characterized by higher susceptibility to the microbial decay. For instance, delayed decay development after UV-C treatments of artificially wounded pears and grapes was observed by Li et al. (2010) and Nigro et al. (1998), respectively.

Table 17.4 Effects of UV treatments on the pathogenic and nonpathogenic microflora present on the surface of fresh commodities

Commodity	UV treatment L/#/P/F	Germicidal effects	References
Apples	UV-C/1/30 W/7.5 kJ•m^{-2}	Enhanced resistance against alternaria rot, brown rot (*Monilinia* spp.), bacterial soft rot (*Erwinia* spp.)	Lu et al. (1991)
	UV-C/1/NA/240 kJ•m^{-2}	3.3 $\log_{10}$ reduction of *E. coli* O157:H7	Yaun et al. (2004)
Blueberries	Pulsed UV/Vis light/60 s (22.6 J•cm^{-2})	4.3 $\log_{10}$ reduction of *E. coli* O157:H7; 2.9 $\log_{10}$ reduction of *Salmonella* spp.	Bialka and Demirci (2007)
Mango fruits	UV-C/NA/15 W/4.93 kJ•m^{-2}	Reduced fungal decay by 60 % after storage for 18 days at 25 °C	González-Aguilar et al. (2001, 2007)
Oranges	UV-C/4/3.8 W/3.0 kJ•m^{-2}	Reduced green mold (*Penicillum digitatum*) decay	Rodov et al. (1992)
Peaches	UV-C/1/30 W/20 kJ•m^{-2}	Reduced brown rot (*Monilinia fructicola*) decay	Lu et al. (1991)
	UV-C/1/NA/4.8 kJ•m^{-2}		Stevens et al. (1998)
Pepper fruits	UV-C/4/30 W/7 kJ•m^{-2}	Reduced grey mold (*Botrytis cinerea*) decay	Vicente et al. (2005)
Raspberries	Pulsed UV/Vis light/60 s (59.4 J•cm^{-2})	3.0 $\log_{10}$ reduction of *E. coli* O157:H7; 3.4 $\log_{10}$ reduction of *Salmonella* spp.	Bialka et al. (2008)
Strawberries	Pulsed UV/Vis light/60 s (59.4 J•cm^{-2})	2.3 $\log_{10}$ reduction of *E. coli* O157:H7; 3.9 $\log_{10}$ reduction of *Salmonella* spp.	Bialka et al. (2008)
	UV-C/3/8 W/2.15 and 4.30 kJ•m^{-2}	Reduced grey mold (*Botrytis cinerea*) by 60 % and 62 %, after 20 days of storage at 10 °C	Erkan et al. (2008)
Tangerines	UV-C/1/NA/1.3 kJ•m^{-2}	Increased resistance against green mold (*Penicillum digitatum*)	Stevens et al. (2005)
Tomatoes	UV-C/NA/30 W/3.7 kJ•m^{-2}	Enhanced resistance against *B. cinerea*	Charles et al. (2008)
	UV-C/1/NA/240 W•m^{-2}	2.19 $\log_{10}$ reduction of *Salmonella* spp.	Yaun et al. (2004)

17.4.5 UV Effects on Shelf Life

Fruits are highly perishable and after harvesting require appropriate handling that will delay their ripening and senescence during storage. The major symptoms of deterioration are quality loss, discoloration, tissue softening, weight loss, increased respiration rate, and ethylene production. Traditionally, through the manipulation of storage conditions, i.e., temperature and atmosphere, attempts were made to prolong the storability of fresh produce. However, these two factors must be optimized to avoid adverse effects. For example, very low temperatures can induce chilling injury in stored commodities. Application of hormetic UV doses can stimulate the expression of defense response genes, and decrease the expression of genes involved in wall degradation, lipid metabolism, and photosynthesis (Pombo et al. 2009; Liu et al. 2011b). These physiological and biochemical changes induced by UV treatments can help to maintain the overall quality and prolong the storability of harvested fresh produce. Better maintenance of nutritional and sensory qualities, delayed ripening, softening and electrolyte leakage, retarded chlorophyll degradation, higher resistance to chilling injury, reduced respiration rate, and weight loss were reported in the case of the variety of UV treated commodities, such as apples (Lu et al. 1991; Hagen et al. 2007), strawberries (Baka et al. 1999; Marquenie et al. 2002; Lammertyn et al. 2004; Allende et al. 2007), peaches (Lu et al. 1991; Gonzalez-Aguilar et al. 2004), limes (Kaewsuksaeng et al. 2011), bananas (Pongprasert et al. 2011), tomatoes (Barka et al. 2000), peppers (Vicente et al. 2005), and broccoli (Costa et al. 2006; Lemoine et al. 2007). Table 17.5 provides several examples of UV effects on the parameters attributed to the shelf life of irradiated fruits.

17.4.6 Factors Affecting the Delivery of UV Dose

Satisfactory microbial reduction can be achieved when the correct UV dose is delivered to the fruit surface. However, delivery of the UV dose to the fruit can be affected by the skin topography and applied procedure and so it needs to be carefully controlled.

Many varieties of fruits are characterized by rough surface and porous veins that allows the bacteria to attach tightly. Moreover, bacteria or pathogens of interest may become incorporated into biofilms with naturally existing microflora (Ukuku et al. 2001). As a consequence, bacteria can be shielded from the UV light and lower microbial reduction might be achieved.

In order to induce the host postharvest resistance to decay and reduce the microbial population, experimental procedures were developed allowing exposure of the entire fruit surface to UV light. This was achieved by the manual rotation of the treated commodities for two or four times during UV treatment (Stevens et al. 2005; Yang et al. 2009). However, as noticed by Stevens et al. (2005) such practices are rather impractical and can seriously affect the commercialization of the postharvest

Table 17.5 UV effects on the parameters attributed to storability of treated commodities

Commodity	UV treatment L/#/P/F	UV effects on storability	References
Limes	UV-B/1/NA/19 kJ•m^{-2}	Retarded chlorophyll degradation; better maintenance of internal fruit quality and antioxidants (ascorbic acid)	Kaewsuksaeng et al. (2011)
Bananas	UV-C/1/8 W/0.03 kJ•m^{-2}	Inhibited PPO activity; delayed yellowing and chlorophyll degradation; reduction of ethylene production, respiration rate, and chilling injury symptoms	Pongprasert et al. (2011)
Mangoes	UV-C/NA/15 W/4.93 kJ•m^{-2}	Maintained better visual appearance and fruit firmness; retarded weight loss; suppressed decay symptoms; developed resistance to chilling injury	González-Aguilar et al. (2001)
Peaches	UV-C/1/NA/20 kJ•m^{-2}	Delayed fruit maturation; increased flesh firmness and acidity; lower pH and soluble solids content	Lu et al. (1991)
Pears	UV-C/2/NA/5 kJ•m^{-2}	Better maintenance of fruit quality and ascorbic acid content, retarded senescence	Li et al. (2010)
Strawberries	UV-C/6/NA/0.25 kJ•m^{-2}	Lower respiration rate; higher titratable acidity and fruit firmness; slower rate of senescence	Baka et al. (1999)
Mature-green tomatoes	UV-C/NA/NA/3.47 kJ•m^{-2}	Retarded tissue softening and color development; delayed climacteric response by 7 days; reduced respiration rate and ethylene production	Maharaj et al. (1999)

UV treatments of fresh produce. Authors verified if fruit rotating can have a major impact on the reduction of bitter rot (*Colletotrichum gloeosporioides*), brown rot (*M. fructicola*), and green mold (*P. digitatum*) in apples, peaches, and tangerines, respectively. Exposure to UV-C light in the stationary position of the stem ends of apples (7.5 kJ•m^{-2}), peaches (7.5 kJ•m^{-2}), and tangerines (1.3 kJ•m^{-2}) resulted in comparable or slightly better resistance to mold decay than when fruits were rotated four different times. The lowest resistance to the spoilage decay was induced when only one or two different sides of fruits were exposed to the UV light. The difference in fruit response to the applied treatment procedures were attributed by Stevens et al. (2005) to the sites of UV-C photoreception and possible transmission mechanisms of the transduction signal within the phloem vascular tissue of fruits. Recently Obande and Shama (2011) applied the biodosimetry in order to measure the UV-C dose delivered to a polystyrene sphere that could mimic the shape of fruits such as apples, peaches, and tomatoes. The spheres were inoculated with spores of *Bacillus subtilis* and exposed to UV-C light with applied static and rotary procedures. Authors reported that under UV irradiation conditions at the theoretical dose of 10.6 J, spore biodosimetry yielded 9.1 ± 0.9 J for a single exposure to UV-C for 80 s, 10.7 ± 1.0 J in case of two rotations by 180° (2 × 40 s), and 6.1 ± 0.6 J for a sphere rotated 4 times by 90° (4 × 20 s). The lowest UV dose, i.e., 3.5 J, was obtained in the case of continuously rotated sphere for 80 s. From the comparison of the results obtained by Stevens et al. (2005) and Obande and Shama (2011) it comes a small contradiction. The highest UV dose for the polystyrene sphere was obtained with the rotation for two times. Application of the same procedure in the case of fruits yielded in the lowest decay inhibition. Certainly, correct determination of the UV dose delivered to the fruits is very important for the future commercialization. However, more work has to be done in order to find the correlation between applied UV dose, its distribution over the fruit surface, and physiological mechanisms induced by the UV hormetic processing.

17.5 UV Preservation of Fruit Products

17.5.1 UV Pasteurization of Fruit Juices

Fresh fruit juices are popular beverages in the world market. They are perceived as wholesome, nutritious, all day beverages. For items such as juices or juice beverages, minimal processing techniques are expected to be used to retain fresh physical, chemical, and nutritional characteristics with extended refrigerated shelf life. The US FDA approval of UV-light as an alternative treatment to thermal pasteurization of fresh juice products (US FDA 2000b) led to the growing interest and research in UV technology. Key factors that influence the efficacy of UV treatment of fruit juices include optical properties, design of UV processing systems, and UV resistance of pathogenic and spoilage organisms. Chemical composition, pH, dissolved solids (°Brix), and water activity have to be considered as hurdles that can modify the efficacy of UV microbial inactivation. There are a number of studies recently published that examined the UV light not only as a potential means of alternative pasteurization by studying effects on microflora but also on enzymes, flavor, color, and nutrient content of fresh juices and nectars (Koutchma 2009).

17.5.1.1 UV Absorption of Fruit Juices

Fruit juices are characterized by a diverse range of chemical, physical, and optical properties. Optical properties (absorbance and scattering) are the major factors impacting UV light transmission and consequently microbial inactivation. UV absorbance and transmittance at 253.7 nm are important parameters to design UV preservation process using LPM or LPHO source. In the case of the broadband continuous UV and pulsed lamps it is important to measure the spectra of the absorbance or transmittance in the UV germicidal region from 200 to 400 nm. In terms of UV transmittance, fruit juices can be characterized as transparent fluids if $10\ \% < \text{UVT} < 100\ \%$, opaque fluids if UVT ~0 %, and semitransparent fluids if $0 < \text{UVT} < 10\ \%$ for anything in between. In a majority of cases, juices will absorb UV radiation. For example, clear or clarified juices (apple, grape, or cranberry juices) can be considered as a case of semitransparent fluids. Juices with suspended solids or particles (apple cider, orange juice) are opaque fluids. Chemical composition such as vitamins content and concentration of dissolved and suspended solids determines the level of juices UVT.

The Beer–Lambert law (Eq. 17.1) is used to describe absorption behavior of fluids. In the case of Lambertian fluids, the relationship between absorbance (A) and concentration of an absorber of UV radiation (c, mol•L^{-1}), extinction coefficient (ε, L•mol^{-1}•cm^{-1}) or molar absorptivity of the absorbing species, and path length of light (d, cm) is linear.

$$A = \varepsilon \times c \times d \tag{17.1}$$

In the case of fruit juices with suspended solids, the function of $A = F(\varepsilon, c, d)$ can be nonlinear, which is typical for non-Lambertian fluids. Examples of the optical characteristics of some clarified fruit juices and opaque juices with particles are shown in Fig. 17.4a, b. Integrated sphere attachment to spectrophotometer and micro-cuvettes was used to measure total transmittance of juice samples due to their low UVT. Total transmittance measurement included both absorptive and scattering properties that contribute to how UV photons travel in juice matrixes.

As it can be noted in Fig. 17.4a, b clear juices including apple, cranberry, and white grape, and juices with particles such as apple cider and coconut water, followed linear behavior as Lambertian fluids, which is typical behavior for category of semitransparent juices. The majority of fruit juices with suspended particles did not follow the Beer–Lambert law. More research has to be done to separate absorptive

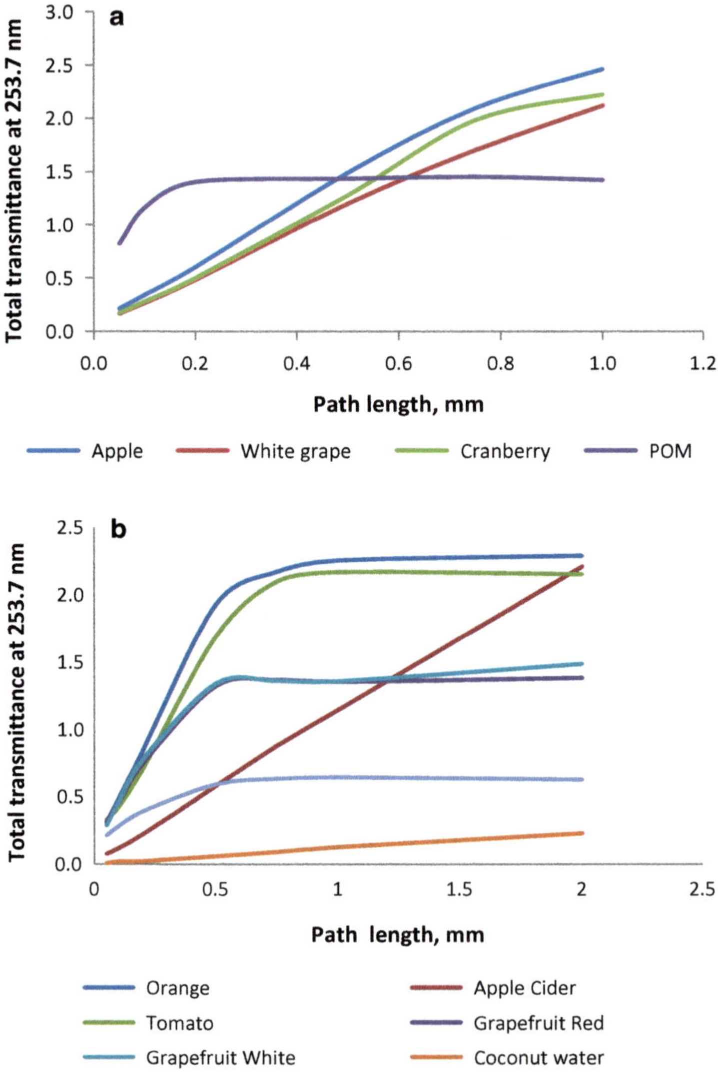

Fig. 17.4 (**a**) Total transmittance of clear fruit juices measured using an integrated sphere. (**b**) Total transmittance of fruit juices with suspended solids measured using an integrated sphere

and light scattering behavior of juices and understand their contribution to microbial inactivation. Knowledge of total absorption coefficients is necessary to calculate absorbed fluence of juices using Eqs. (17.2) and (17.5) from Sect. 17.2. The absorption coefficients of a few brands of freshly squeezed and commercial juices that are Lambertian liquids are summarized in Table 17.6.

Table 17.6 Absorption and UV transmittance of Lambertian fresh juices at 253.7 nm

		UV transmittance, %	
Juice	Absorption coefficient (cm^{-1})	0.1 (cm)	1 (cm)
Apple	26.4	0.2	0.00
Cranberry	22	0.6	0.00
White grape	22.1	0.6	0.00
Apple cider	11.2	7.6	0.00
Coconut water	1.15	76.7	7.08
Coconut liquid	5.2	30.2	0.00

Coconut water and coconut liquid were transparent at 0.1 cm liquid and semitransparent at 1 cm. Apple cider was a semitransparent fluid in 0.1 cm and opaque at 1 cm. All other clear juices were opaque at both path lengths. The absorption coefficient of fresh non-treated apple cider that contained suspended particles was approximately of 12 cm^{-1} which is lower than other fruit juices with particles as well as clarified brands. The higher absorbance of the clarified commercial brands can be probably due to contribution of added preservatives and vitamin C. From this prospective, the UV treatment of freshly pressed fruit juices looks more favorable.

17.5.1.2 UV Processing Systems for Juices

A number of continuous flow UV systems were developed and validated for a variety of fruit juices or other fruit beverages ranging from exotic tropical juices and nectars, to the more common apple cider and apple juice. The reactor designs include traditional annular, thin film, static and dynamic mixers (Taylor-Couette UV reactor), and coiled tube devices. Annular type laminar reactors were used for the treatment of apple juice and cider (Worobo 1998) and mango nectar (Guerrero-Beltran and Barbosa-Canovas 2006). The length and gap size can vary depending on the type of treated juice or flow rate. Thin film reactors are characterized by laminar flow with a parabolic velocity profile. Extensive research of the application of UV-light for fresh apple cider by Worobo (1998) yielded a design and production model of a thin film with 0.8 mm gap "CiderSure" UV reactor that was approved for a safe use to reduce microbial load of apple cider. UV treatment of orange juice was reported by Tran and Farid (2004) using a vertical single UV lamp thin film reactor. The thickness of the film was approximately 0.21–0.48 mm. Another commercial thin film reactor is the PureUV/SurePure reactor that was used for treatment of apple juice, guava-and-pineapple juice, mango nectar, strawberry nectar and two different orange and tropical juices (Keyser et al. 2008). This reactor is a single-lamp system with a thin fluid film formed between the lamp surface and a surrounding rippled or undulating outer wall. The reactor consisted of inlet, outlet chambers and a corrugated spiral tube between the chambers. Another type of static mixers is coiled tube UV reactors that are used to increase liquid delivery to UV source by more mixing due to Dean effect (Dean 1927). Salcor Inc. has promoted a UV reactor in which juice is pumped through the Teflon tubes coiled in a helix, with 12 LPM lamps inside and 12 lamps outside the helix (Anonymous 1999; Koutchma et al. 2007). The curved flow path can result in a pair of counter-rotating vortices with their axis along the length of the coil. Koutchma et al. (2007) validated the performance of a coiled UV module 420 model (Salcor Inc., Fallbrook, CA) for fresh tropical juices pasteurization. Geveke (2005) processed apple cider with a single lamp UV system surrounded by a coil of UV transparent Chemfluor tubing. Forney et al. (2004) used dynamic mixer Taylor-Coutte design to improve UV inactivation efficiency in apple juice.

17.5.1.3 Inactivation of Pathogenic, Nonpathogenic, and Spoilage Organisms

Table 17.7 summarizes the results of several reports on inactivation of pathogenic and nonpathogenic bacteria in fruit juices using continuous UV light sources. These data were obtained using static (collimated beam device) and continuous flow UV systems. The approaches to determine UV fluence also differed so reported results are not directly comparable.

Bobe et al. (2007) studied the presence and concentrations of pathogenic and indicator microorganisms in apple cider processed in Michigan. Neither *E. coli* O157:H7 nor *Salmonella* were detected in any tested cider samples, suggesting a very low frequency of pathogens in apple cider. The persistent and relatively high frequency of generic *E. coli* observed in samples indicated a continued risk of pathogen contamination in apple cider, especially when it is untreated. Basaran et al. (2004) compared log reductions among the *E. coli* strains in the apple cider made of different cultivars. The result failed to show any statistically significant relationship. However, the results of this study indicate that regardless of the apple cultivar used, a minimum 5-log reduction is achieved for all of the strains of *E. coli* O157:H7 tested. Gabriel and Nakano (2009) examined the UV resistance of strains of *E. coli* (K-12 and O157:H7), *Salmonella* (enteritidis and typhimurium), and *Listeria monocytogenes* (AS-1 and M24-1) that were individually suspended in phosphate-buffered saline (PBS) and apple juice prior exposure to UV radiation (220–300 nm). The AS-1 and M24-1 strains of *L. monocytogenes* were found to be most resistant to UV in PBS (0.28–0.29 min) while the AS-1 strain was most resistant in juice (1.26 min). The AS-1 strain of *L. monocytogenes* and *E. coli* O157:H7 were most heat resistant when suspended in PBS (4.41 min) and juice (4.43 min), respectively. Ye et al. (2007) reported that *Yersinia pseudotuberculosis* was less resistant to UV light than *E. coli* K12.

Table 17.8 summarizes results of reported studies in terms of inactivation of spoilage microorganisms in fresh juices. Variations in UV fluence levels can be accounted for due to limitations in dosimetry and fluid absorbance measurements. Mold spores are considered to be very UV resistant, with the resistance higher than

Table 17.7 UV inactivation of pathogenic and nonpathogenic microorganisms in fresh juices

Juice	Type of UV reactor			Fluence, (mJ•cm^{-2})	Test organism	Log (No./*N*)	Reference
	Flow regime	Number/UV lamp/ power	Gap size (mm)				
Apple cider	Thin film laminar	10/LPM	NA	9–61	*E. coli O157:H7*	3.8	Wright (2000)
Apple cider	Laminar	8/LPM/39 W	0.8	14.32	*C. parvum* Oocyst	5	Hanes et al. (2002)
Apple cider	Laminar	8/LPM/39 W	0.8	14	*E. coli* O157:H7 (933, ATCC 43889, and ATCC 43895)	5	Basaran et al. (2004)
Apple juice	Petri dish	220–300 nm/15 W	*d*=5	At 50 cm up to 0–33 min	*Escherichia coli (K-12 and O157:H7)*		Gabriel and Nakano (2009)
					Salmonella (enteritidis and typhimurium)		
					Listeria monocytogenes (AS-1, M24-1)		
Orange juice	Petri dish	4/LPM/30 W		2.19 J•cm^{-2}	*E. coli O157:H7*	5	Oteiza et al. 2010
Apple cider	Laminar	8/LPM/39 W	0.8	NA	*E. coli* ATCC 25922	5–6	Worobo (1998)
Apple juice	Thin Laminar	8/LPM/39 W	0.8	14.5	*E. coli* K12	3–4	Koutchma et al. (2004)
Apple cider	Turbulent	12/LPM/42 W	5–10	0.75	*E. coli* K12	<1	Koutchma et al. (2004)
Apple juice	Dean flow	1/LPM/15 W	Id 3.6	34 J•mL^{-1}	*E. coli* K12	3.4	Geveke (2005)
					L. innocua	2.5	
Apple juice	Taylor-Coutte	4/MPM/0.684	5.5	21.7	*E. coli 15597*	3–5	Forney et al. 2004
			2				
Apple juice	Thin film laminar	1/LPM/15	5		*Yersinia pseudotuberculosis*	1	Ye et al. (2007)
					E. coli K 12	1	

Table 17.8 UV inactivation of spoilage microorganisms in fresh juices

Juice	Type of UV reactor			Fluence	Test organism	Log (No./*N*)	Reference
	Flow regime	Number/UV lamp/ power	Gap size (mm)	($mJ \cdot cm^{-2}$)			
Orange	Thin film laminar vertical	1/LPM/30 W	0.21–0.48	74	APC	0.53	Tran and Farid (2004)
					Yeasts	0.36	
Apple	Laminar	2/LPM/25 W	NA	45,000	*E. coli*	1.34	Guerrero-Beltran and Barbosa-Canovas (2005)
					APC*	4.29	
					Y&M**	5.10	
Mango nectar	Laminar	2/LPM/25 W	NA	45,000	APC	2.94	Guerrero-Beltran and Barbosa-Canovas (2006)
					Yeasts	2.71	
Model of tropical juices	Turbulent, Dean Flow	24/LPM/65 W	ID 10–12	21.5	Yeasts	Up to 6	Koutchma et al. (2007)
Orange					Molds	1.5	
Guava					Molds	1.2	
Carrot					APC	3.2	
Pineapple					Y&M	1.0	
Apple	Turbulent, Re >7500	1–10/LPM/100 W	NA	234	APC	>3.50	Keyser et al. (2008)
					Y&M	>2.99	
Guava-and-pineapple				1404	APC	3.31	
				468	Y&M	2.23	
Mango nectar				702	APC	0.40	
					Y&M	0.44	
Strawberry nectar				1404	APC	1.32	
					Y&M	2.45	

APC* – aerobic plate count; Y&M** – yeasts and molds

of *B. subtilis* spores, followed by yeasts and lactic bacteria (Warriner et al. 2004, unpublished proprietary data). However, data on UV effectiveness against food borne pathogenic and spoilage microorganisms of high importance are limited or available in confidential reports and need to be generated. Data generated in air or water cannot be used for the calculation of UV process of low UVT food liquids. The results should be considered by juice processors in selecting appropriate surrogate organisms for UV light process lethality validations.

17.5.2 UV Surface Treatment of Fresh Fruit and Fresh-Cut Produce

cUV and PL treatments result in various levels of inactivation of spoilage and pathogenic microflora on the surface of a wide variety of foods. Comprehensive reviews of the literature in this field have been compiled by the US FDA (2000b) and by Woodling and Moraru (2005). The variability of the results (a 2- to 8-log reduction was generally reported) is most likely due to the different challenge microorganisms used in various studies, the intensity of the treatment, and the different properties of the treated substrates. Woodling and Moraru (2005) demonstrated that the efficacy of PL is affected by substrate properties such as topography and hydrophobicity, which affect both the distribution of microbial cells on the substrate surface and the interaction between light and the substrate (i.e., reflection and absorption of light). Surface disinfection of fresh and cut fruit products is a basis for longer shelf life. In designing a PL treatment for fruit items, both source (as light wavelength, energy density, duration and number of the pulses, interval between pulses) and target (as product transparency, color, size, smoothness, and cleanliness of surface) parameters are critical for process optimization, in order to maximize the effectiveness of product microbial inactivation and to minimize product alteration. Such alteration can be mainly determined by an excessive increase of temperature causing thermal damage to fruits but also by an excessive content of UV-C light which could result in some undesired photochemical damage to fruit itself or packaging materials.

17.5.2.1 Fresh-Cut Produce

Fresh-cut fruits became popular among consumers due to an increased preference for minimally processed fresh-like and ready-to-eat products. Mechanical operations of fresh-cut fruits production, such as peeling, slicing, and shredding, often result in enzymatic browning, off-flavors, texture breakdown, and lower resistance of fresh-cut produce to microbial spoilage in comparison with the unprocessed commodities (Lemoine et al. 2007) because of the presence of natural microflora on the surface of raw commodities. Therefore during operations of cutting and shredding, cross contamination may occur that might increase the risks of foodborne outbreaks.

To improve the hygiene and safety during the mechanical processing, sanitizing and dripping treatments are commonly applied. During washing and dipping steps, raw or fresh-cut material is immersed into tap water containing sanitizing agents (chlorine, sodium hypochlorite) to remove spoilage microorganisms, pesticide residues, and plant debris from product surface (Martin-Belloso et al. 2006). To reduce the usage of sanitizing chemicals, UV light alone or in combination with ozone or another preservative agent was explored as novel processing alternative. Fonseca and Rushing (2006) examined the effects of UV-C light (1.4–13.7 kJ•m^{-2} at 253.7 nm) on the quality of fresh-cut watermelon compared to the common sanitizing solutions. Dipping cubes in chlorine (40 μL•L^{-1}) and ozone (0.4 μL•L^{-1}) was not effective in reducing microbial populations and cubes quality was lower after these aqueous treatments compared to UV-irradiated cubes or control. In commercial trials, exposure of packaged watermelons cubes to UV-C at 4.1 kJ•m^{-2} produced more than 1-log reduction in microbial populations by the end of the product's shelf life without affecting juice leakage, color, and overall visual quality. Higher UV doses did not show differences neither in microbial populations nor in quality deterioration (13.7 kJ•m^{-2}). Spray applications of hydrogen peroxide (2 %) and chlorine (40 μL•L^{-1}), without subsequent removal of excess water, failed to further decrease microbial load of cubes exposed to UV-C light at 4.1 kJ•m^{-2}. It was concluded that when properly utilized, UV-C light is the only method tested that could be potentially used for sanitizing fresh-cut watermelon. Similarly, exposure of sliced apples to UV-C resulted in higher (~1 log) reduction of *Listeria innocua* ATCC 33090, *E. coli* ATCC 11229, and *Saccharomyces cerevisiae* KE 162 in comparison to the apples pre-treated with anti-browning and sanitizing agent (1 % w/v ascorbic acid – 0.1 % w/v calcium chloride). The combination of UV-C with anti-browning pre-treatment better preserved color of sliced apples during storage at 5 °C for 7 days (Gómez et al. 2010). Other studies have shown that UV-C treatment applied alone was efficient in the reduction of a number of microbiological organisms present on the surface of fresh-cut crops. The examples of successful applications of UV-C light are given in Table 17.9.

Similarly to raw crops, the effectiveness of UV treatment on the reduction of microbial deterioration and quality retention was defined by the delivered UV dose and overall characteristics of the surface exposed to the UV light. Lamikanra et al. (2005) stressed out that the moment of application of UV light during the fruit processing is an important factor. In their studies the authors exposed the cantaloupe melon to UV-C at 254 nm during cutting and after cut of the fruits. Cutting of cantaloupe melon under the UV-C light was as effective as post-cut treatment in reduction of yeast, molds, and *Pseudomonas* spp. populations. However fruit cutting during simultaneous exposure to UV-C resulted in improved product quality, i.e., reduced rancidity and respiration rate, and also increased firmness retention, when compared to post-cut and control samples. Better preservation of fruits processed

Table 17.9 Summary of studies of the effect of UV-C light on reduction of microorganisms in fresh-cut produce

Fresh-cut Commodity	Microbiological organism	Number/UV lamp/ power Fluence	Reference
Watermelon	Mesophilic, psychrophilic, and enterobacteria	15/LPM/36 W 1.6, 2.8, 4.8, 7.2 kJ•m^{-2}	Artés-Hernández et al. (2010)
Cantaloupe melon	Yeast, mold, *Pseudomonas* spp., mesophilic aerobes, Lactic acid bacteria	1/LPM/N/A 0.0118 kJ•m^{-2}	Lamikanra et al. (2005)
Apple	*Listeria innocua* ATCC 33090; *Escherichia coli* ATCC 11229 and *Saccharomyces cerevisiae* KE 162	2/LPM/15 W 5.6 ± 0.3, 8.4 ± 0.5, and 14.1 ± 0.9 kJ•m^{-2}	Gómez et al. (2010)
Pear	*Listeria innocua* ATCC 33090, *Listeria monocytogenes* ATCC 19114 D, *Escherichia coli* ATCC 11229, and *Zygosaccharomyces bailii* NRRL 7256	2/LPM/15 W 15, 31, 35, 44, 56, 66, 79, and 87 kJ•m^{-2}	Schenk et al. (2007)

during the UV exposure can be related to the defence response of the wounded plant enhanced by the UV. Mechanical injury of the plant tissues activates the expression of wound-inducible genes. UV radiation is capable to induce the expression of plant defence-related proteins that are normally activated during wounding. For example, Lamikanra et al. (2005) reported a significant increase in ascorbate peroxidase enzyme activity during storage of cantaloupe melon processed under UV-C light. Peroxidases protect plant cells against oxidation. Higher levels of terpenoids (β-cyclocitral, *cis*- and *trans*- β-ionone, terpinyl acetate, geranylacetone, and dihydroactinidiolide) that can play important roles as phytoalexins in the disease resistance of a variety of plant families were found in cantaloupe tissues (Lamikanra et al. 2005; Beaulieu 2007). Significant increase of anti-oxidative compounds, such as phenolics and flavonoids, was also observed by Alothman et al. (2009) in UV treated fresh-cut banana, pineapple, and guava fruits. However a decrease in vitamin C was observed in all fruits.

In terms of UV effects on fruits flavor, Beaulieu (2007) and Lamikanra et al. (2005) reported that fruits processed with the UV light preserved their aroma to the same extent as non-treated control samples. Detailed studies of volatile compounds in thin-sliced cantaloupe tissues revealed that UV treatment is not responsible for the chemical transformations to ester bonds, esterase, and lipase decrease. However Beaulieu (2007) indicated that improper cutting, handling, sanitation treatment, and storage can radically alter the desirable volatile aroma profile in cut cantaloupe, and potentially leads to decreased consumer acceptance.

17.6 UV Effects on Chemicals in Fruit Products

17.6.1 Degradation of Patulin

Patulin [4-hydroxy-4H-furo (3, 2-c)-pyran-2-(6H)-one] is a mycotoxin produced by a wide range of molds involved in fruit spoilage. *Penicillium expansum* is the predominant patulin producing fungus in naturally rotted apples (Lovett et al. 1974). Although cases of contamination were reported in various peaches, cherries, berries, and strawberries, patulin occurs most frequently in rot lesions of apples. Beretta et al. (2000) reported 21 patulin positive samples of rotten areas of apples in a total of 26 samples. The concentration of patulin has been detected up to 130 mg•kg^{-1}. As with the majority of mycotoxins, patulin is stable and can persist in juice over extended time periods. Although the washing and removal of rotten apples may reduce 90 % of the original patulin concentration (Leggott et al. 2000), patulin contamination in apple juice was detected up to 733 μg•L^{-1} and reported by Ehlers (1986), Gökmen and Acar (1998), and Yurdun et al. (2001). Patulin is a health concern for both consumers and manufactures, which may cause acute but more frequently, chronic intoxications leading to nervousness, convulsion, lung congestion, oedema, hyperaemia, immunotoxic, immunosuppressive, and teratogenic effect (Roll et al. 1990). Because of the prevalence of patulin and possible accumulation of the toxin within the body over time, the Codex Alimentarius Commission (2003) and the US FDA (2005) have recommended a limit for patulin content on apple products intended for human consumption of 50 μg•L^{-1} (50 ppb). The European Union has gone further and imposed a maximum limit of 10 μg•L^{-1} (10 ppb) for baby food and formulae.

Although several methods for control and elimination of patulin have been proposed, there is no unifying method being commercially successful for reducing patulin while keeping produce quality. A few recent studies evaluated feasibility of UV radiation as a possible commercially alternative for the reduction of patulin and patulin producing *Penicillium* spores in fresh apple juice. Dong et al. (2010) used the CiderSure 3500 commercial UV system equipped with the 8 LPM lamps for patulin destruction. It was reported that UV exposure of 14.2–99.4 mJ•cm^{-2} resulted in a significant and nearly linear decrease in patulin levels while producing no quantifiable changes in the chemical composition (i.e., pH, Brix, and total acids) or organoleptic properties of the cider.

Zhu et al. (2012) investigated UVC-light to control patulin content in model solution, apple cider, and apples juice by using R-52G MINERALIGHT® UV Lamp and studied the kinetics of degradation of patulin. It was shown that 56.5 %, 87.5 %, 94.8 %, and 98.6 % reduction of patulin can be achieved in the model solution,

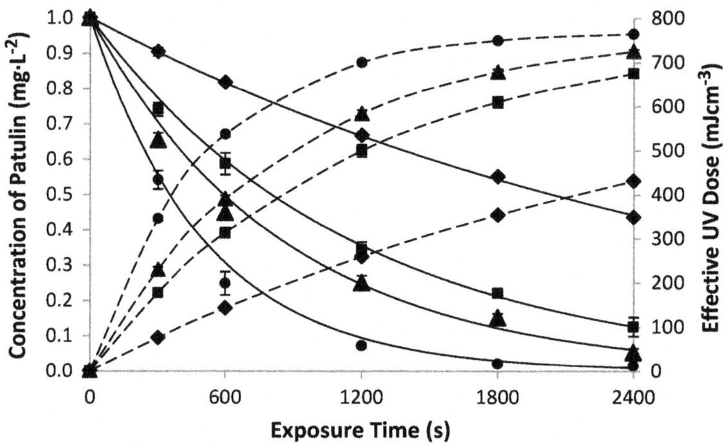

Fig. 17.5 Degradation of patulin in 4 kinds of media during 40 min of UV exposure (0.2 cm of sample thickness and 3.0 mW · cm^{-2} of incident intensity)

apple cider, apple juice without vitamin C addition, and apple juice with vitamin C addition, respectively. Sample (2-mm length) was initially spiked with 1 mg•L^{-1} of patulin after UV exposure for 40 min at UV intensity of 3.00 mW•cm^{-2}. The effective UV doses which were directly absorbed by patulin for photochemical reaction were 430, 674, 724, and 763 mJ · cm^{-3}, respectively (Fig. 17.5). Similar applied UV fluence of 7064 mJ•cm^{-2} was adopted for all samples. The decimal reduction time (D-value) was estimated at 112.6, 44.2, 32.6, and 19.4 min, respectively. Degradation of patulin complied with the first-order reaction model. Both time-based and fluence-based reaction rate constants were determined for predict of patulin degradation. The fluence-based model should be more beneficial given that the uniform degradation rate constant in the same media can be obtained from one specific experiment but consequently to be adopted for further prediction with different UV intensity and sample thickness (UV path length). Yan's work also compared the patulin degradation rate in dynamic system with well stirring during UV radiation and in static system without mixing. The study revealed the reaction rate constant of dynamic samples (model solution: 2.95E-4 s^{-1}, juice: 4.31E-4 s^{-1}) were significantly higher than static ones (model solution: 2.79E-4 s^{-1}, juice: 3.49E-4 s^{-1}, $P<0.05$) when applied UV intensity and sample length were identical. Although the patulin solution is homogeneous, the intensity of UV light is not uniform along the volume of the solution. Based on Beer–Lambert Law, the UV intensity decreases exponentially when IV light enter the liquid sample. The stirring applied in the dynamic system increased the collision chance between patulin molecular and photons and consequently increased the reaction rate. The patulin degradation rate constant in apple juice was significantly higher than in model solution ($P<0.05$). This suggests that apple juice constituents enhanced the degradation of patulin. Polyphenols and ascorbic acids contained in apple juice can be activated by UV light and produce free radicals that react with patulin molecules. However, further work will be required to confirm this hypothesis. This study provided strong evidence that UV radiation can become an effective method of reducing the patulin level in apple cider and apple juice.

17.6.2 Inactivation of Enzymes

Enzymatic activity actually depends on the native structure of the protein which, by principle, can be modified following photo-oxidation promoted by exposure to UV and visible light. Photo-oxidation of enzymes can occur via two major routes: (1) direct photo-oxidation arising from the absorption of radiation by the protein structure or bound chromophore and (2) indirect protein oxidation mediated by singlet oxygen generated by energy transfer by either protein bound, or other chromophores (Davies and Truscott 2001). The effect of UV light on the activity and structure of fruit enzymes is still a matter of speculation. Limited and controversial information is available in the literature.

Color is a very important quality parameter in fruit juices. It is related to non-enzymatic and enzymatic browning, due to polyphenol oxidase (PPO) activity. The effect of UV light on the inactivation of enzymes related to food quality is diverse. While Noci et al. (2008) reported no effect of UV on apple PPO activity, Manzocco et al. (2009) reported about 80 % inactivation of PPO at approximately of 1250 mJ•cm^{-2} of UV fluence. Guerrero-Beltran and Barbosa-Canovas (2006) found that after UV treatment of mango nectar at 44,633 mJ•cm^{-2} PPO reduced its activity to 19 %. Falguera et al. (2011) irradiated apple juices made from four different varieties (Golden, Starking, Fuji, and King David) during 120 min with a polychromatic mercury lamp of 400 W in a range of 250 and 740 nm with an incident energy of 3.88×10^{-1} Einstein•min^{-1}. The treatment was effective in the inactivation of PPO after 100 min, while peroxidase was completely destroyed in 15 min in all the four varieties. It should be noted that the major absorbance peak of PPO enzyme matched with the largest peak of the emission spectrum of the lamp.

One important factor in orange juice appearance is the "cloud" formed by pectin. Pectin methylesterase (PME) is an enzyme that tends to de-esterify pectin, and which inactivation is consequently pursued. Tran and Farid (2004) reported the results of UV treatment of reconstituted orange juice. In addition to the decimal reduction dose for the standard aerobic plate count, effects on shelf life, pH, color, vitamin C, and destruction of PME enzyme were studied. The shelf life of freshly squeezed orange juice was extended to 5 days as a result of limited exposure of UV light of 73.8 mJ•cm^{-2}. No destruction of PME (5 %), which is a major cause of cloud loss of juices, was reported whereas the activity of this enzyme was significantly decreased (70 %) by mild heat treatment at 70 °C for 2 s.

17.6.3 Effects on Essential Vitamins

Even though vitamins may be present in small amounts in fresh juices they are of concern because some vitamins are considered light sensitive. Water soluble light sensitive vitamins include C (ascorbic acid), B12 (cobalamin), B6 (pyridoxine), B2 (riboflavin), and folic acid. Fat soluble, light sensitive vitamins include A, K, E (alpha-tocopherol), and carotene. Most studies were conducted on the effects of light on vitamins in the wavelength range of 290–700 nm, which includes both UV and visible light. They have involved exposure to fluorescent lamps, but there are limited data available at 253.7 nm. Since vitamin C is characterized by high UV absorbance within the germicidal wavelength range (peak at approximately of 260 nm) but does not absorb light significantly above 300 nm, the content of vitamin C also affected the magnitude of absorption coefficient. The destruction of vitamin C during exposure to UV light may alter the absorption properties of the treated juice. Ye et al. (2007) measured vitamin C content before and after UV treatment. Two brands of packaged apple juice (pasteurized, no preservatives), Sahara Burst and Gordon Food Service, were enriched with Vitamin C. The UV system consisted of four chambers with varied lengths and a single LPM bulb at output power of 25 W at 253.7 nm. Approximately 50 % destruction of vitamin C was observed after one complete pass through the system at the slowest flow rate. The effect of vitamin C destruction on the value of the absorption coefficient in apple juice enriched with this vitamin was also measured. After three passes through the UV system at the flow rate of 4 $mL{\bullet}s^{-1}$ the absorption coefficient of apple juice reduced to approximately 20 % of initial value. It was concluded that juices enriched with vitamin C require significantly higher doses of UV irradiation for pasteurization purposes. A comparison of vitamin C destruction and inactivation of *E. coli* K12, in commercial apple juice (Motts) exposed to UV at the fluence rate of 1.0 $mW{\bullet}cm^{-2}$ showed that *E. coli* bacteria were more sensitive to UV light exposure with a destruction rate almost of 2.5 times higher compared to samples containing vitamin C. When destruction of vitamin C in apple juice was measured after processing using a commercial multiple lamp UV unit CiderSure1500, it was found that after three consecutive passes through the system at the slowest flow rate of 57 $mL{\bullet}s^{-1}$, approximately 50–60 % of the initial concentration of vitamin C (25 mg/100 g) remained. Comparison of the destruction of vitamin C in clarified apple juice with absorption coefficient of 15 cm^{-1} and orange juice of 54 cm^{-1} after exposing both juices to the identical levels of UV fluence of 1.0 $mW{\bullet}cm^{-2}$ in a Petri dish demonstrated that the destruction rate was 8 times faster in clarified apple juice due to greater levels of available absorbed energy (Koutchma et al. 2008). Falguera et al. (2011) studied the effect of a mercury lamp of 400 W in a range of 250 and 740 nm at incident energy of 3.88×10^{-1} Einstein min^{-1} on the content of vitamin C in juices from Golden, Starking, and Fuji. The loss in Golden juice after 120 min of UV irradiation was 5.7 %, while in Starking one was 5.6 %, and in Fuji one 4.0 %. In the juice from King David the loss was 70.0 %. This significant difference was attributed to the lack of pigmentation of this juice. In the three first cases, more vitamin C was damaged in the first 60 min than in the second hour, meaning that as pigments were degraded (and the juice color was lighter) its protective effect was less important. In the King David juice the loss after 0 min was 62.4 % of the initial content, and after 60 min it was 69.8 %. In recent years pulsed UV sources gained interest for their application for food processing due to potentially greater germicidal effectiveness and depth penetration. Orlowska et al. (2012) compared the effects of continuous (LPM and MPM) and pulsed UV (PUV) sources on the vitamin C content of fortified apple juice and milk. Applied PUV lamps were characterized by different emission spectra in the range of 200–350 nm, energy per pulse, and frequency (PUV-1: 31 J/pulse, 8 Hz; PUV-2: 344 J/pulse, 0.75 Hz; PUV-3: 644 J/pulse, 0.5 Hz). Comparison was made at the UV fluence that was determined based on 5-log microbial reduction requirement, i.e., 10 $mJ{\bullet}cm^{-2}$ for LPM and MPM, and 5 $mJ{\bullet}cm^{-2}$ for the PUV sources. The UV treatments with the MPM and PUV-2 induced significant ($P<0.05$) reduction of vitamin C by -5.45 ± 0.27 % and -8.52 ± 0.50 % in apple juice, -61.73 ± 3.08 % and -35.80 ± 1.79 % in milk, respectively. The other two pulsed UV lamps didn't affect significantly ($P>0.05$) vitamin C in apple juice, and its reduction was on the same level as in the case of LPM, i.e., $-1.30\pm0.$ 07 %. Similarly PUV-1 and PUV-3 caused least changes in ascorbic acid content in milk, i.e., -12.31 ± 0.62 % and -21.66 ± 1.08 %, respectively, whereas treatment with the LPM lamp resulted in reduction of vitamin C by -35.13 ± 1.56 %. Results have shown that PUV-3 source can constitute a promising alternative for UV treatments as it offers deeper penetration in opaque liquids due to broader emission spectrum in comparison to LPM, and about 10 times shorter exposure times when compared with PUV-1. Authors also stressed out the importance of knowledge of the optical properties of ingredients and their chemical interactions in UV treated beverage and the emission spectra of applied UV sources. For instance, a significantly higher reduction of vitamin C in milk was observed, in comparison to apple juice (<10 %), which can be associated with the riboflavin, also known as vitamin B2. Riboflavin is a photosensitive compound characterized by four absorption peaks in the UV range (222, 266, 373 nm) and in visible light range (445 nm). As it can be seen in Fig. 17.6 the peaks of MPM emission spectrum overlap the broad riboflavin peak with its maximum of absorbance at 266 nm. This can lead to the occurrence of photochemical reactions if sufficient energy is delivered to the UV exposed system. From the literature (Gilmore and Dimick 1979; Bender 2003) it is known that riboflavin photolysis leads to the formation of lumiflavin and lumichrome, which catalyze the oxidation of other milk ingredients, such as vitamin C. Therefore in order to explore the full potential and applications of pulsed UV sources for specific food systems more studies have to be conducted.

Vitamin A is another vitamin of great importance in fresh juices because it contributes to more than 2 % of the nutritional value of the Recommended Daily Allowance (RDA). After exposure of vitamin A in malate buffer to UV light at the fluence of 200 $mJ{\bullet}cm^{-2}$, approximately 50 % of vitamin A initial concentration remained. Orange juice is an essential source of vitamin C and A. One 8 fluid ounce (3.69 mL) serving of orange juice contributes approximately to 210 % of RDA of vitamin C and 10 % RDA of vitamin A in the diet. The destruction of essential

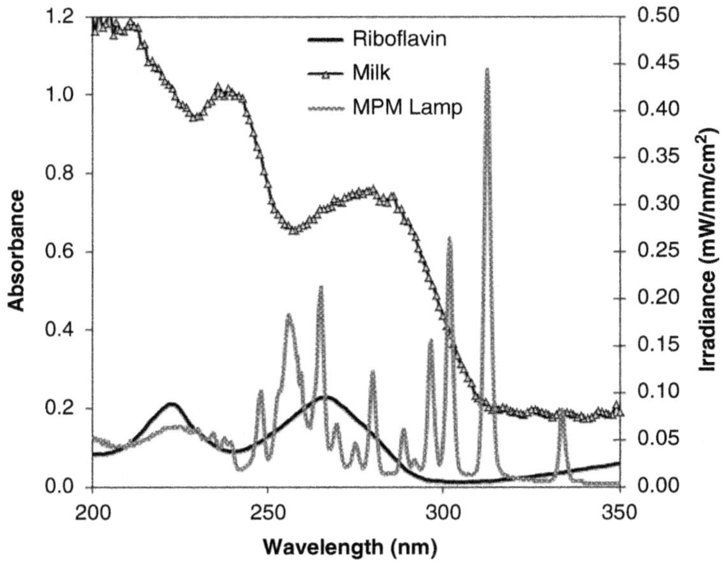

Fig. 17.6 Absorbance of milk (0.2 mm quartz cuvette) and riboflavin (0.08 mg•mL^{-1}; 0.5 mm quartz cuvette) with light output of MPM lamp

vitamins in orange juice was reported by Anonymous (1999) after treatment in the commercial Salcor UV module (Salcor Co, CA) at a flow rate of 7.5 gpm (28.39 L•min^{-1}) when total accumulative UV dose was 298.9 mJ•cm^{-2}. The highest destruction of riboflavin and beta carotene (~50 %) was observed. However, in terms of vitamins C, B6, and A only 16.6–11 % of those vitamins were destroyed after exposure to UV light.

17.6.4 Degradation of Herbicides

The use of agricultural pesticides has increased dramatically and has consequently led to increasing concerns related to their toxicity, stability, and pollution of soil, water, and air. Triazine herbicides are among the most commonly used herbicides in the world. A maximum admissible concentration of 0.1 µg•L^{-1} per individual pesticide was set in the EEC Directive on the Quality of Water Intended for Human Consumption. Evgenidou and Fytianos (2002) studied the photodegradation of three triazines, atrazine, simazine, and prometryn, in aqueous solutions and natural waters using UV radiation ($\lambda > 290$ nm). Experimental results showed the rate of photodecomposition in aqueous solutions depends on the nature of the triazines and follows first-order kinetics. The half-lives of triazines in distilled water and surface waters ranged from 2.7 to 11.6 h with exposure of high-pressure mercury UV lamp. The work demonstrated the effects of photodegradation of triazines during direct

UV exposure and indirect (UV with H_2O_2) irradiation and suggested the existence of various degradation routes resulting in complex and interconnected pathways.

17.7 Sustainability of UV Technology

Expected increase of world population up to 9 billion by 2050 brings the necessity to implement sustainable practices that will allow meeting the needs of the present without compromising the ability of future generations to meet their own needs. These include wiser management of the natural resources use, product stewardship, strengthening energy efficiency, development of new technologies that reduce the consumption of resources, and eradication of poverty.

UV light is an emerging nonthermal technology that has much to offer for the sustainable development of the society. Its application for the food processing is energy and cost-effective, and also was proven to yield fresh-like, safe and highly nutritional fruits and fruit products, such as juices. Moreover, UV light applied as a postharvest technology can significantly reduce the loss of fresh produce, which in the developed countries is of the order of 20 % and as high as 50 % in developing countries (Obande and Shama 2011). It was shown by many researchers that UV technology might be used as alternative method to control postharvest diseases caused by fungi. This in turn may substantially reduce the usage of fungicides as well as other chemicals that pose serious health hazard and environmental risks (Lu et al. 1991).

The major disadvantage of UV technology is the mercury content in UV sources. The potential mercury exposure due to lamp sleeve breakage is a health concern. Breakage of lamps can occur when lamps are in operation and during maintenance. The mercury contained within a UV lamp is isolated from exposure by the lamp envelope and surrounding lamp sleeve. For the mercury to be released, both the lamp and lamp sleeve must break. The mercury content in a single UV lamp used for water treatment typically ranges from 0.005 to 0.4 g (5–400 mg). LPM lamps have less mercury (5–50 mg/lamp) compared to LPHO (26–150 mg/lamp) and MPM lamps (200–400 mg/lamp). The EPA established a maximum contaminant level (MCL) for mercury at 0.002 mg•L^{-1}. The EPA has found mercury to potentially cause kidney damage from short-term exposures at levels above the 0.002 mg•L^{-1} MCL (EPA 1995). The concern over the impact of mercury release into the food plant environment stimulated the development and validation of mercury-free special technologies lamps and LEDs.

17.8 Conclusions and Future Trends

Ultraviolet (UV) light technology using continuous and pulsed modes is a viable nonthermal alternative for fruits and fruit products processing. A large number of reviewed studies reported successful applications of UV light for eliminating or reducing the levels of undesirable pathogenic, nonpathogenic and spoilage microorganisms on the surfaces of fresh fruits and fruit products like juices. In order to achieve the required microbial reduction along with color, texture, and flavor preservation, optimal UV processing conditions and proper UV source have to be found for a given product. Moreover, UV light can be recommended as an effective mean to control microbial loads in the air, water, nonfood, and food contact surfaces in fruit processing facilities. A variety of UV sources are commercially available or currently under development that can be applied for specific fruit processing purposes whereas LPM lamps and xenon PL are currently the dominant sources for UV treatment of fruits since they were approved by the US FDA and Health Canada. A number of UV-light continuous flow systems that included annular laminar and turbulent flow reactors, thin film devices, static and dynamic mixers were developed and validated for a variety of fruit juices for pasteurization purposes. The correct UV design can reduce the interference of low UVT and viscosity associated with some juices and therefore improves the UV inactivation efficiency. More work is needed in regards of design of UV systems capable of delivering sufficient UV doses to all parts of the treated liquid with low UVT such as fruit juices.

Recent studies reported a potential of UV light for enhancement of health promoting compounds such as antioxidants, polyphenols, and flavonoids. Numerous studies cited here have shown the beneficial effects of the UV treatment on the preservation of many fruits, both raw and fresh-cut. However on the basis of the available literature data the mechanism that underlies the hormetic response in fresh produce is still under debate. In response to the exposure of UV light, plants activate different enzymes peroxidases, reductases, chitinases which differ by chemical structure and absorptive properties in UV-A, UV-B, and UV-C ranges. Therefore plant response varies depending on applied UV emission spectrum and UV dose. To improve the state of the current knowledge on UV processing on fresh produce, further studies are necessary that will measure and report conditions and parameters of the UV treatment, such as lamp characteristic, emitted wavelength, and UV fluence levels.

The effect of UV-light on quality of fruits and their products requires further studies. Despite the fact that UV is a pure nonthermal treatment, possible undesirable effects may include damage to vitamins and proteins, destruction of the antioxidants, changes in color and formation of off-flavors and aromas depending on UV spectra and applied dose. In addition, the effects of UV light on the potential formation of chemical compounds in foods that may present a health threat should be evaluated to determine if there is any toxicological or chemical safety concerns associated with products that have undergone UV treatment. Closer examination of UV light potential to destroy undesirable compounds or pollutants also deserves more attention. Due to low penetration of UV light, the combinations with other postharvest technologies (ozone, ultrasound, modified packaging atmosphere, sanitizing and anti-browning agents) might be attractive for processors and also more efficient. Limited data are available on UV processing combined with other treatments and further studies are necessary to undertake.

References

Allende, A., A. Marín, B. Buendía, F. Tomás-Barberán, and M.I. Gil. 2007. Impact of combined postharvest treatments (UV-C light, gaseous O_3, superatmospheric O_2 and high CO_2) on health promoting compounds and shelf-life of strawberries. *Postharvest Biology and Technology* 46: 201–211.

Alothman, M., R. Bhat, and A.A. Karim. 2009. UV radiation-induced changes of antioxidant capacity of fresh-cut tropical fruits. *Innovative Food Science and Emerging Technologies* 10: 512–516.

Anonymous. 1999. *A food additive petition for the use of ultraviolet light in the reduction of microorganisms on juice products.* Submitted to FDA regarding CFR 21 179. Glendore, CA: California Day-Fresh Foods Inc. pp. 1–117.

———. 2000. Kinetics of microbial inactivation for alternative food processing technologies. Institute of Food Technologists. J. Food Sci. Supplement http//vm.cfsan.fda.gov/~comm/ift-pref.html. Accessed: 7 February, 2012

Arakawa, O. 1991. Effect of temperature on anthocyanin accumulation in apple fruit as affected by cultivar, stage of fruit ripening and bagging. *The Journal of Horticultural Science & Biotechnology* 66: 763–768.

Arcas, M.C., J.M. Botía, A.M. Ortuño, and J.A. Del Río. 2000. UV irradiation alters the levels of flavonoids involved in the defence mechanism of *Citrus aurantium* fruits against *Penicillium digitatum. European Journal of Plant Pathology* 106: 617–622.

Artés-Hernández, F., P.A. Robles, P.A. Gómez, A. Tomás-Callejas, and F. Artés. 2010. Low UV-C illumination for keeping overall quality of fresh-cut watermelon. *Postharvest Biology and Technology* 55: 114–120.

Baka, M., J. Mercier, R. Corcuff, F. Castaigne, and J. Arul. 1999. Photochemical treatment to improve storability of fresh strawberries. *Journal of Food Science* 64: 1068–1072.

Bakhshi, D., and O. Arakawa. 2006. Induction of phenolic compounds biosynthesis with light irradiation in the flesh of Red and yellow apples. *Journal of Applied Horticulture* 8: 101–104.

Barka, E.A., S. Kalantari, J. Makhlouf, and J. Arul. 2000. Impact of UV-C irradiation on the cell wall-degrading enzymes during ripening of tomato (*Lycopersicon esculentum L.*) fruit. *Journal of Agriculture and Food Chemistry* 48: 667–671.

Basaran, N., A. Quintero-Ramos, M.M. Moake, J.J. Churey, and R.W. Worobo. 2004. Influence of apple cultivars on inactivation of different strains of *Escherichia coli* O157:H7 in apple cider by UV irradiation. *Applied and Environmental Microbiology* 70: 6061–6065.

Beaulieu, J.C. 2007. Effect of UV irradiation on cut cantaloupe: terpenoids and esters. *Journal of Food Science* 72: S272–S281.

Bender, D.A. 2003. *Nutritional biochemistry of the vitamins.* Cambridge: Cambridge University Press.

Ben-Yehosua, S., V. Rodov, J. Kim, and S. Carmeli. 1992. Preformed and induced antifungal materials of citrus fruit in relation to the enhancement of decay resistance by heat and ultraviolet treatment. *Journal of Agricultural Food Chemistry* 40: 1217–1221.

Beretta, B., A. Gaiaschi, C.L. Galli, and P. Restani. 2000. Patulin in apple-based foods: occurrence and safety evaluation. *Food Additives & Contaminants* 5: 399–406.

Bialka, K.L., and A. Demirci. 2007. Decontamination of *Escherichia coli* O157:H7 and *salmonella enterica* on blueberries using ozone and pulsed UV-light. *Journal of Food Science* 72: M391–M396.

Bialka, K.L., A. Demirci, and V.M. Puri. 2008. Efficacy of pulsed UV-light for the decontamination of *Escherichia coli* O157:H7 and *salmonella* spp. on raspberries and strawberries. *Journal of Food Science* 73: M201–M207.

Bobe, G., D. Thede, T.A.T. Eyck, and L.D. Bourquin. 2007. Microbial levels in Michigan apple cider and their association with manufacturing practices. *Journal of Food Protection* 70: 1187–1193.

Bolton, J.R., and K.G. Linden. 2003. Standardization of methods for fluence UV dose determination in bench-scale UV experiments. *Journal of Environmental Engineering* 129: 209–215.

Boue, S.M., T.E. Cleveland, C. Carter-Wientjes, B.Y. Shih, D. Bhatnagar, J.M. McLachlan, and M.E. Burow. 2009. Phytoalexin-enriched functional foods. *Journal of Agricultural and Food Chemistry* 57: 2614–2622.

Charles, M.T., J. Mercier, J. Makhlouf, and J. Arul. 2008. Physiological basis of UV-C-induced resistance to *botrytis cinerea* in tomato fruit I. Role of pre- and post-challenge accumulation of the phytoalexin-rishitin. *Postharvest Biology and Technology* 47: 10–20.

Codex Alimentarius Commission. 2003. Code of Practice for the Prevention and Reduction of Patulin Contamination in Apple Juice and Apple Juice Ingredients in Other Beverages, www.codexalimentarius.net/download/standards/405/CXC_050e.pdf (accessed: February 5, 2012)

Costa, L., A.R. Vicente, P.M. Civello, A.R. Chaves, and G.A. Martínez. 2006. UV-C treatment delays postharvest senescence in broccoli florets. *Postharvest Biology and Technology* 39: 204–210.

D'hallewin, G., M. Schirra, E. Manueddu, A. Piga, and S. Ben-Yehoshua. 1999. Scoparone and scopoletin accumulation and ultraviolet-C induced resistance to postharvest decay in oranges as influenced by harvest date. *Journal of the American Society for Horticultural Science* 124: 702–707.

D'hallewin, G., M. Schirra, M. Pala, and S. Ben-Yehoshua. 2000. Ultraviolet C irradiation at 0.5 $kJ \cdot m^{-2}$ reduces decay without causing damage or affecting postharvest quality of star ruby grapefruit (*C. paradisi* Macf.). *Journal of Agricultural and Food Chemistry* 48: 4571–4575.

Davies, M.J., and R.J.W. Truscott. 2001. Photo-oxydation of proteins and its role in cataractogenesis. *Journal of Photochemistry and Photobiology B: Biology* 63: 114–125.

Dean, W.R. 1927. Note on the motion of fluid in a curved pipe. *Philosophical Magazine Journal of Science* 4: 208–223.

Dong, Q., D.C. Manns, G. Feng, T. Yue, J.J. Churet, and R.W. Worobo. 2010. Reduction of patulin in apple cider by UV radiation. *Journal of food Protection* 73: 69–74.

Ehlers, D. 1986. HPLC-Bestimmung von patulin in obstsäften-probenaufarbeitung mit einem modifizierten extraktions-und reinigungsverfahren. *Lebensmittelchemie und Gerichtliche Chemie* 40: 1–5.

El Ghaouth, A., C.L. Wilson, and A.M. Callahan. 2003. Induction of chitinase, β-1,3-glucanase and phenylalanine ammonia lyase in peach fruit by UV-C treatment. *Phytopathology* 93: 349–355.

EPA, Office of Water, 1995. National Primary Drinking Water Regulations Contaminant Fact Sheets Inorganic Chemicals—Technical Version. EPA 811-F-95-002-T, Washington, D.C.

Erkan, M., S.Y. Wang, and C.Y. Wang. 2008. Effect of UV treatment on antioxidant capacity, antioxidant enzyme activity and decay in strawberry fruit. *Postharvest Biology and Technology* 48: 163–171.

Evgenidou, E., and K. Fytianos. 2002. Photodegradation of triazine herbicides in aqueous solutions and natural waters. *Journal of Agricultural and Food Chemistry* 50: 6423–6427.

Falguera, V., J. Pagán, and A. Ibarz. 2011. Effect of UV irradiation on enzymatic activities and physicochemical properties of apple juices from different varieties. *LWT - Food Science and Technology* 44: 115–119.

Fonseca, J.M., and J.W. Rushing. 2006. Effect of ultraviolet-C light on quality and microbial population of fresh-cut watermelon. *Postharvest Biology and Technology* 40: 256–261.

Forney, L., J.A. Pierson, and Z. Ye. 2004. Juice irradiation with Taylor-Coutte flow: UV inactivation of *Escherichia coli*. *Journal of Food Protection* 67: 2410–2415.

Gabriel, A.B., and H. Nakano. 2009. Inactivation of *Salmonella*, *Escherichia coli* and *Listeria monocytogenes* in phosphate-buffered saline and apple juice by ultraviolet and heat treatments. *Food Control* 20: 443–446.

Gardner, P.T., T.A.C. White, D.B. McPhail, and G.G. Duthie. 2000. The relative contributions of vitamin C, carotenoids and phenolics to the antioxidant potential of fruit juices. *Food Chemistry* 68: 471–474.

Geveke, D. 2005. UV inactivation of bacteria in apple cider. *Journal of Food Protection* 68: 1739–1742.

Gilmore, T.M., and P.S. Dimick. 1979. Photochemical changes in major whey proteins of cow's milk. *Journal of Dairy Science* 62: 189–194.

Gökmen, V., and J. Acar. 1998. Incidence of patulin in apple juice concentrates produced in Turkey. *Journal of Chromatography A* 815: 99–102.

Gómez, P.L., S.M. Alzamora, M.A. Castro, and D.M. Salvatori. 2010. Effect of ultraviolet-C light dose on quality of cut-apple: microorganism, color and compression behavior. *Journal of Food Engineering* 98: 60–70.

González-Aguilar, G.A., C.Y. Wang, J.G. Buta, and D.T. Krizek. 2001. Use of UV-C irradiation to prevent decay and maintain postharvest quality of ripe `Tommy Atkins' mangoes. *International Journal of Food Science and Technology* 36: 767–773.

Gonzalez-Aguilar, G., C.Y. Wang, and G.J. Buta. 2004. UV-C irradiation reduces breakdown and chilling injury of peaches during cold storage. *Journal of the Science of Food and Agriculture* 84: 415–422.

González-Aguilar, G.A., R. Zavaleta-Gatica, and M.E. Tiznado-Hernández. 2007. Improving postharvest quality of mango 'Haden' by UV-C treatment. *Postharvest Biology and Technology* 45: 108–116.

Gonzalez-Aguilar, G.A., J.A. Villa-Rodriguez, J.F. Ayala-Zavala, and E.M. Yahia. 2010. Improvement of the antioxidant status of tropical fruits as a secondary response to some postharvest treatments. *Trends in Food Science & Technology* 21: 475–482.

Guerrero-Beltran, J.A., and G.V. Barbosa-Canovas. 2005. Reduction of *Saccharomyces cerevisiae*, *Escherichia coli* and *Listeria innocua* in apple juice by ultraviolet light. *Journal of Food Process Engineering* 28: 437–452.

———. 2006. Inactivation of *Saccharomyces cerevisiae* and polyphenoloxidase in mango nectar treated with UV Light. *Journal of Food Protection* 69: 362–368.

Guerrero, R.F., B. Puertas, M.I. Fernández, M. Palma, and E. Cantos-Villar. 2010. Induction of stilbenes in grapes by UV-C: comparison of different subspecies of *Vitis*. *Innovative Food Science and Emerging Technologies* 11: 231–238.

Hagen, S.F., G.I.A. Borge, G.B. Bengtsson, W. Bilger, A. Berge, K. Haffner, and K.A. Solhaug. 2007. Phenolic contents and other health and sensory related properties of apple fruit (*Malus domestica* Borkh., cv. Aroma): effect of postharvest UV-B irradiation. *Postharvest Biology and Technology* 45: 1–10.

Hanes, D.E., P.A. Orlandi, D.H. Burr, M.D. Miliotis, M.G. Robi, J.W. Bier, G.J. Jackson, M.J. Arrowood, J.J. Churey, and R.W. Worobo. 2002. Inactivation of *Crytosporidium parvum* oocysts in fresh apple cider using ultraviolet irradiation. *Applied and Environmental Microbiology* 68: 4168–4172.

Hansen, S.L., S. Purup, and L.P. Christensen. 2003. Bioactivity of falcarinol and its content in carrots. *Journal of the Science of Food and Agriculture* 83: 1010–1017.

Hijnen, W.A.M., E.F. Beerendonk, and G.J. Medema. 2006. Inactivation credit of UV radiation for viruses, bacteria and protozoan (oo)cysts in water: a review. *Water Research* 40: 3–22.

Interdonato, R., M. Rosa, C.B. Nieva, J.A. González, M. Hilal, and F.E. Prado. 2011. Effects of low UV-B doses on the accumulation of UV-B absorbing compounds and total phenolics and carbohydrate metabolism in the peel of harvested lemons. *Environmental and Experimental Botany* 70: 204–211.

Jasinghe, V.J., and C.O. Perera. 2006. Ultraviolet irradiation: the generator of vitamin D2 in edible mushrooms. *Food Chemistry* 95: 638–643.

Kaewsuksaeng, S., Y. Urano, S. Aiamla-or, M. Shigyo, and N. Yamauchi. 2011. Effect of UV-B irradiation on chlorophyll-degrading enzyme activities and postharvest quality in stored lime (*Citrus latifolia* Tan.) fruit. *Postharvest Biology and Technology* 61: 124–130.

Keyser, M., I. Műllera, F.P. Cilliersb, W. Nelb, and P.A. Gouwsa. 2008. UV radiation as a non-thermal treatment for the inactivation of microorganisms in fruit juices. *Innovative Food Science and Emerging Technologies* 9: 348–354.

Konczak, I., and W. Zhang. 2004. Anthocyanins—More than nature's colours. *Journal of Biomedicine and Biotechnology* 5: 239–240.

Koutchma, T., S. Keller, B. Parisi, and S. Chirtel. 2004. Ultraviolet disinfection of juice products in laminar and turbulent flow reactors. *Innovative Food Science & Emerging Technologies* 5: 179–189.

Koutchma, T., B. Parisi, and E. Patazca. 2007. Validation of UV coiled tube reactor for fresh fruit juices. *Journal of Environmental Science and Engineering* 6: 319–328.

Koutchma, T., L. Forney, and C. Moraru. 2008. *Ultraviolet light in food technology: principles and applications*. Boca Raton: CRS Press, Taylor and Francis Group.

Koutchma, T. 2009. Advances in UV light technology for non-thermal processing of liquid foods. *Food and Bioprocess Technology* 2: 138–155.

Kowalski, W.J., and C.E. Dunn. 2002. Current trends in UVGI air and surface disinfection. *INvironment Professional* 8: 4–6.

Kowalski, W.J. 2006. *Aerobiological engineering handbook: A guide to airborne disease control technologies*. New York: McGraw-Hill.

Labas, M.D., R.J. Brandi, C.A. Martin, and A.E. Cassano. 2006. A contribution to the UV dose concept for bacteria disinfection in well mixed photoreactors. *Chemical Engineering Journal* 116: 197–202.

Lamikanra, O., D. Kueneman, D. Ukuku, and K.L. Bett-Garber. 2005. Effect of processing under ultraviolet light on the shelf life of fresh-cut cantaloupe melon. *Journal of Food Science* 70: C534–C538.

Lammertyn, J., B. De Ketelaere, D. Marquenie, G. Molenberghs, and B.M. Nicolaï. 2004. Mixed models for multicategorical repeated response: modelling the time effect of physical treatments on strawberry sepal quality. *Postharvest Biology and Technology* 30: 195–207.

Leggott, N.L., H.F. Vismer, E.W. Sydenham, G.S. Shephard, P. Rheeder, and W.F.O. Marasas. 2000. Occurrence of patulin in the commercial processing of apple juice. *South Africa Journal of Science* 5: 241–243.

Lemoine, M.L., P.M. Civello, G.A. Martínez, and A.R. Chaves. 2007. Influence of postharvest UV-C treatment on refrigerated storage of minimally processed broccoli (*Brassica oleracea var. Italica*). *Journal of the Science of Food and Agriculture* 87: 1132–1139.

Lers, A., S. Burd, E. Lomaniec, S. Droby, and E. Chalutz. 1998. The expression of a grapefruit gene encoding an isoflavone reductaselike protein is induced in response to UV irradiation. *Plant Molecular Biology* 36: 847–856.

Li, X., X. Zheng, S. Yan, and S. Li. 2008. Effects of salicylic acid (SA), ultraviolet radiation (UV-B and UV-C) on trans-resveratrol inducement in the skin of harvested grape berries. *Frontiers of Agriculture in China* 2: 77–81.

Li, J., Q. Zhang, Y. Cui, J. Yan, J. Cao, Y. Zhao, and W. Jiang. 2010. Use of UV-C treatment to inhibit the microbial growth and maintain the quality of Yali pear. *Journal of Food Science* 75: M503–M507.

Liu, C., X. Han, L. Cai, X. Lu, T. Ying, and Z. Jiang. 2011a. Postharvest UV-B irradiation maintains sensory qualities and enhances antioxidant capacity in tomato fruit during storage. *Postharvest Biology and Technology* 59: 232–237.

Liu, C., L. Cai, X. Han, and T. Ying. 2011b. Temporary effect of postharvest UV-C irradiation on gene expression profile in tomato fruit. *Gene* 486: 56–64.

Lovett, J., B. Boutin, and R.G. Thompson. 1974. Patulin production in cherries by *Penicillium* and *Aspergillus* species. *Journal of Milk Food Technology* 37: 530.

Lu, J.Y., C. Stevens, V.A. Khan, M. Kabwe, and C.L. Wilson. 1991. The effect of ultraviolet irradiation on shelflife and ripening of peaches and apples. *Journal of Food Quality* 14: 299–305.

Maharaj, R. 1995. The effect of ultraviolet radiation (UV-C) on the postharvest storage behaviour of tomato (*Lycopersicon esculentum* Mill, *cv. Capello*). Ph.D. Dissertation. Québec, Canada: Université Laval.

Maharaj, R., J. Arul, and P. Nadeau. 1999. Effect of photochemical treatment in the preservation of fresh tomato (*Lycopersicon esculentum cv.* Capello) by delaying senescence. *Postharvest Biology and Technology* 15: 13–23.

Manzocco, L., B. Quarta, and A. Dri. 2009. Polyphenoloxidase inactivation by light exposure in model systems and apple derivatives. *Innovative Food Science and Emerging Technologies* 10: 506–511.

Marais, E., G. Jacobs, and D.M. Holcroft. 2001. Postharvest irradiation enhances anthocyanin synthesis in apples but not in pears. *HortScience* 36: 639–644.

Marquenie, D., C.W. Michiels, A.H. Geeraerd, A. Schenk, C. Soontjens, J.F. Van Impe, and B.M. Nicolaï. 2002. Using survival analysis to investigate the effect of UV-C and heat treatment on storage rot of strawberry and sweet cherry. *International Journal of Food Microbiology* 73: 187–196.

Martin-Belloso, O., R. Soliva-Fortuny, and G. Oms-Oliu. 2006. Fresh-cut fruits. In *Handbook of fruits and fruit processing*, ed. Y.H. Hui, 129–144. Ames: Blackwell Publishing.

Mau, J.-L., P.-R. Chen, and J.-H. Yang. 1998. Ultraviolet irradiation increased vitamin D_2 content in edible mushrooms. *Journal of Agricultural and Food Chemistry* 46: 5269–5272.

Nigro, F., A. Ippolito, and G. Lima. 1998. Use of UV-C light to reduce *botrytis* storage rot of table grapes. *Postharvest Biology and Technology* 13: 171–181.

Noci, F., J. Riener, M. Walkling-Ribeiro, D.A. Cronin, D.J. Morgan, and J.G. Lyng. 2008. Ultraviolet irradiation and pulsed electric fields (PEF) in a hurdle strategy for the preservation of fresh apple juice. *Journal of Food Engineering* 85: 141–146.

Obande, M.A., and G. Shama. 2011. The use of biodosimetry to measure the UV-C dose delivered to a sphere, and implications for the commercial treatment of fruit. *Journal of Food Engineering* 104: 1–5.

Obande, M.A., G.A. Tucker, and G. Shama. 2011. Effect of preharvest UV-C treatment of tomatoes (*Solanum lycopersicon* Mill.) on ripening and pathogen resistance. *Postharvest Biology and Technology* 62: 188–192.

Orlowska, M., Koutchma, T., Grapperhaus, M., Gallagher, J., Schaefer, R., and Defelice, C. 2012. Continuous and pulsed ultraviolet light for non-thermal treatment of liquid foods. Part 1: Effects on quality of fructose solution, apple juice and milk. *Food and Bioprocess Technology*. Online first. DOI: 10.1007/s11947-012-0779-8.

Oteiza, J.M., L. Giannuzzi, and N. Zaritzky. 2010. Ultraviolet treatment of orange juice to inactivate *E. coli* O157:H7 as affected by native microflora. *Food and Bioprocess Technology* 3: 603–614.

Pombo, M.A., M.C. Dotto, G.A. Martínez, and P.M. Civello. 2009. UV-C irradiation delays strawberry fruit softening and modifies the expression of genes involved in cell wall degradation. *Postharvest Biology and Technology* 51: 141–148.

Pombo, M.A., H.G. Rosli, G.A. Martínez, and P.M. Civello. 2011. UV-C treatment affects the expression and activity of defense genes in strawberry fruit (*Fragaria × Ananassa,* Duch.). *Postharvest Biology and Technology* 59: 94–102.

Pongprasert, N., Y. Sekozawa, S. Sugaya, and H. Gemma. 2011. A novel postharvest UV-C treatment to reduce chilling injury (membrane damage, browning and chlorophyll degradation) in banana peel. *Scientia Horticulturae* 130: 73–77.

Reay, P.F., and J.E. Lancaster. 2001. Accumulation of anthocyanins and quercetin glycosides in 'Gala' and 'Royal Gala' apple fruit skin with UV-B—Visible irradiation: modifying effects of fruit maturity, fruit side, and temperature. *Scientia Horticulturae* 90: 57–68.

Rodov, V., S. BenYehoshua, J.J. Kim, B. Shapiro, and Y. Ittah. 1992. Ultraviolet illumination induces scoparone production in kumquat and orange fruits and improves decay resistance. *Journal of the American Society for Horticultural Science* 117: 788–791.

Roll, R., G. Matthiaschk, and A. Korte. 1990. Embryotoxicity and mutagenicity of mycotoxins. *Journal of Environmental Pathology, Toxicology and Oncology* 10: 1–7.

Schenk, M., S. Guerrero, and S. Maris Alzamora. 2007. Response of some microorganisms to ultraviolet treatment on fresh-cut pear. *Food and Bioprocess Technology* 1: 384–392.

Shama, G., and P. Alderson. 2005. UV hormesis in fruits: A concept ripe for commercialisation. *Trends in Food Science and Technology* 16: 128–136.

Shama, G. 2007. *A new role for UV? extensions to the shelf life of plant foods by UV-induced Effects*. Paper presented at the IOA-IUVA Joint World Congress, Los Angeles, California, USA, August 27–29.

Soda, K. 2011. Polyamines—The principal candidate substance of soybean-induced health. In *Soybean and health*, ed. H.A. El-Shemy, 489–502. Croatia: InTech.

Srilaong, V., S. Aiamla-or, A. Soontornwat, M. Shigyo, and N. Yamauchi. 2011. UV-B irradiation retards chlorophyll degradation in lime (*Citrus latifolia* Tan.) fruit. *Postharvest Biology and Technology* 59: 110–112.

Stevens, C., V.A. Khan, J.Y. Lu, C.L. Wilson, P.L. Pusey, M.K. Kabwe, E.C.K. Igwegbe, E. Chalutz, and S. Droby. 1998. The germicidal and hormetic effects of UV-C light on reducing brown rot disease and yeast microflora of peaches. *Crop Protection* 17: 75–84.

Stevens, C., V.A. Khan, C.L. Wilson, J.Y. Lu, E. Chalutz, and S. Droby. 2005. The effect of fruit orientation of postharvest commodities following low dose ultraviolet light-c treatment on host induced resistance to decay. *Crop Protection* 24: 756–759.

Tran, M.T., and M. Farid. 2004. Ultraviolet treatment of orange juice. *Innovative Food Science and Emerging Technologies* 5: 495–502.

Tsormpatsidis, E., M. Ordidge, R.G.C. Henbest, A. Wagstaffe, N.H. Battey, and P. Hadley. 2011. Harvesting fruit of equivalent chronological age and fruit position shows individual effects of UV radiation on aspects of the strawberry ripening process. *Environmental and Experimental Botany* 74: 178–185.

Ubi, B.E., C. Honda, H. Bessho, S. Kondo, M. Wada, S. Kobayashi, and T. Moriguchi. 2006. Expression analysis of anthocyanin biosynthetic genes in apple skin: effect of UV-B and temperature. *Plant Science* 170: 571–578.

Ukuku, D.O., V. Pilizota, and G.M. Sapers. 2001. Bioluminescence ATP assay for estimating total plate counts of surface microflora of whole cantaloupe and determining efficacy of washing treatment. *Journal of Food Protection* 64: 813–819.

U.S. Food and Drug Administration. 2000a. 21 CFR Part 179. Irradiation in the production, processing and handling of food. *Federal Register* 65:71056–71058.

———. 2000b. Kinetics of Microbial Inactivation for Alternative Food Processing Technologies: Pulsed Light Technology, http://www.fda.gov/Food/ScienceResearch/ResearchAreas/SafePracticesforFoodProcesses/ucm103058.htm (accessed March 1, 2011)

———. 2005. CPG Sec.510.150. Apple juice, apple juice concentrates, and apple juice products—adulteration with patulin, http://www.fda.gov/ICECI/ComplianceManuals/CompliancePolicyGuidanceManual/ucm074427.htm (accessed December 10, 2011).

Vicente, A.R., C. Pineda, L. Lemoine, P.M. Civello, G.A. Martinez, and A.R. Chaves. 2005. UV-C treatments reduce decay, retain quality and alleviate chilling injury in pepper. *Postharvest Biology and Technology* 35: 69–78.

Wang, C.Y., C.-T. Chen, and S.Y. Wang. 2009. Changes of flavonoid content and antioxidant capacity in blueberries after illumination with UV-C. *Food Chemistry* 117: 426–431.

Warriner, K., S. Movahedi, and W.M. Waites. 2004. Laser-based packaging sterilization in aseptic processing. In *Improving the thermal processing of foods*, ed. P. Richardson, 277–303. Cambridge: Woodhead Publishing Limited.

Woodling, S.E., and C.I. Moraru. 2005. Influence of surface topography on the effectiveness of pulsed light treatment for the reduction of *Listeria innocua* on stainless steel surfaces. *Journal of Food Science* 70: 245–351.

Worobo, R.W., J.J. Churey, and O. Padilla-Zakour. 1998. Apple cider: Treatment options to comply with new regulations. *Journal of the Association of Food and Drug Officials* 62: 19–26.

Wright, H.B. 2000. Comparison and validation of UV dose calculations for low- and medium-pressure mercury arc lamps. *Water Environment Research* 72: 439–443.

Yang, D.S., R.R. Balandrán-Quintana, C.F. Ruiz, R.T. Toledo, and S.J. Kays. 2009. Effect of hyperbaric, controlled atmosphere, and UV treatments on peach volatiles. *Postharvest Biology and Technology* 51: 334–341.

Yaun, B.R., S.S. Sumner, J.D. Eifert, and J.E. Marcy. 2004. Inhibition of pathogens on fresh produce by ultraviolet energy. *International Journal of Food Microbiology* 90: 1–8.

Ye, Z., T. Koutchma, B. Parisi, J. Larkin, and L.J. Forney. 2007. Ultraviolet inactivation kinetics of *E. coli* and *Y. pseudotuberculosis* in annular reactors. *Journal of Food Science* 72: E271–E278.

Yurdun, T., G.Z. Omurtag, and Ö. Ersoy. 2001. Incidence of patulin in apple juices marketed in Turkey. *Journal of Food Protection* 11: 1851–1853.

Zhang, D., B. Yu, J. Bai, M. Qian, Q. Shu, J. Su, and Y. Teng. 2012. Effects of high temperatures on UV-B/visible irradiation induced postharvest anthocyanin accumulation in 'Yunhongli No. 1' (*Pyrus pyrifolia* Nakai) pears. *Scientia Horticulturae* 134: 53–59.

Zhu, Y., Wu, F., Koutchma, T., Warriner, K., and Zhou, T. 2012. Validation and Kinetics of Patulin Degradation in Model Solution, Apple Cider and Apple Juice by Ultraviolet Radiation. (manuscript was submitted.)

Chapter 18
Ozone Antimicrobial Effects on Fruits and Fruit Juices

David R. Kasler and Ahmed E. Yousef

18.1 Introduction

Agricultural commodities, including fresh produce, are exposed to microbial contamination during production, harvesting, and post-harvest processes. Many of these commodities are treated thermally before consumption, but fresh produce does not receive such treatments. Similar to other fresh produce items, fruits are consumed raw or minimally processed; hence, the risk to consumers needs to be addressed. At the outset, it may be prudent to explain some of the terms as used in this chapter. A fruit remains "raw" or "fresh" if its physical characteristics at the time of harvest have not been altered considerably. Freezing, blanching, irradiating, or cooking often alters fruit's raw tissues characteristics sufficiently to render it "processed," thus it can no longer be described as fresh. Trimming, peeling, and cutting change the physical characteristics of a fruit, but these processes mostly maintain the integrity of edible tissues, thus the product is often described as "fresh cut." It is understandable that a fruit is washed and often sanitized before it is transformed into a "fresh-cut" product. Processing (e.g., juice extraction, pasteurization, or sterilization) alters fresh fruit's physical characteristics and may also change its chemical, microbiological, and nutritional qualities; thus the end product is described as "processed." When processing is done carefully enough to maintain fruit's chemical and nutritional qualities, this processing is described as "minimal." Minimal processing is not intended to pasteurize or sterilize the product, thus, considerable microbial population could remain in the minimally processed fruit. Minimal processes designed to decontaminate fresh produce may decrease the population of viable microbes on a fruit by 1–3 log cfu/g.

D.R. Kasler • A.E. Yousef (✉)
The Ohio State University, Columbus, OH, USA
e-mail: yousef.1@osu.edu

A. Rosenthal et al. (eds.), *Fruit Preservation*, Food Engineering Series,
https://doi.org/10.1007/978-1-4939-3311-2_18

Several agricultural commodities are considered fruits. The fruit groups emphasized in this chapter include pomes (e.g., apples), berries (e.g., grapes), citrus (e.g., oranges), melons (e.g., cantaloupes), and aggregate fruits (e.g., strawberries). In preparation for retail distribution and consumption, fruits are commonly washed and sanitized. Washing removes extraneous materials and visible contaminants. Sanitization could be combined with washing, if the product is barely soiled, or applied subsequent to the washing step. The most commonly used sanitizer is aqueous hypochlorous acid (HOCl), and the process may be described as chlorine treatment (or chlorination), even though elemental chlorine (Cl_2) is not the active agent applied directly to fresh produce during the sanitization process. Use of electrolyzed water may be considered a chlorine treatment. Alternatives to chlorine have been studied and sometimes implemented commercially; these include ozone, chlorine dioxide, and peracetic acid.

Ozone has been approved by the United States Food and Drug Administration (FDA) as a food additive (FDA 2001). Many researchers and food processors have tested ozone for fruit decontamination with varying degrees of success. Results of this research have been implemented in some packing facilities. Organic producers allow ozone use, as a potent alternative to chlorine (Extension 2011). Ozone is a gas at atmospheric conditions and the molecule is the triatomic form of the oxygen atom. Ozone decomposes rapidly at room temperature to diatomic oxygen, and rate of decomposition increases as gas temperature increases. A powerful oxidizer, ozone has an oxidation potential of 2.07 V, which is higher than that of chlorine (1.36 V) (Horvath et al. 1985). Ozone is unique that it can be applied to food products in both gaseous form and aqueous solutions. With the gaseous form, ozone can be applied at concentration up to 14 % by weight using oxygen as the carrier gas (Chawla et al. 2012). With aqueous solutions, gaseous ozone is bubbled and dissolved in water, reaching easily to concentrations up to 6 ppm (Suslow 2004); however, in one study, 25 ppm ozone in water (on weight bases) was achieved (Achen and Yousef 2001).

18.2 Ozone Processing Consideration

Before ozone is used to inactivate microbes on fruits, the following factors should be considered.

- Ozone is very reactive with most organic and inorganic compounds, regardless of their source. If the reaction happens between ozone and bacterial cell components, this leads to microbial death. However, organic matter, extraneous to microbial cell, also reacts with ozone and eliminates its antimicrobial characteristics.
- Accessibility of ozone to microbial contaminant is hindered by surface roughness and entrapment of microorganisms in microscopic crevices.
- Similar to all water-soluble gases, ozone in liquids diffuses at much smaller rate than it does when the molecule is in the gaseous state.

18.3 Ozone Application Along the Fruit Supply Chain

As the case with most food, fruits are subject to microbial contamination throughout the supply chain; this has been well-documented in published literature (e.g., OMAF Food Inspection Branch 2001). Measures to protect fruits against harmful microbes should be started at the field, continued during harvest, intensified during post-harvest operations, and maintained during distribution, warehousing, retailing, and preparation at consumer's home. A weakness in any link of this long chain could lead to contamination with harmful organisms or proliferation of disease-causing pathogens. Many of these measures fall under good production and manufacturing practices, whereas others are deliberate intervention steps. Ozone, and similar antimicrobial agents, could be applied at various stages of the fruit supply chain with the goal of decreasing product microbial load (Fig. 18.1).

18.3.1 Field Application

Many plant diseases are caused by microorganisms which render fruits unsuitable for human consumption. Additionally, infestation of fruits with pests often leads to contamination with microorganisms that can spoil the product or cause diseases to humans. Several field treatments of fruit-bearing plants may decrease these hazards. One of these treatments involves spraying plants with ozone solution as a potential biological control agent. Although the treatment is not specifically designed for fruit-bearing plants, or optimized to reduce microbial load on fruits, it is potentially useful in improving the microbiological qualities of the product.

Recently, researchers reported that ozonated water spray seems to help in controlling insects and diseases of grapes and vines in the field. When commercial winery and vineyards were sprayed with ozonated water, 2 % of vines showed signs of insects while 38 % of the control group showed insect damage (Wood 2013). Disease on the vines was reduced during the growing season, but by the end, disease was similar between the two groups of vines. Despite many reports of the detrimental effect of ozone pollution on plants (e.g., Wang et al. 2014), ozone dissolved in water was not reported to have any adverse effects on the grapes or grape vines.

Application of ozone in soybean fields has been tested by Steffen and Rice (2008a) in the 2006–2007 growing season in Brazil. A three-step sequential application of electrolyzed water, ozonated water, and ultraviolet (UV) light was used to control insects on a soybean field of 6500 acres. According to the authors, the crop did not require any other pesticide or insecticide treatments during the growing season and the ozone treatment produced economic benefits to the farmer. Although the tests just described were applied to soybeans, the concept is likely applicable to orchard trees and other fruit-bearing plants. The same authors (Steffen and Rice 2008b) described a similar process using 8 mg/L ozonated water in sequence with the electrolyzed water and UV light. Mechanistically, application of ozone solution

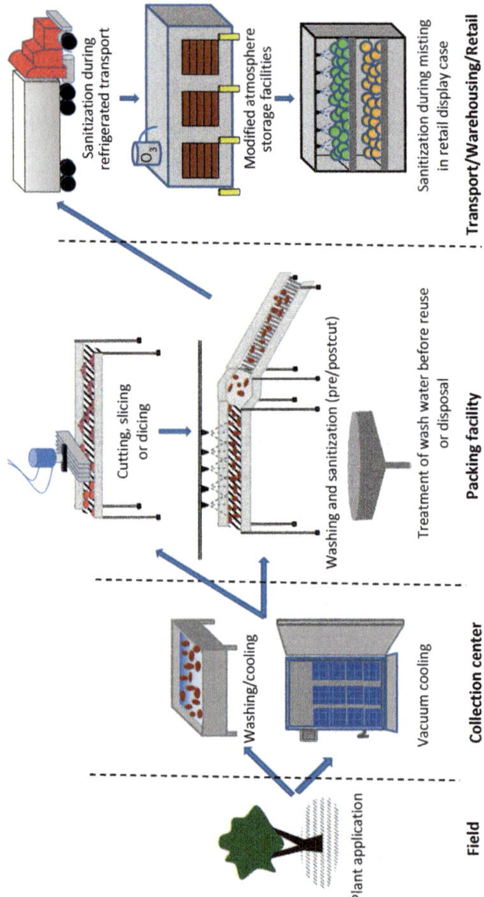

Fig. 18.1 Current and potential ozone applications across the fruit supply chain

is believed to stimulate plant defenses against microbial invasion (e.g., fungal infections). Stressing the growing plants with solutions that generate reactive oxygen species (e.g., ozonated water) may trigger systemic-acquired resistance response that helps plants fight insect infestation and microbial diseases.

18.3.2 Freshly Harvested Fruits

Washing, chilling, and vacuum cooling are some of the processes applied, individually or in combination, to freshly harvested fruits. These processes are often applied at a collection facility near the fields. Ozone can be applied in conjunction with these operations, but variable outcome is expected. Washing fruits near the field is not a common practice by major fruit producers, but it is implemented particularly by some small producers. The washing process often generates considerable amount of organic matter in the wash water, rendering any ozone treatment useless. Ozone is expected to be consumed while reacting with water's organic load. Therefore, it is not recommended to apply ozone in conjunction with initial washing processes.

Removing field heat and minimizing post-harvest cell respiration that generates additional heat are critical for maintaining the quality of many fruits. Cooling freshly harvested fruits can accomplish these goals. Application of forced-air, hydro, or vacuum cooling (Li 2014) is a common practice in the fresh produce industry. Microbial cross-contamination during hydrocooling may be minimized by chlorination of the cooling water. Alternatively, ozone may be dissolved in cooling water before application to fruits. Vacuum cooling is done for some fruits such as cantaloupe. The process involves applying vacuum to evaporate free water on fruits, thus cooling the product rapidly. A process that combines vacuum cooling and sanitization with gaseous ozone has been developed (Vurma et al. 2009). When tested on spinach leaves inoculated with *Escherichia coli* O157:H7, the process decreased the population of the inoculated bacterium by up to 2.4 log cfu/g. Decontamination efficacy depended on ozone concentration and pressure as well as treatment time. The researchers optimized the process so that maximum microbial inactivation is achieved without causing damage to the delicate spinach leaves. The optimized ozonation-vacuum cooling process inactivated 1.8 log of *E. coli* O157:H7 population on spinach leaves. Although the process was tested on leafy green only (Vurma 2009; Yesil 2012), it is likely applicable to several types of fruits.

18.3.3 Packinghouse

Some fruits such as strawberries are not washed at any point in the supply chain, whereas others are trimmed, washed, sanitized, cut, and packaged at specialized packinghouses. Washing in the packinghouse is often followed with sanitization using chlorine or alternative sanitizers. Application of these sanitizers during washing generally results in modest decrease in microbial load of fresh produce (Kim et al. 1999; Sapers 2001). Sanitizers are likely consumed by organics (including microbes) released from the surface of fresh produce during washing. Thus, efficacy of these sanitizers against microbes attached or embedded into fresh produce surface is limited. Despite these limitations, aqueous sanitizers decrease microbial load of wash water, which allows its reuse in the continuous washing line, throughout a production shift. In the absence of a sanitizer, microbial load would increase as wash water continues to be used; thus product microbial quality worsens progressively during the production shift.

Aqueous chlorine is the most used sanitizers on fresh produce at packinghouses. The active ingredient, HOCl, is known to react with many organic and inorganic compounds during treatment of water (Debordea and von Guntena 2008). Reaction of chlorine with organic matter in water creates a variety of chlorinated organic by-products; these include trihalomethanes (mainly chloroform) and haloacetic acids, with smaller amounts of haloaldehydes, haloacetonitriles, and haloketones (Euro Chlor Communications 2013). Potential adverse effects on health from these by-products, particularly trihalomethanes, led to the tightening of chlorine treatments regulations. It is generally assumed that chlorine treatment of fresh produce generates chlorinated organic by-products; hence, alternatives to chlorine have been investigated. Ozone, as an alternative sanitizer, also reacts with organic and inorganic compounds in water (Debordea and von Guntena 2008), but hazardous reaction by-products are not expected.

For fruits washed and sanitized in the packinghouse, it would be possible to use ozone in lieu of chlorine. Aqueous ozone is created by thoroughly mixing ozone gas with water through a venturi in a dissolution column, or through a small porous filter (Perry and Yousef 2011). The process of preparing aqueous ozone is influenced by three factors, (1) incoming ozone gas concentration, (2) gas pressure, and (3) water temperature. Increasing the concentration of ozone in water could be accomplished by increasing the ozone concentration and pressure of incoming gas and lowering water temperature (Chawla et al. 2012).

Decontamination of apples was attempted using aqueous ozone (Achen and Yousef 2001). Apples were inoculated with *E. coli* O157:H7 and dipped for 3 min in an aqueous solution containing high concentration of ozone (22–24 mg/L). The treatment decreased the pathogen population of apple surface by 2.6 log cfu/g. Dipping apples in an ozone solution (22 mg/mL) while bubbling ozone in the mixture decreased pathogen population by 3.7 log cfu/g. Greater efficacy was observed when the researchers used inorganic surfactant in combination with the ozone treatment. Based on these findings, it may be concluded that use of aqueous ozone improves the safety of fresh fruits. The study, however, illustrated the limitations of aqueous ozone as a sanitizer. Levels of ozone used in the study were higher than what is expected in an industrial setting. The decontamination of apple's stem-calyx region was difficult to accomplish. The study also emphasized the need for ozone-compatible inorganic surfactant to help in releasing entrapped microorganisms on apple surface.

Effluent of packinghouse may also be subjected to ozone treatment. Although no specific detail about its efficacy was given, a process for ozone use to treat fruit-washing waste water is available (Ozone Technologies Group 2014). Similar waste water treatment technology has been explored in the poultry industry (Praxair 1999).

18.3.4 Application During Transportation or Cold Storage

Based on research findings by Vurma et al. (2009), it was suggested that quality and safety of fresh produce could be improved by application of ozone during transportation of these products. The researchers envisioned that ozone gas could be applied, in a continuous mode, to fresh produce held in moderately sealed chambers such as transportation containers or storage coolers. A stream of gas containing low concentrations of ozone may be introduced into these chambers to replace existing atmosphere. Recently, Cullen and Tiwari (2012) reported the commercial application of this technology for storage of apples, cherries, kiwi, peaches, plums, strawberries, grapes, tomatoes, and blackberries.

Ozone has also been used to disinfect cold storage rooms and water and air treatment equipment associated with fruit handling. Ozone was used in produce-holding cold rooms to decrease *Listeria monocytogenes* contamination and to minimize cross-contamination (Suslow 2004). Use of ozone during storage may lower the level of ethylene, which is produced by fruits to induce ripening (Skog and Chu 2001). Uncontrolled ripening could lead to fruit microbial spoilage during storage.

18.3.5 Ozone at Grocer and Retail Establishments

Grocers and retailers often spray fresh produce in display racks with tap water to maintain product freshness. Tap water has been replaced gradually with sanitizers, such as ozone solutions and electrolyzed water, to mist fresh produce on retail racks. The US national grocery chain, Whole Food, recently installed ozone generators into existing misting systems in fresh produce racks and washing systems in the grocery stores in California (Eco Safe Systems Food Safety 2010). The treatment potentially decreases microbial load (including pathogens) remaining on products that were contaminated in the field and minimizes accidental microbial contaminants from consumer handling in grocery stores.

18.4 Ozone-Treated Fruits

Fruits may be treated with ozone to inactivate pathogens and to delay or prevent spoilage during distribution and storage. Additionally, ozone application has been suggested to decrease residues of pesticides that have been applied to fruits in the field. Although information about fruits is somewhat limited, Wu et al. (2007) found up to 61 % removal of residual pesticides on leafy green vegetables by ozone. Suitability of ozone application depends on fruit type and stage of processing. Treatments suitable for whole fruits could be ineffective or undesirable for cut fruits.

18.4.1 Pomes

Pome is a fruit consisting of an outer thickened fleshy layer and a central core that contains the seeds. Examples of pomes are apples and pears. The surface of these fruits is relatively smooth and sometimes covered with natural waxy layer. This surface characteristic makes these fruits well suited for decontamination by aqueous ozone. Microbial contaminants that gain access to internal tissues through skin bruises or cuts are difficult to eliminate by aqueous sanitization technologies.

It is conceivable that apples and similar fruits could be contaminated, in the field, with human pathogens. Consumption of unpasteurized apple juice or cider has been linked to disease outbreaks resulting from infection with pathogenic *E. coli* (Besser et al. 1993; CDC 1996). These outbreaks prompted US-FDA to recommend thermal pasteurization of apple cider and other juice products, or use of alternative processing steps to decrease the counts of the pathogen in question by 5 log cfu/mL (FDA 1998). Decontamination of apples before consumption (or use in juice production) should improve safety of both fresh and processed products. Aqueous ozone has been investigated as a sanitizer for treatment of fresh apples that were contaminated with a high inoculum of *E. coli* O157:H7 (Achen and Yousef 2001). The researchers concluded the following: (1) dipping contaminated apple in a solution containing high concentration ozone (21–28 mg/L) decreased pathogen population by 2.6 log cfu/g, (2) bubbling ozone gas in the ozone solution further decreased the population by additional 1.1 log cfu/g, (3) pathogens population in the stem-calyx region of the apple decreased by less than 1 log cfu/g, regardless the ozone delivery method, (4) cutting the apple proved to be detrimental to the antimicrobial efficacy of the ozone solution, and (5) there is a need for a wetting agent that can be used in conjunction with apple washing and sanitization, provided that the agent does not react or interfere with ozone.

Rodgers et al. (2004) used aqueous ozone solution (3 ppm) to treat apples that were inoculated with *E. coli and L. monocytogenes*. Despite this relatively low ozone concentration, the researchers showed that pathogens populations decreased on whole apples by more than 5 log cfu/g after a 5 min wash with the ozone solution. Variability in ozone efficacy between the two studies just reviewed is likely caused by differences in methodological details.

In addition to apple decontamination, ozone has been tested to remove residues of pesticides applied to fruits in the field. In a study by Ong et al. (1996), ozonated water at 0.25 ppm was used to wash apples contaminated with selected pesticides that control diseases, mites, and insects. Effect of temperatures (21 °C and 44 °C) and pH (4.5–10.7) on the treatment also was investigated. It was concluded that ozone at reduced pH and temperature can be effective at removing pesticides from fruit surfaces.

Ozone treatment of apples and pears has also been documented as reducing ethylene, a phytohormone responsible for fruit ripening, in storage rooms containing these fruits. After 107 days storage and use of 0.4 ppm (v/v) gaseous ozone at 0 °C, ethylene content was maintained lower than 2 ppm while air-stored fruits had 25 ppm. Using this small amount of ozone caused no significant change in scald index, titratable acidity, firmness, or total soluble solids in treated fruits (Skog and Chu 2001).

18.4.2 Berries

A berry is defined botanically as a fleshy fruit produced from a single ovary. Examples of berries are blueberry, cranberry, and grape. Intact smooth surface layer of these fruits provides protection against ingress of microorganisms to internal tissues. Despite this natural protection, berries could be contaminated, throughout the supply chain, with microorganisms that cause fruit spoilage or diseases to humans. Contamination could be associated with soil at the pre-production stage, agricultural practices during production (e.g., contaminated irrigation water or organic fertilizer), harvesting and packing (e.g., poor workers hygiene), storage, transportation, wholesale, and retail (OMAF Food Inspection Branch 2001).

Table grapes were treated with gaseous ozone to prevent spoilage during storage (Cayuela et al. 2009). Gaseous ozone (2 ppm by volume) was applied continuously or intermittently (12 h per day) during storage of table grapes at 5 °C for 72 days. Ozone treatments considerably decreased decay of stored grapes, compared to those stored in air, with continuous ozone treatment being the most effective for controlling losses during storage. Interestingly, the intermittent ozone treatment resulted in the highest resveratrol content in stored grapes; resveratrol is a nutritionally desired ingredient in food.

Gaseous ozone, at much higher levels than in the previous study, was applied to control post-harvest gray mold infestation of grapes caused by *Botrytis cinerea* (Gabler et al. 2010). The grapes were placed in a small chamber and subjected to moderate vacuum (33 kPa) and gaseous ozone treatment (5000 ppm by volume) at 5 °C. The treatment lasted for 60 min and treated grapes were kept in cold storage for 1 month. Gray mold incidence was reduced among inoculated grapes from 92 to 19 % by 60 min of the gaseous ozone treatment.

Blueberries were inoculated with *Salmonella enterica* or *E. coli* O157:H7 and treated with ozone (Bialka and Demirci 2007). Berries were treated with ozone gas (5 % by weight) for 64 min under pressure (83 kPa) or in a continuous exposure mode. These two treatments inactivated the pathogens by 3.0 log cfu/g and 2.2 log cfu/g, respectively. The researchers also treated inoculated berries with aqueous ozone at 8.9 ppm for 64 min. The treatment inactivated the two pathogens by 4.9 log and 4.7 log, respectively.

18.4.3 Melons

Regardless the botanical definition, melons addressed here include cantaloupe and watermelon. These fruits have relatively thick rind that protects internal edible tissues. The surface layer varies in roughness depending on the variety. Contaminants on rough surfaces are expected to be difficult to remove by washing or even during sanitization processes. When melons are cut, pathogen on the surface may contaminate the edible portion of the fruit.

A 2012 multistate outbreak of infections in the US, caused by *Salmonella enterica* serovars Typhimurium and Newport, was linked to cantaloupe (CDC 2012b). The source of contaminated cantaloupe was Chamberlain Farms Produce, Inc. of Owensville, Indiana. The disease outbreak resulted in 261 illnesses and three deaths. Consumption of cantaloupes also has been linked to listeriosis outbreak in 2011 (CDC 2012a). The infection was caused by five subtypes of *L. monocytogenes* and resulted in 147 illnesses and 33 deaths. The cantaloupes came from Jensen Farms, and outbreak subtypes were detected in samples from equipment and cantaloupe from the Jensen Farms' packing facility in Granada, Colorado, USA. In Europe, sliced watermelon was linked to an outbreak of salmonellosis in 2011 (Food Safety News 2012). Infection with *Salmonella* Newport was the cause of the outbreak which results 54 illnesses and one death.

Safety and microbiological quality of melons could be improved by application of a suitable sanitizer such as ozone. Rodgers and colleagues (2004) inoculated cantaloupes with *L. monocytogenes* and *E. coli* O157:H7 and immersed the product for 5 min in solution containing 5 ppm ozone. The treatment decreased pathogens populations by more than 5 log cfu/g. Ozone was less effective against the natural microbiota of cantaloupe. Populations of mold, yeast, and mesophilic bacterial were initially reduced by 1.6 log cfu/g, 1.4 log cfu/g, and 4.2 log cfu/g, respectively, but increased during the refrigerated storage at 4 °C.

In another study (Selma et al. 2008), cantaloupe was subjected to gaseous ozone treatments ranging from 5000 to 20,000 ppm, in a closed vessel with vacuum pretreatments, in order to eliminate inoculated *Salmonella* and natural spoilage microorganisms such as coliforms, *Pseudomonas fluorescens*, yeast, and lactic acid bacteria. When cantaloupe rind disks (12.6 cm^2) were inoculated with *Salmonella* and treated with 10,000 ppm ozone gas for 30 min, population of the pathogen decreased by 4.2 log cfu/rind-disk for unripe cantaloupe and 2.8 cfu/rind-disk for ripe cantaloupe. Treatment of fresh-cut cantaloupe with 5000 ppm ozone for 30 min reduced natural microbiota only by 0.5–1.1 log cfu/cube. Increasing the ozone concentration to 20,000 ppm for 30 min increased lethality of all natural microbiota, particularly *Pseudomonas fluorescens,* which was reduced by 1.9 log cfu/cube. Shelf-life of ozone-treated cantaloupe showed that all natural microbiota maintained a 1–2 log cfu/cube lower than untreated fresh cut cantaloupe. Quality of treated fresh cut cantaloupe was also found to be affected minimally by ozone processing.

Sliced watermelon was examined in a study by Fonseca and Rushing (2006). Watermelon cubes were dipped in 0.4 ppm ozonated water and the treatment was found to have limited antimicrobial effect. It was also reported that both the color and overall quality of the watermelon were affected negatively by the ozone treatments. Considering the studies on cantaloupe, if whole watermelons are to be treated with ozone, it is likely that the treatment would decrease microbial population on the rind considerably.

18.4.4 Citrus

Citrus fruits are produced by evergreen trees of the genus *Citrus*. The fruit has thick rind and juicy pulp. Healthy rind can protect edible portion of the fruit from microbial contaminants. Additionally, acidity of the juicy pulp hinders the growth of many bacteria. Despite these natural safeguards, citrus fruits can support growth of molds and allow survival of some bacterial pathogens. Oranges and tangerines are the citrus fruits covered in this chapter.

18.4.4.1 Oranges

Development of molds on cold-stored oranges, with or without ozone treatment, was investigated by Palou et al. (2001). Ozone gas was applied at two levels under distinctly different storage conditions; (1) 0.3 ppm (by volume) ozone for 4 weeks at 5 °C and (2) 1 ppm (by volume) ozone at 10 °C for 2 weeks. Mold growth was examined throughout storage and it was found that ozonated storage at 0.3 ppm increased time of fruit disease onset. At 14 days of storage, fruit disease incidents were lower with ozone storage, but after 21 days ozone-stored and untreated fruits had the same disease incidents. Despite the disease incidents being high, the disease was less severe in fruits stored with ozone. Similar effects of delayed incidence were seen at elevated levels of ozone (1 ppm) and temperature (10 °C). It was reported in both instances that ozone storage did not damage the fruit.

The same research group examined the efficacy of gaseous ozone against *Penicillium digitatum* and *P. italicum* during storage of oranges (Palou et al. 2003). Oranges were packaged in different packaging materials and stored for 14 days in a room held at 12.8 °C. Ozone level was maintained at 0.72 ppm (by volume). Effectiveness of ozone during storage was influenced greatly by the packing material used, but the treatment resulted in satisfactory inhibition of mold sporulation when oranges were stored in vented returnable plastic containers. Bagged fruit inside cartons showed no significant change in mold sporulation even though the packaging was porous.

18.4.4.2 Tangerines

Whangchai et al. (2010) examined the effect of electrolyzed oxidizing (EO) water, when followed by gaseous ozone treatment, on the decay of tangerines by *Penicillium digitatum*. Fruits were dipped in EO water for 4–16 min and stored at 5 °C for up to 28 days. After EO water treatment, some of the fruit samples were subjected to 200 mg/L gaseous ozone for 2 h a day, for 28 days. Tangerines that were EO water-dipped and ozone-stored showed a decrease in disease incidence of up to 10 %, compared to the EO water-dipped and air-stored control after the entire 28-day storage.

Tangerines that were EO-dipped for at least 8 min and treated with gaseous ozone produced no disease incidence after 28 days. The researchers reported no effects of the treatment on weight loss, total soluble solids, titratable acidity, or peel color.

18.4.5 Aggregate Fruits

In contrast to a simple fruit which develops from one ovary, an aggregate fruit develops from the merger of several ovaries that were separate in a single flower. The flower produces many tiny fruits clustered tightly together, like raspberries, blackberries, and strawberries. These fruits caused numerous disease outbreaks (e.g., CDC 2002; Food Safety News 2011). Washing aggregate fruits is avoided throughout the supply chain to minimize product contamination with microorganisms and ensuing spoilage. To overcome this limitation, treatment of these fruits with gaseous ozone could protect consumers against microbial contaminants.

Rodgers et al. (2004) studied the decontamination of a variety of fruit products using various sanitizers, including ozone. Strawberries were inoculated with *E. coli* and *L. monocytogenes* and treated with ozonated water (3 ppm) for 5 min. The treatment decreased pathogens populations on strawberries by more than 5 log. Populations of pathogens remained less than 1 log cfu/g after 9 days of storage at 4 °C.

Effect of ozone-enriched atmosphere on the growth of *Botrytis cinerea* on stored strawberries was investigated by Nadas et al. (2006). Strawberries were inoculated with the mold and stored at 2 °C for 3 days with or without ozone gas at 1.5 ppm (by volume). After storage, fruits were transferred to a 20 °C-room and stored for 4 days, and decay incidences were observed. Results showed that ozone storage significantly decreased decay incidence and that ozone-treated fruit had less weight loss and fruit softening than their air-treated counterparts. The only trait negatively affected by ozone was the aroma of the fruit, which was recovered later when the fruits were removed from the ozone storage.

18.4.6 Kiwi Fruit

Kiwi fruits were tested for spoilage control by a combination of titanium dioxide (TiO_2) photocatalytic oxidation with aqueous ozone (Hur et al. 2005). Processing was performed by dipping kiwifruit in the aqueous solution of ozone and TiO_2 for 30 min. Concentration of ozone was maintained at 0.5 ppm (by weight). Although spores of *Diaporthe actinidiae* suspension, a common agent causing kiwi diseases, had a germination rate reduction of more than 60 % when treated with TiO_2 and O_3 simultaneously, reduction in spore germination on kiwi fruit was only 13 %, compared to untreated fruit. Despite the high incidence of decay on the fruit, the disease affected only 11.5 % of the treated fruit surface, as opposed to 63 % of the control. The ozone and TiO_2 treatment was found to be better than a common chemical fungicide, flusilazole, used on the fruit. The researchers also found that the ozone and

TiO_2 treatment could be used to decompose the flusilazole if it had been applied pre-harvest to the fruit.

18.4.7 Dried Fruits

Although drying of food prevents microbial growth, dormant microorganisms in these products are difficult to inactivate. When food processors search for methods to decontaminate dry food, they face the reality that limited options are available. Dried fruits are no exception being difficult to decontaminate in the event of incidence of pathogens. Ozone gas has strong antimicrobial properties and thus may be considered by researchers and processors who develop new decontamination methods. Ozone's high oxidative power could be detrimental to the quality of treated dry food, but process outcome should depend on the type of food treated. Little information is available in published literature on the use of ozone to decontaminate dried fruits.

18.4.7.1 Figs

Dried figs have been examined for ozone efficacy against natural microbiota in the product (Oztenkin et al. 2006). Gaseous ozone at 5 and 10 ppm was applied to figs for 3–5 h and changes in populations of aerobic mesophilic microorganisms, coliforms, and yeast/molds were measured. Initial populations of these microbial groups in figs were low (1.5–2.6 log cfu/g), and thus modest change in these populations (1.0–1.5 log reductions) was observed following the ozone treatment. The authors concluded that reduction of figs microbiota requires a minimum of 5 ppm ozone treatment for 3 h.

Control of bacterial vegetative cells and spores on dried figs by gaseous ozone was examined by Akbas and Ozdemir (2008). Dried figs were inoculated with *E. coli*, *Bacillus cereus*, and *B. cereus* spores and treated with 0.1–9 ppm ozone gas for 6 h. Ozone, at 1 ppm, decreased the populations of *E coli* and *B. cereus* by 3.5 log cfu/g. Viable *B. cereus* spore population decreased by 2 log using ozone at greater than 1 ppm for 6 h. Ozone-processed figs were highly palatable according to results of the sensory analysis conducted in the study that featured both visual analysis and taste testing. Scores for sweetness, rancidity, flavor, appearance, and overall palatability were not statistically different between ozone-treated and untreated figs.

18.4.7.2 Date Fruits

Dates have been tested for gaseous ozone effectiveness against product's microbial load. Najafi and Haddad Khodaparast (2009) found gaseous ozone (5 ppm) to be effective at reducing mesophylic, coliforms, yeast and molds, and *Staphylococcus aureus* on dates after 60 min of treatment.

18.5 Ozone-Treated Fruit Juices

Fruit juices have been implicated in several diseases outbreaks, as described in a previous section. Thermal processing of fruit juices is a common practice that improves product shelf-life and eliminates potentially harmful microorganisms. However, there is a market for minimally processed juices that appeal to consumers who prefers products' natural flavors and potential nutritional benefits. Ozone has been sought as an alternative to heat pasteurization of fruit juices. Attempts to accomplish this goal resulted in controversial outcomes. Organic components of the juice could consume ozone and thus diminish the antimicrobial efficacy of the treatment. Oxidation of these components by ozone also could decrease the nutritional value of the juice. Despite these obvious drawbacks, many researchers studied juice processing by ozone; results of these studies are reported herein.

18.5.1 Apple Juice

Potential pasteurization of apple juice by ozone was examined in a study by Patil et al. (2010a). Ozone gas was bubbled through the juice that had been previously inoculated with two strains of *E. coli*. Treatment setup included ozone flow at 0.12 L/min, ozone concentration of 0.048 mg/min/mL, and a treatment time of 18 min or less. Results showed that pasteurization (reduction of test microorganism count by 5 log cfu/mL) could be achieved using ozone gas that was bubbled directly into the juice. This goal was accomplished using the treatment conditions just described when juice pH was 5.0 and treatment time was 18 min, or when juice pH was 3.0 and treatment time was 3 min. In a similar study, Patil et al. (2010b) measured the changes in the quality of apple juice as affected by the ozone treatment. Results showed that the ozone treatment sufficient to pasteurize the apple juice significantly changed color and phenol content of the product.

Steenstrup and Floros (2004) inoculated apple cider with *E. coli* O157:H7 and treated the product with gaseous ozone at different concentrations (860 and greater than 1000 ppm) and temperatures (5 and 20 °C). The authors noticed a time lag of 3.5–6.7 min between application of ozone to the cider and onset of microbial inactivation. Additionally, there was a critical concentration of dissolved ozone in the cider that had to be reached before the onset of *E. coli* inactivation. The two observations probably are related to the ozone demand of the juice. D-value and total processing time (lag time plus 5 D-values) were shortest at higher ozone concentration and higher treatment temperature. The authors concluded that inactivation of *E. coli* OI57:H7 by ozone generally was fast enough to allow practical applications in cider production, and the process should be considered as an alternative to thermal pasteurization.

Choi and Nielsen (2005) compared thermal and nonthermal processes, including ozone treatment, in terms of apple cider quality. Gaseous ozone was applied to apple cider (200 mL aliquots) at a flow rate of 0.5 L/min, a concentration of 860 ppm, and a treatment time of 28 min. Treated cider was tested, during 21 days of storage, for changes in these quality parameters: pH, titratable acidity, turbidity, sedimentation, color, Brix, and sugar content. Treatment of cider with ozone increased sedimentation, lowered sucrose content, and decreased soluble solids by day 21.

18.5.2 Orange Juice

Patil et al. (2009) used ozone gas to inactive *E. coli* in orange juice. Ozone gas (75–78 ppm) was bubbled into the juice at a flow rate of 0.12 L/min for up to 30 min. Various orange juice media were examined including unfiltered juice, filtered juice, and juice with reduced pulp. With unfiltered juice, an ozone treatment of 15–18 min was needed to inactivate 5 log of the targeted microorganisms. Low and no pulp orange juices took 6 min to achieve 5 log reduction. Effects of ozone on the juice quality were not examined in this study. However, quality of ozone-treated orange juice was investigated in a study by Tiwari et al. (2008). Ozone gas at flow rates up to 0.25 L/min was bubbled through the juice for up to 10 min. The study was designed such that ozone treatment would produce 5 log reduction, at least, of *E. coli* 0157:H7 population. No significant differences were found in pH, Brix, titratable acid, cloud value, and browning between the control and the ozone-treated juice. However, it was reported that sedimentation and color were affected by ozone treatment; juice became lighter with ozone-treatment.

18.6 Summary

Ozone can be applied at almost any step in the fruit supply chain, from the orchard to the display case at local grocery stores. The gas can be easily produced commercially and used in either gaseous or aqueous states. In addition to improving fruit safety and extending product shelf-life, treatments may also be selected to enhance the nutritional quality of food or remove residues of pesticide applied to fruits in the field.

Ozone processing of fruits to eliminate spoilage and pathogenic microorganisms is very case-sensitive and subjected to a number of factors. For example, ozone application at certain links of fruit supply chain (e.g., storage) produced positive effects, but results at other stages (e.g., juice processing) are largely inconclusive. It is also difficult to make decisive recommendation on the optimum treatment conditions that fit many situations in the fruit industry. Despite this difficulty, it may be concluded that efficacy of ozone on fruits depends on these critical factors: fruit being processed and its state (i.e., whole or cut), target microorganism, ozone production and application method, ozone concentration and pressure, processing time, and treatment temperature.

Lack of agreement between results of different laboratories could be attributed to methodological variances. Researcher and processors who develop ozone decontamination technologies need to document carefully all experimental factors, including ozone measurement units, treatment conditions, product inoculation and microbial enumeration methods, and result-reporting format.

Use of this potent antimicrobial agent and strong oxidizer requires careful consideration and process optimization so that the goal of the treatment is accomplished with minimal shortcomings. The potential benefits of ozone in fruit processing seem very vast. Adoption of ozone is likely to continue in the future by the fresh fruit industry; however, the rate at which this occurs will depend on how well researchers provide consistent measurements and results, and how well publications help us understand details that affect ozone processing.

References

Achen, M., and A.E. Yousef. 2001. Efficacy of ozone against *Escherichia coli* 0157:H7 on apples. *Journal of Food Science* 66: 1380–1384.

Akbas, M., and M. Ozdemir. 2008. Application of gaseous ozone to control populations of *Escherichia coli, Bacillus cereus* and *Bacillus cereus* spores in dried figs. *Food Microbiology* 25: 386–391.

Besser, R.E., S.M. Lett, J. Todd Weber, M.P. Doyle, T.J. Barrett, J.G. Wells, and P.M. Griffin. 1993. An outbreak of diarrhea and hemolytic uremic syndrome from *Escherichia coli* O157:H7 in fresh-pressed apple cider. *Journal of the American Medical Association* 269: 2217–2220.

Bialka, K., and A. Demirci. 2007. Decontamination of *Escherichia coli* O157:H7 and *Salmonella enterica* on blueberries using ozone and pulsed UV-light. *Journal of Food Science* 72: M391–M396.

Cayuela, J.A., A. Vazquez, A.G. Perez, and J.M. Garcia. 2009. Control of table grapes postharvest decay by ozone treatment and resveratrol induction. *Food Science and Technology International* 15: 495–502.

CDC. 1996. Outbreak of *Escherichia coli* O157:H7 infections associated with drinking unpasteurized commercial apple juice. British Columbia, California, Colorado, and Washington, October 1996. *Morbidity and Mortality Weekly Report* 45: 975–982.

———. 2002. *Outbreak of cyclosporiasis associated with imported raspberries, Philadelphia, Pennsylvania, 2000.* http://wwwnc.cdc.gov/eid/article/8/8/02-0012_article (accessed June 20, 2014).

———. 2012a. *Multistate outbreak of listeriosis linked to whole cantaloupes from Jensen farms, Colorado.* http://www.cdc.gov/listeria/outbreaks/cantaloupes-jensen-farms/082712/index.html (accessed June 20, 2014).

———. 2012b. *Multistate outbreak of* Salmonella Typhimurium *and* Salmonella *Newport infections linked to cantaloupe (final update).* http://www.cdc.gov/salmonella/typhimurium-cantaloupe-08-12/index.html (accessed June 20, 2014).

Chawla, A.S., D.R. Kasler, S.K. Sastry, and A.E. Yousef. 2012. Food decontamination using ozone. In *Food decontamination: Novel methods and applications*, ed. A. Demirci and M. Ngadi, 495–532. Cambridge, England: Woodhead.

Choi, L.H., and S.S. Nielsen. 2005. The effects of thermal and nonthermal processing methods on apple cider quality and consumer acceptability. *Journal of Food Quality* 28: 13–29.

Cullen, P., and B. Tiwari. 2012. Applications of ozone in fruit processing. In *Contemporary food engineering: Advances in fruit processing technologies*, ed. S. Rodrigues, F. Andre, and N. Fernande, 185–202. London: CRC.

Debordea, M., and U. von Guntena. 2008. Reactions of chlorine with inorganic and organic compounds during water treatment-kinetics and mechanisms: A critical review. *Water Research* 42: 13–51.

Eco Safe Systems Food Safety. 2010. *Eyewittness news at 11pm. ABC7.* Los Angeles: ABC7. https://www.youtube.com/watch?v=Swye5WhTJ-0 (accessed June 20, 2014).

Euro Chlor Communications. 2013. *Chlorination by-products.* http://www.eurochlor.org/media/40538/01_chlorination_by-products.pdf (accessed June 20, 2014).

Extension. 2011. *Approved chemicals for use in organic postharvest systems*. http://www.extension.org/pages/18355/approved-chemicals-for-use-in-organic-postharvest-systems#.U6Y2Fhsg9Mt (accessed June 20, 2014).

FDA. 1998. Food labeling: Warning and notice statements; labeling of juice products. *Federal Register* 63: 20468–20493.

———. 2001. Secondary direct food additives permitted in food for human consumption. *Federal Register* 66: 33829–33830

Fonseca, J., and J. Rushing. 2006. Effect of ultraviolet-C light on quality and microbial population of fresh-cut watermelon. *Postharvest Biology and Technology* 40: 256–261.

Food Safety News. 2011. *Did deer cause Oregon's strawberry outbreak?* http://www.foodsafetynews.com/2011/08/epis-pinpoint-strawberries-in-or-e-coli-outbreak/#.U6uKO00g99A (accessed June 20, 2014).

———. 2012. *Watermelon* Salmonella *outbreak in UK, Europe*. http://www.foodsafetynews.com/2012/02/watermelon-salmonella-outbreak-in-uk-europe/#.U6tif00g99A (accessed June 20, 2014).

Gabler, F.M., J. Mercier, J.I. Jimenez, and J.L. Smilanick. 2010. Integration of continuous biofumigation with *Muscodor albus* with pre-cooling fumigation with ozone or sulfur dioxide to control postharvest gray mold of table grapes. *Postharvest Biology and Technology* 55: 78–84.

Horvath, M., L. Bilitzky, and J. Huttner. 1985. *Ozone*. New York: Elsevier.

Hur, J.-S., O. Soon-Ok, K.-M. Lim, J.S. Jung, J.-W. Kim, and Y.J. Koh. 2005. Novel effects of TiO_2 photocatalytic ozonation on control of postharvest fungal spoilage of kiwifruit. *Postharvest Biology and Technology* 35: 109–113.

Kim, G.-G., A.E. Yousef, and G.W. Chism. 1999. Use of ozone to inactivate microorganisms on lettuce. *Journal of Food Safety* 19: 17–34.

Li, C. 2014. *Precooling fruits and vegetables in Georgia*. http://extension.uga.edu/publications/detail.cfm?number=C1004 (accessed June 20).

Nadas, A., M. Olmo, and J.M. Garcia. 2006. Growth of *Botrytis cinerea* and strawberry quality in ozone-enriched atmospheres. *Journal of Food Science* 68: 1798–1802.

Najafi, M.B.H., and M.H. Haddad Khodaparast. 2009. Efficacy of ozone to reduce microbial populations in date fruits. *Food Control* 20: 27–30.

OMAF Food Inspection Branch. 2001. *Fresh grapes food safety risk assessment*. http://www.omafra.gov.on.ca/english/food/inspection/fruitveg/risk_assessment_pdf/grapes/50ra.pdf (accessed June 20, 2014).

Ong, K.C., J.N. Cash, M.J. Zabik, M. Siddiq, and A.L. Jones. 1996. Chlorine and ozone washes for pesticide removal from apples and processed apple sauce. *Food Chemistry* 55: 153–160.

Ozone Technologies Group. 2014. *Fresh fruits and vegetables*. http://www.ozonetechnologiesgroup.com/applications_FoodBeverage_FruitsVegetables.php (accessed June 10).

Oztenkin, S., B. Zorlugenc, and F. Zorlugenc. 2006. Effects of ozone treatment on microflora of dried figs. *Journal of Food Engineering* 75: 369–399.

Palou, L., J.L. Smilanick, C.H. Crisosto, and M. Mansour. 2001. Effect of gaseous ozone exposure on the development of green and blue molds on cold stored citrus fruit. *Plant Disease* 85: 632–638.

Palou, L., J.L. Smilanick, C.H. Crisosto, M. Mansour, and P. Plaza. 2003. Ozone gas penetration and control of the sporulation of *Penicillium digitatum* and *Penicillium italicum* within commercial packages of oranges during cold storage. *Crop Protection* 22: 1131–1134.

Patil, S., P. Bourke, J.M. Frias, B.K. Tiwari, and P.J. Cullen. 2009. Inactivation of *Escherichia coli* in orange juice using ozone. *Innovative Food Science & Emerging Technologies* 10: 551–557.

Patil, S., V.P. Valdramidis, P.J. Cullen, J. Frias, and P. Bourke. 2010a. Inactivation of *Escherichia coli* by ozone treatment of apple juice at different pH levels. *Food Microbiology* 27: 835–840.

Patil, S., B. Torres, B.K. Tiwari, H.H. Wijngaard, P. Bourke, P.J. Cullen, C.P. O'Donnell, and V.P. Valdramidis. 2010b. Safety and quality assessment during the ozonation of cloudy apple juice. *Journal of Food Science* 75: M437–M443.

Perry, J.J., and A.E. Yousef. 2011. Decontamination of raw foods using ozone-based sanitization techniques. *Annual Review of Food Science and Technology* 2: 281–298.

Praxair. 1999. *Praxair and Zentox to provide water reuse systems to poultry industry, increase productivity, reduce costs*. http://www.praxair.com/news/1999/praxair-and-zentox-to-provide-water-reuse-systems-to-poultry-industry-alliance-to-help-industry (accessed June 20, 2014).

Rodgers, S.L., J.N. Cash, M. Siddiq, and E. Ryser. 2004. A comparison of different chemical sanitizers for inactivating *Escherichia coli* O157:H7 and *Listeria monocytogenes* in solution and on apples, lettuce, strawberries, and cantaloupe. *Journal of Food Protection* 67: 721–731.

Sapers, G.M. 2001. Efficacy of washing and sanitizing methods for disinfection of fresh fruit and vegetable products. *Food Technology and Biotechnology* 39: 305–311.

Selma, M., A.M. Ibanez, M. Cantwell, and T. Suslow. 2008. Reduction by gaseous ozone of *Salmonella* and microbial flora associated with fresh-cut cantaloupe. *Food Microbiology* 25: 558–565.

Skog, L.J., and C.L. Chu. 2001. Effect of ozone on qualities of fruits and vegetables in cold storage. *Canadian Journal of Plant Science* 81: 773–778.

Steenstrup, L.D., and J.D. Floros. 2004. Inactivation of *E. coli* 0157:H7 in apple cider by ozone at various temperatures and concentrations. *Journal of Food Processing and Preservation* 28: 103–116.

Steffen, H., and R. Rice. 2008. The PhytO3 tech crop protection technology-trial results in a 2700 ha (6500 Acre) soy farm in Brazil. *Ozone Science and Engineering* 30: 210–215.

———. 2008b. The PhytO3 tech crop protection technology for microorganisms and insect control using ozone, UV, and dipole-electrical air jet spray technologies- technical basis and possible chemistries involved. *Ozone Science and Engineering* 30: 216–227

Suslow, T. 2004. *Ozone application for postharvest disinfection of edible horticulture crops*. http://anrcatalog.ucdavis.edu/pdf/8133.pdf (accessed June 20, 2014).

Tiwari, B.K., K. Muthukumarappan, C.P. O'Donnell, and P.J. Cullen. 2008. Kinetics of freshly squeezed orange juice quality changes during ozone processing. *Journal of Agricultural and Food Chemistry* 56: 6416–6422.

Vurma, M. 2009. *Development of ozone-based processes for decontamination of fresh produce to enhance safety and extend shelf life*. Ph.D. Dissertation. Columbus, OH: The Ohio State University.

Vurma, M., R.B. Pandit, S.K. Sastry, and A.E. Yousef. 2009. Inactivation of *Escherichia coli* O157:H7 and natural microbiota on spinach leaves using gaseous ozone during vacuum cooling and simulated transportation. *Journal of Food Protection* 72: 1538–1546.

Wang, Y., Q. Song, M. Frei, Z. Shao, and L. Yang. 2014. Effects of elevated ozone, carbon dioxide and the combination of both on the grain quality of Chinese hybrid rice. *Environmental Pollution* 189: 9–17.

Whangchai, K., K. Saengnil, C. Singkamanee, and J. Uthaibutra. 2010. Effect of electrolyzed oxidizing water and continuous ozone exposure on the control of *Penicillium digitatum* on Tangerine cv. 'Sai Nam Pung' during storage. *Crop Protection* 29: 386–389.

Wood, D. 2013. *Ozonated water: Vineyard pest control without chemicals*. http://midwestwine-press.com/2013/03/31/ozonated-water-vineyard-pest-control-without-chemicals/ (accessed June 20, 2014).

Wu, J.G., T.G. Luan, C.Y. Lan, W.H. Lo, and G.Y.S. Chan. 2007. Efficacy evaluation of low-concentration of ozonated water in removal of residual diazinon, parathion, methyl-parathion and cypermethrin on vegetable. *Journal of Food Engineering* 79: 803–809.

Yesil, M. 2012. *Efficacy of gaseous ozone in combination with vacuum cooling and pre-washing for the inactivation of* Escherichia coli *o157:H7 on fresh produce*. Master Thesis. Columbus, OH: The Ohio State University.

Index

A. Rosenthal et al. (eds.), *Fruit Preservation*, Food Engineering Series,
https://doi.org/10.1007/978-1-4939-3311-2

Printed by Printforce, the Netherlands